Probability and Its Applications

A series published in association with the Applied Probability Trust

Editors: J. Gani, C.C. Heyde, T.G. Kurtz

D1416045

Olav Kallenberg

Probabilistic Symmetries and Invariance Principles

 Springer

at times that were meant to be spent with the family. On a personal note, I adore her for being such a wonderful mother to our children, and I owe her special thanks for tolerating my daily piano practice and for never complaining, as I am filling our house with piles and piles of books on every aspect of cultural history.

Olav Kallenberg
January 2005

Contents

Preface   v

Introduction   1

1. The Basic Symmetries   24

1.1. Infinite sequences   24
1.2. Finite sequences   30
1.3. Continuous-time symmetries   35
1.4. Infinite-interval processes   43
1.5. Measures on a finite interval   46
1.6. Simple or diffuse random measures   52
1.7. Rotations and L^p-symmetries   57
1.8. Miscellaneous complements   66

2. Conditioning and Martingales   69

2.1. Contractable sequences   69
2.2. Continuous-time symmetries   75
2.3. Semi-martingale criteria   83
2.4. Further criteria and representation   90
2.5. Norm relations and regularity   95
2.6. Path properties   103
2.7. Palm measure invariance   111

3. Convergence and Approximation   125

3.1. Discrete-time case   125
3.2. Random measures   129
3.3. Exchangeable processes   136
3.4. Approximation and representation   143
3.5. Restriction and extension   149
3.6. Coupling and path properties   155
3.7. Sub-sequence principles   162

4. Predictable Sampling and Mapping   169

4.1. Skipping and sampling   169
4.2. Gauss and Poisson reduction   173
4.3. Predictable mapping   176
4.4. Predictable contraction   185
4.5. Brownian and stable invariance   189
4.6. Mapping of optional times   200

5. Decoupling Identities   209

   5.1. Integrability and norm estimates   209
   5.2. Exchangeable sums   216
   5.3. Martingale representations   223
   5.4. Exchangeable integrals   229
   5.5. Lévy integrals   233
   5.6. Contractable sums and integrals   240
   5.7. Predictable sampling revisited   249

6. Homogeneity and Reflections   255

   6.1. Symmetries and dichotomies   255
   6.2. Local homogeneity   261
   6.3. Reflection invariance   268
   6.4. Local time and intensity   271
   6.5. Exponential and uniform sampling   279
   6.6. Hitting points and intervals   285
   6.7. Markov properties   290
   6.8. Homogeneity and independence   294

7. Symmetric Arrays   300

   7.1. Notation and basic symmetries   300
   7.2. Coupling, extension, and independence   305
   7.3. Coding and inversion   310
   7.4. Contractable arrays   318
   7.5. Exchangeable arrays   325
   7.6. Equivalence criteria   328
   7.7. Conditional distributions   335
   7.8. Symmetric partitions   342

8. Multi-variate Rotations   350

   8.1. Rotational symmetries   350
   8.2. Gaussian and rotatable processes   356
   8.3. Functionals on a product space   359
   8.4. Preliminaries for rotatable arrays   363
   8.5. Separately rotatable arrays and functionals   370
   8.6. Jointly rotatable functionals   378
   8.7. Jointly rotatable arrays   384
   8.8. Separately exchangeable sheets   391
   8.9. Jointly exchangeable or contractable sheets   395

9. Symmetric Measures in the Plane   401

   9.1. Notions of invariance   401
   9.2. General prerequisites   405
   9.3. Symmetries on a square   411
   9.4. Symmetry on a strip   416

9.5. Technical preparation 422
9.6. Symmetries on a quadrant 432

Appendices 440

A1. Decomposition and selection 440
A2. Weak convergence 445
A3. Multiple stochastic integrals 449
A4. Complete monotonicity 457
A5. Palm and Papangelou kernels 459

Historical and Bibliographical Notes 464

Bibliography 477

Author Index 497

Subject Index 501

Symbol Index 509

Introduction

The hierarchy of distributional symmetries considered in this book, along with the associated classes of transformations, may be summarized as follows:

invariant objects	transformations
stationary	shifts
contractable	sub-sequences
exchangeable	permutations
rotatable	isometries

where each invariance property is clearly stronger than the preceding one. All four symmetries may be considered in discrete or continuous time (or space) and in one or several dimensions. There is also the distinction between bounded and unbounded index sets.

The most obvious problem in this area is to characterize the class of objects of a given type with a specified symmetry property. Thus, for example, de Finetti's theorem describes the infinite, exchangeable sequences of random variables as mixed i.i.d. This classical result is rather easy to prove by modern martingale or weak convergence arguments. Other characterizations may be a lot harder. Thus, it takes about 30 pages of tight mathematical reasoning to derive the characterization of contractable arrays of arbitrary dimension, and for the multi-variate rotatable case another 40 pages may be required. In other cases again, no simple representation seems to exist. Thus, for example, stationary sequences are unique mixtures of ergodic ones, but there is no (known) representation of a (strictly) stationary and ergodic sequence in terms of simpler building blocks. The situation for finite, contractable sequences is even worse, since here the integral representation in terms of extreme points is not even unique.

The next step might be to explore the relationship between the various symmetries. For example, Ryll-Nardzewski's theorem shows that every infinite, contractable sequence of random variables is even exchangeable, so that, for infinite sequences, the two symmetries are in fact equivalent. (The equivalence fails for finite sequences.) A higher-dimensional counterpart is the much deeper fact that every contractable array on a tetrahedral index set can be extended (non-uniquely) to an exchangeable array on the corresponding product set. For a connection with stationarity, it is easy to show that an infinite sequence is contractable iff (if and only if) it is strongly stationary, in the sense of invariance in distribution under optional shifts. Let us finally mention the fundamental and nontrivial fact that every continuous and contractable (in the sense of the increments) process on $\mathbb{R}_+ = [0, \infty)$ with zero drift is also rotatable.

Connections can also take the form of limit theorems or approximation properties. Thus, for example, we may approximate a finite exchangeable sequence by infinite ones. This, of course, is nothing else than the familiar asymptotic equivalence between sampling with or without replacement from a finite population, explored in scores of statistical papers. Similar approximation theorems can be established in continuous time, where modern weak convergence and coupling methods play a prominent role.

Our investigation may continue with a more detailed study of the various symmetric random objects. For example, though on $\mathbb{R}_+$ the exchangeable (increment) processes are just mixtures of the familiar Lévy processes—this is the continuous-time counterpart of de Finetti's theorem, first noted by Bühlmann—on $[0, 1]$ one gets a much broader class of exchangeable processes, and it becomes interesting to explore the path and other properties of the latter. It may then be natural to relate the various symmetries to a filtration of σ-fields and to employ the powerful machinery of modern martingale theory and stochastic calculus.

This new dynamical approach has led to some startling new discoveries, opening up entirely new domains of study. We have already mentioned the elementary connection between exchangeability and strong stationarity. Further instances are given by the wide range of characterizations involving direct or reverse martingales in discrete or continuous time. Less obvious are the predictable sampling and mapping theorems, where the defining properties of contractable or exchangeable sequences and processes are extended to suitably predictable random transformations. Apart from their intrinsic interest, those results also serve as valuable general tools, providing short and streamlined proofs of the arcsine and related theorems for random walks and Lévy processes. A connected, indeed even stronger, class of theorems are the decoupling identities for sums and integrals with respect to exchangeable processes, discussed in further detail below.

All the mentioned problems continue to make sense in higher dimensions, and the last third of the book deals with multi-variate symmetries of various kind. As already noted, already the basic characterization problems here become surprisingly hard, and the picture is still incomplete.

$$- \ - \ -$$

We turn to a more detailed summary of the contents of the book, introducing at the same time some crucial definitions and notation. Let us first define the basic symmetries, as they appear already in Chapter 1. Given an infinite sequence of random elements $\xi = (\xi_1, \xi_2, \ldots)$, we say that ξ is *contractable* (sometimes even called *spreading invariant* or *spreadable*) if every sub-sequence has the same distribution, so that

$$(\xi_{k_1}, \xi_{k_2}, \ldots) \stackrel{d}{=} (\xi_1, \xi_2, \ldots) \tag{1}$$

for any positive integers $k_1 < k_2 < \cdots$, where $\overset{d}{=}$ denotes equality in distribution. It is clearly enough to consider sub-sequences obtained by omitting a single element. (Our term comes from the fact that we may form a sub-sequence by omitting some elements and then *contracting* the resulting sequence to fill in the resulting gaps. This clearly corresponds to a *spreading* of the associated index set.) A finite sequence $\xi_1, \ldots, \xi_n$ is said to be contractable if all sub-sequences of equal length have the same distribution.

For the stronger property of *exchangeability*, we require (1) to hold for any distinct (not necessarily increasing) elements $k_1, k_2, \ldots$ of the index set. Thus, for infinite sequences, the k_n are required to form an injective (but not necessarily surjective) transformation on $\mathbb{N} = \{1, 2, \ldots\}$. However, it is clearly enough to require that (1) be fulfilled for any *finite permutation* of $\mathbb{N}$, defined as a bijective transformation $n \mapsto k_n$ such that $k_n = n$ for all but finitely many n. (Indeed, it suffices to consider transpositions of pairs of adjacent elements.) For technical purposes, it is useful to note that the finite permutations form a countable group of transformations of the index set. (By contrast, the classes of shifts or sub-sequences of $\mathbb{N}$ are only semigroups, which leads occasionally to some technical problems.)

Finally, when the ξ_k are real-valued (or take values in a suitable linear space), we may define *rotatability* of ξ by requiring that every finite sub-sequence be invariant in distribution under arbitrary orthogonal transformations. Thus, for every n, the distribution μ_n of $(\xi_1, \ldots, \xi_n)$ is assumed to be spherically symmetric on $\mathbb{R}^n$. It is then clear, at least intuitively, that each μ_n is a mixture of uniform distributions over concentric spherical shells around the origin. The situation for infinite sequences may be less obvious. Since permutations are special cases of rotations, we note that every rotatable sequence is exchangeable. Similarly, by the injectivity of contractions, every exchangeable sequence is clearly contractable. Finally, shifts are special contractions, so every contractable sequence is stationary.

In continuous time, we may define the corresponding symmetries in terms of the increments. The initial value playing no role, we may then restrict our attention to processes starting at 0, and by suitable scaling and shifting we may assume that the index set I is either $[0, 1]$ or $\mathbb{R}_+$. Considering any disjoint intervals $I_1, I_2, \ldots \subset I$ of equal length, listed in the order from left to right, we may say that a process X on I is contractable, exchangeable, or rotatable if the increments of X over $I_1, I_2, \ldots$ have the corresponding property (for all choices of intervals). A problem with this definition is that the underlying transformations are only applied to the increments of X (rather than to X itself).

A preferable approach would be to base the definition on suitable pathwise transformations. Thus, for any points $a < b$ in I, we may form the *contraction* $C_{a,b} X$ by deleting the path on (a, b) and attaching the continued path starting at (b, X_b) to the loose end at (a, X_a), employing a suitable parallel displacement in space and time. By a similar construction, for any three points $a < b < c$ in I, we may form the *transposition* $T_{a,b,c}X$ by swapping the paths over the intervals $[a, b]$ and $[b, c]$. Then we say that X is *contractable* if $C_{a,b}X \overset{d}{=} X$ for all $a < b$ and *exchangeable* if $T_{a,b,c}X \overset{d}{=} X$ for all $a < b < c$. For rotations, a Hilbert space approach often seems more appropriate, as explained below.

$$- - -$$

Chapter 1 begins with a proof of de Finetti's theorem. In its original form, the theorem says that every infinite, exchangeable sequence of random variables $\xi = (\xi_1, \xi_2, \ldots)$ is *mixed i.i.d.* In other words, there exists a probability measure ν on the set of distributions m on $\mathbb{R}$ such that

$$\mathcal{L}(\xi) \equiv P\{\xi \in \cdot\} = \int m^\infty \nu(dm). \tag{2}$$

(Here m^∞ denotes the distribution of an i.i.d. sequence based on the measure m.) The condition is clearly even sufficient, so it characterizes the class of exchangeable sequences.

A more sophisticated version of the theorem is in terms of conditioning. Thus, an infinite sequence ξ as above is exchangeable iff it is *conditionally i.i.d.* In other words, there exists a random probability measure ν on $\mathbb{R}$ such that, conditionally on ν, the ξ_k are i.i.d. with the common distribution ν. This we can write conveniently as

$$P[\xi \in \cdot | \nu] = \nu^\infty \quad \text{a.s.} \tag{3}$$

Even this characterization is superseded by the stronger statement of Ryll-Nardzewski, the fact that every infinite, contractable sequence of random variables is conditionally i.i.d. Thus, for infinite sequences of random variables, the four stated properties—contractable, exchangeable, mixed i.i.d., and conditionally i.i.d.—are all equivalent. This is the modern statement of de Finetti's theorem, proved in Section 1.1.

de Finetti's theorem suggests a corresponding result in continuous time, characterizing exchangeable processes on $\mathbb{R}_+$ as mixtures of Lévy processes. The proposed statement, first noted by Bühlmann, requires some qualifications, owing to some technical difficulties associated with the uncountable nature of the index set. One way to make the claim precise is to consider only exchangeable processes defined on the rationals, and assert that any such process X can be extended a.s. to a mixture of Lévy processes. Another way is to require X to be continuous in probability and claim that X has then a version with the stated property. A third option is to assume from the outset that X is right-continuous with left-hand limits (rcll), which ensures the validity of the original claim. Whatever the approach, the conclusion may be stated in either a mixing form, in the format of (2), or a conditional form, akin to (3). We finally note that, by Ryll-Nardzewski's theorem, even Bühlmann's result remains true in a stronger contractable version.

The results of de Finetti and Bühlmann are no longer true for bounded index sets. The obvious counterpart in discrete time is the fact that finite exchangeable sequences are mixtures of so-called *urn sequences*, obtained by sampling without replacement from a finite collection (that may be represented by tickets in an urn). The general result in continuous time is much harder and will be discussed later. However, the special case of random measures is accessible by elementary means, and we may prove that a random measure ξ on $[0, 1]$ is exchangeable iff it can be represented in the form

$$\xi = \alpha\lambda + \sum_k \beta_k \delta_{\tau_k} \text{ a.s.,} \tag{4}$$

where α and $\beta_1, \beta_2, \ldots$ are non-negative random variables and $\tau_1, \tau_2, \ldots$ is an independent sequence of i.i.d. $U(0, 1)$ random variables, $U(0, 1)$ being the uniform distribution on $[0, 1]$. Here λ denotes Lebesgue measure on $[0, 1]$, and δ_t is the measure assigning a unit mass to the point t.

In particular, we see that a simple point process ξ on $[0, 1]$ is exchangeable iff it is a *mixed binomial process* of the form $\xi = \sum_{k \leq \kappa} \delta_{\tau_k}$, where $\tau_1, \tau_2, \ldots$ are i.i.d. $U(0, 1)$ and κ is an independent random variable. This may be contrasted with the infinite-interval case, where ξ is mixed Poisson by Bühlmann's theorem. Those rather elementary results, of considerable importance in their own right, are interesting also because of their connections with classical analysis, as they turn out to be essentially equivalent to Bernstein's characterizations of completely monotone functions. Likewise, de Finetti's original theorem for exchangeable events is equivalent to Hausdorff's celebrated moment representation of completely monotone sequences.

We turn to the rotatable case. Here Freedman's theorem states that an infinite sequence of random variables $\xi = (\xi_1, \xi_2, \ldots)$ is rotatable iff it is mixed i.i.d. centered Gaussian. In its conditional form the condition says that, given a suitable random variable $\sigma \geq 0$, the ξ_k are conditionally i.i.d. $N(0, \sigma^2)$. The latter description corresponds to the a.s. representation $\xi_k = \sigma\zeta_k$, $k \in \mathbb{N}$, where $\zeta_1, \zeta_2, \ldots$ are i.i.d. $N(0, 1)$ and independent of σ. In either form, the result is equivalent to Schoenberg's theorem in classical analysis— the remarkable fact that a continuous function φ on $\mathbb{R}_+$ with $\varphi(0) = 1$ is completely monotone, hence a Laplace transform, iff for every $n \in \mathbb{N}$ the function $f_n(x) = \varphi(|x|^2) = \varphi(x_1^2 + \cdots + x_n^2)$ on $\mathbb{R}^n$ is non-negative definite, hence a characteristic function.

The rotational invariance may be expressed in the form $\sum_k c_k \xi_k \overset{d}{=} \xi_1$, where $c_1, c_2, \ldots$ are arbitrary constants satisfying $\sum_k c_k^2 = 1$. It is natural to consider the more general case of l^p-*invariance*, where we require instead that

$$\sum_k c_k \xi_k \overset{d}{=} \|c\|_p \xi_1, \quad c = (c_1, c_2, \ldots) \in l^p.$$

This makes sense for arbitrary $p \in (0, 2]$, and the property is equivalent to the a.s. representation $\xi_k = \sigma\zeta_k$, where the ζ_k are now i.i.d. symmetric p-

stable and σ is an independent random variable. This is another classical result, due to Bretagnolle, Dacunha-Castelle, and Krivine.

$$- - -$$

The main purpose of Chapter 2 is to introduce and explore the basic martingale connections in discrete and continuous time. Given a finite or infinite random sequence $\xi = (\xi_1, \xi_2, \ldots)$ adapted to a filtration $\mathcal{F} = (\mathcal{F}_n)$, we say that ξ is $\mathcal{F}$-*contractable* or $\mathcal{F}$-*exchangeable* if for every $n \in \mathbb{N}$ the shifted sequence $\theta_n \xi = (\xi_{n+1}, \xi_{n+2}, \ldots)$ is conditionally contractable or exchangeable, respectively, given $\mathcal{F}_n$. When $\mathcal{F}$ is the filtration induced by ξ, we note that the two properties reduce to the corresponding elementary versions. However, the added generality is often useful for applications. For infinite sequences, either property holds iff ξ is *strongly stationary*, in the sense that $\theta_\tau \xi \overset{d}{=} \xi$ for every finite optional (or stopping) time τ. It is also equivalent that the *prediction sequence*

$$\mu_n = P[\theta_n \xi \in \cdot | \mathcal{F}_n], \quad n \geq 0,$$

be a measure-valued martingale on $\mathbb{Z}_+ = \{0, 1, \ldots\}$. Similar results hold in continuous time.

Even more striking are perhaps the reverse martingale connections. To state the basic discrete-time version, let $\xi = (\xi_1, \xi_2, \ldots)$ be a finite or infinite random sequence in an arbitrary measurable space, and consider the associated sequence of *empirical distributions*

$$\eta_n = n^{-1} \sum\nolimits_{k \leq n} \delta_{\xi_k}, \quad n \geq 1.$$

Then ξ turns out to be exchangeable iff the η_n form a reverse, measure-valued martingale. In continuous time, we consider any integrable semi-martingale X on $I = [0, 1]$ or $\mathbb{R}_+$ with jump point process ξ and let $[X^c]$ denote the quadratic variation of the continuous component X^c. Then X is exchangeable iff the process

$$Y_t = t^{-1}(X_t, \xi[0, t], [X^c]_t), \quad t \in I \setminus \{0\},$$

is a reverse martingale.

Returning to the associated forward martingales, we show that any integrable and contractable process X on $\mathbb{Q}_{[0,1]} = \mathbb{Q} \cap [0, 1]$ can be extended to a special semi-martingale on $[0, 1)$ with associated jump point process ξ, such that $[X^c]$ is linear and X and ξ have compensators of the form

$$\hat{X} = M \cdot \lambda, \qquad \hat{\xi} = \lambda \otimes \eta, \tag{5}$$

where M and η denote the martingales

$$M_t = \frac{E[X_1 - X_t | \mathcal{F}_t]}{1 - t}, \qquad \eta_t = \frac{E[\xi_1 - \xi_t | \mathcal{F}_t]}{1 - t}.$$

In particular, this allows us to extend X to a process with rcll paths. We may henceforth add the latter property to the defining characteristics of a contractable or exchangeable process.

By a much deeper analysis, it can be shown that every integrable and exchangeable process X on $[0, 1]$ has an a.s. representation

$$X_t = \alpha t + \sigma B_t + \sum_k \beta_k (1\{\tau_k \leq t\} - t), \quad t \in [0, 1], \tag{6}$$

for some i.i.d. $U(0, 1)$ random variables τ_k, an independent Brownian bridge B, and an independent collection of random coefficients α, σ, and $\beta_1, \beta_2, \ldots$, where $1\{\cdot\} = 1_{\{\cdot\}}$. This clearly generalizes the representation (4) of exchangeable random measures on $[0, 1]$.

The martingale description involving (5) has a surprising partial converse. Here we consider any uniformly integrable, special semi-martingale X on $[0, 1]$ with jump point process ξ such that the end values X_1, $[X^c]_1$, and ξ_1 are a.s. non-random. Then X is exchangeable iff the compensating processes $\hat{X}$, $[X^c]$, and $\hat{\xi}$ are absolutely continuous and admit densities that form martingales on $(0, 1)$. This result is related to Grigelionis' characterization of mixed Lévy processes—hence of exchangeable processes on $\mathbb{R}_+$—as semi-martingales whose local characteristics are a.s. linear.

The various martingale descriptions enable us to prove some powerful norm relations for contractable and related processes. For example, for any L^p-valued, exchangeable processes X on $[0, 1]$, we have the relations

$$\|X_t\|_p \asymp \|X_1^*\|_p \asymp \left\| ([X]_1 + X_1^2)^{1/2} \right\|_p,$$

uniformly in X and for $t \in (0, 1)$ and $p > 0$ in compacts. (Here $a \asymp b$ means that the ratio a/b is bounded above and below by positive constants, and $X_t^* = \sup_{s \leq t} |X_s|$.) We can also use martingale methods to estimate the local growth rates of arbitrary exchangeable processes.

The chapter closes with a discussion of Palm measures. For simple point processes ξ on S, the *Palm distribution* Q_s at a point $s \in S$ can be thought of as the conditional distribution of ξ, given that ξ has a point at s. (For the existence, we need to assume that the intensity measure $E\xi$ is σ-finite.) Each measure Q_s is again the distribution of a simple point process ξ_s on S, which is clearly such that one of the points lies at s. The *reduced Palm distribution* Q_s' can be defined as the law of the point process $\xi_s' = \xi_s - \delta_s$, obtained from ξ_s by omitting the trivial point at s. Now a central result says that Q_s' is (or, rather, can be chosen to be) independent of s iff ξ is a mixed binomial or Poisson process based on $E\xi$. Recall that the latter are precisely the exchangeable point processes on S, except that exchangeability is now defined in the obvious way with respect to the measure $E\xi$. As a special case, one recovers a celebrated result of Slivnyak—the fact that ξ is Poisson iff the measures Q_s' all agree with the original distribution $\mathcal{L}(\xi)$.

$- - -$

Chapter 3 deals with weak convergence and related approximation properties. To introduce the subject, recall that by de Finetti's theorem every infinite exchangeable sequence $\xi = (\xi^1, \xi^2, \ldots)$ can be described in terms of a *directing random measure* ν, such that conditionally on ν, the ξ^k are i.i.d. with the common distribution ν. It is easy to see that the distributions of ξ and ν determine each other uniquely. Now consider a whole sequence of such infinite exchangeable sequences $\xi_n = (\xi_n^1, \xi_n^2, \ldots)$, $n \in \mathbb{N}$, with associated directing random measures $\nu_1, \nu_2, \ldots$. We may then ask for conditions ensuring the distributional convergence $\xi_n \xrightarrow{d} \xi$, where ξ is again exchangeable and directed by ν. Here one naturally expects that $\xi_n \xrightarrow{d} \xi$ iff $\nu_n \xrightarrow{d} \nu$, which is indeed true for a suitable choice of topology in the space of measures.

In a similar way, we can go on and consider exchangeable sequences, measures, or processes on any bounded or unbounded index set. The first step is then to identify the corresponding *directing random elements*, which may be summarized as follows:

	bounded	unbounded
sequences	ν	ν
measures	(α, β)	(α, ν)
processes	(α, σ, β)	(α, σ, ν)

For unbounded sequences, ν is the directing measure in de Finetti's theorem, and for bounded sequences $\xi = (\xi^1, \ldots, \xi^m)$ we may choose ν to be the empirical distribution $m^{-1} \sum_k \delta_{\xi^k}$. For exchangeable random measures ξ on $[0, 1]$, we have a representation (4) in terms of some non-negative random variables α and $\beta_1, \beta_2, \ldots$, and we may choose the directing random elements to be α and β, where β denotes the point process $\sum_k \delta_{\beta_k}$ on $(0, \infty)$. Similarly, exchangeable processes X on $[0, 1]$ have a representation as in (6), which suggests that we choose the directing elements α, σ, and $\beta = \sum_k \delta_{\beta_k}$. For exchangeable processes on $\mathbb{R}_+$, we may derive the directing triple (α, σ, ν) from the characteristics of the underlying Lévy processes, and for exchangeable random measures on $\mathbb{R}_+$ the choice of directing pair (α, ν) is similar.

In each of the mentioned cases, there is a corresponding limit theorem, similar to the one for exchangeable sequences. For example, if X and $X_1, X_2, \ldots$ are exchangeable processes on $[0, 1]$ with directing triples (α, σ, β) and $(\alpha_n, \sigma_n, \beta_n)$, respectively, then

$$X_n \xrightarrow{d} X \quad \Longleftrightarrow \quad (\alpha_n, \sigma_n, \beta_n) \xrightarrow{d} (\alpha, \sigma, \beta), \tag{7}$$

for carefully chosen topologies on the appropriate function and measure spaces. But much more is true. To indicate the possibilities, consider some finite, exchangeable sequences $\xi_n = (\xi_n^k; k \le m_n)$ with associated summation processes

$$X_n(t) = \sum_{k \le m_n t} \xi_n^k, \quad t \in [0, 1],$$

and introduce the random triples $(\alpha_n, \sigma_n, \beta_n)$, where

$$\alpha_n = \sum_k \xi_n^k, \qquad \sigma_n = 0, \qquad \beta_n = \sum_k \delta_{\xi_n^k}.$$

Then (7) remains true for any exchangeable process X on $[0, 1]$ directed by (α, σ, β), provided only that $m_n \to \infty$. The corresponding result for summation processes on $\mathbb{R}_+$ generalizes Skorohod's functional limit theorem for i.i.d. random variables. We can also use a similar approach to establish the representation (6) of exchangeable processes on $[0, 1]$, now in full generality, without imposing the previous integrability condition.

The processes in (6) are similar to, but more general than Lévy processes, which leads to the obvious challenge of extending the wide range of path properties known for the latter to the broader class of general exchangeable processes. The problems in the general case are often much harder, owing to the lack of simple independence and Markov properties. One powerful method for obtaining such results is by *coupling*. Given an exchangeable process X on $[0, 1]$ with constant directing triple (α, σ, β), we may then try to construct a Lévy process Y approximating X in a suitable path-wise sense, to ensure that at least some of the path properties of Y will carry over to X. In particular, such an approach allows us to extend some delicate growth results for Lévy processes, due to Khinchin, Fristedt, and Millar, to a general exchangeable setting.

The chapter concludes with a discussion of sub-sequence principles. Here we note that, given any tight sequence of random elements $\xi = (\xi_1, \xi_2, \ldots)$ in an arbitrary Polish space S, we can extract an asymptotically exchangeable sub-sequence $\xi \circ p = (\xi_{p_1}, \xi_{p_2}, \ldots)$, in the sense that the shifted sequences $\theta_n(\xi \circ p)$ tend in distribution, as $n \to \infty$, toward a fixed exchangeable sequence ζ. A stronger result, established in various forms by several people, including Dacunha-Castelle and Aldous, is the *weak sub-sequence principle*, where the asserted limit holds in the sense of the weak convergence

$$E[\eta; \theta_n(\xi \circ p) \in \cdot] \xrightarrow{w} E\eta\nu^\infty, \quad \eta \in L^1,$$

ν being the directing random measure of ζ. In other words, the previously noted convergence $\theta_n(\xi \circ p) \xrightarrow{d} \zeta$ is *stable*, in the sense of Rényi. A related and even more powerful result is the *strong sub-sequence principle*, due to Berkes and Péter, ensuring that for any $\varepsilon > 0$ we can choose the sub-sequence $\xi \circ p$ and an approximating exchangeable sequence ζ such that

$$E[\rho(\xi_{p_n}, \zeta_n) \wedge 1] \leq \varepsilon, \quad n \in \mathbb{N},$$

where ρ denotes a fixed, complete metrization of S.

– – –

Chapter 4 deals with various properties of invariance under predictable transformations in discrete or continuous time. To explain the discrete-time results, recall that a finite or infinite random sequence $\xi = (\xi_1, \xi_2, \ldots)$ is exchangeable if its distribution is invariant under non-random permutations of the elements, as in (1). The *predictable sampling* theorem extends the distributional invariance to certain random permutations. More precisely, letting $\tau_1, \tau_2, \ldots$ be a.s. distinct predictable times, taking values in the (finite or infinite) index set of ξ, we have

$$(\xi_{\tau_1}, \xi_{\tau_2}, \ldots) \stackrel{d}{=} (\xi_1, \xi_2, \ldots).$$

(Recall that a random time τ is said to be *predictable* if $\tau - 1$ is optional, hence an ordinary stopping time.) If ξ is only assumed to be contractable, then the same property holds for any strictly increasing sequence of predictable times $\tau_1, \tau_2, \ldots$. The latter result is a version of the *optional skipping* theorem, first established for i.i.d. sequences by Doob. The more general result for exchangeable sequences yields simple proofs of some classical arcsine laws and fluctuation identities for random walks and Lévy processes.

The corresponding continuous-time results are much harder. In the exchangeable case, we may define the random transformations in terms of predictable processes V on the interval $I = [0, 1]$ or $\mathbb{R}_+$, taking values in the same set I and such that $\lambda \circ V^{-1} = \lambda$ a.s., where λ denotes Lebesgue measure on I. In other words, we assume the paths of V to be measure-preserving transformations on I, corresponding to the permutations considered in discrete time. Given a suitable process X on I, we now define the transformed process $X \circ V^{-1}$ by

$$(X \circ V^{-1})_t = \int_I 1\{V_s \le t\} \, dX_s, \quad t \in I, \tag{8}$$

in the sense of stochastic integration of predictable processes with respect to general semi-martingales. To motivate the notation, we note that if $X_t = \xi[0, t]$ for some random measure ξ, then $X \circ V^{-1}$ is the distribution function of the transformed measure $\xi \circ V^{-1}$. The *predictable mapping* theorem states that if X is an exchangeable process on I, then $X \circ V^{-1} \stackrel{d}{=} X$.

In the contractable case, we need to consider predictable subsets A of the index set I with $\lambda A \ge h$. The corresponding time-change process τ is given by

$$\tau_t = \inf\{s \in I; \ \lambda(A \cap [0, s]) > t\}, \quad t \le h,$$

and we may define the associated *contraction* $C_A X$ of X by

$$(C_A X)_t = X(A \cap [0, \tau_t]) = \int_0^{\tau_t} 1_A(s) \, dX_s, \quad t \le h,$$

again in the sense of general stochastic integration. (A technical difficulty is to establish the existence of the integral on the right, since in general contractable processes are not known to be semi-martingales.) We can then show that if X is contractable, then $C_A X \stackrel{d}{=} X$ on $[0, h)$.

Even stronger invariance properties can be established when X is a stable Lévy process. In the simplest case, we assume that X is strictly p-stable for some $p \in (0, 2]$ and consider a predictable process $U \geq 0$ such that U^p is locally integrable. Let V be another predictable process such that

$$(U^p \cdot \lambda) \circ V^{-1} = \lambda \text{ a.s.,} \tag{9}$$

where the left-hand side is defined as in (8) in terms of the integral process $(U^p \cdot \lambda)_t = \int_0^t U_s^p ds$. Then we can show that

$$(U \cdot X) \circ V^{-1} \stackrel{d}{=} X,$$

where $U \cdot X$ denotes the stochastic integral $\int_0^t U dX$ and the mapping by V is again defined as in (8). In particular, the result leads to time-change representations for stable integrals. If X is symmetric p-stable, we can drop the condition $U \geq 0$, provided we replace U by $|U|$ in (9). Even stronger results are obtainable when X is a Brownian motion or bridge.

The quoted results may be regarded as far-reaching extensions of the classical time-change reductions of continuous local martingales to Brownian motion, due to Dambis, Dubins and Schwarz, and Knight, and the corresponding reductions of quasi-left-continuous, simple point processes to Poisson, due to Papangelou and Meyer. In fact, an abstract version that combines all those classical results into a single theorem plays the role of a universal tool in this chapter.

Time-change reductions of optional times are closely related to the theory of exchangeable processes. To see the connection, let ξ be an exchangeable random measure on $[0, 1]$, given by (4) for some i.i.d. $U(0, 1)$ random variables $\tau_1, \tau_2, \ldots$, and assume for simplicity that the coefficients are a.s. non-random with $\beta_1 > \beta_2 > \cdots$. Then for any random mapping V of $[0, 1]$ into itself, we have

$$\xi \circ V^{-1} = \alpha\lambda \circ V^{-1} + \sum_k \beta_k \, \delta_{V(\tau_k)},$$

and we see that $\xi \circ V^{-1} \stackrel{d}{=} \xi$ iff V is a.s. λ-preserving and satisfies

$$(V_{\tau_1}, V_{\tau_2}, \ldots) \stackrel{d}{=} (\tau_1, \tau_2, \ldots). \tag{10}$$

In fact, the predictable mapping theorem shows that (10) holds automatically as soon as V is predictable with $\lambda \circ V^{-1} = \lambda$ a.s.

Now consider an arbitrary sequence of optional times $\tau_1, \tau_2, \ldots$ with associated marks $\kappa_1, \kappa_2, \ldots$ in some measurable space K, and introduce the compensators $\eta_1, \eta_2, \ldots$ of the random pairs (τ_j, κ_j). The corresponding *discounted compensators* $\zeta_1, \zeta_2, \ldots$ are random sub-probability measures on $\mathbb{R}_+ \times K$, defined as the unique solutions to the Doléans differential equations

$$d\zeta_j = -Z_-^j \, d\eta_j, \quad Z_0^j = 1, \qquad j \in \mathbb{N},$$

where $Z_t^j = 1 - \zeta_j([0, t] \times K)$. Consider any predictable mappings $T_1, T_2, \dots$ on $\mathbb{R}_+ \times K$ taking values in a space S, and fix some probability measures $\mu_1, \mu_2, \dots$ on S. Then the conditions

$$\zeta_j \circ T_j^{-1} \le \mu_j \text{ a.s.}, \quad j \in \mathbb{N}, \tag{11}$$

ensure that the images $\gamma_j = T_j(\tau_j, \kappa_j)$ will be independent with distributions μ_j. This may be surprising, since the inequalities in (11) are typically strict and should allow considerable latitude for tinkering with any initial choice of mappings T_j.

$$- \; - \; -$$

Chapter 5 deals with the closely related family of *decoupling identities*. To explain the background, consider a sequence of i.i.d. random variables $\xi_1, \xi_2, \dots$ with finite mean $E\xi_1$ and a predictable sequence of random variables $\eta_1, \eta_2, \dots$. Under suitable integrability conditions on the η_k, we note that

$$E \sum_k \xi_k \eta_k = E(\xi_1) E \sum_k \eta_k.$$

Similarly, assuming $E\xi_1 = 0$ and $E\xi_1^2 < \infty$, we have

$$E\left(\sum_k \xi_k \eta_k\right)^2 = E(\xi_1^2) E \sum_k \eta_k^2,$$

under appropriate integrability conditions. The remarkable thing about these formulas is that the right-hand side is the same regardless of the dependence between the sequences (ξ_k) and (η_k). In other words, if we choose $(\tilde{\eta}_k) \overset{d}{=} (\eta_k)$ to be independent of (ξ_k), then (under the previous conditions)

$$\begin{aligned}
E \sum_k \xi_k \eta_k &= E \sum_k \xi_k \tilde{\eta}_k, \\
E\left(\sum_k \xi_k \eta_k\right)^2 &= E\left(\sum_k \xi_k \tilde{\eta}_k\right)^2.
\end{aligned}$$

This is the idea of *decoupling:* Whatever the dependence may be between the variables ξ_k and η_k, we can evaluate the expressions on the left as if the two sequences were independent.

The situation in continuous time is similar. Thus, assuming X to be a Lévy process with associated filtration $\mathcal{F} = (\mathcal{F}_t)$, letting V be an $\mathcal{F}$-predictable process on $\mathbb{R}_+$, and choosing $\tilde{V} \overset{d}{=} V$ to be independent of X, we have (again under suitable integrability conditions)

$$E \int_0^\infty V \, dX = E(X_1) E \int_0^\infty V_s \, ds = E \int_0^\infty \tilde{V} \, dX,$$

$$E\left(\int_0^\infty V \, dX\right)^2 = E(X_1^2) E \int_0^\infty V_s^2 \, ds = E\left(\int_0^\infty \tilde{V} \, dX\right)^2,$$

where, for the latter equations, we need to assume that $EX_1 = 0$. The formulas follow from standard facts in stochastic calculus. Specializing to the case where $\eta_k = 1\{\tau \ge k\}$ or $V_t = 1\{\tau \ge t\}$ for some optional time $\tau < \infty$ and writing $X_k = \sum_{j \le k} \xi_j$, we get the classical *Wald identities*

$$E(X_\tau) = E(X_1) E(\tau), \qquad E(X_\tau^2) = E(X_1^2) E(\tau),$$

where the latter formula requires $EX_1 = 0$ and both equations are valid, in discrete and continuous time, under suitable integrability conditions.

The relations mentioned so far are quite elementary and rather obvious. With some further effort, one can derive higher-dimensional formulas of the same kind, such as

$$E\Big(\sum_k \xi_k \eta_k\Big)^d = E\Big(\sum_k \xi_k \tilde{\eta}_k\Big)^d,$$

$$E\Big(\int_0^\infty V\, dX\Big)^d = E\Big(\int_0^\infty \tilde{V}\, dX\Big)^d,$$

which, in addition to appropriate integrability conditions, require that the sums or integrals

$$S_m = \sum_k \eta_k^m, \qquad I_m = \int_0^\infty V_s^m\, ds,$$

be non-random (or at least $\mathcal{F}_0$-measurable) for $1 \le m < d$. From here on, we may easily proceed to general product moments, which leads to decoupling identities of the form

$$E\prod_{j\le d}\sum_k \xi_{jk}\,\eta_{jk} = E\prod_{j\le d}\sum_k \xi_{jk}\,\tilde{\eta}_{jk}, \tag{12}$$

$$E\prod_{j\le d}\int_0^\infty V_j\, dX_j = E\prod_{j\le d}\int_0^\infty \tilde{V}_j\, dX_j, \tag{13}$$

requiring non-randomness of the sums or integrals

$$S_J = \sum_k \prod_{j\in J}\eta_{jk}, \qquad I_J = \int_0^\infty \prod_{j\in J}V_j(s)\, ds, \tag{14}$$

for all nonempty, proper subsets $J \subset \{1,\dots,d\}$.

Up to this point, we have assumed the sequence $\xi = (\xi_{jk})$ to be i.i.d. (in the second index k) or the process $X = (X_j)$ to be Lévy. The truly remarkable fact is that moment formulas of similar type remain valid for finite exchangeable sequences (ξ_{jk}) and for exchangeable processes (X_j) on $[0,1]$. Thus, in the general exchangeable case we can still prove decoupling identities such as (12) or (13), the only difference being that the sums or integrals in (14) are now required to be non-random even for $J = \{1,\dots,d\}$.

These innocent-looking identities are in fact quite amazing, already when $d = 1$. For a simple gambling illustration, suppose that the cards of a well-shuffled deck are drawn one by one. You are invited to bet an amount η_k on the kth card, based on your knowledge of previous outcomes, and the bank will return the double amount if the card is red, otherwise nothing. Also assume that, before entering the game, you must fix your total bet $\sum_k \eta_k$. Then (12) shows that your expected total gain is 0. This is surprising, since you might hope to improve your chances by betting most of your money when the proportion of red cards in the remaining deck is high. If $\sum_k \eta_k^2$ is also fixed in advance, then even the variance of your total gain is independent of your strategy, and so on for higher moments.

Though the methods of proof in Chapter 5 are very different from those in the previous chapter, the results are actually closely related. Indeed, it is not hard to derive the predictable sampling theorem in Chapter 4 from the corresponding decoupling identities, and similarly in continuous time. We can even prove contractable versions of the decoupling identities that are strong enough to imply the optional skipping theorem for contractable sequences, and similarly for contractable processes on $[0, 1]$.

– – –

In Chapter 6 we consider exchangeable random sets and related processes on $[0, 1]$ or $\mathbb{R}_+$. To motivate the basic ideas, consider an rcll process X on $\mathbb{R}_+$ with $X_0 = 0$ such that the random set $\Xi = \{t \geq 0; \ X_t = 0\}$ is a.s. unbounded. Then X is said to be *regenerative* at 0 if for any optional time τ taking values in Ξ, the shifted process $\theta_\tau X$ is independent of $\mathcal{F}_\tau$ with the same distribution as X, where $\mathcal{F} = (\mathcal{F}_t)$ denotes the right-continuous filtration induced by X. *Local homogeneity* is the same property without the independence condition. Thus, X is said to be locally homogeneous at 0 if $\theta_\tau X \overset{d}{=} X$ for any optional time τ in Ξ.

In view of de Finetti's theorem and the characterization of infinite exchangeable sequences by strong stationarity, it is easy to believe (but not quite so easy to prove) that X is locally homogeneous iff it is a mixture of regenerative processes. Since the latter can be described in terms of a local time random measure ξ supported by Ξ, along with a homogeneous Poisson point process η of excursions, as specified by Itô's excursion law, in the general locally homogeneous case we obtain a conditional representation of the same type. In particular, the cumulative local time process $L_t = \xi[0, t]$ is a mixture of inverse subordinators, hence the inverse of a non-decreasing, exchangeable process on $\mathbb{R}_+$.

For processes on $[0, 1]$, we need to replace the local homogeneity by a suitable reflection property. To make this precise, let us first consider the case of a random, closed subset $\Xi \subset [0, 1]$ containing 0. For any optional time τ in Ξ, we may construct a reflected set $R_\tau \Xi$ by reversing the restriction of Ξ to $[\tau, 1]$. The *strong reflection property* is defined by the condition $R_\tau \Xi \overset{d}{=} \Xi$ for every optional time τ in Ξ. The definition is similar for processes X on $[0, 1]$, except that the initial reflection needs to be combined with a reversal of each excursion.

With reflection invariance defined in this way, the theory becomes analogous to the one for processes on $\mathbb{R}_+$. Thus, under the stated condition, we have again a local time random measure ξ supported by Ξ, along with an exchangeable point process η of excursions, such that X admits a conditional Itô-type representation in terms of ξ and η. In particular, the cumulative local time L, normalized such that $L_1 = 1$, now becomes the inverse of a non-decreasing exchangeable process on $[0, 1]$.

We proceed to describe an interesting sampling property of the local time process L. First suppose that Ξ is regenerative. Combining the regenerative property with the loss-of-memory property characterizing the exponential distribution, we note that if the time τ is exponentially distributed and independent of Ξ, then $\sigma = L_\tau$ has again an exponential distribution. Iterating this result, we see that if $\tau_1 < \tau_2 < \cdots$ form a homogeneous Poisson process on $\mathbb{R}_+$ independent of Ξ, then the variables $L_{\tau_1}, L_{\tau_2}, \ldots$ form a compound Poisson process. (Multiplicities occur when several times τ_k hit the same excursion interval.) Ignoring repetitions in the sequence (L_{τ_k}), we get another homogeneous Poisson process $\sigma_1, \sigma_2, \ldots$, and by suitably normalizing L we can arrange that

$$(\sigma_1, \sigma_2, \ldots) \stackrel{d}{=} (\tau_1, \tau_2, \ldots). \tag{15}$$

As stated in this form, the result clearly carries over to the general locally homogeneous case.

So far things are quite simple and elementary. Now consider instead a random subset Ξ of $[0, 1]$, satisfying the strong reflection property and admitting a normalized local time process L, as described above. Here we take the variables $\tau_1, \tau_2, \ldots$ to be i.i.d. $U(0, 1)$, independently of Ξ, and let $\sigma_1, \sigma_2, \ldots$ be the distinct elements of the sequence $L_{\tau_1}, L_{\tau_2}, \ldots$. Then, surprisingly, the σ_k are again i.i.d. $U(0, 1)$, and so (15) remains fulfilled.

Still considering an exchangeable random set Ξ in $[0, 1]$, as described above, we now assume that $\lambda\Xi = 0$ a.s. Continuing Ξ periodically and shifting by an independent $U(0, 1)$ random variable, we obtain a stationary random set $\tilde{\Xi}$ in $\mathbb{R}$. Under suitable regularity conditions, the distributions of Ξ and $\tilde{\Xi}$ can be shown to resemble each other locally, apart from a normalizing factor. Here only some simple aspects of this similarity are discussed. For any fixed interval $I \subset [0, 1]$, we may consider the probabilities that I intersects Ξ or $\tilde{\Xi}$, which will typically tend to 0 as I shrinks to a single point $t \in (0, 1)$. Under mild restrictions on the underlying parameters, we can show that the local time intensity $E\xi$ is absolutely continuous with a nice density p, and that for almost every $t \in (0, 1)$

$$P\{\Xi \cap I \neq \emptyset\} \sim p_t \, P\{\tilde{\Xi} \cap I \neq \emptyset\}, \quad I \downarrow \{t\}.$$

Similar relations hold for regenerative sets in $\mathbb{R}_+$. To appreciate these formulas, we note that the probabilities on the left are very difficult to compute. For those on the right the computation is easy.

If X is locally homogeneous at several states, it is clearly conditionally regenerative at each of them. It may be less obvious that these properties can be combined, under suitable conditions, into a conditional strong Markov property on the corresponding part of the state space. Strengthening the local homogeneity into a property of *global homogeneity*, we may even deduce the Markov property in its usual, unconditional form. This leads us naturally to regard the strong Markov property of a process X at an optional time $\tau < \infty$ as the combination of two properties,

$$P[\theta_\tau X \in \cdot | \mathcal{F}_\tau] = P[\theta_\tau X \in \cdot | X_\tau] = \mu(X_\tau, \cdot) \quad \text{a.s.},$$

a conditional independence and a global homogeneity condition. We have already indicated how the latter property at every τ implies the strong Markov property, hence also the conditional independence. Under certain regularity conditions, we can even obtain an implication in the opposite direction. Thus, again quite surprisingly, the homogeneity and independence components of the strong Markov property are then equivalent.

$$- - -$$

The last three chapters deal with certain multi-variate symmetries. Crucial for our subsequent developments is the discussion of exchangeable and contractable arrays in Chapter 7. Here our main aim is to derive representations of separately or jointly contractable or exchangeable arrays of arbitrary dimension. For motivation we note that, by de Finetti's theorem, any infinite sequence of exchangeable random variables $X_1, X_2, \ldots$ has a representation

$$X_n = f(\alpha, \xi_n) \text{ a.s.,} \quad n \in \mathbb{N},$$

in terms of a measurable function f on $[0,1]^2$ and some i.i.d. $U(0,1)$ random variables α and $\xi_1, \xi_2, \ldots$.

A two-dimensional array $X = (X_{ij}; i, j \in \mathbb{N})$ is said to be *separately exchangeable* if

$$X \circ (p, q) \equiv (X_{p_i, q_j}; i, j \in \mathbb{N}) \stackrel{d}{=} (X_{ij}; i, j \in \mathbb{N}) \equiv X,$$

for any (finite) permutations $p = (p_i)$ and $q = (q_j)$ of $\mathbb{N}$, and *jointly exchangeable* if the same property holds when $p = q$, so that for any permutation p of $\mathbb{N}$,

$$X \circ p \equiv (X_{p_i, p_j}; i, j \in \mathbb{N}) \stackrel{d}{=} (X_{ij}; i, j \in \mathbb{N}) \equiv X.$$

Restricting p and q to sub-sequences $p_1 < p_2 < \cdots$ and $q_1 < q_2 < \cdots$ yields the corresponding properties of *separate* or *joint contractability*. However, since any separately contractable array is also separately exchangeable, by Ryll-Nardzewski's theorem, it is enough to consider the jointly contractable case.

For jointly exchangeable arrays it is often more natural to consider the index set N_2, consisting of all pairs $i, j \in \mathbb{N}$ with $i \neq j$. This is because an array $X = (X_{ij})$ on $\mathbb{N}^2$ is jointly exchangeable iff the same property holds for the array of pairs (X_{ii}, X_{ij}) indexed by N_2. Similarly, for (jointly) contractable arrays, we may prefer the index set T_2 of pairs $i, j \in \mathbb{N}$ with $i < j$, since an array $X = (X_{ij})$ on $\mathbb{N}^2$ is contractable iff the same property holds for the array of triples (X_{ii}, X_{ij}, X_{ji}) on T_2. It is also convenient to think of T_2 as the class of sets $\{i, j\} \subset \mathbb{N}$ of cardinality 2.

The first higher-dimensional representation theorems were obtained, independently, by Aldous and Hoover, who proved that an array $X = (X_{ij})$ on $\mathbb{N}^2$ is separately exchangeable iff

$$X_{ij} = f(\alpha, \xi_i, \eta_j, \zeta_{ij}) \text{ a.s.,} \quad i, j \in \mathbb{N},$$

for some measurable function f on $[0,1]^4$ and some i.i.d. $U(0,1)$ random variables α, ξ_i, η_j, ζ_{ij}, $i,j \in \mathbb{N}$. Hoover also settled the more general case of jointly exchangeable arrays of arbitrary dimension. In particular, he showed that a two-dimensional array $X = (X_{ij})$ is jointly exchangeable iff

$$X_{ij} = f(\alpha, \xi_i, \xi_j, \zeta_{ij}) \text{ a.s.}, \quad i,j \in \mathbb{N}, \tag{16}$$

for some measurable function f as above and some i.i.d. $U(0,1)$ random variables α, ξ_i, and $\zeta_{ij} = \zeta_{ji}$. These results and their higher-dimensional counterparts are quite deep, and their proofs occupy much of Chapter 7.

In the same chapter, we also derive the corresponding representations of (jointly) contractable arrays. In two dimensions we have the same representation as in (16), except that i and j should now be restricted to the triangular index set T_2. In higher dimensions, it is natural to choose as our index set the class $\tilde{\mathbb{N}}$ of all finite subsets $J \subset \mathbb{N}$. The representation can then be stated compactly in the form

$$X_J = f(\xi_I; I \subset J), \quad J \in \tilde{\mathbb{N}}, \tag{17}$$

where $\xi = (\xi_J)$ is an array of i.i.d. $U(0,1)$ random variables, also indexed by $\tilde{\mathbb{N}}$, and f is a measurable function on a suitable space. This formula can be used to extend X in a natural way to a jointly exchangeable array on the index set $\overline{\mathbb{N}}$, consisting of all finite sequences $(k_1, \ldots, k_d)$ in $\mathbb{N}$ with distinct entries. The surprising conclusion is that *an array on $\tilde{\mathbb{N}}$ is contractable iff it admits an exchangeable extension to $\overline{\mathbb{N}}$.* Note that this extension is far from unique, owing to the non-uniqueness of the representing function in (17).

After a detailed study of some matters of uniqueness and conditioning, too technical to describe here, we conclude the chapter with a discussion of symmetric partitions. Informally, a random partition of $\mathbb{N}$ into disjoint subsets $A_1, A_2, \ldots$ is said to be exchangeable if an arbitrary permutation of $\mathbb{N}$ yields a partition with the same distribution. To formalize this, we may introduce a random array X on $\mathbb{N}$ with entries 0 and 1, such that $X_{ij} = 1$ iff i and j belong to the same set A_k. The partition $\{A_k\}$ is then defined to be exchangeable if X is jointly exchangeable, in the sense of the previous discussion. The classical result, due to Kingman, states that $\{A_k\}$ is exchangeable iff it admits a *paint-box* representation

$$X_{ij} = 1\{\xi_i = \xi_j\} \text{ a.s.}, \quad i,j \in \mathbb{N}, \tag{18}$$

in terms of a sequence of exchangeable random variables $\xi_1, \xi_2, \ldots$. The term comes from the interpretation of the variables ξ_j as colors, chosen at random from a possibly infinite paint box, which determine a partition of $\mathbb{N}$ into subsets of different colors.

There is nothing special about exchangeable partitions. Letting $\mathcal{T}$ be an arbitrary family of injective maps $p: \mathbb{N} \to \mathbb{N}$, we can show that a random partition $\{A_k\}$ is $\mathcal{T}$-invariant (in distribution) iff it admits a representation as in (18) in terms of a $\mathcal{T}$-invariant sequence of random variables. To indicate the possibilities, we can apply the stated result, along with the previous representations of arrays, to obtain representations of exchangeable or contractable partitions of $\widetilde{\mathbb{N}}$ or $\tilde{\mathbb{N}}$, respectively.

$$- \ - \ -$$

In Chapter 8 we turn our attention to multi-variate rotatability. For arrays, the definitions are analogous to those in the contractable and exchangeable cases. Thus, for suitable two-dimensional arrays X, U, and V indexed by $\mathbb{N}^2$, we define the transformed array $Y = (U \otimes V)X$ by

$$Y_{ij} = \sum\nolimits_{h,k} U_{ih} V_{jk} X_{hk}, \quad i,j \in \mathbb{N}.$$

To rotate X in the first index, we choose U to be an orthogonal matrix on a finite index set $I^2 \subset \mathbb{N}^2$, and extend U to $\mathbb{N}^2$ by putting $U_{ij} = \delta_{ij} \equiv 1_{\{i=j\}}$ when either i or j lies in I^c. Similarly, we get a rotation in the second index by choosing V to be an orthogonal matrix of the same type. Then X is said to be *separately rotatable* if

$$(U \otimes V)X \stackrel{d}{=} X, \quad U, V \in \mathcal{O}, \tag{19}$$

where $\mathcal{O}$ denotes the class of infinite, orthogonal matrices as above, rotating only finitely many coordinates. It is *jointly rotatable* if (19) holds whenever $U = V \in \mathcal{O}$. The definitions carry over immediately to arrays of arbitrary dimension.

We have already seen that, by Freedman's theorem, an infinite sequence $X = (X_1, X_2, \ldots)$ is rotatable iff it can be represented in the form $X_j = \sigma \zeta_j$, where $\zeta_1, \zeta_2, \ldots$ are i.i.d. $N(0,1)$ and σ is an independent random variable. For separately rotatable arrays $X = (X_{ij})$ on $\mathbb{N}^2$, we get the a.s. representation

$$X_{ij} = \sigma \zeta_{ij} + \sum\nolimits_k \alpha_k \xi_{ki} \eta_{kj}, \quad i,j \in \mathbb{N}, \tag{20}$$

in terms of some i.i.d. $N(0,1)$ random variables ξ_{ki}, η_{kj}, and ζ_{ij} and an independent collection of random coefficients σ and $\alpha_1, \alpha_2, \ldots$, where the latter need to be such that $\sum_k \alpha_k^2 < \infty$ a.s. This is a quite deep result, first conjectured by Dawid, and then proved (under a moment condition) by Aldous. For jointly rotatable arrays on $\mathbb{N}^2$, we get instead an a.s. representation of the form

$$X_{ij} = \rho \delta_{ij} + \sigma \zeta_{ij} + \sigma' \zeta_{ji} + \sum\nolimits_{h,k} \alpha_{hk} (\xi_{ki} \xi_{kj} - \delta_{ij} \delta_{hk}), \quad i,j \in \mathbb{N}, \tag{21}$$

for some i.i.d. $N(0,1)$ random variables ξ_{ki} and ζ_{ij} and an independent collection of random coefficients ρ, σ, σ', and α_{hk}, where the latter must satisfy $\sum_{h,k} \alpha_{hk}^2 < \infty$ a.s., to ensure convergence of the double series in (21).

The representations (20) and (21) of separately or jointly rotatable arrays are easily extended to the continuous parameter case, as follows. Here we consider processes X on $\mathbb{R}_+^2$ with $X(s,t) = 0$ when $s = 0$ or $t = 0$, and we may define rotatability in the obvious way in terms of the two-dimensional increments

$$\Delta_{h,k} X(s,t) = X(s+h, t+k) - X(s+h, t) - X(s, t+k) + X(s,t),$$

where $s, t, h, k \geq 0$ are arbitrary. Assuming X to be continuous in probability, we get in the separately rotatable case an a.s. representation

$$X(s,t) = \sigma Z(s,t) + \sum_k \alpha_k \, B_k(s) \, C_k(t), \quad s, t \geq 0,$$

for a Brownian sheet Z, some independent Brownian motions $B_1, B_2, \ldots$ and $C_1, C_2, \ldots$, and an independent collection of random coefficients σ and $\alpha_1, \alpha_2, \ldots$ with $\sum_k \alpha_k^2 < \infty$ a.s. In the jointly rotatable case, we get instead an a.s. representation of the form

$$\begin{aligned} X(s,t) &= \rho(s \wedge t) + \sigma Z(s,t) + \sigma' Z(t,s) \\ &\quad + \sum_{h,k} \alpha_{hk} \left(B_h(s) \, B_k(t) - \delta_{hk} \, (s \wedge t) \right). \end{aligned} \tag{22}$$

The last two representation formulas are best understood in their measure-valued or functional versions

$$X = \sigma Z + \sum_k \alpha_k \, (B_k \otimes C_k), \tag{23}$$

$$X = \rho \lambda_D + \sigma Z + \sigma' \tilde{Z} + \sum_{h,k} \alpha_{hk} \, (B_h \otimes B_k), \tag{24}$$

where λ_D denotes Lebesgue measure along the main diagonal D in $\mathbb{R}_+^2$, $\tilde{Z}$ is the reflection of Z given by $\tilde{Z}_{s,t} = Z_{t,s}$, and $B_h \otimes B_k$ denotes the double stochastic integral formed by the processes B_h and B_k. In particular, we note that

$$(B_k \otimes B_k)([0,s] \times [0,t]) = B_k(s) B_k(t) - s \wedge t, \quad s, t \geq 0,$$

by the expansion of multiple Wiener–Itô integrals in terms of Hermite polynomials, which explains the form of the last term in (22).

The representations in the discrete and continuous parameter cases can be unified by a Hilbert-space approach, which also clears the way for extensions to higher dimensions. Here we consider any *continuous linear random functional (CLRF)* X on an infinite-dimensional, separable Hilbert space H, where linearity means that

$$X(ah + bk) = aXh + bXk \text{ a.s.}, \quad h, k \in H, \quad a, b \in \mathbb{R},$$

and continuity is defined by $Xh \xrightarrow{P} 0$ (or $E[|Xh| \wedge 1] \to 0$) as $h \to 0$ in H. Rotatability of X means that $X \circ U \overset{d}{=} X$ for any unitary operator U on H, where $(X \circ U)h = X(Uh)$, and Freedman's theorem shows that X is rotatable iff $X = \sigma \eta$ for some isonormal Gaussian process (G-process) η on H and an independent random variable σ.

For CLRFs X on $H^{\otimes 2} = H \otimes H$ we may define separate and joint rotatability in the obvious way in terms of tensor products of unitary operators, and we get the a.s. representations

$$Xh = \sigma\zeta h + (\xi \otimes \eta)(\alpha \otimes h), \tag{25}$$

$$Xh = \sigma\zeta h + \sigma'\tilde{\zeta} h + \eta^{\otimes 2}(\alpha \otimes h), \tag{26}$$

in terms of some independent G-processes ξ, η, and ζ on $H^{\otimes 2}$ and an independent pair (σ, α) or triple $(\sigma, \sigma', \alpha)$, where σ and σ' are real random variables and α is a random element of $H^{\otimes 2}$. Here (25) is just an abstract version of (20) and (23). However, (26) is less general than (21) or (24), since there is no term in (26) corresponding to the diagonal terms $\rho\delta_{ij}$ in (21) or $\rho(s \wedge t)$ in (24). This is because these terms have no continuous extension to $H^{\otimes 2}$.

Those considerations determine our strategy: First we derive representations of separately or jointly rotatable CLRFs on the tensor products $H^{\otimes d}$, which leads to simple and transparent formulas, independent of the choice of ortho-normal basis in H. Applying the latter representations to the various diagonal terms, we can then deduce representation formulas for rotatable arrays of arbitrary dimension.

It may come as a surprise that the representations of rotatable random functionals on the tensor products $H^{\otimes d}$ can also be used to derive representation formulas for separately or jointly exchangeable or contractable random sheets of arbitrary dimension. (By a *random sheet* on $\mathbb{R}_+^d$ or $[0, 1]^d$ we mean a continuous random process X such that $X = 0$ on each of the d coordinate hyper-planes.) To see the connection, recall that a continuous process X on $\mathbb{R}_+$ or $[0, 1]$ with $X_0 = 0$ is exchangeable iff $X_t = \alpha t + \sigma B_t$ a.s., where B is a Brownian motion or bridge, respectively, independent of the random coefficients α and σ. Omitting the drift term αt gives $X_t = \sigma B_t$, which we recognize, for processes on $\mathbb{R}_+$, as the general form of a rotatable process. When X is defined on $[0, 1]$, we may apply the scaling transformation

$$Y(t) = (1 + t) X\left(\frac{t}{1 + t}\right), \quad t \geq 0, \tag{27}$$

to convert X into a rotatable process on $\mathbb{R}_+$.

In higher dimensions, we can decompose X into drift terms associated with the different coordinate subspaces of $\mathbb{R}_+^d$, and then apply transformations of type (27), if necessary, to get a description of X in terms of rotatable processes of different dimension. The previously established representations in the rotatable case can then be used to yield the desired representation formulas for exchangeable random sheets. For contractable random sheets on $\mathbb{R}_+^d$, we may easily reduce to the exchangeable case by means of the general extension property from Chapter 7. The indicated approach and resulting formulas provide a striking confirmation of the close relationship between the various symmetries in our general hierarchy.

$- - -$

The final Chapter 9 deals with separately or jointly exchangeable random measures ξ on a finite or infinite rectangle R in the plane, where the exchangeability may be defined in terms of the increments over a regular square grid in R. By suitable shifting and scaling, we may reduce to the cases where R is one of the sets $[0,1]^2$, $[0,1] \times \mathbb{R}_+$, and $\mathbb{R}_+^2$, and since joint exchangeability makes no sense for the infinite strip $[0,1]^2 \times \mathbb{R}_+$, there are precisely five remaining cases to consider. In each of them, our main objective is to derive a general representation formula for ξ. (Though most of our methods seem to extend rather easily to higher dimensions, the complexity of the argument, already in dimension two, discourages us from attacking the general multi-variate representation problem.)

To introduce the representations of ξ, it may be helpful first to consider the special case where ξ is a simple point process. Since every exchangeable distribution can be shown to be a unique mixture of extreme or ergodic distributions, where the latter are such that no further decomposition is possible, it is also enough to consider the extreme case.

Beginning with the unit square, we get for any extreme, separately exchangeable, simple point process ξ on $[0,1]^2$ the a.s. representation

$$\xi = \sum_{i,j} a_{ij} (\delta_{\sigma_i} \otimes \delta_{\tau_j}), \tag{28}$$

where $\sigma_1, \sigma_2, \ldots$ and $\tau_1, \tau_2, \ldots$ are i.i.d. $U(0,1)$ and the a_{ij} are constants taking values 0 or 1. (Since a_{ij} can be 1 for at most finitely many pairs (i,j), the double sum in (28) is in fact finite.) In the jointly exchangeable case, we have instead the representation

$$\xi = \sum_{i,j} a_{ij} (\delta_{\tau_i} \otimes \delta_{\tau_j}), \tag{29}$$

in terms of a single sequence $\tau_1, \tau_2, \ldots$ of i.i.d. $U(0,1)$ random variables.

The representations in the remaining three cases are much more complicated. For extreme, separately exchangeable, simple point processes ξ on $\mathbb{R}_+ \times [0,1]$, we get an a.s. representation of the form

$$\xi = \sum_{i,j} \alpha_{ij} (\delta_{\sigma_i} \otimes \delta_{\tau_j}) + \sum_i (\delta_{\sigma_i} \otimes \eta_i), \tag{30}$$

where $\sigma_1, \sigma_2, \ldots$ form a homogeneous Poisson process on $\mathbb{R}_+$, the variables $\tau_1, \tau_2, \ldots$ are independent of the σ_i and i.i.d. $U(0,1)$, and the sums

$$\hat{\eta}_i = \eta_i + \sum_j \alpha_{ij} \delta_{\tau_j}, \quad i \in \mathbb{N}, \tag{31}$$

are conditionally independent and identically distributed point processes on $[0,1]$. The points σ_i and τ_j can be chosen such that

$$
\begin{aligned}
\{\sigma_1, \sigma_2, \ldots\} &= \{s \geq 0;\, \xi(\{s\} \times [0,1]) \geq 1\}, \\
\{\tau_1, \tau_2, \ldots\} &= \{t \in [0,1];\, \xi(\mathbb{R}_+ \times \{t\}) = \infty\}.
\end{aligned}
$$

(To appreciate the latter condition, we note that with probability one $\xi(\mathbb{R}_+ \times \{t\}) \in \{0, 1, \infty\}$ for all t.) Then (31) gives the total mass along each line segment $\{\sigma_i\} \times [0, 1]$, divided into contributions to the lines $\mathbb{R}_+ \times \{\tau_j\}$ and to the complementary set. Finally, the components η_i outside the lines need to be conditionally independent binomial processes on $[0, 1]$ independent of the coefficients α_{ij}.

The difficulties of clearly specifying the mutual dependence between the various components in (30) is evident from this attempt at a verbal description. A more precise and transparent formulation can be achieved by means of a suitable *coding*, similar to the one employed for the symmetric arrays in Chapter 7. We may then write (30) in the form

$$\xi = \sum_{i,j} f_j(\vartheta_i)(\delta_{\sigma_i} \otimes \delta_{\tau_j}) + \sum_i \sum_{k \le g(\vartheta_i)} (\delta_{\sigma_i} \otimes \delta_{\rho_{ik}}),$$

for some measurable functions $f_j : \mathbb{R}_+ \to \{0, 1\}$ and $g : \mathbb{R}_+ \to \mathbb{Z}_+$, where the variables τ_j and ρ_{ik} are i.i.d. $U(0, 1)$, and the pairs (σ_i, ϑ_i) form an independent, unit rate Poisson process on $\mathbb{R}_+^2$.

For separately exchangeable, simple point process ξ on $\mathbb{R}_+^2$, we get a representation of the form

$$\xi = \zeta + \sum_{i,j} \alpha_{ij}(\delta_{\sigma_i} \otimes \delta_{\tau_j}) + \sum_i (\delta_{\sigma_i} \otimes \eta_i) + \sum_j (\eta'_j \otimes \delta_{\tau_j}),$$

where ζ is a homogeneous Poisson process on $\mathbb{R}_+^2$, independent of all remaining terms, and the η_i and η'_j are conditionally independent Poisson processes on $\mathbb{R}_+$, independent of the variables σ_i and τ_j, the latter being such that

$$\begin{aligned}
\{\sigma_1, \sigma_2, \ldots\} &= \{s \ge 0; \ \xi(\{s\} \times \mathbb{R}_+) = \infty\}, \\
\{\tau_1, \tau_2, \ldots\} &= \{t \ge 0; \ \xi(\mathbb{R}_+ \times \{t\}) = \infty\}.
\end{aligned}$$

A complete verbal description of the underlying joint distribution, though still possible, would now be too complicated to be useful. This is even more true in the jointly exchangeable case, where the corresponding representation can be written as

$$\xi = \zeta + \tilde{\zeta}' + \sum_{i,j} \alpha_{ij}(\delta_{\tau_i} \otimes \delta_{\tau_j}) + \sum_j (\delta_{\tau_j} \otimes \eta_j + \eta'_j \otimes \delta_{\tau_j}).$$

The difficulties are again resolved through the device of coding. Then for separately exchangeable point processes ξ, we may write

$$\begin{aligned}
\xi &= \sum_k \delta_{a\rho_k} + \sum_{i,j} f(\vartheta_i, \vartheta'_j, \zeta_{ij})(\delta_{\tau_i} \otimes \delta_{\tau'_j}) \\
&\quad + \sum_{i,k} (\delta_{\tau_i} \otimes \delta_{g(\vartheta_i)\sigma_{ik}}) + \sum_{j,k} (\delta_{g'(\vartheta'_j)\sigma'_{jk}} \otimes \delta_{\tau'_j}),
\end{aligned}$$

for some measurable functions $f \in \{0, 1\}$ and $g, g' \ge 0$, a constant $a \ge 0$, some independent, unit rate Poisson processes $\{(\tau_i, \vartheta_i)\}$, $\{(\tau'_j, \vartheta'_j)\}$, and $\{\rho_k\}$ on $\mathbb{R}_+^2$, and $\{\sigma_{ik}\}$ and $\{\sigma'_{jk}\}$, $k \in \mathbb{N}$, on $\mathbb{R}_+$, and some independent i.i.d. $U(0, 1)$ random variables ζ_{ij}.

For jointly exchangeable processes ξ, one can derive a somewhat more complicated representation of a similar type. However, it may now be preferable to extend the Poisson processes to a higher dimension and write the representation in the alternative form

$$
\begin{aligned}
\xi \;=\; & \sum_{k} \left(l(\eta_k)\, \delta_{\rho_k, \rho'_k} + l'(\eta_k)\, \delta_{\rho'_k, \rho_k} \right) \\
& + \sum_{i,j} f(\vartheta_i, \vartheta_j, \zeta_{\{i,j\}})\, \delta_{\tau_i, \tau_j} \\
& + \sum_{j,k} \left(g(\vartheta_j, \chi_{jk})\, \delta_{\tau_j, \sigma_{jk}} + g'(\vartheta_j, \chi_{jk})\, \delta_{\sigma_{jk}, \tau_j} \right),
\end{aligned}
$$

in terms of some independent, unit rate Poisson processes $\{(\tau_j, \vartheta_j)\}$ and $\{(\sigma_{jk}, \chi_{jk})\}$, $k \in \mathbb{N}$, on $\mathbb{R}_+^2$ and $\{(\rho_k, \rho'_k, \eta_k)\}$ on $\mathbb{R}_+^3$, along with some i.i.d. $U(0,1)$ random variables $\zeta_{\{i,j\}}$, $i \leq j$, where f, g, g', and l, l' are measurable functions between suitable spaces. The latter representation has the further advantage of being extendable to the general random measure case.

— — —

This is where the book (but certainly not the subject) ends. Much more could be said about symmetric distributions, and some of the additional material might fit naturally into our account. Other important aspects of the theory are omitted since they seem to require very different methods. For example, the fundamental theory of extremal and ergodic decompositions is reviewed only briefly in an appendix, with some of the main proofs omitted. Important areas of application—to statistics, genetics, and other fields— are not even mentioned, this time even for another reason: These areas are already covered by a huge literature, and as a non-expert, I would have little more of interest to contribute.

Yet, what remains is a comprehensive and, as far as one can tell, reasonably complete theory involving the basic notions of distributional symmetry, comprising an abundance of interesting connections and ideas, unified by both methods and results. One may hope that the present exposition will provide a sound basis for further developments in both theory and applications. Indeed, a perceptive reader will surely identify some unexplored territory and find a wealth of challenging open problems.

Chapter 1

The Basic Symmetries

The purpose of this chapter is to introduce the basic notions of distributional symmetry and to prove some fundamental results that are accessible by elementary methods. Already in Section 1.1, dealing with infinite exchangeable sequences, we prove Ryll-Nardzewski's extension of de Finetti's celebrated theorem, along with some technical facts about uniqueness, independence, and measurability. Section 1.2 deals with the corresponding theory for finite sequences, where it becomes interesting to compare the notions of contractable, exchangeable, and mixed i.i.d. sequences.

Continuous-time versions of the basic symmetries are introduced and compared in Section 1.3. In Section 1.4 we characterize exchangeable and contractable random processes and measures defined on an infinite interval. The corresponding results for a finite interval are accessible, at this stage, only for random measures, where the general representation is derived in Section 1.5. More can be said in the special cases of simple point processes and diffuse random measures, treated in further detail in Section 1.6.

Rotatable sequences and their representation in terms of Gaussian random variables are studied in Section 1.7, which also contains a discussion of the more general L^p-symmetries, where the basic representations are in terms of general stable noise processes. The chapter concludes with a collection of miscellaneous complements and exercises.

1.1 Infinite Sequences

A finite or infinite random sequence $\xi = (\xi_1, \xi_2, \ldots)$ in a measurable space $(S, \mathcal{S})$ is said to be *exchangeable* if

$$(\xi_{k_1}, \ldots, \xi_{k_m}) \overset{d}{=} (\xi_1, \ldots, \xi_m) \tag{1}$$

for any collection $k_1, \ldots, k_m$ of distinct elements in the index set of ξ. We also say that ξ is *contractable* if (1) holds whenever $k_1 < \cdots < k_m$. Informally, ξ is exchangeable if its distribution $\mathcal{L}(\xi) = P \circ \xi^{-1}$ is invariant under finite permutations and contractable if all sub-sequences of equal length have the same distribution. Note that, trivially, every exchangeable sequence is also contractable.

We shall relate the mentioned notions of symmetry to some basic structural properties. Then say that a random sequence $\xi = (\xi_j)$ in S is *conditionally i.i.d.* if

$$P[\xi \in \cdot | \mathcal{F}] = \nu^\infty \quad \text{a.s.} \tag{2}$$

for some σ-field $\mathcal{F}$ and some *random probability measure* ν on S. In other words, ν is a probability kernel from the basic probability space $(\Omega, \mathcal{A})$ to S, or, equivalently, a random element in the space $\mathcal{M}_1(S)$ of probability measures on S, endowed with the σ-field generated by all projection maps $\pi_B : \mu \mapsto \mu B$, $B \in \mathcal{S}$. Since the measure ν in (2) is clearly a.s. $\mathcal{F}$-measurable (measurable with respect to the P-completion $\overline{\mathcal{F}}$), the relation remains valid with $\mathcal{F} = \sigma(\nu)$. We may also take expected values of both sides to see that ξ is *mixed i.i.d.*, in the sense that

$$P\{\xi \in \cdot\} = E\nu^\infty = \int_{\mathcal{M}_1(S)} m^\infty P\{\nu \in dm\}. \tag{3}$$

The following fundamental result shows that all the mentioned properties are essentially equivalent, as long as the sequence ξ is infinite. (For finite sequences, we shall see below that all equivalences may fail.) The result requires a regularity condition on the state space S of ξ. Recall that a measurable space $(S, \mathcal{S})$ is said to be *Borel* if there exists a measurable bijection f of S onto a Borel set of $\mathbb{R}$ such that the inverse mapping f^{-1} is again measurable. In particular (FMP A1.2),[1] any Polish space S has this property when endowed with its Borel σ-field $\mathcal{S} = \mathcal{B}(S)$.

Theorem 1.1 *(infinite exchangeable sequences, de Finetti, Ryll-Nardzewski)*
Let ξ be an infinite sequence of random elements in a measurable space S. Then these three conditions are equivalent when S is Borel:

  (i) ξ *is contractable,*

  (ii) ξ *is exchangeable,*

  (iii) ξ *is conditionally i.i.d.*

For general S, we still have (i) $\Leftrightarrow$ (ii) $\Leftarrow$ (iii).

We shall give several proofs of this result, beginning with a simple argument based on the mean ergodic theorem (FMP 10.6).

First proof: For Borel spaces S, it suffices to prove that (i) implies (iii), the implications (iii) $\Rightarrow$ (ii) $\Rightarrow$ (i) being obvious. Then let ξ be a contractable sequence in S. Put $\mathcal{I}_\xi = \xi^{-1}\mathcal{I}$, where $\mathcal{I}$ denotes the shift-invariant σ-field in $(S, \mathcal{S})^\infty$, and note that the conditional distribution $\nu = P[\xi_1 \in \cdot | \mathcal{I}_\xi]$ exists since S is Borel (FMP 6.3). Fix any set $I \in \mathcal{I}$ and some bounded, measurable functions $f_1, \ldots, f_m$ on S. Noting that $\{\xi \in I\} = \{\theta_k \xi \in I\}$ for all k,

---

[1] Throughout, FMP refers to my book *Foundations of Modern Probability*, 2nd ed.

using the contractability of ξ, and applying the mean ergodic and dominated convergence theorems, we get as $n \to \infty$

$$
\begin{aligned}
E\,1_I(\xi) \prod_{k \leq m} f_k(\xi_k) &= n^{-m} \sum_{j_1,\ldots,j_m \leq n} E\,1_I(\xi) \prod_{k \leq m} f_k(\xi_{kn+j_k}) \\
&= E\,1_I(\xi) \prod_{k \leq m} n^{-1} \sum_{j \leq n} f_k(\xi_{kn+j}) \\
&\to E\,1_I(\xi) \prod_{k \leq m} \nu f_k,
\end{aligned}
$$

and so the extreme members agree. Hence, by a monotone-class argument (FMP 1.1) and the definition of conditional probabilities,

$$
P[\xi \in B | \mathcal{I}_\xi] = \nu^\infty B \quad \text{a.s.,} \quad B \in \mathcal{S}^\infty,
$$

which proves (2) with $\mathcal{F} = \mathcal{I}_\xi$.

It remains to extend the implication (i) $\Rightarrow$ (ii) to general spaces S. Then let ξ be an infinite, contractable sequence in S. Fix any bounded, measurable functions $f_1, \ldots, f_m$ on S, and note that the random vectors

$$
\eta_j = (f_1(\xi_j), \ldots, f_m(\xi_j)), \quad j \in \mathbb{N},
$$

form a contractable sequence in $\mathbb{R}^m$. Since the latter space is Borel, it follows that (η_j) is also exchangeable. In particular, we get for any permutation $k_1, \ldots, k_m$ of $1, \ldots, m$

$$
E \prod_{j \leq m} f_k(\xi_{k_j}) = E \prod_{j \leq m} f_k(\xi_j),
$$

and (1) follows by a monotone-class argument. This shows that ξ is exchangeable. $\qquad\square$

To avoid reliance on the mean ergodic theorem, we may use instead the following elementary estimate.

Lemma 1.2 (*L^2-bound*) *Let $\xi_1, \ldots, \xi_n \in L^2$ with $E\xi_j = m$, $\mathrm{var}(\xi_j) = \sigma^2 < \infty$, and $\mathrm{cov}(\xi_i, \xi_j) = \sigma^2 \rho$ for all $i \neq j$, and fix any distributions $(p_1, \ldots, p_n)$ and $(q_1, \ldots, q_n)$ on $\{1, \ldots, n\}$. Then*

$$
E\Big(\sum_i p_i \xi_i - \sum_i q_i \xi_i\Big)^2 \leq 2\sigma^2(1 - \rho) \sup_j |p_j - q_j|.
$$

Proof: Put $c_j = p_j - q_j$, and note that $\sum_j c_j = 0$ and $\sum_j |c_j| \leq 2$. Hence,

$$
\begin{aligned}
E\Big(\sum_i c_i \xi_i\Big)^2 &= E\Big(\sum_i c_i(\xi_i - m)\Big)^2 \\
&= \sum_{i,j} c_i c_j \,\mathrm{cov}(\xi_i, \xi_j) \\
&= \sigma^2 \rho \Big(\sum_i c_i\Big)^2 + \sigma^2(1 - \rho) \sum_i c_i^2 \\
&\leq \sigma^2(1 - \rho) \sum_i |c_i| \sup_j |c_j| \\
&\leq 2\sigma^2(1 - \rho) \sup_j |c_j|. \qquad\square
\end{aligned}
$$

Second proof of Theorem 1.1: Let ξ be contractable, and fix any bounded, measurable function f on S. By Lemma 1.2 and the completeness of L^2 (FMP 1.31), there exists a random variable α_f such that

$$\left\| n^{-1} \sum_{k \le n} f(\xi_{m+k}) - \alpha_f \right\|_2^2 \lesssim n^{-1} \|f\|^2, \quad m \ge 0, \ n \ge 1, \tag{4}$$

where $\|f\| = \sup_x |f(x)|$ and $f \lesssim g$ means that $f \le cg$ for some constant $c < \infty$. Thus, for fixed $m_1, m_2, \ldots \in \mathbb{Z}_+$ we have $n^{-1} \sum_{k \le n} f(\xi_{m_n + k}) \to \alpha_f$ a.s. as $n \to \infty$ along a subsequence $N' \subset \mathbb{N}$ (FMP 4.2). In particular, α_f is a.s. $\mathcal{I}_\xi$-measurable (FMP 10.4). By the contractability of ξ and dominated convergence we get, a.s. along N' for any $I \in \mathcal{I}$,

$$E[f(\xi_1); \xi \in I] = E\left[n^{-1} \sum_{j \le n} f(\xi_j); \xi \in I \right] \to E[\alpha_f; \xi \in I],$$

which implies

$$\alpha_f = E[f(\xi_1) | \mathcal{I}_\xi] = \nu f \quad \text{a.s.}$$

The proof can now be completed as before. $\qquad\square$

We can also prove Theorem 1.1 by a simple martingale argument, based on the following useful observation. Recall that $\perp\!\!\!\perp_\eta$ denotes conditional independence given η.

Lemma 1.3 *(contraction and independence)* *Let ξ, η, and ζ be random elements such that $(\xi, \eta) \overset{d}{=} (\xi, \zeta)$ and $\sigma(\eta) \subset \sigma(\zeta)$. Then $\xi \perp\!\!\!\perp_\eta \zeta$.*

Proof: Fix any measurable set B in the range space of ξ, and define

$$\mu_1 = P[\xi \in B | \eta], \qquad \mu_2 = P[\xi \in B | \zeta].$$

Then (μ_1, μ_2) is a bounded martingale with $\mu_1 \overset{d}{=} \mu_2$, and so $E(\mu_2 - \mu_1)^2 = E\mu_2^2 - E\mu_1^2 = 0$, which implies $\mu_1 = \mu_2$ a.s. Hence, $\xi \perp\!\!\!\perp_\eta \zeta$ by FMP 6.6. $\qquad\square$

Third proof of Theorem 1.1 (Aldous): If ξ is contractable, then

$$(\xi_m, \theta_m \xi) \overset{d}{=} (\xi_k, \theta_m \xi) \overset{d}{=} (\xi_k, \theta_n \xi), \quad k \le m \le n.$$

Now write $\mathcal{T}_\xi = \bigcap_n \sigma(\theta_n \xi)$ and fix any $B \in \mathcal{S}$. Using Lemma 1.3 followed by reverse martingale convergence (FMP 7.23) as $n \to \infty$, we get a.s.

$$P[\xi_m \in B | \theta_m \xi] = P[\xi_k \in B | \theta_m \xi] = P[\xi_k \in B | \theta_n \xi] \to P[\xi_k \in B | \mathcal{T}_\xi],$$

which shows that the extreme members agree a.s. In particular,

$$P[\xi_m \in B | \theta_m \xi] = P[\xi_m \in B | \mathcal{T}_\xi] = P[\xi_1 \in B | \mathcal{T}_\xi] \quad \text{a.s.}$$

Here the first relation yields $\xi_m \perp\!\!\!\perp_{\mathcal{T}_\xi} \theta_m \xi$ for all $m \in \mathbb{N}$, and so by iteration $\xi_1, \xi_2, \ldots$ are conditionally independent given $\mathcal{T}_\xi$. The second relation shows that the conditional distributions agree a.s. This proves (2) with $\mathcal{F} = \mathcal{T}_\xi$ and $\nu = P[\xi_1 \in \cdot | \mathcal{T}_\xi]$. $\qquad \Box$

We proceed to establish some uniqueness and extremality properties associated with equations (2) and (3). Say that an exchangeable or contractable sequence ξ is *extreme* if its distribution μ cannot be expressed as a nontrivial mixture $p\mu_1 + (1-p)\mu_2$ of exchangeable or contractable distributions. The relation $\mathcal{F} = \mathcal{G}$ a.s. between two σ-fields means that the corresponding P-completions agree.

Proposition 1.4 *(uniqueness and extremality) Let ξ be an infinite, exchangeable sequence in a Borel space $(S, \mathcal{S})$, such that $P[\xi \in \cdot | \mathcal{F}] = \nu^\infty$ a.s. for some σ-field $\mathcal{F}$ and some random probability measure ν on S. Then*

(i) *ν is a.s. unique, ξ-measurable, and given by*

$$n^{-1} \sum\nolimits_{k \leq n} 1_B(\xi_k) \to \nu B \quad a.s., \quad B \in \mathcal{S};$$

(ii) *$\mathcal{F} \perp\!\!\!\perp_\nu \xi$, and $\mathcal{F} \subset \sigma(\xi)$ implies $\mathcal{F} = \sigma(\nu)$ a.s.;*
(iii) *$\mathcal{L}(\xi) = \int m^\infty \mu(dm)$ iff $\mu = \mathcal{L}(\nu)$;*
(iv) *$\mathcal{L}(\xi)$ is extreme iff ν is a.s. non-random.*

Justified by the uniqueness in (i), we shall henceforth refer to the random distribution ν in (2) as the *directing random measure* of ξ and say that ξ is *directed* by ν.

Proof: (i) Fix any measurable function $f \geq 0$ on S, and conclude from the disintegration theorem (FMP 6.4) and the law of large numbers (FMP 4.23) that

$$P\Big\{ n^{-1} \sum\nolimits_{k \leq n} f(\xi_k) \to \nu f \Big\} = E\nu^\infty \Big\{ x; \, n^{-1} \sum\nolimits_{k \leq n} f(x_k) \to \nu f \Big\} = 1.$$

This proves the asserted convergence, and the a.s. uniqueness of ν follows by FMP 1.17 since S is Borel.

(ii) This is clear by FMP 6.6 and 6.7.

(iii) Let $\tilde{\nu}$ be a random probability measure with distribution μ. By FMP 6.9 we may construct a random sequence $\eta = (\eta_j)$ in S satisfying $P[\eta \in \cdot | \tilde{\nu}] = \tilde{\nu}^\infty$ a.s. Then

$$P \circ \eta^{-1} = E\tilde{\nu}^\infty = \int m^\infty \mu(dm),$$

and comparing with (2) we see that $\nu \overset{d}{=} \tilde{\nu}$ implies $\xi \overset{d}{=} \eta$. Conversely, assuming $\xi \overset{d}{=} \eta$ and applying (i) to both ξ and η, we get $\nu f \overset{d}{=} \tilde{\nu} f$ for all measurable functions $f \geq 0$ on S, and so $\tilde{\nu} \overset{d}{=} \nu$ by FMP 12.1.

(iv) Use part (iii) and the fact that a probability measure on $\mathcal{M}_1(S)$ is extreme iff it is degenerate. $\qquad \Box$

The next result refers to the case where a random sequence $\xi = (\xi_j)$ in a measurable space S is exchangeable or contractable *over a random element η* in some space T, in the sense that the pairs $\zeta_j = (\xi_j, \eta)$ form an exchangeable or contractable sequence in $S \times T$.

Corollary 1.5 *(conditional independence) For any Borel spaces S and T, let the infinite random sequence $\xi = (\xi_j)$ in S be exchangeable over a random element η in T with directing random measure ν. Then $\xi \perp\!\!\!\perp_\nu \eta$, and the sequence of pairs $\zeta_j = (\xi_j, \eta)$ is directed by $\nu \otimes \delta_\eta$.*

Proof: The last assertion follows from Proposition 1.4 (i). In particular, $P[\xi \in \cdot | \nu, \eta] = \nu^\infty$ a.s., and so $\eta \perp\!\!\!\perp_\nu \xi$ by the same proposition. $\qquad\square$

To state the next result, we introduce in $(S, \mathcal{S})^\infty$ the tail σ-field $\mathcal{T}$, the shift-invariant σ-field $\mathcal{I}$, and the *exchangeable σ-field $\mathcal{E}$*. The latter consists of all sets $B \in \mathcal{S}^\infty$ that are invariant under every finite permutation of $\mathbb{N}$. Given a random sequence ξ in S, we denote the corresponding induced σ-fields in Ω by $\mathcal{T}_\xi = \xi^{-1}\mathcal{T}$, $\mathcal{I}_\xi = \xi^{-1}\mathcal{I}$, and $\mathcal{E}_\xi = \xi^{-1}\mathcal{E}$.

Corollary 1.6 *(equivalent σ-fields, Hewitt and Savage, Olshen) Let ξ be an infinite, exchangeable sequence in a Borel space S with directing random measure ν. Then*

$$\sigma(\nu) = \mathcal{I}_\xi = \mathcal{T}_\xi = \mathcal{E}_\xi \quad a.s.$$

In particular, all four σ-fields are trivial when ξ is i.i.d.

The last assertion, in the case of the exchangeable σ-field $\mathcal{E}_\xi$, is the celebrated *Hewitt–Savage zero-one law* (FMP 3.15).

Proof: From the first and third proofs of Theorem 1.1 we see that

$$P[\xi \in \cdot | \mathcal{I}_\xi] = P[\xi \in \cdot | \mathcal{T}_\xi] = \nu^\infty \quad a.s.,$$

and so the first two equalities follow by Proposition 1.4 (ii). Next let $B \in \mathcal{E}$ be arbitrary, and note that ξ remains exchangeable over $1_B(\xi)$. Hence, $1_B(\xi) \perp\!\!\!\perp_\nu \xi$ by Corollary 1.5, and so $\xi^{-1}B \in \sigma(\nu)$ a.s. by FMP 6.7. This shows that $\mathcal{E}_\xi \subset \sigma(\nu)$ a.s., and the converse relation is obvious since ν is a.s. invariant under finite permutations of ξ. $\qquad\square$

Let us write $\xi \circ p = (\xi_{p_1}, \xi_{p_2}, \ldots)$ for sequences $\xi = (\xi_1, \xi_2, \ldots)$ in S and $p = (p_1, p_2, \ldots)$ in $\mathbb{N}$. We say that the random sequences $\xi_i = (\xi_i^1, \xi_i^2, \ldots)$, $i \in I$, are *separately exchangeable or contractable*, if for any permutations or sub-sequences $p_1, p_2, \ldots$ of $\mathbb{N}$ we have

$$(\xi_1 \circ p_1, \xi_2 \circ p_2, \ldots) \overset{d}{=} (\xi_1, \xi_2, \ldots).$$

Corollary 1.7 *(separate exchangeability and independence, de Finetti) For any Borel space S and index set I, let $\xi_i = (\xi_i^k)$, $i \in I$, be infinite, separately exchangeable sequences in S with directing random measures ν_i. Then the elements ξ_i^k are conditionally independent with distributions ν_i, given the family $\nu = \{\nu_i; i \in I\}$.*

Proof: For convenience, we may write

$$\xi = \{\xi_i\}, \quad \xi \setminus \xi_i = \{\xi_j; j \neq i\}, \quad \xi_i \setminus \xi_i^k = \{\xi_i^h; h \neq k\}.$$

By Theorem 1.1 and Corollary 1.5, we have for any $i \in I$ and $k \in \mathbb{N}$

$$\xi_i^k \perp\!\!\!\perp_{\nu_i} (\xi_i \setminus \xi_i^k), \qquad \xi_i \perp\!\!\!\perp_{\nu_i} (\xi \setminus \xi_i), \tag{5}$$

and so by combination $\xi_i^k \perp\!\!\!\perp_{\nu_i} (\xi \setminus \xi_i^k)$. The asserted conditional independence now follows by a conditional version of FMP 3.8. Furthermore, we see from (5) and Proposition 1.4 (i) that $\xi_i \perp\!\!\!\perp_{\nu_i} \nu$, and so by FMP 6.6

$$P[\xi_i^k \in \cdot | \nu] = P[\xi_i^k \in \cdot | \nu_i] = \nu_i \quad \text{a.s.} \qquad \square$$

1.2   Finite Sequences

Though all three equivalences in Theorem 1.1 may fail for finite sequences $\xi = (\xi_1, \ldots, \xi_n)$, we may still derive a general representation formula for exchangeable distributions that resembles the conditional i.i.d. property in the infinite case. To state the result, consider any non-random sequence $s = (s_1, \ldots, s_n)$ in a measurable space $(S, \mathcal{S})$, and introduce the corresponding counting measure $\mu = \sum_{k \leq n} \delta_{s_k}$. The associated *factorial measure* $\mu^{(n)}$ on S^n is given by

$$\mu^{(n)} B = \sum_p \delta_{s \circ p} B = \sum_p 1_B(s \circ p), \quad B \in \mathcal{S}^n,$$

where the summation extends over all permutations $p = (p_1, \ldots, p_n)$ of $\{1, \ldots, n\}$. Both μ and $\mu^{(n)}$ are clearly independent of the order of the elements $s_1, \ldots, s_n$, and we note that μ is a measurable function of s whereas $\mu^{(n)}$ is a measurable function of μ. The measure $\mu^{(n)}/n!$ arises naturally as the distribution of the so-called *urn sequence* $\xi = (\xi_1, \ldots, \xi_n)$, obtained by successive drawing without replacement from the finite set $\{s_1, \ldots, s_n\}$. In particular, $\mu^{(n)}$ has the one-dimensional marginals μ.

The following counterpart of Theorem 1.1 shows that every finite exchangeable sequence is a mixture of urn sequences.

Proposition 1.8 *(finite exchangeable sequences) Let $\xi = (\xi_1, \ldots, \xi_n)$ be a finite random sequence in a measurable space S, and put $\beta = \sum_k \delta_{\xi_k}$. Then ξ is exchangeable iff $P[\xi \in \cdot | \beta] = \beta^{(n)}/n!$ a.s., in which case ξ is extreme iff β is a.s. non-random.*

Proof: Suppose that ξ is exchangeable. Since β is an invariant function of ξ, we have $(\xi \circ p, \beta) \stackrel{d}{=} (\xi, \beta)$ for any permutation p of $\{1, \ldots, n\}$. By Fubini's theorem the relation extends to any random permutation $\pi \perp\!\!\!\perp \xi$. In particular, we may choose π to be exchangeable, so that $P\{\pi = p\} = 1/n!$ for all p. Then for any measurable function $f: S^n \to \mathbb{R}_+$ and set $A \in \sigma(\beta)$, we obtain

$$E[f(\xi); A] = E[f(\xi \circ \pi); A] = E[E^\xi f(\xi \circ \pi); A] = E[\beta^{(n)} f/n!; A],$$

which shows that $P[\xi \in \cdot | \beta]$ has the stated form. The converse assertion is obvious from the definition of $\beta^{(n)}$. Given the uniqueness of β, we may prove the last assertion in the same way as parts (iii) and (iv) of Proposition 1.4. $\square$

In particular, we note that the distribution of a finite exchangeable sequence $\xi = (\xi_1, \ldots, \xi_n)$ is determined by the associated *empirical distribution* $\nu = n^{-1} \sum_k \delta_{\xi_k}$. Though a finite exchangeable sequence may not be extendible to an infinite sequence with the same property, we have the following simple approximation of finite exchangeable sequences by infinite ones. Here we write $\|\mu\|$ for the total variation of the signed measure μ.

Proposition 1.9 *(approximation)* *Let $\xi = (\xi_1, \ldots, \xi_n)$ be an exchangeable sequence in S with empirical distribution ν, and consider an infinite, exchangeable sequence η directed by ν. Then*

$$\|\mathcal{L}(\xi_1, \ldots, \xi_k) - \mathcal{L}(\eta_1, \ldots, \eta_k)\| \leq \frac{k(k-1)}{n}, \quad k \leq n.$$

Our proof relies on a comparison of sampling with or without replacement from a finite population, as made precise by the following statement. Here $U\{1, \ldots, n\}$ denotes the uniform distribution on the set $\{1, \ldots, n\}$.

Lemma 1.10 *(sampling equivalence)* *Let $\xi_1, \xi_2, \ldots$ be i.i.d. $U\{1, \ldots, n\}$, and let $\eta_1, \ldots, \eta_n$ be the numbers $1, \ldots, n$ listed in their order of first appearance in the sequence (ξ_j). Then the η_j are exchangeable, and*

$$P \bigcap_{j \leq k} \{\xi_j = \eta_j\} \geq 1 - \frac{k(k-1)}{2n}, \quad 1 \leq k \leq n. \tag{6}$$

Proof: Fix any permutation $p = (p_1, \ldots, p_n)$ of $1, \ldots, n$. Then $p_{\eta_1}, \ldots, p_{\eta_n}$ lists the numbers $1, \ldots, n$ in their order of first appearance in the sequence $p_{\xi_1}, p_{\xi_2}, \ldots$. Since the latter variables are again i.i.d. $U\{1, \ldots, n\}$, we get $(p_{\eta_1}, \ldots, p_{\eta_n}) \stackrel{d}{=} (\eta_1, \ldots, \eta_n)$, and the asserted exchangeability follows as we take the average over all permutations p. To prove (6), we note that

$$P\bigcap_{j \leq k} \{\xi_j = \eta_j\} = P\{\xi_1, \ldots, \xi_k \text{ distinct}\}$$

$$= \prod_{j \leq k} \frac{n - j + 1}{n} \geq 1 - \frac{k(k-1)}{2n},$$

where the last relation may be verified by induction. $\square$

Proof of Proposition 1.9: Consider some i.i.d. $U\{1, \ldots, n\}$ random variables $\tau_1, \tau_2, \ldots$ independent of ξ, and let $\sigma_1, \ldots, \sigma_n$ be the numbers $1, \ldots, n$ enumerated in their order of first appearance in the sequence (τ_k). Introduce the sequences $\tilde{\xi}_j = \xi_{\sigma_j}$, $1 \leq j \leq n$, and $\tilde{\eta}_j = \xi_{\tau_j}$, $j \geq 1$, and note that $\tilde{\xi} \overset{d}{=} \xi$ and $\tilde{\eta} \overset{d}{=} \eta$, the former relation by Lemma 1.10. Using the estimate of the same lemma, we get for any $k \leq n$

$$
\begin{aligned}
\|\mathcal{L}(\xi_1, \ldots, \xi_k) - \mathcal{L}(\eta_1, \ldots, \eta_k)\| &\leq 2P\bigcup\nolimits_{j \leq k} \{\tilde{\xi}_j \neq \tilde{\eta}_j\} \\
&\leq 2P\bigcup\nolimits_{j \leq k} \{\sigma_j \neq \tau_j\} \\
&\leq k(k-1)/n. \qquad \square
\end{aligned}
$$

A finite contractable sequence ξ need not be exchangeable. However, we may construct an exchangeable sequence η that shares with ξ some basic features. The result is often useful to extend estimates for exchangeable sequences or processes to a contractable setting.

Lemma 1.11 *(one-dimensional coupling) For any contractable sequence $\xi = (\xi_1, \ldots, \xi_n)$ in a measurable space S, there exists an exchangeable sequence $\eta = (\eta_1, \ldots, \eta_n)$ in S such that*

$$
\sum\nolimits_{j \leq k} \delta_{\xi_j} \overset{d}{=} \sum\nolimits_{j \leq k} \delta_{\eta_j}, \quad k = 1, \ldots, n.
$$

Proof: Let $\pi = (\pi_1, \ldots, \pi_n)$ be an exchangeable permutation of $1, \ldots, n$ independent of ξ, and define

$$
\eta = \xi \circ \pi = (\xi_{\pi_1}, \ldots, \xi_{\pi_n}).
$$

Note that η is exchangeable by Fubini's theorem. Fixing any $k \in \{1, \ldots, n\}$, we may enumerate the set $\{\pi_1, \ldots, \pi_k\}$ in increasing order as $\tau_1 < \cdots < \tau_k$. Using the contractability of ξ, the independence $(\tau_j) \perp\!\!\!\perp \xi$, and Fubini's theorem, we get

$$
\sum\nolimits_{j \leq k} \delta_{\xi_j} \overset{d}{=} \sum\nolimits_{j \leq k} \delta_{\xi \circ \tau_j} = \sum\nolimits_{j \leq k} \delta_{\xi \circ \pi_j} = \sum\nolimits_{j \leq k} \delta_{\eta_j}. \qquad \square
$$

For an application of the last result, we consider the following contractable version of a classical inequality.

Proposition 1.12 *(moment comparison, Hoeffding) Let $\xi_1, \ldots, \xi_n$ form a contractable sequence in $\mathbb{R}^d$ with empirical distribution ν, and let $\eta_1, \eta_2, \ldots$ be conditionally i.i.d. with distribution ν. Define $X_k = \sum_{j \leq k} \xi_j$ and $Y_k = \sum_{j \leq k} \eta_j$. Then for any convex function f on $\mathbb{R}^d$, we have*

$$
Ef(X_k) \leq Ef(Y_k), \quad k = 1, \ldots, n,
$$

whenever either side exists.

Proof: By Lemma 1.11 we may assume that the ξ_j are exchangeable. We may also assume that the ξ_j are a.s. distinct, since we can otherwise introduce an independent, exchangeable permutation $\pi_1, \ldots, \pi_n$ of $1, \ldots, n$ and consider the sequence of pairs $\zeta_j = (\xi_j, \pi_j)$ in $\mathbb{R}^{d+1}$. Next, we can use Proposition 1.8 to reduce to the case where ν is non-random. Finally, in view of Lemma 1.10, we can take $\xi_1, \ldots, \xi_n$ to be the first n distinct elements of the sequence $\eta_1, \eta_2, \ldots$.

Now introduce for every $k \le n$ the tail σ-field $\mathcal{T}_k = \sigma\{\xi_j; \, j > k\}$, and note that the sequences $\xi_1, \ldots, \xi_k$ and $\eta_1, \ldots, \eta_k$ are conditionally exchangeable given $\mathcal{T}_k$. In particular,

$$X_k = E[X_k | \mathcal{T}_k] = kE[\xi_1 | \mathcal{T}_k] = kE[\eta_1 | \mathcal{T}_k] = E[Y_k | \mathcal{T}_k],$$

and so, by the conditional version of Jensen's inequality in $\mathbb{R}^d$ (FMP 3.5), we get

$$Ef(X_k) = Ef(E[Y_k | \mathcal{T}_k]) \le EE[f(Y_k) | \mathcal{T}_k] = Ef(Y_k). \qquad \square$$

Though the notions of exchangeability and contractability fail to be equivalent for finite sequences, equivalence does hold under additional hypotheses. Here we say that a sequence ξ is contractable or exchangeable *over a σ-field* $\mathcal{F}$ if it is conditionally contractable or exchangeable, given any set $A \in \mathcal{F}$ with $PA > 0$.

Theorem 1.13 *(contractable and exchangeable sequences, Ivanoff and Weber, Kallenberg) Let the sequence $\xi = (\xi_1, \ldots, \xi_n)$ in $(S, \mathcal{S})$ be contractable over a σ-field $\mathcal{F}$. Then each of these conditions implies that ξ is even exchangeable over $\mathcal{F}$:*

(i) *S is a measurable, Abelian group, and $\alpha = \sum_j \xi_j$ is $\mathcal{F}$-measurable;*
(ii) *$\beta = \sum_j \delta_{\xi_j}$ is $\mathcal{F}$-measurable;*
(iii) *card $S \le 2$.*

Proof: (i) The result is clearly true for $n = 1$. Proceeding by induction, we assume that the statement holds for sequences of length $< n$, and turn to a contractable sequence $\xi = (\xi_1, \ldots, \xi_n)$ with $\mathcal{F}$-measurable sum $\alpha = \sum_j \xi_j$. Then $\theta\xi = (\xi_2, \ldots, \xi_n)$ is contractable over $\mathcal{G} = \mathcal{F} \vee \sigma(\xi_1)$ and $\sum_{j>1} \xi_j = \alpha - \xi_1$ is $\mathcal{G}$-measurable. By the induction hypothesis we conclude that $\theta\xi$ is even exchangeable over $\mathcal{G}$, which means that

$$(\mathcal{F}, \xi_1, \xi_{p_2}, \ldots, \xi_{p_n}) \stackrel{d}{=} (\mathcal{F}, \xi_1, \ldots, \xi_n), \qquad (7)$$

for any permutation $(p_2, \ldots, p_n)$ of $(2, \ldots, n)$, in the sense of the same equality with $\mathcal{F}$ replaced by 1_A for an arbitrary $A \in \mathcal{F}$.

Next we see from the contractability of ξ that

$$(\mathcal{F}, \xi_2, \ldots, \xi_n) \stackrel{d}{=} (\mathcal{F}, \xi_1, \ldots, \xi_{k-1}, \xi_{k+1}, \ldots, \xi_n), \qquad k \le n.$$

Since $\xi_k = \alpha - \sum_{j \neq k} \xi_j$ and α is $\mathcal{F}$-measurable, we obtain

$$(\mathcal{F}, \xi_1, \ldots, \xi_n) \overset{d}{=} (\mathcal{F}, \xi_k, \xi_1, \ldots, \xi_{k-1}, \xi_{k+1}, \ldots, \xi_n), \quad k \leq n. \tag{8}$$

Now fix any permutation $(k_1, \ldots, k_n)$ of $(1, \ldots, n)$, take $k = k_1$, and let U be the permutation matrix that transforms the right-hand side of (8) into $(\mathcal{F}, \xi_{k_1}, \ldots, \xi_{k_n})$. Applying U to both sides of (8), we get a relation of the form

$$(\mathcal{F}, \xi_1, \xi_{p_2}, \ldots, \xi_{p_n}) \overset{d}{=} (\mathcal{F}, \xi_{k_1}, \ldots, \xi_{k_n}),$$

for a suitable permutation $(p_2, \ldots, p_n)$ of $(2, \ldots, n)$. Combining this with (7) gives

$$(\mathcal{F}, \xi_{k_1}, \ldots, \xi_{k_n}) \overset{d}{=} (\mathcal{F}, \xi_1, \ldots, \xi_n),$$

which shows that ξ is exchangeable over $\mathcal{F}$. This completes the induction.

(ii) Arguing as in the proof of Theorem 1.1, we may reduce to the case where S is Euclidean. Then $\alpha = \sum_j \xi_j = \int x \beta(dx)$ is $\mathcal{F}$-measurable, and the result follows by part (i). We can also prove the result directly, as follows:

Let $\pi \perp\!\!\!\perp (\mathcal{F}, \xi)$ be an exchangeable permutation of $1, \ldots, n$, and put $\tilde{\xi}_k = \xi \circ \pi_k$ for $k \leq n$. Introduce the filtrations $(\mathcal{G}_k)$ and $(\tilde{\mathcal{G}}_k)$ induced by the sequences $(\mathcal{F}, \xi_1, \ldots, \xi_n)$ and $(\mathcal{F}, \tilde{\xi}_1, \ldots, \tilde{\xi}_n)$, and define $\beta_k = \sum_{j \leq k} \delta_{\xi_j}$ and $\tilde{\beta}_k = \sum_{j \leq k} \delta_{\tilde{\xi}_j}$. We claim that

$$(\mathcal{F}, \xi_1, \ldots, \xi_k) \overset{d}{=} (\mathcal{F}, \tilde{\xi}_1, \ldots, \tilde{\xi}_k), \quad 0 \leq k \leq n. \tag{9}$$

This is trivially true for $k = 0$. Proceeding by induction, we assume that (9) holds for some fixed $k < n$. Using the contractability of ξ, we get for any measurable function $g \geq 0$ on S

$$
\begin{aligned}
E[g(\xi_{k+1})|\mathcal{G}_k] &= (n-k)^{-1} E\Big[\sum_{m>k} g(\xi_m)\,\Big|\,\mathcal{G}_k\Big] \\
&= (n-k)^{-1}(\beta - \beta_k)g.
\end{aligned}
$$

Since $\tilde{\xi}$ is exchangeable over $\mathcal{F}$ and based on the same $\mathcal{F}$-measurable random measure β, a similar relation holds for $\tilde{\xi}_{k+1}$, $\tilde{\mathcal{G}}_k$, and $\tilde{\beta}_k$. Using the induction hypothesis, we get for any measurable function $f \geq 0$ on S^k

$$
\begin{aligned}
E[f(\xi_1, \ldots, \xi_k)g(\xi_{k+1})|\mathcal{F}] &= E[f(\xi_1, \ldots, \xi_k)E[g(\xi_{k+1})|\mathcal{G}_k]|\mathcal{F}] \\
&= (n-k)^{-1} E[f(\xi_1, \ldots, \xi_k)(\beta - \beta_k)g|\mathcal{F}] \\
&= (n-k)^{-1} E[f(\tilde{\xi}_1, \ldots, \tilde{\xi}_k)(\beta - \tilde{\beta}_k)g|\mathcal{F}] \\
&= E[f(\tilde{\xi}_1, \ldots, \tilde{\xi}_k)E[g(\tilde{\xi}_{k+1})|\tilde{\mathcal{G}}_k]|\mathcal{F}] \\
&= E[f(\tilde{\xi}_1, \ldots, \tilde{\xi}_k)g(\tilde{\xi}_{k+1})|\mathcal{F}],
\end{aligned}
$$

which extends to (9) with k replaced by $k+1$. This completes the induction, and the assertion follows for $k = n$.

(iii) By conditioning we may assume that $\mathcal{F}$ is trivial. Let $S = \{a, b\}$. For $n = 2$, the contractability of ξ yields

$$
\begin{aligned}
P\{\xi_1 = a, \xi_2 = b\} &= P\{\xi_1 = a\} - P\{\xi_1 = \xi_2 = a\} \\
&= P\{\xi_2 = a\} - P\{\xi_1 = \xi_2 = a\} \\
&= P\{\xi_2 = a, \xi_1 = b\},
\end{aligned}
$$

which implies $(\xi_1, \xi_2) \stackrel{d}{=} (\xi_2, \xi_1)$. When $n > 2$, fix any $k \in \{1, \ldots, n-1\}$, and note that the pair (ξ_k, ξ_{k+1}) is conditionally contractable, given the remaining elements $\xi_1, \ldots, \xi_{k-1}$ and $\xi_{k+2}, \ldots, \xi_n$. By the result for $n = 2$, the pair (ξ_k, ξ_{k+1}) is even conditionally exchangeable, and therefore

$$
\xi \stackrel{d}{=} (\xi_1, \ldots, \xi_{k-1}, \xi_{k+1}, \xi_k, \xi_{k+2}, \ldots, \xi_n).
$$

Since every permutation of ξ is a product of such transpositions, the asserted exchangeability follows. □

Since an exchangeable sequence is also contractable, extremality can be understood in the sense of either property. We show that the two notions are equivalent.

Corollary 1.14 (*extremality*) *For any finite, exchangeable sequence $\xi = (\xi_1, \ldots, \xi_n)$ in a space S, these two conditions are equivalent:*

(i) *ξ is extreme in the exchangeable sense,*

(ii) *ξ is extreme in the contractable sense.*

Proof: First assume (ii), and let $\mu = \mathcal{L}(\xi) = p\mu_1 + (1 - p)\mu_2$ for some exchangeable distributions μ_1, μ_2 and some $p \in (0, 1)$. Since the μ_i are also contractable, condition (ii) yields $\mu_1 = \mu_2$, which proves (i).

Next assume (i), and let $\mu = p\mu_1 + (1 - p)\mu_2$ for some contractable distributions μ_1, μ_2 and some $p \in (0, 1)$. By Proposition 1.8 we note that μ is restricted to permutations of some fixed sequence $a_1, \ldots, a_n \in S$, and so the same thing is true for μ_1 and μ_2. The latter are then exchangeable by Theorem 1.13 (ii), and so $\mu_1 = \mu_2$ by condition (i). This proves (ii). □

1.3 Continuous-Time Symmetries

In continuous time we may identify three natural symmetry properties, each of which is described in terms of some simple transformations of $[0, 1]$ or $\mathbb{R}_+$. Assuming $0 \le a \le b$, we introduce the *reflections*

$$
R_a(t) = \begin{cases} a - t, & t \le a, \\ t, & t > a, \end{cases}
$$

contractions

$$C_{a,b}(t) = \begin{cases} t, & t \leq a, \\ \infty, & t \in (a, b], \\ t - b + a, & t > b, \end{cases}$$

and *transpositions*

$$T_{a,b}(t) = \begin{cases} t + b - a, & t \leq a, \\ t - a, & t \in (a, b], \\ t, & t > b. \end{cases}$$

An $\mathbb{R}^d$-valued process X on $I = [0, 1]$, $\mathbb{R}_+$, $\mathbb{Q}_{[0,1]}$, or $\mathbb{Q}_+ = \mathbb{Q} \cap [0, \infty)$ with $X_0 = 0$ is said to be *reflectable* if $X \circ R_a^{-1} \overset{d}{=} X$ for all $a \in I$, *contractable* if $X \circ C_{a,b}^{-1} \overset{d}{=} X$ for all $a < b$ in I, and *exchangeable* if $X \circ T_{a,b}^{-1} \overset{d}{=} X$ for all $a \leq b$ in I, where the second relation is understood to hold on $[0, 1 - b + a]$ when $I = [0, 1]$. Here we are using the notation

$$(X \circ f^{-1})_t = \int_I 1\{s \in I; \, f(s) \leq t\} \, dX_s, \quad t \in I,$$

where the integral over a finite union of disjoint intervals is defined by

$$\int_U dX_s = \sum_j (X_{b_j} - X_{a_j}), \quad U = \bigcup_j (a_j, b_j],$$

and similarly for intervals of types $[a_j, b_j]$, $[a_j, b_j)$, or (a_j, b_j). In particular, we note that

$$(X \circ R_a^{-1})_t = X_{t \vee a} - X_{(a-t)_+}, \tag{10}$$
$$(X \circ C_{a,b}^{-1})_t = X_{t \wedge a} + X_{b + (t-a)_+} - X_b. \tag{11}$$

It is sometimes more convenient to write

$$X \circ R_a^{-1} = R_a X, \qquad X \circ C_{a,b}^{-1} = C_{a,b} X, \qquad X \circ T_{a,b}^{-1} = T_{a,b} X.$$

The definitions of reflectable, contractable, and exchangeable random measures ξ on a product space $I \times S$ are similar, except that the mappings $\xi \circ R_a^{-1}$, $\xi \circ C_{a,b}^{-1}$, and $\xi \circ T_{a,b}^{-1}$ should now be understood in the sense of measure theory, with the added convention $\xi(\{0\} \times S) = 0$ a.s. Note that the definitions for processes and measures are consistent when $X_t = \xi(0, t]$ for some random measure ξ on $[0, 1]$ or $\mathbb{R}_+$ with $\xi\{s\} = 0$ a.s. for all s.

For random processes X or measures ξ on I, we may also consider some elementary notions of exchangeability or contractability, defined in terms of the increments. Then define in the two cases

$$\xi_{nj} = X_{j/n} - X_{(j-1)/n},$$
$$\xi_{nj} = \xi((j-1)/n, j/n], \quad n \in \mathbb{N}, \, j \leq n \text{ or } j \in \mathbb{N},$$

respectively, and say that X or ξ has *exchangeable* or *contractable increments* if the sequence $\xi_{n1}, \xi_{n2}, \ldots$ is exchangeable or contractable for every $n \in \mathbb{N}$.

The following result exhibits some relations between the various notions of symmetry and invariance.

Theorem 1.15 *(equivalent symmetries) Let X be an $\mathbb{R}^d$-valued process on $I = \mathbb{R}_+$ or $\mathbb{Q}_+$ with $X_0 = 0$. Then these three conditions are equivalent:*

(i) *X is contractable,*

(ii) *X is exchangeable,*

(iii) *X is reflectable.*

If instead $I = [0,1]$ or $\mathbb{Q}_{[0,1]}$, then (i) $\Leftarrow$ (ii) $\Leftrightarrow$ (iii), with equivalence throughout when X_1 is non-random. If $I = \mathbb{Q}_+$ or $\mathbb{Q}_{[0,1]}$, or if $I = \mathbb{R}_+$ or $[0,1]$ and X is right-continuous, then (i) and (ii) are equivalent to the statements

(i′) *X has contractable increments,*

(ii′) *X has exchangeable increments.*

All assertions remain true for random measures on $(0,\infty) \times S$ or $(0,1] \times S$, where S is an arbitrary Borel space.

Proof: For $a \leq b$ it is easy to verify the relation

$$R_{b-a} \circ R_b \circ R_a = T_{a,b} \quad \text{on } (0,\infty),$$

which shows that (iii) implies (ii). Letting $a \leq b \leq c$, we next define

$$T_{a,b,c}(t) = \begin{cases} t + c - b, & t \in (a,b], \\ t - b + a, & t \in (b,c], \\ t, & t \notin (a,c], \end{cases}$$

and note that

$$T_{a,b,c} = T_{c-a,c} \circ T_{b-a,c-a} \circ T_{a,c}. \tag{12}$$

Since every permutation is a product of transpositions, we see in particular that (ii) implies (ii′). Next we may check that, for any $a \leq b \leq c$,

$$\{s \geq 0;\, T_{a,b,c}(s) \leq t\} = \{s \geq 0;\, C_{a,b}(s) \leq t\}, \quad t \in [0, c - b + a].$$

Taking $c = 1$ and using (12), we conclude that (ii) implies (i) for processes on $[0,1]$ or $\mathbb{Q}_{[0,1]}$. Letting $c \to \infty$, we get the same implication for processes on $\mathbb{R}_+$ or $\mathbb{Q}_+$. We also note that, trivially, (i) implies (i′).

Let us now write $X_I = X_t - X_s$ for $I = (s,t]$, $[s,t]$, $[s,t)$, or (s,t), and note that by (10)

$$(R_a X)_I = \begin{cases} X_{a-I}, & I \subset [0,a], \\ X_I, & I \subset [a,\infty). \end{cases} \tag{13}$$

Assuming (ii), we may iterate (12) to obtain

$$(X_{I_1}, \ldots, X_{I_n}) \overset{d}{=} (X_{J_1}, \ldots, X_{J_n}), \tag{14}$$

for any sets of disjoint intervals $I_1, \ldots, I_n$ and $J_1, \ldots, J_n$ with lengths $|I_k| = |J_k|$ for all k. Fixing any times $0 = t_0 < t_1 < \cdots < t_n$, where $a = t_m$ for some

$m \leq n$, we may apply (14) with $I_k = t_k - t_{k-1}$ for all $k \leq n$, $J_k = a - I_k$ for $k \leq m$, and $J_k = I_k$ for $k > m$, and conclude from (13) that

$$((R_a X)_{I_1}, \ldots, (R_a X)_{I_n}) \stackrel{d}{=} (X_{I_1}, \ldots, X_{I_n}).$$

Hence,

$$((R_a X)_{t_1}, \ldots, (R_a X)_{t_n}) \stackrel{d}{=} (X_{t_1}, \ldots, X_{t_n}),$$

and since $t_1, \ldots, t_n$ were arbitrary, we obtain $R_a X \stackrel{d}{=} X$, which shows that (ii) implies (iii).

If $I = \mathbb{Q}_+$ or $\mathbb{Q}_{[0,1]}$, then clearly (ii') implies

$$((T_{a,b} X)_{t_1}, \ldots, (T_{a,b} X)_{t_n}) \stackrel{d}{=} (X_{t_1}, \ldots, X_{t_n})$$

for any $a < b$ and $t_1, \ldots, t_n$ in I. Hence, $T_{a,b} X \stackrel{d}{=} X$, which shows that (ii') implies (ii). The same argument shows that (i') implies (i) in this case, except that $t_1, \ldots, t_n$ should be restricted to $[0, 1 - b + a]$ when $I = \mathbb{Q}_{[0,1]}$. If instead $I = \mathbb{R}_+$ or $[0, 1]$, we may argue that (ii') implies (14) for any sets of disjoint intervals $I_1, \ldots, I_n$ and $J_1, \ldots, J_n$ with rational endpoints such that $|I_k| = |J_k|$ for all k. Using the right continuity of X, we obtain the same formula without restrictions on the endpoints, which is clearly equivalent to (ii). Thus, the implication (ii') $\Rightarrow$ (ii) remains valid in this case. A similar argument shows that (i') $\Rightarrow$ (i).

To see that (i) implies (ii) when $I = \mathbb{R}_+$ or $\mathbb{Q}_+$, define $\mathcal{T} = \bigcap_t \sigma(\theta_t X - X_t)$. Using (i) and proceeding as in the third proof of Theorem 1.1, we get for any $s < t$ in I

$$P[X_t - X_s \in \cdot | \theta_t X - X_t] = P[X_t - X_s \in \cdot | \mathcal{T}] = P[X_{t-s} \in \cdot | \mathcal{T}]. \qquad (15)$$

Here the first relation yields

$$(X_t - X_s) \perp\!\!\!\perp_{\mathcal{T}} (\theta_t X - X_t), \quad s < t \text{ in } I,$$

which shows that the increments of X are conditionally independent given $\mathcal{T}$ (FMP 3.8, 6.6). Next we see from the second relation in (15) that the conditional distribution of $X_t - X_s$ depends only on $t - s$. Hence, (14) holds for any intervals $I_1, \ldots, I_n$ and $J_1, \ldots, J_n$ as before, and (ii) follows.

To see that (i) implies (ii) when $I = [0, 1]$ or $\mathbb{Q}_{[0,1]}$ and X_1 is non-random, we consider the more general case where X is contractable over a σ-field $\mathcal{F}$ and X_1 is $\mathcal{F}$-measurable. We need to prove that $T_{a,b} X \stackrel{d}{=} X$ for any $a \leq b$ in $[0, 1]$. It is then enough to consider an arbitrary partition of $[0, b - a]$ into sub-intervals $I_1, \ldots, I_n$ and to show that

$$(X_{I_1 + a}, \ldots, X_{I_n + a}, C_{a,b} X) \stackrel{d}{=} (X_{I_1}, \ldots, X_{I_n}, C_{0, b-a} X). \qquad (16)$$

This is trivially true for $n = 0$, since in that case $a = b$.

Now assume (16) to be true for partitions into $n - 1$ intervals I_k, and turn to the case of n intervals $I_1, \ldots, I_n$. Writing $I_1 = [0, c]$, we have

$(\mathcal{F}, C_{a,a+c}X) \stackrel{d}{=} (\mathcal{F}, C_{0,c}X)$ by the assumed contractability, and since X_1 is $\mathcal{F}$-measurable it follows that

$$(X_{I_1+a}, C_{a,a+c}X) \stackrel{d}{=} (X_{I_1}, C_{0,c}X). \tag{17}$$

Next we note that $C_{0,c}X$ is contractable over $\mathcal{G} = \mathcal{F} \vee \sigma(X_c)$ with the $\mathcal{G}$-measurable final value $X_1 - X_c$. Invoking the induction hypothesis, we may apply (16) to the process $C_{0,c}X$, transformation $C_{a,b-c}$, and partition $I_2 - c$, $\ldots, I_n - c$ of $[0, b - a - c]$, and then use (17) to obtain

$$
\begin{aligned}
(X_{I_1}, &\ldots, X_{I_n}, C_{0,b-a}X) \\
&= (X_{I_1}, (C_{0,c}X)_{I_2-c}, \ldots, (C_{0,c}X)_{I_n-c}, C_{0,b-a-c} \circ C_{0,c}X) \\
&\stackrel{d}{=} (X_{I_1}, (C_{0,c}X)_{I_2-c+a}, \ldots, (C_{0,c}X)_{I_n-c+a}, C_{a,b-c} \circ C_{0,c}X) \\
&\stackrel{d}{=} (X_{I_1+a}, (C_{a,a+c}X)_{I_2-c+a}, \ldots, (C_{a,a+c}X)_{I_n-c+a}, C_{a,b-c} \circ C_{a,a+c}X) \\
&= (X_{I_1+a}, \ldots, I_{I_n+a}, C_{a,b}X).
\end{aligned}
$$

This completes the induction, and the assertion follows.

To prove the last assertion, we may apply the previous arguments to the measure-valued process $X_t = \xi((0, t] \times \cdot)$. The only difficulty is with the proof of the implication (ii) $\Rightarrow$ (iii), since (13) may fail when X is discontinuous at an endpoint of I. Now this does not affect the subsequent argument, since X has no fixed discontinuities. In fact, if $P\{\xi(\{t\} \times S) \neq 0\} > 0$ for some $t = t_0 > 0$, then by (ii) the same relation holds at every $t > 0$, which is impossible since ξ has at most countably many fixed discontinuities. $\qquad \square$

We may use the last result to prove an interesting and useful closure property for exchangeable and contractable processes. The result for Lévy processes and subordinators is of course classical.

Theorem 1.16 (*composition*) *Let X and Y be independent processes on $[0, 1]$ or $\mathbb{R}_+$ with $X_0 = Y_0 = 0$, where X is $\mathbb{R}^d$-valued and measurable, and Y is non-decreasing with values in the domain of X. Then*

(i) *if X and Y are exchangeable, so is $X \circ Y$;*

(ii) *if X and Y are contractable, so is $X \circ Y$.*

Proof: (i) Applying (10) to the processes X, Y, and $X \circ Y$, we get for any t in the domain I of Y

$$
\begin{aligned}
(R_{Y_t}X) \circ (R_tY)_s &= (R_{Y_t}X) \circ (Y_{s\vee t} - Y_{(t-s)_+}) \\
&= X \circ \left(Y_t \vee (Y_{s\vee t} - Y_{(t-s)_+}) \right) \\
&\quad - X \circ (Y_t - Y_{s\vee t} + Y_{(t-s)_+})_+ \\
&= X \circ Y_{s\vee t} - X \circ Y_{(t-s)_+} = (R_t(X \circ Y))_s,
\end{aligned}
$$

which shows that

$$R_t(X \circ Y) = (R_{Y_t}X) \circ (R_tY), \quad t \in I. \tag{18}$$

By the exchangeability and independence of X and Y, we have for any $t \in I$

$$P[R_{Y_t} X \in \cdot | Y] = P[X \in \cdot | Y],$$
$$P[R_t Y \in \cdot | X] = P[Y \in \cdot | X],$$

and so by (18)

$$R_t(X \circ Y) = (R_{Y_t} X) \circ (R_t Y) \stackrel{d}{=} X \circ (R_t Y) \stackrel{d}{=} X \circ Y.$$

By Theorem 1.15 it follows that $X \circ Y$ is exchangeable.

(ii) Here we may apply (11) to the processes X, Y, and $X \circ Y$ to get for any $a \le b$ in I

$$
\begin{aligned}
(C_{Y_a, Y_b} X) \circ (C_{a,b} Y)_s &= (C_{Y_a, Y_b} X) \circ (Y_{s \wedge a} + Y_{b+(s-a)_+} - Y_b) \\
&= X \circ \left((Y_{s \wedge a} + Y_{b+(s-a)_+} - Y_b) \wedge Y_a \right) - X \circ Y_b \\
&\quad + X \circ \left(Y_b + (Y_{s \wedge a} + Y_{b+(s-a)_+} - Y_b - Y_a)_+ \right) \\
&= (X \circ Y)_{s \wedge a} + (X \circ Y)_{b+(s-a)_+} - X \circ Y_b \\
&= (C_{a,b}(X \circ Y))_s,
\end{aligned}
$$

which shows that

$$C_{a,b}(X \circ Y) = (C_{Y_a, Y_b} X) \circ (C_{a,b} Y), \quad a \le b. \tag{19}$$

By the contractability and independence of X and Y, we have for any $a \le b$

$$P[C_{Y_a, Y_b} X \in \cdot | Y] = P[X \in \cdot | Y],$$
$$P[C_{a,b} Y \in \cdot | X] = P[Y \in \cdot | X],$$

and so by (19)

$$C_{a,b}(X \circ Y) = (C_{Y_a, Y_b} X) \circ (C_{a,b} Y) \stackrel{d}{=} X \circ (C_{a,b} Y) \stackrel{d}{=} X \circ Y,$$

which shows that $X \circ Y$ is contractable. □

We turn to another kind of operation that preserves the properties of contractability or exchangeability. Here the relation $A < B$ between two subsets $A, B \subset \mathbb{R}$ means that $a < b$ for any $a \in A$ and $b \in B$. For processes X on an interval I and for any sub-interval $J = (a, b] \subset I$, we define the *restriction* of X to J by $X^J = X^{a,b} = \theta_a X^b - X_a$, or

$$X_t^J = X_t^{a,b} = X_{(a+t) \wedge b} - X_a, \quad t \in I.$$

Theorem 1.17 *(interval sampling) Let X be an $\mathbb{R}^d$-valued, measurable process on $[0, 1]$ or $\mathbb{R}_+$, and consider an independent sequence of disjoint, random sub-intervals $I_1, \ldots, I_n$. Then*

(i) *for $I_1 < \cdots < I_n$, if X and $(\lambda I_1, \ldots, \lambda I_n)$ are contractable, so is $(X^{I_1}, \ldots, X^{I_n})$;*

(ii) *if X and $(\lambda I_1, \ldots, \lambda I_n)$ are exchangeable, so is $(X^{I_1}, \ldots, X^{I_n})$.*

Proof: (i) By the contractability and independence of the I_k it suffices to prove that, whenever the intervals $I_1 < \cdots < I_n$ and $J_1 < \cdots < J_n$ are independent of X with

$$(\lambda I_1, \ldots, \lambda I_n) \stackrel{d}{=} (\lambda J_1, \ldots, \lambda J_n),$$

we have

$$(X^{I_1}, \ldots, X^{I_n}) \stackrel{d}{=} (X^{J_1}, \ldots, X^{J_n}).$$

By the independence and contractability of X, we may then assume that $\bigcup_k I_k$ and $\bigcup_k J_k$ are intervals starting at 0. But then the assertion is an immediate consequence of Fubini's theorem.

(ii) Here it is clearly enough to show that the distribution of $X^{I_1}, \ldots, X^{I_n}$ is invariant under transpositions of adjacent elements. By conditioning we may then reduce to the case $n = 2$, and so it is enough to consider only two disjoint intervals I and J. Since X is measurable, we may next invoke Fubini's theorem to reduce to the case where the lengths λI and λJ take only two values, a and b. By contractability we may finally assume that $I \cup J = (0, a + b]$. Noting that

$$(X^{0,a}, X^{a,a+b}) \stackrel{d}{=} (X^{b,b+a}, X^{0,b}),$$
$$P\{\lambda I = a\} = P\{\lambda I = b\} = \tfrac{1}{2},$$

by the exchangeability of X and $(\lambda I, \lambda J)$, we get

$$
\begin{aligned}
\mathcal{L}(X^I, X^J) &= P\{I = (0, a]\}\, \mathcal{L}(X^{0,a}, X^{a,a+b}) \\
&\quad + P\{I = (0, b]\}\, \mathcal{L}(X^{0,b}, X^{b,b+a}) \\
&\quad + P\{I = (a, a + b]\}\, \mathcal{L}(X^{a,a+b}, X^{0,a}) \\
&\quad + P\{I = (b, b + a]\}\, \mathcal{L}(X^{b,b+a}, X^{0,b}) \\
&= \tfrac{1}{2} \mathcal{L}(X^{0,a}, X^{a,a+b}) + \tfrac{1}{2} \mathcal{L}(X^{0,b}, X^{b,b+a}) \\
&= P\{I = (0, a]\}\, \mathcal{L}(X^{a,a+b}, X^{0,a}) \\
&\quad + P\{I = (0, b]\}\, \mathcal{L}(X^{b,b+a}, X^{0,b}) \\
&\quad + P\{I = (a, a + b]\}\, \mathcal{L}(X^{0,a}, X^{a,a+b}) \\
&\quad + P\{I = (b, b + a]\}\, \mathcal{L}(X^{0,b}, X^{b,b+a}) \\
&= \mathcal{L}(X^J, X^I),
\end{aligned}
$$

which shows that indeed $(X^I, X^J) \stackrel{d}{=} (X^J, X^I)$. $\square$

The next result shows how the conditions defining the exchangeability or contractability of a random measure ξ on a product space $S \times I$ can be extended in a natural way to more general sets or functions. Here $I = [0, 1]$ or $\mathbb{R}_+$, and for any $B \in \mathcal{B}(I)$ we define the general *contraction* $C_B\xi$ by

$$(C_B\xi)f = \int\int_{S \times B} f(s, \lambda(B \cap [0, t])) \, \xi(ds \, dt), \quad f \geq 0.$$

Theorem 1.18 *(extended invariance) Let ξ be a random measure on $S \times I$, where S is Borel and $I = [0, 1]$ or $\mathbb{R}_+$. Then*

(i) *ξ is contractable iff $C_B\xi \overset{d}{=} \xi$ on $[0, \lambda B]$ for every $B \in \mathcal{B}(I)$;*

(ii) *ξ is exchangeable iff $\xi \circ f^{-1} \overset{d}{=} \xi$ for every λ-preserving map f on I.*

Proof: (i) Let ξ contractable on $S \times I$. For finite interval unions B, the asserted property then holds by Theorem 1.15. To prove the general result, we may assume that $\xi(S \times \cdot)$ is a.s. locally finite and $\lambda B < \infty$. By a monotone-class argument it suffices to show that, for any $t_1 < \cdots < t_m$,

$$((C_B\xi)_{t_1}, \ldots, (C_B\xi)_{t_m}) \overset{d}{=} (\xi_{t_1}, \ldots, \xi_{t_m}), \tag{20}$$

where $\xi_t = \xi(\cdot \times [0, t])$ and $(C_B\xi)_t = (C_B\xi)(\cdot \times [0, t])$. By the regularity of λ, we may then choose some finite interval unions $U_1, U_2, \ldots$ such that $\lambda(B \Delta U_n) \to 0$. Since (20) holds for each U_n, it is enough to show that $\|(C_{U_n}\xi)_t - (C_B\xi)_t\| \overset{P}{\to} 0$ as $n \to \infty$ for fixed $t \geq 0$. This can be written as $\|\xi(\cdot \times U_n^t) - \xi(\cdot \times B^t)\| \overset{P}{\to} 0$, where we define

$$B^t = \{s \in B; \, \lambda(B \cap [0, s]) \leq t\},$$

and similarly for U_n^t in terms of U_n. Since clearly $\lambda(B^t \Delta U_n^t) \to 0$, we need only show that $\lambda A_n \to 0$ implies $\xi(\cdot \times A_n) \overset{P}{\to} 0$. By the regularity of λ we may then assume that the sets A_n are open, and by the continuity of ξ we may take them to be finite interval unions. Since ξ is contractable, we may finally choose $A_n = (0, a_n]$ for some constants $a_n \to 0$, in which case the stated convergence holds a.s. by the continuity of ξ.

(ii) Suppose that ξ is exchangeable on $S \times I$ and that f is λ-preserving on I. For any disjoint intervals $I_1, \ldots, I_m \subset I$ with inverse images $B_k = f^{-1}I_k$, $k \leq m$, we need to show that

$$(\xi_{B_1}, \ldots, \xi_{B_m}) \overset{d}{=} (\xi_{I_1}, \ldots, \xi_{I_m}), \tag{21}$$

where $\xi_B = \xi(\cdot \times B)$. As before, we may then choose some finite interval unions $U_1^n, \ldots, U_m^n$, $n \in \mathbb{N}$, such that $\lambda(U_k^n \Delta B_k) \to 0$ for each k. Since the B_k are disjoint, we may assume that $U_1^n, \ldots, U_m^n$ have the same property for each n, and since $\lambda B_k = \lambda I_k$ for all k, we may also assume that $\lambda U_k^n \leq \lambda I_k$ for any n and k. Then there exist some intervals $I_k^n \subset I_k$ with $\lambda I_k^n = \lambda U_k^n$, and we note that

$$\lambda(I_k^n \Delta I_k) = \lambda I_k - \lambda I_k^n = \lambda B_k - \lambda U_k^n \leq \lambda(U_k^n \Delta B_k) \to 0.$$

As before, we conclude that $\xi_{U_k^n} \xrightarrow{P} \xi_{B_k}$ and $\xi_{I_k^n} \xrightarrow{P} \xi_{I_k}$ for all k. Furthermore, the exchangeability of ξ yields

$$(\xi_{U_1^n}, \dots, \xi_{U_m^n}) \stackrel{d}{=} (\xi_{I_1^n}, \dots, \xi_{I_m^n}).$$

Relation (21) now follows as we let $n \to \infty$. $\qquad\qquad\square$

Part (ii) of the last result can also be obtained as an easy consequence of the representations in Proposition 1.21 and Theorem 1.25 below. Much more general results of this type will be established in Chapter 4.

1.4 Infinite-Interval Processes

In analogy with Theorem 1.1, we may expect a contractable or exchangeable process X on $\mathbb{R}_+$ to have conditionally stationary and independent (or i.i.d.) increments. Just as in the unconditional case, there will then exist a convolution semigroup of *random* probability measures μ_r on the range space $\mathbb{R}^d$, such that the increments $X_t - X_s$ are conditionally independent with distributions μ_{t-s}, given the whole family $\{\mu_r\}$.

To make all this precise, recall that for any non-random convolution semigroup $\{\mu_r\}$ on $\mathbb{R}^d$, there exists an $\mathbb{R}^d$-valued *Lévy process* X (i.e., an rcll process with stationary, independent increments and $X_0 = 0$) such that $\mathcal{L}(X_t - X_s) = \mu_{t-s}$ for all $s < t$ (FMP 15.12). Furthermore, the distribution of X determines and is determined by the *characteristics* (α, ρ, ν), where $\alpha \in \mathbb{R}^d$ is the drift coefficient of X, $\rho = (\rho_{ij})$ denotes the covariance matrix of the diffusion component, and ν is the *Lévy measure* on $\mathbb{R}^d \setminus \{0\}$ governing the distribution of jumps (FMP 15.7). More precisely, X has the *Lévy–Itô representation*

$$X_t = \alpha t + \sigma B_t + \int_0^t \int x(\eta - 1\{|x| \le 1\}\lambda \otimes \nu)(ds\,dx), \quad t \ge 0, \qquad (22)$$

where η is a Poisson process on $\mathbb{R}_+ \times (\mathbb{R}^d \setminus \{0\})$ with intensity measure $\lambda \otimes \nu$, B is an independent Brownian motion in $\mathbb{R}^d$, and σ is a $d \times d$-matrix such that $\sigma\sigma' = \rho$ (FMP 15.4). The integrability condition $\int (|x|^2 \wedge 1)\nu(dx) < \infty$ ensures the existence of the compensated Poisson integral in (22) (FMP 12.13). Indeed, the convergence is a.s. uniform on bounded intervals, as may be seen from Lemma 2.19 below.

Theorem 1.19 *(exchangeable processes on* $\mathbb{R}_+$*, Bühlmann) Let X be an $\mathbb{R}^d$-valued process on $\mathbb{Q}_+$ with $X_0 = 0$. Then these conditions are equivalent:*

(i) *X is contractable,*

(ii) *X is exchangeable,*

(iii) *X has conditionally i.i.d. increments.*

In that case, X extends a.s. to an rcll process $\tilde{X}$ on $\mathbb{R}_+$, where $\tilde{X}$ is conditionally Lévy with some random characteristics α, ρ, and ν. The latter are then a.s. unique, and X is extreme iff (α, ρ, ν) is a.s. non-random.

An exchangeable process X as above is said to be *directed* by the random elements α, ρ, and ν, and the latter are often referred to, collectively, as the *directing triple* of X.

First proof: For the first assertion, it is enough to show that (i) implies (iii), the remaining implications being obvious. Thus, assume that X is contractable. In the proof of Theorem 1.15 we saw that X has conditionally stationary and independent increments, given a suitable σ-field $\mathcal{T}$. Hence, the conditional distributions $\mu_t = P[X_t \in \cdot | \mathcal{T}]$, $t \in \mathbb{Q}_+$, a.s. satisfy the semigroup property $\mu_s * \mu_t = \mu_{s+t}$, $s, t \in \mathbb{Q}_+$. We may then introduce the associated random characteristics α, ρ, and ν, and from FMP 15.16 we know that the latter are a.s. unique and measurably determined by the family $\{\mu_t\}$.

Now let η be a Cox process directed by ν with $\eta \perp\!\!\!\perp_\nu (\alpha, \rho)$ (cf. FMP 12.7 for existence), and introduce an independent Brownian motion B in $\mathbb{R}^d$. Writing σ for the square root of the nonnegative definite matrix ρ, we may use (22) to construct a *mixed Lévy process* Y. Then clearly $X \overset{d}{=} Y$ on $\mathbb{Q}_+$. Since Y is a.s. rcll and hence may be regarded as a random element in the Borel space $D(\mathbb{R}_+, \mathbb{R}^d)$, the transfer theorem (FMP 6.10) ensures the existence of an rcll process $\tilde{X} \overset{d}{=} Y$ satisfying $X = \tilde{X}$ a.s. on $\mathbb{Q}_+$. Finally, we see from (22) and Fubini's theorem that $\tilde{X}$ is conditionally Lévy with characteristics (α, ρ, ν). The last assertion follows as before from the uniqueness of (α, ρ, ν).   □

It is again illuminating to consider alternative proofs. Here we show how the result can also be deduced as a corollary of Theorem 1.1.

Second proof of Theorem 1.19: Suppose that X is contractable. Introduce the processes

$$Y_n^k(t) = X(t + (k-1)/n) - X((k-1)/n), \quad t \in [0, n^{-1}], \quad k, n \in \mathbb{N},$$

and note that the sequence $Y_n^1, Y_n^2, \dots$ is again contractable for each n. By Theorem 1.1 it is then conditionally i.i.d. with a common distribution ν_n. Since the ν_n are a.s. measurably determined by X and hence by every sequence Y_n^k, $k \in \mathbb{N}$, we may apply Proposition 1.4 (ii) twice to obtain

$$\sigma(\nu_m) = \sigma(\nu_{mn}) = \sigma(\nu_n) \text{ a.s.}, \quad m, n \in \mathbb{N},$$

which shows that the σ-field $\mathcal{I} = \sigma(\nu_n)$ is a.s. independent of n. In particular, X has conditionally stationary, independent increments, given $\mathcal{I}$. The proof can now be completed as before. □

The last result leads immediately to a characterization of contractable or exchangeable random measures on $\mathbb{R}_+$. We may consider the more general case of random measures on a product space $S \times \mathbb{R}_+$, where S is an arbitrary Borel space and contractability is understood in the sense of transformations of $\mathbb{R}_+$. By a *marked* point process on $S \times \mathbb{R}_+$ we mean a simple point process ξ such that $\xi(S \times \{t\}) \leq 1$ for all $t \geq 0$.

Lemma 1.20 *(exchangeable point processes) Let ξ be a marked point process on $S \times \mathbb{R}_+$, where S is Borel. Then ξ is contractable iff it is Cox and directed by $\nu \otimes \lambda$ for some random measure ν on S. In that case, ν is a.s. unique, and ξ is extreme iff ν is a.s. non-random.*

Proof: Suppose that ξ is contractable. Arguing as in the proof of the last theorem, we see that ξ has stationary, independent increments over $\mathbb{Q}_+$, given the tail σ-field $\mathcal{T}_\xi$. By continuity, the statement extends immediately to the increments over $\mathbb{R}_+$. Hence, by the Erlang–Lévy characterization in FMP 12.10, we conclude that ξ is conditionally Poisson with intensity measure $\nu \otimes \lambda$, where $\nu = E[\xi(\cdot \times [0,1])|\mathcal{T}_\xi]$. Thus, ξ is a Cox process directed by $\nu \otimes \lambda$, and we note that ν is a.s. unique by the law of large numbers. The last assertion now follows as in the proofs of Proposition 1.4 (iii) and (iv). □

The last result is easily extended to a characterization of general exchangeable or contractable random measures on $S \times \mathbb{R}_+$. Write $\mathcal{M}'(S) = \mathcal{M}(S) \setminus \{0\}$, where $\mathcal{M}(S)$ denotes the class of σ-finite (or locally finite) measures on S.

Proposition 1.21 *(exchangeable random measures) Let ξ be a random measure ξ on $S \times \mathbb{R}_+$, where S is Borel. Then ξ is contractable iff*

$$\xi = \alpha \otimes \lambda + \iint (\mu \otimes \delta_t)\, \eta(d\mu\, dt) \quad a.s.,$$

for some random measures α on S and ν on $\mathcal{M}'(S)$ and a Cox process $\eta \perp\!\!\!\perp_\nu \alpha$ directed by $\nu \otimes \lambda$. In that case, α and ν are a.s. unique, and ξ is extreme iff (α, ν) is a.s. non-random.

Proof: Suppose again that ξ is contractable. On $\mathcal{M}'(S) \times \mathbb{R}_+$ we may introduce the marked point process

$$\eta = \sum\nolimits_{t \geq 0} \delta_{\xi(\cdot \times \{t\})} \otimes \delta_t,$$

which determines the purely discontinuous component ξ_d of ξ, via the formula

$$\xi_d = \sum\nolimits_{t \geq 0} \xi(\cdot \times \{t\}) \otimes \delta_t = \iint (\mu \otimes \delta_t)\, \eta(d\mu\, dt).$$

Putting $\xi_c = \xi - \xi_d$, we see from Theorem 1.15 that the pair (ξ_c, ξ_d) is again contractable on the space $\{1, 2\} \times S \times \mathbb{R}_+$, which implies that (ξ_c, η) is contractable on $(S \cup \mathcal{M}(S)) \times \mathbb{R}_+$. Proceeding as in Lemma 1.20, we conclude that $\xi_c = \alpha \otimes \lambda$ a.s. for some random measure α on S and that η is conditionally Poisson with intensity measure of the form $\nu \otimes \lambda$, given the pair (α, ν). This shows that η is a Cox process with directing random measure $\nu \otimes \lambda$, and it also implies the asserted conditional independence. The stated uniqueness is obvious for α, and for ν it follows from the corresponding statement in Lemma 1.20. The last assertion is an immediate consequence. $\qquad \Box$

1.5   Measures on a Finite Interval

The characterization problem for exchangeable processes on $[0, 1]$ is more difficult and will not be solved completely until Chapter 3. Here we shall only consider the special case of exchangeable random measures on a product space $S \times [0, 1]$, where S is an arbitrary Borel space. We begin with a characterization of contractability for simple point processes and diffuse random measures on $[0, 1]$. A more detailed discussion of this case appears in the next section.

We say that ξ is a *binomial process* on $[0, 1]$ based on λ and $k \in \mathbb{Z}_+$ if $\xi = \sum_{j \le k} \delta_{\tau_j}$ a.s. for some i.i.d. $U(0, 1)$ random variables $\tau_1, \dots, \tau_k$. The term is motivated by the fact that the variables ξB are binomially distributed with parameters k and λB. We also consider mixtures of such processes, where $\tau_1, \tau_2, \dots$ are i.i.d. $U(0, 1)$ and the constant k is replaced by a random variable $\kappa \perp\!\!\!\perp (\tau_j)$.

Proposition 1.22 *(simple or diffuse random measures, Davidson, Matthes et al., Kallenberg)*

    (i) *A simple point process ξ on $[0, 1]$ is contractable iff it is a mixed binomial process.*

    (ii) *A diffuse random measure ξ on $[0, 1]$ is contractable iff $\xi = \alpha \lambda$ a.s. for some random variable $\alpha \ge 0$.*

In both cases, ξ is also exchangeable.

We give an elementary proof based on Propositions 1.8 and 1.13 and the law of large numbers. A shorter but more sophisticated argument is presented in the next section.

Proof: (i) Let $\tilde{\xi}$ be a mixed binomial process on $[0, 1]$ based on $\xi[0, 1]$ and λ. Writing $I_{nj} = n^{-1}(j - 1, j]$ for $1 \le j \le n$, we introduce for every $n \in \mathbb{N}$ the point processes

$$\eta_n = \sum_j (\xi I_{nj} \wedge 1)\, \delta_{j/n}, \qquad \tilde{\eta}_n = \sum_j (\tilde{\xi} I_{nj} \wedge 1)\, \delta_{j/n},$$

and define

$$\kappa = \xi[0,1] = \tilde{\xi}[0,1], \quad \kappa_n = \eta_n[0,1], \quad \tilde{\kappa}_n = \tilde{\eta}_n[0,1].$$

The sequences of weights $\xi I_{nj} \wedge 1$ and $\tilde{\xi} I_{nj} \wedge 1$ are exchangeable, the former by Proposition 1.13 (iii). Hence, by Proposition 1.8, both are mixed urn sequences with values in $\{0,1\}$. Letting $\|\cdot\|_n$ denote half the total variation on the σ-field $\mathcal{U}_n = \sigma\{I_{n1}, \ldots, I_{nn}\}$, we get

$$\|\mathcal{L}(\xi) - \mathcal{L}(\tilde{\xi})\|_n$$
$$\leq \quad \|\mathcal{L}(\xi) - \mathcal{L}(\eta_n)\|_n + \|\mathcal{L}(\eta_n) - \mathcal{L}(\tilde{\eta}_n)\| + \|\mathcal{L}(\tilde{\eta}_n) - \mathcal{L}(\tilde{\xi})\|_n$$
$$\leq \quad P\{\kappa \neq \kappa_n\} + P\{\kappa_n \neq \tilde{\kappa}_n\} + P\{\tilde{\kappa}_n \neq \kappa\}$$
$$\leq \quad 2P\{\kappa_n < \kappa\} + 2P\{\tilde{\kappa}_n < \kappa\} \to 0.$$

Hence, $\mathcal{L}(\xi) = \mathcal{L}(\tilde{\xi})$ on every $\mathcal{U}_n$, which extends to $\xi \stackrel{d}{=} \tilde{\xi}$ by a monotone-class argument.

(ii) For each $n \in \mathbb{N}$, let η_n be a Cox process directed by $n\xi$, and note that the random variables $\eta_n B$ are mixed Poisson with means $n\xi B$. Since η_n is also a simple, contractable point process on $[0,1]$ (FMP 12.5), part (i) shows that it is a mixed binomial process based on $\eta_n[0,1]$ and λ. Noting that the Cox property is preserved by independent thinnings (FMP 12.3), we conclude that the variables $\eta_n B$ are also mixed Poisson with conditional means $n\xi[0,1]\lambda B$. Applying the law of large numbers twice, we obtain

$$\xi B \stackrel{P}{\leftarrow} (\eta_n B/n) \stackrel{P}{\to} \xi[0,1]\lambda B, \quad B \in \mathcal{B}[0,1],$$

which implies $\xi B = \xi[0,1]\lambda B$ a.s. for all $B \in \mathcal{B}[0,1]$. This extends by a monotone-class argument to $\xi = \xi[0,1]\lambda$ a.s.

The last statement is obvious from the invariance of Lebesgue measure λ and the form of the binomial processes. $\square$

We continue with a useful relationship between contractable sequences and processes.

Theorem 1.23 *(contractable sequences and processes) Let ξ be a marked point process on $S \times [0,1]$, where S is Borel and $\xi(S \times [0,1]) = n$ is a.s. finite and non-random. Put $\bar{\xi} = \xi(S \times \cdot)$, and let $\eta = (\eta_1, \ldots, \eta_n)$ be the associated sequence of marks. Then*

(i) *ξ is contractable iff $\bar{\xi}$ and η are independent and contractable.*

In that case,

(ii) *ξ and η are simultaneously extreme,*

(iii) *ξ and η are simultaneously exchangeable.*

Proof: (i) Suppose that $\bar{\xi}$ and η are independent and contractable. Fix any finite interval union $U \subset [0,1]$, and put $I = [0, \lambda U]$. Write ξ^U and $\bar{\xi}^U$ for the contractions of ξ to $S \times I$ and $\bar{\xi}$ to I, respectively, and let $\eta_1^U, \ldots, \eta_{\bar{\xi}U}^U$ be the marks of ξ^U, enumerated in the order from left to right. Using Fubini's theorem twice, along with the contractability of η and $\bar{\xi}$, and Theorem 1.15, we get for $k \leq n$ and any measurable function $f \geq 0$ on $\mathcal{M}(I) \times S^k$

$$
\begin{aligned}
E[f(\bar{\xi}^U, \eta_1^U, \ldots, \eta_k^U);\ \bar{\xi}U = k] \\
= \ & \int Ef(\mu, \eta_1, \ldots, \eta_k)\, P\{\bar{\xi}^U \in d\mu,\ \bar{\xi}U = k\} \\
= \ & \int Ef(\mu, \eta_1, \ldots, \eta_k)\, P\{\bar{\xi}^I \in d\mu,\ \bar{\xi}I = k\} \\
= \ & E[f(\bar{\xi}^I, \eta_1, \ldots, \eta_k);\ \bar{\xi}I = k],
\end{aligned}
$$

which shows that $\xi^U \overset{d}{=} \xi^I$. Hence, ξ is contractable.

Conversely, suppose that ξ is contractable. Let U and I be as before, and fix any $B \in \mathcal{S}^n$. Since $\xi^{U^c} \overset{d}{=} \xi^{I^c}$ by Theorem 1.15, we obtain

$$
\begin{aligned}
P\{\bar{\xi}U = 0,\ \eta \in B\} &= P\{\bar{\xi}U^c = n,\ \eta^{U^c} \in B\} \\
&= P\{\bar{\xi}I^c = n,\ \eta^{I^c} \in B\} \\
&= P\{\bar{\xi}I = 0,\ \eta \in B\}.
\end{aligned}
$$

Applying this to subsets U of a fixed interval union V with associated interval $J = [0, \lambda V]$ and using a version of FMP 12.8, we obtain

$$
P\{\bar{\xi}^V \in \cdot,\ \eta \in B\} = P\{\bar{\xi}^J \in \cdot,\ \eta \in B\}.
$$

Thus, $\bar{\xi}$ is conditionally contractable given $\eta \in B$, whenever the latter event has positive probability. By Proposition 1.22 it follows that $\bar{\xi}$ is conditionally a binomial process based on n and λ, and since the conditional distribution is independent of B, we conclude that $\bar{\xi}$ is contractable and independent of η.

To see that even η is contractable, fix any numbers $a < b$ in $(0,1)$, and put

$$
I = [0, a], \qquad U = [0, a] \cup (b, 1], \qquad V = [0, 1 - b + a].
$$

Then $\xi^U \overset{d}{=} \xi^V$ by the contractability of ξ and Theorem 1.15. Combining this with the independence of $\bar{\xi}$ and η, we get for any $B \in \mathcal{S}^{n-1}$ and $k \in \{0, \ldots, n-1\}$

$$
\begin{aligned}
P\{(\eta_1, \ldots, \eta_k, \eta_{k+2}, \ldots, \eta_n) \in B\}\, P\{\bar{\xi}I = k,\ \bar{\xi}U = n - 1\} \\
= \ & P\{(\eta_1^U, \ldots, \eta_{n-1}^U) \in B,\ \bar{\xi}I = k,\ \bar{\xi}U = n - 1\} \\
= \ & P\{(\eta_1^V, \ldots, \eta_{n-1}^V) \in B,\ \bar{\xi}I = k,\ \bar{\xi}V = n - 1\} \\
= \ & P\{(\eta_1, \ldots, \eta_{n-1}) \in B\}\, P\{\bar{\xi}I = k,\ \bar{\xi}U = n - 1\}.
\end{aligned}
$$

Dividing by the common factor $P\{\bar{\xi}I = k,\ \bar{\xi}U = n - 1\} > 0$, we obtain

$$
(\eta_1, \ldots, \eta_k, \eta_{k+2}, \ldots, \eta_n) \overset{d}{=} (\eta_1, \ldots, \eta_{n-1}), \quad k = 0, \ldots, n - 1,
$$

which extends by iteration to the required contractability condition.

(ii) The measure ξ is clearly a measurable function of the pair $(\bar{\xi}, \eta)$, and conversely, $\bar{\xi}$ and η can be measurably recovered from ξ. Hence, the corresponding distributions $\mathcal{L}(\xi)$ and $\mathcal{L}(\bar{\xi}) \otimes \mathcal{L}(\eta)$ determine each other uniquely. Since $\bar{\xi}$ is a fixed binomial process, it follows that $\mathcal{L}(\xi)$ and $\mathcal{L}(\eta)$ are uniquely determined by each other. Furthermore, the bi-linearity of the product measure $\mathcal{L}(\bar{\xi}) \otimes \mathcal{L}(\eta)$ implies that the mapping $\mathcal{L}(\eta) \mapsto \mathcal{L}(\xi)$ is linear. Hence, for any $p \in (0,1)$, the relations

$$\mathcal{L}(\eta) = p\mu_1 + (1-p)\mu_2, \qquad \mathcal{L}(\xi) = p\tilde{\mu}_1 + (1-p)\tilde{\mu}_2$$

are equivalent whenever μ_1 and μ_2 are contractable probability measures on S^n and $\tilde{\mu}_1$ and $\tilde{\mu}_2$ are the corresponding contractable distributions on $\mathcal{M}[0,1]$. If η is extreme, we have $\mu_1 = \mu_2$, and then also $\tilde{\mu}_1 = \tilde{\mu}_2$, which shows that even ξ is extreme. By the same argument, extremality of ξ implies the same property for η.

(iii) Since $\bar{\xi}$ is a binomial process based on n and λ, the transfer theorem (FMP 6.10) guarantees the existence of an a.s. representation $\bar{\xi} = \sum_j \delta_{\sigma_j}$, where $\sigma_1, \ldots, \sigma_n$ are i.i.d. $U(0,1)$ random variables with $\sigma \equiv (\sigma_j) \perp\!\!\!\perp_{\bar{\xi}} \eta$. Since also $\eta \perp\!\!\!\perp \bar{\xi}$, we have in fact $\sigma \perp\!\!\!\perp \eta$ by the chain rule for conditional independence (FMP 6.8). Enumerating the σ_j in increasing order as $\sigma_{\pi_1}, \ldots, \sigma_{\pi_n}$, we note that $\pi_1, \ldots, \pi_n$ form a σ-measurable permutation of $1, \ldots, n$. Introducing the inverse permutation $\pi' = (\pi'_1, \ldots, \pi'_n)$ and writing $\eta_{\pi'_j} = \eta \circ \pi'_j$, for convenience, we obtain

$$\xi = \sum_i \delta_{\eta_i} \otimes \delta_{\sigma \circ \pi_i} = \sum_j \delta_{\eta \circ \pi'_j} \otimes \delta_{\sigma_j}. \tag{23}$$

Now suppose that η is exchangeable. By Fubini's theorem we see that $(\eta \circ \pi', \sigma) \overset{d}{=} (\eta, \sigma)$, where $\eta \circ \pi' = (\eta_{\pi'_1}, \ldots, \eta_{\pi'_n})$. Hence, $(\eta \circ \pi') \perp\!\!\!\perp \sigma$, and the asserted exchangeability of ξ follows from (23) by another application of Fubini's theorem. Conversely, suppose that ξ is exchangeable. Then this remains conditionally true, given the invariant point process $\beta = \sum_j \delta_{\eta_j}$ on S, and so we may assume the latter to be non-random. But then η is exchangeable by Theorem 1.13 (ii). $\qquad\square$

We can use the last theorem to characterize the exchangeable random measures on a product space $S \times [0,1]$, where S is Borel. For convenience, we begin with the case of marked point processes. Recall that if $\beta = \sum_j \delta_{\beta_j}$ is an arbitrary point process on S, then a *uniform* or *λ-randomization* of β is defined as a point process on $S \times [0,1]$ of the form $\xi = \sum_j \delta_{\beta_j, \tau_j}$, where the τ_j are i.i.d. $U(0,1)$ and independent of $\beta_1, \beta_2, \ldots$.

Lemma 1.24 *(exchangeable point processes)* *Let ξ be a marked point process on $S \times [0,1]$, where S is Borel. Then ξ is exchangeable iff it is a uniform randomization of the point process $\beta = \xi(\cdot \times [0,1])$. In that case, ξ is extreme iff β is a.s. non-random.*

Proof: If ξ is exchangeable, it remains so under conditioning on β. Since the class of uniform randomizations is closed under mixing, we may henceforth assume that β is non-random. Next we note that ξ remains exchangeable on any product set $A \times [0,1]$ with $A \in \mathcal{S}$. Since the randomization property for every such set with $\beta A < \infty$ implies the same property on $S \times [0,1]$, we may further assume that $\beta S = n < \infty$.

Under those additional hypotheses, we introduce the time and mark sequences $\tau = (\tau_1, \ldots, \tau_n)$ and $\eta = (\eta_1, \ldots, \eta_n)$ of ξ and note that, by Theorem 1.23, both $\bar{\xi} = \sum_j \delta_{\tau_j}$ and η are exchangeable with $\tau \perp\!\!\!\perp \eta$. Fixing any enumeration $b_1, \ldots, b_n$ of the points of β, we may conclude from Proposition 1.8 and the transfer theorem that $\eta = b \circ \pi \equiv (b_{\pi_1}, \ldots, b_{\pi_n})$ a.s. for some exchangeable permutation $\pi \perp\!\!\!\perp \tau$ of $1, \ldots, n$. The same theorem ensures that $\tau = \sigma \circ \kappa \equiv (\sigma_{\kappa_1}, \ldots, \sigma_{\kappa_n})$ for some i.i.d. $U(0,1)$ random variables $\sigma_1, \ldots, \sigma_n$ with $\sigma \equiv (\sigma_j) \perp\!\!\!\perp \pi$ and a σ-measurable permutation $\kappa = (\kappa_1, \ldots, \kappa_n)$ of $1, \ldots, n$. Writing π' for the inverse of the permutation π, we obtain

$$\xi = \sum_j \delta_{\eta_j} \otimes \delta_{\tau_j} = \sum_j \delta_{b \circ \pi_j} \otimes \delta_{\sigma \circ \kappa_j} = \sum_i \delta_{b_i} \otimes \delta_{\sigma \circ \kappa \circ \pi'_i}.$$

We may now use Fubini's theorem to see that the permutation $\kappa \circ \pi'$ of $1, \ldots, n$ remains exchangeable, conditionally on σ. Hence, $(\kappa \circ \pi') \perp\!\!\!\perp \sigma$, and since even σ is exchangeable, it follows that $\sigma \circ \kappa \circ \pi' \overset{d}{=} \sigma$. This shows that ξ has the required form. The last assertion may be proved by the same arguments as for Proposition 1.4. $\square$

We turn to the corresponding characterization for general random measures on $S \times [0,1]$. It is interesting to compare with the characterization of contractable random measures on $S \times \mathbb{R}_+$ in Proposition 1.21.

Theorem 1.25 *(exchangeable random measures)* *Let ξ be a random measure on $S \times [0,1]$, where S is Borel. Then ξ is exchangeable iff*

$$\xi = \alpha \otimes \lambda + \sum_j \beta_j \otimes \delta_{\tau_j} \quad a.s.,$$

for some i.i.d. $U(0,1)$ random variables $\tau_1, \tau_2, \ldots$ and an independent collection of random measures α and $\beta_1, \beta_2, \ldots$ on S. In that case, ξ is extreme iff α and $\beta = \sum_j \delta_{\beta_j}$ are a.s. non-random.

First proof: As in case of Proposition 1.21, we introduce the marked point process

$$\eta = \sum_{t \in [0,1]} \delta_{\xi(\cdot \times \{t\})} \otimes \delta_t,$$

now defined on the product space $\mathcal{M}'(S) \times [0,1]$, and construct the purely discontinuous component ξ_d of ξ through the formula

$$\xi_d = \sum_t \xi(\cdot \times \{t\}) \otimes \delta_t = \int\!\!\int (\mu \otimes \delta_t)\, \eta(d\mu\, dt). \tag{24}$$

Putting $\xi_c = \xi - \xi_d$, we see from Theorem 1.15 that the pair (ξ_c, η) is again exchangeable. Since the same condition holds conditionally, given the invariant measures $\alpha = \xi_c(\cdot \times [0,1])$ and $\beta = \eta(\cdot \times [0,1])$, we may assume that the latter are non-random. Then $\xi_c = \alpha \otimes \lambda$ a.s. by Proposition 1.22, and Lemma 1.24 shows that $\eta = \sum_j \delta_{b_j, \tau_j}$ a.s. for some i.i.d. $U(0,1)$ random variables $\tau_1, \tau_2, \ldots$, where $b_1, b_2, \ldots$ is an arbitrary enumeration of the points of β. The desired representation is now immediate from (24). The last assertion follows by the usual arguments, given the fact that α and β are a.s. unique, measurable functions of ξ. □

It is interesting to see how the underlying lemmas can also be proved directly, by an entirely different method involving moment measures.

Second proof of Theorem 1.25: By Theorem 1.15 it is enough to consider the cases where $\bar{\xi} = \xi(S \times \cdot)$ is a.s. diffuse or ξ is a marked point process on $S \times [0,1]$, in either case with a non-random projection $\beta = \xi(\cdot \times [0,1])$. In the point process case, we may also assume that β is simple, since we can otherwise consider a uniform randomization of ξ, which is again exchangeable with a simple projection onto the extended mark space.

In the diffuse case, it suffices to show that $\xi(A \times \cdot)$ is a.s. invariant for every $A \in \mathcal{S}$ with $\beta A < \infty$. We may then assume ξ to be a diffuse, exchangeable random measure on $[0,1]$. Using the exchangeability of ξ, we get for any disjoint, rational sub-intervals I and J

$$E\xi I = a\lambda I, \qquad E\xi^2(I \times J) = b\lambda^2(I \times J),$$

for some constants $a, b \geq 0$. Noting that $E\xi^2$ and λ^2 both vanish on the main diagonal of $[0,1]^2$, we may use a monotone-class argument to conclude that $E\xi = a\lambda$ and $E\xi^2 = b\lambda^2$. Since $\xi[0,1] = c$ is non-random, we obtain $a = c$ and $b = c^2$, and so for any $B \in \mathcal{B}[0,1]$

$$\mathrm{var}(\xi B) = E(\xi B)^2 - (E\xi B)^2 = c^2\lambda^2 B^2 - (c\lambda B)^2 = 0,$$

which implies $\xi B = c\lambda B$ a.s. Here the exceptional null set can be chosen to be independent of B, which leads to the a.s. relation $\xi = c\lambda$.

Turning to the case of marked point processes, we may fix an enumeration $b_1, b_2, \ldots$ of the marks and define $\xi_j = \delta_{\tau_j} = \xi(\{b_j\} \times \cdot)$ for all j. Considering the contributions to disjoint, rational intervals and extending by a monotone-class argument, we see as before that $E(\xi_1 \otimes \cdots \otimes \xi_n) = \lambda^n$. Noting that each product measure $\xi_j^k = \delta_{\tau_j}^k$ is supported by the main diagonal D_k in $[0,1]^k$ with the one-dimensional projections $\delta_{\tau_j} = \xi_j$, we obtain more generally

$$E \bigotimes_{j \leq n} \xi_j^{k_j} = \bigotimes_{j \leq n} \lambda_{k_j},$$

where λ_k denotes the measure on D_k with one-dimensional projections λ. Since the joint distribution of $\xi_1, \xi_2, \ldots$ is uniquely determined by the product measures of arbitrary order, we conclude that β determines the distribution of ξ. We may now compare with a uniform randomization $\tilde{\xi}$ of β, which is again an exchangeable, marked point process with the same projection onto S. Using the mentioned uniqueness, we obtain the required equality $\xi \overset{d}{=} \tilde{\xi}$. $\square$

1.6 Simple or Diffuse Random Measures

The simple point processes and diffuse random measures deserve special attention for at least two reasons: Their distributions are uniquely determined by the *avoidance function* $\varphi(B) = P\{\xi B = 0\}$ (FMP 12.9), and they exhibit some special features that are not present in the general case. Though some of the quoted results are familiar already from earlier sections, the following unified approach may be illuminating.

Let us first describe the exchangeable sequences and processes in the four fundamental cases of discrete or continuous, finite or infinite time intervals. We may begin with the *binomial process* on $[0,1]$ based on λ and k, which is a simple point process of the form $\xi = \sum_{j \leq k} \delta_{\tau_j}$ (or any point process with the same distribution), where $\tau_1, \tau_2, \ldots$ are independent $U(0,1)$ random variables. Replacing k by a $\mathbb{Z}_+$-valued random variable $\kappa \perp\!\!\!\perp (\tau_j)$, we obtain a *mixed binomial process* ξ on $[0,1]$, which satisfies

$$P\{\xi B = 0\} = E(1 - \lambda B)^\kappa = \psi_\kappa(1 - \lambda B), \quad B \in \mathcal{B}[0,1], \qquad (25)$$

where $\psi_\kappa(s) = Es^\kappa$ denotes the probability generating function of κ. In particular, we get a homogeneous Poisson process with rate $r \geq 0$ by choosing κ to be Poisson distributed with mean r. Extending by independence to the interval $\mathbb{R}_+$ and mixing with respect to r (so that ξ becomes conditionally Poisson with the random rate $\rho \geq 0$), we obtain a *mixed Poisson process* ξ on $\mathbb{R}_+$, satisfying

$$P\{\xi B = 0\} = Ee^{-\rho \lambda B} = \varphi_\rho(\lambda B), \quad B \in \mathcal{B}(\mathbb{R}_+), \qquad (26)$$

where $\varphi_\rho(u) = Ee^{-u\rho}$ denotes the Laplace transform of ρ.

We may also consider point processes on the discrete intervals $\{1, \ldots, n\}$ or $\mathbb{N}$, where the simple point processes ξ may be identified with finite or infinite sequences of $\{0,1\}$-valued random variables $\xi_1, \xi_2, \ldots$. If the ξ_j are i.i.d. with $E\xi_j = r$, then $\xi = (\xi_1, \xi_2, \ldots)$ becomes a *Bernoulli sequence* with rate $r \in [0,1]$. Randomizing the parameter r, as before, we obtain a *mixed Bernoulli sequence* with a random rate ρ in $[0,1]$, which satisfies

$$E[\xi_{k_1} \cdots \xi_{k_m}] = E\rho^k, \quad k_1 < \cdots < k_m. \qquad (27)$$

For completeness, we may also consider the pure or mixed *hyper-geometric sequences* ξ on $\{1, \ldots, n\}$, where the former may be defined as urn sequences with values in $\{0, 1\}$. Writing $n^{(k)} = n!/(n-k)!$ for $0 \le k \le n$, we get in the general case

$$E[\xi_{k_1} \cdots \xi_{k_m}] = E[\kappa^{(m)};\, \kappa \ge m]/n^{(m)}, \quad k_1 < \cdots < k_m, \tag{28}$$

for some random variable κ in $\{0, \ldots, n\}$.

The following theorem summarizes the characterizations of contractable sequences or processes in the mentioned four cases, along with the corresponding results for diffuse random measures.

Theorem 1.26 (*simple or diffuse random measures*) *A simple point process ξ on an interval I is contractable, hence also exchangeable, iff*

(i) *ξ is a mixed hyper-geometric sequence when $I = \{1, \ldots, n\}$,*

(ii) *ξ is a mixed Bernoulli sequence when $I = \mathbb{N}$,*

(iii) *ξ is a mixed binomial process when $I = [0, 1]$,*

(iv) *ξ is a mixed Poisson process when $I = \mathbb{R}_+$.*

A diffuse random measure ξ on $I = [0, 1]$ or $\mathbb{R}_+$ is contractable iff $\xi = \alpha\lambda$ a.s. for some random variable $\alpha \ge 0$.

All the listed assertions are special cases of previously established results. Thus, part (i) is an immediate consequence of Proposition 1.8 and Theorem 1.13 (iii), part (ii) is a special case of Theorem 1.1, part (iii) appears in Proposition 1.22, and part (iv) was proved in Lemma 1.20. Finally, the diffuse case was established in Proposition 1.22. Some of those earlier proofs could be readily simplified, if we were only willing to use the (nontrivial) fact that contractability implies exchangeability in the simple and diffuse cases. The latter statement, recorded for point processes already in Proposition 1.13 and Corollary 1.21, is an easy consequence of the fundamental uniqueness result for simple and diffuse random measures in FMP 12.8.

It is interesting to note that the last theorem is essentially equivalent to the celebrated Hausdorff–Bernstein theorem in analysis, which characterizes the completely monotone sequences or functions on a finite or infinite interval. A statement of that result, along with the relevant definitions, is given in Theorem A4.1.

Second proof of Theorem 1.26: If ξ is a simple point process, then by FMP 12.8 it suffices to establish formulas (25)–(28) for suitable random variables κ or ρ. In cases (i) and (ii), we may then introduce the moments $c_k = E[\xi_1 \cdots \xi_k]$ and verify, by induction on m, that

$$(-1)^m \Delta^m c_k = E[\xi_1 \cdots \xi_k \,(1 - \xi_{k+1}) \cdots (1 - \xi_{k+m})] \ge 0, \tag{29}$$

for appropriate k and m. This shows that the sequence (c_k) is completely monotone, and so by Theorem A4.1 we have $c_k = E[\kappa^{(k)};\, \kappa \ge k]/n^{(k)}$ or $c_k = E\rho^k$, respectively, for some random variable κ in $\{0, \ldots, n\}$ or ρ in $[0, 1]$. Formulas (27) and (28) now follow by the contractability of ξ.

In cases (iii) and (iv), define $f(t) = P\{\xi(0,t] = 0\}$, put $I_k = h(k-1,k]$ for fixed $h > 0$, and conclude from (29) for the variables $\xi_k = 1\{\xi I_k = 0\}$ that f is completely monotone on $[0,1]$ or $\mathbb{R}_+$, respectively. Next consider any disjoint intervals $J_1, J_2, \ldots \subset [0,1]$ of lengths $h_n > 0$. By Fatou's lemma and the contractability of ξ, we have

$$0 \leq 1 - \liminf_n f(h_n) = \limsup_n P\{\xi J_n > 0\}$$
$$\leq P\{\xi J_n > 0 \text{ i.o.}\} \leq P\{\xi[0,1] = \infty\} = 0,$$

which shows that $f(0+) = 1$. Hence, Theorem A4.1 yields $f(t) = E(1-t)^\kappa$ or $f(t) = Ee^{-t\rho}$, respectively, for some random variable κ in $\mathbb{Z}_+$ or ρ in $\mathbb{R}_+$. Formulas (25) and (26) now follow by the contractability of ξ.

Next assume that ξ is diffuse and contractable on $[0,1]$ or $\mathbb{R}_+$. Let η be a Cox process directed by ξ, and note that η is simple (FMP 12.5) and contractable. Applying part (iii) to η, we get for any $t \in [0,1]$

$$Ee^{-\xi[0,t]} = P\{\eta[0,t] = 0\} = E(1-t)^{\eta[0,1]} = Ee^{-t\xi[0,1]}.$$

By the contractability of ξ and FMP 12.8 (ii) we obtain $\xi \overset{d}{=} \alpha\lambda$ on $[0,1]$ with $\alpha = \xi[0,1]$, which implies $\xi = \alpha\lambda$ a.s. on the same interval. By scaling we obtain the same result for measures on $\mathbb{R}_+$. $\qquad\square$

The next result shows how suitable restrictions of an exchangeable point process are again exchangeable and satisfy appropriate independence relations. Given a random element ξ, a σ-field $\mathcal{F}$, and an event $A \in \mathcal{F}$, we say that ξ is independent of $\mathcal{F}$ with distribution μ on a set A, if $P[\xi \in \cdot | \mathcal{F}] = \mu$ a.s. on A. Here, clearly, ξ need not even be defined on A^c. Note that for binomial processes on a real interval, the underlying measure is assumed to be λ, unless otherwise specified.

Proposition 1.27 (*restrictions of exchangeable processes*) *Consider a simple, contractable point process $\xi = \sum_j \delta_{\tau_j}$ on $I = [0,1]$ or $\mathbb{R}_+$, where $\tau_1 < \tau_2 < \cdots$. Then*

(i) *for any finite interval $J \subset I$, the restriction $\xi_J = 1_J \cdot \xi$ is a mixed binomial process on J satisfying $\xi_J \perp\!\!\!\perp_{\xi J} \xi_{J^c}$;*

(ii) *for every $n \in \mathbb{N}$, the ratios $\tau_1/\tau_n, \ldots, \tau_{n-1}/\tau_n$ form a binomial process on $[0,1]$, independent of $\tau_n, \tau_{n+1}, \ldots$ on the set $\{\xi I \geq n\}$.*

Proof: (i) First assume that ξ is a contractable process on a finite interval I. Then Theorem 1.26 shows that ξ is a mixed binomial process on I and hence admits a representation $\xi = \sum_{j \leq \kappa} \delta_{\sigma_j}$, where the σ_j are i.i.d. $U(I)$ and independent of $\kappa = \xi I$. For any sub-interval $J \subset I$, we note that ξ_J and ξ_{J^c} are conditionally independent binomial processes on J and J^c, respectively, given the random variables κ and $\iota_k = 1\{k \leq \kappa, \sigma_k \in J\}$, $k \in \mathbb{N}$. Thus, ξ_J is conditionally a binomial process on J with parameter ξJ, given the process

ξ_{J^c} and the variables ξJ and $\iota_1, \iota_2, \ldots$. Since the conditional distribution is a function of ξJ alone, we obtain $\xi_J \perp\!\!\!\perp_{\xi J} (\xi_{J^c}, \iota_1, \iota_2, \ldots)$, and the required independence follows. The result extends to processes on $I = \mathbb{R}_+$, by a martingale or monotone-class argument.

(ii) It is enough to prove the assertion on $\{\tau_n < t\}$ for a fixed $t \in I$, and by Theorem 1.26 we may then assume that ξ is a mixed binomial process on $[0, t]$. By conditioning and scaling, we may next reduce to the case where ξ is a binomial process on $[0, 1]$ based on $m \geq n$ and λ. Then $\xi = \sum_{j \leq m} \delta_{\sigma_j}$ a.s. for some i.i.d. $U(0, 1)$ random variables $\sigma_1, \ldots, \sigma_m$. Since the latter have joint distribution λ^m on $[0, 1]^m$, we note that $(\tau_1, \ldots, \tau_m)$ has distribution $m! \lambda^m$ on the tetrahedral subset $\Delta_m = \{t_1 < \cdots < t_m\}$. By elementary conditioning it follows that $(\tau_1, \ldots, \tau_{n-1})$ has conditional distribution $(n-1)! \lambda^n / \tau_n^{n-1}$ on $\tau_n \Delta_{n-1} = \{t_1 < \ldots < t_{n-1} < \tau_n\}$, given the remaining variables $\tau_n, \ldots, \tau_m$. Using the disintegration theorem (FMP 6.4), we conclude that the ratios $\tau_1 / \tau_n, \ldots, \tau_{n-1} / \tau_n$ have distribution $(n-1)! \lambda^{n-1}$ on Δ_{n-1}, conditionally on $\tau_n, \ldots, \tau_m$, which is equivalent to our assertion. $\quad\square$

This leads to some useful characterizations of mixed Poisson processes. Given a simple point process on $\mathbb{R}_+$ with points $\tau_1 < \tau_2 < \cdots$, we define the associated *spacing variables* as $\gamma_k = \tau_k - \tau_{k-1}$, $k \in \mathbb{N}$, where $\tau_0 = 0$.

Proposition 1.28 *(mixed Poisson processes, Nawrotzki, Freedman, Matthes et al., Kallenberg) Let ξ be a simple point process on $\mathbb{R}_+$ with infinitely many points $\tau_1 < \tau_2 < \cdots$. Then these conditions are equivalent:*

(i) *ξ is contractable, hence also exchangeable;*

(ii) *ξ is a mixed Poisson process;*

(iii) *ξ is a mixed binomial process on $[0, t]$ for every $t > 0$;*

(iv) *$\tau_1, \ldots, \tau_{n-1}$ form a binomial process on $[0, \tau_n]$, conditionally on τ_n for every n;*

(v) *ξ is stationary with exchangeable spacing variables $\gamma_1, \gamma_2, \ldots$.*

Proof: Conditions (i)–(iii) are equivalent by Theorem 1.26, and Proposition 1.27 yields (i) $\Rightarrow$ (iv). Conversely, (iv) implies that ξ is exchangeable on $[0, t]$, conditionally on $\{\tau_n \geq t\}$ for any $t > 0$ and $n \in \mathbb{N}$. Condition (i) now follows as we let $n \to \infty$ and then $t \to \infty$.

Condition (v) follows from (ii) by FMP 12.15. Conversely, assume (v) and let (γ_k) be directed by μ. Then μ is clearly shift-invariant, and by conditioning we may reduce to the case where ξ is a stationary renewal process. Writing $f(t) = \mu(t, \infty)$ and $c = (E\gamma_k)^{-1}$, we see from FMP 9.18 that $f'(t) = -cf(t)$, which implies $f(t) = e^{-ct}$. By FMP 12.15 it follows that ξ is Poisson with constant rate c. $\quad\square$

A sequence $\xi = (\xi_1, \ldots, \xi_\kappa)$ of *random* length κ is said to be exchangeable, if for every $n \in \mathbb{N}$ the elements $\xi_1, \ldots, \xi_n$ are exchangeable, conditionally on $\{\kappa = n\}$. Sequences of this type arise naturally as the sets of excursions of exchangeable processes. Since excursions are handled most naturally by the martingale methods of subsequent chapters, we restrict our present attention to some simple special cases.

The spacing variables of a simple point process ξ on $[0, 1]$ with points $\tau_1 < \cdots < \tau_\kappa$ are given by $\gamma_k = \tau_k - \tau_{k-1}$, $k = 1, \ldots, \kappa + 1$, where $\tau_0 = 0$ and $\tau_{\kappa+1} = 1$. If ξ is a random sequence in an arbitrary space S, with a specified element 0 such that $\xi_k = 0$ at times $\tau_1 < \tau_2 < \cdots$, we define the first *excursion* $\eta^1 = (\eta_1^1, \eta_2^1, \ldots)$ of ξ by taking $\eta_j^1 = \xi_j$ for $j \leq \tau_1$ and $\eta_j^1 = 0$ for $j > \tau_1$. If $\tau_{n-1} < \infty$, then the nth excursion η^n is similarly defined in terms of the shifted sequence $\theta_{\tau_{n-1}}\xi$. We say that an excursion η is *complete* if it ends with a string of zeros.

Proposition 1.29 *(spacings and excursions)*

   (i) *For any simple, exchangeable point process ξ on $[0, 1]$ or $\mathbb{R}_+$, the induced spacing variables $\gamma_1, \gamma_2, \ldots$ are exchangeable.*

   (ii) *For any finite or infinite, exchangeable sequence $\xi_1, \xi_2, \ldots$ in a space S, the excursions $\eta_1, \eta_2, \ldots$ from a fixed state $0 \in S$ are exchangeable.*

   (iii) *For any finite or infinite sequence $\xi_0, \xi_1, \ldots$ in $\mathbb{R}^d$ with exchangeable increments and $\xi_0 = 0$, the associated complete excursions $\eta_1, \eta_2, \ldots$ from 0 are exchangeable.*

Proof: (i) Since mixtures of exchangeable sequences are again exchangeable, it suffices, by Theorem 1.26, to assume that ξ is a Poisson process on $\mathbb{R}_+$ or a binomial process on $[0, 1]$. In the former case the assertion is obvious, since the γ_j are i.i.d. exponentially distributed random variables. For every $n \in \mathbb{N}$ it follows that $\gamma_1, \ldots, \gamma_n$ are conditionally exchangeable given τ_n, which yields the exchangeability of the ratios $\gamma_1/\tau_n, \ldots, \gamma_n/\tau_n$. Now according to Proposition 1.27 (ii), the latter variables have the same joint distribution as the spacing variables induced by a binomial process based on $n - 1$ and λ, and the result in the binomial case follows.

(ii) In the infinite case we may assume that the state space S is Borel, since we can otherwise consider the image of ξ under an arbitrary measurable mapping $f \colon S \to \mathbb{Z}_+$ with $f(x) = 0$ iff $x = 0$, and then extend by a monotone-class argument. By Theorem 1.1 we may then take the ξ_j to be i.i.d., in which case the result follows from the strong Markov property.

In the finite case, Proposition 1.8 allows us to assume that ξ is an urn sequence. In particular, the number of zeros is then fixed, and so is the number k of excursions. It remains to prove that

$$P \bigcap_j \{\eta_j = y_j\} = P \bigcap_j \{\eta_{p_j} = y_j\}, \tag{30}$$

for any excursions $y_1, \ldots, y_k$ in S and permutation $p_1, \ldots, p_k$ of $1, \ldots, k$. Expressing (30) in terms of the ξ_j, we see that the relation is an immediate consequence of the exchangeability of ξ.

(iii) The argument for (ii) applies with only minor changes. Since the number of excursions is no longer fixed in general, we need to supplement the events in (30) by requiring that ξ have exactly n complete excursions. Since the latter condition can be expressed in terms of the increments $\Delta\xi_j$ following the first n excursions $\eta_1, \ldots, \eta_n$, we may proceed as before to complete the proof. $\square$

1.7   Rotations and L^p-Symmetries

Here we take as our starting point the standard *Gaussian* or *normal* distribution $N(0,1)$ on $\mathbb{R}$ with Lebesgue density $(2\pi)^{-1/2}e^{-x^2/2}$. It has the remarkable property that, if $\eta_1, \eta_2, \ldots$ are i.i.d. $N(0,1)$, then the distribution of $(\eta_1, \ldots, \eta_n)$ is spherically symmetric for every $n > 0$. This is clear from the form of the joint density $(2\pi)^{-n/2}e^{-|x|^2/2}$, where $|x| = (x_1^2 + \cdots + x_n^2)^{1/2}$. In Corollary 1.32 below we shall see how the stated symmetry property characterizes the normal distribution.

If $\eta_1, \ldots, \eta_n$ are i.i.d. $N(0,1)$, then the normalized vector with components

$$\xi_k = \frac{\eta_k}{(\eta_1^2 + \cdots + \eta_n^2)^{1/2}}, \quad k = 1, \ldots, n,$$

is uniformly (or symmetrically) distributed on the unit sphere in $\mathbb{R}^n$. We need the intuitively obvious (but less elementary) fact (FMP 2.29) that the latter property determines uniquely the distribution of $(\xi_1, \ldots, \xi_n)$. In particular, the distribution of a spherically symmetric random vector $\zeta = (\zeta_1, \ldots, \zeta_n)$ is then determined by $\mathcal{L}(\zeta_1^2 + \cdots + \zeta_n^2) = \mathcal{L}(|\zeta|^2)$.

The previous construction of the uniform distribution on the unit sphere in $\mathbb{R}^n$ yields a short proof of the following classical result, where the low-dimensional projections of this distribution are shown to be approximately Gaussian for large n. Here we write $\|\mu\|$ for the total variation of the signed measure μ.

Lemma 1.30 *(Gaussian approximation, Maxwell, Borel)   For each $n \in \mathbb{N}$, let the random vector $(\xi_{n1}, \ldots, \xi_{nn})$ be uniformly distributed on the unit sphere in $\mathbb{R}^n$, and let $\eta_1, \eta_2, \ldots$ be i.i.d. $N(0,1)$. Then*

$$\lim_{n\to\infty} \left\| \mathcal{L}(n^{1/2}(\xi_{n1}, \ldots, \xi_{nk})) - \mathcal{L}(\eta_1, \ldots, \eta_k) \right\| = 0, \quad k \in \mathbb{N}. \tag{31}$$

Proof: Writing $\rho_n^2 = \sum_{j \le n} \eta_j^2$, we may take $\xi_{nk} = \eta_k/\rho_n$ for all $k \le n \in \mathbb{N}$, so that

$$\frac{1}{n(\xi_{n1}^2 + \cdots + \xi_{nk}^2)} = \frac{\sigma_n^2}{n\rho_k^2} = \frac{1}{n} + \frac{\rho_n^2 - \rho_k^2}{n\rho_k^2}.$$

Letting f_k denote the density of ρ_k^{-2}, putting $\zeta_{nk} = n/(\rho_n^2 - \rho_k^2)$, and using the spherical symmetry and the independence $\rho_k^2 \perp\!\!\!\perp \zeta_{nk}$, we obtain

$$\|\mathcal{L}(n^{1/2}(\xi_{n1}, \ldots, \xi_{nk})) - \mathcal{L}(\eta_1, \ldots, \eta_k)\|$$
$$= \|\mathcal{L}(n(\xi_{n1}^2 + \cdots + \xi_{nk}^2)) - \mathcal{L}(\rho_k^2)\|$$
$$= \int |E\zeta_{nk} f_k(\zeta_{nk}(x - n^{-1})) - f_k(x)| \, dx$$
$$\leq E \int |\zeta_{nk} f_k(\zeta_{nk}(x - n^{-1})) - f_k(x)| \, dx.$$

Here $\zeta_{nk} \to 1$ a.s. as $n \to \infty$ by the law of large numbers, and so the integrand tends to 0 by the continuity of f_k. Hence, the right-hand side tends to 0 by mean and dominated convergence (FMP 1.32). □

The last result leads easily to the following fundamental characterization of rotatable sequences. Here a sequence of random variables $\xi = (\xi_1, \xi_2, \ldots)$ is said to be *rotatable* if every finite subsequence $(\xi_1, \ldots, \xi_n)$ has a spherically symmetric distribution.

Theorem 1.31 *(rotatable sequences, Freedman) An infinite sequence of random variables $\xi_1, \xi_2, \ldots$ is rotatable iff $\xi_j = \sigma\eta_j$ a.s., $j \in \mathbb{N}$, for some i.i.d. $N(0, 1)$ random variables $\eta_1, \eta_2, \ldots$ and an independent random variable $\sigma \geq 0$. The latter is then a.s. unique.*

Proof: Suppose that ξ is rotatable, and introduce an independent sequence of i.i.d. $N(0, 1)$ random variables $\eta = (\eta_1, \eta_2, \ldots)$. Putting $\sigma_n^2 = (\xi_1^2 + \cdots + \xi_n^2)/n$, we see from Lemma 1.30 that $\sigma_n \eta \xrightarrow{d} \xi$. In particular, the sequence (σ_n) is tight, and therefore $\sigma_n \xrightarrow{d} \sigma$ along a subsequence $N' \subset \mathbb{N}$ for some random variable $\sigma \geq 0$. Assuming $\sigma \perp\!\!\!\perp \eta$, we get $\sigma_n \eta \xrightarrow{d} \sigma\eta$ along N', and so $\xi \overset{d}{=} \sigma\eta$, by the uniqueness of the limiting distribution. The transfer theorem (FMP 6.10) then yields the corresponding a.s. relation, and we get $\sigma_n^2 \to \sigma^2$ a.s. by the law of large numbers, which proves the asserted uniqueness. □

The last result yields another classical characterization.

Corollary 1.32 *(rotatability and independence, Maxwell) Let $\xi_1, \ldots, \xi_n$ be independent random variables, where $n \geq 2$. Then $(\xi_1, \ldots, \xi_n)$ is rotatable iff the ξ_k are i.i.d. centered Gaussian.*

Proof: Assume the stated symmetry. In particular, the ξ_k are i.i.d. and may be extended to an infinite i.i.d. sequence $\xi = (\xi_1, \xi_2, \ldots)$. The rotatability yields $r_1\xi_1 + r_2\xi_2 \overset{d}{=} \xi_1$ for any $r_1, r_2 \in \mathbb{R}$ with $r_1^2 + r_2^2 = 1$, and so by iteration $\sum_{k \leq n} r_k\xi_k \overset{d}{=} \xi_1$ for any constants $r_1, \ldots, r_n \in \mathbb{R}$ with $\sum_k r_k^2 = 1$. By the Cramér–Wold theorem (FMP 5.5) it follows that $(\xi_1, \ldots, \xi_n)$ is rotatable for every $n \in \mathbb{N}$, and so by Theorem 1.31 we have $\xi = \sigma\eta$ a.s. for some i.i.d. $N(0, 1)$ sequence $\eta = (\eta_k)$ and an independent random variable $\sigma \geq 0$. But then $n^{-1} \sum_{k \leq n} \xi_k^2 \to \sigma^2$ a.s. by the law of large numbers applied to η, and

Kolmogorov's zero-one law shows that σ is a.s. a constant. This means that the ξ_k are i.i.d. $N(0, \sigma^2)$. $\square$

Conversely, we can prove the last corollary directly, and then derive Theorem 1.31 as an easy corollary. This approach will be pursued below, in the more general context of stable distributions. Yet another approach is to base the proof of Theorem 1.31 on Schoenberg's celebrated Theorem A4.3 in classical analysis. In fact, the two theorems are essentially equivalent.

Second proof of Theorem 1.31: Let ξ be rotatable, write φ for the characteristic function of ξ_1, and put $f(t) = \varphi(\sqrt{t})$ for $t \geq 0$. For any $u \in \mathbb{R}^d$ we have $u\xi \equiv \langle u, \xi \rangle \overset{d}{=} |u|\xi_1$ by rotatability, and so

$$Ee^{iu\xi} = Ee^{i|u|\xi_1} = \varphi(|u|) = f(|u|^2), \quad u \in \mathbb{R}^d. \tag{32}$$

By Theorem A4.2 (the easy direction), $f(|u|^2)$ is then nonnegative definite on $\mathbb{R}^d$ for every $d \in \mathbb{N}$, and so by Theorem A4.3 the function f is completely monotone on $\mathbb{R}_+$. Since it is also continuous with $f(0) = 1$, Theorem A4.1 (iv) yields $f(t) = Ee^{-t\sigma^2/2}$ for some random variable σ. By (32) and Fubini's theorem we obtain

$$Ee^{iu\xi} = E \exp(-\tfrac{1}{2}\sigma^2|u|^2) = Ee^{-iu(\sigma\eta)}, \quad u \in \mathbb{R}^d.$$

Hence, the uniqueness theorem for characteristic functions yields $\xi \overset{d}{=} \sigma\eta$, which can be strengthened as before to the required a.s. equality $\xi = \sigma\eta$. $\square$

The rotational symmetry of $(\xi_1, \ldots, \xi_n)$ is equivalent, by the Cramér–Wold theorem (FMP 5.5), to the condition

$$r_1\xi_1 + \cdots + r_n\xi_n \overset{d}{=} (r_1^2 + \cdots + r_n^2)^{1/2}\xi_1, \quad r_1, \ldots, r_n \in \mathbb{R}.$$

We turn to the more general case where all linear combinations have the same distribution, apart from a scaling. A real or complex random variable ξ is said to be *symmetric stable*, if for any independent random variables $\xi_1, \xi_2, \ldots \overset{d}{=} \xi$ and constants $r_1, \ldots, r_n \in \mathbb{R}$ or $\mathbb{C}$, respectively, we have

$$r_1\xi_1 + \cdots + r_n\xi_n \overset{d}{=} r\xi_1 \tag{33}$$

for some constant $r \in \mathbb{R}$ or $\mathbb{C}$. Similarly, we say that a random variable $\xi \geq 0$ is *positive stable*, if (33) holds for any $r_1, \ldots, r_n \geq 0$ with a suitable choice of $r \geq 0$. If there is no risk for confusion, we may often omit the qualifications "symmetric" or "positive" when referring to p-stable distributions.

The following result shows that, in all three cases, there exists a constant $p \in (0, 2]$ such that the coefficients r and $r_1, \ldots, r_n$ are related by

$$|r_1|^p + \cdots + |r_n|^p = |r|^p, \tag{34}$$

or, in the positive case, simply by $r_1^p + \cdots + r_n^p = r^p$. The corresponding *p-stable* distributions are then unique up to a normalization. In particular, Corollary 1.32 shows that the symmetric 2-stable distributions on $\mathbb{R}$ are precisely the centered Gaussian laws.

Proposition 1.33 *(p-stable distributions, Lévy) Let the random variable ξ be symmetric stable in $\mathbb{C}$ or $\mathbb{R}$, or positive stable in $\mathbb{R}_+$. Then the coefficients $r, r_1, \ldots, r_n$ in (33) are related by (34) for some constant $p \in (0, 2]$, or in the positive case, for some $p \in (0, 1]$. All such values of p may occur, and for each p the distribution of ξ is unique up to a normalization.*

The result may be derived from the representation of Lévy processes or infinitely divisible distributions (FMP 15.9). Here we give a direct, elementary proof. Our argument requires the following simple extension of the classical Cramér–Wold theorem (FMP 5.5).

Lemma 1.34 *(one-dimensional projections) Let ξ and $\xi_1, \xi_2, \ldots$ be random elements in $\mathbb{C}^d$, where $d \in \mathbb{N}$ is arbitrary. Then $\xi_n \xrightarrow{d} \xi$ iff*

$$\Re\langle u, \xi_n \rangle \xrightarrow{d} \Re\langle u, \xi \rangle, \quad u \in \mathbb{C}^d. \tag{35}$$

Proof: Writing $u = s + it$ with $s, t \in \mathbb{R}^d$, we get from (35)

$$\langle s, \Re\xi_n \rangle - \langle t, \Im\xi_n \rangle \xrightarrow{d} \langle s, \Re\xi \rangle - \langle t, \Im\xi \rangle, \quad s, t \in \mathbb{R}^d.$$

Hence, the Cramér–Wold theorem in $\mathbb{R}^{2d}$ yields

$$(\Re\xi_n, \Im\xi_n) \xrightarrow{d} (\Re\xi, \Im\xi),$$

which is clearly equivalent to $\xi_n \xrightarrow{d} \xi$. $\qquad \square$

Proof of Proposition 1.33: We may clearly exclude the case where $\xi = 0$ a.s. If $\xi_1, \xi_2, \ldots$ are real and i.i.d. symmetric stable, then $\xi_1 - \xi_2 \stackrel{d}{=} r\xi$ for some $r \in \mathbb{R} \setminus \{0\}$, which implies $-\xi \stackrel{d}{=} \xi$. In the complex case, we may put $\varepsilon_n = e^{2\pi i/n}$ and note that

$$\varepsilon_n^1 \xi_1 + \cdots + \varepsilon_n^n \xi_n \stackrel{d}{=} r\xi,$$

for some $r \in \mathbb{C} \setminus \{0\}$. This implies $\varepsilon_n^k \xi \stackrel{d}{=} \xi$, and so by continuity $r\xi \stackrel{d}{=} |r|\xi$. Hence, (33) remains true with $r_1, \ldots, r_n$ and r replaced by their absolute values, and so it suffices to prove (34) for positive $r_1, \ldots, r_n$ and r.

In all three cases, define $s_1, s_2, \ldots \geq 0$ by

$$\xi_1 + \cdots + \xi_n \stackrel{d}{=} s_n\xi, \quad n \in \mathbb{N}.$$

Then by iteration $s(n^k) = s_n^k$ for all $k, n \in \mathbb{N}$, where $s(n) = s_n$, and also

$$s_m\xi_1 + s_{n-m}\xi_2 \stackrel{d}{=} s_n\xi, \quad 1 \leq m < n.$$

If s_m/s_n is unbounded, we may divide by $s_m \vee s_{n-m}$ and then take limits along a subsequence to obtain $\xi_1 + c\xi_2 = 0$ a.s. for some $c \in [0,1]$, which is impossible. Thus, s_m/s_n is bounded for $m < n$. Applying this to the ratios $s_m^k/s_n^k = s(m^k)/s(n^k)$, we see that the sequence s_n is even non-decreasing.

Now fix any $k, m, n \in \mathbb{N}$, and choose $h \in \mathbb{N}$ such that $m^h \leq n^k \leq m^{h+1}$. Then

$$h \log s_m \ \leq \ k \log s_n \ \leq \ (h+1) \log s_m,$$
$$(h+1) \log m \ \geq \ k \log n \ \geq \ h \log m.$$

Dividing the two formulas and letting $k \to \infty$ for fixed m and n, we get

$$\frac{\log s_m}{\log m} = \frac{\log s_n}{\log n}, \quad m, n \in \mathbb{N},$$

which implies $\log s_n = c \log n = \log n^c$ for some $c \geq 0$, and hence $s_n = n^c$. Here in fact $c > 0$, since $c = 0$ would imply $\xi = 0$ a.s. In particular, we get for any $t_1, t_2 \in \mathbb{N}$

$$t_1^c \xi_1 + t_2^c \xi_2 \stackrel{d}{=} (t_1 + t_2)^c \xi, \tag{36}$$

which extends by division and continuity to arbitrary $t_1, t_2 \geq 0$. This is clearly equivalent to (33) and (34) with $p = c^{-1}$.

Letting φ denote the characteristic function or Laplace transform of $\Re\xi$, we see from (36) that

$$\varphi(t_1^c)\varphi(t_2^c) = \varphi((t_1 + t_2)^c), \quad t_1, t_2 \geq 0,$$

which shows that the function $f(t) = \varphi(t^c)$ satisfies the Cauchy equation $f(t_1)f(t_2) = f(t_1 + t_2)$. Since f is bounded and continuous, we obtain $f(t) = e^{-at}$ for some $a \geq 0$, and so $\varphi(t) = e^{-at^p}$ for all $t \geq 0$. This proves that $\mathcal{L}(\Re\xi)$ is unique up to a normalization. In the complex case, we get

$$\Re(r\xi) \stackrel{d}{=} \Re(|r|\xi) = |r|\Re\xi, \quad r \in \mathbb{C},$$

and so the uniqueness extends to $\mathcal{L}(\xi)$ by Lemma 1.34.

If $\xi \geq 0$ a.s., then $\varphi(t) = Ee^{-t\xi}$ is convex on $\mathbb{R}_+$, and we get $p \leq 1$. In the symmetric case, we put $\eta = \Re\xi$ and note that

$$\frac{1 - e^{-at^p}}{t^2} = \frac{1 - \varphi(t)}{t^2} = E \frac{1 - \cos t\eta}{t^2 \eta^2} \eta^2, \quad t > 0.$$

If $p > 2$, then as $t \to 0$ we get $0 = \frac{1}{2}E\eta^2$, which implies $\eta = 0$ a.s. and therefore $\xi = 0$ a.s. Hence, in this case $p \in (0,2]$.

To prove the existence, we note that any constant $\xi \geq 0$ is trivially positive 1-stable. Further note that if $\eta \stackrel{d}{=} \zeta$ are independent, centered, Gaussian, then η and $\xi = \eta + i\zeta$ are real and complex, symmetric 2-stable.

Next let η be a Poisson process on $(0, \infty)$ with intensity measure $cx^{-p-1}dx$, and put $\xi = \int x\eta(dx)$. Then $\xi < \infty$ for $p \in (0, 1)$, and by FMP 12.2 we have

$$
\begin{aligned}
\log Ee^{-t\xi} &= -c \int_0^\infty (1 - e^{-tx}) x^{-p-1} dx \\
&= -ct^p \int_0^\infty (1 - e^{-x}) x^{-p-1} dx,
\end{aligned}
$$

which shows that ξ is positive p-stable. Finally, let η be a Poisson process on $\mathbb{C} \setminus \{0\}$ whose intensity measure $E\eta$ has Lebesgue density $c|z|^{-p-2}$, and put $\xi = \int z\eta(dz)$, where the integral is defined as the limit along the regions $B_\varepsilon^c = \{|z| > \varepsilon\}$. This limit exists a.s. for $p \in (0, 2)$, and we have

$$
\begin{aligned}
\log E \exp(i\Re(t\xi)) &= -c \int_0^\infty r^{-p-1} dr \int_0^{2\pi} (1 - \exp(ir|t| \cos \theta)) \, d\theta \\
&= -4c|t|^p \int_0^\infty (1 - \cos r) \, r^{-p-1} dr \int_0^{\pi/2} (\cos \theta)^p \, d\theta,
\end{aligned}
$$

which shows that $\Re\xi$ and ξ are real and complex, symmetric p-stable, respectively. $\qquad\square$

The previous results lead easily to an extension of Theorem 1.31. Here we say that an infinite random sequence $\xi = (\xi_1, \xi_2, \dots)$ in $\mathbb{C}$, $\mathbb{R}$, or $\mathbb{R}_+$ is l^p-*symmetric* if (33) holds for all $n \in \mathbb{N}$ and r and $r_1, \dots, r_n$ in $\mathbb{C}$, $\mathbb{R}$, or $\mathbb{R}_+$, respectively, satisfying (34).

Theorem 1.35 (l^p-*symmetry, Bretagnolle, Dacunha-Castelle, and Krivine*) *An infinite random sequence $\xi = (\xi_j)$ in $\mathbb{C}$, $\mathbb{R}$, or $\mathbb{R}_+$ is l^p-symmetric iff $\xi = \sigma\eta$ a.s. for some i.i.d., p-stable sequence $\eta = (\eta_j)$ and an independent random variable $\sigma \geq 0$. In that case, σ is a.s. unique up to a normalization.*

Proof: Let ξ be l^p-symmetric. It is then exchangeable by Lemma 1.34, and so by Theorem 1.1 and Corollary 1.6 it is conditionally i.i.d., given the tail σ-field $\mathcal{T}_\xi$. Using Lemma 1.34 again, we note that (33) remains true with coefficients r and r_j related by (34), conditionally on $\xi_{n+1}, \dots, \xi_m$ for arbitrary $m > n$. By direct and reverse martingale convergence (FMP 7.23), ξ is then conditionally l^p-symmetric given $\mathcal{T}_\xi$, hence conditionally i.i.d. p-stable. Fixing a p-stable i.i.d. sequence $\zeta \perp\!\!\!\perp \xi$, we see from Proposition 1.33 that ξ is conditionally distributed as a positive multiple $\sigma\zeta$, where σ is $\mathcal{T}_\xi$-measurable by the law of large numbers. Since $\zeta \perp\!\!\!\perp \sigma$, we obtain $\xi \overset{d}{=} \sigma\zeta$, and the corresponding a.s. relation follows by the transfer theorem in FMP 6.10. $\qquad\square$

We turn to a general, abstract version of the last result. Given a linear space L over $F = \mathbb{C}$, $\mathbb{R}$, or $\mathbb{R}_+$, we define a *linear random functional* on L as an F-valued process ξ on L such that

$$
\xi(af + bg) = a\xi f + b\xi g \text{ a.s.}, \quad a, b \in F, \ f, g \in L.
$$

In the special case where L is a function space, we say that ξ has *independent increments* if $\xi f_1, \ldots, \xi f_n$ are independent whenever $f_1, \ldots, f_n$ have disjoint supports. In particular, we may fix a σ-finite measure $(S, \mathcal{S}, \mu)$ and a constant $p > 0$, and choose $L = L^p(\mu) = L_F^p(\mu)$, defined as the class of F-valued, measurable functions f on S with $\|f\|_p < \infty$. By a *p-stable noise* on S with *control measure* μ we mean a linear random functional ξ on $L^p(\mu)$ such that

$$\xi f \stackrel{d}{=} \|f\|_p \zeta, \quad f \in L^p(\mu), \tag{37}$$

for some p-stable random variable ζ in F. We need to establish the existence and uniqueness of ξ.

Lemma 1.36 *(p-stable noise) Fix a σ-finite measure space $(S, \mathcal{S}, \mu)$, let $F = \mathbb{C}$, $\mathbb{R}$, or $\mathbb{R}_+$, and let $p \in (0, 1]$ when $F = \mathbb{R}_+$ and $p \in (0, 2]$ otherwise. Then there exists a linear random functional ξ on $L_F^p(\mu)$ satisfying (37) for some p-stable random variable $\zeta \neq 0$ in F. Here ξ has independent increments, and $\mathcal{L}(\xi)$ is unique up to a normalization.*

Proof: For $p = 1$ and $F = \mathbb{R}_+$, we can simply take $\xi = \mu$. For $p = 2$ and $F = \mathbb{R}$, we may choose ξ to be an *isonormal Gaussian process* on the Hilbert space $L^2(\mu)$, defined as a centered Gaussian process on L^2 with covariance function $E\xi(f)\xi(g) = \langle f, g \rangle$. Finally, when $p = 2$ and $F = \mathbb{C}$, we may take $\xi = \eta + i\zeta$, where η and ζ are independent copies of the Gaussian process in the previous case.

To prove the existence for $p \in (0, 2)$, or for $p \in (0, 1)$ when $F = \mathbb{R}_+$, let η be a Poisson process on $S \times (F \setminus \{0\})$ with intensity measure $\mu \otimes \nu$, where ν has Lebesgue density $|z|^{-p-2}$ when $F = \mathbb{C}$ and $|x|^{-p-1}$ when $F = \mathbb{R}$ or $\mathbb{R}_+$, and put

$$\xi f = \iint x f(s) \, \eta(ds \, dx), \quad f \in L^p.$$

Proceeding as in the proof of Proposition 1.33, we can easily verify that ξ satisfies (37) for a suitable ζ.

By Lemma 1.34 we note that the distribution of ξ is uniquely determined by that of ζ. The asserted uniqueness up to normalizations then follows from the corresponding statement in Proposition 1.33. Utilizing the uniqueness, it is enough to prove the independence of the increments for the special processes constructed above. But for those, the statement follows easily from basic properties of Poisson processes and Gaussian distributions. $\square$

It is now easy to establish a version of Proposition 1.35 for L^p-symmetric linear random functionals ξ on L^p, defined by the requirement that $\|f\|_p = \|g\|_p$ implies $\xi f \stackrel{d}{=} \xi g$. To allow a wider range of applications, we consider the more general case when ξ is defined on an abstract linear space L, mapped into $L^p(\mu)$ by a linear operator A. Here we need to assume that the image AL is an infinite-dimensional subset of L^p, in the sense that the closure $\overline{AL}$ contains infinitely many functions $f_1, f_2, \ldots \in L^p$ with positive norms and disjoint supports. When this is true, we say that AL is a _separating_ subspace of L^p. We also say that $\mathcal{L}(\xi f)$ depends only on $\|Af\|_p$ if $\xi f \overset{d}{=} \xi g$ whenever $f, g \in L$ are such that $\|Af\|_p = \|Ag\|_p$.

Theorem 1.37 _(L^p-symmetric random functionals) Consider a linear random functional ξ on a linear space L over $F = \mathbb{C}$, $\mathbb{R}$, or $\mathbb{R}_+$, a σ-finite measure μ on a Borel space S, and a linear operator $A \colon L \to L^p_F(\mu)$ such that AL is a separating subspace of L^p. Then $\mathcal{L}(\xi f)$ depends only on $\|Af\|_p$ iff_

$$\xi f = \sigma \eta(Af) \quad a.s., \quad f \in L, \tag{38}$$

for some p-stable noise η on S with control measure μ and an independent random variable $\sigma \geq 0$. The latter is then a.s. unique up to a normalization.

Proof: We may define a linear random functional ζ on AL by

$$\zeta(Af) = \xi f \quad a.s., \quad f \in L. \tag{39}$$

To see that ζ is well-defined, suppose that $Af = Ag$ for some $f, g \in L$. By the linearity of A we have for any $a, b \in F$

$$A(af + bg) = aAf + bAg = (a + b)Af = A((a + b)f),$$

and so by the linearity and invariance of ξ

$$a\xi f + b\xi g = \xi(af + bg) \overset{d}{=} \xi((a + b)f) = a\xi f + b\xi f.$$

Hence, Lemma 1.34 yields $(\xi f, \xi g) \overset{d}{=} (\xi f, \xi f)$, and so $\xi f = \xi g$ a.s. To verify the a.s. linearity of ζ, let $f, g \in L$ and $a, b \in F$ be arbitrary, and conclude from the linearity of A and ξ that a.s.

$$
\begin{aligned}
\zeta(aAf + bAg) &= \zeta(A(af + bg)) = \xi(af + bg) \\
&= a\xi f + b\xi g = a\zeta(Af) + b\zeta(Ag).
\end{aligned}
$$

To extend ζ to the closure $\overline{AL}$, let $f_1, f_2, \ldots \in L$ with $\|Af_n\|_p \to 0$. Fixing any $g \in L$ with $\|Ag\|_p = 1$ and using the linearity of A and the homogeneity of the norm, we get

$$\|Af_n\|_p = \|Af_n\|_p \|Ag\|_p = \|A(\|Af_n\|_p g)\|_p,$$

and so by the linearity and invariance of ξ

$$\zeta(Af_n) = \xi f_n \overset{d}{=} \xi(\|Af_n\|_p g) = \|Af_n\|_p \xi g \overset{P}{\to} 0.$$

Hence, ζ is uniformly L^p/L^0-continuous on AL, and so it extends a.s. uniquely to an L^0-continuous process on $\overline{AL}$. The linearity on AL extends by continuity to $\overline{AL}$, and we note that $\mathcal{L}(\zeta f)$ still depends only on $\|f\|_p$.

Since AL is separating, we may choose $f_1, f_2, \ldots \in \overline{AL}$ with disjoint supports and norm 1. Then for any $r_1, \ldots, r_n \in F$ we have

$$\|\textstyle\sum_k r_k f_k\|_p^p = \sum_k \|r_k f_k\|_p^p = \sum_k |r_k|^p = \left\|(\textstyle\sum_k |r_k|^p)^{1/p} f_1\right\|_p^p,$$

and so

$$\textstyle\sum_k r_k \zeta f_k = \zeta \sum_k r_k f_k \overset{d}{=} \left(\sum_k |r_k|^p\right)^{1/p} \zeta f_1.$$

By Proposition 1.35 there exists a p-stable sequence (η_j) and an independent random variable $\sigma \geq 0$ such that $\zeta f_j = \sigma \eta_j$ a.s. for all j. Letting $\eta \perp\!\!\!\perp \sigma$ be a p-stable noise on $L^p(\mu)$ with the same normalization, we get for any $f \in \overline{AL}$

$$\zeta f \overset{d}{=} \zeta(\|f\|_p f_1) = \|f\|_p \sigma \eta_1 \overset{d}{=} \sigma \eta f,$$

and so by Lemma 1.34 we have $\zeta \overset{d}{=} \sigma \eta$ a.s. on $\overline{AL}$. The corresponding a.s. relation is then obtained from FMP 6.10, and (38) follows by means of (39). To prove the a.s. uniqueness of σ, we may use the law of large numbers, as before. □

By a *linear isometry* of $L^p = L_F^p(\mu)$ we mean a linear operator U on L^p such that $\|Uf\|_p = \|f\|_p$ for all $f \in L^p$. Given a linear random functional ξ on L^p, we say that $\mathcal{L}(\xi)$ is invariant under linear isometries of L^p if $\xi U \overset{d}{=} \xi$ for all such operators U, where $(\xi U)f = \xi(Uf)$. By Lemma 1.34 this is equivalent to the condition $\xi(Uf) \overset{d}{=} \xi f$ for all $f \in L^p$. We consider the following strengthened version of Theorem 1.37, in the special case where A is the identity mapping on L^p and μ is diffuse.

Theorem 1.38 *(invariance under L^p-isometries) Let ξ be a linear random functional on $L^p = L_F^p(\mu)$, where μ is a diffuse, σ-finite measure on a Borel space S, $F = \mathbb{C}$, $\mathbb{R}$, or $\mathbb{R}_+$, and $p > 0$. Then $\mathcal{L}(\xi)$ is invariant under linear isometries of L^p iff $\xi = \sigma \eta$ a.s. for some p-stable noise on S with control measure μ and an independent random variable $\sigma \geq 0$.*

To deduce the result from Theorem 1.37, we need the following technical result.

Lemma 1.39 *(extension) Let $f, g \in L^p = L_F^p(\mu)$ with $\|f\|_p = \|g\|_p$, where μ is a diffuse, σ-finite measure on a Borel space S, $F = \mathbb{C}$, $\mathbb{R}$, or $\mathbb{R}_+$, and $p > 0$. Then $f = Ug$ or $g = Uf$ for some linear isometry U on L^p.*

Proof: Since S is Borel, we may assume that $S = [0, 1]$. First note that if μ_1 and μ_2 are diffuse measures on S with $\mu_1 1 = \mu_2 1 < \infty$ and distribution functions $F_i(t) = \mu_i[0, t]$, then the formula $Vf = f \circ F_1^{-1} \circ F_2$ defines a linear isometry $V : L^p(\mu_1) \to L^p(\mu_2)$ satisfying $V1 = 1$. If fact, writing

$\varphi = F_1^{-1} \circ F_2$ and and noting that $\mu_1 = \mu_2 \circ \varphi^{-1}$ since μ_2 is diffuse, we get for any $f \in L^p(\mu_1)$

$$\mu_2 |Vf|^p = \mu_2 |f \circ \varphi|^p = (\mu_2 \circ \varphi^{-1}) |f|^p = \mu_1 |f|^p.$$

Now fix any $f, g \in L^p(\mu)$ with $\|f\|_p = \|g\|_p$, and let A and B denote the corresponding supports. We may assume that either $\mu B^c > 0$ or $\mu A^c = \mu B^c = 0$, since we can otherwise interchange the roles of f and g. By the previously mentioned result there exists a linear isometry $V_1 \colon L^p(|f|^p \cdot \mu) \to L^p(|g|^p \cdot \mu)$ satisfying $V_1 1 = 1$, and we may define

$$U_1 h = g V_1(h/f), \quad h \in L^p(1_A \cdot \mu).$$

Then clearly $U_1 f = g$ on B, and moreover

$$
\begin{aligned}
(1_B \cdot \mu) |U_1 h|^p &= (1_B \cdot \mu) |g V_1(h/f)|^p \\
&= (|g|^p \cdot \mu) |V_1(h/f)|^p \\
&= (|f|^p \cdot \mu) |h/f|^p = (1_A \cdot \mu) |h|^p,
\end{aligned}
$$

which shows that U_1 is a linear isometry from $L^p(1_A \cdot \mu)$ to $L^p(1_B \cdot \mu)$.

Next we may choose some measurable functions $f', g' \in L^p$ with supports A^c and $B' \subset B^c$, respectively, such that $\|f'\|_p = \|g'\|_p$. Proceeding as before in case of f and g, we may construct a linear isometry $U_2 \colon L^p(1_{A^c} \cdot \mu) \to L^p(1_{B_c} \cdot \mu)$ satisfying $U_2 f' = g'$. Now define

$$Uh = 1_B U_1(1_A h) + 1_{B^c} U_2(1_{A^c} h), \quad h \in L^p(\mu).$$

Then $Uf = g$, and for any $h \in L^p(\mu)$ we have

$$
\begin{aligned}
\mu |Uh|^p &= (1_B \cdot \mu) |U_1(1_A h)|^p + (1_{B^c} \cdot \mu) |U_2(1_{A^c} h)|^p \\
&= (1_A \cdot \mu) |h|^p + (1_{A^c} \cdot \mu) |h|^p = \mu |h|^p,
\end{aligned}
$$

which shows that U is a linear isometry on $L^p(\mu)$. $\qquad\square$

Proof of Theorem 1.38: Assume that $\mathcal{L}(\xi)$ is invariant under linear L^p-isometries, and fix any $f, g \in L^p$ with $\|f\|_p = \|g\|_p$. By Lemma 1.39 there exists a linear isometry U on L^p such that either $Uf = g$ or $Ug = f$. Since $\xi U \overset{d}{=} \xi$ by hypothesis, we have $\xi f \overset{d}{=} \xi g$. The desired representation now follows by Theorem 1.37. $\qquad\square$

1.8   Miscellaneous Complements

Here we are listing, with details omitted, some interesting problems or statements related to the previous material. The motivated reader is invited to work through the list as a set of exercises.

1. Derive part (ii) of Theorem 1.26 from part (i) by a limiting argument. (*Hint:* Let $\xi = (\xi_k)$ be an infinite, contractable sequence in $\{0, 1\}$, and define $\rho_n = \sum_{k \leq n} \xi_k / n$. Then $\rho_n \xrightarrow{d} \rho$ along a subsequence for some random variable ρ in $[0, 1]$. Using (i), let $n \to \infty$ for fixed m to check that $E \prod_{k \leq m} \xi_k = E \rho^m$. Finally, use the contractability and a monotone-class argument to verify that ξ has the required distribution.)

2. Derive part (iv) of Theorem 1.26 from part (iii) by a limiting argument. (*Hint:* Let ξ be a simple, exchangeable point process on $\mathbb{R}_+$. Then (iii) yields $P\{\xi B = 0\} = E(1 - \lambda B/n)^{\kappa_n}$, $B \in \mathcal{B}[0, n]$, $n \in \mathbb{N}$, where $\kappa_n = \xi[0, n]$. Show that $\kappa_n / n \xrightarrow{d} \rho$ along a subsequence for some random variable $\rho \geq 0$, and conclude that $P\{\xi B = 0\} = E e^{-\rho \lambda B}$.)

3. Let φ be a random closed set in $[0, 1]$ or $\mathbb{R}_+$, and assume that φ is exchangeable, in the sense that the events $A_{nk} = \{\varphi \cap [(k-1)/n, k/n)\}$, $k \leq n$ or $k \in \mathbb{N}$, are exchangeable for fixed $n \in \mathbb{N}$. Show that φ is a.s. locally finite on $\{\varphi^c \neq \emptyset\}$. Then use Theorem 1.26 to derive a representation of $P\{\varphi \cap [0, t] = \emptyset\}$. (*Hint:* In the $[0, 1]$-case, define $\nu_n = \sum_k 1_{A_{nk}}$, and note that $\nu_n \uparrow \nu$. Show that $\varphi = [0, 1]$ a.s. on $\{\nu = \infty\}$ and that φ is otherwise locally finite. In the $\mathbb{R}_+$-case, consider the random sets $\varphi_n = \varphi \cap [0, n]$ for arbitrary n.)

4. Use the spherical symmetry of the standard Gaussian distribution in $\mathbb{R}^n$ to compute the area A_n of the n-dimensional unit sphere. Then derive the volume V_n of the unit ball in $\mathbb{R}^n$. (*Hint:* The area of a sphere of radius $r > 0$ equals $A_n r^{n-1}$.)

5. Let $\xi_1, \ldots, \xi_4$ be i.i.d. $N(0, 1)$, and put $\rho_k = (\xi_1^2 + \cdots + \xi_k^2)^{1/2}$. Show that the random variables ρ_1 / ρ_3 and ρ_2^2 / ρ_4^2 are $U(0, 1)$.

6. For $n \in \mathbb{N}$, let $(\xi_n^1, \ldots, \xi_n^n)$ be uniformly distributed over the unit sphere in $\mathbb{R}^n$. Show that the ξ_n^j are asymptotically independent, in the sense that $\|\mathcal{L}(\xi_n^1, \ldots, \xi_n^k) - \bigotimes_{j \leq k} \mathcal{L}(\xi_n^j)\| \to 0$ as $n \to \infty$ for fixed k. Prove the same statement when the ξ_n^j are extreme, exchangeable in j for fixed n.

7. For $n \in \mathbb{N}$, let ξ_n be an extreme, exchangeable, marked point process on $[0, n]$ with κ_n points, where the sequence (κ_n / n) is tight. Show that the processes ξ_n have asymptotically independent increments, in the sense of the previous problem. Show also that the result may fail in general, without the stated tightness condition.

8. Let ξ be a mixed binomial process on $[0, 1]$ such that $\xi I \perp\!\!\!\perp \xi J$ for some disjoint, non-degenerate intervals $I, J \subset [0, 1]$. Show that ξ is Poisson.

9. Let ξ be a mixed binomial process on $[0, 1]$ such that $(\xi I \wedge 1) \perp\!\!\!\perp (\xi J \wedge 1)$ for any two disjoint intervals $I, J \subset [0, 1]$. Show that ξ is Poisson.

10. Let ξ be a mixed binomial process on $[0,1]$ such that $\xi \overset{d}{=} \xi_p$ for every $p \in (0,1)$, where ξ_p is a p-thinning of some point process η_p on $[0,1]$. Show that ξ is a mixed Poisson process.

11. Let $\tau_1, \tau_2, \ldots$ form a mixed binomial or Poisson process ξ on some interval I, and fix any integers $m < n$. Show that, on the set $\{\xi I \geq n\}$, the points $(\tau_k - \tau_m)/(\tau_n - \tau_m)$, $m < k < n$, form a binomial process independent of $\tau_1, \ldots, \tau_m$ and $\tau_n, \tau_{n+1}, \ldots$.

12. (*Rényi*) Let $\tau_1 < \cdots < \tau_n$ form a binomial process on $\mathbb{R}_+$ based on an exponential distribution μ, and put $\tau_0 = 0$. Show that the variables $(n - k)(\tau_{k+1} - \tau_k)$, $0 \leq k < n$, are i.i.d. μ.

13. (*Kendall*) Let ξ be a mixed Poisson process on $\mathbb{R}_+$ with directing measure $\alpha\mu(dt) = \alpha e^t dt$, where $P\{\alpha > x\} = e^{-x}$. Show that $Z_t = \xi[0,t] + 1$ is a birth process on $\mathbb{N}$ with rates $\lambda_k = k$. (*Hint:* Let Z be a birth process, as stated, with jumps at $\tau_1 < \tau_2 < \cdots$. Compute the joint density of $\tau_1, \ldots, \tau_n$ and check that, conditionally on τ_n, the variables $\tau_1, \ldots, \tau_{n-1}$ form a binomial process on $[0, \tau_n]$ based on μ. Conclude from Corollary 1.28 that ξ is mixed Poisson with directing measure of the form $\alpha\mu$. It remains to check that $Ee^{-\alpha\mu[0,t]} = P\{\tau_1 > t\} = e^{-t}$ when $P\{\alpha > x\} = e^{-x}$.)

14. Derive the Hausdorff–Bernstein Theorem A4.1 from Theorem 1.26. (*Hint:* Given a completely monotone sequence $a = (a_n)$ or function f with $a_0 = 1$ or $f(0+) = 1$, use the Daniell–Kolmogorov theorem to construct an associated exchangeable random sequence or point process ξ, and apply Theorem 1.26 to find the form of a or f.)

15. Derive Schoenberg's Theorem A4.3 from Theorem 1.31. (*Hint:* Let f be continuous with $f(0) = 1$ and such that $f_n(u) = f(|u|^2)$ is nonnegative definite on $\mathbb{R}^n$ for every n. Then the f_n are characteristic functions by Bochner's Theorem A4.2, and by the Daniell–Kolmogorov theorem there exists an infinite random sequence $\xi = (\xi_k)$ such that $(\xi_1, \ldots, \xi_n)$ has characteristic function f_n for every n. Here ξ is rotatable by the Cramér–Wold theorem, and the required form of f follows from Theorem 1.31. For the converse assertion, use Bernstein's Theorem A4.1.)

Chapter 2

Conditioning and Martingales

In this chapter we apply the powerful machinery of modern martingale theory and stochastic calculus to the study of contractable or exchangeable random sequences and processes, where the basic symmetries are now related to an underlying filtration. We begin in Section 2.1 with the discrete-time theory, where infinite contractable sequences are characterized by strong stationarity and by the martingale property of the associated prediction sequence. Here we also characterize finite or infinite, exchangeable sequences by the reverse martingale property of their empirical distributions. The corresponding results in continuous time are considered in Section 2.2.

A deeper analysis is presented in Section 2.3, where exchangeable processes on $\mathbb{R}_+$ or $[0, 1]$ are characterized as special semi-martingales, whose local characteristics are absolutely continuous with martingale densities. Similar methods allow us in Section 2.4, under a suitable moment condition, to derive the missing representation formula for exchangeable processes on $[0, 1]$. In Section 2.5 we use martingale techniques to establish some powerful norm inequalities for exchangeable processes, and in Section 2.6 we use a suitable super-martingale to estimate the rates of local growth.

Our final Section 2.7 deals with invariance properties of reduced Palm distributions, characterizing exchangeable random measures and point processes on an abstract space, in the spirit of Slivnyak's celebrated result for Poisson processes on the line.

2.1 Contractable Sequences

The notions of exchangeability and contractability may be related to a filtration in a natural way. Thus, a finite or infinite random sequence $\xi = (\xi_1, \xi_2, \ldots)$ is said to be exchangeable or contractable with respect to a filtration $\mathcal{F} = (\mathcal{F}_0, \mathcal{F}_1, \ldots)$ if it is $\mathcal{F}$-adapted and such that, for $n = 0, 1, \ldots$, the shifted sequence $\theta_n \xi = (\xi_{n+1}, \xi_{n+2}, \ldots)$ is conditionally exchangeable or contractable given $\mathcal{F}_n$. To avoid relying on the existence of regular conditional distributions, we may take the conditional statement to mean that the stated property holds for $\theta_n \xi$, conditionally on A, for every $A \in \mathcal{F}_n$ with $PA > 0$. The new conditions of exchangeability or contractability clearly reduce to the previous, unqualified ones when $\mathcal{F}$ is the filtration induced by ξ.

For infinite sequences, it is useful to prove an extended version of Theorem 1.1 involving the stronger versions of exchangeability and contractability. Then say that ξ is *conditionally $\mathcal{F}$-i.i.d.* if there exists a random probability measure ν on S such that

$$P[\xi_{k+1} \in \cdot \,|\, \mathcal{F}_k, \nu] = \nu \text{ a.s.}, \quad k \in \mathbb{Z}_+. \tag{1}$$

This will be seen to be equivalent to the seemingly stronger relation

$$P[\theta_k \xi \in \cdot \,|\, \mathcal{F}_k, \nu] = \nu^\infty \text{ a.s.}, \quad k \in \mathbb{Z}_+. \tag{2}$$

We can also establish some less obvious equivalences with properties involving optional times and martingales. Let us say that the sequence ξ is *strongly stationary* or *$\mathcal{F}$-stationary* if $\theta_\tau \xi \stackrel{d}{=} \xi$ for every optional (or stopping) time $\tau < \infty$. For finite sequences $\xi = (\xi_1, \ldots, \xi_n)$, the condition is interpreted as

$$(\xi_{\tau+1}, \ldots, \xi_{\tau+k}) \stackrel{d}{=} (\xi_1, \ldots, \xi_k) \text{ whenever } \tau + k \le n \text{ a.s.}$$

The *$\mathcal{F}$-prediction sequence* $\mu = (\mu_k)$ of ξ is defined by $\mu_k = P[\theta_k \xi \in \cdot \,|\, \mathcal{F}_k]$, and we say that μ is an *$\mathcal{F}$-martingale* if $\mu_k f = E[f(\theta_k \xi)|\mathcal{F}_k]$ is a martingale in the usual sense for every bounded, measurable function f on the appropriate space. In this form, the definition makes sense even when S is not Borel and μ may have no measure-valued version. For finite sequences $\xi = (\xi_1, \ldots, \xi_n)$, we may interpret the martingale condition as

$$(1_A, \xi_{k+2}, \ldots, \xi_n) \stackrel{d}{=} (1_A, \xi_{k+1}, \ldots, \xi_{n-1}), \quad A \in \mathcal{F}_k, \ 0 \le k \le n-2.$$

The mentioned properties are related as follows.

Proposition 2.1 *(strong stationarity and prediction) Let ξ be an infinite, $\mathcal{F}$-adapted random sequence in a Borel space S, with $\mathcal{F}$-prediction sequence μ. Then these conditions are equivalent:*

(i) *ξ is conditionally $\mathcal{F}$-i.i.d.,*
(ii) *ξ is $\mathcal{F}$-exchangeable,*
(iii) *ξ is $\mathcal{F}$-contractable,*
(iv) *$\theta_\tau \xi \stackrel{d}{=} \xi$ for every $\mathcal{F}$-optional time $\tau < \infty$,*
(v) *μ is a measure-valued $\mathcal{F}$-martingale.*

Conditions (iii)–(v) *remain equivalent for finite sequences ξ.*

Proof: First assume (i), in the form of condition (1). Letting $f_1, \ldots, f_n \ge 0$ be measurable functions on S and iterating (1), we obtain

$$
\begin{aligned}
E[f_1(\xi_{k+1}) &\cdots f_n(\xi_{k+n})|\mathcal{F}_k, \nu] \\
&= E[f_1(\xi_{k+1}) \cdots f_{n-1}(\xi_{k+n-1})\, E[f_n(\xi_{k+n})|\mathcal{F}_{k+n-1}, \nu]|\mathcal{F}_k, \nu] \\
&= E[f_1(\xi_{k+1}) \cdots f_{n-1}(\xi_{k+n-1})|\mathcal{F}_k, \nu]\, \nu f_n \\
&= \cdots = \nu f_1 \cdots \nu f_n,
\end{aligned}
$$

which extends to (2) by a monotone-class argument. In particular, ξ satisfies conditions (ii) and (iii), which are equivalent by Theorem 1.1.

Conversely, (ii) implies that the sequence $\theta_n \xi$ is exchangeable over $\mathcal{F}_n$ for every $n \in \mathbb{Z}_+$, and so by Corollary 1.5 there exist some random probability measures ν_n on S such that a.s.

$$P[\theta_n \xi \in \cdot \,|\, \nu_n, \mathcal{F}_n] = P[\theta_n \xi \in \cdot |\nu_n] = \nu_n^\infty, \quad n \in \mathbb{Z}_+.$$

In particular, we obtain $P[\theta_m \xi \in \cdot |\nu_n] = \nu_n^\infty$ a.s. for all $m > n$, and so by Proposition 1.4 the random measures ν_n are a.s. independent of n. Hence, (2) holds with $\nu = \nu_0$, which proves (i). This shows that (i)–(iii) are equivalent.

To prove the equivalence of (iii)–(v), we may restrict our attention to infinite sequences ξ, the proof in the finite case requiring only obvious changes. Assuming (iii), we get for $k \in \mathbb{Z}_+$ and any bounded, measurable function $f \colon S^\infty \to \mathbb{R}$

$$E[\mu_{k+1} f | \mathcal{F}_k] = E[f(\theta_{k+1} \xi) | \mathcal{F}_k] = E[f(\theta_k \xi) | \mathcal{F}_k] = \mu_k f,$$

which proves (v). Conversely, for any k and f as before, we get from (v)

$$E[f(\theta_{k+1} \xi) | \mathcal{F}_k] = E[\mu_{k+1} f | \mathcal{F}_k] = \mu_k f = E[f(\theta_k \xi) | \mathcal{F}_k],$$

which implies

$$(1_A, \xi_{h+1}, \dots, \xi_k, \xi_{k+2}, \dots) \overset{d}{=} (1_A, \xi_{h+1}, \dots), \quad A \in \mathcal{F}_h, \ h \le k.$$

Iterating this relation for fixed h and A, we see that $\theta_h \xi$ is conditionally contractable given $\mathcal{F}_h$, and (iii) follows.

Next we see from FMP 6.2 that $\mu_\tau = P[\theta_\tau \xi \in \cdot | \mathcal{F}_\tau]$ for every $\mathcal{F}$-optional time $\tau < \infty$. By FMP 7.13 we conclude that (v) is equivalent to

$$Ef(\theta_\tau \xi) = E\mu_\tau f = E\mu_0 f = Ef(\xi)$$

for any bounded measurable function f, which is in turn equivalent to (iv). $\square$

For finite sequences $\xi = (\xi_1, \dots, \xi_n)$, we proceed to characterize exchangeability by the *strong reflection property*

$$(\xi_{\tau+1}, \dots, \xi_n) \overset{d}{=} (\xi_n, \dots, \xi_{\tau+1}),$$

where τ is an arbitrary optional time in $\{0, \dots, n-1\}$. We may abbreviate the condition as $Q_\tau \xi \overset{d}{=} \theta_\tau \xi$, where the *reflection operators* Q_k are defined for fixed n by $Q_k(a_1, \dots, a_n) = (a_n, \dots, a_{k+1})$. We also say that ξ is a *conditional $\mathcal{F}$-urn sequence* with *occupation measures* $\beta_k = \sum_{j \le k} \delta_{\xi_j}$, $k \le n$, if a.s.

$$P[\xi_{k+1} \in \cdot \,|\, \beta_n, \mathcal{F}_k] = \frac{\beta_n - \beta_k}{n - k}, \quad k < n. \tag{3}$$

This will be seen to be equivalent to the seemingly stronger relation

$$P[\theta_k \xi \in \cdot \mid \beta_n, \mathcal{F}_k] = \frac{(\beta_n - \beta_k)^{(n-k)}}{(n-k)!}, \quad k < n, \tag{4}$$

where $(\beta_n - \beta_k)^{(n-k)}$ denotes the factorial measure of $\beta_n - \beta_k$, first introduced in connection with Proposition 1.8.

Proposition 2.2 *(reflection invariance) Let ξ be a finite, $\mathcal{F}$-adapted random sequence in a Borel space S. Then these conditions are equivalent:*

(i) *ξ is $\mathcal{F}$-exchangeable,*

(ii) *ξ has the $\mathcal{F}$-reflection property,*

(iii) *ξ is a conditional $\mathcal{F}$-urn sequence.*

Proof: First assume (iii), in the form of condition (3). Then for any measurable functions $f_{k+1}, \ldots, f_n \geq 0$ on S, we have

$$
\begin{aligned}
E[f_{k+1}&(\xi_{k+1}) \cdots f_n(\xi_n)|\beta_n, \mathcal{F}_k] \\
&= E[f_{k+1}(\xi_{k+1}) \cdots f_{n-1}(\xi_{n-1})\, E[f_n(\xi_n)|\beta_n, \mathcal{F}_{n-1}]|\beta_n, \mathcal{F}_k] \\
&= E[f_{k+1}(\xi_{k+1}) \cdots f_{n-1}(\xi_{n-1})|\beta_n, \mathcal{F}_k]\,(\beta_n - \beta_{n-1})f_n \\
&= \cdots = \prod_{j=k+1}^{n} \frac{(\beta_n - \beta_{j-1})f_j}{n-j} = \frac{(\beta_n - \beta_k)^{(n-k)}}{(n-k)!} \bigotimes_{j=k+1}^{n} f_j,
\end{aligned}
$$

which extends to (4) by a monotone-class argument. This shows in particular that ξ satisfies (i). Since reflections are special cases of permutations, the implication (i) $\Rightarrow$ (ii) is obvious.

Now assume (ii). Considering optional times in the set $\{k, n\}$, we obtain

$$P[Q_{n-k}\theta_k\xi \in \cdot|\mathcal{F}_k] = P[\theta_k\xi \in \cdot|\mathcal{F}_k] \quad \text{a.s.,} \quad k \leq n-2.$$

Hence, for any $k \leq n-2$ and $A \in \mathcal{F}_k$,

$$(1_A, \xi_{k+1}, \ldots, \xi_m, \xi_n, \ldots, \xi_{m+1}) \overset{d}{=} (1_A, \xi_{k+1}, \ldots, \xi_n), \quad k < m < n.$$

Noting that the reflections on the left generate all permutation of $\theta_k\xi$, we conclude that the latter sequence is exchangeable over 1_A, and (i) follows since A was arbitrary.

Next assume (i). Then for any permutation p of $(k+1, \ldots, n)$,

$$(1_A, \xi_1, \ldots, \xi_k, \theta_k\xi \circ p) \overset{d}{=} (1_A, \xi), \quad A \in \mathcal{F}_k.$$

Since β_n is invariant under permutations of ξ, we obtain

$$(1_A, \beta_n, \theta_k\xi \circ p) \overset{d}{=} (1_A, \beta_n, \theta_k\xi), \quad A \in \mathcal{F}_k,$$

which shows that $\theta_k \xi$ is exchangeable over $(\beta_n, \mathcal{F}_k)$. Hence, for any measurable function $f \geq 0$ on S,

$$
\begin{aligned}
(n-k)E[f(\xi_{k+1})| \, \beta_n, \mathcal{F}_k] &= \sum_{j>k} E[f(\xi_j)| \, \beta_n, \mathcal{F}_k] \\
&= E\Big[\sum_{j>k} f(\xi_j)\Big| \, \beta_n, \mathcal{F}_k\Big] \\
&= (\beta_n - \beta_k)f,
\end{aligned}
$$

which proves (iii). $\qquad\square$

To state the next result, recall that an integer-valued random time τ is said to be *predictable* with respect to the discrete filtration $\mathcal{F}$ if $\tau - 1$ is $\mathcal{F}$-optional. The following result shows that, under additional hypotheses, the strong stationarity in Proposition 2.1 can be replaced by the weaker condition $\xi_\tau \overset{d}{=} \xi_1$ for any finite, predictable time $\tau \geq 1$. This is equivalent to a weaker version of the previous martingale property.

Theorem 2.3 *(local prediction) Let ξ be an infinite, stationary random sequence with induced filtration $\mathcal{F}$, taking values in a Borel space S. Then these conditions are equivalent:*

(i) *ξ is $\mathcal{F}$-exchangeable,*

(ii) *ξ is $\mathcal{F}$-contractable,*

(iii) *$\xi_\tau \overset{d}{=} \xi_1$ for every finite, $\mathcal{F}$-predictable time $\tau \geq 1$,*

(iv) *$\mu_k = P[\xi_{k+1} \in \cdot | \mathcal{F}_k]$ is a measure-valued $\mathcal{F}$-martingale.*

Conditions (i)–(iv) *remain equivalent for any finite sequence ξ adapted to a filtration $\mathcal{F}$, provided that the occupation measure β is $\mathcal{F}_0$-measurable.*

Proof: In both the finite and infinite cases we have trivially (i) $\Rightarrow$ (ii), and Proposition 2.1 yields (ii) $\Rightarrow$ (iii). Arguing as in the proof of the latter result, we see that also (iii) $\Leftrightarrow$ (iv). Thus, it remains to show that (iv) implies (i).

First consider a finite sequence ξ of length n, adapted to a filtration $\mathcal{F}$ such that β is $\mathcal{F}_0$-measurable. Assuming (iv), we get for $k < n$ and any bounded, measurable function f on S

$$
\begin{aligned}
E[f(\xi_{k+1})| \, \beta, \mathcal{F}_k] &= E[f(\xi_{k+1})|\mathcal{F}_k] = \cdots = E[f(\xi_n)|\mathcal{F}_k] \\
&= (n-k)^{-1}\sum_{j>k} E[f(\xi_j)|\mathcal{F}_k] \\
&= (n-k)^{-1} E[\beta f - \beta_k f|\mathcal{F}_k] \\
&= (n-k)^{-1}(\beta - \beta_k)f,
\end{aligned}
$$

which proves condition (iii) of Proposition 2.2. Even condition (i) of the same result is then fulfilled, which means that ξ is $\mathcal{F}$-exchangeable.

Now suppose instead that ξ is infinite and stationary with induced filtration $\mathcal{F}$. Define $\nu = P[\xi_1 \in \cdot | \mathcal{I}_\xi]$, where $\mathcal{I}_\xi$ denotes the shift-invariant σ-field of ξ. Assuming (iv) and using the point-wise ergodic theorem (FMP 10.6) and dominated convergence, we get for any bounded, measurable function f on S

$$
\begin{aligned}
E[f(\xi_{k+1})|\mathcal{F}_k] &= E[f(\xi_{k+2})|\mathcal{F}_k] = \cdots \\
&= n^{-1}\sum_{j\leq n}E[f(\xi_{k+j})|\mathcal{F}_k] \\
&= E\left[n^{-1}\sum_{j\leq n}f(\xi_{k+j})\middle|\mathcal{F}_k\right] \to E[\nu f|\mathcal{F}_k],
\end{aligned}
$$

which implies

$$
E[f(\xi_{k+1})|\mathcal{F}_k] = E[\nu f|\mathcal{F}_k] \text{ a.s.,} \quad k \in \mathbb{Z}_+. \tag{5}
$$

We now extend ξ to a stationary sequence $\hat{\xi}$ indexed by $\mathbb{Z}$ (FMP 10.2) and define $\hat{\mathcal{F}}_k = \sigma\{\hat{\xi}_j; j \leq k\}$ for all $k \in \mathbb{Z}$. Using the stationarity of $\hat{\xi}$ and martingale convergence, and noting that ν is $\hat{\mathcal{F}}_k$-measurable for every k, we may strengthen (5) to

$$
E[f(\xi_{k+1})|\,\nu, \hat{\mathcal{F}}_k] = E[f(\xi_{k+1})|\hat{\mathcal{F}}_k] = E[\nu f|\hat{\mathcal{F}}_k] = \nu f,
$$

and so $P[\xi_{k+1} \in \cdot|\nu, \mathcal{F}_k] = \nu$, which implies (i). $\qquad\square$

Finally, we consider a reverse martingale that plays a basic role in subsequent sections. Recall that, for any finite or infinite sequence ξ, the associated *empirical distributions* are given by

$$
\eta_n = n^{-1}\sum_{k\leq n}\delta_{\xi_k}, \quad n \geq 1.
$$

Theorem 2.4 *(empirical distributions) Let ξ be a finite or infinite random sequence with empirical distributions $\eta_1, \eta_2, \ldots$. Then ξ is exchangeable iff the η_k form a reverse, measure-valued martingale.*

Proof: We introduce the tail filtration

$$
\mathcal{T}_k = \sigma(\theta_{k-1}\eta) = \sigma(\eta_k, \theta_k\xi), \quad k \geq 1,
$$

where the second equality follows from the relations

$$
k\eta_k = (k-1)\eta_{k-1} + \delta_{\xi_k}, \quad k \geq 1. \tag{6}
$$

Suppose that ξ is exchangeable. Then $(\xi_1, \ldots, \xi_m)$ is exchangeable over $\mathcal{T}_m$ for every m in the index set of ξ, and so for $k = 1, \ldots, m$ and any measurable function $f \geq 0$ on S, we have a.s.

$$
\begin{aligned}
E[f(\xi_1)|\mathcal{T}_m] &= \cdots = E[f(\xi_k)|\mathcal{T}_m] \\
&= k^{-1}\sum_{j\leq k}E[f(\xi_j)|\mathcal{T}_m] \\
&= E\left[k^{-1}\sum_{j\leq k}f(\xi_j)\middle|\mathcal{T}_m\right] = E[\eta_k f|\mathcal{T}_m].
\end{aligned}
$$

Taking $k = m - 1$ and m gives

$$E[\eta_{m-1}f|\mathcal{T}_m] = E[\eta_m f|\mathcal{T}_m] = \eta_m f \quad \text{a.s.,}$$

which proves that $\eta = (\eta_k)$ is a reverse martingale.

Conversely, suppose that η is a reverse martingale. Then by (6) we have for $k \geq 1$ and any bounded, measurable function f on S

$$E[f(\xi_k)|\mathcal{T}_k] = k\eta_k f - (k - 1)E[\eta_{k-1}f|\mathcal{T}_k] = \eta_k f. \tag{7}$$

Fixing $m \geq 1$, we define for $k = 1, \ldots, m$

$$\zeta_k = \xi_{m-k+1}, \qquad \beta_k = \sum_{j \leq k} \delta_{\zeta_j} = m\eta_m - (m - k)\eta_{m-k}.$$

The ζ_k are clearly adapted to the filtration $\mathcal{F}_k = \mathcal{T}_{m-k}$, $0 \leq k \leq m$, and by (7) we have for $k = 0, \ldots, m - 1$

$$
\begin{aligned}
E[f(\zeta_{k+1})|\beta_m, \mathcal{F}_k] &= E[f(\xi_{m-k})|\mathcal{T}_{m-k}] = \eta_{m-k}f \\
&= (m - k)^{-1}(\beta_m - \beta_k)f.
\end{aligned}
$$

Hence, the sequence $(\zeta_1, \ldots, \zeta_m) = (\xi_m, \ldots, \xi_1)$ is exchangeable by Proposition 2.2, which shows that even ξ is exchangeable. □

2.2 Continuous-Time Symmetries

Given a filtration $\mathcal{F}$, we say that a process X is $\mathcal{F}$-*exchangeable* or $\mathcal{F}$-*contractable* if it is $\mathcal{F}$-adapted and such that, for every $t \geq 0$, the shifted process $\theta_t X - X_t$ is conditionally exchangeable or contractable given $\mathcal{F}_t$. Next we say that an $\mathcal{F}$-adapted process X is conditionally $\mathcal{F}$-*Lévy* if there exists a random triple $\gamma = (\alpha, \rho, \nu)$ such that, for every $t \geq 0$, the process $\theta_t X - X_t$ is conditionally Lévy with characteristics γ, given $(\mathcal{F}_t, \gamma)$. As in the discrete-time case, we note that these notions reduce to the unqualified ones when $\mathcal{F}$ is the filtration induced by X.

The strong stationarity of X is now defined in terms of the increments. Thus, we say that X has $\mathcal{F}$-*stationary increments* if $\theta_\tau X - X_\tau \overset{d}{=} X$ for every optional time $\tau < \infty$. Finally, we consider the martingale property of the $\mathcal{F}$-*prediction process*

$$\mu_t = P[\theta_t X - X_t \in \cdot|\mathcal{F}_t], \quad t \geq 0.$$

The definitions for processes on $[0, 1]$ require only some obvious modifications, similar to those needed in discrete time.

The case of random measures ξ on a product space $I \times S$ is similar, where $I = \mathbb{R}_+$, $\mathbb{Q}_+$, $[0, 1]$, or $\mathbb{Q}_{[0,1]}$, and S is an arbitrary Borel space. Thus, we say that ξ is $\mathcal{F}$-*exchangeable* or $\mathcal{F}$-*contractable* if $\theta_t \xi$ is exchangeable or contractable over $\mathcal{F}_t$ for every t, where $\theta_t \xi$ denotes the shifted measure defined by $(\theta_t \xi) B = \xi(B + t)$ for every measurable subset $B \subset I \times S$. The *strong stationarity* of ξ is now defined by the condition $\theta_\tau \xi \overset{d}{=} \xi$ for every optional time $\tau < \infty$, and the *prediction process* is given by $\mu_t = P[\theta_t \xi \in \cdot | \mathcal{F}_t]$.

We begin with a continuous-time version of Proposition 2.1.

Proposition 2.5 *(strong stationarity and prediction)   Let X be an $\mathbb{R}^d$-valued, $\mathcal{F}$-adapted process on $\mathbb{Q}_+$ with $\mathcal{F}$-prediction process μ and with $X_0 = 0$. Then these statements are equivalent:*

(i) X *is conditionally $\mathcal{F}$-Lévy,*
(ii) X *is $\mathcal{F}$-exchangeable,*
(iii) X *is $\mathcal{F}$-contractable,*
(iv) $\theta_\tau X - X_\tau \overset{d}{=} X$ *for every optional time $\tau < \infty$,*
(v) μ *is a measure-valued $\mathcal{F}$-martingale.*

The equivalence of (iii)–(v) extends to processes on $\mathbb{Q}_{[0,1]}$. All statements remain true for right-continuous processes on $\mathbb{R}_+$ or $[0, 1]$, and they extend with obvious changes to random measures on $(0, \infty) \times S$ or $(0, 1] \times S$, for any Borel space S.

Proof: For processes on $\mathbb{Q}_+$ or $\mathbb{Q}_{[0,1]}$, the arguments for Proposition 2.1 apply with obvious changes. Thus, instead of using Theorem 1.1 and the associated uniqueness assertions in Proposition 1.4, we need to employ the continuous-time versions of the same statements from Theorem 1.19. Assuming (v), we now obtain

$$P[\theta_t X - X_t \in \cdot | \mathcal{F}_s] = P[\theta_s X - X_s \in \cdot | \mathcal{F}_s], \quad 0 \le s \le t,$$

and it follows as before that, conditionally on $\mathcal{F}_t$ for fixed $t \ge 0$, the shifted process $\theta_t X - X_t$ has contractable increments, which proves (iii). Next we note as before that (v) is equivalent to $\theta_\tau X - X_\tau \overset{d}{=} X$, for any optional time $\tau < \infty$ that takes only countably many values.

The result extends with the same proof to processes on $\mathbb{R}_+$ or $[0, 1]$, except that (iv) is only obtained in the weaker form with the optional time τ restricted to a finite index set. To deduce the general version of (iv), we may approximate τ from the right by such elementary optional times τ_n (FMP 7.4) and conclude that by right continuity

$$(\theta_{\tau_n} X)_s - X_{\tau_n} \to (\theta_\tau X)_s - X_\tau, \quad s \ge 0. \tag{8}$$

The relation $\theta_{\tau_n} X - X_{\tau_n} \overset{d}{=} X$ then extends to the limit. The same argument proves the result for random measures on $(0, \infty) \times S$ or $(0, 1] \times S$. $\square$

For processes X on $[0, 1]$, we define the *$\mathcal{F}$-reflection property* by the condition $Q_\tau X \overset{d}{=} \theta_\tau X - X_\tau$, in the sense of finite-dimensional distributions and for any $\mathcal{F}$-optional time τ in $[0, 1]$, where

$$\begin{aligned} (Q_t X)_s &= X_1 - X_{(1-s) \vee t}, \\ (\theta_t X)_s &= X_{(t+s) \wedge 1}, \end{aligned} \qquad s, t \in [0, 1].$$

Note that the processes $Q_\tau X$ and $\theta_\tau X - X_\tau$ are both defined on $[0, 1]$, start at 0, and are constantly equal to $X_1 - X_\tau$ on $[1 - \tau, 1]$. For random measures ξ on a product space $(0, 1] \times S$, the reflection $Q_t \xi$ is defined instead as a random measure on $[0, 1 - t) \times S$, given by

$$(Q_t \xi)(B \times C) = \xi((1 - B) \times C), \quad B \in \mathcal{B}[0, 1 - t), \ C \in \mathcal{S}.$$

We may now state a partial extension of Proposition 2.2 to continuous time. The representation problem for exchangeable processes on $[0, 1]$ is more difficult and will not be addressed until Theorem 2.18.

Proposition 2.6 *(reflection invariance) Let X be an $\mathbb{R}^d$-valued, $\mathcal{F}$-adapted process on $\mathbb{Q}_{[0,1]}$ with $X_0 = 0$. Then X is $\mathcal{F}$-exchangeable iff it has the $\mathcal{F}$-reflection property. The result remains true for right-continuous processes on $[0, 1]$, and also for random measures on $(0, 1] \times S$, where S is Borel.*

Proof: Recall from Theorem 1.15 that a process on $\mathbb{Q}_{[0,1]}$ is exchangeable iff it has exchangeable increments. We may then apply Proposition 2.2 to the increments of X on an arbitrary rational interval $[t, 1]$, and conclude that X is $\mathcal{F}$-exchangeable iff the reflection property $Q_\tau X \stackrel{d}{=} \theta_\tau X - X_\tau$ holds for any optional time $\tau \leq 1$ taking only finitely many rational values. This extends immediately to the general case.

For right-continuous processes X on $[0, 1]$, the same argument applies at any real time $t \in [0, 1]$ and shows that X is $\mathcal{F}$-exchangeable iff $Q_\tau X \stackrel{d}{=} \theta_\tau X - X_\tau$ for every optional time τ taking values in some set $A_h = \{1 - jh; \ j \geq 0\}$ with $h > 0$. For general τ, we may approximate from the right by such times τ_n and infer from the right-continuity of X that (8) holds and $(Q_{\tau_n} X)_s \to (Q_\tau X)_s$ for all s. Hence, the reflection property at τ_n carries over to τ.

For random measures ξ on $(0, 1] \times S$, we note that the measure-valued process $X_t = \xi((0, t] \times \cdot)$ has no fixed discontinuities under either condition. We may then proceed as before to see that ξ is $\mathcal{F}$-exchangeable iff $Q_\tau \xi \stackrel{d}{=} \theta_\tau \xi$ for every optional time τ that takes at most countably many values. To extend the latter condition to arbitrary optional times τ, we may approximate by such special times $\tau_n \downarrow \tau$. Assuming $S = \mathbb{R}$, we get $Q_{\tau_n} \xi \stackrel{v}{\to} Q_\tau \xi$ and $\theta_{\tau_n} \xi \stackrel{v}{\to} \theta_\tau \xi$, where $\stackrel{v}{\to}$ denotes vague convergence in the measure space $\mathcal{M}((0, 1] \times \mathbb{R})$. This yields the required reflection property at τ. $\quad\square$

In order to apply the more advanced results of martingale theory and stochastic calculus to exchangeable or contractable processes X, we need the underlying filtration $\mathcal{F}$ to be right-continuous and complete, where the latter property is the requirement that each $\mathcal{F}_t$ should contain all null sets in the completion of $\mathcal{F}_\infty$. The following result shows that this can always be assumed, as long as X is right-continuous. For a precise statement, recall (FMP 7.8) that any filtration $\mathcal{F}$ has a smallest extension $\overline{\mathcal{F}}$ with the mentioned properties, often referred to as the *usual augmentation* of $\mathcal{F}$.

Lemma 2.7 *(augmented filtration)* *Consider a right-continuous,* $\mathbb{R}^d$*-valued process* X *on* I *or a random measure* ξ *on* $I \times S$*, where* S *is Borel and* $I = \mathbb{R}_+$ *or* $[0,1]$*, and suppose that* X *or* ξ *is exchangeable or contractable with respect to a filtration* $\mathcal{F}$*. Then the same property holds for the right-continuous and complete extension of* $\mathcal{F}$*.*

Proof: If $\theta_t X - X_t$ is exchangeable or contractable over $\mathcal{F}_t$, it remains so over $\mathcal{F}_{s+}$ for every $s < t$. Letting $t \downarrow s$ and using the right continuity of X, we conclude that $\theta_s X - X_s$ is exchangeable or contractable over $\mathcal{F}_{s+}$. The last statement extends immediately to any completion of $\mathcal{F}_{s+}$, and it remains to note that $\overline{\mathcal{F}}_s$ agrees with the P-completion of $\mathcal{F}_{s+}$ with respect to $\mathcal{F}_\infty$. The proof for random measures is similar.   □

The remainder of this section is devoted to a study of exchangeable or contractable, marked point processes on $\mathbb{R}_+$ or $[0,1]$. The results, of interest in their own right, prepare for our treatment of the general semi-martingale case in subsequent sections.

Recall that the *compensator* of a marked point process ξ on $S \times I$, where $I = [0,1]$ or $\mathbb{R}_+$, is defined as the a.s. unique, predictable random measure $\hat{\xi}$ on the same space such that $\xi - \hat{\xi}$ is a measure-valued, local martingale (FMP 25.22). By a *density* of $\hat{\xi}$ we mean a measurable, measure-valued process $\eta = (\eta_t)$ on S, hence a kernel from $\Omega \times I$ to S, such that $\hat{\xi}[0,t] = (\lambda \otimes \eta)_t = \int_0^t \eta_s ds$ a.s. for all $t \in I$. A *conditional martingale* on an interval I is defined as a process M on I such that $E[|M_t| \| \mathcal{F}_s] < \infty$ and $E[M_t | \mathcal{F}_s] = M_s$ a.s. for all $s < t$ in I.

Theorem 2.8 *(compensator density)* *Let* ξ *be an* $\mathcal{F}$*-contractable, marked point process on* $S \times [0,1]$*, where* S *is Borel. Then* ξ *has* $\mathcal{F}$*-compensator* $\hat{\xi} = \lambda \otimes \eta$*, where* $\eta = (\eta_t)$ *is a conditional, measure-valued* $\mathcal{F}$*-martingale on* $(0,1)$*, given by*

$$\eta_t = \frac{E[\xi_1 - \xi_t | \mathcal{F}_t]}{1 - t}, \quad t \in [0,1).$$

Our proof is based on a couple of technical lemmas.

Lemma 2.9 *(conditional integrability)* *Let* ξ *be a simple,* $\mathcal{F}$*-contractable point process on* $[0,1]$ *or* $\mathbb{R}_+$*. Then*

$$E[\xi[0,t] | \mathcal{F}_s] < \infty \quad a.s., \quad 0 < s < t.$$

Proof: Fix any $t > s > 0$, and note that ξ is a mixed binomial process on $[0,t]$ by Theorem 1.26. Writing binomial coefficients as $(n /\!\!/ k)$ and putting $r = s/t$ and $\xi_s = \xi[0,s]$, we get for any $k \in \mathbb{Z}_+$

$$
\begin{aligned}
E[E[\xi_t | \mathcal{F}_s]; \, \xi_s = k] &= E[\xi_t; \, \xi_s = k] = E\,\xi_t P[\xi_s = k | \xi_t] \\
&= \sum_{n \geq k} n\, P\{\xi_t = n\}\, (n /\!\!/ k)\, r^k\, (1 - r)^{n-k} \\
&\leq \sum_{n \geq k} n\, (n /\!\!/ k)\, (1 - r)^{n-k} < \infty,
\end{aligned}
$$

where the last series converges by the ratio test, since its terms a_n satisfy

$$\frac{a_n}{a_{n-1}} = \frac{n^2(1-r)}{(n-1)(n-k)} \to 1 - r < 1.$$

Hence, $E[\xi_t|\mathcal{F}_s] < \infty$ a.s. on the set $\bigcup_k \{\xi_s = k\} = \{\xi_s < \infty\} = \Omega$. □

Lemma 2.10 *(conditional martingales) Any conditional martingale M on $\mathbb{R}_+$ is a local martingale.*

Proof: Put $\alpha_t = E[\|M_t\| \mid \mathcal{F}_0]$, and note that for $t \leq 1$

$$\alpha_t = E[\|E[M_1|\mathcal{F}_t]\| \mid \mathcal{F}_0] \leq E[E[\|M_1\| \mid \mathcal{F}_t] \mid \mathcal{F}_0] = \alpha_1,$$

by the conditional form of Jensen's inequality. Defining

$$M_t^n = M_t 1\{\alpha_1 \leq n\}, \quad t \in [0,1], \ n \in \mathbb{N},$$

we get for any $t \leq 1$

$$E|M_t^n| = E[|M_t|; \alpha_1 \leq n] = E[\alpha_t; \alpha_1 \leq n] \leq E[\alpha_1; \alpha_1 \leq n] \leq n.$$

Furthermore, we have for any $s \leq t$ in $[0,1]$

$$E[M_t^n|\mathcal{F}_s] = E[M_t; \alpha_1 \leq n|\mathcal{F}_s] = M_s 1\{\alpha_1 \leq n\} = M_s^n,$$

which shows that M^n is a martingale on $[0,1]$. Thus, M is a local martingale on the same interval. It is now easy to continue inductively and show that M is a local martingale on $\mathbb{R}_+$. □

Proof of Theorem 2.8: Put $\xi_t = \xi(\cdot \times [0,t])$, and introduce the processes

$$\zeta_t = E[\xi_1|\mathcal{F}_t], \quad \eta_t = \frac{\zeta_t - \xi_t}{1-t}, \quad t \in [0,1). \tag{9}$$

By Lemmas 2.9 and 2.10 we note that, for any $B \in \mathcal{S}$ with $\xi_1 B < \infty$, the process ζB is a local martingale on $(0,1)$ and hence admits a right-continuous version on the same interval. Letting $s < t$ in $(0,1)$ be such that $1 - s$ and $1 - t$ are rationally dependent, we conclude from the $\mathcal{F}$-contractability of ξ that a.s.

$$E[\eta_t B|\mathcal{F}_s] = \frac{E[\xi_1 B - \xi_t B|\mathcal{F}_s]}{1-t} = \frac{E[\xi_1 B - \xi_s B|\mathcal{F}_s]}{1-s} = \eta_s B.$$

The formula extends to arbitrary $s < t$ in $(0,1)$, by approximation of s from the right. This shows that η is a conditional martingale on $(0,1)$.

Now $\xi_t = \zeta_t - (1-t)\eta_t$ by (9), and integration by parts gives

$$d\xi_t B = d\zeta_t B - (1-t)d\eta_t B + \eta_t B \, dt.$$

Here the first two terms on the right are local martingales on $(0, 1)$, and the last term $\eta_t B\,dt$ is continuous and hence predictable. By the definition of the compensator $\hat{\xi}$ it follows that $d\hat{\xi}_t B = \eta_t B\,dt$ or $\hat{\xi}_t = \int_0^t \eta_t dt$. Thus, η has the stated property, and it remains to note that η has a measure-valued version since S is Borel. $\square$

We turn to a continuous-time version of Theorem 2.3.

Theorem 2.11 *(density criterion) Let ξ be a marked point process on $S \times I$, where S is Borel and $I = [0, 1]$ or $\mathbb{R}_+$. When $I = [0, 1]$, assume that ξ is $\mathcal{F}$-adapted and $\xi_1 = \xi(\cdot \times [0, 1])$ is $\mathcal{F}_0$-measurable, and when $I = \mathbb{R}_+$, let ξ be stationary with induced filtration $\mathcal{F}$. Then these conditions are equivalent:*

(i) *ξ is $\mathcal{F}$-exchangeable,*
(ii) *ξ is $\mathcal{F}$-contractable,*
(iii) *$\hat{\xi} = \lambda \otimes \eta$ a.s. for a conditional $\mathcal{F}$-martingale η on I°.*

Proof: Since (i) $\Rightarrow$ (ii) is obvious and (ii) $\Rightarrow$ (iii) by Theorem 2.8, it is enough to show that (iii) $\Rightarrow$ (i). Beginning with the case where $I = [0, 1]$, let ξ be $\mathcal{F}$-adapted on $S \times [0, 1]$ with an $\mathcal{F}_0$-measurable projection ξ_1 on S, and suppose that $\hat{\xi}$ admits a conditional $\mathcal{F}$-martingale density $\eta = (\eta_t)$ on $(0, 1)$. For any $A \in \mathcal{S}$ with $\xi_1 A < \infty$ a.s., we have

$$
\begin{aligned}
\xi_1 A - \xi_t A &= E[\xi_1 A - \xi_t A | \mathcal{F}_t] = E[\hat{\xi}_1 A - \hat{\xi}_t A | \mathcal{F}_t] \\
&= E\left[\int_t^1 \eta_s A\,ds \,\Big|\, \mathcal{F}_t\right] = \int_t^1 E[\eta_s A | \mathcal{F}_t]\,ds \\
&= (1 - t)\,\eta_t A,
\end{aligned}
$$

which shows that $\xi_1 - \xi_t = (1 - t)\eta_t$ for all $t \in (0, 1)$. Hence,

$$
\hat{\xi}_t = \int_0^t \frac{\xi_1 - \xi_s}{1 - s}\,ds, \quad t \in [0, 1). \tag{10}
$$

Now fix any disjoint sets $A_1, \ldots, A_n \in \mathcal{S}$ such that $\xi_1 A_k < \infty$ for all k, put $\kappa_k = \xi_1 A_k$, and let $\tau_1^k < \cdots < \tau_{\kappa_k}^k$ be the points of the process ξA_k. Define $\hat{\tau}_j^k = \hat{\xi} A_k(\tau_j^k)$ and put $\gamma_j^k = \hat{\tau}_j^k - \hat{\tau}_{j-1}^k$, where $\hat{\tau}_0^k = 0$. By (10) we have

$$
\gamma_j^k = (\kappa_k - j + 1) \log\left(\frac{1 - \tau_{j-1}^k}{1 - \tau_j^k}\right), \quad j \le \kappa_k, \ k \le n,
$$

and we may solve recursively for the τ_j^k to obtain

$$
\tau_j^k = 1 - \exp\left(-\sum_{i \le j} \frac{\gamma_i^k}{\kappa_k - i + 1}\right), \quad j \le \kappa_k, \ k \le n. \tag{11}
$$

Using the basic time-change reduction of multivariate point processes (FMP 25.26), we see that the random variables γ_i^k are conditionally independent and exponentially distributed with unit mean, given the σ-field $\mathcal{F}_0$. Hence, by (11), the distribution of $(\xi A_1, \ldots, \xi A_n)$ is uniquely determined by that of ξ_1.

Now consider a uniform randomization $\tilde{\xi}$ of ξ_1, regarded as a marked point process on $S \times [0, 1]$, and let $\tilde{\mathcal{F}}$ denote the right-continuous and complete filtration induced by $\mathcal{F}_0$ and $\tilde{\xi}$. Then by Theorem 2.8 the pair $(\tilde{\xi}, \tilde{\mathcal{F}})$ satisfies the same conditions as $(\xi, \mathcal{F})$, and so

$$(\xi A_1, \ldots, \xi A_n) \stackrel{d}{=} (\tilde{\xi} A_1, \ldots, \tilde{\xi} A_n).$$

Since $A_1, \ldots, A_n$ were arbitrary, we get $\xi \stackrel{d}{=} \tilde{\xi}$, and in particular ξ is exchangeable. Applying the same argument, for arbitrary $t \in [0, 1)$, to the conditional distribution of $\theta_t \xi$ given $\mathcal{F}_t$, we conclude that ξ is indeed $\mathcal{F}$-exchangeable.

Next suppose that ξ is stationary on $S \times \mathbb{R}_+$ with induced filtration $\mathcal{F}$, and that the compensator $\hat{\xi}$ of ξ has a conditional martingale density $\eta = (\eta_t)$ on $(0, \infty)$. Using Fubini's theorem and the martingale properties of $\xi - \hat{\xi}$ and η, we get a.s., for any measurable function $f \geq 0$ on S and times $s < t$ and $h > 0$,

$$\begin{aligned}
E[\xi_{t+h} f - \xi_t f | \mathcal{F}_s] &= E\left[\hat{\xi}_{t+h} f - \hat{\xi}_t f \,\middle|\, \mathcal{F}_s\right] = E\left[\int_t^{t+h} \eta_r f \, dr \,\middle|\, \mathcal{F}_s\right] \\
&= \int_t^{t+h} E[\eta_r f | \mathcal{F}_s] \, dr = \int_t^{t+h} \eta_s f \, dr = h \, \eta_s f,
\end{aligned}$$

where $\xi_{t+h} f - \xi_t f = \xi(\cdot \times (t, t+h]) f$. Since the right-hand side is independent of t, we may use Fubini's theorem and the mean ergodic theorem to obtain

$$\begin{aligned}
E[\xi_{s+h} f - \xi_s f | \mathcal{F}_s] &= n^{-1} \int_s^{s+n} E[\xi_{t+h} f - \xi_t f | \mathcal{F}_s] \, dt \\
&= E\left[n^{-1} \int_s^{s+n} (\xi_{t+h} f - \xi_t f) \, dt \,\middle|\, \mathcal{F}_s\right] \\
&\to E[E[\xi_h f | \mathcal{I}_\xi] | \mathcal{F}_s] = h \, E[\nu f | \mathcal{F}_s],
\end{aligned}$$

where ν denotes the sample intensity of ξ, given by $E[\xi | \mathcal{I}_\xi] = \nu \otimes \lambda$ a.s. (FMP 10.19). Since the left-hand side is independent of n, we conclude that

$$E[\xi_{s+h} f - \xi_s f | \mathcal{F}_s] = h \, E[\nu f | \mathcal{F}_s] \quad \text{a.s.}, \quad s, h \geq 0. \tag{12}$$

Let us now extend ξ to a stationary, marked point process $\hat{\xi}$ on $S \times \mathbb{R}$ with induced filtration $\hat{\mathcal{F}} = (\hat{\mathcal{F}}_s)$. By stationarity and martingale convergence, we get from (12)

$$E[\hat{\xi}_{s+h} f - \hat{\xi}_s f | \hat{\mathcal{F}}_s] = h \nu f \quad \text{a.s.}, \quad s \in \mathbb{R}, \quad h \geq 0.$$

Since f was arbitrary, it follows that $\hat{\xi}$ has $\hat{\mathcal{F}}$-compensator $\nu \otimes \lambda$. Using the general Poisson reduction in FMP 25.24, we conclude that $\hat{\xi}$ is a Cox process on $S \times \mathbb{R}$ directed by $\nu \otimes \lambda$. In particular, this implies the asserted exchangeability of ξ. $\qquad\square$

We may also prove a continuous-time version of Theorem 2.4.

Theorem 2.12 *(empirical measures) Let ξ be a marked point process on $S \times I$, where S is Borel and $I = (0, 1]$ or $(0, \infty)$, and define $\eta_t = t^{-1}\xi_t$, $t > 0$. Then ξ is exchangeable iff $\eta = (\eta_t)$ is a reverse, conditional, measure-valued martingale.*

Proof: Assume that ξ is exchangeable. Introduce the tail filtration

$$\mathcal{T}_t = \sigma(\theta_t \eta) = \sigma(\xi_t, \theta_t \xi), \quad t > 0,$$

where the second equality follows from the relation

$$(t + h)\eta_{t+h} = \xi_t + \theta_t \xi(\cdot \times (0, h]), \quad t > 0.$$

Let $s < t$ be rationally dependent in $(0, \infty)$, so that $s = mh$ and $t = nh$ for some $h > 0$ and $m, n \in \mathbb{N}$. Noting that ξ is exchangeable over $\mathcal{T}_t$ on $[0, t]$ for every $t \in I$, we get for any measurable function $f \geq 0$ on S

$$
\begin{aligned}
E[\xi_h f | \mathcal{T}_t] &= m^{-1} \sum_{k \leq m} E[\xi_{kh} f - \xi_{(k-1)h} f | \mathcal{T}_t] \\
&= m^{-1} E\Big[\sum_{k \leq m} (\xi_{kh} f - \xi_{(k-1)h} f) \Big| \mathcal{T}_t \Big] \\
&= m^{-1} E[\xi_s f | \mathcal{T}_t] = h\, E[\eta_s f | \mathcal{T}_t].
\end{aligned}
$$

Combining this with the corresponding relation for η_t gives

$$E[\eta_s f | \mathcal{T}_t] = h^{-1} E[\xi_h f | \mathcal{T}_t] = E[\eta_t f | \mathcal{T}_t] = \eta_t f, \tag{13}$$

which shows that η is a conditional, reverse martingale on $t\mathbb{Q}_+$ for every $t > 0$. To extend (13) for fixed $t > 0$ to arbitrary $s < t$, it suffices to approximate s from the right by times in $t\mathbb{Q}_+$, and then use the right-continuity of $\eta_s f$ and dominated convergence.

Now suppose instead that η is a reverse, conditional martingale. Since the jumps of $\eta_t f$ are positive for every measurable function $f \geq 0$ on S, we note in particular that η is continuous in probability. By Lemma 2.10 we may apply the martingale regularization theorem (FMP 7.27) and conclude that η remains a reverse, conditional martingale with respect to the left-continuous, complete extension $\overline{\mathcal{T}}$ of $\mathcal{T}$. Also note that the continuity of η implies the corresponding property for ξ, so that $\xi(\cdot \times \{t\}) = 0$ a.s. for all $t > 0$.

Using Fubini's theorem and the definition and martingale property of η, we get for any times $s \leq t$ and measurable functions $f \geq 0$ on S

$$
\begin{aligned}
E[\xi_t f - \xi_s f | \overline{\mathcal{T}}_t] &= E[t\eta_t f - s\eta_s f | \overline{\mathcal{T}}_t] = (t - s)\eta_t f \\
&= \int_s^t \eta_t f\, dr = \int_s^t E[\eta_r f | \overline{\mathcal{T}}_t]\, dr \\
&= E\Big[\int_s^t \eta_r f\, dr \Big| \overline{\mathcal{T}}_t \Big],
\end{aligned}
$$

which shows that ξ has $\overline{\mathcal{T}}$-compensator $\hat{\xi} = \lambda \otimes \eta$. The exchangeability of ξ now follows by Theorem 2.11. $\square$

2.3   Semi-Martingale Criteria

Our further development of the continuous-time theory is more difficult, as it depends on the sophisticated concepts and results of semi-martingale theory and general stochastic calculus. The purpose of this section is to extend any contractable process on $\mathbf{Q}_{[0,1]}$ to a semi-martingale on the real interval $[0, 1]$, and to characterize the exchangeable processes on $\mathbb{R}_+$ or $[0, 1]$ in terms of their semi-martingale characteristics.

Recall that an $\mathbb{R}^d$-valued process X is called a *special semi-martingale* if it admits a decomposition $X = M + \hat{X}$, where M is a local martingale and $\hat{X}$ is a predictable process of locally finite variation, starting at 0. The decomposition is then a.s. unique (FMP 17.2, 25.16), and $\hat{X}$ is called the *compensator* of X. We have also an a.s. unique decomposition $X = X^c + X^d$ into a continuous local martingale X^c starting at 0, called the *continuous martingale component* of X, and a purely discontinuous semi-martingale X^d (FMP 26.14). With X^c we may associate the matrix-valued *covariation process* ρ with components $\rho_t^{ij} = [X_i^c, X_j^c]_t$, $i, j \leq d$. We finally introduce the *jump point process* ξ on $(\mathbb{R}^d \setminus \{0\}) \times \mathbb{R}_+$ given by $\xi(B \times I) = \sum_{t \in I} 1_B(\Delta X_t)$, along with the associated compensator $\hat{\xi}$. The three processes $\hat{X}$, ρ, and $\hat{\xi}$ are often referred to, collectively, as the *local characteristics* of X.

For technical reasons, we need to impose an integrability condition on X. Say that a process X is *(uniformly) $\mathcal{F}_0$-integrable* if it is a.s. (uniformly) integrable with respect to the conditional distribution $P[X \in \cdot | \mathcal{F}_0]$.

Theorem 2.13 *(regularization and martingale properties) Let X be an $\mathcal{F}_0$-integrable, $\mathcal{F}$-contractable, $\mathbb{R}^d$-valued process on $\mathbf{Q}_{[0,1]}$, and let $\overline{\mathcal{F}}$ denote the right-continuous, complete augmentation of $\mathcal{F}$. Then X extends to a uniformly $\overline{\mathcal{F}}_0$-integrable, $\overline{\mathcal{F}}$-contractable, special $\overline{\mathcal{F}}$-semi-martingale on $[0, 1)$ with jump point process ξ, such that $[X^c]$ is a.s. linear, $\hat{X} = M \cdot \lambda$, and $\hat{\xi} = \lambda \otimes \eta$, for some conditional $\overline{\mathcal{F}}$-martingales M on $[0, 1)$ and η on $(0, 1)$ given by*

$$M_t = \frac{E[X_1 - X_t | \mathcal{F}_t]}{1 - t}, \quad \eta_t = \frac{E[\xi_1 - \xi_t | \mathcal{F}_t]}{1 - t}, \qquad t \in [0, 1).$$

Proof: Introduce the processes

$$N_t = E[X_1 | \mathcal{F}_t], \quad M_t = \frac{N_t - X_t}{1 - t}, \qquad t \in \mathbf{Q}_{[0,1)}, \tag{14}$$

and note that N is a conditional $\mathcal{F}$-martingale. By the $\mathcal{F}$-contractability of X, we have for any $s \leq t$ in $\mathbf{Q}_{[0,1)}$

$$E[M_t | \mathcal{F}_s] = \frac{E[X_1 - X_t | \mathcal{F}_s]}{1 - t} = \frac{E[X_1 - X_s | \mathcal{F}_s]}{1 - s} = M_s,$$

which shows that even M is a conditional $\mathcal{F}$-martingale. By the martingale regularization theorem (FMP 7.27) and Lemma 2.10, the right-hand limits M_{t+} and N_{t+} exist outside a fixed P-null set and form conditional $\overline{\mathcal{F}}$-martingales on $[0,1)$. Noting that by (14)

$$X_t = N_t - (1-t)M_t, \quad t \in \mathbf{Q}_{[0,1)}, \tag{15}$$

and using Lemma 2.10, we see that X_{t+} exists in the same sense and defines an $\overline{\mathcal{F}}$-semi-martingale on $[0,1)$. The $\mathcal{F}$-contractability of X extends to $\overline{\mathcal{F}}$ and X_+ by approximation from the right. Furthermore, for rational $t \in [0,1)$ and $h_n > 0$ with $h_n \to 0$, the contractability of X yields a.s.

$$X_{t+} - X_t \leftarrow X_{t+h_n} - X_t \stackrel{d}{=} X_{t+2h_n} - X_{t+h_n} \to X_{t+} - X_{t+} = 0,$$

which shows that $X_{t+} = X_t$ a.s. for every $t \in \mathbf{Q}_{[0,1)}$. Thus, X_+ is a.s. an extension of X to $[0,1]$, and we may henceforth write X instead of X_+.

Equation (15) extends immediately to $[0,1)$. In particular, X is uniformly $\overline{\mathcal{F}}_0$-integrable on $[0,\frac{1}{2}]$, and the same property holds on $[0,1]$ by the contractability of X. Integration by parts (FMP 26.6) in (15) gives

$$dX_t = dN_t - (1-t)dM_t + M_t dt, \quad t \in [0,1),$$

which shows that X is a special $\overline{\mathcal{F}}$-semi-martingale on $[0,1)$ with compensator $d\hat{X}_t = M_t dt$. By Theorem 1.15 the point process ξ on $(\mathbf{R}^d \setminus \{0\}) \times [0,1]$ inherits the $\overline{\mathcal{F}}$-contractability from X, and so by Theorem 2.8 it has $\overline{\mathcal{F}}$-compensator $\hat{\xi} = \lambda \otimes \eta$, where η is the stated conditional, measure-valued $\overline{\mathcal{F}}$-martingale on $(0,1)$. Finally, Theorem 1.15 ensures that the co-variation processes $\rho_t^{ij} = [X_i^c, X_j^c]_t$ are again $\overline{\mathcal{F}}$-contractable as well as continuous and of locally finite variation (FMP 17.5), properties that carry over to the increasing and decreasing Jordan components (FMP 2.18) of ρ. Hence, Theorem 1.26 yields $\rho_t^{ij} = \rho_1^{ij}t$ a.s. for some $\overline{\mathcal{F}}_0$-measurable random variables ρ_1^{ij}. $\square$

The necessary conditions in Theorem 2.13 are not sufficient to guarantee exchangeability, and further hypotheses are needed. Since different sets of conditions are required for processes on $[0,1]$ and $\mathbf{R}_+$, we treat the two cases separately. We begin with two martingale criteria for exchangeable processes on $\mathbf{R}_+$. For general semi-martingales X we define the local characteristics as before, except that the jumps of X need to be suitably truncated before we form the compensator $\hat{X}$. The precise way of truncation is inessential.

Theorem 2.14 *(martingale criteria on $\mathbf{R}_+$, Grigelionis, Kallenberg)* Let X be an $\mathbf{R}^d$-valued $\mathcal{F}$-semi-martingale on $\mathbf{R}_+$ with $X_0 = 0$, where $\mathcal{F}$ is right-continuous and complete. Then each of these conditions implies that X is $\mathcal{F}$-exchangeable:

(i) *The local characteristics of X are a.s. linear;*

(ii) *X is a special semi-martingale, $\mathcal{F}$ is the filtration induced by X, the local characteristics of X admit martingale densities, and X is locally L^1-bounded with stationary increments.*

In subsequent proofs we shall often use the basic reduction Theorem 4.5, whose proof is independent of the results in this chapter.

Proof: (i) Letting $M = (M^1, \ldots, M^d)$ denote the continuous martingale component of X, we have $[M^i, M^j]_t \equiv t\rho_{ij}$ a.s. for some $\mathcal{F}_0$-measurable random variables ρ_{ij}. Writing $U_{t,r} = 1\{r \leq t\}$ and noting that, for $i, j \leq d$ and $s, t \geq 0$,

$$M_t^i = \int_0^\infty U_{t,r} dM_r^i,$$

$$\int_0^\infty U_{s,r} U_{t,r} d[M^i, M^j]_r = \rho_{ij} \int_0^\infty 1\{r \leq s \wedge t\} \, dr = (s \wedge t) \rho_{ij},$$

we see from Theorem 4.5 that, conditionally on $\mathcal{F}_0$, the process $M = (M_t^i)$ on $\mathbb{R}_+ \times \{1, \ldots, d\}$ is centered Gaussian with covariance function

$$E[M_s^i M_t^j | \mathcal{F}_0] = (s \wedge t) \rho_{ij}, \quad s, t \geq 0, \quad i, j \leq d.$$

In other words, M is conditionally a Brownian motion in $\mathbb{R}^d$ with covariance matrix $\rho = (\rho_{ij})$.

Next we consider the jump point process ξ of X and note that $\hat{\xi}_t \equiv t\nu$ a.s. for some $\mathcal{F}_0$-measurable random measure ν on $\mathbb{R}^d \setminus \{0\}$. Using Theorem 4.5 with V as the identity mapping on $\mathbb{R}_+ \times (\mathbb{R}^d \setminus \{0\})$, we see that, conditionally on $\mathcal{F}_0$, the process ξ is independent of M and Poisson with intensity measure $\lambda \otimes \nu$. Furthermore, FMP 26.6 (iv) yields

$$\int_0^t \int |x|^2 \xi(ds\,dx) < \infty \text{ a.s.}, \quad t \geq 0,$$

and so, by FMP 12.13 (i) and (iii), the integrals

$$N_t = \int_0^t \int_{|x| \leq 1} x \, (\xi - \lambda \otimes \nu)(ds\,dx), \quad t \geq 0,$$

converge a.s. and define a local martingale.

We may finally introduce the process J of jumps of modulus > 1, and note that the residual $\hat{X}_t = X_t - M_t - N_t - J_t$ is a.s. linear by hypothesis and hence of the form $t\alpha$ for some $\mathcal{F}_0$-measurable random vector α in $\mathbb{R}^d$. Comparing with the general representation in FMP 15.4, we conclude that X is conditionally a Lévy process with characteristic triple (α, ρ, ν), given the σ-field $\mathcal{F}_0$. In particular, X is then exchangeable over $\mathcal{F}_0$. The same argument shows that, more generally, the shifted process $\theta_t X - X_t$ is exchangeable over $\mathcal{F}_t$ for every $t \geq 0$. Thus, X is $\mathcal{F}$-exchangeable.

(ii) Here the argument is similar to that for random measures in Theorem 2.11. Writing $\hat{X}$ for the compensator of X, we have $\hat{X}_t = \int_0^t M_s ds$ for some martingale M. Noting that $X - \hat{X}$ is a true martingale and using Fubini's theorem, we get for any $s \leq t$

$$E[X_t - X_s | \mathcal{F}_s] = E[\hat{X}_t - \hat{X}_s | \mathcal{F}_s] = E\left[\int_s^t M_r dr \middle| \mathcal{F}_s\right]$$

$$= \int_s^t E[M_r | \mathcal{F}_s] \, dr = (t - s) M_s.$$

Dividing by $t - s$ and applying the mean ergodic theorem (FMP 10.9), we obtain $E[\overline{X}|\mathcal{F}_s] = M_s$ a.s. with $\overline{X} = E[X_1|\mathcal{I}_X]$, where $\mathcal{I}_X = X^{-1}\mathcal{I}$ denotes the σ-field of X-measurable events, invariant under the combined shifts $\theta_t X - X_t$, $t \geq 0$. Hence,

$$E[X_t - X_s|\mathcal{F}_s] = (t - s)E[\overline{X}|\mathcal{F}_s], \quad 0 \leq s \leq t. \tag{16}$$

We now extend X to a process on $\mathbb{R}$ with stationary increments (FMP 10.2), and write $\mathcal{F}_s^t$ for the σ-field induced by the increments of X on $(s, t]$. Then (16) holds with $\mathcal{F}_s$ replaced by $\mathcal{F}_0^s$, and by stationarity

$$E[X_t - X_s|\mathcal{F}_r^s] = (t - s)E[\overline{X}|\mathcal{F}_r^s], \quad r \leq s \leq t.$$

The formula extends by martingale convergence, first to $r = -\infty$, and then to the right-continuous and complete filtration $\mathcal{G}$ induced by $(\mathcal{F}_{-\infty}^s)$. Since $\overline{X}$ is $\mathcal{G}_s$-measurable for every s, e.g. by the ergodic theorem, we obtain

$$E[X_t - X_s|\mathcal{G}_s] = (t - s)\overline{X} \text{ a.s.}, \quad s \leq t,$$

which shows that $X_t - t\overline{X}$ is a $\mathcal{G}$-martingale.

Next consider the matrix-valued process $\rho_t = [X^c]_t = [X]_t^c$. By hypothesis, ρ admits a matrix-valued martingale density, and in particular $E|\rho_t| < \infty$ for all t. Writing $\bar{\rho} = E[\rho_1|\mathcal{I}_X]$, we see as before that $\rho_t - t\bar{\rho}$ is a $\mathcal{G}$-martingale. Since the components of ρ are continuous and of locally finite variation, we conclude (FMP 17.2) that $\rho_t \equiv t\bar{\rho}$ a.s. We finally consider the jump point process ξ on $\mathbb{R}^d \setminus \{0\}$, and note as before that $\xi_t B - t\bar{\xi} B$ is a $\mathcal{G}$-martingale for every Borel set B with $0 \notin \overline{B}$, where the random measure $\bar{\xi}$ is a.s. defined by $\bar{\xi} B = E[\xi_1 B|\mathcal{I}_X]$.

The preceding argument shows that X remains a special semi-martingale with respect to $\mathcal{G}$ and that the associated local characteristics are a.s. linear. By part (i) it follows that X is $\mathcal{G}$-exchangeable. In particular, X is exchangeable on $\mathbb{R}_+$. $\square$

We turn to a similar martingale characterization of exchangeable processes on $[0, 1]$.

Theorem 2.15 *(martingale criterion on $[0, 1]$)* *Let X be a uniformly $\mathcal{F}_0$-integrable, $\mathbb{R}^d$-valued, special $\mathcal{F}$-semi-martingale on $[0, 1)$ with $X_0 = 0$, where $\mathcal{F}$ is right-continuous and complete, and let ξ denote the jump point process of X. Suppose that $X_1 = X_{1-}$, ξ_1, and $[X^c]_1$ exist and are $\mathcal{F}_0$-measurable, and that $(X, [X^c])$ and $\hat{\xi}$ admit conditional martingale densities on $[0, 1)$ and $(0, 1)$, respectively. Then X is $\mathcal{F}$-exchangeable.*

Our proof will be based on two lemmas.

Lemma 2.16 *(local characteristics) Let X and ξ be such as in Theorem 2.15. Then $[X^c]$ is a.s. linear, and for every $t \in [0, 1)$ we have a.s.*

$$\hat{\xi}_t = \int_0^t \frac{\xi_1 - \xi_s}{1 - s} \, ds, \tag{17}$$

$$\hat{X}_t = tX_1 - \int_0^t ds \int_0^s \frac{d(X_r - \hat{X}_r)}{1 - r}. \tag{18}$$

Proof: Formula (17) agrees with (10), which was established under more general hypotheses in the proof of Theorem 2.11. Next let N denote the conditional martingale density of the matrix-valued process $\rho = [X^c]$. Using the $\mathcal{F}_0$-measurability of ρ_1, Fubini's theorem, the martingale property of N, and the integration-by-parts formula for general semi-martingales (FMP 26.6), we get for any $t \in [0, 1]$

$$\begin{aligned}
\int_t^1 N_u \, du &= \rho_1 - \rho_t = E[\rho_1 - \rho_t | \mathcal{F}_t] = E\left[\int_t^1 N_u \, du \,\middle|\, \mathcal{F}_t\right] \\
&= \int_t^1 E[N_u | \mathcal{F}_t] \, du = (1 - t) N_t \\
&= N_0 + \int_0^t (1 - s) \, dN_s - \int_0^t N_s \, ds.
\end{aligned}$$

Solving for the second term on the right gives a.s.

$$\int_0^t (1 - s) \, dN_s = \rho_1 - N_0, \quad t \in [0, 1],$$

which implies $(1 - s)dN_s = 0$ and hence $N_t \equiv N_0 = \rho_1$ a.s. This shows that $\rho_t \equiv t\rho_1$ a.s.

Let us finally introduce the conditional martingale density M of $\hat{X}$. Since $\hat{X}$ is continuous, we may use a BDG (Burkholder–Davis–Gundy) inequality (FMP 26.12) and the representation of co-variation processes in terms of the jumps (FMP 26.15) to obtain

$$E[(X - \hat{X})_1^{*2} | \mathcal{F}_0] \lesssim \text{tr}\,[X]_1 = \text{tr}\,\rho_1 + \int |x|^2 \xi_1(dx) < \infty,$$

where $X_t^* = \sup_{s \le t} |X_s|$. In particular, $X - \hat{X}$ extends to a conditional martingale on $[0, 1]$. Since $|M|$ is a conditional sub-martingale on $[0, 1)$ (FMP 7.11), we may use Fubini's theorem to get for any $t \in [0, 1)$

$$E\left[\int_0^t |M_s| ds \,\middle|\, \mathcal{F}_0\right] = \int_0^t E[|M_s| | \mathcal{F}_0] \, ds \le tE[|M_t| | \mathcal{F}_0] < \infty.$$

This justifies that we employ Fubini's theorem to write, for any $s \le t < 1$,

$$\begin{aligned}
E[X_t - X_s | \mathcal{F}_s] &= E[\hat{X}_t - \hat{X}_s | \mathcal{F}_s] = E\left[\int_s^t M_u \, du \,\middle|\, \mathcal{F}_s\right] \\
&= \int_s^t E[M_u | \mathcal{F}_s] \, du = (t - s) M_s.
\end{aligned}$$

The last formula extends to $t = 1$ by uniform integrability, and since $X_1 - X_s$ is $\mathcal{F}_s$-measurable, we obtain a.s.

$$X_1 - X_s = (1 - s)M_s, \quad s \in [0, 1]. \tag{19}$$

Integrating by parts yields

$$-(1 - s)\, dM_s + M_s ds = dX_s = d(X_s - \hat{X}_s) + d\hat{X}_s,$$

and so, by the uniqueness of the canonical decomposition,

$$d\hat{X}_s = M_s ds, \qquad dM_s = -\frac{d(X_s - \hat{X}_s)}{1 - s}.$$

Integrating both relations gives

$$
\begin{aligned}
\hat{X}_t &= \int_0^t M_s ds = \int_0^t \left(M_0 - \int_0^s \frac{d(X_r - \hat{X}_r)}{1 - r} \right) ds \\
&= tM_0 - \int_0^t ds \int_0^s \frac{d(X_r - \hat{X}_r)}{1 - r},
\end{aligned}
$$

and (18) follows since $M_0 = X_1$ a.s. in view of (19).   □

Lemma 2.17 *(uniqueness) For processes X as in Theorem 2.15, the conditional distribution $P[X \in \cdot | \mathcal{F}_0]$ is an a.s. unique, measurable function of X_1, $[X^c]_1$, and ξ_1.*

Proof: By Lemma 2.16 we have $[X^c]_t = t[X^c]_1 \equiv t\rho$ a.s., and so by Theorem 4.5 the continuous martingale component X^c is conditionally a Brownian motion with covariance matrix ρ, given the σ-field $\mathcal{F}_0$. Turning to the distribution of ξ, fix any disjoint Borel sets $A_1, \ldots, A_n$ in $\mathbb{R}^d$ that are bounded away from 0, and put $\kappa_r = \xi_1 A_r$ for $r = 1, \ldots, n$. By the proof of Theorem 2.11, we can express the points $\tau_1^r < \tau_2^r < \cdots$ of the processes $\xi_t A_r$ as

$$\tau_j^r = 1 - \exp\left(-\sum_{i \le j} \frac{\gamma_i^r}{\kappa_r - i + 1}\right), \quad j \le \kappa_r, \ r \le n,$$

where the variables γ_i^r are conditionally independent and exponentially distributed with mean 1, given $\mathcal{F}_0$. In view of Theorem 4.5, the γ_i^r are also conditionally independent of X^c, which specifies completely the joint conditional distribution of the processes X_t^c and $\xi_t A_1, \ldots, \xi_t A_n$ in terms of the random matrix ρ and the variables $\kappa_1, \ldots, \kappa_n$. By a monotone-class argument, the conditional distribution of the pair (X^c, ξ) is then determined a.s. by ρ and ξ_1. It is easy to check that this specification is also measurable.

To complete the proof, it is enough to show that X can be expressed as a measurable function of X^c and ξ. Then note that the purely discontinuous component X^d of $X - \hat{X}$ is given a.s. by

$$X_t^d = \int_0^t \int_{x \ne 0} x(\xi - \hat{\xi})(dx\, ds), \quad t \in [0, 1],$$

where the integral converges in probability since $\text{tr}\,[X]_1 < \infty$ a.s. (FMP 26.12). Since also (17) exhibits the compensator $\hat{\xi}$ in terms of ξ, we conclude that $X - \hat{X} = X^c + X^d$ is determined a.s. by (X^c, ξ). Next (18) expresses $\hat{X}$ as a function of $X - \hat{X}$, and so even $X = (X - \hat{X}) + \hat{X}$ is specified a.s. by (X^c, ξ). Again it is easy to verify that all steps in the construction are measurable. □

Proof of Theorem 2.15: Put $\alpha = X_1$, $\beta = \xi_1$, and $\rho = [X^c]_1$, let σ denote the non-negative definite square root of ρ, and write $\beta = \sum_j \delta_{\beta_j}$, where the random vectors $\beta_1, \beta_2, \ldots$ in $\mathbb{R}^d$ can be chosen to be $\mathcal{F}_0$-measurable. Introduce a Brownian bridge B in $\mathbb{R}^d$ and an independent sequence of i.i.d. $U(0, 1)$ random variables $\tau_1, \tau_2, \ldots$, all independent of $\mathcal{F}_0$. Define

$$Y_t = \alpha t + \sigma B_t + \sum_j \beta_j (1\{\tau_j \le t\} - t), \quad t \in [0, 1]. \tag{20}$$

To prove that the series converges, we note that

$$\sum_j \beta_j^2 = \int |x|^2 \xi_1(dx) = \int |x|^2 \xi_{1/2}(dx) + \int |x|^2 (\xi_1 - \xi_{1/2})(dx) < \infty,$$

since the first integral on the right is bounded by $\text{tr}[X]_{1/2} < \infty$ and the second one has the same distribution by Theorem 2.11. By the three-series criterion (FMP 4.18) it follows that, conditionally on $\mathcal{F}_0$ and for fixed $t \in [0, 1]$, the partial sums S_t^n converge a.s. and in L^2 toward some limit S_t. Since the processes $M_t^n = S_t^n / (1 - t)$ are conditional martingales on $[0, 1)$, the same thing is true for $M_t = S_t / (1 - t)$, and so by symmetry and Doob's inequality (FMP 7.16)

$$
\begin{aligned}
E[(S^n - S)_1^{*2} | \mathcal{F}_0] &\le 2E[(M^n - M)_{1/2}^{*2} | \mathcal{F}_0] \\
&\le 2^3 E[(M_{1/2}^n - M_{1/2})^2 | \mathcal{F}_0] \\
&= 2^5 E[(S_{1/2}^n - S_{1/2})^2 | \mathcal{F}_0] \to 0.
\end{aligned}
$$

The individual terms in (20) being rcll and exchangeable over $\mathcal{F}_0$, we note that the sum Y has the same properties. Furthermore, we have $Y_1 = \alpha = X_1$, and the jump sizes of Y are given by the same point process $\beta = \xi_1$ as for X. Since Y is $\mathcal{F}_0$-integrable, we conclude from Theorem 2.13 that Y is a special $\mathcal{G}$-semi-martingale, where $\mathcal{G}$ denotes the right-continuous and complete filtration generated by $\mathcal{F}_0$ and Y. Noting that the sum S is a purely discontinuous semi-martingale (FMP 26.14), we have also a.s.

$$[Y^c]_t = [\sigma B]_t = t\sigma\sigma' = t\rho = [X^c]_t.$$

We may now apply Lemma 2.17 to see that $P[X \in \cdot | \mathcal{F}_0] = P[Y \in \cdot | \mathcal{F}_0]$. In particular, since Y is exchangeable over $\mathcal{F}_0$, the same thing is true for X. Applying this result to the shifted processes $\theta_t X - X_t$ and the associated σ-fields $\mathcal{F}_t$, we see more generally that X is $\mathcal{F}$-exchangeable. □

2.4   Further Criteria and Representation

The previous arguments can be used to derive an explicit representation of exchangeable processes on $[0,1]$. For technical reasons, we impose an extra moment condition, which will be removed in Chapter 3.

Theorem 2.18 *(representation on $[0,1]$) Let X be an integrable, $\mathbb{R}^d$-valued process on $\mathbb{Q}_{[0,1]}$. Then X is exchangeable iff*

$$X_t = \alpha t + \sigma B_t + \sum_j \beta_j (1\{\tau_j \leq t\} - t) \quad a.s., \quad t \in \mathbb{Q}_{[0,1]}, \tag{21}$$

for some i.i.d. $U(0,1)$ random variables $\tau_1, \tau_2, \ldots$, an independent Brownian bridge B in $\mathbb{R}^d$, and an independent array of random elements σ in $\mathbb{R}^{d \times d}$ and $\alpha, \beta_1, \beta_2, \ldots$ in $\mathbb{R}^d$ with $\sum_j |\beta_j|^2 < \infty$ a.s. In that case, α, $\rho = \sigma \sigma'$, and $\beta = \sum_j \delta_{\beta_j}$ are a.s. unique, the series in (21) converges a.s., uniformly on $[0,1]$, and (21) defines an exchangeable, rcll extension of X to $[0,1]$.

The following lemma is needed to prove the asserted uniformity. It also yields the corresponding convergence in Theorem 1.19.

Lemma 2.19 *(uniform convergence) Let $X = (X_t^r)$ be a real-valued process on $\mathbb{Q}_+ \times I$ for an interval $I \subset \mathbb{R}$, such that X_t^r is rcll in $t \in I$ with independent increments in $r \in \mathbb{Q}_+$ and satisfies $(X^r - X^s)^* \xrightarrow{P} 0$ as $s, r \to \infty$. Then there exists an rcll process $\tilde{X}$ on I such that $(X^r - \tilde{X})^* \to 0$ a.s.*

Proof: Choosing $r_n \to \infty$ in $\mathbb{Q}_+$ with

$$E[(X^{r_n} - X^{r_{n-1}})^* \wedge 1] \leq 2^{-n}, \quad n \in \mathbb{N},$$

we note that $(X^{r_m} - X^{r_n})^* \to 0$ a.s. as $m, n \to \infty$. Hence, there exists a right-continuous process $\tilde{X}$ on I satisfying $(X^{r_n} - \tilde{X})^* \to 0$ a.s., and we get $(X^r - \tilde{X})^* \xrightarrow{P} 0$ as $r \to \infty$ along $\mathbb{Q}_+$. Write $Y^r = X^r - \tilde{X}$. Fixing any $\varepsilon > 0$ and $r \in \mathbb{Q}_+$ and a finite subset $A \subset [r, \infty) \cap \mathbb{Q}$, and putting $\sigma = \sup\{s \in A; (Y^s)^* > 2\varepsilon\}$, we get as in FMP 16.8

$$
\begin{aligned}
P\{(Y^r)^* > \varepsilon\} &\geq P\Big\{(Y^r)^* > \varepsilon, \ \max_{s \in A}(Y^s)^* > 2\varepsilon\Big\} \\
&\geq P\{\sigma < \infty, \ (Y^r - Y^\sigma)^* \leq \varepsilon\} \\
&\geq P\Big\{\max_{s \in A}(Y^s)^* > 2\varepsilon\Big\} \min_{s \in A} P\{(Y^r - Y^s)^* \leq \varepsilon\},
\end{aligned}
$$

which extends immediately to $A = [r, \infty) \cap \mathbb{Q}$. Solving for the first factor on the right gives

$$P\Big\{\sup_{s \geq r}(Y^s)^* > 2\varepsilon\Big\} \leq \frac{P\{(Y^r)^* > \varepsilon\}}{1 - \sup_{s \geq r} P\{(Y^r - Y^s)^* > \varepsilon\}} \to 0,$$

which shows that $\sup_{s \geq r}(Y^s)^* \xrightarrow{P} 0$ and hence $(X^r - \tilde{X})^* = (Y^r)^* \to 0$ a.s. $\square$

Proof of Theorem 2.18: Suppose that $\sum_j |\beta_j|^2 < \infty$ a.s. To prove the asserted uniform convergence, we may clearly assume that α, σ, and $\beta_1, \beta_2, \ldots$ are non-random. Then introduce the martingales

$$M_t^n = (1-t)^{-1} \sum_{j \leq n} \beta_j (1\{\tau_j \leq t\} - t), \quad t \in [0,1),$$

and note that by Doob's inequality

$$E(M^m - M^n)_{1/2}^{*2} \lesssim E(M^m - M^n)_{1/2}^2 \lesssim \sum_{j > m \wedge n} \beta_j^2 \to 0,$$

as $m, n \to \infty$. By Lemma 2.19 we conclude that M^n, and hence also the series in (21), converges a.s., uniformly on the interval $[0, \frac{1}{2}]$. By symmetry we have the same result on $[\frac{1}{2}, 1]$, and by combination we get the required uniform a.s. convergence on $[0, 1]$. In particular, X is a.s. rcll, and the coefficients β_j may be recovered from X as the magnitudes of the jumps. Also note that $\alpha = X_1$ and $\rho = [\sigma B, \sigma B]_1$ a.s. The expression in (21) clearly defines an exchangeable process on $[0, 1]$.

Now suppose instead that X is integrable and exchangeable on $\mathbb{Q}_{[0,1]}$. By Theorem 2.13 it extends a.s. to an exchangeable special semi-martingale on $[0, 1)$ with respect to the right-continuous and complete filtration $\mathcal{F}$ induced by X. The extended process being rcll, we may introduce the associated jump point process ξ, which satisfies $\int |x|^2 \xi_1(dx) \leq \text{tr}\,[X, X]_1 < \infty$ a.s. From the cited theorem we see that $[X^c]$ is a.s. linear and that $\hat{X}$ and $\hat{\xi}$ are a.s. absolutely continuous, with densities that are conditional martingales on $[0, 1)$ and $(0, 1)$, respectively. Since X remains exchangeable over (X_1, ξ_1) by Theorem 1.15, the stated properties remain valid with $\mathcal{F}$ replaced by the right-continuous and complete filtration $\mathcal{G}$ induced by X and (X_1, ξ_1). But then $\alpha = X_1 = X_{1-}$, $\beta = \xi_1$, and $\rho = [X^c]_1$ are $\mathcal{G}_0$-measurable, and we may argue as in the proof of Theorem 2.15 to see that $X \overset{d}{=} Y$ for the process Y in (20), where σ denotes the non-negative definite square root of ρ. We may finally use the transfer theorem (FMP 6.10) to obtain a corresponding a.s. representation of X. $\qquad\square$

Our next aim is to consider reverse martingale criteria for a process X on $I = [0, 1]$ or $\mathbb{R}_+$ to be exchangeable. The following result extends Theorem 2.4 for random sequences and Theorem 2.12 for marked point processes.

Theorem 2.20 *(reverse martingale criterion) Let X be an integrable process in $D_0(I, \mathbb{R}^d)$ with jump point process ξ, where $I = [0, 1]$ or $\mathbb{R}_+$. Then X is exchangeable iff the process $t^{-1}(X, \xi, [X]^c)_t$ is a reverse, conditional martingale on $I \setminus \{0\}$.*

Note that if X_t/t is a reverse, conditional martingale, then X itself is a reverse semi-martingale, which ensures the existence of the process $[X]^c$.

Proof: First assume that X is exchangeable. Then X is a semi-martingale by Theorem 2.13, and so by Theorem 1.15 the triple $(X, \xi, [X^c])$ exists and is again exchangeable on I. Introducing the tail filtration

$$\mathcal{T}_t = \sigma\{(X, \xi, [X^c])_u,\ u \geq t\}, \quad t \in I,$$

we conclude that $(X, \xi, [X^c])$ is exchangeable over $\mathcal{T}_t$ on $[0, t]$ for every $t \in I$. Writing $M_t = t^{-1}X_t$ and letting $s \leq t$ be rationally dependent, so that $s = mh$ and $t = nh$ for some $m, n \in \mathbb{N}$ and $h > 0$, we get

$$
\begin{aligned}
E[X_h | \mathcal{T}_t] &= m^{-1}\sum_{k \leq m} E[X_{kh} - X_{(k-1)h} | \mathcal{T}_t] \\
&= m^{-1} E\Big[\sum_{k \leq m}(X_{kh} - X_{(k-1)h})\Big| \mathcal{T}_t\Big] \\
&= m^{-1} E[X_s | \mathcal{T}_t] = h E[M_s | \mathcal{T}_t].
\end{aligned}
$$

Combining with the same relation for M_t gives

$$E[M_s | \mathcal{T}_t] = h^{-1} E[X_h | \mathcal{T}_t] = E[M_t | \mathcal{T}_t] = M_t,$$

which shows that M is a reverse, conditional $\mathcal{T}$-martingale on $t\mathbb{Q}\cap(I\setminus\{0\})$ for every $t > 0$. The result extends by right continuity and uniform integrability to the entire interval $I \setminus \{0\}$. A similar argument applies to the processes $t^{-1}\xi_t$ and $t^{-1}[X^c]_t$.

Now suppose instead that $M_t = t^{-1}X_t$ is a reverse, conditional $\mathcal{T}$-martingale. Using Fubini's theorem and the definition and martingale property of M, we get for any $s \leq t$ in $(0, 1]$

$$
\begin{aligned}
E[X_t - X_s | \mathcal{T}_t] &= E[tM_t - sM_s | \mathcal{T}_t] \\
&= (t - s)M_t = \int_s^t M_t dr \\
&= \int_s^t E[M_r | \mathcal{T}_t]\, dr = E\Big[\int_s^t M_r dr\Big| \mathcal{T}_t\Big],
\end{aligned}
$$

which shows that X is a special, reverse semi-martingale with respect to $\mathcal{T}$ with compensator $\hat{X} = M \cdot \lambda$. A similar argument shows that if $\eta_t = t^{-1}\xi_t$ and $\rho_t = t^{-1}[X^c]_t$ are reverse, conditional $\mathcal{T}$-martingales, then ξ and $[X^c]$ have $\mathcal{T}$-compensators $\hat{\xi} = \lambda \otimes \eta$ and $[X^c]^\wedge = \rho \cdot \lambda$, respectively. Since $[X^c]$ is continuous with locally finite variation, we have in fact $[X^c] = [X^c]^\wedge = \rho \cdot \lambda$ a.s.

The continuity of $\hat{\xi}$ implies that ξ is continuous in probability, and so the same thing is true for X. Hence, the previous statements remain true for the left-continuous versions of the various processes, and Doob's regularization theorem (FMP 7.27) allows us to replace $\mathcal{T}$ by the generated left-continuous and complete filtration $\overline{\mathcal{T}}$. For fixed $u \in I$, the martingale conditions of Theorem 2.15 are then fulfilled for the right-continuous process Y on $[0, u]$ with associated right-continuous and complete filtration $\mathcal{F}$, given by

$$Y_t = X_u - X_{u-t-}, \quad \mathcal{F}_t = \overline{\mathcal{T}}_{u-t}, \qquad t \in [0, u].$$

We also note that the terminal values

$$Y_u = Y_{u-} = X_u, \qquad \xi_u, \qquad [Y^c]_u = [Y]_u^c = [X]_u^c = [X^c]_u$$

are all measurable with respect to $\mathcal{F}_0 = \overline{\mathcal{T}}_u$. Hence, the quoted theorem shows that Y is $\mathcal{F}$-exchangeable on $[0, u]$. In particular, X is then exchangeable on $[0, u]$, and u being arbitrary, it follows that X is exchangeable on I. $\square$

We conclude this section with two technical propositions that will be useful in subsequent chapters. They will be proved here under the additional assumption that the exchangeable processes considered on $[0, 1]$ are representable as in Theorem 2.18. The general results will then follow from the developments in Chapter 3.

To introduce the first of those results, we say that the processes $X_1, X_2, \ldots$ on $I = [0, 1]$ or $\mathbb{R}_+$ are *jointly $\mathcal{F}$-exchangeable* if the process $(X_1, \ldots, X_n)$ is $\mathcal{F}$-exchangeable for every $n \in \mathbb{N}$. The stronger notion of *separate $\mathcal{F}$-exchangeability* is defined by the requirement that the shifted process $\theta_t X_i - X_i(t)$ be exchangeable over $\mathcal{F}_t$ and $(X_j, j \neq i)$ for any $i \in \mathbb{N}$ and $t \in I$. In either case, the family (X_j) is said to be *$\mathcal{F}$-extreme* if the $\mathcal{F}_t$-conditional distribution is a.s. extreme for every $t \in I$. By *$\mathcal{F}_0$-extremality* we mean the same property for $t = 0$ only. Let us also say that the X_j are *$\mathcal{F}$-independent* if they are adapted and conditionally independent given $\mathcal{F}_t$ for every t.

Lemma 2.21 *(separate exchangeability and independence) Let $X_1, X_2, \ldots$ be $\mathcal{F}$-exchangeable and $\mathcal{F}_0$-extreme processes on $[0, 1]$ or $\mathbb{R}_+$. Then the X_k are separately $\mathcal{F}$-exchangeable iff they are $\mathcal{F}$-independent, in which case the whole family is $\mathcal{F}$-extreme. The corresponding statements hold in discrete time.*

Proof: We prove the result for representable processes X in continuous time only, the discrete-time case being similar. First we claim that an $\mathcal{F}$-exchangeable and $\mathcal{F}_0$-extreme process X of this type is even $\mathcal{F}$-extreme. Then note that, by the uniqueness of the representation, a process X as in Theorem 1.19 or 2.18 is extreme iff the characteristics (α, ρ, ν) or (α, ρ, β) are a.s. non-random. The claim now follows from the fact that the characteristics of the shifted process $\theta_t X - X_t$ are measurably determined by those for X and for the stopped process X^t.

Now assume that the X^j are separately exchangeable and individually extreme. Then the conditional distribution $P[X_i \in \cdot | X_j, j \neq i]$ is a.s. exchangeable for every $i \in \mathbb{N}$, and so the extremality yields a.s.

$$P[X_i \in \cdot | X_j, j \neq i] = P\{X_i \in \cdot\}, \quad i = 1, \ldots, d. \tag{22}$$

Thus, $X_i \perp\!\!\!\perp (X_j, j \neq i)$ for all i, which shows that the X_j are independent. Conversely, the independence of $X_1, X_2, \ldots$ implies (22), and so their separate exchangeability follows from the corresponding individual property.

To see that the whole family (X_j) is extreme, suppose that the joint distribution μ is a nontrivial convex combination of some separately exchangeable distributions μ_1 and μ_2. By the individual extremality we note that μ_1 and μ_2 have the same marginal distributions as μ, and so, by the previous proof, the X_j remain independent under μ_1 and μ_2. This shows that $\mu_1 = \mu_2 = \mu$, and the required extremality follows.

Let us now suppose that the X_j are individually $\mathcal{F}$-exchangeable and $\mathcal{F}_0$-extreme. As before they are even $\mathcal{F}$-extreme, and so the unconditional statements apply to the conditional distribution of $\theta_t X - X_t$ given $\mathcal{F}_t$, which proves the required equivalence and extremality. □

To motivate the next result, we note that the representation in Theorem 2.18 may require an extension of the original probability space, to accommodate the random variables $\tau_1, \tau_2, \ldots$. Similarly, for an $\mathcal{F}$-exchangeable process with a representation as in the cited theorem, the individual terms may not be $\mathcal{F}$-exchangeable or even $\mathcal{F}$-adapted. Here we show how to achieve the desired properties by a suitable extension of $\mathcal{F}$. The result will be useful in Chapter 5.

Given a filtration $\mathcal{F}$ on some index set I, we say (following FMP 18.4) that $\mathcal{G}$ is a *standard extension* of $\mathcal{F}$ if

$$\mathcal{F}_t \subset \mathcal{G}_t \perp\!\!\!\perp_{\mathcal{F}_t} \mathcal{F}, \quad t \in I.$$

The condition ensures that all adaptedness and martingale properties for the original filtration $\mathcal{F}$ carry over to $\mathcal{G}$.

Lemma 2.22 *(term-wise exchangeability) Let X be an $\mathcal{F}$-exchangeable process on $[0, 1]$, admitting a representation as in Theorem 2.18. Then the individual terms can be chosen to be separately $\mathcal{G}$-exchangeable and $\mathcal{G}$-extreme for a suitable extension $\mathcal{G}$ of $\mathcal{F}$. If X is $\mathcal{F}_0$-extreme, we can choose $\mathcal{G}$ to be a standard extension of $\mathcal{F}$.*

Again we note that the representability requirement is redundant, since the stated property holds automatically for every exchangeable process on $[0, 1]$, by Theorem 3.15 below.

Proof: Suppose that X has characteristics (α, ρ, β), and let ξ denote the point process of jump times and sizes. To resolve the multiplicities of the latter, we may introduce a uniform randomization $\tilde{\xi} \perp\!\!\!\perp_{\xi} \mathcal{F}$ of ξ on the space $\mathbb{R}^d \times [0, 1]^2$, by independently attaching some i.i.d. $U(0, 1)$ labels ϑ_j to the original points (β_j, τ_j). (The precise definition and construction are described in FMP 12.7.) The points $\tilde{\beta}_j = (\beta_j, \vartheta_j)$ are a.s. distinct and may be enumerated measurably in terms of the projection $\tilde{\beta} = \tilde{\xi}(\cdot \times [0, 1])$, which ensures that the associated times τ_j will be i.i.d. $U(0, 1)$ and independent of $(\alpha, \rho, \tilde{\beta})$ and B. We choose $\mathcal{G}$ to be the filtration induced by $\mathcal{F}$, $\tilde{\xi}$, and $(\alpha, \rho, \tilde{\beta})$.

Through this construction, the pair $(X, \tilde{\xi})$ will be $\mathcal{G}$-exchangeable and $\mathcal{G}_0$-extreme, and with the chosen enumeration, the individual terms

$$X_t^j = \beta_j(1\{\tau_j \le t\} - t), \quad t \in [0,1], \quad j \in \mathbb{N},$$

become $\mathcal{G}$-exchangeable and $\mathcal{G}$-independent. This extends by subtraction to include even the continuous component $X_t^0 = \sigma B_t$. All those terms are also seen to be $\mathcal{G}_0$-extreme. Hence, by Lemma 2.21 they are even separately $\mathcal{G}$-exchangeable and $\mathcal{G}$-extreme.

Now assume that X is already $\mathcal{F}_0$-extreme, so that α, ρ, and β are $\mathcal{F}_0$-measurable. Then we need to show that

$$(\mathcal{F}_t, \tilde{\xi}^t, \alpha, \rho, \tilde{\beta}) \perp\!\!\!\perp_{\mathcal{F}_t} \mathcal{F}, \quad t \in [0,1],$$

where $\tilde{\xi}^t$ denotes the restriction of $\tilde{\xi}$ to $\mathbb{R}^d \times [0,t] \times [0,1]$. Omitting the $\mathcal{F}_t$-measurable components α, ρ, and $\mathcal{F}_t$ and subtracting the projection $\tilde{\beta}_t = \tilde{\xi}^t(\cdot \times [0,t])$ from $\tilde{\beta}$, we may write the last relations in the equivalent form

$$(\tilde{\xi}^t, \tilde{\beta} - \tilde{\beta}_t) \perp\!\!\!\perp_{\mathcal{F}_t} \mathcal{F}, \quad t \in [0,1].$$

But here the left-hand side is a uniform randomization of the corresponding $\mathcal{F}_t$-measurable pair $(\xi^t, \beta - \beta_t)$, where ξ^t is the restriction of ξ to $\mathbb{R}^d \times [0,t])$ with projection β_t onto $\mathbb{R}^d$. The result is then a consequence of the conditional independence in the original construction. $\square$

2.5   Norm Relations and Regularity

The martingale properties of exchangeable and contractable processes can be used to establish some basic norm relations. Here we consider continuous-time, exchangeable or contractable processes on $[0,1]$, as well as *summation processes* of the form

$$X_t = \sum_{j \le nt} \xi_j, \quad t \in [0,1],$$

based on finite, exchangeable or contractable sequences of random variables $\xi_1, \dots, \xi_n$. The results can be used to establish some local growth properties of exchangeable and contractable processes. They will also play an instrumental role for the weak convergence theory in the next chapter.

Theorem 2.23 *(norm comparison)* *Let X be a real, exchangeable semi-martingale on $[0,1)$, and define $\gamma = ([X]_1 + X_1^2)^{1/2}$. Then uniformly in X and (t,p), we have*

(i) *for $(t,p) \in (0,1) \times (0,\infty)$ in compacts,*

$$\|X_t\|_p \asymp \|X_1^*\|_p \asymp \|\gamma\|_p \,;$$

(ii) *for $(t,p) \in [0,1) \times (0,\infty)$ in compacts,*

$$t^{1/(p\wedge 1)}\|\gamma\|_p \lesssim \|X_t\|_p \asymp \|X_t^*\|_p \lesssim t^{1/(p\vee 2)}\|\gamma\|_p \,.$$

The same relations hold for summation processes based on exchangeable n-sequences, as long as $t \geq n^{-1}$. The bounds are sharp, though the upper rate in (ii) can be improved to $t^{1/(p\vee 1)}$ when X is non-decreasing. All estimates remain valid for contractable processes in L^p, possibly except for those involving X^* when $p < 1$.

An elementary inequality will be helpful for the proof.

Lemma 2.24 *(hyper-contraction)* *Let ξ be a random variable satisfying $\|\xi\|_4 \leq c\|\xi\|_2 < \infty$ for some constant $c > 0$. Then*

$$(3c^4)^{-1/p}\|\xi\|_2 \leq \|\xi\|_p \leq \|\xi\|_2, \quad p \in (0,2].$$

Proof: By scaling we may assume that $\|\xi\|_2 = 1$. Then $E\xi^4 \leq c^4$, and so the Paley–Zygmund inequality (FMP 4.1) yields

$$P\{\xi^2 > t\} \geq c^{-4}(1-t)^2, \quad t \in [0,1].$$

Hence, by FMP 3.4 we get for any $r \in (0,1]$

$$
\begin{aligned}
E|\xi|^{2r} &= r\int_0^\infty P\{\xi^2 > t\}t^{r-1}dt \\
&\geq rc^{-4}\int_0^1 (1-t)^2 t^{r-1}dt \\
&= \frac{2}{c^4(r+1)(r+2)} \geq \frac{1}{3c^4}.
\end{aligned}
$$

To obtain the asserted lower bound, it remains to take $r = p/2$ and raise the extreme sides to the power $1/p$. The upper bound holds by Jensen's inequality.   □

Proof of Theorem 2.23: In view of the length and complexity of the argument, we divide the proof into five parts:

1. *Part* (i) *for exchangeable semi-martingales:* Consider any process X as in Theorem 2.18 with constant coefficients, and put $Y_t = X_t - \alpha t$. Then

$M_t = Y_t/(1-t)$ is a martingale on $[0,1)$, and we may integrate by parts to get

$$dY_t = (1-t)dM_t - M_t dt,$$
$$d[X]_t = d[Y]_t = (1-t)^2 d[M]_t.$$

Hence, for fixed $t \in [0,1)$ and $p \geq 1$, the BDG-inequalities (FMP 26.12) yield

$$\|Y_t^*\|_p \leq \|M_t^*\|_p \asymp \|[M]_t^{1/2}\|_p \asymp \|[X]_t^{1/2}\|_p \leq [X]_1^{1/2},$$

and so by Minkowski's inequality

$$\|X_t\|_p \leq \|X_t^*\|_p \lesssim \|[X]_t^{1/2}\|_p + t|\alpha| \lesssim \gamma. \tag{23}$$

On the other hand, we see from (21) and Jensen's inequality that, for any $p \geq 2$ and for fixed $t \in (0,1)$,

$$\|X_t\|_p^2 \geq E X_t^2 = t(1-t)[X]_1 + t^2 \alpha^2 \gtrsim \gamma^2. \tag{24}$$

Finally, by the exchangeability of X and Minkowski's inequality,

$$\|X_t^*\|_p \leq \|X_1^*\|_p \leq [t^{-1}]\|X_t^*\|_p \asymp \|X_t^*\|_p. \tag{25}$$

Combining (23)–(25), we obtain for any $p \geq 2$

$$\|X_t\|_p \asymp \|X_t^*\|_p \asymp \|X_1^*\|_p \asymp \gamma, \tag{26}$$

which extends by Lemma 2.24 to arbitrary $p > 0$.

When the coefficients α, β, and σ are random, we see from Fubini's theorem and (26) that, for fixed t and p,

$$E[|X_t|^p | \alpha, \beta, \sigma] \asymp E[X_t^{*p} | \alpha, \beta, \sigma] \asymp E[X_1^{*p} | \alpha, \beta, \sigma] \asymp \gamma^p \text{ a.s.},$$

and all bounds being uniform, we may take expected values to obtain

$$\|X_t\|_p \asymp \|X_t^*\|_p \asymp \|X_1^*\|_p \asymp \|\gamma\|_p. \tag{27}$$

It is easy to check that the underlying estimates are uniform for t bounded away from 0 and 1, as well as for p bounded away from 0 and ∞.

2. *Part* (ii) *for exchangeable semi-martingales:* Fix any $u \in (0,1)$, and let $t \in [0,u]$ be arbitrary. Applying (27) to the exchangeable process $Y_s = X_{st/u}$, $s \in [0,1]$, we obtain

$$\|X_t\|_p = \|Y_u\|_p \asymp \|Y_u^*\|_p = \|X_t^*\|_p, \tag{28}$$

uniformly in $t \leq u$. This proves the second relation in (ii).

To prove the upper bound in (ii), it suffices as before to consider processes X with constant coefficients α, β, and σ. Noting that $[X]$ is again exchangeable by Theorem 1.15, we get for any $t < 1$ and $p \geq 2$

$$\|[X]_t^{1/2}\|_p^p = E[X]_t^{p/2} \leq [X]_1^{p/2-1} E[X]_t = t[X]_1^{p/2},$$

and so by (23)

$$\|X_t\|_p \lesssim \|[X]_t^{1/2}\|_p + t|\alpha| \leq t^{1/p}[X]_1^{1/2} + t|\alpha| \lesssim t^{1/p}\gamma.$$

If instead $p \leq 2$, then by (24) and Jensen's inequality

$$\|X_t\|_p^2 \leq EX_t^2 = t(1-t)[X]_1 + \alpha^2 t^2 \lesssim t\gamma^2.$$

Finally, when X is non-decreasing, we have for $p \geq 1$

$$\|X_t\|_p^p = EX_t^p \leq X_1^{p-1}EX_t = t\alpha^p \leq t\gamma^p,$$

while for $p \leq 1$ we get by Jensen's inequality

$$\|X_t\|_p \leq \|X_t\|_1 = EX_t = t\alpha \leq t\gamma.$$

It is easy to check that these estimates are uniform for t bounded away from 1 and for p bounded away from 0 and ∞.

To establish the lower bound, we may use part (i) and proceed as in (25) to get for any $t \in (0,1)$ and $p > 0$

$$\|\gamma\|_p \asymp \|X_1^*\|_p \lesssim [t^{-1}]^{1/(p\wedge1)}\|X_t^*\|_p \leq t^{-1/(p\wedge1)}\|X_t^*\|_p,$$

again with the desired uniformity in t and p. This completes the proof of (ii). To see that the three bounds are sharp, it suffices to consider the special processes

$$X_t^1 = t, \qquad X_t^2 = 1\{\tau \leq t\}, \qquad X_t^3 = B_t,$$

where τ is $U(0,1)$ and B is a standard Brownian motion.

3. *Exchangeable summation processes:* Consider a summation process X based on an exchangeable sequence $\xi_1, \ldots, \xi_n$. Introduce some independent i.i.d. $U(0,1)$ random variables $\tau_1, \ldots, \tau_n$, and put $Y_t = \sum_j \xi_j 1\{\tau_j \leq t\}$. Then Y is exchangeable with $Y_1 = X_1$, and we note that

$$Y_1^* \overset{d}{=} X_1^*, \qquad [Y]_1 + Y_1^2 = [X]_1 + X_1^2 = \gamma^2.$$

Hence, the continuous-time result yields

$$\|X_t\|_p \leq \|X_1^*\|_p = \|Y_1^*\|_p \asymp \|\gamma\|_p,$$

uniformly for p bounded away from 0 and ∞.

To get a reverse estimate, we may assume that $n \geq 3$, since trivially $|X_t| = X_t^*$ when $n \leq 2$ and $t < 1$. Fixing any $t \in [2/n, 1)$ and writing $m = [nt]$, we get in the extreme case and for $p \geq 2$

$$\begin{aligned}
\|X_t\|_p^2 &\geq EX_t^2 = mE\xi_1^2 + m(m-1)E\xi_1\xi_2 \\
&= \frac{m}{n}\sum_j \xi_j^2 + \frac{m(m-1)}{n(n-1)}\sum_{i\neq j}\xi_i\xi_j \\
&= \frac{m}{n}\left(1 - \frac{m-1}{n-1}\right)\sum_j \xi_j^2 + \frac{m(m-1)}{n(n-1)}\left(\sum_j \xi_j\right)^2 \\
&= t'(1-t'')[X]_1 + t't''X_1^2 \gtrsim \gamma^2,
\end{aligned}$$

where $t' = m/n$ and $t'' = (m - 1)/(n - 1)$. Proceeding as before, we see that (i) remains true in this case, uniformly for t bounded away from 0 and 1, and for p bounded away from 0 and ∞. As for part (ii), the previous argument applies when t is a multiple of n^{-1}, and the general result follows by interpolation.

4. *Contractable summation processes:* If the sequence $\xi_1, \dots, \xi_n$ is contractable, then by Lemma 1.11 we may choose an exchangeable sequence $\eta_1, \dots, \eta_n$ such that

$$\sum\nolimits_{j \le k} \delta_{\xi_j} \stackrel{d}{=} \sum\nolimits_{j \le k} \delta_{\eta_j}, \quad k \le n.$$

For the associated summation process X and Y we get

$$X_t \stackrel{d}{=} Y_t, \quad t \in [0, 1]; \qquad \gamma^2 \equiv [X]_1 + X_1^2 \stackrel{d}{=} [Y]_1 + Y_1^2,$$

and so by the previous case we have for suitable t and p

$$\|X_t^*\|_p \ge \|X_t\|_p \asymp \|\gamma\|_p, \tag{29}$$
$$t^{1/(p \wedge 1)} \|\gamma\|_p \lesssim \|X_t\|_p \lesssim t^{1/(p \vee 2)} \|\gamma\|_p,$$

with the stated improvement when the ξ_j are non-negative.

To derive the reverse estimate in (29), we may assume that the ξ_j are integrable, since otherwise $\|\gamma\|_p = \infty$ for $p \ge 1$ and there is nothing to prove. By Theorem 2.13 the process $X - \hat{X}$ is then a martingale on $[0, 1)$ with quadratic variation $[X]$, where $\hat{X}$ admits the martingale density

$$M_t = \frac{E[X_1 - X_t | \mathcal{F}_t]}{1 - t}, \quad t \in [0, 1).$$

By a BDG-inequality we have for any $t \in [0, 1]$ and $p \ge 1$

$$\|(X - \hat{X})_t^*\|_p \lesssim \|[X]_t^{1/2}\|_p \le \|\gamma\|_p.$$

Next we may use the continuous-time version of Minkowski's inequality (FMP 1.30), the sub-martingale property of $|M_t|^p$ (FMP 7.11), Jensen's inequality, and the equivalence in (29) to get for any $t \in (0, 1)$ and $p \ge 1$

$$\|\hat{X}_t^*\|_p \le \left\| \int_0^t |M_s| ds \right\|_p \le \int_0^t \|M_s\|_p ds$$
$$\le \|M_t\|_p \lesssim \|X_{1-t}\|_p \lesssim \|\gamma\|_p.$$

Combining the last two estimates and using Minkowski's inequality, we obtain

$$\|X_t^*\|_p \le \|(X - \hat{X})_t^*\|_p + \|\hat{X}_t^*\|_p \lesssim \|\gamma\|_p,$$

and we note as in (25) that the bound extends to $\|X_1^*\|_p$. This proves (i) when $t \in (0, 1)$ is fixed, uniformly for bounded $p \ge 1$. We may finally use the scaling argument in (28) to see that the relation $\|X_t\|_p \asymp \|X_t^*\|_p$ remains uniform even for t bounded away from 1.

5. *Contractable semi-martingales:* For fixed $t \in (0,1)$, we may approximate X on $[0, t]$ by the step processes

$$X_s^n = X([ns/t]t/n), \quad s \in [0,t], \ n \in \mathbb{N}, \tag{30}$$

and note that $X_t^n \equiv X_t$ and $(X^n)_t^* \uparrow X_t^*$ as $n = 2^k \to \infty$. Using the result in the previous case, employing monotone convergence, and arguing as in (25), we obtain

$$\|X_t\|_p = \|X_t^n\|_p \asymp \|(X^n)_t^*\|_p \to \|X_t^*\|_p \asymp \|X_1^*\|_p \,.$$

Next we define X^n as in (30) with $t = 1$ and put $\gamma_n = [X^n]_1 + (X_1^n)^2$. As a contractable semi-martingale, X is continuous in probability, and therefore $X_t^n \xrightarrow{P} X_t$ for all $t \in [0,1]$. Furthermore, by a standard approximation property for semi-martingales (cf. FMP 17.18), we have $[X^n]_t \xrightarrow{P} [X]_t$ for all $t < 1$, which extends to $t = 1$ by contractability. Thus, $\gamma_n \xrightarrow{P} \gamma$. Applying the result in the previous case to the processes X^n, we see that

$$\|X_t^n\|_p \asymp \|\gamma_n\|_p, \qquad t^{1/(p \wedge 1)}\|\gamma_n\|_p \lesssim \|X_t^n\|_p \lesssim t^{1/(p \vee 2)}\|\gamma_n\|_p, \tag{31}$$

with the appropriate improvement when X is non-decreasing. If $X_t \in L^p$ for some $t \in \mathbb{Q}_{(0,1)}$, we conclude that the sequences $(X^n)_1^*$ and γ_n are L^p-bounded, and so by uniform integrability we have for any $q < p$ and $t \in [0,1]$

$$\|X_t^n\|_q \to \|X_t\|_q, \qquad \|\gamma_n\|_q \to \|\gamma\|_q \,.$$

Since also $\|X_t\|_q \to \|X_t\|_p$ and $\|\gamma\|_q \to \|\gamma\|_p$ as $q \to p$ by monotone convergence, relations (31) extend in the limit to X and γ. □

The estimates of Theorem 2.23 allow us to improve on Theorem 2.13.

Theorem 2.25 *(regularization) Any $\mathcal{F}$-contractable process X on $\mathbb{Q}_{[0,1]}$ has an a.s. rcll extension $\tilde{X}$ to $[0,1]$, which remains contractable with respect to the augmented filtration $\overline{\mathcal{F}}$. If $E[|X_t||\overline{\mathcal{F}}_0] < \infty$ a.s. for some $t \in (0,1)$, then $\tilde{X}$ is a uniformly $\overline{\mathcal{F}}_0$-integrable, special semi-martingale on $[0,1]$.*

In Theorem 3.15 we shall use this result to show that every $\mathcal{F}$-exchangeable process on $[0,1]$ or $\mathbb{R}_+$ is a semi-martingale. Our present proof requires the following standard criterion for regularity.

Lemma 2.26 *(optional continuity and regularity, Aldous) Let X be an $\mathbb{R}^d$-valued process on $\mathbb{Q}_{[0,1]}$ satisfying*

$$X_{\tau_n + h_n} - X_{\tau_n} \xrightarrow{P} 0, \tag{32}$$

for any optional times τ_n and positive constants $h_n \to 0$ such that $\tau_n + h_n \leq 1$. Then X extends a.s. to an rcll process on $[0,1]$.

Proof: By the corresponding tightness criterion (FMP 16.11) and its proof we see that $\tilde{w}(X, h) \xrightarrow{P} 0$ as $h \to 0$, where $\tilde{w}$ is the modified modulus of continuity associated with the Skorohod topology on $D = D([0, 1], \mathbb{R}^d)$, and so by monotonicity we have $\tilde{w}(X, h) \to 0$ a.s. Arguing directly from definitions or approximating by step processes, we conclude that the right and left limits $X^{\pm}$ exist outside a fixed P-null set (FMP A2.2), and also that $X^+ \in D$ a.s. It remains to note that $X^+ = X$ a.s. on $\mathbb{Q}_{[0,1]}$, since X is continuous in probability in view of (32). $\square$

Proof of Theorem 2.25: Fix any $u \in (0, 1)$, and let X be the summation process based on an extreme, exchangeable n-sequence with $n \geq u^{-1}$. Then Theorem 2.23 yields $\|X_u\|_1 \asymp \|X_u\|_2 \asymp \gamma$, and so by the Paley–Zygmund inequality (FMP 4.1) there exists a constant $c > 0$ such that

$$P\{|X_u| > rc\gamma\} \geq c(1 - r)_+^2, \quad r > 0.$$

Substituting $s = r\gamma$ and passing to the non-extreme case, we obtain

$$P\{|X_u| > cs\} \geq cE(1 - s/\gamma)_+^2, \quad s > 0,$$

where $(1 - s/\gamma)_+$ is taken to be 0 when $\gamma = 0$. Using Chebyshev's inequality gives

$$
\begin{aligned}
P\{\gamma > 2s\} &= P\left\{1 - s/\gamma > \tfrac{1}{2}\right\} \\
&= P\left\{4(1 - s/\gamma)_+^2 > 1\right\} \\
&\leq 4E(1 - s/\gamma)_+^2 \leq 4c^{-1}P\{|X_u| > cs\}. \quad (33)
\end{aligned}
$$

By another application of Theorem 2.23, we get in the extreme case $E|X_t| \leq bt^{1/2}\gamma$ for $t \in [n^{-1}, \tfrac{1}{2}]$, where $b > 0$ is an absolute constant. Hence, by Chebyshev's inequality and (33), we have for any $\varepsilon > 0$

$$
\begin{aligned}
P\{|X_t| > \varepsilon\} &\leq 2sbt^{1/2}\varepsilon^{-1} + P\{\gamma > 2s\} \\
&\leq 2bst^{1/2}\varepsilon^{-1} + 4c^{-1}P\{|X_u| > cs\}.
\end{aligned}
$$

The last estimate extends to contractable summation processes by Lemma 1.11, and since n was arbitrary, it remains true for any $t \leq \tfrac{1}{2}$ when X is a contractable process on $\mathbb{Q}_{[0,1]}$. The right-hand side tends to 0 as $t \to 0$ and then $s \to \infty$, which shows that $X_t \xrightarrow{P} 0$ as $t \to 0$.

Now consider any optional times τ_n and positive constants $h_n \to 0$ such that $\tau_n + h_n \leq 1$, and conclude from Proposition 2.5 that

$$X_{\tau_n + h_n} - X_{\tau_n} \stackrel{d}{=} X_{h_n} \xrightarrow{P} 0.$$

By Lemma 2.26 we can then extend X to an a.s. rcll process on $[0, 1]$, and by right continuity we note that X remains contractable with respect to the augmented filtration $\overline{\mathcal{F}}$ of $\mathcal{F}$.

If $E[|X_t||\overline{\mathcal{F}}_0] < \infty$ a.s., then Theorem 2.23 yields $E[X^*|\overline{\mathcal{F}}_0] < \infty$, and so by Theorem 2.13 we see that X is a special $\overline{\mathcal{F}}$-semi-martingale on $[0, 1)$. Furthermore, the compensator $\hat{X}$ is absolutely continuous with a conditional martingale density M as in (14). Using Jensen's inequality, the contractability of X, and Theorem 2.23, we get for any $t \in [0, 1)$

$$E[|M_t||\overline{\mathcal{F}}_0] \leq \frac{E[|X_1 - X_t||\overline{\mathcal{F}}_0]}{1-t} = \frac{E[|X_{1-t}||\overline{\mathcal{F}}_0]}{1-t} \lesssim \frac{E[X^*|\overline{\mathcal{F}}_0]}{(1-t)^{1/2}}.$$

Thus, by Fubini's theorem,

$$\begin{aligned} E\left[\int_0^1 |d\hat{X}_t| \,\Big|\, \overline{\mathcal{F}}_0\right] &= \int_0^1 E[|M_t||\overline{\mathcal{F}}_0]\,dt \\ &\lesssim E[X^*|\overline{\mathcal{F}}_0] \int_0^1 (1-t)^{-1/2}\,dt < \infty, \end{aligned}$$

which shows that $\hat{X}$ has $\overline{\mathcal{F}}_0$-integrable variation. This in turn implies that $X - \hat{X}$ is a uniformly $\overline{\mathcal{F}}_0$-integrable local martingale on $[0, 1]$. $\qquad\square$

We may also use the previous methods to examine the local martingales occurring in the semi-martingale decomposition of a contractable process, first considered in Theorem 2.13. The following result is useful in connection with stochastic integration.

Proposition 2.27 *(martingale norms) Let X be an $\mathcal{F}$-contractable process on $[0, 1]$ such that $\|X_t\|_p < \infty$ for some $t \in (0, 1)$ and $p \geq 1$. Then $X = M \cdot \lambda + N$ a.s. for some L^p-martingales M and N on $[0, 1)$ satisfying*

$$\||\lambda|M|\|_p \vee \|N^*\|_p \lesssim \|X^*\|_p < \infty.$$

Proof: Writing $p'' = p \vee 2$ and using Jensen's inequality, the contractability of X, and Theorems 2.13 and 2.23, we get for $s \in (0, 1)$ the uniform estimate

$$\begin{aligned} (1-s)\|M_s\|_p &= \|E[X_1 - X_s|\mathcal{F}_s]\|_p \\ &\leq \|X_1 - X_s\|_p = \|X_{1-s}\|_p \\ &\lesssim (1-s)^{1/p''}\|X_t\|_p. \end{aligned}$$

Hence, by the extended Minkowski inequality (FMP 1.30),

$$\||\lambda|M|\|_p \leq \lambda \|M\|_p \lesssim \|X_t\|_p \int_0^1 (1-s)^{1/p''-1}ds \lesssim \|X_t\|_p < \infty.$$

Since $N = X - M \cdot \lambda$ a.s., we may finally use Minkowski's inequality and Theorem 2.23 to obtain

$$\|N^*\|_p \leq \|X^*\|_p + \|(M \cdot \lambda)^*\|_p \lesssim \|X_t\|_p < \infty. \qquad\square$$

2.6   Path Properties

Here we use martingale methods to study some path properties of exchange-able processes, which requires us first to develop some general tools. We begin with a basic super-martingale in discrete time.

Proposition 2.28 *(discrete-time super-martingale, Dubins and Freedman, Kallenberg) Let $X = (X_0, X_1, \ldots)$ be an $\mathcal{F}$-adapted sequence of random variables with $X_0 = 0$, and let ξ denote the jump point process of X. Fix an even function $f \geq 0$ on $\mathbb{R}$ with $f(0) = 0$ and a non-increasing, convex function $g \geq 0$ on $\mathbb{R}_+$, and put $A_n = \hat{\xi}_n f$. Suppose that either X is a local martingale, f is convex, and f' is concave on $\mathbb{R}_+$, or else that f is concave on $\mathbb{R}_+$. Then we can define a super-martingale $Y \geq 0$ by*

$$Y_n = 2g(A_n) - f(X_n)g'(A_n), \quad n \in \mathbb{Z}_+.$$

Note that A is allowed to be infinite. This causes no problem since both g and g' have finite limits $g(\infty) \geq 0$ and $g'(\infty) = 0$ at ∞. We may also allow $g'(0) = -\infty$, with the understanding that $0 \cdot \infty = 0$.

Proof: First assume that f is convex and f' is concave on $\mathbb{R}_+$. For any $a \in \mathbb{R}$ and $x \geq 0$, we have

$$
\begin{aligned}
f(a + x) - f(a) - xf'(a) &= \int_0^x \{f'(a+s) - f'(a)\}\, ds \\
&\leq \int_0^x \{f'(s/2) - f'(-s/2)\}\, ds \\
&= 2\int_0^x f'(s/2)\, ds = 4\int_0^{x/2} f'(t)\, dt \\
&= 4f(x/2) \leq 2f(x),
\end{aligned}
$$

where the second relation holds by the concavity of f' and the last one is due to the convexity of f. Hence, for $x \geq 0$

$$f(a + x) \leq f(a) + xf'(a) + 2f(x). \tag{34}$$

Since f is even and f' is odd, it follows that

$$
\begin{aligned}
f(a - x) &= f(-a + x) \leq f(-a) + xf'(-a) + 2f(x) \\
&= f(a) - xf'(a) + 2f(-x),
\end{aligned}
$$

which extends (34) to arbitrary $a, x \in \mathbb{R}$.

Now consider any martingale $X = (X_n)$ such that the sequences $A = (A_n)$ and $Y = (Y_n)$ are both a.s. finite. Using (34) and the martingale property of X, we get for any $n \in \mathbb{N}$

$$
\begin{aligned}
E[f(X_n)|\mathcal{F}_{n-1}] &\leq f(X_{n-1}) + 2E[f(\Delta X_n)|\mathcal{F}_{n-1}] \\
&= f(X_{n-1}) + 2\Delta A_n,
\end{aligned}
\tag{35}
$$

and so, by the convexity of g,

$$
\begin{aligned}
E[Y_n | \mathcal{F}_{n-1}] &\leq 2g(A_n) - \{f(X_{n-1}) + 2\Delta A_n\}g'(A_n) \\
&= 2g(A_{n-1}) - f(X_{n-1})g'(A_n) \\
&\quad + 2\{g(A_n) - g(A_{n-1}) - \Delta A_n g'(A_n)\} \\
&\leq 2g(A_{n-1}) - f(X_{n-1})g'(A_{n-1}) = Y_{n-1}, \qquad (36)
\end{aligned}
$$

which shows that Y is a super-martingale. The martingale property of X is not required when f is concave on $\mathbb{R}_+$, since (35) then follows trivially from the subadditivity of f.

Let us now examine the cases of infinite values. First we note that Y is a.s. finite. This is clear if we can only show that $f(X_n) = 0$ a.s. on the set $\{A_n = 0\}$. Then write

$$
\begin{aligned}
E[f(\Delta X_n); \; \Delta A_n = 0] &= E[E[f(\Delta X_n) | \mathcal{F}_{n-1}]; \; \Delta A_n = 0] \\
&= E[\Delta A_n; \; \Delta A_n = 0] = 0,
\end{aligned}
$$

and conclude that $f(\Delta X_n) = 0$ a.s. on $\{\Delta A_n = 0\}$. Excluding the trivial case where $f \equiv 0$, we obtain $\Delta X_1 = \cdots = \Delta X_n = 0$ a.s. on $\{A_n = 0\}$, which implies $X_n = 0$ a.s. on the same set. Hence,

$$
Y_n = 2g(0) - f(0)g'(0) = 2g(0) < \infty \quad \text{a.s. on } \{A_n = 0\}.
$$

Next we consider the possibility that A_n may be infinite. Then (36) remains true with the same proof on the set $\{A_n < \infty\}$, and on $\{A_n = \infty\}$ we have trivially

$$
\begin{aligned}
E[Y_n | \mathcal{F}_{n-1}] &= E[2g(\infty) | \mathcal{F}_{n-1}] = 2g(\infty) \\
&\leq 2g(A_{n-1}) - f(X_{n-1})g'(A_{n-1}) = Y_{n-1}.
\end{aligned}
$$

It remains to extend the result for convex functions f to local martingales X. Then consider any optional time τ such that the stopped sequence $X_n^\tau = X_{\tau \wedge n}$ is a true martingale. Noting that

$$
E[f(\Delta X_k^\tau) | \mathcal{F}_{k-1}] = E[f(\Delta X_k) | \mathcal{F}_{k-1}] 1\{\tau \geq k\} = \Delta A_k 1\{k \leq \tau\},
$$

we obtain

$$
A_n^\tau = \sum_{k \leq n \wedge \tau} \Delta A_k = \sum_{k \leq n} E[f(X_k^\tau) | \mathcal{F}_{k-1}],
$$

and so the sequence

$$
Y_n^\tau = 2g(A_n^\tau) - f(X_n^\tau)g'(A_n^\tau), \quad n \in \mathbb{Z}_+,
$$

is a positive super-martingale. Applying this result to a localizing sequence $\tau_1, \tau_2, \ldots$ and using Fatou's lemma, we obtain the same property for the original sequence Y_n. $\qquad\square$

We proceed with a continuous-time version of the last result.

Proposition 2.29 *(continuous-time super-martingale) Let X be a quasi-left-continuous semi-martingale with $X_0 = 0$, and let ξ denote the jump point process of X. Fix an even function $f \geq 0$ on $\mathbb{R}$ with $f(0) = 0$ and a non-increasing, convex function $g \geq 0$ on $\mathbb{R}_+$, and let A be a predictable process with $A_0 = 0$ and $dA_t \geq d\hat{\xi}_t f$. Suppose that either X is a purely discontinuous local martingale, f is convex, and f' is concave on $\mathbb{R}_+$, or else that X is of pure jump type and f is concave on $\mathbb{R}_+$. Then we can define a super-martingale $Y \geq 0$ by*

$$Y_t = 2g(A_t) - f(X_t)g'(A_t), \quad t \geq 0.$$

When $f(x) \equiv x^2$, the statement remains true with $A_t = \langle X \rangle_t$ for any local L^2-martingale X. If instead $f(x) \equiv |x|$ and X has locally finite variation, we may take $A = \hat{V}$, where V is the total variation process of X.

Here g' is right-continuous by convention. As before, we may allow A to take infinite values, and we may also have $g'(0) = -\infty$, with the understanding that $0 \cdot \infty = 0$. For convenience, we begin with a couple of lemmas.

Lemma 2.30 *(convex functions and super-martingales) Let $X \geq 0$ be a special semi-martingale with compensator $\hat{X}$ and $X_0 = 0$, consider a non-decreasing, predictable process A with $A_0 = 0$ and $dA_t \geq d\hat{X}$, and fix a non-increasing, convex function $g \geq 0$ on $\mathbb{R}_+$. Then we can define a super-martingale $Y \geq 0$ by*

$$Y_t = g(A_t) - X_t g'(A_t), \quad t \geq 0.$$

Proof: By suitable approximation we may assume that $g'(0)$ is finite. Integrating by parts, as in FMP 26.10, and using the convexity of g, the positivity of X, and the monotonicity of A, we obtain

$$\begin{aligned} dY_t &= dg(A_t) - d(X_t g'(A_t)) \\ &= dg(A_t) - X_{t-} dg'(A_t) - g'(A_t) dX_t \\ &\leq dg(A_t) - g'(A_t) dX_t. \end{aligned}$$

Hence, by FMP 26.4 and the monotonicity of g and $A - \hat{X}$,

$$\begin{aligned} d\hat{Y}_t &\leq dg(A_t) - g'(A_t) d\hat{X}_t \\ &\leq dg(A_t) - g'(A_t) dA_t. \end{aligned} \tag{37}$$

To estimate the right-hand side, we may assume that g is smooth on $(0, \infty)$, since we can otherwise approximate g' from above by a suitable convolution with the desired properties. Using the general substitution rule in FMP 26.7, we get for any $s < t$

$$\begin{aligned} g(A_t) - g(A_s) &= \int_{s+}^{t+} g'(A_{r-}) dA_r + \sum_{r \in (s,t]} (\Delta g(A_r) - g'(A_{r-})\Delta A_r) \\ &= \int_{s+}^{t+} g'(A_r) dA_r + \sum_{r \in (s,t]} (\Delta g(A_r) - g'(A_r)\Delta A_r). \end{aligned}$$

Noting that
$$\Delta g(A_r) \le g'(A_r)\Delta A_r, \quad r \ge 0,$$
by the convexity of g and the monotonicity of A, we get $dg(A_t) \le g'(A_t)dA_t$, and so by (37) the process $\hat{Y}$ is non-increasing, which shows that Y is a local super-martingale. Since $Y \ge 0$, we conclude that Y is even a true super-martingale.   □

The following lemma gives the required estimate for convex functions f and purely discontinuous martingales X. Here we write $\int_s^t$ for integration over $(s, t]$.

Lemma 2.31 *(domination)* *Let X be a quasi-left-continuous, purely discontinuous, local martingale with jump point process ξ, and consider an even, convex function $f \ge 0$ on $\mathbb{R}$ such that f' is concave on $\mathbb{R}_+$ and $f(0) = f'(0) = 0$. Then*
$$f(X_t) \le f(X_s) + \int_s^t f'(X_{r-})dX_r + 2(\xi_t - \xi_s)f, \quad s < t.$$

Proof: Fix any $s < t$ in $\mathbb{R}_+$ and a partition $s = t_0 < t_1 < \cdots < t_n = t$. We claim that
$$\begin{aligned} f(X_t) &\le f(X_s) + \sum_k f'(X_{t_{k-1}})(X_{t_k} - X_{t_{k-1}}) \\ &\quad + 2\sum_k f(X_{t_k} - X_{t_{k-1}}). \end{aligned} \tag{38}$$
Indeed, the formula for $n = 1$ agrees with (34) for $a = X_s$ and $x = X_t - X_s$. Proceeding by induction, we assume the result to be true for partitions into $n - 1$ intervals. Turning to the case of n intervals, we obtain
$$\begin{aligned} f(X_t) &\le f(X_{t_1}) + \sum_{k>1} f'(X_{t_{k-1}})(X_{t_k} - X_{t_{k-1}}) \\ &\quad + 2\sum_{k>1} f(X_{t_k} - X_{t_{k-1}}), \end{aligned}$$
and by (34)
$$f(X_{t_1}) \le f(X_s) + f'(X_s)(X_{t_1} - X_s) + 2f(X_{t_1} - X_s).$$

Adding the two inequalities gives (38).

The first sum in (38) may be regarded as a stochastic integral with respect to X of the predictable step process $f'(X^n)$, where
$$X_t^n = \sum_k X_{t_{k-1}} 1\{t \in (t_{k-1}, t_k]\}, \quad t \ge 0.$$
Choosing successively finer partitions with mesh size $\max_k (t_k - t_{k-1}) \to 0$, we get $f'(X_t^n) \to f'(X_{t-})$ for all $t \ge 0$, and since the processes $f'(X^n)$ are uniformly locally bounded, we conclude from FMP 26.4 that
$$\sum_k f'(X_{t_{k-1}})(X_{t_k} - X_{t_{k-1}}) \xrightarrow{P} \int_s^t f'(X_{r-})dX_r.$$

To deal with the second sum in (38), we first assume that X has finite variation and finitely many jumps in $(s, t]$, say at $\tau_1, \ldots, \tau_m$. Letting $\tau_j \in (t_{k-1}, \tau_k]$ for $k = \kappa_j$, $j = 1, \ldots, m$, and writing $K = \{\kappa_1, \ldots, \kappa_m\}$, we note that

$$\sum_{k \in K} f(X_{t_k} - X_{t_{k-1}}) \to \sum_j f(\Delta X_{\tau_j}) = \xi_t f - \xi_s f.$$

Since $f(x) = o(x)$ as $x \to 0$, the contribution from the complement K^c tends to 0, and so the entire sum tends to the same limit $\xi_t f - \xi_s f$.

Proceeding to martingales X with the stated properties, we may first reduce by localization to the case where f' is bounded. In fact, let f_n be the even, convex function satisfying $f_n(x) = f(x)$ when $|f'(x)| \leq n$ and $f'_n(x) = \pm n$ when $|f'(x)| > n$. Writing $\tau_n = \inf\{t \geq 0; |f'(x)| > n\}$ and using the result in the special case, we get for any $m \leq n$

$$f_n(X_t^{\tau_m}) \leq f_n(X_s^{\tau_m}) + \int_s^t f'(X_{r-}^{\tau_m}) dX_r^{\tau_m} + 2(\xi_t^{\tau_m} - \xi_s^{\tau_m}) f_n,$$

and the desired formula follows as we let $n \to \infty$ and then $m \to \infty$. By a further localization, we may assume that $EX^* < \infty$.

Since X is quasi-left-continuous, we can choose a sequence of totally inaccessible times $\tau_1, \tau_2, \ldots$ enumerating the jump times of X. For every $n \in \mathbb{N}$ we introduce the process X^n of compensated jumps at $\tau_1, \ldots, \tau_n$, and note that each X^n has finite variation and finitely many jumps. Hence, the result in the special case yields

$$f(X_t^n) \leq f(X_s^n) + \int_s^t f'(X_{r-}^n) dX_r^n + 2(\xi_t^n - \xi_s^n) f, \qquad (39)$$

where ξ^n is the jump point process associated with X^n. Since X is purely discontinuous, we have $(X - X^n)^* \xrightarrow{P} 0$, and therefore $f(X_t^n) \xrightarrow{P} f(X_t)$ and $f(X_s^n) \xrightarrow{P} f(X_s)$ by the continuity of f. We also note that $(\xi_t^n - \xi_s^n) f \to (\xi_t - \xi_s) f$ by monotone convergence.

To deal with the stochastic integral in (39), we may write $Y = f'(X)$ and $Y^n = f'(X^n)$, and conclude from a BDG-inequality (FMP 26.12) that

$$\begin{aligned} E(Y_-^n &\cdot X^n - Y_- \cdot X)_t^* \\ &\leq E(Y_-^n \cdot (X^n - X))_t^* + E((Y_-^n - Y_-) \cdot X)_t^* \\ &\leq E((Y_-^n)^2 \cdot [X^n - X])_t^{1/2} + E((Y_-^n - Y_-)^2 \cdot [X])_t^{1/2}. \end{aligned}$$

Noting that the processes Y and Y^n are uniformly bounded with $(Y - Y^n)^* \xrightarrow{P} 0$ and that $E[X - X^n]_\infty^{1/2} \to 0$, we see that the right-hand side tends to 0 by dominated convergence as $n \to \infty$. Hence,

$$\int_0^t f'(X_{s-}^n) dX_s^n \xrightarrow{P} \int_0^t f'(X_{s-}) dX_s, \quad t \geq 0,$$

and the desired relation follows as we let $n \to \infty$ in (39). $\qquad \square$

Proof of Proposition 2.29: Define $U = f(X)$, with f and X as stated. By Lemma 2.30 it suffices to show that U is a special semi-martingale with compensator bounded by $2A$ for the specified processes A. This is obvious when X is a local L^2-martingale and $f(x) \equiv x^2$, since the compensator of X^2 equals $\langle X \rangle$ by definition (FMP 26.1). It is equally obvious when $f(x) \equiv |x|$ and X has locally finite variation. Next let f be non-decreasing and concave on $\mathbb{R}_+$ with $f(0) = 0$, and let X be a pure jump type process. Then, by subadditivity, we have for any $s < t$

$$f(X_t) - f(X_s) \leq \sum_{r \in (s,t]} f(\Delta X_r) = \xi_t f - \xi_s f,$$

and so the compensator of $U = f(X)$ is bounded by $A = \hat{\xi} f$. The case of convex functions f and purely discontinuous martingales follows from Lemma 2.31, since the integral $f'(X_-) \cdot X$ is a local martingale by FMP 26.13. $\square$

We may use the previous results to study the local growth of exchangeable processes on $[0, 1]$. A different approach based on coupling is exhibited in connection with Corollaries 3.29 and 3.30. Given an even, convex function $f \geq 0$ with $f(0) = 0$, we write f^{-1} for the non-negative inverse of f.

Theorem 2.32 *(local growth, Fristedt, Millar, Kallenberg)* *Let X be an exchangeable process on $[0, 1]$ with characteristics $(\alpha, 0, \beta)$, and fix an even, continuous function $f \geq 0$ on $\mathbb{R}$ with $f(0) = 0$ and $\beta f < \infty$ a.s. Then the following statements hold as $t \to 0$:*

(i) *If f is convex, f' is concave on $\mathbb{R}_+$ with $f'(0) = 0$, and $c > 1$, then*

$$\frac{X_t}{f^{-1}(t|\log t|^c)} \to 0 \quad a.s., \qquad \frac{X_t}{f^{-1}(t)} \xrightarrow{P} 0.$$

(ii) *If $\sum_j |\beta_j| < \infty$ and $\alpha = \sum_j \beta_j$, and if f is concave on $\mathbb{R}_+$, then*

$$\frac{X_t}{f^{-1}(t)} \to 0 \quad a.s.$$

Proof: (i) We may clearly assume that α and β are non-random, and since $t/f^{-1}(t|\log t|^c) \to 0$ as $t \to 0$, we may even take $\alpha = 0$. Then Theorem 2.13 shows that $M_t = X_t/(1-t)$ is a purely discontinuous, quasi-left-continuous martingale on $[0, 1)$, and that the compensator $\hat{\mu}_t$ of the associated jump point process μ_t satisfies

$$\hat{\mu}_t - \hat{\mu}_s = \int_s^t \frac{d\hat{\xi}_r}{1-r} \leq \int_s^t \frac{\beta \, dr}{(1-r)^2} \leq \frac{(t-s)\beta}{(1-t)^2}, \quad s < t < 1.$$

Hence, by Proposition 2.29 and FMP 7.18,

$$f(M_t)g'(2t\beta f) \to a \leq 0 \quad a.s., \quad t \to 0,$$

for any non-increasing and convex function $g \colon \mathbb{R}_+ \to \mathbb{R}_+$. In particular, we may take $g'(t) = -(t|\log t|^c)^{-1}$ for small $t > 0$ and arbitrary $c > 1$ to get as $t \to 0$

$$\frac{f(M_t)}{t|\log t|^c} \to a_c \geq 0 \text{ a.s.}, \quad c > 1.$$

Comparing with the limit for exponents $c' \in (1, c)$, we see that in fact $a_c = 0$. Since $f(rx) \leq r^2 f(x)$ for $r \geq 1$ by the concavity of f', we obtain

$$\frac{|X_t|}{f^{-1}(t|\log t|^c)} = \frac{(1-t)|M_t|}{f^{-1}(t|\log t|^c)} \lesssim \left(\frac{f(M_t)}{t|\log t|^c} \right)^{1/2} \to 0 \text{ a.s.},$$

as required.

Next we see from Lemma 2.31 that

$$f(M_t) \leq \int_0^t f'(M_{s-})dM_s + 2\mu_t f, \quad t \in [0, 1).$$

Putting $\tau_n = \inf\{t > 0; |M_t| > n\}$ and noting that the stopped integral process $(f'(M-) \cdot M)^{\tau_n}$ is a true martingale, we get

$$Ef(M_{t \wedge \tau_n}) \leq 2E\mu_t f = 2E\hat{\mu}_t f \leq 2t(1-t)^{-2}\beta f,$$

and so by Fatou's lemma

$$Ef(X_t) \leq Ef(M_t) \leq \liminf_{n \to \infty} Ef(M_{t \wedge \tau_n}) \leq 2t(1-t)^{-2}\beta f.$$

Using this relation and noting that $f(x+y) \leq 2f(x)+2f(y)$ for any $x, y \in \mathbb{R}$, we get for $t \in (0, 1]$ and $n \in \mathbb{N}$

$$
\begin{aligned}
t^{-1}Ef(X_t) &\leq t^{-1}Ef\left(\sum_{j \leq n} \beta_j (1\{\tau_j \leq t\} - t) \right) \\
&\quad + t^{-1}Ef\left(\sum_{j > n} \beta_j (1\{\tau_j \leq t\} - t) \right) \\
&\lesssim t^{-1}f\left(t \sum_{j \leq n} |\beta_j| \right) + \sum_{j > n} f(\beta_j),
\end{aligned}
$$

which tends to 0 as $t \to 0$ and then $n \to \infty$. This gives $f(X_t)/t \xrightarrow{P} 0$, and we conclude as before that $X_t/f^{-1}(t) \xrightarrow{P} 0$.

(ii) Here the subadditivity of f yields

$$f(X_t) = f\left(\sum_j \beta_j 1\{\tau_j \leq t\} \right) \leq \sum_j f(\beta_j) 1\{\tau_j \leq t\}, \quad t \in [0, 1].$$

Now Theorem 2.12 or 2.20 shows that the processes

$$M_t^n = t^{-1} \sum_{j > n} f(\beta_j) 1\{\tau_j \leq t\}, \quad t \in (0, 1], \ n \in \mathbb{Z}_+, \tag{40}$$

are reverse, L^1-bounded martingales, and therefore converge a.s. as $t \to 0$ toward some finite limits M_0^n. Since trivially $M_t^0 - M_t^n \to 0$ a.s., Fatou's lemma gives

$$EM_0^0 = EM_0^n \le \liminf_{t \to 0} EM_t^n = \sum_{j>n} f(\beta_j),$$

and n being arbitrary, we conclude that $M_0^0 = 0$ a.s. Hence, $f(X_t)/t \to 0$ a.s., and the assertion follows as before, since $f(rt) \le rf(t)$ for $r > 1$ by the concavity of f. $\qquad \square$

We may also use the previous methods to study the convergence of the series of centered jumps in the representation formula for exchangeable processes exhibited in Theorem 2.18.

Proposition 2.33 *(uniform convergence)* Let $\tau_1, \tau_2, \ldots$ be i.i.d. $U(0,1)$, and fix an even, continuous function $f \ge 0$ on $\mathbb{R}$ with $f(0) = 0$ and $\beta f < \infty$. Then the following statements hold as $n \to \infty$:

(i) If f is convex, f' is concave on $\mathbb{R}_+$ with $f'(0) = 0$, and $p \in (0,1)$, then

$$\sup_{t \in (0,1)} \frac{\left| \sum_{j>n} \beta_j (1\{\tau_j \le t\} - t) \right|}{f^{-1}(t^p(1-t)^p)} \to 0 \quad a.s.$$

(ii) If f is concave on $\mathbb{R}_+$, then

$$\sup_{t \in (0,1]} \frac{\left| \sum_{j>n} \beta_j 1\{\tau_j \le t\} \right|}{f^{-1}(t)} \to 0 \quad a.s.$$

Proof: (i) Introduce the martingales $M_t^n = X_t^n/(1-t)$, where

$$X_t^n = \sum_{j>n} \beta_j (1\{\tau_j \le t\} - t), \quad t \in [0,1], \; n \in \mathbb{Z}_+,$$

and note as before that the jump point processes μ^n of M^n satisfy $d\hat{\mu}_t^n \le 4\beta^n f \, dt$ on $[0, \frac{1}{2}]$, where $\beta^n f = \sum_{j>n} f(\beta_j)$. Hence, we may form a super-martingale Y^n as in Proposition 2.29, based on the processes M_t^n and $A_t^n = 4t\beta^n f$ and for any non-increasing, convex function $g \ge 0$ on $\mathbb{R}_+$. By the Bernstein–Lévy maximum inequalities (FMP 7.15) we get for any $r > 0$

$$rP\{\sup_{t \le 1/2}(-f(X_t^n)g'(4t\beta^n f)) \ge r\} \le rP\{\sup_t Y_t^n \ge r\}$$
$$\le 3\sup_t EY_t^n \le 6g(0) < \infty,$$

which shows that the sequence of suprema on the left is tight. Fixing any $p \in (0,1)$, we may choose $g'(t) = -t^{-p}$ for small $t > 0$ and conclude from FMP 4.9 that as $n \to \infty$

$$\sup_{t \in (0,1/2]} t^{-p} f(X_t^n) \xrightarrow{P} 0.$$

By symmetry the same result holds for the processes X^n_{1-t}, and so by combination

$$\sup_{t\in(0,1)} \frac{f(X^n_t)}{t^p(1-t)^p} \xrightarrow{P} 0.$$

Since $f(rt) \le r^2 f(t)$ for all $r \ge 1$, we conclude that

$$\sup_{t\in(0,1)} \frac{|X^n_t|}{f^{-1}(t^p(1-t)^p)} \xrightarrow{P} 0,$$

and the corresponding a.s. convergence follows by Lemma 2.19.

(ii) Here we introduce the processes

$$X^n_t = \sum_{j>n} \beta_j 1\{\tau_j \le t\}, \quad t \in [0,1], \ n \in \mathbb{Z}_+,$$

and note as before that $f(X^n_t) \le tM^n_t$, where the M^n are given by (40). Since the latter processes are reverse martingales on $(0,1]$ with $EM^n_t = \beta^n f < \infty$, we may apply the Bernstein–Lévy inequalities to get as $n \to \infty$

$$rP\{(M^n)^* > r\} \le \beta^n f \to 0, \quad r > 0,$$

which shows that $(M^n)^* \xrightarrow{P} 0$. Since $(M^n)^*$ is non-increasing in n, the last result remains true in the a.s. sense, and we obtain

$$\sup_t t^{-1} f(X^n_t) \le (M^n)^* \to 0 \ \text{ a.s.}$$

The assertion now follows since $f(rx) \le rf(x)$ for any $r \ge 1$ by the concavity of f. □

2.7 Palm Measure Invariance

Fix a σ-finite measure λ on a measurable space $(S, \mathcal{S})$. If ξ is a mixed Poisson process on S directed by $\rho\lambda$ for some random variable $\rho \ge 0$, then clearly

$$P\{\xi B = 0\} = \varphi(\lambda B), \quad B \in \mathcal{S},$$

where $\varphi(t) = Ee^{-t\rho}$ denotes the Laplace transform of ρ. When $0 < \lambda S < \infty$, we note that the same formula holds for a mixed binomial process ξ based on the probability measure $\lambda/\lambda S$ and a $\mathbb{Z}_+$-valued random variable κ, provided that we choose $\varphi(t) = E(1 - t/\lambda S)^\kappa$ for $t \in [0, \lambda S]$. In either case, ξ has Laplace functional

$$Ee^{-\xi f} = \varphi(\lambda(1 - e^{-f})), \quad f \ge 0, \tag{41}$$

where the function f is understood to be measurable. This shows that the distribution of ξ is uniquely given by the pair (λ, φ), and justifies that we write $\mathcal{L}(\xi) = M(\lambda, \varphi)$.

In the context of Palm measures, it is convenient to allow the basic "probability" measure P to be σ-finite, provided that the associated "distribution" of ρ or κ is such that $\varphi(t) < \infty$ for all $t > 0$. Then the "distribution" of ξ is still determined by (λ, φ), and the notation $M(\lambda, \varphi)$ continues to make sense. In fact, by an extension of the Hausdorff–Bernstein Theorem A4.1, we may choose φ to be any completely monotone function on $(0, \lambda S)$. In particular, given a pair (λ, φ) as above with $\varphi(0) = 1$, we may consider the σ-finite measure $M(\lambda, -\varphi')$.

The following result characterizes the mixed Poisson and binomial processes in terms of their reduced Palm measures. For the definitions and elementary properties of Palm and supporting measures, we refer to Appendix A5.

Theorem 2.34 *(reduced Palm measures, Papangelou, Kallenberg) Let ξ be a point process on a Borel space S with reduced Palm measures Q'_s, $s \in S$. Then we can choose the latter to be independent of s iff ξ is a mixed Poisson or binomial process, in which case $\mathcal{L}(\xi) = M(\lambda, \varphi)$ iff $Q'_s = M(\lambda, -\varphi')$ a.e. λ.*

This leads in particular to the following classical characterization of the general Poisson process.

Corollary 2.35 *(Poisson criterion, Slivnyak) Let ξ be a point process on a Borel space S with reduced Palm distributions Q'_s, $s \in S$. Then $Q'_s = \mathcal{L}(\xi)$ a.e. $E\xi$ iff ξ is Poisson.*

Proof: Assume the stated condition. Then Theorem 2.34 yields

$$\mathcal{L}(\xi) = M(\lambda, \varphi) = M(\lambda, -\varphi'),$$

for some measure λ and function φ. Solving the equation $\varphi = -\varphi'$ gives $\varphi(t) = e^{-t}$, which means that ξ is Poisson with intensity λ. Reversing the argument yields the result in the opposite direction. $\qquad\square$

Our proof of Theorem 2.34 is based on a simple connection between Palm measures and regular conditional distributions.

Lemma 2.36 *(Palm measures and conditioning) Let ξ be a random measure on a Borel space $(S, \mathcal{S})$ with Palm measures Q_s, $s \in S$, fix a set $C \in \mathcal{S}$ with $\xi C < \infty$ a.s., and consider a random element τ in C satisfying*

$$P[\tau \in \cdot|\xi] = (1_C \cdot \xi)/\xi C \quad a.s. \text{ on } \{\xi C > 0\}. \tag{42}$$

Then for any measurable function $f \geq 0$, we have

$$E[f(\xi, \tau)|\xi C, \tau] = \int f(\mu, \tau)\, q(\tau, \xi C, d\mu) \quad a.s. \text{ on } \{\xi C > 0\}, \tag{43}$$

where q is a probability kernel from $C \times (0, \infty)$ to $\mathcal{M}(S)$ given by

$$q(s, x, \cdot) = Q_s[\,\cdot\,|\mu C \in dx], \quad s \in C,\ x > 0. \tag{44}$$

Proof: By the definition of the Palm measures Q_s and the a.s. finiteness of ξC, we note that the measures $R_s = Q_s\{\mu C \in \cdot\}$ are σ-finite for $s \in C$ a.e. λ, where λ denotes the supporting measure associated with the kernel Q. This ensures the existence, for λ-almost every $s \in C$, of a probability kernel $q(s, \cdot)$ from $(0, \infty)$ to $\mathcal{M}(S)$ satisfying (44). Furthermore, we see from FMP 7.26 that $q(s, x, \cdot)$ has a product measurable version on $C \times (0, \infty)$. It is also clear that q is independent of the choice of λ.

To prove (43), we may clearly assume that $\lambda C > 0$. Using (42), (44), the definition of Palm and supporting measures, and the disintegration theorem (FMP 6.4), we get for any product measurable function $f \geq 0$ on $\mathcal{M}(S) \times C$

$$
\begin{aligned}
Ef(\xi, \tau) &= E \int_C f(\xi, s)\, \xi(ds)/\xi C \\
&= \int_C \lambda(ds) \int f(\mu, s)\, Q_s(d\mu)/\mu C \\
&= \int_C \lambda(ds) \int x^{-1} R_s(dx) \int f(\mu, s)\, q(s, x, d\mu).
\end{aligned}
$$

Writing $r_s(dx) = R_s(dx)/x$ for $x > 0$, we obtain

$$
\begin{aligned}
\mathcal{L}(\tau, \xi C, \xi) &= \lambda \otimes r \otimes q &&\text{on} \quad C \times (0, \infty) \times \mathcal{M}(S), \\
\mathcal{L}(\tau, \xi C) &= \lambda \otimes r &&\text{on} \quad C \times (0, \infty),
\end{aligned}
$$

in the sense of composition of kernels (FMP 1.41). Hence, for any measurable subset $B \subset C \times (0, \infty)$,

$$
\begin{aligned}
E[f(\xi, \tau);\ (\tau, \xi C) \in B] &= \iint_B (\lambda \otimes r)(dx\, ds) \int f(\mu, s)\, q(s, x, d\mu) \\
&= E\left[\int f(\mu, \tau)\, q(\tau, \xi C, d\mu);\ (\tau, \xi C) \in B\right],
\end{aligned}
$$

and (43) follows. $\qquad\square$

Proof of Theorem 2.34: First suppose that $\mathcal{L}(\xi) = M(\lambda, \varphi)$. In particular, we note that λ is then a supporting measure for ξ. Fixing a measurable function $f \geq 0$ on S with $\lambda f > 0$ and a set $B \in \mathcal{S}$ with $\lambda B < \infty$, we get by (41)

$$
Ee^{-\xi f - t\xi B} = \varphi(\lambda(1 - e^{-f - t1_B})), \quad t \geq 0.
$$

Taking right derivatives at $t = 0$ gives

$$
E\xi B e^{-\xi f} = -\varphi'(\lambda(1 - e^{-f}))\, \lambda(1_B e^{-f}),
$$

where the formal differentiation on each side is justified by dominated convergence. Writing Q_s for the Palm measures of ξ associated with the supporting measure λ, we obtain

$$
\int_B \lambda(ds) \int e^{-\mu f}\, Q_s(d\mu) = -\varphi'(\lambda(1 - e^{-f})) \int_B e^{-f(s)}\, \lambda(ds),
$$

and since B was arbitrary, we get for $s \in S$ a.e. λ

$$\int e^{-\mu f} Q_s(d\mu) = -\varphi'(\lambda(1 - e^{-f})) e^{-f(s)},$$

which implies

$$\int e^{-\mu f} Q'_s(d\mu) = \int e^{-(\mu f - \delta_s)f} Q_s(d\mu) = -\varphi'(\lambda(1 - e^{-f})).$$

This extends by monotone convergence to arbitrary $f \geq 0$, and we may conclude that $Q'_s = M(\lambda, -\varphi')$ a.e. λ. In particular, Q'_s is independent of $s \in S$ a.e. λ.

Conversely, suppose that we can choose the supporting measure λ and the associated reduced Palm measures Q'_s such that $Q'_s = Q'$ is independent of s. Let us first assume that $\xi S < \infty$ a.s. and $P\{\xi \neq 0\} > 0$. We may then introduce a random element τ in S satisfying

$$P[\tau \in \cdot |\xi] = \xi/\xi S \quad \text{a.s. on } \{\xi S > 0\}. \tag{45}$$

For $n \in \mathbb{N}$ with $P\{\xi S = n\} > 0$, we get by Lemma 2.36

$$\begin{aligned} P[\xi - \delta_\tau \in M | \xi S = n, \tau \in ds] &= Q_s[\mu - \delta_s \in M | \mu S = n] \\ &= Q'[M | \mu S = n - 1], \end{aligned}$$

which implies

$$\tau \perp\!\!\!\perp_{\xi S} (\xi - \delta_\tau) \quad \text{on } \{\xi S > 0\}. \tag{46}$$

Next we get, for any $B \in \mathcal{S}$ and for $n \in \mathbb{N}$ with $P\{\xi S = n\} > 0$,

$$\begin{aligned} P[\tau \in B | \xi S = n] \\ = E[\xi B | \xi S = n]/n &= \frac{E[\xi B; \, \xi S = n]}{n P\{\xi S = n\}} \\ = \frac{\int_B Q_s\{\mu S = n\} \lambda(ds)}{n P\{\xi S = n\}} &= \frac{Q'\{\mu S = n - 1\} \lambda B}{n P\{\xi S = n\}}. \end{aligned}$$

In particular, we may take $B = S$ to see that $0 < \lambda S < \infty$. Dividing the expressions for B and S, and using (46), we obtain

$$P[\tau \in \cdot | \xi S, \xi - \delta_\tau] = P[\tau \in \cdot | \xi S] = \lambda/\lambda S \quad \text{a.s. on } \{\xi S > 0\}. \tag{47}$$

Since S is Borel, we may write $\xi = \sum_{j \leq \kappa} \delta_{\sigma_j}$ for some random elements $\sigma_1, \ldots, \sigma_\kappa$ of S, where $\kappa = \xi S$. Now introduce some mutually independent random variables $\pi_1, \pi_2, \ldots \perp\!\!\!\perp \xi$ with distributions

$$P\{\pi_k = j\} = k^{-1}, \quad j = 1, \ldots, k, \quad k \in \mathbb{N}.$$

Define $\tau_1 = \sigma \circ \pi_\kappa$, select τ_2 from the remaining set $\{\sigma_j; \, j \neq \pi_\kappa\}$ as variable number $\pi_{\kappa-1}$, and continue recursively until the whole sequence $\tau_1, \ldots, \tau_\kappa$

has been constructed, by means of $\pi_\kappa, \ldots, \pi_1$, as a random permutation of $\sigma_1, \ldots, \sigma_\kappa$. Then $(\tau_1, \ldots, \tau_\kappa)$ is clearly conditionally exchangeable given $\kappa = \xi S$, and (45) follows with $\tau = \tau_1$. Using (47) and invoking the independence of the variables π_k, we obtain

$$P[\tau_1 \in \cdot \,|\, \kappa,\, \tau_2, \ldots, \tau_\kappa] = \lambda/\lambda S \quad \text{a.s. on } \{\kappa > 0\},$$

which extends by exchangeability to

$$P[\tau_i \in \cdot \,|\, \kappa;\, \tau_j,\, j \neq i] = \lambda/\lambda S, \quad i \leq \kappa, \ \text{a.s. on } \{\kappa > 0\}.$$

The τ_i are then conditionally i.i.d. $\lambda/\lambda S$ given $\kappa = \xi S$, which means that ξ is a mixed binomial process based on $\lambda/\lambda S$. In other words, $\mathcal{L}(\xi) = M(\lambda, \varphi)$ for some completely monotone function φ on $[0, \lambda S]$.

Let us finally allow ξ to be unbounded. Choosing sets $B_n \uparrow S$ in $\mathcal{S}$ with $\xi B_n < \infty$ a.s., we see from the previous argument that

$$\mathcal{L}(1_{B_n}\xi) = M(1_{B_n}\lambda, \varphi_n), \quad n \in \mathbb{N},$$

for some completely monotone functions φ_n on $[0, \lambda B_n]$, $n \in \mathbb{N}$. The uniqueness of the representation yields $\varphi_m = \varphi_n$ on the common interval $[0, \lambda B_m \wedge \lambda B_n]$, and so there exists a common extension φ on $[0, \lambda S)$, which is again completely monotone on the new interval. For any measurable function $f \geq 0$ on S, we may put $f_n = 1_{B_n}f$ and note that

$$Ee^{-\xi f_n} = \varphi(\lambda(1 - e^{-f_n})), \quad n \in \mathbb{N}.$$

Equation (41) now follows as we let $n \to \infty$, by monotone and dominated convergence together with the continuity of φ. This shows that again $\mathcal{L}(\xi) = M(\lambda, \varphi)$. $\qquad\square$

We turn to a similar but more sophisticated result for general random measures. Here we write $\xi \in S_1(\lambda, \alpha, \beta)$ if ξ is symmetrically distributed with respect to the positive and bounded measure λ on S with diffuse component $\xi_d = \alpha\lambda/\lambda S$, with atom sizes β_j given by the point process $\beta = \sum_j \delta_{\beta_j}$, and with associated atom positions τ_j that are independent random variables with the common distribution $\hat{\lambda} = \lambda/\lambda S$. Note that this forces λ to be diffuse, unless $\alpha = 0$ and $\beta(0, \infty) \leq 1$ a.s. In the exceptional case, we have $\xi = \beta_1\delta_{\tau_1}$ for some independent random elements β_1 and τ_1, where the distribution of the latter is arbitrary.

Similarly, by $\xi \in S_\infty(\lambda, \alpha, \nu)$ we mean that ξ is again λ-symmetric, but now with conditionally independent increments directed by the pair (α, ν). Thus, ξ has again diffuse component $\xi_d = \alpha\lambda$, and the atoms of ξ are given by a Cox process $\eta \perp\!\!\!\perp_\nu \alpha$ on $S \times (0, \infty)$ directed by $\lambda \otimes \nu$. Given the Palm measures Q_s of a random measure ξ, we now define the associated *reduced* versions Q'_s by

$$Q'_s = Q_s\{\mu;\, (\mu\{s\},\, \mu - \mu\{s\}\delta_s) \in \cdot\}, \quad s \in S.$$

Theorem 2.37 *(invariant Palm measures) Let ξ be a random measure on a Borel space S with supporting measure λ and associated reduced Palm measures Q'_s. Then $Q'_s = Q'$ is a.e. independent of s iff ξ is λ-symmetric, in which case even Q' is λ-symmetric, and λ is diffuse unless ξ is a.s. degenerate. Furthermore, letting $\lambda = E\xi$ be σ-finite and $Q'_s \equiv Q' = \mathcal{L}(\eta, \zeta)$, we have $\eta \perp\!\!\!\perp \zeta$ iff either of these conditions holds:*

(i) *$\xi \in S_1(\lambda, 0, \beta)$ for some mixed binomial process β on $(0, \infty)$,*

(ii) *$\xi \in S_\infty(\lambda, \rho\alpha, \rho\nu)$ for a fixed pair (α, ν) and some random $\rho \geq 0$.*

The last assertion contains the remarkable fact that, in the case of independence $\eta \perp\!\!\!\perp \zeta$, the atomic structure of ξ, regarded as a point process on the product space $S \times (0, \infty)$, is separately exchangeable with respect to the measures $E\xi$ and $E\beta$ or $E\nu$, respectively. It is only in the mixed Poisson case that we can also have a diffuse component, which is then proportional to ν.

A lemma will again be helpful for the proof. Here and below, we write $l\beta = l \cdot \beta$ for $l(x) \equiv x$, and when $0 < \xi S < \infty$ we put $\hat{\xi} = \xi/\xi S$.

Lemma 2.38 *(symmetry and independence) Let ξ be a random measure on S with $\xi S < \infty$ a.s., let $P[\tau \in \cdot | \xi] = \hat{\xi}$ a.s. on $\{\xi \neq 0\}$, and define*

$$(\eta, \zeta) = (\xi\{\tau\}, \, \xi - \xi\{\tau\}\delta_\tau) \quad on \ \{\xi \neq 0\}.$$

Then $\xi \in$ some $S_1(\lambda, \alpha, \beta)$ iff $\tau \perp\!\!\!\perp (\eta, \zeta)$ conditionally on $\{\xi \neq 0\}$, where λ is diffuse unless ξ is a.s. degenerate. In that case, (η, ζ) is again conditionally λ-symmetric, say with $\zeta \in S_1(\lambda, \alpha', \beta')$ on $\{\xi \neq 0\}$, and writing $\gamma = \alpha\delta_0 + l\beta$ and $\gamma' = \alpha'\delta_0 + l\beta'$, we have for any measurable function $f \geq 0$

$$E[f(\eta, \gamma') | \xi \neq 0] = E\left[\int f(x, \gamma - x\delta_x)\hat{\gamma}(dx) \Big| \gamma \neq 0\right].$$

Proof: Define $\lambda = P[\tau \in \cdot | \xi \neq 0]$, let α be the total diffuse mass of ξ, write $\beta = \sum_j \delta_{\beta_j}$ for the point process of atom sizes of ξ, and put $\gamma = \alpha\delta_0 + l\beta$. Assuming $\tau \perp\!\!\!\perp (\eta, \zeta)$ and noting that γ is a measurable function of (η, ζ), we obtain

$$P[\tau \in \cdot | \gamma, \eta, \zeta] = \lambda \quad \text{a.s. on } \{\gamma \neq 0\}. \tag{48}$$

Since also $P[\tau \in \cdot | \xi, \gamma] = \hat{\xi}$ a.s. on $\{\gamma \neq 0\}$, and since conditional λ-symmetry implies the corresponding unconditional property, we may henceforth assume that γ is non-random and nonzero.

Now let $\tau_1, \tau_2, \ldots$ be the positions of the atoms of sizes $\beta_1, \beta_2, \ldots$, enumerated in exchangeable order in case of multiplicities, as in the proof of Theorem 2.34. Considering (48) on the set $\{\eta = \beta_k\}$ and arguing as in the previous proof, we obtain

$$P[\tau_k \in \cdot | \xi\{\tau_k\}, \, \xi - \xi\{\tau_k\}\delta_{\tau_k}; \, \tau_j, \, j \neq k] = \lambda \quad \text{a.s. }, \quad k \leq \beta(0, \infty),$$

which shows that the τ_j are i.i.d. λ. When $\alpha > 0$, we note that also a.s.

$$
\begin{aligned}
\xi_d/\alpha &= P[\tau \in \cdot \,|\, \xi\{\tau\} = 0,\, \xi] \\
&= P[\tau \in \cdot \,|\, \eta = 0,\, \zeta] = P\{\tau \in \cdot\} = \lambda,
\end{aligned}
$$

which implies $\xi_d = \alpha\lambda$ a.s. In particular, λ is diffuse when $\alpha > 0$. The same thing is true when $\beta(0,\infty) > 1$, since otherwise we would have $P\{\tau_1 = \tau_2\} > 0$, contradicting the definition of the τ_j. The λ-symmetry of ξ then follows under either condition. In the degenerate case, the required symmetry is obvious from the representation $\xi = \beta_1\delta_\tau$.

By the disintegration theorem (FMP 6.4) and the definitions of τ, η, and γ, we get for any measurable function $f \geq 0$ on $\mathbb{R}_+$

$$
E[f(\eta)|\xi] = E[f(\xi\{\tau\})|\xi] = \int_S f(\xi\{s\})\,\hat{\xi}(ds) = \int_0^\infty f(x)\,\hat{\gamma}(dx),
$$

a.s. on $\{\xi \neq 0\}$. Using this fact and applying the disintegration theorem once more, we conclude that

$$
\begin{aligned}
E[f(\eta, \gamma')|\,\xi \neq 0] &= E[f(\eta,\, \gamma - \eta\delta_\eta)|\,\gamma \neq 0] \\
&= E\left[\int_0^\infty f(x, \gamma - x\delta_x)\,\hat{\gamma}(dx)\,\Big|\,\gamma \neq 0\right],
\end{aligned}
$$

as asserted.

Now suppose instead that ξ is λ-symmetric for some probability measure λ on S, assumed to be diffuse unless ξ is a.s. degenerate. By an easy extension of Theorem 1.25 or its proof, ξ has then an a.s. representation

$$
\xi = \alpha\lambda + \sum_j \beta_j \delta_{\tau_j},
$$

for some i.i.d. random elements $\tau_1, \tau_2, \ldots$ with distribution λ and an independent sequence of random variables $\alpha, \beta_1, \beta_2, \ldots$. To establish the required independence $\tau \perp\!\!\!\perp (\eta, \zeta)$ and λ-symmetry of (η, ζ), it is enough to prove the corresponding statements for the conditional distribution, given $\gamma = \alpha\delta_0 + l\beta$ with $\beta = \sum_j \delta_{\beta_j}$, and to verify that τ has conditional distribution λ. For convenience, we may then assume that γ is non-random.

We may construct η and τ by a randomization in two steps, as follows. First we choose η to be independent of $\tau_1, \tau_2, \ldots$ with distribution

$$
\begin{aligned}
P\{\eta = 0\} &= \alpha/\gamma\mathbb{R}_+, \\
P\{\eta = \beta_k\} &= m_k\beta_k/\gamma\mathbb{R}_+, \quad k \geq 1,
\end{aligned}
$$

where $m_k = \beta\{\beta_k\}$. When $\eta = \beta_k$, we may next choose $\tau = \tau_{j_k}$, where $j_k = \min\{j \geq 1;\, \beta_j = \beta_k\}$. If instead $\eta = 0$, we may choose τ to be an independent random variable with distribution λ. It is easy to check that τ satisfies the required relation $P[\tau \in \cdot|\xi] = \hat{\xi}$.

Since $\zeta = \xi$ when $\eta = 0$, we have a.s.

$$\begin{aligned} P[\tau \in \cdot \,|\, \zeta, \eta = 0] &= P[\tau \in \cdot \,|\, \xi, \eta = 0] \\ &= P[\tau \in \cdot \,|\, \eta = 0] = \lambda. \end{aligned}$$

Similarly, by the independence of η and $\tau_1, \tau_2, \ldots$,

$$\begin{aligned} P[\tau \in \cdot \,|\, \zeta, \eta = \beta_k] &= P[\tau_{j_k} \in \cdot \,|\, \tau_i, i \neq j_k; \eta = \beta_k] \\ &= P\{\tau_{j_k} \in \cdot\} = \lambda. \end{aligned}$$

Combining the two results gives $P[\tau \in \cdot \,|\, \eta, \zeta] = \lambda$ a.s., which means that τ is independent of (η, ζ) with distribution λ.

Next we see from the independence $\eta \perp\!\!\!\perp \xi$ that

$$P[\zeta \in \cdot \,|\, \eta = 0] = P[\xi \in \cdot \,|\, \eta = 0] = P\{\xi \in \cdot\},$$

which is λ-symmetric by hypothesis. Similarly, writing

$$\zeta_k = \xi - \beta_k \delta_{\tau_{j_k}}, \quad k \geq 1,$$

and noting that $\zeta_k \perp\!\!\!\perp \eta$, we get

$$P[\zeta \in \cdot \,|\, \eta = \beta_k] = P[\zeta_k \in \cdot \,|\, \eta = \beta_k] = P\{\zeta_k \in \cdot\},$$

which is again λ-symmetric. Hence, ζ is conditionally λ-symmetric given η, which implies the unconditional λ-symmetry of the pair (η, ζ). $\quad\square$

Proof of Theorem 2.37: In proving the first assertion, it suffices to consider the restrictions of ξ to sets $B \in \mathcal{S}$ with $\xi B < \infty$ a.s., and to simplify the notation we may then assume that $\xi S < \infty$ a.s. Letting τ be such that $P[\tau \in \cdot \,|\, \xi] = \xi / \xi S$ on $\{\xi S > 0\}$, we get by Lemma 2.36 for any $x > 0$ and $s \in S$

$$\begin{aligned} P[(\xi\{\tau\}, \xi - \xi\{\tau\}\delta_\tau) &\in \cdot \,|\, \xi S \in dx, \tau \in ds] \\ &= Q_s[(\mu\{s\}, \mu - \mu\{s\}\delta_s) \in \cdot \,|\, \mu S \in dx], \end{aligned} \tag{49}$$

in the sense of the quoted lemma. If Q'_s is independent of s, then so is the right-hand side of (49), which implies

$$\tau \perp\!\!\!\perp_{\xi S} (\xi\{\tau\}, \xi - \xi\{\tau\}\delta_\tau) \text{ on } \{\xi S > 0\}. \tag{50}$$

Since the defining property of τ remains conditionally valid given ξS, Lemma 2.38 applies to the conditional distributions, and we may conclude that, a.s. for given $\xi S > 0$, ξ is either degenerate at τ or conditionally symmetrically distributed with respect to the diffuse random measure

$$P[\tau \in \cdot \,|\, \xi S] = E[\xi / \xi S \,|\, \xi S] = E[\xi \,|\, \xi S] / \xi S \text{ a.s. on } \{\xi S > 0\}. \tag{51}$$

Now for any $B \in \mathcal{S}$ and $A \in \mathcal{B}$,

$$
\begin{aligned}
E[\xi B; \, \xi S \in A] &= \int_B Q_s\{\mu S \in A\} \, \lambda(ds) \\
&= Q'\{(x, \mu); \, x + \mu S \in A\} \, \lambda B,
\end{aligned}
$$

and combining this with the same equation for $B = S$ gives

$$
\begin{aligned}
E[E[\xi B | \xi S]; \, \xi S \in A] &= E[\xi B; \, \xi S \in A] \\
&= E[\xi S; \, \xi S \in A] \, \lambda B / \lambda S.
\end{aligned}
$$

Since A was arbitrary, we obtain

$$
E[\xi B | \xi S] = (\lambda B / \lambda S) \, \xi S \quad \text{a.s.}, \quad B \in \mathcal{S},
$$

and so by (51)

$$
P[\tau \in \cdot | \xi S] = E[\xi | \xi S] / \xi S = \lambda / \lambda S \quad \text{a.s. on } \{\xi S > 0\}.
$$

This shows that ξ remains unconditionally symmetric with respect to λ. We also note that λ is diffuse by Lemma 2.38, unless ξ is a.s. degenerate.

Conversely, suppose that ξ is λ-symmetric, where λ is diffuse except when ξ is a.s. degenerate. To show that $Q'_s = Q'$ is a.e. independent of s, it is enough to consider the restrictions of ξ to any sets $B \in \mathcal{S}$ with $0 < \lambda B < \infty$, and so we may assume that $\xi \in S_1(\lambda, \alpha, \beta)$ with $0 < \lambda S < \infty$. Then clearly $E[\xi | \xi S] = (\xi S) \lambda / \lambda S$ a.s., and so for any $B \in \mathcal{S}$ we have

$$
\begin{aligned}
\int_B Q_s\{\mu S \in \cdot\} \, \lambda(ds) &= E[\xi B; \, \xi S \in \cdot] \\
&= E[E[\xi B | \xi S]; \, \xi S \in \cdot] \\
&= E[\xi S; \, \xi S \in \cdot] \, \lambda B / \lambda S,
\end{aligned}
$$

which shows that a.e.

$$
Q_s\{\mu S \in \cdot\} \, \lambda S = E[\xi S; \, \xi S \in \cdot], \tag{52}
$$

independently of s. On the other hand, Lemma 2.38 yields (50) with τ defined as before, and so by (49) we see that even the conditional distributions

$$
Q_s[(\mu\{s\}, \, \mu - \mu\{s\}\delta_s) \in \cdot | \mu S], \quad s \in S, \tag{53}
$$

are a.e. independent of s. This, in combination with (52), shows that $Q'_s = Q'$ is independent of s a.e. λ.

From Lemma 2.38 and (49) we see that the conditional distributions in (53) are λ-symmetric, and so the same thing is true for the σ-finite measure Q'. Writing $\gamma = \alpha\delta_0 + l\beta$ as before, and letting $\gamma(\mu)$ denote the corresponding measure formed from a general μ, we get by (49) and Lemma 2.38 for any $r > 0$

$$
\begin{aligned}
Q_s\big[f(\mu\{s\},\,\gamma(\mu) - \mu\{s\}\delta_{\mu\{s\}})\big|\,\mu S \in dr\big] \\
= E\big[f(\xi\{\tau\},\,\gamma - \xi\{\tau\}\delta_{\xi\{\tau\}})\big|\,\xi S \in dr\big] \\
= E\Big[\int_0^\infty f(x,\gamma - x\delta_x)\hat{\gamma}(dx)\Big|\,\xi S \in dr\Big],
\end{aligned}
$$

where $\hat{\gamma} = \gamma/\gamma \mathbb{R}_+$. Combining with (52), we conclude that

$$
\begin{aligned}
Q_s f(\mu\{s\},\,\gamma(\mu) - \mu\{s\}\delta_{\mu\{s\}}) \\
= \int_0^\infty Q_s\big[f(\mu\{s\},\,\gamma(\mu) - \mu\{s\}\delta_{\mu\{s\}})\big|\,\mu S \in dr\big]\,Q_s\{\mu S \in dr\} \\
= \int_0^\infty E\Big[\int_0^\infty f(x,\gamma - x\delta_x)\hat{\gamma}(dx)\Big|\,\xi S \in dr\Big]\,E[\xi S;\,\xi S \in dr]\,/\lambda S \\
= \int_0^\infty E\Big[\int_0^\infty f(x,\gamma - x\delta_x)\gamma(dx)\Big|\,\xi S \in dr\Big]\,P\{\xi S \in dr\}/\lambda S \\
= E\int_0^\infty f(x,\gamma - x\delta_x)\,\gamma(dx)/\lambda S.
\end{aligned}
$$

Writing $Q' = \mathcal{L}(\eta,\zeta)$ and $\gamma_\zeta = \gamma(\zeta)$, we obtain

$$
Q'f(\eta,\gamma_\zeta)\,\lambda S = E\int_0^\infty f(x,\gamma - x\delta_x)\,\gamma(dx). \tag{54}
$$

Now suppose that $\lambda = E\xi$ is σ-finite and $Q' = \mathcal{L}(\eta,\zeta)$ with $\eta \perp\!\!\!\perp \zeta$. To prove (i) or (ii), it suffices as before to assume that $\xi \in S_1(\lambda,\alpha,\beta)$ with $0 < \lambda S < \infty$. Writing $\gamma' = \gamma_\zeta$, we get by (54) for any scalars $t \geq 0$ and measurable functions $f \geq 0$

$$
\begin{aligned}
Ee^{-\eta t}Ef(\gamma') &= Ee^{-\eta t}f(\gamma') \\
&= E\int_0^\infty e^{-xt}f(\gamma - x\delta_x)\gamma(dx)/E\xi S \\
&= \int_0^\infty e^{-xt}Ef(\gamma_x - x\delta_x)E\gamma(dx)/E\xi S,
\end{aligned}
$$

where $\gamma_x = \alpha_x\delta_0 + l\beta_x$ is such that $\mathcal{L}(\gamma_x)$ equals the Palm distribution of γ at $x \geq 0$. Assuming $0 < Ef(\gamma') < \infty$, we obtain

$$
Ee^{-\eta t} = \int_0^\infty e^{-xt}\,\frac{Ef(\gamma_x - x\delta_x)}{Ef(\gamma')}\,\frac{E\gamma(dx)}{E\xi S}.
$$

By the uniqueness theorem for Laplace transforms (FMP 5.3) we conclude that, for $x \geq 0$ a.e. $E\gamma$, the ratio $Ef(\gamma_x - x\delta_x)/Ef(\gamma')$ is independent of f. Comparing with the value for $f \equiv 1$ gives

$$
Ef(\gamma_x - x\delta_x) = Ef(\gamma'), \qquad x \geq 0 \text{ a.e. } E\gamma, \ f \geq 0.
$$

Considering separately the cases $x > 0$ and $x = 0$, we see that

$$
\begin{aligned}
Ef(\alpha', \beta') &= \frac{Ef(\alpha, \beta - \delta_x)\gamma(dx)}{E\gamma(dx)} = \frac{Ef(\alpha, \beta - \delta_x)\beta(dx)}{E\beta(dx)} \\
&= Ef(\alpha_x, \beta_x - \delta_x), \quad x > 0 \text{ a.e. } E\beta, \quad (55) \\
Ef(\alpha', \beta') &= E\alpha f(\alpha, \beta)/E\alpha \quad \text{when } E\alpha > 0,
\end{aligned}
$$

and so by combination

$$
Ef(\alpha_x, \beta_x - \delta_x)\, E\alpha = E\alpha f(\alpha, \beta), \quad x > 0 \text{ a.e. } E\beta. \quad (56)
$$

From this point on, it is enough to consider the restrictions of β to any compact subsets of $(0, \infty)$, and so we may assume that $E\kappa < \infty$, where $\kappa = \beta \mathbb{R}_+$. Arguing as in the proof of Theorem 2.34, though now with τ satisfying $P[\tau \in \cdot | \alpha, \beta] = \beta/\kappa$, we see from (55) that β is a mixed binomial process with $\beta \perp\!\!\!\perp_\kappa \alpha$. In particular,

$$
E[\beta | \alpha, \kappa] = E[\beta | \kappa] = (\kappa/E\kappa)E\beta. \quad (57)
$$

Condition (i) or (ii) holds trivially when $\alpha = 0$ a.s., and since (ii) is also trivially fulfilled when $\kappa = 0$ a.s., we may henceforth assume that $E\alpha > 0$ and $E\kappa > 0$. Then for any $t \geq 0$ and $s \in [0, 1]$, we get by (56) and (57)

$$
\begin{aligned}
\frac{E\alpha e^{-\alpha t}s^\kappa}{E\alpha} &= \frac{Ee^{-\alpha t}s^{\kappa-1}\beta(dx)}{E\beta(dx)} = \frac{Ee^{-\alpha t}s^{\kappa-1}E[\beta(dx)|\alpha, \kappa]}{E\beta(dx)} \\
&= \frac{Ee^{-\alpha t}s^{\kappa-1}\kappa\, E\beta(dx)}{E\kappa\, E\beta(dx)} = \frac{Ee^{-\alpha t}\kappa s^{\kappa-1}}{E\kappa},
\end{aligned}
$$

and so by conditioning on α

$$
\frac{Ee^{-\alpha t}\alpha E[s^\kappa | \alpha]}{E\alpha} = \frac{Ee^{-\alpha t}E[\kappa s^{\kappa-1} | \alpha]}{E\kappa}, \quad t \geq 0, \ s \in [0, 1]. \quad (58)
$$

Now choose some measurable functions $p_0, p_1, \ldots$ satisfying

$$
P[\kappa = k | \alpha] = p_k(\alpha), \quad k \in \mathbb{Z}_+,
$$

and note that, a.s. for any $s \in [0, 1)$,

$$
\begin{aligned}
E[s^\kappa | \alpha] &= \sum_{k \geq 0} s^k p_k(\alpha) \equiv \psi(\alpha, s), \\
E[\kappa s^{\kappa-1} | \alpha] &= \sum_{k \geq 0} k s^{k-1} p_k(\alpha) = \psi'(\alpha, s),
\end{aligned}
$$

where ψ' denotes the derivative of ψ in the second argument. Putting $c = E\kappa/E\alpha$, we may then write (58) in the form

$$
cEe^{-\alpha t}\alpha\psi(\alpha, s) = Ee^{-\alpha t}\psi'(\alpha, s), \quad t \geq 0, \ s \in [0, 1).
$$

Invoking once again the uniqueness theorem for Laplace transforms, we conclude that

$$c\alpha\psi(\alpha, s) = \psi'(\alpha, s) \text{ a.s.}, \quad s \in [0, 1),$$

where the exceptional null set can be chosen by continuity to be independent of s. Noting that $\psi(\alpha, 1) = 1$, we get the a.s. unique solution

$$E[s^\kappa|\alpha] = \psi(\alpha, s) = \exp(-c\alpha(1 - s)), \quad s \in [0, 1],$$

which shows that κ is conditionally Poisson with mean $c\alpha$. Thus, β is a mixed Poisson process directed by $\alpha E\beta/E\alpha$, which shows that (ii) holds in the form

$$\xi \in S_\infty(\lambda, \rho E\alpha, \rho E\beta), \qquad \rho = \frac{\alpha}{E\alpha\,\lambda S}.$$

To prove the sufficiency of (i) and (ii), we may assume again that $0 < E\xi S < \infty$. When β is a mixed binomial process and $\alpha = 0$, we see from Theorem 2.34 that $\beta_x - \delta_x \overset{d}{=} \hat\beta$ a.e. independently of x, and so by (54) we get for any measurable functions $f, g \geq 0$

$$
\begin{aligned}
Ef(\eta)g(\beta') &= E\int_0^\infty f(x)g(\beta - \delta_x)\,x\beta(dx)/E\xi S \\
&= \int_0^\infty xf(x)\,Eg(\beta_x - \delta_x)\,E\beta(dx)/E\xi S \\
&= Eg(\hat\beta)\int_0^\infty xf(x)\,E\beta(dx)/E\xi S \\
&= Ef(\eta)\,Eg(\beta'),
\end{aligned}
$$

where the last equality follows as we take $f \equiv 1$ and $g \equiv 1$, respectively. This shows that $\eta \perp\!\!\!\perp \beta'$, and since also $\eta \perp\!\!\!\perp_{\beta'} \zeta$, we obtain $\eta \perp\!\!\!\perp \zeta$ by the chain rule for conditional independence (FMP 6.8).

Next assume β to be a Poisson process with $E\beta = \nu$ and let $\alpha \geq 0$ be a constant. Then by Corollary 2.35 we have for any measurable function $f \geq 0$ and set $B \subset (0, \infty)$

$$
\begin{aligned}
E\int_B f(\alpha, \beta - \delta_x)\,\beta(dx) &= \int_B Ef(\alpha, \beta_x - \delta_x)\,E\beta(dx) \\
&= Ef(\alpha, \beta)\,\nu B.
\end{aligned}
$$

Assuming next that β is a Cox process directed by $\rho\nu$, where (α, ν) is again fixed, we get by the previous result together with the disintegration theorem (FMP 6.4)

$$
\begin{aligned}
E\int_B f(\rho\alpha, \beta - \delta_x)\,\beta(dx) &= EE[f(\rho\alpha, \beta)|\rho]\,\rho\nu B \\
&= E\rho f(\rho\alpha, \beta)\,\nu B.
\end{aligned}
$$

Writing $\gamma = \rho\alpha\delta_0 + l\beta$, we obtain

$$
\begin{aligned}
\int_B Ef(\gamma_x - x\delta_x)\,E\gamma(dx) &= E\int_B f(\gamma - x\delta_x)\,\gamma(dx) \\
&= (E\rho f(\gamma)/E\rho)\,E\gamma B.
\end{aligned}
$$

Since this is also trivially true for $B = \{0\}$, we obtain $\gamma_x - x\delta_x \overset{d}{=} \hat{\gamma}$ a.e. independently of $x \geq 0$. We may now apply (54) to obtain

$$
\begin{aligned}
Ef(\eta)g(\gamma_\zeta) &= E\int_0^\infty f(x)g(\gamma - x\delta_x)\,\gamma(dx)/E\xi S \\
&= \int_0^\infty f(x)\,Eg(\gamma_x - x\delta_x)\,E\gamma(dx)/E\xi S \\
&= Eg(\hat{\gamma})\int_0^\infty f(x)\,E\gamma(dx)/E\xi S \\
&= Ef(\eta)\,Eg(\gamma_\zeta),
\end{aligned}
$$

where the last equality follows again for the special choices $f \equiv 1$ and $g \equiv 1$. Hence, in this case $\eta \perp\!\!\!\perp \gamma_\zeta$, and since also $\eta \perp\!\!\!\perp_{\gamma_\zeta} \zeta$, we conclude as before that $\eta \perp\!\!\!\perp \zeta$. $\qquad\square$

The characterizations of symmetric point processes in terms of Palm distributions imply the corresponding characterizations in terms of the dual object of *Papangelou kernel*, defined as in Appendix A5. In this connection, given a simple point process ξ on a Borel space $(S, \mathcal{S})$, we say that ξ satisfies condition (Σ) if

$$
P[\xi B = 0 | 1_{B^c}\xi] > 0 \quad \text{a.s. on } \{\xi B = 1\}, \quad B \in \mathcal{S}.
$$

Theorem 2.39 *(conditional invariance, Papangelou, Kallenberg) Let ξ be a simple point process on a Borel space $(S, \mathcal{S})$ satisfying condition (Σ), and fix a σ-finite measure λ on S. Then $\mathcal{L}(\xi) \in M(\lambda, \varphi)$ for some function φ iff there exist some random variables $\rho_B \geq 0$, $B \in \mathcal{S}$, such that*

$$
E[1_B\xi; \xi B = 1 | 1_{B^c}\xi] = \rho_B 1_B \lambda \quad \text{a.s. on } \{\xi B = 0\}, \quad B \in \mathcal{S}. \tag{59}
$$

Proof: First assume condition (59). Since S is Borel, we may take $S = \mathbb{R}$. We may also assume that both ξ and λ are locally finite, in the sense that $\xi I < \infty$ a.s. and λI for every bounded interval I. Let $\mathcal{U}$ denote the countable class of finite interval unions with rational endpoints. Writing η for the Papangelou measure of ξ, we see from Theorem A5.1 that

$$
1_U \eta = \alpha_U 1_U \lambda \quad \text{a.s. on } \{\xi U = 0\}, \quad U \in \mathcal{U},
$$

for some random variables $\alpha_U \geq 0$. Since $\mathcal{U}$ is countable, we can choose the exceptional null set to be independent of U, and since clearly $\alpha_{U_1} = \alpha_{U_2}$ a.s. whenever $\lambda(U_1 \cap U_2) > 0$, we may also assume that $\alpha_U = \alpha$ is independent of U. Noting that

$$
(\operatorname{supp} \xi)^c = \bigcup\{U \in \mathcal{U}; \xi U = 0\} \quad \text{a.s.}
$$

we conclude that $\eta = \alpha\lambda$ a.s. on $(\operatorname{supp} \xi)^c$. Since ξ is simple, we also see from Theorem A5.1 that $\eta(\operatorname{supp} \xi) = 0$ a.s., and therefore $\eta = \alpha\lambda$ a.s. on S.

Invoking condition (Σ), we next conclude from Theorem A5.1 and Fubini's theorem that, for any measurable function $f \geq 0$ on $S \times \mathcal{M}(S)$,

$$
\begin{aligned}
C'f &= E \int f(s, \xi) \, \eta(ds) \\
&= E \alpha \int f(s, \xi) \, \lambda(ds) \\
&= \int \lambda(ds) \, E\alpha f(s, \xi) \\
&= \int \lambda(ds) \int f(s, \mu) \, E[\alpha; \, \xi \in d\mu].
\end{aligned}
$$

This shows that the reduced Palm measures of ξ associated with the supporting measure λ have versions

$$
Q'_s(M) = E[\alpha; \, \xi \in M], \quad s \in S.
$$

Since the latter measures are independent of s, we conclude from Theorem 2.34 that ξ is a mixed Poisson or binomial process based on λ.

Conversely, suppose that $\mathcal{L}(\xi) \in M(\lambda, \varphi)$ for some function φ. Then for any $B \in \mathcal{S}$ we note that $1_B \xi$ is conditionally a mixed Poisson or binomial process based on $1_B \lambda$, given $1_{B^c} \xi$, and so the measure $E[1_B \xi; \, \xi B = 1 | 1_{B^c} \xi]$ is a.s. proportional to λ, which proves (46).

We can also base our proof of the converse assertion on the theory of Palm measures and Papangelou kernels. By Theorem 2.34 we note that the reduced Palm measures $Q'_s = Q'$ associated with the supporting measure λ can be chosen to be independent of s, and so the reduced Campbell measure of ξ equals $C' = \lambda \otimes Q'$. Condition (Σ) then yields $Q' \ll \mathcal{L}(\xi)$, say with the Radon–Nikodým density g on $\mathcal{M}(S)$. Writing $\alpha = g(\xi)$, we get for any measurable function $f \geq 0$

$$
\begin{aligned}
C'f &= \int Q'(d\mu) \int f(s, \mu) \, \lambda(ds) \\
&= \int g(\mu) \, P\{\xi \in d\mu\} \int f(s, \mu) \, \lambda(ds) \\
&= E \alpha \int f(s, \xi) \, \lambda(ds),
\end{aligned}
$$

which shows that $\eta = \alpha \lambda$ a.s. Condition (46) now follows by Theorem A5.1. $\qquad \square$

Chapter 3

Convergence and Approximation

This chapter deals primarily with the theory of convergence in distribution and related approximation properties for exchangeable sequences and processes. The basic convergence criteria for exchangeable and related sequences are treated in Section 3.1, which also contains a limit theorem for asymptotically invariant sampling from a stationary process. The corresponding continuous-time theory is initiated in Section 3.2 with a discussion of random measures on $[0, 1]$ and $\mathbb{R}_+$. The convergence theory for exchangeable processes on $[0, 1]$ and $\mathbb{R}_+$ is presented in Section 3.3, and the corresponding approximation theorems for summation processes based on exchangeable sequences appear in Section 3.4. As a by-product of those developments, we can now establish the representation theorem for exchangeable processes on $[0, 1]$ in full generality.

In Section 3.5 we use complex variable theory to show that the convergence of a sequence of exchangeable processes is determined by conditions on an arbitrarily short time interval. The basic convergence assertions, originally established in the sense of Skorohod's J_1-topology, are strengthened in Section 3.6 to the sense of the uniform topology. In the same section we prove some coupling theorems that enable us to extend path properties for Lévy processes to the general exchangeable case.

The final Section 3.7 provides a weak and a strong version of the subsequence principle—the remarkable fact that every tight sequence of random variables contains a sub-sequence that is approximately exchangeable.

3.1 Discrete-Time Case

We begin with a limit theorem for contractable sequences $\xi_n = (\xi_n^j)$ of finite lengths $m_n \to \infty$, taking values in a Polish space S. If $\xi = (\xi^j)$ is an infinite random sequence in S, we define the convergence $\xi_n \xrightarrow{d} \xi$ by the finite-dimensional conditions

$$(\xi_n^1, \ldots, \xi_n^k) \xrightarrow{d} (\xi^1, \ldots, \xi^k), \quad k \in \mathbb{N}.$$

Let $\mathcal{M}_1(S)$ denote the set of probability measures on S, equipped with the topology of weak convergence, and recall that $\mathcal{M}_1(S)$ is again Polish. For any random probability measures ν_n and ν on S, we write $\nu_n \xrightarrow{wd} \nu$ for convergence in distribution with respect to the weak topology.

Theorem 3.1 *(convergence of contractable sequences) Let $\xi_1, \xi_2, \ldots$ be contractable sequences in a Polish space S, of finite lengths $m_n \to \infty$ and with empirical distributions ν_n. Then $\xi_n \xrightarrow{d}$ some ξ in S^∞ iff $\nu_n \xrightarrow{wd}$ some ν in $\mathcal{M}_1(S)$, in which case $\mathcal{L}(\xi) = E\nu^\infty$.*

Note that the result contains a version of Ryll-Nardzewski's theorem—the fact that any infinite contractable sequence in a Polish space S is mixed i.i.d. From this result we can easily deduce the stronger version in Theorem 1.1. In particular, the present argument provides a weak-convergence approach to de Finetti's theorem.

Proof: First assume that $\nu_n \xrightarrow{wd} \nu$. Fix any bounded, continuous functions $f_1, \ldots, f_r$ on S, and let $m, n \in \mathbb{N}$ be arbitrary with $mr \leq m_n$. Put $\|f_k\| = \sup_s |f_k(s)|$. Using the contractability of the sequences $\xi_n = (\xi_n^k)$ and applying Lemma 1.2, we get

$$
\begin{aligned}
&\left| E \prod_{k \leq r} f_k(\xi_n^k) - E \prod_{k \leq r} \nu_n f_k \right| \\
&\quad = \left| E \prod_{k \leq r} m^{-1} \sum_{j \leq m} f_k(\xi_n^{km-j}) - E \prod_{k \leq r} \nu_n f_k \right| \\
&\quad \leq \sum_{k \leq r} E \left| m^{-1} \sum_{j \leq m} f_k(\xi_n^{km-j}) - \nu_n f_k \right| \prod_{h \neq k} \|f_h\| \\
&\quad \lesssim m^{-1/2} r \prod_{h \leq r} \|f_h\|,
\end{aligned}
$$

which tends to 0 as $n \to \infty$ and then $m \to \infty$. Since also

$$
E \prod_{k \leq r} \nu_n f_k \to E \prod_{k \leq r} \nu f_k
$$

by the weak continuity of the mapping $\mu \mapsto \prod_{k \leq r} \mu f_k$, we obtain

$$
E \prod_{k \leq r} f_k(\xi_n^k) \to E \prod_{k \leq r} \nu f_k = E \prod_{k \leq r} f_k(\eta^k),
$$

where the sequence $\eta = (\eta^k)$ has distribution $E\nu^\infty$. By the general criterion in FMP 4.29 it follows that $\xi_n \xrightarrow{d} \eta$ in S^∞.

Conversely, suppose that $\xi_n \xrightarrow{d} \xi$ in S^∞. By the contractability of each ξ_n we obtain the weak convergence

$$
E\nu_n = P\{\xi_n^1 \in \cdot\} \xrightarrow{w} P\{\xi^1 \in \cdot\},
$$

and so by Prohorov's theorem (FMP 16.3) the sequence $(E\nu_n)$ is tight in S. By Theorem A2.2 it follows that the sequence of random measures (ν_n) is tight in $\mathcal{M}_1(S)$. Using Prohorov's theorem again, we conclude that any sub-sequence $N' \subset \mathbb{N}$ contains a further sub-sequence N'' such that $\nu_n \xrightarrow{wd} \nu$

along N'', for some random probability measure ν on S. But then the previous part of the proof yields $\xi_n \xrightarrow{d} \eta$ along N'', where $\mathcal{L}(\eta) = E\nu^\infty$. Since also $\xi_n \xrightarrow{d} \xi$, we obtain $\mathcal{L}(\xi) = E\nu^\infty$. By Proposition 1.4 the distribution of ν is then unique, and so the convergence $\nu_n \xrightarrow{wd} \nu$ extends to the entire sequence. □

We turn to the basic convergence criteria for exchangeable sequences of finite or infinite length. To allow for a unified statement, we may refer to the directing random measure of an infinite exchangeable sequence ξ as the *empirical distribution* of ξ.

Theorem 3.2 *(convergence of exchangeable sequences) Let $\xi, \xi_1, \xi_2, \ldots$ be exchangeable sequences in a Polish space S, of common length $m \leq \infty$ and with empirical distributions $\nu, \nu_1, \nu_2, \ldots$. Then $\xi_n \xrightarrow{d} \xi$ in S^m iff $\nu_n \xrightarrow{wd} \nu$ in $\mathcal{M}_1(S)$, in which case even $(\xi_n, \nu_n) \xrightarrow{wd} (\xi, \nu)$ in $S^m \times \mathcal{M}_1(S)$. The statement remains true for sequences ξ_n of finite lengths $m_n \to m = \infty$.*

Proof: First let $m < \infty$, and put $\beta = m\nu$ and $\beta_n = m\nu_n$ for convenience. If $\xi_n \xrightarrow{d} \xi$, then $(\xi_n, \beta_n) \xrightarrow{wd} (\xi, \beta)$ by continuous mapping. Conversely, suppose that $\beta_n \xrightarrow{wd} \beta$. Then the sequence of pairs (ξ_n, β_n) is weakly tight by Theorem A2.2 and Prohorov's theorem, and the latter theorem shows that every sub-sequence $N' \subset \mathbb{N}$ has then a further sub-sequence N'' satisfying

$$(\xi_n, \beta_n) \xrightarrow{wd} (\eta, \tilde{\beta}) \text{ along } N'', \tag{1}$$

for suitable η and $\tilde{\beta}$. In particular, the same convergence holds for the marginals, which implies that η is exchangeable and $\tilde{\beta} \stackrel{d}{=} \beta$. Furthermore, the continuous mapping theorem yields

$$(\beta_n, \beta_n) \xrightarrow{wd} \left(\sum_j \delta_{\eta_j}, \tilde{\beta} \right) \text{ in } (\mathcal{M}_1(S))^2,$$

and the diagonal in $(\mathcal{M}_1(S))^2$ being closed, we conclude from the Portmanteau theorem (FMP 4.25) that $\tilde{\beta} = \sum_j \delta_{\eta_j}$ a.s. Using the disintegration theorem (FMP 6.4) and Theorem 1.8, we obtain for any measurable function $f \geq 0$

$$\begin{aligned} Ef(\eta, \tilde{\beta}) &= E \int f(x, \tilde{\beta})\, \tilde{\beta}^{(m)}(dx)/m! \\ &= E \int f(x, \beta)\, \beta^{(m)}(dx)/m! = Ef(\xi, \beta), \end{aligned}$$

which shows that $(\eta, \tilde{\beta}) \stackrel{d}{=} (\xi, \beta)$. Hence, (1) reduces to $(\xi_n, \beta) \xrightarrow{wd} (\xi, \beta)$ along N'', which remains true along $\mathbb{N}$ since the limit is independent of sub-sequence.

Next let $m = \infty$. Assuming the ν_n to be non-random with $\nu_n \xrightarrow{w} \nu$, we see from FMP 4.29 that $\nu_n^\infty \xrightarrow{w} \nu^\infty$ in $\mathcal{M}_1(S^\infty)$, which means that $\xi_n \xrightarrow{d} \xi$ in S^∞.

Hence, by FMP 4.28 we have $(\xi_n, \nu_n) \xrightarrow{d} (\xi, \nu)$, which extends by Lemma A2.4 to random measures ν_n with $\nu_n \xrightarrow{wd} \nu$. Now assume instead that $\xi_n \xrightarrow{d} \xi$ in S^∞. The sequence (ν_n) is then weakly tight in $\mathcal{M}_1(S)$ by Theorem A2.2 and Prohorov's theorem, and so the latter theorem shows that any sub-sequence $N' \subset \mathbb{N}$ has a further sub-sequence N'' such that $\nu_n \xrightarrow{wd} \tilde{\nu}$ along N'', for some random probability measure $\tilde{\nu}$ on S. As before, we get $\xi_n \xrightarrow{d} \tilde{\xi}$ along N'', where $\tilde{\xi}$ is an exchangeable sequence directed by $\tilde{\nu}$. But then $\tilde{\xi} \stackrel{d}{=} \xi$, and so by Proposition 1.4 we have $\tilde{\nu} \stackrel{d}{=} \nu$. Hence, $\nu_n \xrightarrow{wd} \nu$ along N'', which remains true along $\mathbb{N}$ since the limit is independent of sub-sequence.

Finally, suppose that the sequences ξ_n have finite lengths $m_n \to \infty$. Then Theorem 3.1 shows that $\xi_n \xrightarrow{d} \xi$ iff $\nu_n \xrightarrow{wd} \nu$, where the limits are related by $\mathcal{L}(\xi) = E\nu^\infty$. If the ν_n are non-random with $\nu_n \xrightarrow{w} \nu$, we see as before that $(\xi_n, \nu_n) \xrightarrow{d} (\xi, \nu)$ in $S^\infty \times \mathcal{M}_1(S)$. By Lemma A2.4, this extends to the general case where the ν_n are random with $\nu_n \xrightarrow{wd} \nu$. $\qquad\Box$

We proceed to show how exchangeable sequences arise asymptotically through random sampling from stationary processes. Here we say that the probability measures $\mu_1, \mu_2, \ldots$ on $\mathbb{R}_+^m$ or $\mathbb{Z}_+^m$ are *asymptotically invariant* if

$$\lim_{n \to \infty} \|\mu_n - \mu_n * \delta_t\| = 0, \quad t \in \mathbb{R}_+^m \text{ or } \mathbb{Z}_+^m.$$

For probability measures μ_n defined on $\mathbb{R}_+^\infty$ or $\mathbb{Z}_+^\infty$, we require the same condition for all finite-dimensional projections $\mu_n \circ \pi_{1,\ldots,m}^{-1}$. The shift-invariant σ-field associated with a process X on $\mathbb{R}_+$ or $\mathbb{Z}_+$ is denoted by $\mathcal{I}_X$.

Proposition 3.3 *(asymptotically invariant sampling) Let X be a stationary and measurable process on $T = \mathbb{R}_+$ or $\mathbb{Z}_+$, taking values in a Polish space S. For every $n \in \mathbb{N}$, consider an independent sequence of random variables τ_n^j in T with joint distribution μ_n, where the μ_n are asymptotically invariant in T^∞, and put $\xi_n = (\xi_n^j)$ with $\xi_n^j = X(\tau_n^j)$. Then $\xi_n \xrightarrow{d} \xi$ in S^∞, where the sequence $\xi = (\xi^j)$ is exchangeable in S and directed by $\nu = P[X_0 \in \cdot | \mathcal{I}_X]$.*

Proof: By FMP 4.29 it suffices to show that $Ef(\xi_n) \to Ef(\xi)$ for any finite tensor product $f(x) = f_1(x_1) \cdots f_m(x_m)$, where $f_1, \ldots, f_m$ are measurable functions on S with $\|f_k\| \le 1$. We may then assume that the μ_n are asymptotically invariant on T^m. Letting λ_r denote the uniform distribution on $[0, r]$ or $\{0, \ldots, r\}$, respectively, and writing $X_t = (X_{t_1}, \ldots, X_{t_m})$ for $t = (t_1, \ldots, t_m)$, we get informally

$$
\begin{aligned}
Ef(\xi_n) &= E \int f(X_t)\, \mu_n(dt) \\
&\approx E \int f(X_t)(\mu_n * \lambda_r^m)(dt) \\
&= \int \mu_n(ds)\, E \int f(X_{s+t})\, \lambda_r^m(dt) \\
&\to \int \mu_n(ds)\, E\nu^m f = Ef(\xi).
\end{aligned}
$$

For a formal justification, we may write

$$
\begin{aligned}
|Ef(\xi_n) &- Ef(\xi)| \\
&\leq \left| E \int f(X_t)(\mu_n - \mu_n * \lambda_r^m)(dt) \right| \\
&\quad + \int \mu_n(ds) \left| E \int f(X_{s+t})\lambda_r^m(dt) - E\nu^m f \right| \\
&\leq \|\mu_n - \mu_n * \lambda_r^m\| + \sup_{s \in T^m} E \left| \int f(X_{s+t})\lambda_r^m(dt) - \nu^m f \right| \\
&\leq \int \|\mu_n - \mu_n * \delta_s\| \lambda_r^m(ds) + \sum_{k \leq m} \sup_{s \in T} E \left| \int f_k(X_{s+t})\lambda_r(dt) - \nu f_k \right|.
\end{aligned}
$$

Using the asymptotic invariance of the μ_n and applying the dominated convergence and mean ergodic theorems, we see that the right-hand side tends to 0 as $n \to \infty$ and then $r \to \infty$. $\qquad\square$

3.2 Random Measures

We begin with some convergence criteria for exchangeable random measures ξ on a product space $S \times [0, 1]$. Recall from Theorem 1.25 that if S is Borel, then ξ has a representation

$$
\xi = \alpha \otimes \lambda + \sum_j \beta_j \otimes \delta_{\tau_j} \quad \text{a.s.}, \tag{2}
$$

where α and $\beta_1, \beta_2, \ldots$ are random measures on S and $\tau_1, \tau_2, \ldots$ is an independent sequence of i.i.d. $U(0, 1)$ random variables. Note that the distribution of ξ is determined by that of the pair (α, β), where β is the point process on $\mathcal{M}'(S) = \mathcal{M}(S) \setminus \{0\}$ given by $\beta = \sum_j \delta_{\beta_j}$. We also note that

$$
\xi(\cdot \times [0, 1]) = \alpha + \sum_j \beta_j \equiv \gamma. \tag{3}
$$

In order to discuss the convergence of such random measures ξ, we need to endow S with a suitable topology. To avoid some technical complications, we consider only the case where S is a compact metric space. Then ξ, α, $\beta_1, \beta_2, \ldots$, and γ are all a.s. bounded, and the spaces $\mathcal{M}(S)$ and $\mathcal{M}'(S)$ are locally compact in the weak topology. The convergence criteria become especially simple when expressed in terms of the random measures β^1 on $\mathcal{M}(S \times \mathcal{M}(S))$ given by

$$
\beta^1 = \alpha \otimes \delta_0 + \sum_j \beta_j \otimes \delta_{\beta_j}. \tag{4}
$$

We may also introduce the non-decreasing, measure-valued process

$$
X_t = \xi(\cdot \times [0, t]), \quad t \in [0, 1], \tag{5}
$$

which is clearly right-continuous with left-hand limits and therefore may be regarded as a random element in the Skorohod space $D([0, 1], \mathcal{M}(S))$ endowed with the J_1-topology.

Theorem 3.4 *(convergence on $S \times [0,1]$)* *Let $\xi, \xi_1, \xi_2, \ldots$ be exchangeable random measures on $S \times [0,1]$ directed by (α, β) and (α_n, β_n), $n \in \mathbb{N}$, where S is compact, and define X, γ, β^1 and X^n, γ_n, β_n^1 as in (3), (4), and (5). Then these conditions are equivalent:*

(i) $\xi_n \xrightarrow{wd} \xi$ *in* $\mathcal{M}(S \times [0,1])$,

(ii) $X^n \xrightarrow{d} X$ *in* $D([0,1], \mathcal{M}(S))$,

(iii) $(\beta_n, \gamma_n) \xrightarrow{vd} (\beta, \gamma)$ *in* $\mathcal{N}(\mathcal{M}'(S)) \times \mathcal{M}(S)$,

(iv) $\beta_n^1 \xrightarrow{wd} \beta^1$ *in* $\mathcal{M}(S \times \mathcal{M}(S))$.

Here $\mathcal{N}(S)$ denotes the class of integer-valued measures in $\mathcal{M}(S)$, and $\xrightarrow{vd}$ means convergence in distribution with respect to the vague topology. In the special case of exchangeable random measures on $[0,1]$, the spaces in (iii) and (iv) clearly reduce to $\mathcal{N}(0, \infty) \times \mathbb{R}_+$ and $\mathcal{M}(\mathbb{R}_+)$, respectively.

Proof: First assume (i), so that $\xi_n \xrightarrow{wd} \xi$ in $\mathcal{M}(S \times [0,1])$. Consider any optional times $\tau_1, \tau_2, \ldots$ in $[0,1)$ with respect to the induced filtrations and some positive constants $h_1, h_2, \ldots \to 0$ such that $\tau_n + h_n \leq 1$ a.s. Writing $\bar{\xi}_n = \xi(S \times \cdot)$ and using the finite-interval version of Proposition 2.5, we get as $n \to \infty$ for fixed $h > 0$

$$
\begin{aligned}
E[\bar{\xi}_n[\tau_n, \tau_n + h_n] \wedge 1] &= E[\bar{\xi}_n[0, h_n] \wedge 1] \\
&\leq E[\bar{\xi}_n[0, h] \wedge 1] \to E[\bar{\xi}[0, h] \wedge 1],
\end{aligned}
$$

which tends to 0 as $h \to 0$. Thus, $\bar{\xi}_n[\tau_n, \tau_n + h_n] \xrightarrow{P} 0$, and so by Aldous' criterion (FMP 16.11) the sequence (X^n) is tight in the Skorohod space $D([0,1], \mathcal{M}(S))$.

Now suppose that $X_n \xrightarrow{d} Y$ along a sub-sequence $N' \subset \mathbb{N}$. Then $\xi_n \xrightarrow{wd} \eta$ along N', where Y and η are related as in (5). Since also $\xi_n \xrightarrow{wd} \xi$, we have $\eta \overset{d}{=} \xi$, and so $X^n \xrightarrow{d} X$ along N', where even X and ξ are related as in (5). Hence, by the tightness of (X^n) and Prohorov's theorem, the convergence $X^n \xrightarrow{d} X$ remains valid along $\mathbb{N}$, which proves (ii). Condition (iii) follows easily from (ii) by means of the continuous mapping theorem.

Conversely, assume (iii). Then in particular,

$$
\xi_n(S \times [0,1]) = \gamma_n S \xrightarrow{d} \gamma S < \infty,
$$

and since $S \times [0,1]$ is compact, the sequence (ξ_n) is weakly tight. If $\xi_n \xrightarrow{wd} \eta$ along a sub-sequence $N' \subset \mathbb{N}$, then η is again exchangeable, and so by the previous proof we have $(\beta_n, \gamma_n) \xrightarrow{vd} (\tilde{\beta}, \tilde{\gamma})$ along N', where η and $(\tilde{\beta}, \tilde{\gamma})$ are related as in (2) and (3). But then $(\tilde{\beta}, \tilde{\gamma}) \overset{d}{=} (\beta, \gamma)$, and therefore $\eta \overset{d}{=} \xi$. Hence, we have $\xi_n \xrightarrow{wd} \xi$ along N', and (i) follows by the tightness of (ξ_n).

To prove the equivalence of (iii) and (iv), we may assume that the α_n and β_n are non-random, since the general case will then follow by the definition of weak convergence. First suppose that $\beta_n^1 \overset{w}{\to} \beta^1$ on $S \times \mathcal{M}(S)$. For any continuous function $f \geq 0$ on S, we note that the function

$$g(s, \mu) = f(s), \quad s \in S, \ \mu \in \mathcal{M}(S),$$

is again bounded and continuous. Hence,

$$\gamma_n f = \beta_n^1 g \to \beta^1 g = \gamma f,$$

which shows that $\gamma_n \overset{w}{\to} \gamma$ on S.

Next let $f \geq 0$ be a continuous function with compact support on $\mathcal{M}'(S)$, and note that

$$\varepsilon \equiv \inf\{\mu S; \ \mu \in \mathcal{M}'(S), \ f(\mu) > 0\} > 0.$$

Introducing the bounded and continuous function

$$g(s, \mu) = (\mu S \vee \varepsilon)^{-1} f(\mu), \quad s \in S, \ \mu \in \mathcal{M}(S),$$

we obtain

$$\beta_n f = \beta_n^1 g \to \beta^1 g = \beta f,$$

which shows that $\beta_n \overset{v}{\to} \beta$ on $\mathcal{M}'(S)$.

Conversely, suppose that $\beta_n \overset{v}{\to} \beta$ on $\mathcal{M}'(S)$ and $\gamma_n \overset{w}{\to} \gamma$ on S. The latter convergence yields

$$\beta_n^1(S \times \mathcal{M}(S)) = \alpha_n S + \sum_j \beta_{nj} S = \gamma_n S \to \gamma S < \infty.$$

Writing $c = \sup_n \gamma_n S < \infty$, it follows that the measures β_n^1 are uniformly bounded by c and restricted to the compact set $S \times \{\mu \in \mathcal{M}(S); \ \mu S \leq c\}$. Hence, by Lemma A2.1 the sequence (β_n^1) is weakly relatively compact in $\mathcal{M}(S \times \mathcal{M}(S))$.

Now assume that $\beta_n^1 \overset{w}{\to} \rho$ on $S \times \mathcal{M}(S)$ along a sub-sequence $N' \subset \mathbb{N}$, for some bounded measure ρ on $S \times \mathcal{M}(S)$. By continuous mapping it follows as before that

$$\gamma \overset{w}{\leftarrow} \gamma_n = \beta_n^1(\cdot \times \mathcal{M}(S)) \overset{w}{\to} \rho(\cdot \times \mathcal{M}(S)). \tag{6}$$

Next consider any continuous function $f \geq 0$ on $S \times \mathcal{M}'(S)$ with compact support, and define

$$g(\mu) = \int f(s, \mu) \, \mu(ds), \quad \mu \in \mathcal{M}'(S).$$

Then g has compact support in $\mathcal{M}'(S)$ and is again continuous, by the extended continuous mapping theorem (FMP 4.27). Hence,

$$\rho f \leftarrow \beta_n^1 f = \beta_n g \to \beta g = \beta^1 f,$$

which shows that $\rho = \beta^1$ on $S \times \mathcal{M}'(S)$. Combining this with (6) gives

$$\rho(\cdot \times \{0\}) = \gamma - \beta^1(\cdot \times \mathcal{M}'(S)) = \beta^1(\cdot \times \{0\}),$$

and so $\rho = \beta^1$. Hence, $\beta_n^1 \xrightarrow{w} \beta^1$ along N', and then also along $\mathbb{N}$, since the limit is independent of sub-sequence. $\square$

We proceed to the case of exchangeable random measures ξ on $S \times \mathbb{R}_+$, where S is a compact metric space. Then by Proposition 1.21 we have the general representation

$$\xi = \alpha \otimes \lambda + \int\!\!\int (\mu \otimes \delta_t)\, \eta(d\mu\, dt) \quad \text{a.s.}, \tag{7}$$

where α is a random measure on S and η is a Cox process on $\mathcal{M}'(S) \times \mathbb{R}_+$, directed by a random measure of the form $\nu \otimes \lambda$ and satisfying $\eta \perp\!\!\!\perp_\nu \alpha$. For the double integral to converge, we need to assume that $\int (\mu S \wedge 1)\nu(d\mu) < \infty$ a.s., by FMP 12.13 or Theorem A3.5. The pair (α, ν) is then a.s. unique, and the distributions of ξ and (α, ν) determine each other uniquely.

To allow for a simple convergence criterion in terms of the pairs (α, ν), we put $\hat{\mu} = \mu/(\mu S \vee 1)$ and introduce on $S \times \mathcal{M}(S)$ the a.s. bounded random measure

$$\hat{\nu}^1 = \alpha \otimes \delta_0 + \int (\hat{\mu} \otimes \delta_\mu)\, \nu(d\mu). \tag{8}$$

Theorem 3.5 (*convergence on $S \times \mathbb{R}_+$*) *Let* $\xi, \xi_1, \xi_2, \ldots$ *be exchangeable random measures on* $S \times \mathbb{R}_+$ *directed by* (α, ν) *and* (α_n, ν_n), $n \in \mathbb{N}$, *where S is compact, and define $X, \hat{\nu}^1$ and $X^n, \hat{\nu}_n^1$ as in (5) and (8). Then these conditions are equivalent:*

  (i) $\xi_n \xrightarrow{vd} \xi$ *in* $\mathcal{M}(S \times \mathbb{R}_+)$,
  (ii) $X^n \xrightarrow{d} X$ *in* $D(\mathbb{R}_+, \mathcal{M}(S))$,
  (iii) $\hat{\nu}_n^1 \xrightarrow{wd} \hat{\nu}^1$ *in* $\mathcal{M}(S \times \mathcal{M}(S))$.

In the special case of exchangeable random measures on $\mathbb{R}_+$, we note that the space in (iii) reduces to $\mathcal{M}(\mathbb{R}_+)$.

Proof: Assuming (i) and noting that $\xi(S \times \{t\}) = 0$ a.s. for every $t > 0$, we get $\xi_n \xrightarrow{wd} \xi$ on $S \times [0, t]$ for all $t > 0$, and so by Theorem 3.4 we obtain $X^n \xrightarrow{d} X$ on $D([0, t], \mathcal{M}(S))$ for all $t > 0$, which implies (ii). The implication (i) $\Rightarrow$ (ii) can also be proved directly by the same argument as before. The converse assertion is obvious by continuity. It is then enough to prove the equivalence of (i) and (iii).

To show that (iii) implies (i), it suffices by Lemma A2.4 to consider the case of non-random α and ν. Then by (7) and FMP 12.2 we have for any continuous function $f > 0$ on S

$$-t^{-1} \log E e^{-X_t f} = \alpha f + \int (1 - e^{-\mu f})\, \nu(d\mu) = \hat{\nu}^1 g, \quad t > 0, \tag{9}$$

and similarly for X^n in terms of $\hat{\nu}_n^1$, where

$$g(s, \mu) = f(s) \frac{1 - e^{-\mu f}}{\mu f} (\mu S \vee 1), \quad s \in S, \ \mu \in \mathcal{M}(S),$$

the second factor on the right being defined as 1 when $\mu f = 0$. Here g is clearly continuous on $S \times \mathcal{M}(S)$. To see that it is also bounded, we note that if $0 < \varepsilon \leq f \leq c$, then $g \leq c$ when $\mu S \leq 1$ and $g \leq c/\varepsilon$ when $\mu S > 1$. Using (iii), we obtain

$$Ee^{-X_t^n f} = \exp(-t\hat{\nu}_n^1 g) \to \exp(-t\hat{\nu}^1 g) = Ee^{-X_t f},$$

and so by FMP 5.3 we have $X_t^n f \xrightarrow{d} X_t f$, which is easily extended to any continuous functions $f \geq 0$ on S. Hence, by FMP 16.16 we have $X_t^n \xrightarrow{wd} X_t$. Since the processes X and X^n have independent increments, we conclude from FMP 4.29 that $X^n \xrightarrow{fd} X$, and a further routine extension yields the required convergence in (i).

Now assume (i). Noting that $1 - e^{-t} \geq c(t \wedge 1)$ with $c = 1 - e^{-1}$, we get from (9) for any $r > 0$

$$
\begin{aligned}
Ee^{-rX_1^n S} &= E \exp\left(-r\alpha_n S - \int (1 - e^{-r\mu S}) \nu_n(d\mu)\right) \\
&\leq E \exp(-r\alpha_n S - cr\nu_n[\mu S; \ r\mu S \leq 1] - c\nu_n\{r\mu S > 1\}).
\end{aligned}
$$

Since $X_1^n S \xrightarrow{d} X_1 S < \infty$, we have $rX_1^n S \xrightarrow{P} 0$ as $r \to 0$ uniformly in n, and so

$$r\alpha_n S + r\nu_n[\mu S; \ \mu S \leq r^{-1}] + \nu_n\{\mu S > r^{-1}\} \xrightarrow{P} 0,$$

in the same sense. By Theorem A2.2 it follows that the sequence $\hat{\nu}_n^1$ is weakly tight on $S \times \mathcal{M}(S)$. Now suppose that $\hat{\nu}_n^1 \xrightarrow{wd} \rho$ along a sub-sequence $N' \subset \mathbb{N}$ for some random measure ρ on $S \times \mathcal{M}(S)$. Using the direct assertion, we conclude that $\xi_n \xrightarrow{vd} \eta$ along N', where η and ρ are related as in (7) and (8). But then $\eta \overset{d}{=} \xi$, and by the uniqueness in Proposition 1.21 we obtain $\rho \overset{d}{=} \hat{\nu}^1$. Hence, $\hat{\nu}_n^1 \xrightarrow{wd} \hat{\nu}^1$ along N', which extends to $\mathbb{N}$ since the limit is independent of sub-sequence. $\qquad\square$

We conclude with the case of exchangeable random measures ξ_n on the product spaces $S \times [0, t_n]$, where S is compact and $t_n \to \infty$. In this case, the pairs (α_n, β_n) refer to the scaled random measure ξ_n' given by $\xi_n' B = \xi_n(t_n B)$, and we need to modify the definition of β^1 from (4) by taking

$$\hat{\beta}^1 = \alpha \otimes \delta_0 + \sum_j \hat{\beta}_j \otimes \delta_{\beta_j}. \tag{10}$$

Theorem 3.6 *(convergence along increasing sets) Let ξ be an exchangeable random measure on $S \times \mathbb{R}_+$ directed by (α, ν), where S is compact, and for each $n \in \mathbb{N}$, let ξ_n be exchangeable on $S \times [0, t_n]$ and directed by (α_n, β_n), where $t_n \to \infty$. Define $X, \hat{\nu}^1$ and $X^n, \hat{\beta}_n^1$ as in (5), (8), and (10). Then these conditions are equivalent:*

(i) $\xi_n \xrightarrow{vd} \xi$ in $\mathcal{M}(S \times \mathbb{R}_+)$,

(ii) $X^n \xrightarrow{d} X$ in $D(\mathbb{R}_+, \mathcal{M}(S))$,

(iii) $\hat{\beta}_n^1/t_n \xrightarrow{wd} \hat{\nu}^1$ in $\mathcal{M}(S \times \mathcal{M}(S))$.

As before, the measure space in part (iii) reduces to $\mathcal{M}(\mathbb{R}_+)$ when the ξ_n are exchangeable random measures on the intervals $[0, t_n]$. We begin with an elementary lemma, where we write $\| \cdot \|_A$ for the total variation on the set A.

Lemma 3.7 *(Poisson approximation) If τ is $U(0, 1)$, then for any $t \in (0, 1)$ there exists a unit rate Poisson process η_t on $[0, 1]$ such that*

$$E\|\delta_\tau - \eta_t\|_{[0,t]}^2 \leq 2t^2, \quad t \in [0, 1].$$

Proof: Fixing any $t \in (0, 1)$, let $\sigma_1, \sigma_2, \ldots$ be i.i.d. $U(0, t)$, and introduce an independent pair of random variables κ and κ' in $\mathbb{Z}_+$ with $\kappa \wedge 1 \leq \kappa' \leq 1$, where κ is Poisson with mean t and $P\{\kappa' = 1\} = t$. Define $\eta = \sum_{k \leq \kappa} \delta_{\sigma_k}$, and put $\sigma = \sigma_1$ when $\kappa' = 1$ and $\sigma = 1$ when $\kappa' = 0$, so that η is unit rate Poisson on $[0, t]$ and $\sigma \stackrel{d}{=} \tau$ on $[0, t]$. By the transfer theorem (FMP 6.10) we may next choose a unit rate Poisson process η_t on $[0, 1]$ such that $(\tau, \eta_t) \stackrel{d}{=} (\sigma, \eta)$ on $[0, t]$. Then clearly

$$\begin{aligned}
E\|\delta_\tau - \eta_t\|_{[0,t]}^2 &= E\|\delta_\sigma - \eta\|_{[0,t]}^2 = E(\kappa' - \kappa)^2 \\
&= e^{-t} - 1 + t + \sum_{k \geq 2}(k-1)^2 \frac{t^k}{k!} e^{-t} \leq 2t^2. \qquad \square
\end{aligned}$$

Proof of Theorem 3.6: To show that (iii) implies (i), it suffices by Lemma A2.4 to consider the case of non-random characteristics (α_n, β_n) and (α, ν). For each $n \in \mathbb{N}$, put $\tilde{\nu}_n = \beta_n/t_n$ and $\tilde{\alpha}_n = \alpha_n/t_n$, and let η_n be a Poisson process on $\mathcal{M}'(S) \times \mathbb{R}_+$ with intensity measure $\tilde{\nu}_n \otimes \lambda$. We may next introduce on $S \times \mathbb{R}_+$ the random measures

$$\tilde{\xi}_n = \tilde{\alpha}_n \otimes \lambda + \iint (\mu \otimes \delta_t)\, \eta_n(d\mu\, dt), \quad n \in \mathbb{N},$$

which are locally finite since

$$\int (\mu S \wedge 1)\, \tilde{\nu}_n(d\mu) \leq t_n^{-1} \sum_j \beta_{nj} S < \infty,$$

and have stationary, independent increments with characteristics $(\tilde{\alpha}_n, \tilde{\nu}_n)$. Condition (iii) yields $\tilde{\nu}_n \xrightarrow{w} \nu^1$, and so by Theorem 3.5 we have $\tilde{\xi}_n \xrightarrow{vd} \xi$.

To extend the convergence to the random measures ξ_n, we may assume
that

$$\xi_n = \tilde{\alpha}_n \otimes \lambda + \sum_j \beta_{nj} \otimes \delta_{\tau_{nj}}, \quad n \in \mathbb{N}, \tag{11}$$

where the τ_{nj} are i.i.d. $U(0, t_n)$ for each n. Fix any $t > 0$, and let n be large
enough that $t_n \geq t$. Then by Lemma 3.7 we may choose some independent
Poisson processes η_{nj} with constant rates t_n^{-1} such that

$$E\|\delta_{\tau_{nj}} - \eta_{nj}\|_{[0,t]}^2 \leq 2(t/t_n)^2, \quad n, j \in \mathbb{N}, \tag{12}$$

and we may define the Poisson processes η_n on $S \times \mathbb{R}_+$ by

$$\eta_n = \sum_j \beta_{nj} \otimes \eta_{nj}, \quad n \in \mathbb{N}.$$

Now fix any $[0, 1]$-valued, continuous functions f on S and g on $[0, t]$.
Using the subadditivity of the function $x \wedge 1$, Fubini's theorem, the estimate
(12), and (iii), we get

$$
\begin{aligned}
E(|(\xi_n - \tilde{\xi}_n)(f \otimes g)| \wedge 1) \\
&\leq E\left(\sum_j \beta_{nj} f \, |g(\tau_{nj}) - \eta_{nj} g| \wedge 1\right) \\
&\leq \sum_j E(\beta_{nj} f \, |g(\tau_{nj}) - \eta_{nj} g| \wedge 1) \\
&\leq \sum_j (\beta_{nj} f \, E|g(\tau_{nj}) - \eta_{nj} g| \wedge P\{(\delta_{\tau_{nj}} - \eta_{nj})[0, t] \neq 0\}) \\
&\leq \sum_j (\beta_{nj} f \wedge 1) \, E\|\delta_{\tau_{nj}} - \eta_{nj}\|_{[0,t]}^2 \\
&\leq (t/t_n)^2 \sum_j (\beta_{nj} f \wedge 1) \\
&\leq (t/t_n)^2 \, \hat{\beta}_n^1(S \times \mathcal{M}(S)) \\
&\leq (t^2/t_n) \, \hat{\nu}^1(S \times \mathcal{M}(S)) \to 0,
\end{aligned}
$$

which shows that $(\xi_n - \tilde{\xi}_n)(f \otimes g) \xrightarrow{P} 0$.

More generally, we have for any continuous functions $f_1, \ldots, f_m$ and
$g_1, \ldots, g_m$ as above

$$(\xi_n - \tilde{\xi}_n) \sum_{i \leq m} (f_i \otimes g_i) \xrightarrow{P} 0.$$

Using the Stone–Weierstrass approximation theorem together with the basic
approximation lemma of weak convergence theory (FMP 4.28), we conclude
that $\xi_n f \xrightarrow{d} \xi f$ for any continuous function $f \geq 0$ on $S \times \mathbb{R}_+$ with compact
support, which yields $\xi_n \xrightarrow{vd} \xi$ on $S \times \mathbb{R}_+$ by FMP 16.16. This shows that (iii)
implies (i). We also note that (i) and (ii) are equivalent by the corresponding
assertion in Theorem 3.4.

Now assume (i). Letting each ξ_n be represented as in (11) in terms of
the random measures α_n and β_{nj} and some independent times τ_{nj}, uniformly

distributed on $[0, t_n]$, we get for any $r \geq 0$

$$
\begin{aligned}
E \exp(-r X_1^n S) &\leq E \exp\left(-r \sum_j \beta_{nj} S \, 1\{\tau_{nj} \leq 1\}\right) \\
&= E \prod_j \left(1 - t_n^{-1}(1 - e^{-r\beta_{nj}S})\right) \\
&\leq E \exp\left(-\sum_j t_n^{-1}(1 - e^{-r\beta_{nj}S})\right) \\
&= E \exp\left(-t_n^{-1} \int (1 - e^{-r\mu S}) \, \beta_n(d\mu)\right),
\end{aligned}
$$

where the third relation follows from the estimate $1 - x \leq e^{-x}$ for $x \geq 0$. Noting that the sequence $(X_1^n S)$ is tight and proceeding as in the proof of Theorem 3.5, we conclude that the sequence $(\hat{\beta}_n^1 / t_n)$ is weakly tight on $S \times \mathcal{M}(S)$. Condition (iii) now follows as before. $\qquad \square$

3.3 Exchangeable Processes

In this section, we derive convergence criteria for $\mathbb{R}^d$-valued, exchangeable processes on $[0, 1]$ or $\mathbb{R}_+$. Motivated by Theorem 2.18, we consider first exchangeable processes on $[0, 1]$ of the form

$$
X_t = \alpha t + \sigma B_t + \sum_j \beta_j (1\{\tau_j \leq t\} - t), \quad t \in [0, 1], \tag{13}
$$

where $\tau_1, \tau_2, \ldots$ are i.i.d. $U(0, 1)$ random variables, B is an independent Brownian bridge in $\mathbb{R}^d$, and the random elements α and $\beta_1, \beta_2, \ldots$ in $\mathbb{R}^d$ and σ in $\mathbb{R}^{d^2}$ are independent of (τ_j) and B and satisfy $\sum_j |\beta_j|^2 < \infty$ a.s. (In Theorem 3.15 below we show that *every* exchangeable process on $\mathbb{Q}_{[0,1]}$ is of this form.) If X is instead an exchangeable process on an interval $[0, u]$, we may introduce the process $Y_t = X_{tu}$ on $[0, 1]$ and assume that Y has a representation as in (13). In this case, characteristics such as α, σ, and β will always refer to the re-scaled process.

It is convenient to express our conditions in terms of the point process $\beta = \sum_j \beta_j$ on $\mathbb{R}^d \setminus \{0\}$, regarded as a random element in the space $\mathcal{N}(\mathbb{R}^d \setminus \{0\})$ endowed with the vague topology, and we say that X is *directed* by the triple $(\alpha, \sigma\sigma', \beta)$. We also need to introduce the co-variation matrix

$$
\gamma = [X, X]_1 = \sigma\sigma' + \sum_j \beta_j \beta_j'. \tag{14}
$$

For random processes on $[0, 1]$ or $\mathbb{R}_+$, we write $\xrightarrow{fd}$ for convergence of the finite-dimensional distributions.

Theorem 3.8 *(convergence on $[0,1]$) Let X and $X_1, X_2, \ldots$ be $\mathbb{R}^d$-valued, exchangeable processes on $[0,1]$, as in (13), directed by the triples $(\alpha, \sigma\sigma', \beta)$ and $(\alpha_n, \sigma_n\sigma'_n, \beta_n)$, $n \in \mathbb{N}$, and define γ and the γ_n as in (14). Then these conditions are equivalent:*

(i) $X_n \xrightarrow{d} X$ in $D([0,1], \mathbb{R}^d)$,

(ii) $X_n \xrightarrow{fd} X$,

(iii) $(\alpha_n, \gamma_n, \beta_n) \xrightarrow{vd} (\alpha, \gamma, \beta)$ in $\mathbb{R}^{d+d^2} \times \mathcal{N}(\mathbb{R}^d \setminus \{0\})$.

Our proof requires some tightness criteria of independent interest, which also apply to summation processes of the form

$$X_t = \sum_{j \leq mt} \xi_j, \quad t \in [0,1],$$

where $\xi_1, \ldots, \xi_m$ are exchangeable random vectors in $\mathbb{R}^d$. To allow for a unified treatment, we introduce in the latter case the quantities

$$\alpha = \sum_j \xi_j, \qquad \gamma = \sum_j \xi_j \xi'_j. \tag{15}$$

Lemma 3.9 *(tightness on $[0,1]$) Let $X_1, X_2, \ldots$ be $\mathbb{R}^d$-valued, exchangeable processes on $[0,1]$, as in (13), directed by $(\alpha_n, \sigma_n\sigma'_n, \beta_n)$, $n \in \mathbb{N}$, and define the γ_n as in (14). Let $t_1, t_2, \ldots \in (0,1)$ be bounded away from 0 and 1. Then these conditions are equivalent:*

(i) (X_n) *is tight in* $D([0,1], \mathbb{R}^d)$,

(ii) $(X_n(t_n))$ *is tight in* $\mathbb{R}^d$,

(iii) (α_n) *is tight in* $\mathbb{R}^d$ *and* $(\operatorname{tr} \gamma_n)$ *in* $\mathbb{R}_+$.

This remains true for summation processes X_n based on exchangeable sequences (ξ_{nj}) in $\mathbb{R}^d$ of lengths $m_n \geq t_n^{-1}$, with α_n and γ_n defined as in (15).

Proof: First we note that (i) implies (ii), since $\sup_t |x_t|$ is a continuous function of $x \in D([0,1], \mathbb{R}^d)$. Next we see from Theorem 2.23 that

$$\left(E[|X_n(t_n)|^4 | \alpha_n, \gamma_n] \right)^{1/2} \asymp E[|X_n(t_n)|^2 | \alpha_n, \gamma_n] \asymp \alpha_n^2 + \operatorname{tr} \gamma_n. \tag{16}$$

By Lemma A2.5 it follows that (ii) implies (iii).

Now assume condition (iii). For exchangeable processes on $[0,1]$ we have

$$E[|X_n(t)|^2 | \alpha_n, \gamma_n] \leq t(\operatorname{tr} \gamma_n) + t^2 |\alpha_n|^2 < \infty, \quad t \in [0,1],$$

and similarly for summation processes based on exchangeable sequences. By Jensen's inequality we get for any $c_n > 0$ and $t_n \in [0,1]$

$$
\begin{aligned}
E\left(|c_n X_n(t_n)|^2 \wedge 1 \right) &\leq E\left(E[|c_n X_n(t_n)|^2 | \alpha_n, \gamma_n] \wedge 1 \right) \\
&\leq E\left(c_n^2 (t_n(\operatorname{tr} \gamma_n) + t_n^2 |\alpha_n|^2) \wedge 1 \right).
\end{aligned}
$$

Letting $c_n \to 0$ and applying FMP 4.9 in both directions, we see from (iii) that (ii) is fulfilled. Similarly, we may take $c_n = 1$ and conclude that $X_n(h_n) \xrightarrow{P} 0$ as $h_n \to 0$. Now consider any optional times τ_n with respect to X_n and some positive constants $h_n \to 0$ such that $\tau_n + h_n \leq 1$ a.s. Using Proposition 2.5, together with the earlier observation about $X_n(h_n)$, we obtain

$$X_n(\tau_n + h_n) - X_n(\tau_n) \stackrel{d}{=} X_n(h_n) \xrightarrow{P} 0.$$

Condition (i) now follows by Aldous' criterion in FMP 16.11. $\square$

Proof of Theorem 3.8: First assume (i). Since the process X in (13) is a.s. continuous at every $t \in [0, 1]$, the evaluation maps $\pi_t \colon x \mapsto x_t$ are a.s. continuous at X. Hence, (ii) follows by continuous mapping (FMP 4.27). Conversely, assuming (ii), we see from Lemma 3.9 that (X_n) is tight in $D([0, 1], \mathbb{R}^d)$. Then by Prohorov's theorem (FMP 16.3) every sub-sequence $N' \subset \mathbb{N}$ contains a further sub-sequence N'' such that $X_n \xrightarrow{d}$ some Y along N''. This implies $X_n \xrightarrow{fd} Y$ on some dense set $T \subset [0, 1]$ containing 1, and so $Y \stackrel{d}{=} X$ on T, which extends by right-continuity to all of $[0, 1]$. Thus, $X_n \xrightarrow{d} X$ along N'', and (i) follows since N' was arbitrary.

To prove that (iii) implies (ii), it suffices by Lemma A2.4 to consider non-random triples $(\alpha_n, \gamma_n, \beta_n)$ and (α, γ, β), so that the X_n and X are representable as in (13) with non-random coefficients. The case where $\gamma_n \equiv 0$ is elementary, as is also the case where $(\alpha_n, \sigma_n) \equiv 0$ and the β_n are uniformly bounded, since we can then arrange that either $\beta_{nj} \to \beta_j$ or $\beta_{nj} \to 0$ for all j. Next assume that $(\alpha, \beta) = 0 \equiv \alpha_n$ and the measures β_n are bounded. Then clearly $\max_j |\beta_{nj}| \to 0$, and also $E\,\beta_{nj}(1\{\tau_j \leq t\} - t) = 0$ for every j and t. Furthermore, for any $s \leq t$, we have

$$\begin{aligned}
\mathrm{cov}(X_s^n, X_t^n) &= \sigma_n \,\mathrm{cov}(B_s, B_t)\sigma_n' + \sum_j \beta_{nj}\beta_{nj}'\mathrm{cov}(1\{\tau_j \leq s\}, 1\{\tau_j \leq t\}) \\
&= s(1-t)\gamma_n \to s(1-t)\gamma \\
&= \sigma\,\mathrm{cov}(B_s, B_t)\sigma' = \mathrm{cov}(X_s, X_t).
\end{aligned}$$

Using Lindeberg's theorem in $\mathbb{R}^d$ (FMP 5.5 and 5.12), we conclude that $X_n \xrightarrow{fd} X$.

By the independence of terms, we can combine the three cases into one (FMP 4.29) to obtain $X_n \xrightarrow{fd} X$, as long as β and all the β_n are bounded. In the general case, we may fix a bounded measure $\beta' \leq \beta$ and choose the measures $\beta_n' \leq \beta_n$ to be bounded with associated matrices γ' and γ_n' such that $(\alpha_n, \gamma_n', \beta_n') \to (\alpha, \gamma', \beta')$. Then the corresponding processes satisfy $X_n' \xrightarrow{fd} X'$, and for every $t \in [0, 1]$ we have

$$E|X_{n,t} - X_{n,t}'|^2 \lesssim t(1-t)|\gamma_n - \gamma_n'| \to t(1-t)|\gamma - \gamma'|.$$

Since $\gamma - \gamma'$ can be made arbitrarily small, we may argue by uniform approximation (FMP 4.28) to obtain $X_n \xrightarrow{fd} X$. This proves that (iii) $\Rightarrow$ (ii).

Conversely, assuming (ii), we see from Lemma 3.9 that the sequence of triples $(\alpha_n, \gamma_n, \beta_n)$ is tight in $\mathbb{R}^{d+d^2} \times \mathcal{N}(\mathbb{R}^d \setminus \{0\})$. Then Prohorov's theorem shows that every sub-sequence $N' \subset \mathbb{N}$ contains a further sub-sequence N'' such that (iii) holds along N'', for a suitable limit $(\tilde{\alpha}, \tilde{\gamma}, \tilde{\beta})$. Here $\tilde{\gamma}$ may be expressed as in (14) for some random matrix $\tilde{\sigma}$, and by randomization we may form an associated process Y as in (13). Then the previous part of the proof yields $X_n \xrightarrow{fd} Y$ along N'', and comparing this with (ii) gives $Y \stackrel{d}{=} X$. The uniqueness in Theorem 2.18 implies $(\tilde{\alpha}, \tilde{\gamma}, \tilde{\beta}) \stackrel{d}{=} (\alpha, \gamma, \beta)$, which means that (iii) holds along N''. The full convergence now follows since N' was arbitrary. $\qquad\square$

We turn to the case of contractable processes on $\mathbb{R}_+$. Recall from Theorem 1.19 that any such process X is also exchangeable with representation

$$X_t = \alpha t + \sigma B_t + \int_0^t \int x \, (\eta - 1\{|x| \le 1\}\lambda \otimes \nu)(ds\,dx), \quad t \ge 0, \qquad (17)$$

for some random triple (α^h, σ, ν) in $\mathbb{R}^{d+d^2} \times \mathcal{M}(\mathbb{R}^d \setminus \{0\})$ satisfying $\int(|x|^2 \wedge 1)\nu(dx) < \infty$ a.s., an independent, $\mathbb{R}^d$-valued Brownian motion B, and a Cox process $\eta \perp\!\!\!\perp_\nu (\alpha, \sigma, B)$ directed by ν. Here we say that X is *directed* by the triple $(\alpha, \sigma\sigma', \nu)$. It is convenient in this case to introduce, for every $h > 0$, the random vector and matrix

$$\begin{aligned}
\alpha^h &= \alpha - \int x\, 1\{h < |x| \le 1\}\, \nu(dx), \\
\gamma^h &= \sigma\sigma' + \int xx'\, 1\{|x| \le h\}\, \nu(dx), \qquad (18)
\end{aligned}$$

with the obvious sign convention when $h > 1$.

Theorem 3.10 (*convergence on* $\mathbb{R}_+$) *Let X and $X_1, X_2, \ldots$ be $\mathbb{R}^d$-valued, contractable processes on $\mathbb{R}_+$ directed by $(\alpha, \sigma\sigma', \nu)$ and $(\alpha_n, \sigma_n\sigma_n', \nu_n)$, $n \in \mathbb{N}$, and define (α^h, γ^h) and (α_n^h, γ_n^h) as in (18). Then for any $h > 0$ with $\nu\{|x| = h\} = 0$ a.s., these conditions are equivalent:*

(i) $X_n \stackrel{d}{\to} X$ *in* $D(\mathbb{R}_+, \mathbb{R}^d)$,

(ii) $X_n \xrightarrow{fd} X$ *in* $\mathbb{R}^d$,

(iii) $(\alpha_n^h, \gamma_n^h, \nu_n) \xrightarrow{vd} (\alpha^h, \gamma^h, \nu)$ *in* $\mathbb{R}^{d+d^2} \times \mathcal{M}(\overline{\mathbb{R}^d} \setminus \{0\})$.

Though the random measures ν and the ν_n are defined on $\mathbb{R}^d \setminus \{0\}$, we note that the appropriate topology is that of the extended space $\overline{\mathbb{R}^d} \setminus \{0\}$, where $\overline{\mathbb{R}^d}$ denotes the one-point compactification of $\mathbb{R}^d$. Again it is useful to establish separately the associated tightness criteria. For convenience, we associate with each triple $(\alpha, \sigma\sigma', \nu)$ the random quantity

$$\rho = |\alpha_n|^2 + \text{tr}(\sigma\sigma') + \int (|x|^2 \wedge 1)\, \nu(dx). \qquad (19)$$

Let us also say that the ν_n are *tight at* ∞ if

$$\lim_{r\to\infty} \limsup_{n\to\infty} E(\nu_n\{x; |x| > r\} \wedge 1) = 0.$$

This is clearly equivalent to the condition $\nu_n\{|x| > r_n\} \xrightarrow{P} 0$, for every sequence $r_n \to \infty$.

Lemma 3.11 *(tightness on* $\mathbb{R}_+$*)* *Let* $X_1, X_2, \dots$ *be* $\mathbb{R}^d$*-valued, contractable processes on* $\mathbb{R}_+$ *directed by* $(\alpha_n, \sigma_n\sigma_n', \nu_n)$, $n \in \mathbb{N}$, *and define the* ρ_n *as in* (19). *Let* $t_1, t_2, \dots \in (0, \infty)$ *be bounded away from* 0 *and* ∞. *Then these conditions are equivalent:*

(i) (X_n) *is tight in* $D(\mathbb{R}_+, \mathbb{R}^d)$,
(ii) $(X_n(t_n))$ *is tight in* $\mathbb{R}^d$,
(iii) (ρ_n) *is tight in* $\mathbb{R}_+$ *and* (ν_n) *is tight at* ∞.

This remains true for exchangeable processes X_n *on* $[0, u_n] \to \mathbb{R}_+$, *with the triples* $(\alpha_n, \sigma_n\sigma_n', \nu_n)$ *replaced by* $u_n^{-1}(\alpha_n, \sigma_n\sigma_n', \beta_n)$, *as well as for exchangeable summation processes* $X_t^n = \sum_{j\le tm_n} \xi_{nj}$ *and constants* $t_n \ge m_n^{-1}$.

Proof: The equivalence of (i) and (ii) in Lemma 3.9 extends immediately to processes defined on $\mathbb{R}_+$ or $[0, u_n] \to \mathbb{R}_+$. In particular, (ii) is equivalent to the same condition for any fixed $t_n = t$. Now let X_n^r denote the process obtained from X_n by omitting all jumps of modulus $> r$, and note that

$$E[|X_n^r(t)|^2 | \alpha_n, \sigma_n, \nu_n] = t\operatorname{tr}(\gamma_n^r) + t^2|\alpha_n^r|^2 \le (t \vee t^2)\rho_n^r,$$

where

$$\rho_n^r = |\alpha_n^r|^2 + \operatorname{tr}(\sigma\sigma') + \int (|x|^2 \wedge r)\nu_n(dx), \quad r > 0.$$

Assuming (iii), we see that (ρ_n^r) is tight for fixed $r > 0$, and so $c_n\rho_n^r \xrightarrow{P} 0$ as $c_n \to 0$. Furthermore, for any $r > 0$ we have

$$\begin{aligned}
E\left(c_n|X_n(t)|^2 \wedge 1\right) &\le E\left(c_n|X_n^r(t)|^2 \wedge 1\right) + P\{X_n^r(t) \ne X_n(t)\} \\
&\le E\left(c_n(t \vee t^2)\rho_n^r \wedge 1\right) + E(t\nu_n\{|x| > r\} \wedge 1).
\end{aligned}$$

Letting $n \to \infty$ and then $r \to \infty$, we see that the right-hand side tends to 0, uniformly for bounded $t > 0$. Hence, $c_n|X_n(t)|^2 \xrightarrow{P} 0$, and so $(X_n(t))$ is uniformly tight for bounded t, which proves (ii).

To prove the reverse implication, we may assume that (ii) holds for fixed $t_n = t \in (0, 1)$. Letting β_n be the jump point process for the restriction of X_n to $[0, 1]$, we conclude from Lemma 3.9 that the sequences $\operatorname{tr}(\sigma_n\sigma_n')$ and $\int |x|^2\beta_n(dx)$ are tight. Noting that β_n is a Cox process directed by ν_n and letting $c_n > 0$, we get by FMP 12.2

$$E\exp\left(-c_n\int |x|^2\beta_n(dx)\right) = E\exp\left(-\int\left(1 - e^{-c_n|x|^2}\right)\nu_n(dx)\right), \tag{20}$$

which tends to 1 as $n \to \infty$ when $c_n \to 0$. Hence,

$$c_n \int (|x|^2 \wedge 1) \, \nu_n(dx) + \nu_n\{c_n|x|^2 > 1\} \le \int \left(1 - e^{-c_n|x|^2}\right) \nu_n(dx) \xrightarrow{P} 0,$$

which implies the required tightness of $\int(|x|^2 \wedge 1)\nu_n(dx)$, as well as the tightness at ∞ of (ν_n). Thus, (iii) holds for the centered processes $X'_n(t) = X_n(t) - \alpha_n t$, and by the direct assertion it follows that $(X'_n(t))$ is tight. The tightness of (α_n) now follows by subtraction, which completes the proof for processes on $\mathbb{R}_+$.

For processes on $[0, u_n] \to \mathbb{R}_+$, let β_n^1 be the jump point processes associated with the restrictions to $[0, 1]$, and note that β_n^1 is a u_n^{-1}-thinning of β_n. Hence, by FMP 12.2 we have instead of (20)

$$E \exp\left(-c_n \int |x|^2 \beta_n^1(dx)\right) = E \exp \int \log\left(1 - u_n^{-1}\left(1 - e^{-c_n|x|^2}\right)\right) \beta_n(dx),$$

which again tends to 1 under (ii) when $c_n \to 0$. Writing $\hat{\beta}_n = \beta_n/u_n$, we get in this case

$$c_n \int (|x|^2 \wedge 1) \, \hat{\beta}_n(dx) + \hat{\beta}_n\{c_n|x|^2 > 1\}$$
$$\le -\int \log\left(1 - u_n^{-1}\left(1 - e^{-c_n|x|^2}\right)\right) \beta_n(dx) \xrightarrow{P} 0,$$

which shows that the integrals $\int(|x|^2 \wedge 1)\hat{\beta}_n(dx)$ are tight and the random measures $\hat{\beta}_n$ are tight at ∞. The proof of (iii) may now be completed as before.

We postpone the discussion of summation processes, where the results follow most easily from the approximation theorems of the next section. $\quad\square$

Proof of Theorem 3.10: The implication (i) $\Rightarrow$ (ii) holds by continuity, and the converse assertion follows from the corresponding statement in Theorem 3.8. To prove that (iii) implies (ii), it suffices by Lemma A2.4 to consider non-random characteristics $(\alpha_n, \sigma_n \sigma'_n, \nu_n)$. The random vectors $X_n(t)$ are then infinitely divisible, and the one-dimensional convergence holds by the classical criteria (FMP 15.14). The general finite-dimensional convergence then follows by the independence of the increments (FMP 4.29).

Now assume (ii). By Lemma 3.11 the random variables ρ_n in (19) are tight in $\mathbb{R}_+$ and the random measures ν_n are tight at ∞. Hence, the ν_n are vaguely tight on $\overline{\mathbb{R}^d} \setminus \{0\}$, and so by Prohorov's theorem any sub-sequence $N' \subset \mathbb{N}$ contains a further sub-sequence N'' such that $\nu_n \xrightarrow{vd}$ some $\tilde{\nu}$ along N''. The tightness at ∞ ensures $\tilde{\nu}\{\infty\} = 0$ a.s., which means that $\tilde{\nu}$ is a.s. a random measure on $\mathbb{R}^d \setminus \{0\}$. It is also clear from the tightness of ρ_n that $\int(|x|^2 \wedge 1)\tilde{\nu}(dx) < \infty$ a.s. Fixing any $k > 0$ with $\tilde{\nu}\{|x| = k\} = 0$ a.s., we may choose yet another sub-sequence $N''' \subset N''$ such that

$$(\alpha_n^k, \gamma_n^k, \nu_n) \xrightarrow{vd} (\tilde{\alpha}^k, \tilde{\gamma}^k, \tilde{\nu}) \quad \text{in} \quad \mathbb{R}^{d+d^2} \times \mathcal{M}(\overline{\mathbb{R}^d} \setminus \{0\}),$$

for suitable $\tilde{\alpha}^k$ and $\tilde{\gamma}^k$. Since the differences

$$\gamma_n^k - \int_{|x| \le k} xx' \nu_n(dx) = \sigma_n \sigma_n', \quad n \in \mathbb{N},$$

are non-negative definite, the corresponding property holds a.s. in the limit, and so we may choose some random vector $\tilde{\alpha}$ and matrix $\tilde{\sigma}$ satisfying (18) with $(\alpha^h, \gamma^h, \nu^h)$ replaced by $(\tilde{\alpha}^k, \tilde{\gamma}^k, \tilde{\nu})$. If $\tilde{X}$ is a mixed Lévy process based on $(\tilde{\alpha}, \tilde{\sigma}, \tilde{\nu})$, the direct assertion yields $X_n \xrightarrow{d} \tilde{X}$ along N''''. But then $\tilde{X} \stackrel{d}{=} X$, and so $(\tilde{\alpha}, \tilde{\sigma}\tilde{\sigma}', \tilde{\nu}) \stackrel{d}{=} (\alpha, \sigma\sigma', \nu)$ by the uniqueness in Theorem 1.19. In particular, we may then choose $k = h$ and obtain (iii) along N''''. The convergence extends to the entire sequence since N' was arbitrary. $\qquad\square$

We conclude with the case of exchangeable processes X_n, as in (13), defined on some intervals $[0, u_n] \to \mathbb{R}_+$. Here we define the corresponding directing triples $(\alpha_n, \sigma_n \sigma_n', \beta_n)$ in terms of the scaled processes Y_n on $[0, 1]$, given by $Y_n(t) = X_n(tu_n)$, $t \in [0, 1]$. With any characteristic triple $(\alpha, \sigma\sigma', \beta)$ and constant $h > 0$, we associate the random vector and matrix

$$\begin{aligned}
\alpha^h &= \alpha - \sum_j \beta_j 1\{|\beta_j| > h\}, \\
\gamma^h &= \sigma\sigma' + \sum_j \beta_j \beta_j' 1\{|\beta_j| \le h\}. \qquad (21)
\end{aligned}$$

Theorem 3.12 *(convergence on increasing intervals) Let X be an $\mathbb{R}^d$-valued, exchangeable process on $\mathbb{R}_+$ directed by $(\alpha, \sigma\sigma', \nu)$, and consider for every $n \in \mathbb{N}$ an exchangeable process X_n on $[0, u_n]$, as in (13), directed by $(\alpha_n, \sigma_n \sigma_n', \beta_n)$, where $u_n \to \infty$. Define (α^h, γ^h) and (α_n^h, γ_n^h) as in (18) and (21). Then for any $h > 0$ with $\nu\{|x| = h\} = 0$ a.s., the following conditions are equivalent:*

(i) $X_n \xrightarrow{d} X$ *in* $D(\mathbb{R}_+, \mathbb{R}^d)$,

(ii) $X_n \xrightarrow{fd} X$ *in* $\mathbb{R}^d$,

(iii) $u_n^{-1}(\alpha_n^h, \gamma_n^h, \beta_n) \xrightarrow{vd} (\alpha^h, \gamma^h, \nu)$ *in* $\mathbb{R}^{d+d^2} \times \mathcal{M}(\overline{\mathbb{R}^d} \setminus \{0\})$.

Proof: The equivalence of (i) and (ii) may be proved as before. To complete the proof, it is enough to show that (iii) implies (ii), since the reverse implication will then follow, as in the previous proof, by means of Lemma 3.11. By Lemma A2.4 we may assume that the characteristics $(\alpha_n, \sigma_n \sigma_n', \beta_n)$ in (iii) are non-random. In that case, we shall prove (ii) by applying Theorem 3.10 to some approximating Lévy processes $\tilde{X}_n$ with characteristics

$$(\tilde{\alpha}_n, \tilde{\sigma}_n \tilde{\sigma}_n', \tilde{\nu}_n) = u_n^{-1}(\alpha_n^1, \sigma_n \sigma_n', \beta_n), \quad n \in \mathbb{N}.$$

To construct the $\tilde{X}_n$, we may write

$$\begin{aligned}
X_n(t) &= \tilde{\alpha}_n t + \tilde{\sigma}_n (B_t^n - (t/u_n) B_{u_n}^n) \\
&\quad + \sum_j \beta_{nj}(1\{\tau_{nj} \le t\} - (t/u_n) 1\{|\beta_{nj}| \le 1\}), \quad t \in [0, u_n],
\end{aligned}$$

where $\tau_{n1}, \tau_{n2}, \ldots$ are i.i.d. $U(0, u_n)$ random variables and B^n is an independent standard Brownian motion. For any $u \in (0, u_n]$, there exist by Lemma 3.7 some Poisson processes η_{nj} with rates u_n^{-1} such that

$$E\|\delta_{\tau_{nj}} - \eta_{nj}\|^2_{[0,u]} \leq 2(u/u_n)^2, \quad n, j \in \mathbb{N}, \tag{22}$$

where the pairs (τ_{nj}, η_{nj}) can be assumed to be i.i.d. for fixed $n \in \mathbb{N}$ and independent of B^n. The point processes $\eta_n = \sum_j \delta_{\beta_{nj}} \otimes \eta_{nj}$ on $(\mathbb{R}^d \setminus \{0\}) \otimes \mathbb{R}_+$ are then Poisson with intensity measures $\tilde{\nu}_n \otimes \lambda$, and so we may define some Lévy processes $\tilde{X}_n$ with the desired characteristics by taking

$$\tilde{X}_n(t) = \tilde{\alpha}_n t + \tilde{\sigma}_n B^n_t + \sum_j \beta_{nj}(\eta_{nj}(0, t] - (t/u_n)1\{|\beta_{nj}| \leq 1\}).$$

Writing $p_n = u/u_n$, we get by (22) and (iii)

$$
\begin{aligned}
E\big((X_n - \tilde{X}_n)^{*2}_u \wedge 1\big) &\leq p_n^2 E|\tilde{\sigma}_n B^n_{u_n}|^2 + \sum_j (\beta_{nj}^2 \wedge 1)E\|\delta_{\tau_{nj}} - \eta_{nj}\|^2_u \\
&\leq up_n \operatorname{tr}(\tilde{\sigma}_n \tilde{\sigma}'_n) + 2p_n^2 \sum_j (\beta_{nj}^2 \wedge 1) \to 0,
\end{aligned}
$$

which shows that $X_n(t) - \tilde{X}_n(t) \xrightarrow{P} 0$ for every $t > 0$. Noting that (iii) implies $\tilde{X}_n \xrightarrow{fd} X$ by Theorem 3.10 and using FMP 4.28, we obtain (ii). $\quad\square$

3.4 Approximation and Representation

The limit theorems of the last section are easily extended to summation processes $X_t = \sum_{j \leq rt} \xi_j$ of *rate* $r > 0$ based on finite or infinite exchangeable sequences $\xi = (\xi_1, \xi_2, \ldots)$ in $\mathbb{R}^d$. When the length m of ξ is finite, we consider X as a process on $[0, u] = [0, m/r]$. In particular, we get a process on $[0, 1]$ by taking $r = m$. Infinite sequences ξ generate summation processes on $\mathbb{R}_+$.

For sequences ξ of finite length m, we introduce the characteristics

$$\alpha = \sum_{j \leq m} \xi_j, \qquad \beta = \sum_{j \leq m} \delta_{\xi_j}, \qquad \gamma = \sum_{j \leq m} \xi_j \xi'_j. \tag{23}$$

For any $h > 0$, we also consider the truncated quantities

$$\alpha^h = \sum_{j \leq m} \xi_j 1\{|\xi_j| \leq h\}, \qquad \gamma^h = \sum_{j \leq m} \xi_j \xi'_j 1\{|\xi_j| \leq h\}. \tag{24}$$

If ξ is instead an infinite exchangeable sequence with directing random measure η, we define $\nu = r\eta$, and introduce for any $h > 0$ the quantities

$$\alpha^h = \int_{|x| \leq h} x\,\nu(dx), \qquad \gamma^h = \int_{|x| \leq h} xx'\,\nu(dx). \tag{25}$$

This is clearly consistent with our earlier definitions, provided that we take $\sigma = 0$.

With this notation, the basic convergence criteria for exchangeable summation processes take on the same form as the continuous-time results of the preceding section.

Theorem 3.13 *(convergence of summation processes, Hájek, Rosén, Billingsley, Hagberg, Kallenberg) The statements of Theorems 3.8, 3.10, and 3.12 remain true with the X_n replaced by summation processes of rates $r_n \to \infty$, based on finite or infinite exchangeable sequences $\xi_n = (\xi_{nj})$ in $\mathbb{R}^d$, and with characteristics given by (23), (24), and (25).*

The result follows easily from the earlier results by means of a random change of scale. For any random sequences (ξ_n) and (η_n) in a metric space S, we define the relation $\xi_n \overset{d}{\sim} \eta_n$ to mean that $\xi_n \overset{d}{\to} \xi$ iff $\eta_n \overset{d}{\to} \xi$ for any random element ξ of S.

Lemma 3.14 *(time-scale comparison) Let $S_1, S_2, \ldots$ be $\mathbb{R}^d$-valued, unit rate summation processes on $\mathbb{R}_+$ or $[0, m_n]$, and let $N_1, N_2, \ldots$ be independent, unit rate Poisson or binomial processes on the same intervals. Define*

$$X_n(t) = S_n(r_n t), \quad Y_n(t) = S_n \circ N_n(r_n t), \qquad t \in [0, u_n],$$

where $r_n \to \infty$ and $u_n = m_n / r_n$. Then $X_n \overset{d}{\sim} Y_n$ holds in $D([0,1], \mathbb{R}^d)$ when $u_n \equiv 1$ and in $D(\mathbb{R}_+, \mathbb{R}^d)$ when $u_n \to \infty$ or $\equiv \infty$.

Proof: Writing $m_n = \infty$ for processes on $\mathbb{R}_+$, we have in either case

$$EN_n(t) = t, \quad \operatorname{var} N_n(t) = t(1 - m_n^{-1}t) \le t, \qquad t \in [0, m_n],$$

and so

$$E\Big(r_n^{-1} N_n(r_n t)\Big) = t, \quad \operatorname{var}\Big(r_n^{-1} N_n(r_n t)\Big) \le r_n^{-1} t \to 0,$$

which implies $r_n^{-1} N_n(r_n t) \overset{P}{\to} t$ for each t. Hence, by monotone interpolation,

$$\sup_{t \le u} \left| r_n^{-1} N_n(r_n t) - t \right| \overset{P}{\to} 0, \quad u \ge 0,$$

which implies $\rho(X_n, Y_n) \overset{P}{\to} 0$ for a suitable metrization ρ of the Skorohod topology on $D([0,1], \mathbb{R}^d)$ or $D(\mathbb{R}_+, \mathbb{R}^d)$ (cf. FMP A2.2). The assertion now follows by FMP 4.28. $\square$

Proof of Theorem 3.13: Choose some processes $Y_n \overset{d}{\sim} X_n$ as in Lemma 3.14. Then by Theorem 1.23 the Y_n are exchangeable in the continuous-time sense with the same characteristics as the X_n. Hence, by Theorem 3.8, 3.10, or 3.12, condition (iii) is equivalent to $Y_n \overset{d}{\to} X$, which shows that (i) and (iii) are equivalent. Next we note that (i) implies (ii) since X has no fixed discontinuities. Conversely, assuming (ii), we see from Lemma 3.9 that (X_n) is tight in $D([0,1], \mathbb{R}^d)$ or $D(\mathbb{R}_+, \mathbb{R}^d)$. By Prohorov's theorem it remains to show that if $X_n \overset{d}{\to} Y$ along a sub-sequence $N' \subset \mathbb{N}$, then $Y \overset{d}{=} X$. But this is clear by comparison with (ii), since the stated condition implies $X_n \overset{fd}{\longrightarrow} Y$ along N' on some dense subset of $[0,1]$ or $\mathbb{R}_+$. $\square$

We now have the tools to establish the general representation of exchangeable processes on $[0,1]$, previously obtained under a moment condition in Theorem 2.18. We may also provide a new proof of the similar but more elementary representation for processes on $\mathbb{R}_+$, originally derived in Theorem 1.19. A martingale approach to the latter result is implicit in the proof of Theorem 2.14. As a by-product of our representations, we obtain a partial extension of the regularity conditions from Theorem 2.25.

Theorem 3.15 *(regularity and representation)* *An $\mathbb{R}^d$-valued process X on $\mathbb{Q}_{[0,1]}$ or $\mathbb{Q}_+$ is exchangeable iff it has an a.s. representation as in* (13) *or* (17), *respectively, for some random variables, vectors, matrices, and processes as in Theorem 2.18 or 1.19. If X is $\mathcal{F}$-exchangeable, the formula provides an extension of X to an $\overline{\mathcal{F}}$-semi-martingale on $[0,1]$ or $\mathbb{R}_+$, respectively, where $\overline{\mathcal{F}}$ denotes the right-continuous and complete augmentation of $\mathcal{F}$.*

Proof: We may assume that X is exchangeable on $\mathbb{Q}_{[0,1]}$, the proof for processes on $\mathbb{Q}_+$ being similar. For every $n \in \mathbb{N}$, we introduce the increments

$$\xi_{nj} = X(j/n!) - X((j-1)/n!), \quad j = 1, \ldots, n!, \tag{26}$$

and consider the associated summation processes

$$X_n(t) = \sum_{j \le tn!} \xi_{nj}, \quad t \in [0,1].$$

We also define the associated triples $(\alpha_n, \gamma_n, \beta_n)$ as in (23).

The tightness conditions of Lemma 3.9 are trivially fulfilled, since $X_n(t) \to X(t)$ for all $t \in \mathbb{Q}_{[0,1]}$. By Prohorov's theorem, condition (iii) of Theorem 3.8 then holds along a sub-sequence $N' \subset \mathbb{N}$ for some limiting triple (α, γ, β). Here γ can be expressed as in (14) in terms of some matrix σ, and we may choose an associated exchangeable process Y on $[0,1]$ with representation as in (13). By Theorem 3.13 we have $X_n \xrightarrow{fd} Y$ along N', and so $X \stackrel{d}{=} Y$ on $\mathbb{Q}_{[0,1]}$. We may finally use the transfer theorem (FMP 6.10) to obtain a process $\tilde{X} \stackrel{d}{=} Y$ on $[0,1]$ with a similar representation such that $\tilde{X} = X$ a.s. on $\mathbb{Q}_{[0,1]}$.

Now assume that X is $\mathcal{F}$-exchangeable on $\mathbb{Q}_{[0,1]}$. Then X remains $\mathcal{G}$-exchangeable, where $\mathcal{G}$ is the filtration generated by $\mathcal{F}$ and (α, β, γ), and so by Theorem 2.25 it extends a.s. to a special semi-martingale $\tilde{X}$ on $[0,1]$ with respect to the augmented filtration $\overline{\mathcal{G}}$. Using the Bichteler–Dellacherie theorem (FMP 26.21), we see that $\tilde{X}$ remains a semi-martingale on $[0,1]$ with respect to $\overline{\mathcal{F}}$. $\qquad\square$

Our next aim is to extend some of the previous results to contractable processes on $[0,1]$. Here the following continuous-time version of Lemma 1.11

will play a key role. For any contractable semi-martingale X on $[0,1]$, we introduce the associated *characteristic processes*

$$\alpha_t = X_t, \quad \beta_t = \sum_{s \le t} \delta_{\Delta X_s}, \quad \gamma_t^{ij} = [X^i, X^j]_t, \quad t \in [0,1].$$

Without the semi-martingale hypothesis, we can only define α_t and β_t.

Lemma 3.16 *(one-dimensional coupling) Let X be an $\mathbb{R}^d$-valued, contractable semi-martingale on $[0,1]$. Then there exists an exchangeable process $\tilde{X}$ on $[0,1]$, such that the associated characteristic processes satisfy*

$$(\alpha_t, \beta_t, \gamma_t) \overset{d}{=} (\tilde{\alpha}_t, \tilde{\beta}_t, \tilde{\gamma}_t), \quad t \in [0,1]. \tag{27}$$

Without the semi-martingale property, we can only assert that $(\alpha_t, \beta_t) \overset{d}{=} (\tilde{\alpha}_t, \tilde{\beta}_t)$ for all t. A similar statement holds for contractable random measures on $S \times [0,1]$, where S is compact.

Proof: The increments ξ_{nj} in (26) being contractable for fixed n, there exist by Lemma 1.11 some exchangeable sequences $\tilde{\xi}_{nj}$ satisfying

$$\sum_{j \le k} \delta_{\xi_{nj}} \overset{d}{=} \sum_{j \le k} \delta_{\tilde{\xi}_{nj}}, \quad k \le n!, \ n \in \mathbb{N}. \tag{28}$$

Let X^n and $\tilde{X}^n$ denote the associated summation processes on $[0,1]$. When $t \in \mathbb{Q}_{[0,1]}$, we have $\tilde{X}_t^n \overset{d}{=} X_t^n = X_t$ for all but finitely many $n \in \mathbb{N}$, and so by Lemma 3.9 the sequence $(\tilde{X}^n)$ is tight in $D([0,1], \mathbb{R}^d)$. Hence, Prohorov's theorem yields $\tilde{X}^n \overset{d}{\to} \tilde{X}$ along a sub-sequence $N' \subset \mathbb{N}$, where the limiting process $\tilde{X}$ is exchangeable on $[0,1]$. By Theorem 3.13 the corresponding characteristic triples satisfy

$$(\tilde{\alpha}_t^n, \tilde{\beta}_t^n, \tilde{\gamma}_t^n) \overset{vd}{\longrightarrow} (\tilde{\alpha}_t, \tilde{\beta}_t, \tilde{\gamma}_t) \text{ in } \mathbb{R}^{d+d^2} \times \mathcal{N}(\mathbb{R}^d), \quad t \in \mathbb{Q}_{[0,1]}, \tag{29}$$

with n restricted to N'. Using (28) together with the discrete approximation property of co-variation processes (cf. FMP 17.17), we obtain

$$(\tilde{\alpha}_t^n, \tilde{\beta}_t^n, \tilde{\gamma}_t^n) \overset{d}{=} (\alpha_t^n, \beta_t^n, \gamma_t^n) \overset{P}{\to} (\alpha_t, \beta_t, \gamma_t), \quad t \in \mathbb{Q}_{[0,1]}. \tag{30}$$

Comparing (29) and (30) yields (27) for $t \in \mathbb{Q}_{[0,1]}$, and the general result follows by the right-continuity on both sides. Though the convergence $\gamma_t^n \overset{P}{\to} \gamma_t$ may fail when X is not a semi-martingale, the previous argument still applies to the first two components in (27).

Next let ξ be a contractable random measure on $S \times [0,1]$. Write

$$t_{nj} = j/n!, \quad I_{nj} = (j-1,j]/n!, \quad \xi_{nj} = \xi(\cdot \times I_{nj}), \quad j \le n!, \ n \in \mathbb{N},$$

and introduce some exchangeable sequences $(\tilde{\xi}_{nj})$ satisfying (28). Consider on $S \times [0,1]$ the random measures

$$\xi_n = \sum_j \xi_{nj} \otimes \delta_{t_{nj}}, \quad \tilde{\xi}_n = \sum_j \tilde{\xi}_{nj} \otimes \delta_{t_{nj}}, \quad n \in \mathbb{N}.$$

Since
$$\tilde{\xi}_n(S \times [0,1]) \overset{d}{=} \xi_n(S \times [0,1]) = \xi(S \times [0,1]) < \infty,$$
the sequence $(\tilde{\xi}_n)$ is weakly tight (FMP 16.15), and so we have convergence $\tilde{\xi}_n \overset{wd}{\longrightarrow} \tilde{\xi}$ along a sub-sequence $N' \subset \mathbb{N}$, for some limit $\tilde{\xi}$. Approximating as in Lemma 3.14 and using Theorem 3.4, we see that $\tilde{\xi}$ is again exchangeable. Moreover, in the notation of the latter result and for $n \in N'$,

$$(\tilde{\beta}_t^n, \tilde{\gamma}_t^n) \overset{vd}{\longrightarrow} (\tilde{\beta}_t, \tilde{\gamma}_t) \quad \text{in} \quad \mathcal{N}(\mathcal{M}'(S)) \times \mathcal{M}(S), \quad t \in \mathbb{Q}_{[0,1]}.$$

The proof may now be completed as before. $\square$

We can now extend some earlier tightness and convergence criteria to the contractable case.

Theorem 3.17 *(tightness and finite-dimensional convergence)* *Let* $X_1, X_2,$ *... be* $\mathbb{R}^d$*-valued, contractable processes on* $[0,1]$ *or* $[0, u_n] \to \mathbb{R}_+$*, or summation processes on the same intervals of rates* $r_n \to \infty$*, based on contractable sequences in* $\mathbb{R}^d$*. Then*

- (I) *conditions* (i) *and* (ii) *of Lemmas 3.9 and 3.11 are equivalent, and the equivalence extends to* (iii) *when the* X_n *are semi-martingales;*
- (II) $X_n \overset{d}{\to} X$ *in* $D([0,1], \mathbb{R}^d)$ *or* $D(\mathbb{R}_+, \mathbb{R}^d)$ *iff* $X_n \overset{fd}{\longrightarrow} X$*, in which case the limit* X *is again contractable.*

Proof: (I) The implication (i) $\Rightarrow$ (ii) is generally true in $D([0,1], \mathbb{R}^d)$ and $D(\mathbb{R}_+, \mathbb{R}^d)$. Conversely, assume condition (ii), and note that the same condition holds for the associated exchangeable processes $\tilde{X}_n$ in Lemma 3.16. By Lemma 3.9 or 3.11 it follows that $(\tilde{X}_n)$ is tight in $D([0,1], \mathbb{R}^d)$ or $D(\mathbb{R}_+, \mathbb{R}^d)$, respectively. Now consider any optional times τ_n for X_n and some positive constants $h_n \to 0$ such that $\tau_n + h_n \leq 1$ or u_n. Using Proposition 2.5 and noting that $x_n \to x$ in $D([0,1], \mathbb{R}^d)$ or $D(\mathbb{R}_+, \mathbb{R}^d)$ implies $x_n(h_n) \to x(0)$, we get in the continuous-time case

$$X_n(\tau_n + h_n) - X_n(\tau_n) \overset{d}{=} X_n(h_n) \overset{d}{=} \tilde{X}_n(h_n) \overset{P}{\to} 0,$$

and similarly for summation processes based on contractable sequences. Condition (i) now follows by Aldous' criterion in FMP 16.11. If the X_n are semi-martingales, then the equivalence (ii) $\Leftrightarrow$ (iii) holds for the $\tilde{X}_n$ by Lemmas 3.9 and 3.11, and the result carries over to the X_n by means of (27).

(II) If $X_n \overset{d}{\to} X$ in $D([0,1], \mathbb{R}^d)$ or $D(\mathbb{R}_+, \mathbb{R}^d)$, then $X_n \overset{fd}{\longrightarrow} X$ on some set $T \subset [0,1]$ or $\mathbb{R}_+$, respectively, with a countable complement. Fixing any disjoint intervals $I_1, \ldots, I_k$ of equal length, we get for almost every $h > 0$

$$(X_n(I_1 + h), \ldots, X_n(I_k + h)) \overset{d}{\to} (X(I_1 + h), \ldots, X(I_k + h)), \qquad (31)$$

where $X(I)$ and $X_n(I)$ denote the increments of X and X_n over I. If the X_n are contractable in the continuous-time sense, then the limit in (31) is contractable for almost every $h > 0$. The property extends by right continuity to $X(I_1), \ldots, X(I_k)$, which proves the contractability of X in the continuous-time case. The same result then holds for summation processes by Lemma 3.14. In particular, X has no fixed discontinuities, and so the convergence $X_n \xrightarrow{fd} X$ extends to the entire time scale.

Conversely, suppose that $X_n \xrightarrow{fd} X$. By (I) it follows that (X_n) is tight in $D([0,1], \mathbb{R}^d)$ or $D(\mathbb{R}_+, \mathbb{R}^d)$, respectively. If $X_n \xrightarrow{d} Y$ along a sub-sequence $N' \subset \mathbb{N}$, then $X_n \xrightarrow{fd} Y$ holds as before along N'. Since also $X_n \xrightarrow{fd} X$, we get $Y \overset{d}{=} X$. The desired convergence $X_n \xrightarrow{d} X$ now follows by Prohorov's theorem. □

We conclude with some continuous-time versions of Theorem 3.1. As noted before, any random measure ξ on a product space $S \times [0, u]$ can be identified with an rcll process $X_t = \xi(\cdot \times [0, t])$ in $\mathcal{M}(S)$.

Proposition 3.18 *(convergence of contractable random measures) For every $n \in \mathbb{N}$, let ξ_n be a contractable random measure on $S \times [0, u_n]$, where S is compact and $u_n \to \infty$, and choose an associated exchangeable random measure $\tilde{\xi}_n$, as in Lemma 3.16. Then $\xi_n \overset{d}{\sim} \tilde{\xi}_n$ in both $\mathcal{M}(S \times \mathbb{R}_+)$ and $D(\mathbb{R}_+, \mathcal{M}(S))$. A similar result holds for summation processes of rates $r_n \to \infty$, based on contractable sequences of random measures on S.*

Proof: If $\xi_n \xrightarrow{d} \xi$ in $\mathcal{M}(S \times \mathbb{R}_+)$ or $D(\mathbb{R}_+, \mathcal{M}(S))$, then by Theorem 3.17 (I) the sequence $(\tilde{\xi}_n)$ is tight in both spaces, and so by Prohorov's theorem every sub-sequence $N' \subset \mathbb{N}$ has a further sub-sequence N'' such that $\tilde{\xi}_n \xrightarrow{d} \tilde{\xi}$ along N'', in the sense of either topology, for some random measure $\tilde{\xi}$ on $S \times \mathbb{R}_+$. Conversely, if $\tilde{\xi}_n \xrightarrow{d} \tilde{\xi}$, then for every sequence $N' \subset \mathbb{N}$ we have $\xi_n \xrightarrow{d} \xi$ along a further sub-sequence N'' for a suitable limit ξ. In each case, it remains to show that $\xi \overset{d}{=} \tilde{\xi}$.

Then note that, by continuous mapping from $D([0, t], \mathbb{R}^d)$ to $\mathcal{N}(\mathcal{M}'(S)) \times \mathcal{M}(S)$, the associated characteristics satisfy

$$(\beta_t, \gamma_t) \xleftarrow{vd} (\beta_t^n, \gamma_t^n) \overset{d}{=} (\tilde{\beta}_t^n, \tilde{\gamma}_t^n) \xrightarrow{vd} (\tilde{\beta}_t, \tilde{\gamma}_t), \quad t \geq 0, \tag{32}$$

along N''. Now ξ and $\tilde{\xi}$ are both contractable by Theorem 3.17 (II) and hence can be represented, as in Proposition 1.21, in terms of some random pairs (α, ν) and $(\tilde{\alpha}, \tilde{\nu})$. Using (32) and the law of large numbers, we get on $\mathcal{M}(S) \times \mathcal{M}(\mathcal{M}'(S))$ as $t \to \infty$

$$(\alpha, \nu) \xleftarrow{v} t^{-1}(\alpha_t, \beta_t) \overset{d}{=} t^{-1}(\tilde{\alpha}_t, \tilde{\beta}_t) \xrightarrow{v} (\tilde{\alpha}, \tilde{\nu}),$$

which implies $\xi \overset{d}{=} \tilde{\xi}$ by the uniqueness in Proposition 1.21. □

We can also use Lemma 3.16 to prove the following continuous-time version of Proposition 1.12.

Proposition 3.19 *(moment comparison) Let X be an $\mathbb{R}^d$-valued, contractable semi-martingale on $[0, 1]$ with characteristic triple (α, β, γ), and let Y be a mixed Lévy process in $\mathbb{R}^d$ with the same characteristics. Then for any convex function f on $\mathbb{R}^d$, we have*

$$Ef(X_t) \leq Ef(Y_t), \quad t \in [0, 1],$$

whenever either side exists.

Proof: By Lemma 3.16 we may assume that X is exchangeable, in which case we may take the coefficients in the representation (13) to be non-random. By monotone convergence we can next assume that $f(x) = O(|x|)$ as $|x| \to \infty$. It is now easy to reduce to the case where the sum in (13) is finite. Changing one term at a time and using a conditioning argument based on Fubini's theorem, we may further reduce to the case of a single representing term. We may also disregard the centering, which is the same for X and Y.

First assume that $X = \sigma B^0$ and $Y = \sigma B$, where B is a Brownian motion and B^0 a Brownian bridge in $\mathbb{R}^d$. Fixing any $t \in [0, 1]$, we note that $X_t \overset{d}{=} \xi$ and $Y_t \overset{d}{=} \xi + \eta$, where ξ and η are independent, centered, Gaussian random vectors with covariance matrices $t(1 - t)\sigma\sigma'$ and $t^2\sigma\sigma'$, respectively. Using Jensen's inequality in $\mathbb{R}^d$ (FMP 3.5), we get

$$\begin{aligned}
Ef(X_t) &= Ef(\xi) = Ef(E[\xi + \eta|\xi]) \\
&\leq EE[f(\xi + \eta)|\xi] = Ef(\xi + \eta) = Ef(Y_t).
\end{aligned}$$

Next assume that $X_t = \beta 1\{\tau \leq t\}$ and $Y_t = \beta N_t$, where N is a unit rate Poisson process on $\mathbb{R}_+$ and τ is $U(0,1)$. For every $n \in \mathbb{N}$, consider a Bernoulli sequence $\eta_1^n, \eta_2^n, \ldots$ with rate n^{-1}, and put $Y_t^n = \beta \sum_{j \leq nt} \eta_1^n$. Then Proposition 1.12 yields

$$Ef(X([nt]/n)) \leq Ef(Y_t^n), \quad t \in [0, 1].$$

Here the left-hand side tends to $Ef(X_t)$ by continuity and dominated convergence as $n \to \infty$. Furthermore, $Y_t^n \overset{d}{\to} Y_t$ by a standard approximation (FMP 5.7), and also $E|Y_t^n|^2 \leq 2|\beta|^2$. Hence, $Ef(Y_t^n) \to Ef(Y_t)$ by continuous mapping and uniform integrability (FMP 4.11), and the desired relation follows. $\square$

3.5 Restriction and Extension

The next result shows how the condition of finite-dimensional convergence $X_n \xrightarrow{fd} X$ in Theorems 3.8 to 3.12 can be relaxed in various ways. Corresponding uniqueness criteria are obtained for the choice of $X_n \equiv Y$.

Theorem 3.20 *(improved convergence criteria) Let $X, X_1, X_2, \ldots$ be $\mathbb{R}^d$-valued, exchangeable processes defined on $I = [0,1]$ or $[0, u_n] \to I = \mathbb{R}_+$. Then $X_n \xrightarrow{d} X$ in $D(I, \mathbb{R}^d)$ iff*

(i) $X_n \xrightarrow{fd} X$ *on $[0, \varepsilon]$ for some $\varepsilon > 0$;*

(ii) $X_n \xrightarrow{fd} X$ *on $c\mathbb{N}$ for some $c > 0$, when $I = \mathbb{R}_+$;*

(iii) $X_n(t) \xrightarrow{d} X(t)$ *for all t, when $X, X_1, X_2, \ldots$ are real and continuous;*

(iv) $X_n(t) \xrightarrow{d} X(t)$ *for some $t \in (0,1) \setminus \{\frac{1}{2}\}$ or $t > 0$, when $X, X_1, X_2, \ldots$ are extreme.*

A few lemmas will be helpful. The following result shows how various properties of the processes $X_1, X_2, \ldots$ carry over to the proposed limit.

Lemma 3.21 *(closure properties) Let $X_1, X_2, \ldots$ be $\mathbb{R}^d$-valued, exchangeable processes on $[0,1]$ or $[0, u_n] \to \mathbb{R}_+$ such that $X_n \xrightarrow{fd}$ some X. Then even X is exchangeable, and*

(i) X *is extreme whenever this holds for every X_n;*

(ii) X *is continuous whenever this holds for every X_n.*

Proof: The first assertion is obvious, part (i) follows from the criteria in Theorems 3.8 to 3.12, and (ii) holds since $X_n \xrightarrow{d} X$ in $D([0,1], \mathbb{R}^d)$ or $D(\mathbb{R}_+, \mathbb{R}^d)$. □

If X is an exchangeable process on $[0,1]$, the restriction X^p to a sub-interval $[0, p]$ is again exchangeable. The next result shows how the distributions of the associated directing triples are related. Here it is convenient to write $\rho = \sigma\sigma'$ and to introduce the *Fourier–Laplace (FL) transform* of the directing triple (α, ρ, β), given by

$$H(u, v, f) = Ee^{iu\alpha - v\rho - \beta f}, \quad u \in \mathbb{R}^d, \ v \in V_d^+, \ f : \mathbb{R}^d \to \mathbb{C}_+,$$

where $\mathbb{C}_+ = \{z \in \mathbb{C}; \Re z \geq 0\}$. Here $v\rho = \sum_{ij} v_{ij}\rho_{ij}$, and V_d^+ denotes the set of real $d \times d$ matrices v such that $v\rho \geq 0$ for all symmetric, non-negative definite matrices ρ. We also define $l(x) \equiv x$ for $x \in \mathbb{R}^d$.

Lemma 3.22 *(restriction) Let X be an $\mathbb{R}^d$-valued, exchangeable process on $[0,1]$ with restriction X^p to $[0,p]$, and let H_p denote the FL-transform of the directing triple $(\alpha_p, \rho_p, \beta_p)$ of X^p. Then for $p < \frac{1}{2}$ and appropriate u, v, and f, we have*

$$H_p(u, v, f) = H_1\left(pu, \ pv + \tfrac{1}{2}p(1-p)uu', \ ipul - \log(1 - p(1 - e^{iul - f}))\right).$$

When $p < \frac{1}{2}$ we have $\Re(1 - p(1 - e^{iul-f})) > 0$, and we may choose the principal branch of the logarithm. Assuming that $f(x) = O(|x|^2)$ at the origin, we then have the same behavior for the function

$$g(x) = ipux - \log\left(1 - p(1 - e^{iux-f(x)})\right), \quad x \in \mathbb{R}^d. \tag{33}$$

Proof: For convenience, we may drop the subscript 1 when $p = 1$. Noting that

$$\alpha_p = X_p, \quad \rho_p = p\rho, \quad \beta_p = \sum_j \delta_{\beta_j} 1\{\tau_j \le p\},$$

we get formally

$$
\begin{aligned}
H_p(u, v, f) &= E \exp(iu\alpha_p - v\rho_p - \beta_p f) \\
&= E \exp\Big(ipu\alpha + iu\sigma B_p + iu\sum_j \beta_j(1\{\tau_j \le p\} - p) \\
&\quad -pv\rho - \sum_j f(\beta_j)1\{\tau_j \le p\}\Big) \\
&= E \exp\Big(ipu\alpha - \tfrac{1}{2}p(1 - p)u'\rho u - pv\rho - ipu\sum_j \beta_j\Big) \\
&\quad \prod_j\Big(1 - p(1 - e^{iu\beta_j - f(\beta_j)})\Big) \\
&= E \exp\Big(ipu\alpha - (pv + \tfrac{1}{2}p(1 - p)uu')\rho - ipu\beta l \\
&\quad + \beta \log(1 - p(1 - e^{iul-f}))\Big) \\
&= H_1\Big(pu,\ pv + \tfrac{1}{2}p(1 - p)uu',\ ipul - \log(1 - p(1 - e^{iul-f}))\Big).
\end{aligned}
$$

To justify the computation, we may first consider the case where X is extreme with finitely many jumps, and then proceed to the general extreme case by dominated convergence. The result in the composite case follows by conditioning on (α, σ, β). □

We proceed to show how the Laplace transform of a random measure can be extended by analytic continuation. Here *lcsc* is short for a locally compact, second countable, Hausdorff topological space.

Lemma 3.23 *(analytic extension 1)* *Let ξ be a random measure on an lcsc space S, let α be a $\mathbb{C}$-valued, integrable random variable, and fix a continuous function $f_0 : S \to \mathbb{C}_+$ with $\xi|f_0| < \infty$ a.s., such that $E(\alpha e^{-\xi f}) = 0$ for every measurable function f in a locally uniform neighborhood of f_0. Then $E[\alpha|\xi] = 0$ a.s.*

Proof: For any bounded Borel set $B \subset S$, there exists an $\varepsilon > 0$ such that $E(\alpha e^{-\xi f}) = 0$ for every measurable function $f : S \to \mathbb{C}_+$ with $|f - f_0| < \varepsilon$ on B and $f = f_0$ on B^c. Since f_0 is continuous, we may next choose $\delta > 0$ so small that $|f_0(x) - f_0(y)| < \varepsilon/2$ whenever $x, y \in B$ with $d(x, y) < \delta$, where d is an arbitrary metrization of S. For any partition of B into Borel sets $B_1, \ldots, B_n$ of diameter $< \delta$, the function

$$f(x) = f_0(x)1_{B^c}(x) + \sum_k c_k 1_{B_k}(x), \quad x \in S,$$

satisfies $|f - f_0| < \varepsilon$ for all vectors $(c_1, \ldots, c_n)$ in some open subset of $\mathbb{C}_+^n$. Hence, $E(\alpha e^{-\xi f}) = 0$ for all such functions f. Since this expected value is an analytic function of each coefficient $c_1, \ldots, c_n \in \mathbb{C}_+$, the relation extends to any $c_1, \ldots, c_n \in \mathbb{R}_+$. We may next extend the formula by dominated convergence, first to any measurable function $f \geq 0$ with $f = f_0$ outside a bounded set, and then to any continuous function $f \geq 0$ with bounded support. Using the uniqueness theorem for Laplace transforms (FMP 5.3), we conclude that $E[\alpha; \xi \in A] = 0$ for any measurable set $A \subset \mathcal{M}(S)$, and the assertion follows. $\qquad\square$

We need also a similar result for random covariance matrices.

Lemma 3.24 *(analytic extension 2) Consider a symmetric, non-negative definite random $d \times d$ matrix ρ, an integrable random variable α in $\mathbb{C}$, and an element $c \in V_d^+$ such that $E(\alpha e^{-v\rho - c\rho}) = 0$ for all $v \in V_d^+$. Then $E[\alpha|\rho] = 0$ a.s.*

Proof: Introduce the non-negative random variables

$$\eta_{ii} = \rho_{ii}, \quad \eta_{ij} = \rho_{ii} + \rho_{jj} - 2\rho_{ij}, \qquad i \neq j.$$

For any symmetric matrix $u = (u_{ij})$ with non-negative entries, we may define a corresponding matrix $v \in V_d^+$ through the identity $v\rho \equiv u\eta$. It is also clear that $c\rho \equiv b\eta$ for some symmetric matrix $b = (b_{ij})$. By hypothesis, we have $E(\alpha e^{-u\eta}) = 0$ for all matrices $u \geq b$. Here the left-hand side is an analytic function of each coefficient $u_{ij} \in \mathbb{C}_+$, and so the relation extends to arbitrary $u_{ij} \geq 0$. Using the uniqueness theorem for Laplace transforms, we conclude that $E[\alpha; \eta \in A] = 0$ for every Borel set $A \subset \mathbb{R}^{d^2}$, and the assertion follows. $\qquad\square$

Proof of Theorem 3.20: The necessity of (i)–(iv) holds by Theorems 3.8, 3.10, and 3.12. Conversely, assume anyone of the four conditions. Then (X_n) is tight by Lemma 3.9, and so by Prohorov's theorem any sub-sequence $N' \subset \mathbb{N}$ has a further sub-sequence N'' such that $X_n \xrightarrow{d}$ some Y along N''. It remains to show that $Y \overset{d}{=} X$. Now $X_n \xrightarrow{fd} Y$ along N'' by the necessity of (i)–(iv), and hence, in the four cases, we have respectively

  (i) $X \overset{d}{=} Y$ on $[0, \varepsilon]$,

  (ii) $X \overset{d}{=} Y$ on $c\mathbb{N}$,

  (iii) $X_t \overset{d}{=} Y_t$ for all t,

  (iv) $X_t \overset{d}{=} Y_t$ for the special value of t.

Furthermore, we note that Y is exchangeable by Lemma 3.21, and also continuous in case (iii) and extreme in case (iv). To show that $X \overset{d}{=} Y$, we need to discuss each case separately.

(i) By scaling it is enough to consider two exchangeable processes X and $\tilde{X}$ on $[0,1]$ such that $X \stackrel{d}{=} \tilde{X}$ on $[0,p]$ for some $p < \frac{1}{2}$. Writing H and $\tilde{H}$ for the FL-transforms of the corresponding characteristic triples (α, ρ, β) and $(\tilde{\alpha}, \tilde{\rho}, \tilde{\beta})$, we have by Lemma 3.22, for appropriate u, v, and f,

$$H(pu,\ pv + \tfrac{1}{2}p(1-p)uu',\ g) = \tilde{H}(pu,\ pv + \tfrac{1}{2}p(1-p)uu',\ g),$$

where g is defined by (33) in terms of f and u. Applying Lemma 3.24 for fixed u and f, we obtain the extended version

$$E \exp(ipu\alpha - v\rho - \beta g) = E \exp(ipu\tilde{\alpha} - v\tilde{\rho} - \tilde{\beta} g). \tag{34}$$

Fixing u and v, we can solve for f in (33) to obtain

$$f(x) = iux - \log\Big(1 - p^{-1}(1 - e^{ipux - g(x)})\Big) + 2\pi n i, \quad x \in \mathbb{R}^d, \tag{35}$$

where the logarithm exists since

$$1 - p^{-1}(1 - e^{ipux - g(x)}) = e^{iux - f(x)} \neq 0, \quad x \in \mathbb{R}^d.$$

Given any continuous, real-valued function $f_0 > 0$ with $f_0(x) = O(|x|^2)$ at the origin, we note that the corresponding function g_0 in (33) is continuous and $\mathbb{C}_+$-valued with $g_0(x) = O(|x|^2)$. Hence, (35) holds for the pair (f_0, g_0), provided that we choose the right-hand side to be continuous and real for $x = 0$. If the function g is sufficiently close to g_0 in the locally uniform topology, we can use (35) with the same branch of the logarithm to construct a corresponding $\mathbb{C}_+$-valued function f. Then (33) remains true for the pair (f, g), and so even (34) holds for such a g. By Lemma 3.23, the latter relation extends to any real-valued functions $g \geq 0$.

As a final step, we may replace pu by u to obtain $H(u, v, g) \equiv \tilde{H}(u, v, g)$ for all u, v, and g. This implies $(\alpha, \rho, \beta) \stackrel{d}{=} (\tilde{\alpha}, \tilde{\rho}, \tilde{\beta})$, and the required relation $X \stackrel{d}{=} \tilde{X}$ follows by the uniqueness in Theorem 3.15.

(ii) Here we may assume that the processes X and $\tilde{X}$ are exchangeable on $\mathbb{R}_+$ with directing triples (α, ρ, ν) and $(\tilde{\alpha}, \tilde{\rho}, \tilde{\nu})$, respectively, such that $X \stackrel{d}{=} \tilde{X}$ on $\mathbb{N}$. Then the exchangeable sequences $\xi_k = X_k - X_{k-1}$ and $\tilde{\xi}_k = \tilde{X}_k - \tilde{X}_{k-1}$ agree in distribution, and so, by the law of large numbers, the same thing is true for the associated directing random measures η and $\tilde{\eta}$. Now the latter are a.s. infinitely divisible with characteristic triples (α, ρ, ν) and $(\tilde{\alpha}, \tilde{\rho}, \tilde{\nu})$, respectively, which are measurably determined by η and $\tilde{\eta}$. In fact, the mapping $\eta \mapsto (\alpha, \rho, \nu)$ is continuous in the topology of Theorem 3.10, by the same result or by the continuity theorem for infinitely divisible distributions in FMP 15.14. Hence, even $(\alpha, \rho, \nu) \stackrel{d}{=} (\tilde{\alpha}, \tilde{\rho}, \tilde{\nu})$, and the required relation $X \stackrel{d}{=} \tilde{X}$ follows by the uniqueness in Theorem 1.19.

(iii) Here we may assume that X and $\tilde{X}$ are real-valued, continuous, exchangeable processes on $[0,1]$ such that $X_t \stackrel{d}{=} \tilde{X}_t$ for all t. Writing H

and $\tilde{H}$ for the FL-transforms of the characteristic pairs (α, σ^2) and $(\tilde{\alpha}, \tilde{\sigma}^2)$, respectively, we note that

$$
\begin{aligned}
Ee^{iuX_t} &= E\exp(iu\alpha t + iu\sigma B_t)\\
&= E\exp\left(iut\alpha - \tfrac{1}{2}u^2 t(1-t)\sigma^2\right) = H\left(ut, \tfrac{1}{2}u^2 t(1-t)\right),
\end{aligned}
$$

and similarly for $\tilde{X}$. Substituting

$$
s = ut, \quad v = \tfrac{1}{2}u^2 t(1-t) = \tfrac{1}{2}s^2(t^{-1} - 1),
$$

and noting that the range of (s, v) contains the set $(\mathbb{R} \setminus \{0\}) \times \mathbb{R}_+$, we get $H(s, v) \equiv \tilde{H}(s, v)$ by continuity, and so $(\alpha, \sigma^2) \overset{d}{=} (\tilde{\alpha}, \tilde{\sigma}^2)$. Thus, $X \overset{d}{=} \tilde{X}$ by the uniqueness in Theorem 3.15.

(iv) The result for processes on $\mathbb{R}_+$ follows as in FMP 15.8 from the uniqueness theorem for characteristic functions. It is then enough to consider some extreme, exchangeable processes X and $\tilde{X}$ on $[0, 1]$ such that $X_t \overset{d}{=} \tilde{X}_t$ for a fixed $t \in (0, 1) \setminus \{\tfrac{1}{2}\}$. Then, as in Lemma 3.22, we have

$$
\begin{aligned}
Ee^{iu'X_t} &= \exp\left(itu'\alpha + iu'\sigma B_t + iu'\textstyle\sum_j \beta_j(1\{\tau_j \le t\} - t)\right)\\
&= \exp(itu'\alpha - \tfrac{1}{2}t(1-t)u'\rho u)\textstyle\prod_j f_t(iu'\beta_j), \tag{36}
\end{aligned}
$$

and similarly for $\tilde{X}_t$, where

$$
f_t(z) = e^{-tz}(1 - t(1 - e^z)) = 1 + O(z^2), \quad z \in \mathbb{C}.
$$

Since $\sum_j \beta_j^2 < \infty$ and $\sum_j \tilde{\beta}_j^2 < \infty$, the expressions in (36) converge, uniformly for bounded u, toward some entire functions $\varphi_t(u)$ and $\tilde{\varphi}_t(u)$ on $\mathbb{C}^d$, and the relation $Ee^{iu'X_t} = Ee^{iu'\tilde{X}_t}$ extends to $\varphi_t(u) \equiv \tilde{\varphi}_t(u)$.

Now f_t has the zeros

$$
z = \log(t^{-1} - 1) + (2n + 1)\pi i, \quad n \in \mathbb{Z},
$$

and so the solutions to the equation $\varphi_t(u) = 0$ satisfy

$$
u'\beta_j = (2n + 1)\pi - i\log(t^{-1} - 1), \quad j \in \mathbb{N}, n \in \mathbb{Z}.
$$

Taking imaginary parts gives

$$
(\Im u)'\beta_j = \Im(u'\beta_j) = -\log(t^{-1} - 1) \ne 0,
$$

which represents a plane in $\mathbb{R}^d$. Since the zeros of φ_t and $\tilde{\varphi}_t$ agree and have the same multiplicities, we conclude that the sequences $\beta_1, \beta_2, \ldots$ and $\tilde{\beta}_1, \tilde{\beta}_2, \ldots$ agree apart from order. Thus, $\beta = \tilde{\beta}$, and we may divide by the corresponding products in (36) to obtain

$$
\exp(itu'\alpha - \tfrac{1}{2}t(1-t)u'\rho u) = \exp(itu'\tilde{\alpha} - \tfrac{1}{2}t(1-t)u'\tilde{\rho}u), \quad u \in \mathbb{R}^d.
$$

Here we may take absolute values to get $u'\rho u = u'\tilde{\rho}u$ for all $u \in \mathbb{R}^d$, which implies $\rho = \tilde{\rho}$ since both matrices are symmetric. We are then left with the equation $e^{itu'\alpha} = e^{itu'\tilde{\alpha}}$ for arbitrary u, which gives $\alpha = \tilde{\alpha}$. $\qquad\square$

3.6   Coupling and Path Properties

Here our first aim is to strengthen the convergence in Theorems 3.8, 3.10, and 3.12 to the sense of the locally uniform topology. Unfortunately, the usual weak convergence theory doesn't apply to this case, since the uniform topology on $D([0,1], \mathbb{R}^d)$ is non-separable and the induced σ-field is different from the one generated by the coordinate projections. Our present approach is through a coupling argument, which provides an even stronger a.s. version of the indicated result.

Letting $x, x_1, x_2, \ldots \in D(I, \mathbb{R}^d)$ with $I = [0,1]$ or $\mathbb{R}_+$, we write $x_n \overset{u}{\to} x$ for the locally uniform convergence $(x_n - x)_t^* \to 0$, where $t \in I$ is arbitrary. If $X, X_1, X_2, \ldots$ are random processes in $D(I, \mathbb{R}^d)$, then by $X_n \overset{ud}{\longrightarrow} X$ we mean that $f(X_n) \overset{d}{\to} f(X)$ for every measurable function f on $D(I, \mathbb{R}^d)$ that is continuous for the uniform topology, in the sense that $f(x_n) \to f(x)$ whenever $x_n \overset{u}{\to} x$ in $D(I, \mathbb{R}^d)$.

Theorem 3.25 *(uniform convergence, Skorohod, Kallenberg) Let X and $X_1, X_2, \ldots$ be $\mathbb{R}^d$-valued, exchangeable processes on $[0,1]$ or $[0, u_n] \to \mathbb{R}_+$. Then these conditions are equivalent:*

(i) $X_n \overset{fd}{\longrightarrow} X$,

(ii) $X_n \overset{d}{\to} X$,

(iii) $X_n \overset{ud}{\longrightarrow} X$,

(iv) $\tilde{X}_n \overset{u}{\to} X$ *a.s. for some* $\tilde{X}_n \overset{d}{=} X_n$.

Here an elementary lemma will be helpful.

Lemma 3.26 *(combination of sequences) Let $\xi_{k,n} \geq 0$, $k, n \in \mathbb{N}$, be random variables such that $\xi_{k,n} \to 0$ a.s. as $n \to \infty$ for fixed k. Then there exist some constants $k_n \to \infty$ such that*

$$\lim_{n \to \infty} \max_{k \leq k_n} \xi_{k,n} = 0 \quad a.s. \tag{37}$$

Proof: We may assume that the variables $\xi_{k,n}$ are non-increasing in n, non-decreasing in k, and uniformly bounded, since otherwise they can be replaced by

$$\eta_{k,n} = \max_{j \leq k} \sup_{m \geq n} \xi_{k,n} \wedge 1, \quad k, n \in \mathbb{N},$$

which have all the stated properties, satisfy the hypotheses of the lemma, and are such that $\eta_{k_n,n} \to 0$ implies (37). Then define

$$
\begin{aligned}
n_k &= \inf\{n \in \mathbb{N};\ E\xi_{k,n} \leq 2^{-k}\}, & k \in \mathbb{N}, \\
k_n &= \max\{k \leq n;\ n_k \leq n\}, & n \in \mathbb{N}.
\end{aligned}
$$

Since $E\xi_{k,n} \to 0$ as $n \to \infty$ by dominated convergence, we have $n_k < \infty$ for all k, which implies $k_n \to \infty$. Furthermore, by Fubini's theorem,

$$E \sum_k \xi_{k,n_k} = \sum_k E\xi_{k,n_k} \leq \sum_k 2^{-k} < \infty,$$

which shows that $\xi_{k,n_k} \to 0$ a.s. as $k \to \infty$. Noting that $n_{k_n} \leq n$ for large n and using the monotonicity in the second index, we get as $n \to \infty$

$$\xi_{k_n,n} \leq \xi_{k_n,n_{k_n}} \to 0 \quad \text{a.s.}$$

Relation (37) now follows by the monotonicity in the first index. □

Proof of Theorem 3.25: It is enough to prove that (ii) $\Rightarrow$ (iv), since (i) $\Leftrightarrow$ (ii) by the previous results, and the implications (iv) $\Rightarrow$ (iii) $\Rightarrow$ (ii) are obvious. First we consider processes X and $X_1, X_2, \ldots$ on $[0, 1]$ with non-random characteristics (α, γ, β) and $(\alpha_n, \gamma_n, \beta_n)$, $n \in \mathbb{N}$, satisfying $X_n \xrightarrow{d} X$ in $D([0, 1], \mathbb{R}^d)$. Fix any $\varepsilon > 0$ with $\beta\{|x| = \varepsilon\} = 0$, and let $\xi^1, \ldots, \xi^m$ denote the jumps in X of modulus $> \varepsilon$, listed in order from left to right. For large enough $n \in \mathbb{N}$, even X_n has m jumps greater than ε, say $\xi_n^1, \ldots, \xi_n^m$, and we note that

$$(\xi_n^1, \ldots, \xi_n^m) \xrightarrow{d} (\xi^1, \ldots, \xi^m).$$

By FMP 4.30 we may replace the X_n by equivalent processes $X_n^\varepsilon \stackrel{d}{=} X_n$ such that the same convergence holds in the a.s. sense. Since the associated times $\tau^1, \ldots, \tau^m$ and $\tau_n^1, \ldots, \tau_n^m$ form independent binomial processes on $[0, 1]$, e.g. by Theorem 1.23, we may assume that also $\tau_n^k = \tau^k$ a.s. for all k. Then the corresponding jump processes J^ε and $J_1^\varepsilon, J_2^\varepsilon, \ldots$ satisfy $\|J_n^\varepsilon - J^\varepsilon\| \to 0$ a.s.

The remaining components $Y^\varepsilon = X - J^\varepsilon$ and $Y_n^\varepsilon = X_n - J_n^\varepsilon$, $n \in \mathbb{N}$, satisfy $Y_n^\varepsilon \xrightarrow{d} Y^\varepsilon$, in the sense of Skorohod's J_1-topology on $D([0, 1], \mathbb{R}^d)$, and so by FMP 4.30 we may assume that $Y_n^\varepsilon \to Y^\varepsilon$ a.s. for the same topology. Then, with probability one, there exist some increasing bijections $\lambda_1, \lambda_2, \ldots$ on $[0, 1]$ such that $\|Y_n^\varepsilon \circ \lambda_n - Y^\varepsilon\| \to 0$ and $\|\lambda_n - \lambda\| \to 0$, where $\lambda(t) \equiv t$. Furthermore, we note that $\tilde{w}(Y^\varepsilon, h) \to 0$ a.s. as $h \to 0$, where $\tilde{w}$ denotes the modified modulus of continuity in $D([0, 1], \mathbb{R}^d)$ (FMP A2.2). Since the jump sizes of Y^ε are bounded by ε, we conclude for the ordinary modulus of continuity w that

$$\limsup_{h \to 0} w(Y^\varepsilon, h) \leq \varepsilon \quad \text{a.s.}, \quad \varepsilon > 0. \tag{38}$$

Writing $X_n^\varepsilon = J_n^\varepsilon + Y_n^\varepsilon \stackrel{d}{=} X_n$ and noting that

$$\begin{aligned}
\|X_n^\varepsilon - X\| &\leq \|J_n^\varepsilon - J^\varepsilon\| + \|Y_n^\varepsilon - Y^\varepsilon\| \\
&\leq \|J_n^\varepsilon - J^\varepsilon\| + \|Y_n^\varepsilon \circ \lambda_n - Y^\varepsilon\| + w(Y^\varepsilon, \|\lambda_n - \lambda\|),
\end{aligned}$$

we get by (38)

$$\limsup_{n \to \infty} \|X_n^\varepsilon - X\| \leq \varepsilon \quad \text{a.s.}, \quad \varepsilon > 0. \tag{39}$$

To extend the last result to the case of random characteristics, we note that, in view of Theorem 3.8,

$$(\alpha_n, \gamma_n, \beta_n) \xrightarrow{vd} (\alpha, \gamma, \beta) \quad \text{in } \mathbb{R}^{d+d^2} \times \mathcal{N}(\mathbb{R}^d \setminus \{0\}).$$

By FMP 6.12 we may assume this to hold a.s., and from the proof of the same result we see that the processes $X, X_1, X_2, \ldots$ can be taken to be conditionally independent, given the family of characteristics (α, γ, β) and $(\alpha_n, \gamma_n, \beta_n)$. Then all processes are conditionally exchangeable, given the latter quantities, and Theorem 3.8 shows that $X_n \overset{d}{\to} X$ remains conditionally true. We may now to apply FMP 4.30, as before, to the conditional distributions of X and $X_1, X_2, \ldots$, to obtain some processes J_n^ε and Y_n^ε with suitable conditional distributions, satisfying the appropriate a.s. conditions. To justify the last step, we need to check that the construction in the proof of FMP 4.30 depends measurably on the underlying distributions, which is quite straightforward. This ensures the existence, even for general characteristics, of some processes $X_n^\varepsilon \overset{d}{=} X_n$ satisfying (39).

To eliminate the ε in (39), we may apply Lemma 3.26 to the random variables

$$\xi_{k,n} = \left(\|X_n^{1/k} - X\| - k^{-1} \right) \vee 0, \quad k, n \in \mathbb{N},$$

which are clearly such that $\xi_{k,n} \to 0$ a.s. as $n \to \infty$ for fixed k. By the lemma we may then choose a sequence $k_n \to \infty$ with $\xi_{k_n,n} \to \infty$ a.s., and we note that the processes $\tilde{X}_n = X_n^{1/k_n} \overset{d}{=} X_n$ satisfy

$$\|\tilde{X}_n - X\| \le \xi_{k_n,n} + k_n^{-1} \to 0 \quad \text{a.s.}$$

We proceed to the case of processes X and X_n on $\mathbb{R}_+$. By the result in the finite-interval case, there exist some processes $X_n^k \overset{d}{=} X_n$ satisfying $(X_n^k - X)_k^* \to 0$ a.s. as $n \to \infty$ for fixed $k \in \mathbb{N}$. Using Lemma 3.26 with $\xi_{k,n} = (X_n^k - X)_k^*$, we may next assert the existence of some constants $k_n \to \infty$ such that $\xi_{k_n,n} \to 0$ a.s. Putting $\tilde{X}_n = X_n^{k_n} \overset{d}{=} X_n$, we get a.s., as $n \to \infty$ for fixed $t > 0$,

$$(\tilde{X}_n - X)_t^* \le (X_n^{k_n} - X)_{k_n}^* = \xi_{k_n,n} \to 0,$$

which shows that $\tilde{X}_n \overset{u}{\to} X$ a.s. The case of processes X_n on increasing intervals $[0, u_n] \to \mathbb{R}_+$ is similar, apart from the need to impose an additional restriction $k_n \le u_n$. $\qquad \square$

To examine the local path properties of an exchangeable process, it is often convenient to compare with a suitable Lévy process, for which the corresponding properties are well-known. Our basic comparison takes the form of two general coupling theorems, each of which provides some very precise error estimates. To measure the rate of decrease of the coefficients β_j in (13), we introduce the *index of regularity*

$$\rho_X = \inf\{c \ge 0; \ \textstyle\sum_j |\beta_j|^c < \infty\}.$$

Clearly $\rho_X \in [0, 2]$, and we also note that $\rho_X = 1$ is the borderline case for the jump component of X to have finite variation. If X is a stable Lévy process of index $p \in (0, 2)$, then $\rho_X = p$. In the general case, we may iterate the following construction to obtain mixed Lévy approximations with arbitrarily small error terms.

Theorem 3.27 *(coupling 1) Let X be a real-valued, exchangeable process on $[0,1]$ directed by $(\alpha, \sigma^2, \beta)$. Then $X = Y + Z$ a.s. for some exchangeable process (Y, Z) in $\mathbb{R}^2$, where Y is mixed Lévy with the same directing triple and Z has a directing triple $(\alpha', 0, \beta')$ satisfying $\rho_Z^{-1} \geq \rho_X^{-1} + \frac{1}{2}$ a.s.*

Proof: Let X be given by (13). Since the order of terms is irrelevant, we may choose the index set of j to be $\pm\mathbb{N} = \{\pm 1, \pm 2, \ldots\}$ and assume that

$$\beta_{-1} \leq \beta_{-2} \leq \cdots \leq 0 \leq \cdots \leq \beta_2 \leq \beta_1,$$

where some of the β_j may be 0. Independently of the random processes and variables in (13), we introduce a random pair (ϑ, ξ), where ϑ is $N(0,1)$ and $\xi = \sum_j \delta_{\xi_j}$ is an independent, unit rate Poisson process on $\pm\mathbb{N}$ with points ξ_j, enumerated in increasing order. Now consider the point process $\eta = \sum_j \delta(\tau_j, \beta_{\xi_j})$ on $[0,1] \times \mathbb{R}$, where $\delta(x) = \delta_x$, and for $t \in [0,1]$ define

$$Y_t = \alpha t + \sigma(B_t + \vartheta t) + \int_0^t \int_{x \neq 0} x \, (\eta - \lambda \otimes \beta^1)(ds \, dx), \tag{40}$$

where β^1 denotes the restriction of β to the set $[-1,0) \cup (0,1]$. Noting that $B_t + \vartheta t$ is a standard Brownian motion on $[0,1]$ and η is a Cox process on $[0,1] \times (\mathbb{R} \setminus \{0\})$ directed by $\lambda \otimes \beta$ (cf. FMP 12.3), we see from Theorem 1.19 that Y is a mixed Lévy process on $[0,1]$ directed by the same triple $(\alpha, \sigma^2, \beta)$.

Since Y is exchangeable with diffusion component σB_t and with jumps β_{ξ_j} occurring at times τ_j, we may rewrite (40) in the form

$$Y_t = \tilde{\alpha} t + \sigma B_t + \sum_j \beta_{\xi_j} (1\{\tau_j \leq t\} - t), \quad t \in [0,1],$$

for a suitable random variable $\tilde{\alpha}$. Then

$$Z_t \equiv X_t - Y_t = (\alpha - \tilde{\alpha})t + \sum_j (\beta_j - \beta_{\xi_j})(1\{\tau_j \leq t\} - t),$$

which shows that Z is exchangeable with directing triple $(\alpha', 0, \beta')$, where $\alpha' = \alpha - \tilde{\alpha}$ and $\beta' = \sum_j \beta'_j$ with $\beta'_j \equiv \beta_j - \beta_{\xi_j}$. To see that β' satisfies the required index relation, we may clearly take β to be non-random. We can also assume that $\beta_1 \geq \beta_2 \geq \cdots > 0$, thus restricting the summation in (13) to the index set $\mathbb{N}$.

First suppose that $\rho_X < 2$, and fix any $p \in (\rho_X, 2)$. Choose any constants $q \in (0,1)$ and $a \in (0, \frac{1}{2})$ with

$$0 < q^{-1} - p^{-1} < a < \tfrac{1}{2},$$

and put $b = 1 - a$ and $r = b/a$. Define $m_n = (an)^{1/a}$, and note that

$$\Delta m_n \sim m_n' = (an)^r = m_n^b. \tag{41}$$

Write $x_n = \beta_{m_n}$, where $\beta_t \equiv \beta_{[t]}$. Noting that $k^{-b}(\xi_k - k) \to 0$ a.s. by the Marcinkiewicz–Zygmund law of large numbers (FMP 4.23), and using Hölder's inequality and (41), we get

$$\sum_k |\beta_k'|^q \; \lesssim \; \sum_k (\beta_{k-k^b} - \beta_{k+k^b})^q \le a^r \sum_n n^r |\Delta x_n|^q$$
$$\lesssim \; a^r \left(\sum_n n^r x_{n-1}^{c-1} |\Delta x_n| \right)^q \left(\sum_n n^r x_n^p \right)^{1-q},$$

where $c = 1 + p - p/q \in (0, p)$. Using the monotonicity of x_n, Fubini's theorem, and Hölder's inequality, we get as in FMP 3.4

$$c \sum_n n^r x_{n-1}^{c-1} |\Delta x_n| \; \le \; \sum_n n^r |\Delta x_n^c| \lesssim r \sum_n n^{r-1} x_n^c$$
$$\le \; r \left(\sum_n n^{-d} \right)^{1-c/p} \left(\sum_n n^r x_n^p \right)^{c/p},$$

where $d = (q^{-1} - p^{-1})^{-1} - r > 1$. Combining the preceding estimates and noting that $\sum_n n^{-d} < \infty$ and $\sum_n n^r x_n^p \lesssim \sum_j \beta_j^p < \infty$ by (41), we conclude that $\sum_k |\beta_k'|^q < \infty$. This gives $\rho_Z \le q$, and the assertion follows as we let $q \to (p^{-1} + \frac{1}{2})^{-1}$ and then $p \to \rho_X$.

Now assume instead that $\rho_X = 2$, and let $q \in (1, 2)$ be arbitrary. Writing $\gamma_k = \beta_k^q$, we get by convexity

$$\sum_k |\beta_k'|^q \le \sum_k \left| \beta_k^q - \beta_{\xi_k}^q \right| = \sum_k |\gamma_k - \gamma_{\xi_k}|. \tag{42}$$

Since $\sum_k \gamma_k^{2/q} = \sum_k \beta_k^2 < \infty$, the previous case applies to the constants γ_k with $p = 2/q$ and proves convergence of the last sum in (42). Thus, we have again $\rho_Z \le q$, and as $q \to 1$ we get $\rho_Z^{-1} \ge 1 = \rho_X^{-1} + \frac{1}{2}$, as required. $\qquad \square$

We turn to a different kind of coupling, where the given exchangeable process X is essentially squeezed between two mixed Lévy processes $X_\pm$, with directing random measures $\nu_\pm$ close to the original directing point process β. Since a weak regularity condition will now be imposed on β, it may be necessary, for future applications, first to make a preliminary reduction by means of Theorem 3.27. For convenience, we may assume that X has positive jumps. Write $\log_2 x = \log \log(x \vee e)$.

Theorem 3.28 (*coupling 2*) *Let X be a real-valued, exchangeable process on $[0, 1]$ directed by $(\alpha, \sigma^2, \beta)$, where $\beta_1 \ge \beta_2 \ge \cdots$ and $\rho_X < 2$ a.s. Then $X = Y_\pm \mp Z_\pm$ for some exchangeable processes $(Y, Z)_\pm$ in $\mathbb{R}^2$, where $Y_\pm$ are mixed Lévy and directed by $(\alpha_\pm, \sigma^2, \nu_\pm)$, with*

$$\nu_\pm = \sum_k \left(1 \pm (k^{-1} \log_2 k)^{1/2} \right) \delta_{\beta_k}, \tag{43}$$

and $Z_\pm$ are a.s. non-decreasing apart from finitely many negative jumps. If X is non-decreasing with drift 0, we may choose $Y_\pm$ to have the same properties.

Proof: Independently of all random objects in (13), we introduce a $N(0,1)$ random variable ϑ and some independent Poisson processes $\xi^\pm = \sum_k \delta(\xi_k^\pm)$ on $\mathbb{N}$ with $E\xi^\pm\{n\} = 1 \pm (n^{-1} \log_2 n)^{1/2}$. Put $\beta^\pm = \sum_k \delta(\beta_k^\pm)$ with $\beta_k^\pm = \beta(\xi_k^\pm)$, write $\eta^\pm = \sum_k \delta(\tau_k, \beta_k^\pm)$, and define

$$Y_\pm(t) = \alpha t + \sigma(B_t + \vartheta t) + \int_0^t \int_{x \neq 0} x\,(\eta^\pm - \lambda \otimes \beta^\pm)(ds\,dx). \qquad (44)$$

The processes $Y_\pm$ are clearly mixed Lévy with directing triples $(\alpha, \sigma^2, \nu_\pm)$, and we note that $Z_\pm = \pm(Y_\pm - X)$ are exchangeable with directing triples $(\tilde{\alpha}^\pm, 0, \tilde{\beta}^\pm)$, where $\tilde{\beta}_k^\pm = \pm(\beta_k^\pm - \beta_k)$. We shall prove that $\tilde{\beta}_k^\pm \geq 0$ for all but finitely many k, and also that $\sum_k \tilde{\beta}_k^\pm < \infty$ a.s. Then the $Z_\pm$ are clearly non-decreasing, apart from a linear drift and finitely many negative jumps, and the required conditions become fulfilled, after a suitable adjustment of the drift coefficient in (44). For convenience, we may henceforth assume that β is non-random.

Using (43) and some elementary estimates, we get

$$\pm(E\xi^\pm[1,n] - n) \sim \sum_{k \leq n} (k^{-1} \log_2 k)^{1/2}$$
$$\sim \int_e^n (x^{-1} \log_2 x)^{1/2} dx \sim 2(n \log_2 n)^{1/2}. \qquad (45)$$

This gives in particular $E\xi^\pm[1,n] \sim n$, and so, by the Hartman–Wintner law of the iterated logarithm (FMP 14.8), we have eventually for large n

$$\left| \xi^\pm[1,n] - E\xi^\pm[1,n] \right| \leq (3n \log_2 n)^{1/2}.$$

Combining this with (45), we get for any $b > \frac{1}{2}$ and for large enough n

$$n - n^b \leq \xi^-[1,n] \leq n \leq \xi^+[1,n] \leq n + n^b,$$

Since $k - n \sim \pm n^b$ iff $n - k \sim \mp k^b$, we conclude that, for $b > \frac{1}{2}$ and large enough k,

$$k - k^b \leq \xi_k^+ \leq k \leq \xi_k^- \leq k + k^b,$$

and hence

$$\beta_{k+k^b} \leq \beta_k^- \leq \beta_k \leq \beta_k^+ \leq \beta_{k-k^b}.$$

Since $\rho_X < 2$, we finally see from the previous proof that

$$\sum_k (\beta_k^+ - \beta_k^-) \leq \sum_k (\beta_{k-k^b} - \beta_{k+k^b}) < \infty,$$

whenever b is close enough to $\frac{1}{2}$. $\qquad \square$

To illustrate the use of the last two theorems, we now extend some classical local growth results for Lévy processes to the more general setting of exchangeable processes on $[0,1]$. Similar results were derived directly, by different methods, in Theorem 2.32.

Corollary 3.29 *(local growth 1, Khinchin, Millar, Kallenberg)* *Let X be an exchangeable process on $[0,1]$ directed by $(\alpha, \sigma^2, \beta)$. Then*

(i)
$$\limsup_{t \to 0} \frac{X_t}{(2t \log |\log t|)^{1/2}} = |\sigma| \quad a.s.;$$

(ii) *for any convex function f on $\mathbb{R}_+$ with $f(0) = f'(0) = 0$ such that $f(x^{1/2})$ is concave and for arbitrary $c > 1$, we have*

$$\lim_{t \to 0} \frac{X_t}{f^{-1}(t|\log t|^c)} = 0 \quad a.s. \ on \ \Big\{\sigma = 0, \ \sum_k f(|\beta_k|) < \infty\Big\}.$$

Proof: (i) Let $X = Y + Z$ be the decomposition in Theorem 3.27. Then the statement holds for the mixed Lévy process Y, and it remains to prove the result for Z, which reduces the discussion to the case where $\rho_X < 2$ and $\sigma = 0$. Next we may utilize the decomposition $X = Y_+ - Z_-$ of Theorem 3.28. Here the result holds for the mixed Lévy process Y_+, and since $X \leq Y_+$ a.s. near 0, the formula remains true for X.

(ii) For convenience, we may assume that β is non-random, restricted to $(0, \infty)$, and such that $\beta f < \infty$. Consider the decomposition $X = Y + Z$ of Theorem 3.27, and note that the asserted formula holds for Y. Letting β' and β'' denote the directing point processes of Y and Z, respectively, we note that even $\beta' f < \infty$ a.s. since $E\beta' f = \beta f < \infty$. The relation $\beta_j'' = \beta_j - \beta_j'$ yields $|\beta_j''| \leq \beta_j \vee \beta_j'$, and so

$$\sum_j f(|\beta_j''|) \leq \sum_j f(\beta_j) + \sum_j f(\beta_j') = \beta f + \beta' f < \infty.$$

This reduces the discussion to the case where $\rho_X < 2$.

Next we use Theorem 3.28 to approximate X by Lévy processes $Y_\pm$ such that $Y_- \leq X \leq Y_+$ near 0. The associated Lévy measures $\nu_\pm$ satisfy $\nu_\pm \leq 2\beta$ near the origin, and so the condition $\beta f < \infty$ implies $\nu_\pm f < \infty$. The result then holds for both Y_+ and Y_-, and so it remains true for X. □

We may also extend some growth results for subordinators to any non-decreasing, exchangeable processes on $[0,1]$.

Corollary 3.30 *(local growth 2, Fristedt, Kallenberg)* *Let X be a non-decreasing, exchangeable process on $[0,1]$ with directing point process β and drift 0. Then a.s.*

(i) *for any concave, increasing function f on $\mathbb{R}_+$ with $f(0) = 0$,*

$$\lim_{t \to 0} \frac{X_t}{f^{-1}(t)} = 0 \quad iff \quad \beta f < \infty;$$

(ii) *for any function $f \geq 0$ on $\mathbb{R}_+$ such that $f(t)/t$ is increasing,*

$$\lim_{t \to 0} \frac{X_t}{f(t)} = 0 \quad iff \quad \int_0^\infty \beta(f(t), \infty) \, dt < \infty.$$

Proof: The corresponding results for subordinators apply to the approximating processes $Y_\pm$ in Theorem 3.28, with similar criteria in terms of the associated directing random measures $\nu_\pm$. Now the latter conditions are equivalent to those for β, since $\frac{1}{2}\beta \le \nu_\pm \le 2\beta$ near the origin. $\qquad\square$

3.7 Sub-sequence Principles

For motivation, we begin with a simple result for exchangeable sequences. If ξ is an infinite, exchangeable sequence directed by ν, then we know that $P\{\xi \in \cdot\} = E\nu^\infty$. The next result shows that the same relation holds asymptotically in a suitably conditional sense.

Proposition 3.31 *(stable convergence, Rényi and Révész) Let $\xi = (\xi_n)$ be an infinite, $\mathcal{F}$-exchangeable sequence directed by ν, taking values in a Polish space S. Then*

  (i) $E[\theta_n\xi \in \cdot|\mathcal{F}_n] \xrightarrow{w} \nu^\infty$ *a.s.,*
  (ii) $E[\eta; \theta_n\xi \in \cdot] \xrightarrow{w} E\eta\nu^\infty$ *for all $\eta \in L^1$.*

Proof: (i) For any bounded, measurable function f on S^∞, we get a.s. by Proposition 2.1 and martingale convergence

$$
\begin{aligned}
E[f(\theta_n\xi)|\mathcal{F}_n] &= E[E[f(\theta_n\xi)|\mathcal{F}_n, \nu]|\mathcal{F}_n] \\
&= E[\nu^\infty f|\mathcal{F}_n] \to E[\nu^\infty f|\mathcal{F}_\infty] = \nu^\infty f.
\end{aligned}
$$

If S is Polish, then so is S^∞, and we can apply this result to a convergence-determining sequence of bounded, continuous functions $f_1, f_2, \ldots$ on S^∞, which yields the asserted statement.

  (ii) If η is $\mathcal{F}_n$-measurable, then for any bounded, measurable function f on S^∞

$$
Ef(\theta_n\xi)\eta = EE[f(\theta_n\xi)|\nu, \mathcal{F}_n]\eta = E(\nu^\infty f)\eta.
$$

For general $\eta \in L^1$, we get by martingale convergence (FMP 7.23)

$$
\begin{aligned}
|Ef(\theta_n\xi)\eta - E(\nu^\infty f)\eta| &= |E(f(\theta_n\xi) - \nu^\infty f)E[\eta|\mathcal{F}_\infty]| \\
&\le |Ef(\theta_n\xi)(E[\eta|\mathcal{F}_\infty] - E[\eta|\mathcal{F}_n])| \\
&\quad + |E(f(\theta_n\xi) - \nu^\infty f)E[\eta|\mathcal{F}_n]| \\
&\quad + |E(\nu^\infty f)(E[\eta|\mathcal{F}_n] - E[\eta|\mathcal{F}_\infty])| \\
&\le 2\|f\|\, \|E[\eta|\mathcal{F}_n] - E[\eta|\mathcal{F}_\infty]\|_1 \to 0,
\end{aligned}
$$

which shows that $Ef(\theta_n\xi)\eta \to E(\nu^\infty f)\eta$. As before, it remains to apply this to a convergence-determining sequence $f_1, f_2, \ldots$ of bounded continuous functions $f_1, f_2, \ldots$ on S^∞. $\qquad\square$

In particular, part (ii) shows that the sequence $\xi_1, \xi_2, \ldots$ converges stably, in the sense of Rényi, with mixing random measure ν. By a similar argument, most classical limit theorems for i.i.d. sequences of random variables carry over, in a mixed form and in the sense of stable convergence, to arbitrary exchangeable sequences.

We now turn our attention to more general random sequences ξ in a Polish space. Our aim is to show that any tight sequence ξ has a sub-sequence that is approximately exchangeable. We begin with an approximation in the sense of stable convergence. For suitable sequences $\xi = (\xi_n)$ and $p = (p_n)$, we write $\xi \circ p = (\xi_{p_1}, \xi_{p_2}, \ldots)$.

Theorem 3.32 *(weak sub-sequence principle, Dacunha-Castelle, Aldous) For any tight random sequence $\xi = (\xi_n)$ in a Polish space S, there exist a sub-sequence p of $\mathbb{N}$ and a random probability measure ν on S such that*

$$E[\eta; \theta_n(\xi \circ p) \in \cdot] \overset{w}{\to} E\eta\nu^\infty, \quad \eta \in L^1. \tag{46}$$

In particular, the shifted sub-sequence $\theta_n(\xi \circ p)$ converges in distribution to an exchangeable sequence directed by ν. Our proof of Theorem 3.32 is based on two lemmas. First we need to prove convergence of the conditional distributions along a sub-sequence, which requires a simplifying assumption.

Lemma 3.33 *(compactness) Let $\xi = (\xi_n)$ be a tight random sequence in a Polish space S, such that each ξ_n takes only finitely many values. Then there exist a sub-sequence $\xi \circ p = (\xi_{p_n})$ of ξ with induced filtration $\mathcal{F} = (\mathcal{F}_n)$ and a random probability measure ν on S such that*

$$P[(\xi \circ p)_{n+1} \in \cdot | \mathcal{F}_n] \overset{w}{\to} \nu \quad a.s.$$

Proof: Let $f_1, f_2, \ldots$ be a convergence-determining sequence of bounded, continuous functions on S. By the weak compactness of the unit ball in L^2, combined with a diagonal argument, we can choose a sub-sequence $\zeta_1, \zeta_2, \ldots$ of ξ such that

$$E[f_j(\zeta_n); A] \to E[\alpha_j; A], \quad A \in \mathcal{A}, \ j \in \mathbb{N}. \tag{47}$$

for some random variables $\alpha_1, \alpha_2, \ldots \in L^2$. Passing to a further sub-sequence, if necessary, we can ensure that

$$|E[f_j(\zeta_{n+1})|A] - E[\alpha_j|A]| \le 2^{-n}, \quad A \in \mathcal{A}_n, \ j \le n, \ n \in \mathbb{N}, \tag{48}$$

where $\mathcal{A}_n$ denotes the set of atoms A in $\mathcal{F}_n = \sigma(\zeta_1, \ldots, \zeta_n)$ with $PA > 0$. We shall prove that the latter sequence $\zeta = (\zeta_n)$ has the required property.

The sequence $P[\zeta_n \in \cdot | A]$ is again tight for every $A \in \bigcup_m \mathcal{A}_m$, and so by Prohorov's theorem and a diagonal argument we have $P[\zeta_n \in \cdot | A] \overset{w}{\to} \mu^A$ along a sub-sequence for some probability measures μ^A on S. Comparing with (47), we conclude that

$$\mu^A f_j = E[\alpha_j | A], \quad A \in \bigcup_n \mathcal{A}_n, \ j \in \mathbb{N}. \tag{49}$$

Next we introduce the random probability measures

$$\mu_n = \sum\nolimits_{A \in \mathcal{A}_n} 1_A \mu^A, \quad n \in \mathbb{N}, \tag{50}$$

and note that the random variables $\mu_n f$ form a bounded $\mathcal{F}$-martingale for every bounded, measurable function f on S. By martingale convergence we obtain $\mu_n f_j \to \beta_j$ a.s. for some random variables β_j.

The probability measures $E\mu_n$ are independent of n and are therefore trivially tight. Furthermore, we see from Doob's inequality that, for any compact $K \subset S$,

$$E(\sup_n \mu_n K^c)^2 \lesssim \sup_n E(\mu_n K^c)^2 \leq \sup_n E\mu_n K^c,$$

which tends to 0 as $K \uparrow S$. By dominated convergence we get $\sup_n \mu_n K^c \to 0$ a.s., which shows that the sequence (μ_n) is a.s. tight. But then, with probability 1, we have convergence $\mu_n \xrightarrow{w} \mu$ along a sub-sequence, where the limiting probability measure μ may depend on $\omega \in \Omega$. Comparing this with the previously noted convergence $\mu_n f_j \to \beta_j$ a.s., we obtain $\mu_n f_j \to \mu f_j$ a.s. along $\mathbb{N}$ for every $j \in \mathbb{N}$, and since the f_j are convergence-determining, we conclude that $\mu_n \xrightarrow{w} \mu$ a.s. In particular, we see from FMP 1.10 that μ has a measurable version ν.

Now define

$$\nu_n = P[\zeta_{n+1} \in \cdot | \mathcal{F}_n] = \sum\nolimits_{A \in \mathcal{A}_n} 1_A P[\zeta_{n+1} \in \cdot | A], \quad n \in \mathbb{N},$$

and conclude from (48), (49), and (50) that

$$|\nu_n f_j - \mu_n f_j| \leq 2^{-n} \quad \text{a.s.}, \quad j \leq n.$$

For any $j \in \mathbb{N}$, we get a.s.

$$|\nu_n f_j - \nu f_j| \leq |\nu_n f_j - \mu_n f_j| + |\mu_n f_j - \nu f_j| \to 0,$$

which implies $\nu_n \xrightarrow{w} \nu$ a.s. $\square$

Our next step is to extend the previous convergence to the infinite product space S^∞. Here we write $\xrightarrow{wP}$ for convergence in probability with respect to the weak topology.

Lemma 3.34 *(iteration) Let $\xi = (\xi_n)$ be an $\mathcal{F}$-adapted random sequence in a Polish space S, satisfying $P[\xi_{n+1} \in \cdot | \mathcal{F}_n] \xrightarrow{wP} \nu$ for some random probability measure ν on S. Then*

$$E[\eta; \, \theta_n \xi \in \cdot] \xrightarrow{w} E\eta\nu^\infty, \quad \eta \in L^1.$$

Proof: Consider in S an exchangeable sequence $\zeta = (\zeta_n)$ directed by ν and such that $\zeta \perp\!\!\!\perp_\nu \mathcal{A}$. We claim that

$$(\eta, \xi_{n+1}, \ldots, \xi_{n+k}) \xrightarrow{d} (\eta, \zeta_1, \ldots, \zeta_k), \quad k \in \mathbb{Z}_+, \tag{51}$$

for any $\mathcal{F}_\infty$-measurable random element η in a Polish space T. This is vacuously true for $k = 0$. Now assume the statement to be true for a given $k \in \mathbb{Z}_+$. Proceeding by induction, suppose that η is $\mathcal{F}_m$-measurable for some $m \in \mathbb{N}$, and fix a bounded, continuous function $f \geq 0$ on $T \times S^{k+1}$. By FMP 4.27 we note that the mapping

$$(y, x_1, \ldots, x_k, \mu) \mapsto \mu f(y, x_1, \ldots, x_k, \cdot) \tag{52}$$

is again continuous from $T \times S^k \times \mathcal{M}_1(S)$ to $\mathbb{R}_+$. Furthermore, noting that ν is $\mathcal{F}_\infty$-measurable, we see from the induction hypothesis that

$$(\eta, \xi_{n+1}, \ldots, \xi_{n+k}, \nu) \xrightarrow{d} (\eta, \zeta_1, \ldots, \zeta_k, \nu).$$

Since $\nu_n \equiv P[\xi_{n+1} \in \cdot | \mathcal{F}_n] \xrightarrow{wP} \nu$, we may use FMP 4.28 to conclude that

$$(\eta, \xi_{n+1}, \ldots, \xi_{n+k}, \nu_{n+k}) \xrightarrow{d} (\eta, \zeta_1, \ldots, \zeta_k, \nu).$$

Combining this with the continuity in (52) and using the disintegration theorem twice, we get for $n > m$ and as $n \to \infty$

$$\begin{aligned}
Ef(\eta, \xi_{n+1}, \ldots, \xi_{n+k+1}) &= E\nu_{n+k} f(\eta, \xi_{n+1}, \ldots, \xi_{n+k}, \cdot) \\
&\to E\nu f(\eta, \zeta_1, \ldots, \zeta_k, \cdot) \\
&= Ef(\eta, \zeta_1, \ldots, \zeta_{k+1}).
\end{aligned}$$

This proves (51) for $k + 1$ when η is $\mathcal{F}_m$-measurable.

To extend the result to any $\mathcal{F}_\infty$-measurable random element η in T, we may choose some $\mathcal{F}_m$-measurable random elements η_m in T such that $\eta_m \xrightarrow{P} \eta$. This is obvious when η is simple, since for any finite partition $A_1, \ldots, A_r \in \mathcal{F}_\infty$ of Ω there exist some approximating partitions $A_1^m, \ldots, A_r^m \in \mathcal{F}_m$ such that $P(A_i \Delta A_i^m) \to 0$ for all $i \leq r$ (FMP 3.16). The associated approximations η_m of η then satisfy $\eta_m \xrightarrow{P} \eta$. It remains to note that any $\mathcal{F}_\infty$-measurable random element η in T is tight and hence can be approximated in probability by some simple, $\mathcal{F}_\infty$-measurable random elements $\eta^1, \eta^2, \ldots$. The required extension now follows by FMP 4.28. This completes the induction and proves (51) for any $\mathcal{F}_\infty$-measurable random element η.

In particular, every bounded, $\mathcal{F}_\infty$-measurable random variable η satisfies

$$E[\eta; \, \theta_n \xi \in \cdot] \xrightarrow{w} E[\eta; \, \zeta \in \cdot] = E\eta\nu^\infty,$$

which extends by a simple approximation to any $\eta \in L^1(\mathcal{F}_\infty)$. Since ξ and ν are a.s. $\mathcal{F}_\infty$-measurable, we may finally extend the result to arbitrary $\eta \in L^1(\mathcal{A})$, by writing

$$\begin{aligned}
E[\eta; \, \theta_n \xi \in \cdot] &= E[E[\eta | \mathcal{F}_\infty]; \, \theta_n \xi \in \cdot] \\
&\xrightarrow{w} EE[\eta | \mathcal{F}_\infty]\nu^\infty = E\eta\nu^\infty.
\end{aligned} \qquad \square$$

Proof of Theorem 3.32: Fix any metric ρ in S. Since the ξ_n are individually tight, we may choose some random elements $\zeta_1, \zeta_2, \ldots$ in S, each of which takes only finitely many values, such that

$$E[\rho(\xi_n, \zeta_n) \wedge 1] \leq 2^{-n}, \quad n \in \mathbb{N}.$$

Then Fubini's theorem yields $E \sum_n (\rho(\xi_n, \zeta_n) \wedge 1) \leq 1$, and so $\rho(\xi_n, \zeta_n) \to 0$ a.s. By Lemma 3.33 we may next choose a sub-sequence $\zeta \circ p = (\zeta_{p_1}, \zeta_{p_2}, \ldots)$ with induced filtration $\mathcal{F} = (\mathcal{F}_n)$ such that

$$P[(\zeta \circ p)_{n+1} \in \cdot | \mathcal{F}_n] \xrightarrow{w} \nu \text{ a.s.}$$

Then Lemma 3.34 yields

$$E[\eta; \, \theta_n(\zeta \circ p) \in \cdot] \to E\eta\nu^\infty, \quad \eta \in L^1,$$

and (46) follows by means of FMP 4.28. $\square$

Choosing $\zeta_1, \zeta_2, \ldots$ to be conditionally i.i.d. given $\mathcal{A}$ with distribution ν, we note that the statement of Theorem 3.32 is equivalent to $\theta_n(\xi \circ p) \to \zeta$, in the sense of stable convergence. A sequence ξ satisfying the condition $E[\eta; \, \theta_n \xi \in \cdot] \xrightarrow{w} E\eta\nu^\infty$ of Theorem 3.32 is said to be *determining with mixing random measure* ν. For determining sequences it is clear that the mixing measure ν is a.s. unique.

Proceeding to a further sub-sequence, if necessary, we may strengthen the conclusion to approximation in a suitable point-wise sense.

Theorem 3.35 *(strong sub-sequence principle, Berkes and Péter) Let* $\xi = (\xi_n)$ *be a tight random sequence in a separable, complete metric space* (S, ρ), *and fix any* $\varepsilon > 0$. *Then there exist a sub-sequence p of* $\mathbb{N}$ *and an exchangeable sequence* $\zeta = (\zeta_n)$ *in S such that*

$$E[\rho(\xi_{p_n}, \zeta_n) \wedge 1] \leq \varepsilon, \quad n \in \mathbb{N}. \tag{53}$$

Here, as always, we assume the underlying probability space to be rich enough to support an independent randomization variable. Using approximations of this type, it can be shown that any classical limit theorem for i.i.d. random elements of S remains true, in a mixed form, for a suitable sub-sequence of any tight sequence ξ in S. The details of this extension lie outside the scope of the present exposition.

To prove the stated theorem, we need again a couple of lemmas. For the first one, we write $\|\mu - \nu\| = \sup_A |\mu A - \nu A|$.

Lemma 3.36 *(sequential coupling) Let* $\xi = (\xi_n)$ *be an $\mathcal{F}$-adapted random sequence in a Borel space S, and put* $\nu_n = P[\xi_n \in \cdot | \mathcal{F}_{n-1}]$, $n \in \mathbb{N}$. *Fix any probability measures* $\mu_1, \mu_2, \ldots$ *on S. Then there exist some independent random elements ζ_n in S with distributions μ_n, $n \in \mathbb{N}$, such that*

$$P\{\zeta_n \neq \xi_n\} = E\|\mu_n - \nu_n\|, \quad n \in \mathbb{N}.$$

Proof: We may assume that $S = [0, 1]$. Introduce some i.i.d. $U(0, 1)$ random variables $\vartheta_1, \vartheta_2, \ldots$ independent of $\mathcal{F}_\infty$, put $\mathcal{G}_n = \mathcal{F}_n \vee \sigma(\vartheta_1, \ldots, \vartheta_n)$, and note that $\nu_n = P[\xi_n \in \cdot | \mathcal{G}_{n-1}]$ for all n. It is enough to choose the ζ_n to be $\mathcal{G}_n$-measurable and independent of $\mathcal{G}_{n-1}$ with distributions μ_n. Proceeding recursively, we need to show that, for any random variable ξ, σ-field $\mathcal{F}$, distribution μ, and $U(0, 1)$ random variable $\vartheta \perp\!\!\!\perp (\xi, \mathcal{F})$, there exists a random variable $\zeta \perp\!\!\!\perp \mathcal{F}$ with distribution μ, measurable with respect to $(\xi, \mathcal{F}, \vartheta)$, such that

$$P\{\zeta \neq \xi\} = E\|\mu - P[\xi \in \cdot | \mathcal{F}]\|. \tag{54}$$

Assuming first that $\mathcal{F}$ is the trivial σ-field and putting $\nu = \mathcal{L}(\xi)$, we need to construct ζ with distribution μ such that $P\{\zeta \neq \xi\} = \|\mu - \nu\|$. Then consider the Lebesgue decomposition $\mu = p \cdot \nu + \mu_s$ of μ with respect to ν (FMP 2.10), where $\mu_s \perp \nu$ and $p \geq 0$, and put $\zeta = \xi$ when $\vartheta \leq p(\xi)$. This gives the partial distribution $(p \wedge 1) \cdot \nu$, and we can easily complete the construction such that ζ gets distribution μ. By FMP 7.26, the density function $p(x)$ can be chosen to be jointly measurable in μ, ν, and x. Then for fixed μ, any reasonable method of construction yields ζ as a measurable function f_μ of (ξ, ϑ, ν).

Turning to the general case, we may still define $\zeta = f_\mu(\xi, \vartheta, \nu)$, but now with $\nu = P[\xi \in \cdot | \mathcal{F}]$. Noting that $P[(\xi, \vartheta) \in \cdot | \mathcal{F}] = \nu \otimes \lambda$ and using the disintegration theorem (FMP 6.4), we get a.s.

$$
\begin{aligned}
P[\zeta \in \cdot | \mathcal{F}] &= P[f_\mu(\xi, \vartheta, \nu) \in \cdot | \mathcal{F}] \\
&= \iint 1\{f_\mu(x, t, \nu) \in \cdot\}\, \nu(dx)\, dt = \mu, \tag{55} \\
P[\zeta \neq \xi | \mathcal{F}] &= P[f_\mu(\xi, \vartheta, \nu) \neq \xi | \mathcal{F}] \\
&= \iint 1\{f_\mu(x, t, \nu) \neq x\}\, \nu(dx)\, dt = \|\mu - \nu\|. \tag{56}
\end{aligned}
$$

Here (55) shows that $\zeta \perp\!\!\!\perp \mathcal{F}$ with distribution μ, and we get (54) by taking expected values in (56).   □

The main result will first be proved in a simple special case.

Lemma 3.37 *(case of finite state space) When S is finite, the conclusion of Theorem 3.35 holds with (53) replaced by*

$$P\{\xi_{p_n} \neq \zeta_n\} \leq \varepsilon, \quad n \in \mathbb{N}. \tag{57}$$

Proof: By Theorem 3.32 we may assume that ξ is determining with mixing measure ν. Fix any $\varepsilon > 0$. Since S is finite, we may choose a measurable partition of $\mathcal{M}_1(S)$ into finitely many sets $M_1, \ldots, M_m$, each of diameter $< \varepsilon$ with respect to the norm $\|\cdot\|$. For every $k \leq m$ we fix a measure $\mu_k \in M_k$. When the set $\Omega_k = \{\nu \in M_k\}$ has positive probability, we may introduce the conditional probability measure $P_k = P[\cdot | \Omega_k]$ on Ω_k.

We shall construct a sub-sequence $p = (p_n)$ of $\mathbb{N}$ such that

$$\|P_k[\xi_{p_n} \in \cdot|\mathcal{F}_{n-1}] - \mu_k\| \leq \varepsilon \text{ a.s.}, \quad k \leq m, \ n \in \mathbb{N}, \tag{58}$$

where $\mathcal{F}_n = \sigma(\xi_{p_1}, \ldots, \xi_{p_n})$ for all n. Then by Lemma 3.36 there exist some random elements $\zeta_1, \zeta_2, \ldots$, i.i.d. under each P_k with distribution μ_k, such that

$$P_k\{\xi_{p_n} \neq \zeta_n\} \leq \varepsilon, \quad k \leq m, \ n \in \mathbb{N}. \tag{59}$$

The sequence $\zeta = (\zeta_n)$ is clearly exchangeable, and (57) follows from (59) by averaging over k.

Suppose that $p_1, \ldots, p_n$ have already been chosen to satisfy (58). For any $k \leq m$ and $A \in \mathcal{F}_n$ with $P_k A > 0$, we get by Theorem 3.32 as $r \to \infty$

$$
\begin{aligned}
\|P_k[\xi_r \in \cdot|A] - \mu_k\| &\leq \|P_k[\xi_r \in \cdot|A] - E_k[\nu|A]\| + \|E_k[\nu|A] - \mu_k\| \\
&\to \|E_k[\nu|A] - \mu_k\| \leq E_k[\|\nu - \mu_k\| \,|A] < \varepsilon.
\end{aligned}
$$

Since $\mathcal{F}_n$ is finite, we may then choose $r > p_n$ so large that

$$\|P_k[\xi_r \in \cdot|\mathcal{F}_n] - \mu_k\| \leq \varepsilon \text{ a.s.}, \quad k \leq m,$$

which yields (58) for $n + 1$ with $p_{n+1} = r$. □

Proof of Theorem 3.35: By Theorem 3.32 we may assume that $\xi = (\xi_n)$ is determining with mixing measure ν. We may also take $\rho \leq 1$. Fix any $\varepsilon > 0$, and put $\varepsilon' = \varepsilon/3$. Since ξ is tight, we may choose a compact set $K \subset S$ such that

$$P\{\xi_n \notin K\} \leq \varepsilon', \quad n \in \mathbb{N}. \tag{60}$$

Next we may cover K by finitely many open balls B_k of diameter $\leq \varepsilon'$ such that $E\nu\partial B_k = 0$. Taking successive differences, we may construct a partition of $B = \bigcup_k B_k$ into finitely many $E\nu$-continuity sets S_k of diameter $\leq \varepsilon'$. To this we may add the set $S_0 = B^c$, to obtain a finite partition of S. Fixing a point $s_k \in S_k$ for every k, we define a mapping f on S by taking $f(s) = s_k$ when $s \in S_k$.

Since $E\nu\partial S_k = 0$ for all k, the sequence $(f(\xi_n))$ is again determining with mixing measure $\nu \circ f^{-1}$. By Lemma 3.37 we may then choose a sub-sequence p of $\mathbb{N}$ and an exchangeable sequence $\zeta = (\zeta_n)$ such that

$$P\{f(\xi_{p_n}) \neq \zeta_n\} \leq \varepsilon', \quad n \in \mathbb{N}. \tag{61}$$

Write $\hat{s} = f(s)$, and note that $\rho(s, \hat{s}) \leq \varepsilon'$ for all $s \in K$. Using (60) and (61), we obtain

$$
\begin{aligned}
E\rho(\xi_{p_n}, \zeta_n) &\leq E\rho(\xi_{p_n}, \hat{\xi}_{p_n}) + E\rho(\hat{\xi}_{p_n}, \zeta_n) \\
&\leq P\{\xi_{p_n} \notin K\} + E[\rho(\xi_{p_n}, \hat{\xi}_{p_n}); \xi_{p_n} \in K] + P\{\hat{\xi}_{p_n} \neq \zeta_n\} \\
&\leq 3\varepsilon' = \varepsilon. \qquad\qquad\qquad\qquad\qquad\qquad\qquad\qquad \square
\end{aligned}
$$

Chapter 4

Predictable Sampling and Mapping

The core of this chapter consists of an account of the optional skipping and predictable sampling theorems, along with their continuous-time counterparts, the predictable contraction and mapping theorems. Those topics also lead us naturally into the area of random time change.

The relatively elementary discrete-time theory is treated already in Section 4.1. After an interlude about Gauss and Poisson reduction of local martingales and marked point processes in Section 4.2, we consider in Section 4.3 the predictable mapping theorem for exchangeable processes on $[0, 1]$ or $\mathbb{R}_+$, and then in Section 4.4 the predictable contraction theorem for contractable processes on $[0, 1]$. The stronger results attainable for Brownian and stable processes are treated in Section 4.5. In particular, we show how stochastic integrals with respect to stable Lévy processes can be represented in terms of time-changed versions of the original processes.

The final section 4.6 deals with predictable mapping of optional times. As a main result we show, under suitable regularity conditions, how a collection of marked, optional times can be reduced to independent $U(0, 1)$ random variables through an appropriate family of predictable transformations.

4.1 Skipping and Sampling

Our first aim is to show that the defining property of a finite or infinite, contractable sequence extends to certain randomly selected sub-sequences. Recall that a discrete random time τ is said to be $\mathcal{F}$-*predictable* for a given filtration $\mathcal{F}$ if $\tau - 1$ is optional or a stopping time for $\mathcal{F}$. This is equivalent to the condition $\{\tau = k\} \in \mathcal{F}_{k-1}$ for every k.

Proposition 4.1 *(optional skipping, Doob, Kallenberg)* Let $\xi = (\xi_j)$ be a finite or infinite, $\mathcal{F}$-contractable random sequence in a measurable space S, and let $\tau_1 < \cdots < \tau_m$ be $\mathcal{F}$-predictable times in the index set of ξ. Then

$$(\xi_{\tau_1}, \ldots, \xi_{\tau_m}) \overset{d}{=} (\xi_1, \ldots, \xi_m). \tag{1}$$

Proof: By a suitable truncation, we may assume that the index set I is finite, say $I = \{1, \ldots, n\}$. We proceed by induction on $m \leq n$, starting with the vacuous statement for $m = 0$. Assuming (1) to be true for fewer than m elements, we turn to the case of m predictable times $\tau_1 < \cdots < \tau_m$. Letting $f: S^m \to \mathbb{R}_+$ be measurable, we get for any $k \leq n - m + 1$

$$
\begin{aligned}
E[f(\xi_{\tau_1}, \ldots, \xi_{\tau_m}); \tau_1 = k] &= E[f(\xi_k, \xi_{\tau_2}, \ldots, \xi_{\tau_m}); \tau_1 = k] \\
&= E[f(\xi_k, \xi_{k+1}, \ldots, \xi_{k+m-1}); \tau_1 = k] \\
&= E[f(\xi_{n-m+1}, \ldots, \xi_n); \tau_1 = k],
\end{aligned}
$$

where the second equality holds by the induction hypothesis, applied to the $\mathcal{F}$-contractable sequence of triples $(1\{\tau_1 = k\}, \xi_k, \xi_j)$, $j = k + 1, \ldots, n$, and the predictable times $\tau_2, \ldots, \tau_m$, and the last equality holds since $\theta_{k-1}\xi$ is contractable over $\mathcal{F}_{k-1}$. Summing over $k = 1, \ldots, n - m + 1$ and using the contractability of ξ, we obtain

$$
(\xi_{\tau_1}, \ldots, \xi_{\tau_m}) \stackrel{d}{=} (\xi_{n-m+1}, \ldots, \xi_n) \stackrel{d}{=} (\xi_1, \ldots, \xi_m),
$$

as required. This completes the induction. $\square$

The monotonicity assumption of the last result can be dropped when ξ is exchangeable, which holds in particular for infinite sequences.

Theorem 4.2 *(predictable sampling) Let $\xi = (\xi_k)$ be a finite or infinite, $\mathcal{F}$-exchangeable random sequence in a measurable space S. Then (1) holds for any a.s. distinct, $\mathcal{F}$-predictable times $\tau_1, \ldots, \tau_m$ in the index set of ξ.*

Note that we are no longer demanding the τ_k to be increasing, nor do we require ξ to be infinite.

Proof: First suppose that ξ has index set $I = \{1, \ldots, n\}$ and $m = n$. Then the τ_k form a random permutation of I, and we note that the inverse permutation $\alpha_1, \ldots, \alpha_n$ is *predictable*, in the sense that

$$
\{\alpha_j = k\} = \{\tau_k = j\} \in \mathcal{F}_{j-1}, \quad j, k \in \{1, \ldots, n\}.
$$

For every $m \in \{0, \ldots, n\}$, we introduce the unique $\mathcal{F}_{m-1}$-measurable random permutation $\alpha_1^m, \ldots, \alpha_n^m$ of $1, \ldots, n$ satisfying

$$
(\alpha_1^m, \ldots, \alpha_m^m) = (\alpha_1, \ldots, \alpha_m), \qquad \alpha_{m+1}^m < \cdots < \alpha_n^m. \tag{2}
$$

Fix any measurable functions $f_1, \ldots, f_n \geq 0$ on S, and let $0 \leq m < n$. Using the first relation in (2), the $\mathcal{F}_m$-measurability of (α_j^{m+1}) and (α_j^m), the $\mathcal{F}$-exchangeability of ξ, and the disintegration theorem, we obtain

$$
\begin{aligned}
E\prod_k f_{\alpha_k^{m+1}}(\xi_k) &= E\,E^{\mathcal{F}_m}\prod_k f_{\alpha_k^{m+1}}(\xi_k) \\
&= E\prod_{k \leq m} f_{\alpha_k}(\xi_k)\,E^{\mathcal{F}_m}\prod_{k > m} f_{\alpha_k^{m+1}}(\xi_k) \\
&= E\prod_{k \leq m} f_{\alpha_k}(\xi_k)\,E^{\mathcal{F}_m}\prod_{k > m} f_{\alpha_k^m}(\xi_k) \\
&= E\,E^{\mathcal{F}_m}\prod_k f_{\alpha_k^m}(\xi_k) = E\prod_k f_{\alpha_k^m}(\xi_k),
\end{aligned}
$$

where $E^{\mathcal{F}} = E[\cdot|\mathcal{F}]$. Summing over m and noting that $\alpha_k^n = \alpha_k$ and $\alpha_k^0 = k$, we get

$$E \prod_k f_k(\xi_{\tau_k}) = E \prod_k f_{\alpha_k}(\xi_k) = E \prod_k f_k(\xi_k),$$

and (1) follows.

Next suppose that $I = \{1, \dots, n\}$ with $n > m$. We may then extend the sequence (τ_k) to a random permutation of I such that $\tau_{m+1} < \cdots < \tau_n$. For any $r > m$ and $k \le n$ we have

$$\{\tau_r \le k\} = \left\{ \sum_{i \le m} 1\{\tau_i > k\} \ge r - k \right\} \in \mathcal{F}_{k-1},$$

which shows that $\tau_{m+1}, \dots, \tau_n$ are again predictable. Hence, the result in the previous case yields $(\xi_{\tau_1}, \dots, \xi_{\tau_n}) \overset{d}{=} \xi$, and (1) follows.

Finally suppose that $I = \mathbb{N}$. Then for every $n \in \mathbb{N}$, we introduce the random times

$$\tau_j^n = \tau_j 1\{\tau_j \le n\} + (n + j)1\{\tau_j > n\}, \quad j \le m,$$

which are bounded by $m + n$ and also predictable, since

$$
\begin{aligned}
\{\tau_j^n \le k\} &= \{\tau_j \le k\} \in \mathcal{F}_{k-1}, \quad k \le n, \\
\{\tau_j^n = n + j\} &= \{\tau_j > n\} \in \mathcal{F}_{n-1} \subset \mathcal{F}_{n+j-1}.
\end{aligned}
$$

Using the result in the previous case and letting $n \to \infty$, we obtain

$$
\begin{aligned}
\|\mathcal{L}(\xi_1, \dots, \xi_m) - \mathcal{L}(\xi_{\tau_1}, \dots, \xi_{\tau_m})\| \\
= \left\| \mathcal{L}(\xi_{\tau_1^n}, \dots, \xi_{\tau_m^n}) - \mathcal{L}(\xi_{\tau_1}, \dots, \xi_{\tau_m}) \right\| \\
\le \|\mathcal{L}(\tau_1^n, \dots, \tau_m^n) - \mathcal{L}(\tau_1, \dots, \tau_m)\| \\
\le P \bigcup_{j \le m} \{\tau_j > n\} \le \sum_{j \le m} P\{\tau_j > n\} \to 0,
\end{aligned}
$$

and the assertion follows since the left-hand side is independent of n. $\qquad\square$

For an interesting application of the last result, we consider a celebrated identity from fluctuation theory.

Corollary 4.3 (*positivity and maximum, Sparre-Andersen*) *Let* $\xi_1, \dots, \xi_n$ *be exchangeable random variables, and put* $S_k = \sum_{j \le k} \xi_j$ *and* $M = \max_{k \ge 0} S_k$. *Then*

$$\sum_{k \le n} 1\{S_k > 0\} \overset{d}{=} \min\{k \ge 0; \; S_k = M\}.$$

Proof: The random variables $\tilde{\xi}_k = \xi_{n-k+1}$, $1 \le k \le n$, clearly remain exchangeable for the filtration

$$\mathcal{F}_k = \sigma\{S_n, \tilde{\xi}_1, \dots, \tilde{\xi}_k\}, \quad k = 0, 1, \dots, n.$$

Writing $\tilde{S}_k = \sum_{j \le k} \tilde{\xi}_j$, we introduce the predictable permutation

$$\alpha_k = \sum_{j \in [0,k)} 1\{\tilde{S}_j < \tilde{S}_n\} + (n - k + 1)1\{\tilde{S}_{k-1} \ge \tilde{S}_n\}, \quad k = 1, \dots, n,$$

along with its inverse $\tau_1, \ldots, \tau_n$, and put $\xi'_k = \tilde{\xi}_{\tau_k}$ and $S'_k = \sum_{j \le k} \xi'_j$. Informally, this amounts to lining up the negative excursions from the left and the inverted positive ones from the right. Then $(\xi'_k) \overset{d}{=} (\tilde{\xi}_k) \overset{d}{=} (\xi_k)$ by Theorem 4.2, and it is easy to check that

$$
\min\{k \ge 0;\ S_k = \max_j S_j\} \overset{d}{=} \min\{k \ge 0;\ S'_k = \max_j S'_j\}
$$
$$
= \sum_{j \in [0,n)} 1\{\tilde{S}_j < \tilde{S}_n\}
$$
$$
= \sum_{k \in [1,n]} 1\{S_k > 0\}. \qquad \Box
$$

We turn to some multi-variate versions of Proposition 4.1 and Theorem 4.2.

Proposition 4.4 *(multi-variate sampling) Let $\xi_j = (\xi_{jk})$, $j = 1, \ldots, d$, be finite or infinite random sequences in S, indexed by I, consider some $\mathcal{F}$-predictable times τ_{jk} in I, and put $\tilde{\xi}_{jk} = \xi_{j,\tau_{jk}}$. Then $(\tilde{\xi}_{jk}) \overset{d}{=} (\xi_{jk})$ under each of these conditions:*

(i) *ξ is separately $\mathcal{F}$-exchangeable and the τ_{jk} are a.s. distinct for each j;*
(ii) *ξ is separately $\mathcal{F}$-contractable and $\tau_{j1} < \tau_{j2} < \cdots$ for each j.*

Proof: (i) First we take $I = \{1, \ldots, n\}$. For every $j \le d$, let $\tau_{j1}, \ldots, \tau_{jn}$ form a random permutation of I. Introduce the inverse permutation $\alpha_{j1}, \ldots, \alpha_{jn}$, and define the intermediate permutations (α_{jk}^m) as in (2). Proceeding as before, we get for any measurable functions $f_{jk} \ge 0$ on S

$$
E \prod_{j,k} f_{j,\alpha_{jk}^{m+1}}(\xi_{jk}) = E \prod_j \prod_{k \le m} f_{j,\alpha_{jk}}(\xi_{jk})\, E^{\mathcal{F}_m} \prod_j \prod_{k > m} f_{j,\alpha_{jk}^{m+1}}(\xi_{jk})
$$
$$
= E \prod_j \prod_{k \le m} f_{j,\alpha_{jk}}(\xi_{jk})\, E^{\mathcal{F}_m} \prod_j \prod_{k > m} f_{j,\alpha_{jk}^m}(\xi_{jk})
$$
$$
= E \prod_{j,k} f_{j,\alpha_{jk}^m}(\xi_{jk}),
$$

and we may sum over m to obtain

$$
E \prod_{j,k} f_{jk}(\xi_{j,\tau_{jk}}) = E \prod_{j,k} f_{j,\alpha_{jk}}(\xi_{jk}) = E \prod_{j,k} f_{jk}(\xi_{jk}).
$$

This completes the proof for predictable permutations of finite sequences, and the general result follows as before.

(ii) Again we may take $I = \{1, \ldots, n\}$. The statement is obvious for $n = 1$. Assuming it is true for $n - 1$, we turn to sequences of length n. Given any predictable times $\tau_{j1} < \cdots < \tau_{j,m_j}$ in I, we introduce the $\mathcal{F}_0$-measurable times

$$
\tau'_{jk} = 1\{\tau_{jk} = 1\} + (k + n - m_j)1\{\tau_{jk} > 1\}, \quad k \le m_j,\ j \le d.
$$

We also define $\kappa_j = 1\{\tau_{j1} = 1\}$ and $J = \{j;\ \tau_{j1} = 1\}$. Applying the induction hypothesis to the array of pairs $(\xi_{j1}, \kappa_j, \xi_{jk})$, $k = 2, \ldots, n$, and then using the joint contractability of (ξ_{jk}) over $\mathcal{F}_0$, we get for any measurable functions $f_{jk} \geq 0$ on S

$$
\begin{aligned}
E \prod_{j,k} f_{jk}(\xi_{j,\tau_{jk}}) &= E \prod_{j \in J} f_{j1}(\xi_{j1}) \, E^{\mathcal{F}_1} \prod_j \prod_{k > \kappa_j} f_{jk}(\xi_{j,\tau_{jk}}) \\
&= E \prod_{j \in J} f_{j1}(\xi_{j1}) \, E^{\mathcal{F}_1} \prod_j \prod_{k > \kappa_j} f_{jk}(\xi_{j,k+n-m_j}) \\
&= E \prod_{j,k} f_{jk}(\xi_{j,\tau'_{jk}}) = E \prod_{j,k} f_{jk}(\xi_{jk}).
\end{aligned}
$$

This completes the induction, and the result follows. $\qquad\square$

4.2 Gauss and Poisson Reduction

The continuous-time theory is based on a general reduction of a continuous local martingales M and a quasi-left-continuous point process ξ to a pair (X, η), consisting of a centered Gaussian process X and an independent Poisson process η, each defined on an abstract space. As special cases, the result contains the classical time-change reductions of a continuous local martingale to a Brownian motion and of a quasi-left-continuous simple point process to a Poisson process. Generalizations, in different directions, of the latter results will appear throughout the chapter.

Though the general reduction theorem was used already in Chapter 2, it is stated here because of its close connections to other results in this chapter. It is then important to notice that the present proof is self-contained and does not utilize any previous results in this book.

Recall that a marked point process ξ on $S \times [0, 1]$ or $S \times \mathbb{R}_+$ is said to be *quasi-left-continuous* if its compensator $\hat{\xi}$ is a.s. continuous, in the sense that $\hat{\xi}(S \times \{t\}) = 0$ a.s. for every t.

Theorem 4.5 *(Gauss and Poisson reduction)* *Consider a continuous, local martingale M in $\mathbb{R}^d$, a quasi-left-continuous, K-marked point process ξ on $(0, \infty)$ with compensator $\hat{\xi}$, a predictable process $V : \mathbb{R}_+ \times K \to \hat{S}$, and a progressively measurable process $U_t : \mathbb{R}_+ \to \mathbb{R}^d$, $t \in T$, where K and S are Borel and $\hat{S} = S \cup \{\Delta\}$ with $\Delta \notin S$. Suppose that the random measure and process*

$$
\mu = \hat{\xi} \circ V^{-1}, \qquad \rho_{s,t} = \sum_{i,j} \int_0^\infty U_{s,r}^i U_{t,r}^j d[M^i, M^j]_r, \quad s, t \in T,
$$

exist and are $\mathcal{F}_0$-measurable, and define

$$
\eta = \xi \circ V^{-1}, \qquad X_t = \sum_i \int_0^\infty U_{t,r}^i dM_r^i, \quad t \in T.
$$

Then conditionally on $\mathcal{F}_0$, the point process η is Poisson on S with intensity measure μ, and X is an independent, centered Gaussian process on T with covariance function ρ.

In applications, we often need to extend the point process ξ, through a suitable randomization, to ensure that the compensator $\hat{\xi}$ will be a.s. unbounded. Using this device, we can transform any quasi-left-continuous, simple point process on $[0,1]$, by a suitable random time-change, to the *beginning* of a Poisson process on $\mathbb{R}_+$. The underlying randomization will then be assumed without further comments.

Proof: Fixing any constants $c_1, \ldots, c_m \in \mathbb{R}$, elements $t_1, \ldots, t_m \in T$, and disjoint Borel sets $B_1, \ldots, B_n \subset S$ with $\mu B_j < \infty$, we consider the processes

$$N_t = \sum_k c_k \sum_i \int_0^t U_{t_k,r}^i \, dM_r^i,$$

$$Y_t^k = \int_S \int_0^{t+} 1_{B_k}(V_{s,x}) \, \xi(ds \, dx), \quad t \geq 0, \ k \leq n.$$

For any $u_1, \ldots, u_n \geq 0$, we further introduce the exponential local martingales

$$Z_t^0 = \exp(iN_t + \tfrac{1}{2}[N,N]_t),$$

$$Z_t^k = \exp\left(-u_k Y_t^k + (1 - e^{-u_k})\hat{Y}_t^k\right), \quad t \geq 0, \ k \leq n.$$

where the local martingale property holds for Z^0 by FMP 18.1 and for $Z^1, \ldots, Z^n$ by FMP 26.8 applied to the processes $A_t^k = (1 - e^{-u_k})(\hat{Y}_t^k - Y_t^k)$. The same property holds for the product $Z_t = \prod_k Z_t^k$ since the Z^k are strongly orthogonal (FMP 26.4, 26.6). Furthermore,

$$N_\infty = \sum_k c_k X_{t_k}, \qquad Y_\infty^k = \eta B_k$$

$$[N,N]_\infty = \sum_{h,k} c_h c_k \sum_{i,j} \int_0^\infty U_{t_h,r}^i U_{t_k,r}^j \, d[M^i, M^j]_r = \sum_{h,k} c_h c_k \rho_{t_h, t_k},$$

$$\hat{Y}_\infty^k = \int_S \int_K 1_{B_k}(V_{s,x}) \, \hat{\xi}(ds \, dx) = \mu B_k, \quad k \leq n.$$

The product Z remains a local martingale with respect to the conditional probability measure $P_A = P[\,\cdot\,|A]$, for any $A \in \mathcal{F}_0$ with $PA > 0$. Choosing A such that ρ_{t_h, t_k} and μB_k are bounded on A, we see that even Z becomes bounded on A and hence is a uniformly integrable P_A-martingale. In particular, we get $E[Z_\infty|A] = 1$ or $E[Z_\infty; A] = PA$, which extends immediately to arbitrary $A \in \mathcal{F}_0$. This shows that $E[Z_\infty|\mathcal{F}_0] = 1$, and we get

$$E\left[\exp\left(i \sum_k c_k X_{t_k} - \sum_k u_k \eta B_k\right) \Big| \mathcal{F}_0\right]$$
$$= \exp\left(-\tfrac{1}{2} \sum_{h,k} c_h c_k \rho_{t_h, t_k} - \sum_k (1 - e^{-u_k}) \mu B_k\right).$$

Invoking the uniqueness theorems for characteristic functions and Laplace transforms (FMP 5.3), we conclude that the variables $X_{t_1}, \ldots, X_{t_m}$ and $\eta B_1, \ldots, \eta B_n$ have the required joint conditional distribution, and the assertion follows by a monotone-class argument. $\qquad\square$

We turn to an application of the preceding result, needed below, where some infinitely divisible random variables are constructed through stochastic integration with respect to some quasi-left-continuous, purely discontinuous, local martingales. Recall that two local martingales M and N are said to be *strongly orthogonal* if $[M, N] = 0$ a.s., and that M is *purely discontinuous* if its continuous martingale component vanishes, so that $[M] = [M, M]$ is a pure jump-type process (FMP 26.14–15).

Lemma 4.6 *(martingale reduction) Let $M_1, \ldots, M_m$ be strongly orthogonal, purely discontinuous, quasi-left-continuous, local martingales with jump point processes $\xi_1, \ldots, \xi_m$, consider some predictable processes $V_1, \ldots, V_m$, and define $W_k(x, t) = xV_k(t)$, $k \leq m$. Assume that the measures $\nu_k = \hat{\xi}_k \circ W_k^{-1}$ are a.s. non-random and satisfy*

$$\int_{-\infty}^{\infty} (x^2 \wedge |x|)\, \nu_k(dx) < \infty, \quad k \leq m. \tag{3}$$

Then the stochastic integrals $\gamma_k = \int_0^\infty V_k dM_k$ exist for all k and define some independent, centered, infinitely divisible random variables with Lévy measures ν_k and vanishing Gaussian components.

Proof: In view of the Cramér–Wold theorem (FMP 5.5), it suffices to prove that the linear combination $\sum_k c_k \gamma_k$ has the appropriate infinitely divisible distribution for arbitrary $c_1, \ldots, c_k \in \mathbb{R}$. By suitable scaling, we may then assume that $c_k = 1$ for all k. Since $\hat{\xi}_k \circ W_k^{-1} = \nu_k$ for all k, we see from Theorem 4.5 that the random measures $\eta_k = \xi_k \circ W_k^{-1}$ are independent Poisson processes on $\mathbb{R} \setminus \{0\}$ with intensity measures ν_k.

Now introduce the process

$$N_t = \sum_k (V_k \cdot M_k)_t, \quad t \geq 0.$$

To prove the existence of the stochastic integrals on the right, we note that

$$\int_0^\infty V_k^2 d[M_k] = \int_0^\infty \int_{-\infty}^\infty W_k^2 d\xi_k = \int_{-\infty}^\infty x^2 \eta_k(dx).$$

Using Jensen's inequality and the subadditivity of $x^{1/2}$, we get from (3)

$$
\begin{aligned}
E\left(\int_{-\infty}^\infty x^2 \eta_k(dx)\right)^{1/2} &\leq E\left(\int_{-1}^1 x^2 \eta_k(dx)\right)^{1/2} + E\left(\int_{|x|>1} x^2 \eta_k(dx)\right)^{1/2} \\
&\leq \left(E \int_{-1}^1 x^2 \eta_k(dx)\right)^{1/2} + E \int_{|x|>1} |x|\, \eta_k(dx) \\
&= \left(\int_{-1}^1 x^2 \nu_k(dx)\right)^{1/2} + \int_{|x|>1} |x|\, \nu_k(dx) < \infty.
\end{aligned}
$$

This shows that N exists as a uniformly integrable martingale on $[0, \infty]$ (cf. FMP 26.13) satisfying

$$E[N]_\infty^{1/2} = \sum_k E\left(V_k^2 \cdot [M_k]\right)_\infty^{1/2} < \infty.$$

We can now approximate N, for every $\varepsilon > 0$, by the martingale

$$N_t^\varepsilon = \sum_k \int_0^t W_k 1\{|W_k| > \varepsilon\}\, d(\xi_k - \hat{\xi}_k), \quad t \geq 0.$$

The random variables N_∞^ε are clearly centered, compound Poisson with Lévy measure $\sum_k \nu_k$ restricted to the complements $[-\varepsilon, \varepsilon]^c$. Furthermore, a BDG-inequality (FMP 26.12) yields $N_\infty^\varepsilon \to N_\infty = \sum_k \gamma_k$ in L^1 as $\varepsilon \to 0$, and so by 15.14 we get the required distribution for the limit. $\qquad\square$

4.3 Predictable Mapping

In Proposition 1.18 we saw that if ξ is an exchangeable random measure on a product space $S \times I$, where S is Borel and $I = [0,1]$ or $\mathbb{R}_+$, then $\xi \circ f^{-1} \overset{d}{=} \xi$ for every measure-preserving transformation f on I. Now we intend to show that the statement remains true if the deterministic function f is replaced by a predictable process V. (When $I = \mathbb{R}_+$, we may allow V to take the value $+\infty$.)

To state the corresponding result for exchangeable processes X on $I = [0,1]$ or $\mathbb{R}_+$, we need to identify the proper interpretation of $X \circ V^{-1}$. Recalling from Theorem 3.15 that X is a semi-martingale with respect to the augmented filtration, we define $X \circ V^{-1}$ as the process

$$(X \circ V^{-1})_t = \int_I 1\{V_s \leq t\}\, dX_s, \quad t \in I, \tag{4}$$

where the right-hand side is a stochastic integral of the predictable process $U_s = 1\{V_s \leq t\}$. In particular, we note that if X is the distribution function of some random measure ξ on I, so that $X_t = \xi[0,t]$ for all $t \in I$, then $X \circ V^{-1}$ agrees with the distribution function of the transformed measure $\xi \circ V^{-1}$. Hence, in this case, the relations $X \circ V^{-1} \overset{d}{=} X$ and $\xi \circ V^{-1} \overset{d}{=} \xi$ are equivalent.

Theorem 4.7 *(predictable mapping) Consider an $\mathcal{F}$-exchangeable, $\mathbb{R}^d$-valued process X on I or an $\mathcal{F}$-exchangeable random measure ξ on $S \times I$, where S is Borel and $I = [0,1]$ or $\mathbb{R}_+$. Then for any $\mathcal{F}$-predictable transformation V on I such that $\lambda \circ V^{-1} = \lambda$ a.s., we have $X \circ V^{-1} \overset{d}{=} X$ or $\xi \circ V^{-1} \overset{d}{=} \xi$, respectively.*

For exchangeable processes X, the random variable or vector $(X \circ V^{-1})_t$ is only defined up to a null set for every fixed $t \in I$. The asserted relation $X \circ V^{-1} \overset{d}{=} X$ signifies implicitly that $X \circ V^{-1}$ has an rcll version with the same distribution as X (FMP 3.24).

Several proofs of Theorem 4.7 are known. The most straightforward approach, chosen here, may be to deduce the result from the corresponding discrete-time Theorem 4.2 by a suitable approximation. Three lemmas will then be required. To state the first one, we say that the random subset $A \subset I$ is a *simple, predictable set* if it is a finite union of predictable intervals of the form $(\sigma, \tau]$, where $\sigma \leq \tau$ are optional times taking values in a fixed, finite subset of $\mathbb{Q}_I = \mathbb{Q} \cap I$.

Lemma 4.8 *(approximation)* *Let $A_1, \ldots, A_m$ be disjoint, $\mathcal{F}$-predictable sets in $I = [0, 1]$ or $\mathbb{R}_+$ with non-random, rational lengths. Then for every $n \in \mathbb{N}$ there exist some simple, disjoint, $\mathcal{F}$-predictable sets $A_{1,n}, \ldots, A_{m,n} \subset I$ such that $\lambda A_{j,n} = \lambda A_j$ a.s. and*

$$\lim_{n \to \infty} \sum_{j \leq m} E\lambda(A_j \Delta A_{j,n}) = 0. \tag{5}$$

Proof: First assume that $I = [0, 1]$. Let $\mathcal{C}$ be the class of predictable intervals $(\sigma, \tau]$, and write $\mathcal{D}$ for the class of predictable sets $A \subset (0, 1]$, admitting approximations $E\lambda(A \Delta A_n) \to 0$ by simple, predictable sets A_n. Then $\mathcal{C}$ is a π-system generating the predictable σ-field $\mathcal{P}$ on $(0, 1]$ (FMP 25.1), and $\mathcal{D}$ is a λ-system containing $\mathcal{C}$. Hence, $\mathcal{D} = \mathcal{P}$ (FMP 1.1), which means that every predictable set A admits the stated approximation. Applying this result to the sets $A \cap (k - 1, k]$ for arbitrary $k \in \mathbb{N}$, we may extend the statement to predictable sets in $\mathbb{R}_+$ of finite length.

Now let $A_1, \ldots, A_m$ be such as stated, and proceed as above to produce some simple, predictable sets $A_{1n}, \ldots, A_{mn}$ satisfying (5). For every $n \in \mathbb{N}$, we may easily adjust the latter sets, to make them disjoint with desired lengths $\lambda A_1, \ldots, \lambda A_m$. Noting that, for distinct $i, j \leq m$,

$$E|\lambda A_j - \lambda A_{j,n}| \leq E\lambda(A_j \Delta A_{j,n}),$$
$$E\lambda(A_{i,n} \cap A_{j,n}) \leq E\lambda(A_i \Delta A_{i,n}) + E\lambda(A_j \Delta A_{j,n}),$$

we see that the total change is of the same order as the original approximation error. Hence, the new sets $A'_{j,n}$ will again satisfy (5). $\qquad\Box$

To justify the approximation of the last lemma, we need to show that the resulting errors in (4) are negligible. This requires the following continuity property for the associated stochastic integrals. When X is a semi-martingale and A is a predictable set such that 1_A is X-integrable, it is often suggestive to write $X(A) = \int_A dX = \int 1_A dX$.

Lemma 4.9 *(existence and continuity)* *Consider an $\mathcal{F}$-exchangeable random measure ξ or process X on $I = [0, 1]$ or $\mathbb{R}_+$, and let $A, A_1, A_2, \ldots$ be $\mathcal{F}$-predictable sets in I. Then ξA or $X(A)$ exist when $\lambda A < \infty$ a.s., and if $\lambda A_n \xrightarrow{P} 0$ on a set F, we have $\xi A_n \xrightarrow{P} 0$ or $X(A_n) \xrightarrow{P} 0$, respectively, on F.*

Proof: If $X_t = \xi[0, t]$, then clearly $X(A_n) = \xi A_n$. Thus, the statement for random measures is essentially a special case of that for processes, and it is enough to consider the latter. Since the semi-martingale integral is independent of filtration by FMP 26.4, we may extend the latter, if required, to ensure that the characteristic triple (α, β, γ) or (α, ν, γ) becomes $\mathcal{F}_0$-measurable.

Beginning with the case where $I = [0, 1]$, we see from Theorems 2.13 and 2.25 that X is a special semi-martingale such that $[X^c]$ is a.s. linear, and both X itself and its jump point process ξ have absolutely continuous compensators $\hat{X}$ and $\hat{\xi}$, respectively. Thus, we may write $\hat{X} = M \cdot \lambda$ and $\hat{\xi} = \eta \cdot \lambda$, where M is a.s. integrable on $[0, 1]$. Since the quadratic variation $[X]_1 = \gamma$ is finite and $\mathcal{F}_0$-measurable, it follows that the associated predictable variation $\langle X - \hat{X} \rangle$ is again absolutely continuous with an a.s. integrable density N. Hence, by dominated convergence,

$$|\hat{X}(A_n)| \leq \int_{A_n} |M_s| ds \xrightarrow{P} 0,$$

$$\langle X - \hat{X} \rangle(A_n) = \int_{A_n} N_s ds \xrightarrow{P} 0,$$

and so by FMP 26.2 we have $X(A_n) \xrightarrow{P} 0$.

In the $\mathbb{R}_+$-case, we know from Theorem 1.19 that X is conditionally a Lévy process given $\mathcal{F}_0$. Hence, we may write $X = M + L + J$, where M is a local martingale with bounded jumps and linear predictable variation $\langle M \rangle$, L is a linear drift component, and J is a process of isolated large jumps, such that the associated jump-time process N is mixed Poisson with linear compensator $\hat{N}$. If $\lambda A < \infty$ a.s., then the variables $\langle M \rangle(A)$, $L(A)$, and $\hat{N}(A)$ are a.s. finite, which implies the existence of $X(A)$. Furthermore, the condition $\lambda A_n \xrightarrow{P} 0$ yields

$$\langle M \rangle(A_n) \xrightarrow{P} 0, \qquad L(A_n) \xrightarrow{P} 0, \qquad \hat{N}(A_n) \xrightarrow{P} 0,$$

and so by FMP 26.2 we have

$$M(A_n) + L(A_n) \xrightarrow{P} 0, \qquad N(A_n) \xrightarrow{P} 0.$$

Here the latter convergence implies $J(A_n) \xrightarrow{P} 0$, since $|J(A_n)| \wedge 1 \leq N(A_n)$, and again it follows that $X(A_n) \xrightarrow{P} 0$.

Finally, suppose that $\lambda A_n \xrightarrow{P} 0$ on F. Then we may choose some positive constants $c_n \to 0$ such that $P[\lambda A_n > c_n; F] \to 0$. For every $n \in \mathbb{N}$ we define

$$\tau_n = \inf\{t \geq 0; \lambda_t A_n > c_n\}, \qquad A'_n = A_n \cap [0, \tau_n],$$

where $\lambda_t A = \lambda(A \cap [0, t])$. Here the A'_n are predictable with $\lambda A'_n \leq c_n \to 0$, and so $X(A'_n) \xrightarrow{P} 0$. Noting that $X(A_n) = X(A'_n)$ a.s. on $\{\tau_n = \infty\} = \{\lambda A_n \leq c_n\}$, we get for any $\varepsilon > 0$

$$P[|X(A_n)| > \varepsilon; F] \leq P\{|X(A'_n)| > \varepsilon\} + P[\lambda A_n > c_n; F] \to 0,$$

which shows that $X(A_n) \xrightarrow{P} 0$ on F. $\square$

Finally, we need to describe the simple, predictable sets of Lemma 4.8 in terms of discrete, predictable times.

Lemma 4.10 (*enumeration*) *Let A be a simple, $\mathcal{F}$-predictable set in $I = [0,1]$ or $\mathbb{R}_+$, with all interval endpoints belonging to $\mathbb{Z}_+/n$, and define $\mathcal{G}_k = \mathcal{F}_{k/n}$, $k \in \mathbb{Z}_+$. Then there exist some $\mathcal{G}$-predictable times $\tau_1 < \tau_2 < \cdots$ such that*

$$A = \bigcup_{j \leq m}(\tau_j - 1, \tau_j]/n.$$

Proof: The time $\sigma = \inf A$ satisfies $\{\sigma < t\} \in \mathcal{F}_t$ for all $t \in I$ and is therefore $\mathcal{F}$-optional by the right continuity of $\mathcal{F}$ (FMP 7.2). Hence, $n\sigma = \tau_1 - 1$ is $\mathcal{G}$-optional, which means that τ_1 is predictable. Furthermore, the stochastic interval $(\sigma, \sigma + n^{-1}]$ is $\mathcal{F}$-predictable, and so the same thing is true for the difference $A' = A \setminus (\sigma, \sigma + n^{-1}]$. The assertion now follows easily by induction. $\square$

Proof of Theorem 4.7: First we consider the case of exchangeable processes X on I. Fix any disjoint, rational intervals $I_1, \ldots, I_m \subset I$ of equal length h, and define $A_j = V^{-1} I_j$ for all $j \leq m$. The A_j are predictable by the predictability of V, and since $\lambda \circ V^{-1} = \lambda$ a.s. by hypothesis, we have also $\lambda A_j = \lambda I_j$ a.s. for all j. Hence, by Lemma 4.8, we may choose some simple, predictable sets $A_{j,n}$, disjoint for fixed n and with $\lambda A_{j,n} = \lambda A_j$ a.s., such that $A_{j,n} \to A_j$ in the sense of (5). Restricting n to a suitable subsequence $N' \subset \mathbb{N}$, we may assume that each $A_{j,n}$ is a finite union of intervals $I_{nk} = n^{-1}(k-1, k]$. Then by Lemma 4.10 we may write

$$A_{j,n} = \bigcup_{k \leq nh} I_{n,\tau_{njk}}, \quad j \leq m, \ n \in N', \tag{6}$$

where the times τ_{njk} are a.s. distinct for fixed n and predictable with respect to the discrete filtration $\mathcal{G}_k^n = \mathcal{F}_{k/n}$, $k \leq n\lambda I$. The increments $X(I_{nk})$ are then $\mathcal{G}^n$-exchangeable for each $n \in N'$, and so by Theorem 4.2

$$\{X(I_{n,\tau_{njk}}); (j,k) \leq (m,nh))\} \overset{d}{=} \{X(I_{nk} + (j-1)h); (j,k) \leq (m,nh)\}.$$

Using (5) and Lemma 4.9, we get

$$\begin{aligned}
(X(I_1), \ldots, X(I_m)) \quad &\overset{d}{=} \quad (X(A_{1,n}), \ldots, X(A_{m,n})) \\
&\overset{P}{\to} \quad (X(A_1), \ldots, X(A_m)) \\
&= \quad (X \circ V^{-1} I_1, \ldots, X \circ V^{-1} I_m),
\end{aligned}$$

which shows that $X \circ V^{-1} \overset{d}{=} X$ on $\mathbb{Q}_I$. The relation extends to I by another application of Lemma 4.9.

In the case of random measures ξ, we may apply the previous result, for arbitrary $B_1, \ldots, B_d \in \hat{\mathcal{S}}$, to the $\mathbb{R}^d$-valued process

$$X_t = (\xi(B_1 \times [0,t]), \ldots, \xi(B_d \times [0,t])), \quad t \in I,$$

to see that $\xi \circ V^{-1} \overset{d}{=} \xi$ on the class of product sets $B \times [0, t]$. The general result then follows by a monotone-class argument. $\square$

In the special case of processes on $I = \mathbb{R}_+$, we can also prove the result by a direct argument based on Theorem 4.5. The more subtle argument required for $I = [0, 1]$ will be given later.

Second proof of Theorem 4.7, for $I = \mathbb{R}_+$: Extending the original filtration, if necessary, we may assume that the characteristics of X are $\mathcal{F}_0$-measurable. First suppose that X has isolated jumps. Then

$$X_t = \alpha t + \sigma B_t + \int_{\mathbb{R}^d} \int_0^t x\, \xi(dx\, ds), \quad t \geq 0,$$

where ξ is a Cox process on $\mathbb{R}^d \times \mathbb{R}_+$ directed by $\nu \otimes \lambda$ and B is an independent Brownian motion. We may also assume that the triple (α, σ, ν) is $\mathcal{F}_0$-measurable and the pair (B, ξ) is $\mathcal{F}$-exchangeable.

Since ξ has compensator $\hat{\xi} = \nu \otimes \lambda$, we obtain $\hat{\xi} \circ V^{-1} = \nu \otimes \lambda$, which yields $\xi \circ V^{-1} \overset{d}{=} \xi$ by Theorem 4.5. Next, we consider for every $t \geq 0$ the predictable process $U_{t,r} = 1\{V_r \leq t\}$, $r \geq 0$, and note that

$$\int_0^\infty U_{s,r} U_{t,r} d[B^i, B^j]_r = \delta_{ij} \int_0^\infty 1\{V_r \leq s \wedge t\}\, dr = (s \wedge t)\, \delta_{ij}.$$

Using Theorem 4.5 again, we conclude that the $\mathbb{R}^d$-valued process

$$(B \circ V^{-1})_t = \int_0^\infty U_{t,r} dB_r, \quad t \geq 0,$$

is Gaussian with the same covariance function as B. From the same result we also note that $\xi \circ V^{-1}$ and $B \circ V^{-1}$ are conditionally independent given $\mathcal{F}_0$. Hence,

$$((\xi, \lambda, B) \circ V^{-1}, \alpha, \sigma, \nu) \overset{d}{=} (\xi, \lambda, B, \alpha, \sigma, \nu),$$

which implies $X \circ V^{-1} \overset{d}{=} X$.

Turning to the general case, we may write

$$X_t = M_t^\varepsilon + (X_t - M_t^\varepsilon), \quad t \geq 0,$$

where M^ε is the purely discontinuous, local martingale formed by all compensated jumps of modulus $\leq \varepsilon$. Then $(X - M^\varepsilon) \circ V^{-1} \overset{d}{=} (X - M^\varepsilon)$ as above, and it suffices to show that $M_t^\varepsilon \overset{P}{\to} 0$ and $(M^\varepsilon \circ V^{-1})_t \overset{P}{\to} 0$, as $\varepsilon \to 0$ for fixed $t \geq 0$. In the one-dimensional case, we may use the isometric property of the stochastic L^2-integral (cf. FMP 26.2), together with the measure-preserving property of V, to see that a.s.

$$
\begin{aligned}
E^{\mathcal{F}_0}(M^\varepsilon \circ V^{-1})_t^2 &= E^{\mathcal{F}_0}\left(\int_0^\infty 1\{V_s \le t\}\, dM_s^\varepsilon\right)^2 \\
&= E^{\mathcal{F}_0}\int_0^\infty 1\{V_s \le t\}\, d\langle M^\varepsilon\rangle_s \\
&= E^{\mathcal{F}_0}\int_0^\infty 1\{V_s \le t\}ds \int_{|x|\le\varepsilon} x^2\nu(dx) \\
&= t\int_{|x|\le\varepsilon} x^2\nu(dx) \to 0.
\end{aligned}
$$

Hence, by Jensen's inequality and dominated convergence,

$$
E\big[(M^\varepsilon \circ V^{-1})_t^2 \wedge 1\big] \le E\big[E^{\mathcal{F}_0}(M^\varepsilon \circ V^{-1})_t^2 \wedge 1\big] \to 0,
$$

which shows that $(M^\varepsilon \circ V^{-1})_t \xrightarrow{P} 0$. The convergence $M_t^\varepsilon \xrightarrow{P} 0$ is obtained in the special case where $V_s \equiv s$. □

In applications, the relation $\lambda \circ V^{-1} = \lambda$ may often be satisfied only up to a random time $\zeta \in I$. By a localization argument based on Lemma 4.9, it is clear that the process $X \circ V^{-1}$ still exists on the random interval $[0, \zeta)$. We proceed to show that the relation $X \circ V^{-1} \overset{d}{=} X$ also remains valid on $[0, \zeta)$, in the sense that $X \circ V^{-1}$ can be extended beyond ζ to a process with the same distribution as X.

Theorem 4.11 (*predictable embedding*) *Given an $\mathbb{R}^d$-valued, $\mathcal{F}$-exchangeable process X on $I = [0, 1]$ or $\mathbb{R}_+$ and an $\mathcal{F}$-predictable process $V : I \to \bar{I}$, there exists a process $Y \overset{d}{=} X$ on I such that $X \circ V^{-1} = Y$ a.s. on $[0, \zeta)$, where*

$$
\zeta = \sup\{t \in I;\ (\lambda \circ V^{-1})_t = t\}.
$$

To be precise, we assert that $(X \circ V^{-1})_t = Y_t$ a.s. on $\{\zeta > t\}$ for every $t \in I$. Our proof requires several lemmas. First we may reduce to the case where $\lambda \circ V^{-1} \le \lambda$ a.s.

Lemma 4.12 (*leveling*) *For $I = [0, 1]$ or $\mathbb{R}_+$, consider a predictable process $V : I \to \bar{I}$ such that $\lambda \circ V^{-1} \le \lambda$ a.s. on some random interval $[0, \zeta)$. Then there exists a predictable process $U \ge V$, such that a.s. $\lambda \circ U^{-1} \le \lambda$ on $[0, 1)$ or $\mathbb{R}_+$ and $U \wedge \zeta = V \wedge \zeta$.*

Proof: By a suitable truncation, we may reduce to the case where $I = \mathbb{R}_+$. Writing $\lambda_t = \lambda(\cdot \cap [0, t])$, we consider on $\mathbb{R}_+$ the processes

$$
\begin{aligned}
Z_t &= \sup\{r \ge 0;\ \lambda_t \circ V^{-1} \le \lambda \text{ on } [0, r]\}, \\
U_t &= V_t + \infty \cdot 1\{V_t \ge Z_t\}.
\end{aligned}
$$

Since V is predictable and hence progressively measurable, Fubini's theorem shows that $\lambda_t \circ V^{-1}I$ is $\mathcal{F}_t$-measurable for every interval I. Assuming $\lambda_t \circ$

$V^{-1}I \le \lambda I$ for every rational interval $I \subset [0, r]$, we get the same inequality for any open set, and hence, by regularity, for every Borel set in $[0, r]$. This implies

$$\{\lambda_t \circ V^{-1} \le \lambda \text{ on } [0, r]\} \in \mathcal{F}_t, \quad r, t \ge 0,$$

and so Z_t is $\mathcal{F}_t$-measurable for every $t \ge 0$, which means that Z is adapted. Since it is also left-continuous, it is then predictable, and so the same property holds for the process U.

Noting that a.s. $U_t \wedge Z_t = V_t \wedge Z_t$ and $Z_t \ge \zeta$ for all $t \ge 0$, we get $U \wedge \zeta = V \wedge \zeta$ a.s. To show that $\lambda \circ U^{-1} \le \lambda$ a.s., we may assume that V and ζ are non-random with $\lambda \circ V^{-1} \le \lambda$ on $[0, \zeta)$. For every $\varepsilon > 0$, we introduce the function

$$U_{\varepsilon, t} = V_t + \infty \cdot 1\{V_t \ge Z_t - \varepsilon\}, \quad t \ge 0.$$

Fixing any $b, \varepsilon > 0$, we next define

$$s = \sup\{t \ge 0; \ \lambda_t \circ U_\varepsilon^{-1} \le \lambda \text{ on } [0, b]\}.$$

If $s < \infty$, we have $Z_{s+} \le b$, and we may choose a $t > s$ such that $Z_t \ge Z_{s+} - \varepsilon$. Using the definitions of U_ε, Z_t, and s, and noting that $U_{\varepsilon, r} \notin [Z_{s+} - \varepsilon, \infty)$ for $r > s$, we get

$$\begin{aligned}
\lambda_t \circ U_\varepsilon^{-1} &\le \lambda_t \circ V^{-1} \le \lambda &\quad \text{on } [0, Z_t) \supset [0, Z_{s+} - \varepsilon), \\
\lambda_t \circ U_\varepsilon^{-1} &= \lambda_s \circ U_\varepsilon^{-1} \le \lambda &\quad \text{on } [Z_{s+} - \varepsilon, b],
\end{aligned}$$

which shows that $\lambda_t \circ U_\varepsilon^{-1} \le \lambda$ on $[0, b]$. This contradicts the definition of s and shows that $s = \infty$, which implies $\lambda \circ U_\varepsilon^{-1} \le \lambda$ on $[0, b]$. Since $b > 0$ was arbitrary, the same relation holds on $\mathbb{R}_+$. We may finally let $\varepsilon \to 0$ and conclude by monotone convergence that $\lambda \circ U^{-1} \le \lambda$. $\square$

In the case where $I = [0, 1]$, we proceed to construct a predictable and measure-preserving modification of V. Such a construction may be impossible when $I = \mathbb{R}_+$.

Lemma 4.13 *(filling) Let V be a predictable transformation of $[0, 1]$ such that $\lambda \circ V^{-1} = \lambda$ a.s. on some random interval $[0, \zeta)$. Then there exists a predictable process U on $[0, 1]$ such that a.s.*

$$\lambda \circ U^{-1} = \lambda, \qquad \lambda\{t \in [0, 1]; \ U_t \wedge \zeta = V_t \wedge \zeta\} = 1.$$

Proof: By Lemma 4.12 we may assume that $\lambda \circ V^{-1} \le \lambda$ holds identically on $[0, 1)$. For $t, x \in [0, 1]$, we introduce the processes

$$\begin{aligned}
R_{t,x} &= x + \lambda\{s \le t; \ V_s > x\}, \\
X_t &= \inf\{x \in [0, 1]; \ R_{t,x} = 1\}, \qquad\qquad (7) \\
U_t &= V_t \wedge X_{t-}.
\end{aligned}$$

Since $\lambda \circ V^{-1} \leq \lambda$ on $[0,1)$, we note that R is jointly continuous on $[0,1] \times [0,1)$ and non-decreasing in x for fixed t with $R_{t,0} = t$ and $R_{t,1-} = 1 + \lambda_t \circ V^{-1}\{1\}$. Thus, the set $\Xi = \{R = 1\}$ is closed in $[0,1]^2$, and so the infimum in (7) is attained. Furthermore, the process X is clearly non-increasing and right-continuous on $[0,1]$ with $X_0 = 1$ and $X_1 = 0$, and by continuity the pairs (t, X_t) and (t, X_{t-}) belong to Ξ for every t. Writing T for the left-continuous inverse of X, it follows that even $(T_x, x) \in \Xi$ for every $x \in [0,1]$.

Noting that $X_t > x$ iff $t < T_x$ for any $x, t \in [0,1]$, we obtain

$$\lambda\{s;\, U_s > x\} = \lambda\{s;\, V_s \wedge X_{s-} > x\}$$
$$= \lambda\{s < T_x;\, V_s > x\} = 1 - x,$$

which shows that $\lambda \circ U^{-1} = \lambda$. Since also $\lambda \circ V^{-1} = \lambda$ on $[0, \zeta)$, we have

$$\lambda \circ (U \wedge \zeta)^{-1} = \lambda \circ (V \wedge \zeta)^{-1},$$

which implies $\lambda(V \wedge \zeta - U \wedge \zeta) = 0$. Noting that $V \geq U$, we obtain $U \wedge \zeta = V \wedge \zeta$ a.e. λ.

The process V being predictable and hence progressively measurable, we note that $R_{t,x}$ is $\mathcal{F}_t$-measurable for any $x, t \in [0,1]$. By the monotonicity in x it follows that X is adapted, and so the left-continuous version X_{t-} is predictable. Combining this with the predictability of V, we conclude that even U is predictable. $\qquad\square$

Next we show that the process $X \circ V^{-1}$ is unaffected on $[0, \zeta)$ by the changes in V described by the last two lemmas.

Lemma 4.14 *(localization) Let X be an $\mathcal{F}$-exchangeable process on $I = [0,1]$ or $\mathbb{R}_+$, and consider some $\mathcal{F}$-predictable processes U and V on I and a random variable $\zeta \in I$, such that $\lambda\{U \wedge \zeta \neq V \wedge \zeta\} = 0$ and $\lambda \circ V^{-1} \leq \lambda$ a.s. on $[0, \zeta)$. Then*

$$(X \circ U^{-1})_t = (X \circ V^{-1})_t \quad \text{a.s. on } \{\zeta > t\}, \quad t \in I.$$

Proof: For fixed $t \in I$, we introduce the optional time

$$\tau = \sup\{s \in I;\, \lambda_s\{U \wedge t \neq V \wedge t\} = 0,\, (\lambda_s \circ V^{-1})_t \leq t\}$$

and the predictable processes

$$\tilde{U}_s = U_s + \infty \cdot 1\{\tau < s\}, \quad \tilde{V}_s = V_s + \infty \cdot 1\{\tau < s\}, \qquad s \in I.$$

Then a.s.

$$\lambda\{\tilde{U} \wedge t \neq \tilde{V} \wedge t\} = 0, \quad (\lambda \circ \tilde{V}^{-1})_t \leq t,$$

and so, on the set

$$\{\zeta > t\} \subset \{\tau = \infty\} \subset \{U = \tilde{U},\, V = \tilde{V}\},$$

we have formally

$$
\begin{aligned}
(X \circ U^{-1})_t &= \int_0^\tau 1\{U_s \le t\} dX_s = \int_0^\tau 1\{\tilde{U}_s \le t\} dX_s \\
&= \int_0^\tau 1\{\tilde{V}_s \le t\} dX_s = \int_0^\tau 1\{V_s \le t\} dX_s \\
&= (X \circ V^{-1})_t.
\end{aligned}
$$

The last calculation is justified by the local property of stochastic integrals in FMP 26.2 (iv), together with absolute continuity of the compensators $\hat{X}$ and $\langle X, X \rangle$ established in Theorem 2.13. Before applying the latter result, we may need to extend the original filtration to ensure that, for processes on $[0, 1]$, the directing triple $(\alpha, \sigma\sigma', \beta)$ becomes $\mathcal{F}_0$-measurable. The result for processes on $\mathbb{R}_+$ follows by truncation. □

We also need the following coupling lemma.

Lemma 4.15 *(coupling) Let the processes $X^1 \overset{d}{=} X^2 \overset{d}{=} \cdots$ in $D(\mathbb{R}_+, \mathbb{R}^d)$ and the random variable ζ in $[0, \infty]$ be such that*

$$
X_t^n \overset{P}{\to} Y_t \quad on \ \{\zeta > t\}, \quad t \ge 0,
$$

where the process Y is continuous in probability. Then there exists a process $X \overset{d}{=} X^1$ such that

$$
Y_t = X_t \ a.s. \ on \ \{\zeta > t\}, \quad t \ge 0.
$$

Proof: Write $D = D(\mathbb{R}_+, \mathbb{R}^d)$. Since the sequence of pairs (X^n, ζ) is trivially tight in $D \times \overline{\mathbb{R}}_+$, we have convergence $(X^n, \zeta) \overset{d}{\to} (X, \tilde{\zeta})$ along a subsequence, where $X \overset{d}{=} X^1$ and $\tilde{\zeta} \overset{d}{=} \zeta$. Defining

$$
\begin{aligned}
x_u(t) &= ((u - t)_+ \wedge 1)\, x(t), \\
f(x, u) &= (x_u, u), \qquad x \in D, \ u \in \overline{\mathbb{R}}_+, \ t \ge 0,
\end{aligned}
$$

we get $f(X^n, \zeta) \overset{d}{\to} f(X, \tilde{\zeta})$ in $D \times \overline{\mathbb{R}}_+$ by continuity, and so

$$
f(X^n, \zeta) \overset{fd}{\longrightarrow} f(X, \tilde{\zeta}) \quad on \ T,
$$

where T is a dense subset of $\mathbb{R}_+$. On the other hand, $f(X^n, \zeta)_t \overset{P}{\to} f(Y, \zeta)_t$ for all $t \ge 0$, and therefore

$$
f(X^n, \zeta) \overset{fd}{\longrightarrow} f(Y, \zeta) \quad on \ \mathbb{R}_+.
$$

Comparing the two formulas, we see that $f(Y, \zeta) \overset{d}{=} f(X, \tilde{\zeta})$ on T, which extends to $\mathbb{R}_+$ by right-continuity on the right and continuity in probability on the left. The transfer theorem (FMP 6.10) then allows us to assume that

$$
f(Y, \zeta)_t = f(X, \zeta)_t \ a.s., \quad t \ge 0,
$$

which yields the required relation. □

Proof of Theorem 4.11: First we take $I = [0, 1]$. Letting U be such as in Lemma 4.13, we get by Lemma 4.14

$$(X \circ V^{-1})_t = (X \circ U^{-1})_t \text{ a.s. on } \{\zeta > t\}, \quad t \in [0, 1].$$

Furthermore, Theorem 4.7 yields $X \circ U^{-1} \stackrel{d}{=} X$. The assertion now follows, with Y chosen as the right-continuous version of $X \circ U^{-1}$.

Now let $I = \mathbb{R}_+$. By Lemmas 4.12 and 4.14 we may assume that $\lambda \circ V^{-1} \leq \lambda$ on $\mathbb{R}_+$. For every $n \in \mathbb{N}$, we may introduce the predictable process

$$U_n(t) = \begin{cases} V_t + \infty \cdot 1\{V_t > n\}, & t \leq n, \\ \inf\{s \geq 0; \ s - \lambda_n \circ V^{-1}(0, s \wedge n] = t - n\}, & t > n, \end{cases}$$

where $\lambda_n = \lambda(\cdot \cap [0, n])$. Then clearly $\lambda \circ U_n^{-1} = \lambda$ a.s. on $\mathbb{R}_+$, and so by Theorem 4.7 we have $Y^n = X \circ U_n^{-1} \stackrel{d}{=} X$. On the other hand, Lemma 4.9 yields $Y_t^n \stackrel{P}{\to} Y_t$ on $\{\zeta > t\}$ for every $t \geq 0$. The assertion now follows by Lemma 4.15.   □

4.4   Predictable Contraction

Since the conclusion of Theorem 4.7 implies that X is exchangeable, it must be false for more general contractable processes on $[0, 1]$. All we can hope for, in general, is then to prove a continuous-time version of the optional skipping property in Proposition 4.1. The obvious continuous-time analogue would be to consider a predictable set $A \subset [0, 1]$ of fixed length $\lambda A = h$ and define the *A-contraction* $C_A X$ of X by

$$(C_A X)_t = X(A \cap [0, \tau_t]) = \int_0^{\tau_t} 1_A(s) dX_s, \quad t \in [0, h], \tag{8}$$

where the random time scale (τ_t) is given by

$$\tau_t = \inf\{s \in [0, 1]; \ \lambda(A \cap [0, s]) > t\}, \quad t \in [0, h]. \tag{9}$$

It is reasonable to expect that $C_A X \stackrel{d}{=} X$ on $[0, h]$. Unfortunately, already the definition of $C_A X$ in terms of stochastic integrals seems to require X to be a semi-martingale (cf. FMP 26.21), a property that has only been established, in Theorems 2.13 and 2.25, under an extra moment condition. Before the general result can even be stated, we therefore need to extend the stochastic integral in (8) to arbitrary contractable processes X.

Then, given an $\mathcal{F}$-contractable process X on $[0, 1]$ and an $\mathcal{F}$-predictable set $A \subset [0, 1]$, we need to construct the process $X_A(t) = X(A_t)$, where $A_t = A \cap [0, t]$. When A is simple, so that $A = \bigcup_{j \leq m} (\sigma_j, \tau_j]$ for some disjoint, predictable intervals $(\sigma_j, \tau_j]$, we may define X_A as the *elementary* integral

$$X_A(t) = \sum_{j \leq m} (X_{t \wedge \tau_j} - X_{t \wedge \sigma_j}), \quad t \in [0, 1]. \tag{10}$$

We proceed to show that X_A has an additive and continuous extension to arbitrary predictable sets $A \subset [0,1]$. Here the mapping $A \mapsto X_A$ is said to be *additive* on the predictable σ-field $\mathcal{P}$ if $X_{A \cup B} = X_A + X_B$ a.s. for any disjoint sets $A, B \in \mathcal{P}$.

Proposition 4.16 *(selection integral) Let X be an $\mathcal{F}$-contractable process on $[0,1]$. Then the elementary integral in (10) extends, a.s. uniquely, to an additive map $A \mapsto X_A$ on the $\mathcal{F}$-predictable σ-field $\mathcal{P}$, such that $\lambda A_n \xrightarrow{P} 0$ implies $(X_{A_n})_t^* \xrightarrow{P} 0$ for all $t \in [0,1)$. Furthermore, X_A is a.s. rcll on $[0,1)$ with $\Delta X_A = 1_A \Delta X$, and we have $X_A = 1_A \cdot X$ a.s. whenever X is a semimartingale on $[0,1)$.*

First we consider the case of simple, predictable sets.

Lemma 4.17 *(elementary integral) Let X be an $\mathcal{F}$-contractable process on $[0,1]$, and consider some simple, $\mathcal{F}$-predictable sets $A, A_1, A_2, \ldots \subset [0,1]$. Then*

(i) $\lambda A \geq h$ *a.s. implies* $C_A X \overset{d}{=} X$ *on $[0,h]$;*

(ii) $\lambda A_n \xrightarrow{P} 0$ *implies* $(X_{A_n})_t^* \xrightarrow{P} 0$ *for all $t \in [0,1)$.*

Proof: (i) We may assume that A has fixed length h and its interval endpoints belong to the set $\{t_{n,k}; \ k \leq n\}$ for a fixed $n \in \mathbb{N}$, where $t_{n,k} = k/n$. Then A is a union of some intervals $I_{n,k} = (t_{n,k-1}, t_{n,k}]$, and so by Lemma 4.10 we may write $A = \bigcup_{k \leq nh} I_{n,\tau_k}$, where the times $\tau_1 < \cdots < \tau_{nh}$ are predictable with respect to the discrete filtration $\mathcal{G}_k = \mathcal{F}_{k/n}$, $k = 0, \ldots, n$. Noting that the processes

$$Y_k(t) = X(t_{n,k-1} + t) - X(t_{n,k-1}), \quad t \in [0, n^{-1}], \ k = 1, \ldots, n,$$

form a $\mathcal{G}$-contractable sequence, we get by Proposition 4.1

$$(Y_{\tau_1}, \ldots, Y_{\tau_{nh}}) \overset{d}{=} (Y_1, \ldots, Y_{nh}),$$

and the assertion follows.

(ii) Fix any rational time $t \in (0,1)$, and put $B_n = A_n \cup (t,1]$. Then $\lambda B_n \geq 1 - t$, and so by (i) we have $C_{B_n} X \overset{d}{=} X$ on $[0, 1-t]$. Noting that $X_A = C_A X \circ \lambda_A$ with $\lambda_A(t) = \lambda(A \cap [0,t])$, we get for any $h \in (0, 1-t)$

$$\begin{aligned} E[(X_{A_n})_t^* \wedge 1] &\leq E[(C_{B_n} X)_h^* \wedge 1] + P\{\lambda A_n > h\} \\ &= E[X_h^* \wedge 1] + P\{\lambda A_n > h\}, \end{aligned}$$

which tends to 0 as $n \to \infty$ and then $h \to 0$ by the right continuity of X. This shows that $(X_{A_n})_t^* \xrightarrow{P} 0$. $\square$

Proof of Proposition 4.16: Fix any predictable set $A \subset [0,1]$. Proceeding as in Lemma 4.8, we may choose some simple, predictable sets $A_1, A_2, \ldots \subset$

$[0, 1]$ such that $\lambda(A \Delta A_n) \xrightarrow{P} 0$. Then also $\lambda(A_m \Delta A_n) \xrightarrow{P} 0$ as $m, n \to \infty$, and so by Lemma 4.17 (ii) we have for any $t < 0$

$$
\begin{aligned}
(X_{A_m} - X_{A_n})_t^* &= (X_{A_m \backslash A_n} - X_{A_n \backslash A_m})_t^* \\
&\leq (X_{A_m \backslash A_n})_t^* + (X_{A_n \backslash A_m})_t^* \xrightarrow{P} 0.
\end{aligned}
$$

Hence, there exists a process X_A satisfying

$$
(X_A - X_{A_n})_t^* \xrightarrow{P} 0, \quad t \in [0, 1),
$$

and we note that X_A is rcll on $[0, 1)$, since this property holds trivially for the processes X_{A_n}. To see that X_A is independent of the choice of approximating sequence $A_1, A_2, \ldots$, consider another such sequence $B_1, B_2, \ldots$, and apply the previous argument to the alternating sequence $A_1, B_1, A_2, B_2, \ldots$ to see that even $(X_A - X_{B_n})_t^* \xrightarrow{P} 0$.

To prove the stated continuity, let $A_1, A_2, \ldots \subset [0, 1]$ be predictable with $\lambda A_n \xrightarrow{P} 0$, and fix any $t \in (0, 1)$. Approximating as before for each $n \in \mathbb{N}$, we may choose some simple, predictable sets $B_1, B_2, \ldots \subset [0, 1]$ such that

$$
\lambda(A_n \Delta B_n) \xrightarrow{P} 0, \qquad (X_{A_n} - X_{B_n})_t^* \xrightarrow{P} 0.
$$

Then even $\lambda B_n \xrightarrow{P} 0$, and so by Lemma 4.17 (ii)

$$
(X_{A_n})_t^* \leq (X_{B_n})_t^* + (X_{A_n} - X_{B_n})_t^* \xrightarrow{P} 0.
$$

To prove the asserted additivity, consider any disjoint, predictable sets $A, B \subset [0, 1]$. As before, we may choose some simple, predictable sets A_n and B_n such that

$$
\lambda(A \Delta A_n) \xrightarrow{P} 0, \qquad \lambda(B \Delta B_n) \xrightarrow{P} 0. \tag{11}
$$

Here the second relation remains true with B_n replaced by $B_n \backslash A_n$, and so we may assume that even A_n and B_n are disjoint. It is also clear from (11) that $\lambda((A \cup B) \Delta (A_n \cup B_n)) \xrightarrow{P} 0$. Noting that trivially $X_{A_n \cup B_n} = X_{A_n} + X_{B_n}$ a.s., we get for any $t \in (0, 1)$

$$
\begin{aligned}
(X_{A \cup B} - X_A - X_B)_t^* &\leq (X_{A \cup B} - X_{A_n \cup B_n})_t^* \\
&\quad + (X_A - X_{A_n})_t^* + (X_B - X_{B_n})_t^* \xrightarrow{P} 0,
\end{aligned}
$$

which shows that $X_{A \cup B} = X_A + X_B$ a.s.

The relation $\Delta X_A = 1_A \Delta X$ is clearly true for simple, predictable sets $A \subset [0, 1]$. To extend the result to general A, we may again choose some simple, predictable sets A_n such that $\lambda(A \Delta A_n) \xrightarrow{P} 0$. Then $(X_A - X_{A_n})_t^* \xrightarrow{P} 0$ for all $t \in [0, 1)$, which implies

$$
(\Delta X_A - \Delta X_{A_n})_t^* \xrightarrow{P} 0, \quad t \in [0, 1).
$$

It remains to show that

$$(1_A \Delta X - 1_{A_n} \Delta X)^*_t \xrightarrow{P} 0, \quad t \in [0,1).$$

Equivalently, given any predictable sets A_n with $\lambda A_n \xrightarrow{P} 0$, we need to show that $(1_{A_n} \Delta X)^*_t \xrightarrow{P} 0$ for every $t < 1$. Letting ξ denote the jump point process of X, and putting $B_\varepsilon = \{x; |x| > \varepsilon\}$ and $A^t_n = A_n \cap [0, t]$, we must prove that $\xi(A^t_n \times B_\varepsilon) \xrightarrow{P} 0$ for every $\varepsilon > 0$. But this is another consequence of the previously established continuity property.

If X is a semi-martingale on $[0, 1)$, then clearly $X_A = 1_A \cdot X$ a.s., as long as the predictable set A is simple. For general A, suppose that X has a semi-martingale decomposition $M + V$. Then choose the approximating sets A_n such that also

$$\int_0^t 1_{A \Delta A_n} d[M] + \int_0^t 1_{A \Delta A_n} |dV| \xrightarrow{P} 0, \quad t \in [0, 1).$$

Letting $n \to \infty$ for fixed $t \in [0, 1)$, we get by FMP 26.13

$$(X_A - 1_A \cdot X)^*_t \leq (X_A - X_{A_n})^*_t + (1_A \cdot X - 1_{A_n} \cdot X)^*_t \xrightarrow{P} 0,$$

which shows that $X_A = 1_A \cdot X$ a.s. $\qquad\square$

Defining $C_A X$ by (8) and (9), in terms of the stochastic integral in Proposition 4.16, we may now state a continuous-time version of the optional skipping property in Proposition 4.1.

Theorem 4.18 *(predictable contraction)* *Let X be an $\mathcal{F}$-contractable process on $[0, 1]$, and consider an $\mathcal{F}$-predictable set $A \subset [0, 1]$ with $\lambda A \geq h$ a.s. Then $C_A X \overset{d}{=} X$ on $[0, h)$.*

By the right continuity of the processes X_A and (τ_t), we note that even $C_A X$ is right-continuous on $[0, h)$. Thus, the relation $C_A X \overset{d}{=} X$ holds in the function space $D([0, h), \mathbb{R}^d)$. Defining $(C_A X)_h$ as the left-hand limit $(C_A X)_{h-}$, we may extend the result to the interval $[0, h]$.

Proof: Fix any rational times $t_1 < \cdots < t_m$ in $[0, h)$, and define

$$A_j = A \cap (\tau_{t_{j-1}}, \tau_{t_j}], \quad j = 1, \ldots, m,$$

where $t_0 = 0$. Since the τ_t are optional times, the sets $A_1, \ldots, A_m$ are again predictable. From the definition of τ_t it is also clear that $\lambda A_j = t_j - t_{j-1}$ for all j. Using Lemma 4.8 we may choose, for every $n \in \mathbb{N}$, some disjoint, simple, predictable sets $A_{j,n}$ such that $\lambda A_{j,n} = \lambda A_j$ a.s. and $\lambda(A_j \Delta A_{j,n}) \xrightarrow{P} 0$. By a suitable re-numbering, we may clearly assume that each $A_{j,n}$ is a union of intervals $I_{nk} = n^{-1}(k-1, k]$. Trading intervals between the sets, if necessary, we may also arrange that $A_{j-1,n} \leq A_{j,n}$ for all j, in the sense that the conditions $s \in A_{j-1,n}$ and $t \in A_{j,n}$ imply $s < t$.

Now Lemma 4.10 yields a representation of the sets $A_{j,n}$ as in (6), where the times τ_{njk} in the second subscript are predictable with respect to the discrete filtration $\mathcal{G}_k^n = \mathcal{F}_{k/n}$, $k \leq n$. For every j, the times associated with $A_{j-1,n}$ are smaller than those for $A_{j,n}$. Since the increments $X(I_{nk})$ are $\mathcal{G}^n$-contractable, Proposition 4.1 yields

$$(X(A_{1,n}), \ldots, X(A_{m,n})) \stackrel{d}{=} (X_{t_1} - X_{t_0}, \ldots, X_{t_m} - X_{t_{m-1}}).$$

Letting $n \to \infty$ and using the continuity property in Proposition 4.16, we obtain

$$\begin{aligned}((C_A X)_{t_1}, \ldots, (C_A X)_{t_m}) &= (X(U_1), \ldots, X(U_m)) \\ &\stackrel{d}{=} (X_{t_1}, \ldots, X_{t_m}),\end{aligned}$$

where $U_k = \bigcup_{j \leq k} A_j$. This shows that $C_A X \stackrel{d}{=} X$ on the set of rational numbers in $[0, h)$, and the general relation follows by the right continuity on each side. $\quad\square$

4.5 Brownian and Stable Invariance

Here we consider some special cases where the predictable mapping Theorem 4.7 can be improved. For any processes U and V on $I = \mathbb{R}_+$ or $[0, 1]$, we define

$$\lambda V = \int_I V_t dt, \qquad \langle U, V \rangle = \int_I U_t V_t dt, \qquad \|V\|_2 = \langle V, V \rangle^{1/2}.$$

Recall from Theorem 2.25 that an $\mathcal{F}$-exchangeable Brownian bridge in $\mathbb{R}^d$ is a continuous semi-martingale on $[0, 1]$. The associated martingale component is of course a standard Brownian motion.

Theorem 4.19 (*Brownian invariance*)

(i) *Let X be a Brownian motion with respect to $\mathcal{F}$, and consider some $\mathcal{F}$-predictable processes V^t on $\mathbb{R}_+$, $t \geq 0$, such that*

$$\langle V^s, V^t \rangle = s \wedge t \quad a.s., \qquad s, t \geq 0.$$

Then the process $Y_t = \int V^t dX$ is again a Brownian motion.

(ii) *Let X be an $\mathcal{F}$-exchangeable Brownian bridge, and consider some $\mathcal{F}$-predictable processes V^t on $[0, 1]$, $t \in [0, 1]$, such that*

$$\lambda V^t = t, \quad \langle V^s, V^t \rangle = s \wedge t \quad a.s., \qquad s, t \in [0, 1].$$

Then the process $Y_t = \int V^t dX$ exists and is again a Brownian bridge.

To deal with case (ii), we need to convert the relevant stochastic integrals with respect to a Brownian bridge into continuous martingales. This is accomplished by the following result. Given a Lebesgue integrable process V on $[0, 1]$, we define

$$\overline{V}_t = (1 - t)^{-1} \int_t^1 V_s ds, \quad t \in [0, 1).$$

Lemma 4.20 *(Brownian bridge integral)*

(i) *Let X be an $\mathcal{F}$-exchangeable Brownian bridge with martingale component B, and let V be an $\mathcal{F}$-predictable process on $[0, 1]$, satisfying $E^{\mathcal{F}_0} \int_0^1 V^2 < \infty$ a.s. and such that λV is $\mathcal{F}_0$-measurable. Then*

$$\int_0^1 V_t \, dX_t = \int_0^1 (V_t - \overline{V}_t) \, dB_t \quad a.s.$$

(ii) *For any processes U and V in $L^2[0, 1]$, we have*

$$\int_0^1 (U_t - \overline{U}_t)(V_t - \overline{V}_t) \, dt = \int_0^1 U_t V_t \, dt - \lambda U \cdot \lambda V.$$

Proof: (i) Recall from Theorem 2.13 that $M_t = X_t / (1 - t)$ is a martingale on $[0, 1)$. Integrating by parts, we get for any $t \in [0, 1)$

$$dX_t = (1 - t)dM_t - M_t dt = dB_t - M_t dt, \tag{12}$$

and also

$$
\begin{aligned}
\int_0^t V_s M_s \, ds &= M_t \int_0^t V_s ds - \int_0^t dM_s \int_0^s V_r dr \\
&= \int_0^t dM_s \int_s^1 V_r dr - M_t \int_t^1 V_s ds \\
&= \int_0^t \overline{V}_s dB_s - X_t \overline{V}_t.
\end{aligned}
$$

Combining this with (12) gives

$$\int_0^t V_s dX_s = \int_0^t (V_s - \overline{V}_s)dB_s + X_t \overline{V}_t, \quad t \in [0, 1). \tag{13}$$

Applying the Cauchy–Buniakovsky inequality twice and using dominated convergence, we get as $t \to 1$

$$
\begin{aligned}
\left(E^{\mathcal{F}_0} |X_t \overline{V}_t| \right)^2 &\leq E^{\mathcal{F}_0}(M_t^2) E^{\mathcal{F}_0} \left(\int_t^1 |V_s| ds \right)^2 \\
&\leq (1 - t) E^{\mathcal{F}_0}(M_t^2) E^{\mathcal{F}_0} \int_t^1 V_s^2 ds \\
&= t E^{\mathcal{F}_0} \int_t^1 V_s^2 ds \to 0,
\end{aligned}
$$

and so by dominated convergence $X_t \overline{V}_t \xrightarrow{P} 0$. Using (13) and invoking the dominated convergence property of stochastic integrals (FMP 17.13), it remains to show that $\overline{V}$ is B-integrable. In other words, we need to prove that $\int_0^1 \overline{V}_t^2 dt < \infty$ a.s. This will be established as part of (ii).

(ii) We may take $U = V$, since the general case will then follow by polarization. Writing $R_t = \int_t^1 V_s ds = (1 - t)\overline{V}_t$ and integrating by parts, we get on $[0, 1)$

$$
\begin{aligned}
\int \overline{V}_t^2 dt &= \int (1 - t)^{-2} R_t^2 dt \\
&= (1 - t)^{-1} R_t^2 + 2 \int (1 - t)^{-1} R_t V_t dt \\
&= (1 - t)\overline{V}_t^2 + 2 \int \overline{V}_t V_t dt.
\end{aligned}
$$

For bounded V, we conclude that

$$
\int_0^1 (V_t - \overline{V}_t)^2 dt = \int_0^1 V_t^2 dt - (\lambda V)^2 = \int_0^1 (V_t - \lambda V)^2 dt, \tag{14}
$$

and so by Minkowski's inequality

$$
\begin{aligned}
\|\overline{V}\|_2 &\leq \|V\|_2 + \|V - \overline{V}\|_2 \\
&= \|V\|_2 + \|V - \lambda V\|_2 \leq 2\|V\|_2,
\end{aligned}
$$

which extends to the general case by monotone convergence. This shows that $\overline{V} \in L^2$, and the asserted relation follows from (14) by dominated convergence. □

Proof of Theorem 4.19: (i) This is immediate from Theorem 4.5.
(ii) By Lemma 4.20 (i) we have a.s.

$$
Y_t = \int_0^1 (V_r^t - \overline{V}_r^t) dB_r, \quad t \in [0, 1],
$$

where B denotes the Brownian motion $X - \hat{X}$. Furthermore, we see from part (ii) of the same result that

$$
\int_0^1 (V_r^s - \overline{V}_r^s)(V_r^t - \overline{V}_r^t) dr = \int_0^1 V_r^s V_r^t dr - \lambda V^s \cdot \lambda V^t = s \wedge t - st.
$$

The assertion now follows by Theorem 4.5. □

We turn to a study of stable processes. Recall that a Lévy process X is said to be *strictly p-stable* if it satisfies the *self-similarity* or scaling relation $X_{rt} \stackrel{d}{=} r^{1/p} X_t$ for every $r > 0$ (FMP 15.9). If the same relation holds apart from a linear drift term $b_r t$, we say that X is *weakly p-stable*. Excluding the trivial case where $X \equiv 0$, we see from Lemma 1.33 that p-stability may occur only for $p \in (0, 2]$.

A strictly 2-stable Lévy process is simply a constant multiple of Brownian motion. For $p \in (0, 2)$, a Lévy process is known to be weakly p-stable iff it is purely discontinuous with Lévy measure $\nu(dx) = c_{\pm}|x|^{-p-1}dx$ on $\mathbb{R}_{\pm}$, for some constants $c_{\pm} \geq 0$. A weakly p-stable Lévy process with $p \neq 1$ can be reduced, by a suitable centering, to a strictly p-stable process. Such a reduction is no longer possible when $p = 1$, since the linear process $X_t = t$ is itself strictly 1-stable. In fact, any strictly 1-stable process is symmetric apart from a linear drift term, hence a *Cauchy process* with drift. In other words, strict 1-stability implies $c_+ = c_-$.

When $p \in (0, 1)$, a strictly p-stable Lévy process X has locally finite variation and its continuous component vanishes. If instead $p \in (1, 2)$, then X is a purely discontinuous martingale. Though a strictly or weakly 1-stable process has neither of these properties, it is still a purely discontinuous semi-martingale. Given a filtration $\mathcal{F}$, we say that a process X is $\mathcal{F}$-*Lévy* if it is both a Lévy process and a Markov process with respect to $\mathcal{F}$. By Theorem 2.25 we may then assume that $\mathcal{F}$ is right-continuous and complete.

Before we can state the basic invariance properties of stable Lévy processes, we need to examine the matter of integrability. Given a semi-martingale X and a predictable process V, we say that V is X-integrable on the closed interval $[0, \infty]$ and write $V \in \overline{L}(X)$ if X is locally integrable in the usual sense, and the integral process $V \cdot X$ extends to a semi-martingale on $[0, \infty]$. By specializing our conditions to processes V supported by finite intervals $[0, t]$, we obtain criteria for the local integrability $V \in L(X)$.

Proposition 4.21 *(stable integrals) Consider a p-stable $\mathcal{F}$-Lévy process X and an $\mathcal{F}$-predictable process V. Then $V \in \overline{L}(X)$ iff*

   (i) $V \in L^p$ *a.s., when X is strictly p-stable;*

  (ii) $V \in L^p \cap L^1$ *a.s., when X is weakly p-stable for some $p \neq 1$;*

 (iii) $V \in L^1 \cap L \log L$ *a.s., when X is weakly 1-stable.*

Here $L \log L$ denotes the class of functions f on $\mathbb{R}_+$ such that $|f \log |f||$ is integrable. Our proof depends on a random time change, based on the following simple invariance property of the underlying Lévy measure, which will also be needed later.

Lemma 4.22 *(invariance of Lévy measure) Consider a p-stable Lévy measure ν on $\mathbb{R} \setminus \{0\}$ and some measurable functions U and $V \geq 0$ on $\mathbb{R}_+$, and put $W_{x,t} = (xU_t, V_t)$. Assume that ν is symmetric or $U \geq 0$. Then*

$$(|U|^p \cdot \lambda) \circ V^{-1} = \lambda$$

implies

$$(\nu \otimes \lambda) \circ W^{-1} = \nu \otimes \lambda.$$

Proof: By Fubini's theorem, we have for any $r, t > 0$

$$
\begin{aligned}
(\nu \otimes \lambda) \circ W^{-1}((r, \infty) \times [0, t]) &= (\nu \otimes \lambda)\{(x, s);\ xU_s > r,\ V_s \leq t\} \\
&= \int_0^\infty \nu\{x;\ xU_s > r\} 1\{V_s \leq t\}\, ds \\
&= \nu(r, \infty) \int_0^\infty 1\{V_s \leq t\} |U_s|^p\, ds \\
&= t\,\nu(r, \infty),
\end{aligned}
$$

and similarly for the set $(-\infty, -r) \times [0, t]$. The assertion now follows by a monotone-class argument. $\qquad\square$

In subsequent proofs we shall often, without further comments, use the device of *trans-finite extension*, which allows us to apply results like Theorem 4.5 when there is only inequality $\hat{\xi} \circ V^{-1} \leq \mu$ a.s. on S. The idea is to extend X and V, along with the underlying filtration $\mathcal{F}$, to a second copy of the time axis $\mathbb{R}_+$, such that the extended version satisfies $\hat{\xi} \circ V^{-1} = \mu$ a.s. The quoted result then ensures that $\xi \circ V^{-1}$ can be extended to a Poisson process η on S with $E\eta = \mu$.

Proof of Proposition 4.21: The case $p = 2$ being classical, we may assume that $p \in (0, 2)$. Considering separately the processes $V^\pm = (\pm V) \vee 0$, we may reduce to the case where $V \geq 0$. First suppose that $V \in \overline{L}(X)$. For any $n \in \mathbb{N}$ we define

$$
V_{n,t} = V_t \wedge n, \qquad T_{n,t} = (V_n^p \cdot \lambda)_t, \qquad W_{x,t} = (xV_{n,t}, T_{n,t}).
$$

Letting ξ denote the jump point process of X, we note that the process $(V_n \cdot X) \circ T_n^{-1}$ has jump point process $\xi \circ W_n^{-1}$. Since ξ has compensator $\hat{\xi} = \nu \otimes \lambda$ and $\hat{\xi} \circ W_n^{-1} = \nu \otimes \lambda$ on $[0, \lambda V_n^p]$ by Lemma 4.22, the extended version of Theorem 4.5 yields the existence of a Poisson process $\eta_n \overset{d}{=} \xi$ on $\mathbb{R} \times \mathbb{R}_+$ satisfying $\xi \circ W_n^{-1} = \eta_n$ a.s. on $[0, \lambda V_n^p]$.

Now introduce the associated quadratic variation processes

$$
Q_{n,t} = \int\!\!\int_0^t x^2 \eta_n(dx\, ds), \quad t \geq 0,
$$

and note that $[V_n \cdot X]_\infty = Q_n \circ \lambda V_n^p$. For any $t, r > 0$ we have

$$
\begin{aligned}
P\{\lambda V_n^p > t\} &\leq P\{Q_{n,t} \leq r\} + P\{\lambda V_n^p > t,\ Q_{n,t} > r\} \\
&\leq P\{[X]_t \leq r\} + P\{[V_n \cdot X]_\infty > r\}.
\end{aligned}
$$

Noting that $\lambda V_n^p \to \lambda V^p$ and $[V_n \cdot X]_\infty \to [V \cdot X]_\infty$ by monotone convergence as $n \to \infty$, we obtain

$$
P\{\lambda V^p > t\} \leq P\{[X]_t \leq r\} + P\{[V \cdot X]_\infty > r\}.
$$

Since $[X]_\infty = \infty$ and $[V \cdot X]_\infty < \infty$ a.s., the right-hand side tends to 0 as $t \to \infty$ and then $r \to \infty$, which implies $\lambda V^p < \infty$ a.s. This proves the necessity of the condition $V \in L^p$ in (i)–(iii).

To prove the sufficiency when $p \neq 1$, we may take X to be strictly p-stable and let $V \in L^p$. Considering separately the positive and negative jumps, we may assume that again $c_- = 0$. Writing $T_t = (V^p \cdot \lambda)_t$ and $W_t = (xV_t, T_t)$, we see as before that $\xi \circ W^{-1} = \tilde\xi$ a.s. on $[0, \lambda V^p]$, where $\tilde\xi \overset{d}{=} \xi$. Assuming first that $p \in (0, 1)$, we get a.s.

$$
\begin{aligned}
(V \cdot X)_\infty &= \int\!\!\int_0^\infty V_s x \, \xi(dx\,ds) \\
&= \int\!\!\int_0^{\lambda V^p} y\, \tilde\xi(dy\,dr) = \tilde X \circ \lambda V^p < \infty
\end{aligned}
$$

for some process $\tilde X \overset{d}{=} X$, which shows that $V \in \overline{L}(X)$. If instead $p \in (1, 2)$, we note that

$$
\int_0^\infty V^2 d[X] = \int\!\!\int_0^\infty V_s^2 x^2 \xi(dx\,ds) = \int\!\!\int_0^{\lambda V^p} x^2 \tilde\xi(dx\,ds).
$$

Writing $\tilde\xi_t = \tilde\xi(\cdot \times [0, t])$ and using Jensen's inequality and subadditivity, we get

$$
\begin{aligned}
E\left(\int_0^\infty x^2 \tilde\xi_t(dx)\right)^{1/2} &\leq E\left(\int_0^1 x^2 \tilde\xi_t(dx)\right)^{1/2} + E\left(\int_1^\infty x^2 \tilde\xi_t(dx)\right)^{1/2} \\
&\leq \left(E\int_0^1 x^2 \tilde\xi_t(dx)\right)^{1/2} + E\int_1^\infty x \tilde\xi_t(dx) \\
&= \left(t\int_0^1 x^2 \nu(dx)\right)^{1/2} + t\int_1^\infty x\nu(dx) < \infty.
\end{aligned}
$$

Hence, $V \in L(X)$ by FMP 26.13. Since $\lambda V^p < \infty$ a.s., we see that again $V \in \overline{L}(X)$. This completes the proof of (i) for $p \neq 1$. It also proves (ii), since if X is weakly p-stable with drift $\hat X$ and $V \in \overline{L}(X)$, we have $\lambda V^p < \infty$ a.s., which implies $V \in \overline{L}(X - \hat X)$. But then also $V \in \overline{L}(\hat X)$, which means that $V \in L^1$ a.s.

It remains to take $p = 1$. By the necessity part we may then take $V \in L^1$, which allows us to choose an arbitrary centering. For convenience, we may assume that $X = M + J$, where M is a purely discontinuous martingale with jumps bounded by ± 1 and J is a pure step process containing all the remaining jumps. Now introduce the processes

$$
\begin{aligned}
A_t &= \sum_{s \leq t} \Delta M_s 1\{|V_s \Delta M_s| > 1\} \\
&= \int\!\!\int_0^t x 1\{|x| \leq 1 < |xV_s|\}\, \xi(dx\,ds), \quad\quad (15) \\
B_t &= \sum_{s \leq t} \Delta J_s 1\{|V_s \Delta J_s| \leq 1\} \\
&= \int\!\!\int_0^t x 1\{|xV_s| \leq 1 < |x|\}\, \xi(dx\,ds), \quad\quad (16)
\end{aligned}
$$

and note that both A and $V \cdot B$ have isolated jumps of modulus ≤ 1. In fact, assuming $V \geq 0$ and $c_- = 0$, we have in case of A

$$\sum_{s \leq t} 1\{\Delta A_s > 0\} \leq \int\!\!\int_0^t 1\{xV_s > 1\}\, \xi(dx\, ds)$$
$$= \tilde{\xi}((1, \infty) \times [0, \lambda V]) < \infty.$$

Recalling that V is predictable and writing $c = c_+ - c_-$, we may evaluate the associated compensators as

$$\hat{A}_t = \int\!\!\int_0^t x 1\{|x| \leq 1 < |xV_s|\}\, \hat{\xi}(dx\, ds)$$
$$= \int_0^t ds \int x 1\{|x| \leq 1 < |xV_s|\}\, \nu(dx)$$
$$= c \int_0^t ds \int 1\{|V_s|^{-1} < x \leq 1\}\, x^{-1}\, dx$$
$$= c \int_0^t \log(|V_s| \vee 1)\, ds,$$

$$\hat{B}_t = \int\!\!\int_0^t x 1\{|xV_s| \leq 1 < |x|\}\, \hat{\xi}(dx\, ds)$$
$$= \int_0^t ds \int x 1\{|xV_s| \leq 1 < |x|\}\, \nu(dx)$$
$$= c \int_0^t ds \int 1\{1 < x \leq |V_s|^{-1}\}\, x^{-1}\, dx$$
$$= -c \int_0^t \log(|V_s| \wedge 1)\, ds.$$

Let us finally introduce the L^2-martingale $N = M - (A - \hat{A})$, and note that X has the semi-martingale decomposition

$$X = N + (J + A - \hat{A})$$
$$= N + (J + A - B) + (B - \hat{B}) - (\hat{A} - \hat{B}). \tag{17}$$

Assuming $V \geq 0$, we obtain

$$\left(V^2 \cdot \langle N + B - \hat{B} \rangle\right)_\infty = \int\!\!\int_0^\infty (xV_s)^2 1\{|xV_s| \leq 1\}\, \hat{\xi}(dx\, ds)$$
$$= \int_{-1}^1 \int_0^{\lambda V} x^2 \hat{\xi} \circ W^{-1}(dx\, ds)$$
$$= \lambda V \int_{-1}^1 x^2 \nu(dx) < \infty.$$

This shows that $V \in \overline{L}(N)$, already in the elementary sense of L^2-integration (FMP 26.2), and also that the process $V \cdot (B - \hat{B})$ extends to a local martingale on $[0, \infty]$. Next we note that

$$(V \cdot (J + A - B))_t = \sum_{s \leq t} V_s \Delta X_s 1\{|V_s \Delta X_s| > 1\}$$
$$= \int\!\!\int_0^t xV_s 1\{|xV_s| > 1\}\, \xi(dx\, ds), \tag{18}$$

which for $V \geq 0$ has total variation

$$\iint |xV_s| 1\{|xV_s| > 1\} \, \xi(dx \, ds) = \int_{|x|>1} \int_0^{\lambda V} |x| \, \tilde{\xi}(dx \, ds) < \infty.$$

Thus, the first three terms in (17) give rise to semi-martingales on $[0, \infty]$, and so the global integrability $V \in \overline{L}(X)$ holds iff the process

$$\left(V \cdot (\hat{A} - \hat{B})\right)_t = c \int_0^t V_s \log |V_s| \, ds, \quad t \geq 0, \tag{19}$$

exists and has finite variation on $\mathbb{R}_+$. For $c \neq 0$, this is precisely the condition in (iii). □

For any $\mathcal{F}$-predictable processes U and $V \geq 0$ satisfying appropriate integrability conditions, we define the transformed process $(U \cdot X) \circ V^{-1}$ by

$$\left((U \cdot X) \circ V^{-1}\right)_t = \int_0^\infty 1\{V_s \leq t\} \, U_s dX_s, \quad t \geq 0. \tag{20}$$

We proceed to show that if X is symmetric p-stable, or if it is strictly p-stable and $U \geq 0$, then the condition

$$(|U|^p \cdot \lambda) \circ V^{-1} = \lambda \quad \text{a.s.} \tag{21}$$

ensures that the transformation in (20) will preserve the distribution of X. The results are easily modified, through a suitable centering, to cover the case of weakly p-stable Lévy processes X, as long as $p \neq 1$. The weakly 1-stable case is more delicate and requires a separate treatment.

Theorem 4.23 *(stable invariance) Let X be a p-stable $\mathcal{F}$-Lévy process, and let U and $V \geq 0$ be $\mathcal{F}$-predictable processes satisfying (21) and such that $(U \cdot X) \circ V^{-1}$ exists.*

(i) *If X is symmetric p-stable, or if X is strictly p-stable and $U \geq 0$ a.s., then*
$$(U \cdot X) \circ V^{-1} \overset{d}{=} X.$$

(ii) *If X is weakly 1-stable with $c_+ - c_- = c$ and $U \geq 0$ a.s., then*
$$(U \cdot X) \circ V^{-1} + c \, (U \log U \cdot \lambda) \circ V^{-1} \overset{d}{=} X.$$

Proof: The result for $p = 2$ follows from Theorem 4.19 (i) with

$$V_r^t = U_r 1\{V_r \leq t\}, \quad r, t \geq 0,$$

since for any $s, t \geq 0$ we have by (21)

$$\begin{aligned}
\langle V^s, V^t \rangle &= \int_0^\infty U_r^2 1\{V_r \leq s \wedge t\} \, dr \\
&= (U^2 \cdot \lambda) \circ V^{-1}[0, s \wedge t] \\
&= \lambda[0, s \wedge t] = s \wedge t.
\end{aligned}$$

When $p \in (0,2)$, the jump point process ξ of X is Poisson with $\mathcal{F}$-compensator $\hat{\xi} = \nu \otimes \lambda$ a.s., where ν is the p-stable Lévy measure of X. Writing $W_{x,t} = (xU_t, V_t)$, we see from (21) and Lemma 4.22 that $\hat{\xi} \circ W^{-1} = \nu \otimes \lambda$ a.s., and so by Theorem 4.5 we have $\eta \equiv \xi \circ W^{-1} \stackrel{d}{=} \xi$. Assuming first that $p \in (0,1)$, we obtain

$$
\begin{aligned}
\left((U \cdot X) \circ V^{-1} \right)_t &= \int_0^\infty 1\{V_s \le t\} U_s dX_s \\
&= \int\int_0^\infty 1\{V_s \le t\} U_s x\, \xi(dx\, ds) \\
&= \int\int_0^t y\, (\xi \circ W^{-1})(dy\, dr) = \int\int_0^t y\, \eta(dy\, dr), \quad (22)
\end{aligned}
$$

which is a process with the same distribution as X.

Now let $p \in (1,2)$, so that X is a quasi-left-continuous and purely discontinuous martingale. Fixing any times $t_1 < \cdots < t_m$ and putting $I_k = (t_{k-1}, t_k]$ with $t_0 = 0$, we may introduce the strongly orthogonal martingales $M_k = 1_{I_k}(V) \cdot X$. Writing $Y = (U \cdot X) \circ V^{-1}$, we note that

$$
\gamma_k \equiv Y_{t_k} - Y_{t_{k-1}} = \int_0^\infty U dM_k, \quad k = 1, \ldots, m. \quad (23)
$$

The M_k have jump point processes $\xi_k = 1_{I_k}(V) \cdot \xi$ with compensators

$$
\hat{\xi}_k = 1_{I_k}(V) \cdot \hat{\xi} = 1_{I_k}(V) \cdot (\nu \otimes \lambda), \quad k \le m.
$$

Writing $H_{x,t} = xU_t$ and $W_{x,t} = (xU_t, V_t)$, we note that

$$
\begin{aligned}
\hat{\xi}_k \circ H^{-1} &= (\hat{\xi} \circ W^{-1})(\cdot \times I_k) \\
&= (\nu \otimes \lambda)(\cdot \times I_k) = \lambda I_k\, \nu.
\end{aligned}
$$

By Lemma 4.6 it follows that $\gamma_1, \ldots, \gamma_m$ are independent, centered, infinitely divisible random variables with Lévy measures $\lambda I_k \nu$ and vanishing Gaussian components. Hence,

$$
(Y_{t_1}, \ldots, Y_{t_m}) \stackrel{d}{=} (X_{t_1}, \ldots, X_{t_m}), \quad (24)
$$

which shows that $Y \stackrel{d}{=} X$.

It remains to take $p = 1$. If $U \ge 0$, then (21) yields $(U \cdot \lambda) \circ V^{-1} = \lambda$ a.s., which shows that the drift component is a.s. mapped into itself. We may then assume that $X = M + J$, where M is a purely discontinuous martingale with jumps in $[-1, 1]$ and J is a pure jump-type process with jumps of modulus > 1. Defining A and B as in (15) and (16) and putting $\tilde{A} = A - \hat{A}$ and $\tilde{B} = B - \hat{B}$, we see from (17) that

$$
X + (\hat{A} - \hat{B}) = (M - \tilde{A} + \tilde{B}) + (J + A - B).
$$

In view of (19), it is then enough to show that

$$
Y + Z \equiv (U \cdot (M - \tilde{A} + \tilde{B})) \circ V^{-1} + (U \cdot (J + A - B)) \circ V^{-1} \stackrel{d}{=} X.
$$

Letting η denote the restriction of $\xi \circ W^{-1}$ to $[-1,1]^c \times \mathbb{R}_+$, we may proceed as in (18) and (22) to obtain

$$Z_t = ((U \cdot (J + A - B)) \cdot V^{-1})_t = \int\!\!\int_0^t x\,\eta(dx\,ds), \quad t \geq 0.$$

Next we note that (23) remains true with $M_k = 1_{I_k} \cdot X'$, where $X' = M - \tilde{A} + \tilde{B}$ and $I_k = (t_{k-1}, t_k]$ as before. Letting ξ' denote the jump point process of X' and putting $\xi_k = 1_{I_k}(V) \cdot \xi'$, we see as before that $\hat{\xi} \circ H^{-1} = \lambda I_k \nu'$, where ν' denotes the restriction of ν to $[-1,1]$. Using a slightly extended version of Lemma 4.6, we get as in (24)

$$(Y_{t_1}, \ldots, Y_{t_m}, Z) \overset{d}{=} (M_{t_1}, \ldots, M_{t_m}, J)$$

for arbitrary times $t_1 < \cdots < t_m$. This implies $(Y, Z) \overset{d}{=} (M, J)$, and so

$$\left(U \cdot (X + \hat{A} - \hat{B})\right) \circ V^{-1} = Y + Z \overset{d}{=} M + J = X.$$

It remains to note that $\hat{A} = \hat{B} = 0$ a.s. when X is strictly 1-stable. $\square$

The last theorem, in a suitably extended version, leads easily to some interesting time-change representations of stable integrals.

Corollary 4.24 *(time-change representations, Rosiński and Woyczyński, Kallenberg) Consider a p-stable $\mathcal{F}$-Lévy process X and an $\mathcal{F}$-predictable process $V \in L(X)$.*

(i) *If X is symmetric p-stable, there exists a process $Y \overset{d}{=} X$ such that*

$$V \cdot X = Y \circ (|V|^p \cdot \lambda) \quad a.s.$$

(ii) *If X is a strictly p-stable, there exist some mutually independent processes $Y, Z \overset{d}{=} X$ such that*

$$V \cdot X = Y \circ (V_+^p \cdot \lambda) - Z \circ (V_-^p \cdot \lambda) \quad a.s.$$

(iii) *If X is weakly 1-stable with $c_+ - c_- = c$, there exist some mutually independent processes $Y, Z \overset{d}{=} X$ such that a.s.*

$$V \cdot X = Y \circ (V_+ \cdot \lambda) - Z \circ (V_- \cdot \lambda) - c\,(V \log |V|) \cdot \lambda.$$

Our proofs of parts (ii) and (iii) are based on the following multi-variate version of Theorem 4.23, which can be deduced as a simple corollary or may be proved directly by similar methods.

Lemma 4.25 *(multi-variate invariance)* Let X be a p-stable $\mathcal{F}$-Lévy process, and let $U_1, \ldots, U_m, V \geq 0$ be $\mathcal{F}$-predictable processes with $U_i U_j \equiv 0$ a.s. for all $i \neq j$, such that the pairs (U_j, V) satisfy (21).

(i) If X is strictly p-stable, then the processes $X_j = (U_j \cdot X) \circ V^{-1}$ are mutually independent with the same distribution as X.

(ii) If X is weakly 1-stable with $c_+ - c_- = c$, then the previous statement holds for the processes

$$X_j = (U_j \cdot X) \circ V^{-1} + c\,(U_j \log U_j \cdot \lambda) \circ V^{-1}, \quad j = 1, \ldots, m.$$

Proof of Corollary 4.24: (i) Randomizing extensions of V and X, as explained before the proof of Proposition 4.21, we may assume that $\lambda|V|^p = \infty$ a.s. Since the predictable process $T = |V|^p \cdot \lambda$ satisfies $(|V|^p \cdot \lambda) \circ T^{-1} = \lambda$ a.s., we conclude from Theorem 4.23 (i) that $Y \equiv (V \cdot X) \circ T^{-1} \stackrel{d}{=} X$. Putting $T_t^{-1} = \inf\{s \geq 0;\ T_s > t\}$, we obtain

$$Y_t = \int_0^\infty V_s 1\{T_s \leq t\} dX_s = (V \cdot X) \circ T_t^{-1} \text{ a.s.}, \quad t \geq 0,$$

and due to the right continuity of Y, T^{-1}, and $V \cdot X$, we may choose the exceptional null set to be independent of t. Writing $R_t = \inf\{u > t;\ T_u > T_t\}$, we get a.s. for fixed $t \geq 0$

$$Y \circ T_t = (V \cdot X) \circ T^{-1} \circ T_t = (V \cdot X) \circ R_t = (V \cdot X)_t,$$

where the last relation follows from the identity $(|V|^p \cdot \lambda) \circ R = |V|^p \cdot \lambda$, combined with the fact that the jump point process ξ of X has compensator $\hat{\xi} = \nu \otimes \lambda$. Invoking the right continuity of Y, T, and $V \cdot X$, we see that the exceptional null set can be chosen again to be independent of t.

(ii) Here we may assume that $\lambda V_\pm^p = \infty$ a.s. Putting $T_\pm = V_\pm^p \cdot \lambda$, we see from Lemma 4.25 (i) that the processes $Y_\pm = (V_\pm \cdot X) \circ T_\pm^{-1}$ are mutually independent with the same distribution as X. As before, we note that $Y_\pm \circ T_\pm = V_\pm \cdot X$ a.s., which implies

$$V \cdot X = V_+ \cdot X - V_- \cdot X = Y_+ \circ T_+ - Y_- \circ T_-.$$

(iii) Assuming $\lambda V_\pm = \infty$ and putting $T_\pm = V_\pm \cdot \lambda$, we see from Lemma 4.25 (ii) that the processes

$$Y_\pm = (V_\pm \cdot X) \circ T_\pm^{-1} + c\,(V_\pm \log V_\pm \cdot \lambda) \circ T_\pm^{-1}$$

are mutually independent and distributed as X. As before, we get a.s.

$$Y_\pm \circ T_\pm = V_\pm \cdot X + c\,V_\pm \log V_\pm \cdot \lambda,$$

which implies

$$V \cdot X = V_+ \cdot X - V_- \cdot X = Y_+ \circ T_+ - Y_- \circ T_- - c\,V \log|V| \cdot \lambda. \qquad \square$$

4.6   Mapping of Optional Times

The predictable mapping Theorem 4.7 shows that any predictable, measure-preserving transformation of the unit interval $[0, 1]$ preserves the joint distribution of a family of independent $U(0, 1)$ random variables. Indeed, the random measure version of the quoted theorem is essentially equivalent to that statement. Our present aim is to study the effect of predictable mappings on more general collections of random elements. This requires some martingale theory, to describe the dynamical properties of a single random variable.

More precisely, we consider a marked point process ξ on $\mathbb{R}_+ \times K$ consisting of a single point (τ, κ) in $(0, \infty) \times K$, where K is a Borel space with associated σ-field $\mathcal{K}$. We assume that $\xi = \delta_{\tau, \kappa}$ is adapted to a right-continuous and complete filtration $\mathcal{F}$ and denote the associated compensator by η (FMP 25.22). The time τ is then optional with compensator $\bar\eta = \eta(\cdot \times K)$. For convenience, we introduce the measure-valued process $\eta_t = \eta([0, t] \times \cdot)$ and its projection $\bar\eta_t = \eta[0, t]$. Similar conventions apply to other random measures on $\mathbb{R}_+ \times K$.

If $\mathcal{F}$ is the filtration induced by ξ, then η reduces to the *natural compensator*, given explicitly, as in FMP 25.28, by

$$\eta_t B = \int_0^{t \wedge \tau} \frac{\mu(ds \times B)}{\mu([s, \infty) \times K)}, \quad B \in \mathcal{K}, \ t \geq 0, \tag{25}$$

where $\int_a^b$ denotes the integral over $(a, b]$. Conversely, (25) shows that η determines the underlying measure μ up to time τ. Though (25) may fail in general, we can always find a *random* sub-probability measure ζ on $(0, \tau] \times K$, called the *discounted compensator* of ξ, such that (25) holds with μ replaced by ζ.

To construct ζ, we note that the *tail process* $Z_t = 1 - \bar\zeta_t$ is related to $\bar\eta$ through the ordinary differential equation

$$dZ_t = -Z_{t-} d\bar\eta_t, \qquad Z_0 = 1, \tag{26}$$

whose solution is unique and given by *Doléans' exponential* (FMP 26.8)

$$Z_t = \exp(-\bar\eta_t^c) \prod_{s \leq t} (1 - \Delta\bar\eta_s), \quad t \geq 0, \tag{27}$$

where $\Delta\bar\eta_s = \bar\eta_s - \bar\eta_{s-}$ and $\bar\eta^c$ denotes the continuous component of $\bar\eta$. It remains to express ζ in terms of Z and η as $\zeta = Z_- \cdot \eta$, or

$$\zeta(A) = \iint_A Z_{t-} \eta(dt\, dx), \quad A \in \mathcal{B}(\mathbb{R}_+) \otimes \mathcal{K}. \tag{28}$$

We list some basic properties of Z.

Lemma 4.26 *(tail process) Consider an adapted pair (τ, κ) in $(0, \infty) \times K$ with compensator η, and define Z by (27). Then a.s.*

(i) *$\Delta \bar{\eta} < 1$ on $[0, \tau)$ and ≤ 1 on $[\tau]$;*

(ii) *Z is non-increasing with $Z \geq 0$ and $Z_{\tau-} > 0$ a.s.;*

(iii) *$Y = Z^{-1}$ satisfies $dY_t = Y_t d\bar{\eta}_t$, as long as $Z_t > 0$.*

Proof: (i) Define $\sigma = \inf\{t \geq 0;\ \Delta \bar{\eta}_t \geq 1\}$, and note that σ is optional by an elementary approximation argument. The random interval (σ, ∞) is then predictable, and so the same thing is true for the graph $[\sigma] = \{\Delta \bar{\eta} \geq 1\} \setminus (\sigma, \infty)$. Hence, by dual predictable projection (FMP 25.13),

$$P\{\tau = \sigma\} = E\bar{\xi}[\sigma] = E\bar{\eta}[\sigma].$$

Since also

$$1\{\tau = \sigma\} \leq 1\{\sigma < \infty\} \leq \bar{\eta}[\sigma]$$

by the definition of σ, we obtain

$$1\{\tau = \sigma\} = \bar{\eta}[\sigma] = \Delta \bar{\eta}_\sigma \quad \text{a.s. on } \{\sigma < \infty\},$$

which implies $\Delta \bar{\eta}_\sigma \leq 1$ and $\tau = \sigma$ a.s. on the same set.

(ii) Since $1 - \Delta \bar{\eta} \geq 0$ a.s. by (i), we see from (27) that Z is a.s. non-increasing with $Z \geq 0$. Since also $\sum_t \Delta \bar{\eta}_t \leq \bar{\eta}_\tau < \infty$ a.s. and $\sup_{t < \tau} \Delta \bar{\eta}_t < 1$ a.s. in view of (i), we have $Z_{\tau-} > 0$.

(iii) By an elementary integration by parts (cf. FMP 26.10), we have on the set $\{t \geq 0;\ Z_t > 0\}$

$$0 = d(Z_t\, Y_t) = Z_{t-}\, dY_t + Y_t\, dZ_t,$$

and so, by the chain rule for Stieltjes integrals (cf. FMP 26.2 (ii)) together with (26),

$$dY_t = -Z_{t-}^{-1} Y_t\, dZ_t = Z_{t-}^{-1} Y_t\, Z_{t-}\, d\bar{\eta} = Y_t\, d\bar{\eta}. \qquad \square$$

Now consider any predictable process V on $\mathbb{R}_+ \times K$ such that a.s.

$$\int_0^\tau \int |V| d\zeta < \infty, \qquad \int_0^\tau \int V d\zeta = 0 \text{ on } \{Z_\tau = 0\}. \tag{29}$$

We introduce the associated processes

$$U_{t,x} = V_{t,x} + Z_t^{-1} \int_0^t \int V d\zeta, \quad t \geq 0,\ x \in K, \tag{30}$$

$$M_t = U_{\tau,\kappa} 1\{\tau \leq t\} - \int_0^t \int U d\eta, \quad t \geq 0, \tag{31}$$

with the understanding that $0/0 = 0$. The following remarkable identities and martingale properties will play a basic role.

Lemma 4.27 *(fundamental martingale) Consider an adapted pair (τ, κ) in $(0, \infty) \times K$ with compensator η and discounted compensator ζ, and put $Z_t = 1 - \bar{\zeta}_t$. Let V be a predictable process on $\mathbb{R}_+ \times K$ satisfying (29), and define U and M by (30) and (31). Then*

(i) *M exists and satisfies $M_\infty = V_{\tau,\kappa}$ a.s.;*

(ii) *if $E|U_{\tau,\kappa}| < \infty$, we have $EV_{\tau,\kappa} = 0$, and M becomes a uniformly integrable martingale satisfying $\|M^*\|_p \leq \|V_{\tau,\kappa}\|_p$ for every $p > 1$.*

Proof: (i) Write $Y = Z^{-1}$. Using (28), (29), and Lemma 4.26 (ii), we get

$$\eta|V| = \zeta(Y_-|V|) \leq Y_{\tau-}\zeta|V| < \infty.$$

Next we see from (29), (30), and Lemma 4.26 (iii) that

$$\eta|U - V| \leq \zeta|V| \int_0^\tau Y \, d\bar{\eta} = (Y_\tau - 1)\,\zeta|V| < \infty,$$

whenever $Z_\tau > 0$. Similarly, when $Z_\tau = 0$, we obtain

$$\eta|U - V| \leq \zeta|V| \int_0^{\tau-} Y \, d\bar{\eta} = (Y_{\tau-} - 1)\,\zeta|V| < \infty.$$

Thus, in either case, the process U is η-integrable and the definition of M makes good sense.

Now let $t \geq 0$ be such that $Z_t > 0$. Using (30) (twice), Lemma 4.26 (iii), Fubini's theorem, and (28), we get for any $x \in K$

$$
\begin{aligned}
\int_0^t\!\!\int (U - V)\, d\eta &= \int_0^t Y_s \, d\bar{\eta}_s \int_0^s\!\!\int V \, d\zeta = \int_0^t dY_s \int_0^s\!\!\int V \, d\zeta \\
&= \int_0^t\!\!\int (Y_t - Y_{s-})\, V_{s,y} \, d\zeta_{s,y} \\
&= U_{t,x} - V_{t,x} - \int_0^t\!\!\int V \, d\eta.
\end{aligned}
$$

Simplifying and combining with (30), we obtain

$$\int_0^t\!\!\int U \, d\eta = U_{t,x} - V_{t,x} = Y_t \int_0^t\!\!\int V \, d\zeta. \tag{32}$$

We claim that this remains true for $t = \tau$, even when $Z_\tau = 0$. Then use (28), (29), and (32) to write in this case

$$
\begin{aligned}
\int_0^\tau\!\!\int U \, d\eta &= \int_0^{\tau-}\!\!\int U \, d\eta + \int_{[\tau]}\!\!\int U \, d\eta \\
&= Y_{\tau-} \int_0^{\tau-}\!\!\int V \, d\zeta + \int_{[\tau]}\!\!\int V \, d\eta \\
&= Y_{\tau-} \int_0^\tau\!\!\int V \, d\zeta = 0 = U_{\tau,x} - V_{\tau,x}.
\end{aligned}
$$

This shows that (32) in generally true. In particular,

$$V_{\tau,\kappa} = U_{\tau,\kappa} - \int_0^\tau \int U \, d\eta = M_\tau = M_\infty.$$

(ii) If $E|U_{\tau,\kappa}| < \infty$, then by dual predictable projection

$$E\eta|U| = E\delta_{\tau,\kappa}|U| = E|U_{\tau,\kappa}| < \infty,$$

which yields the asserted uniform integrability of M. Applying a similar calculation to the predictable process $1_{[0,\sigma]}U$ for an arbitrary optional time σ, we obtain $EM_\sigma = 0$, which shows that M is a martingale (FMP 7.13). Hence, in view of (i),

$$EV_{\tau,\kappa} = EM_\infty = EM_0 = 0.$$

Furthermore, by Doob's inequality (FMP 7.16) and (i), we get for any $p > 1$

$$\|M^*\|_p \lesssim \|M_\infty\|_p = \|V_{\tau,\kappa}\|_p. \qquad \square$$

The last result is easily extended to higher dimensions. Let us then say that two optional times are *orthogonal*, if they are a.s. distinct and such that their compensators have a.s. no discontinuity points in common. For more general collections of optional times, orthogonality should be understood in the pair-wise sense.

Corollary 4.28 *(product moments)* *For $1 \leq j \leq m$, consider an adapted pair (τ_j, κ_j) in $(0,\infty) \times K_j$ and a predictable process V_j on $\mathbb{R}_+ \times K_j$, and define Z_j, ζ_j, and U_j as in (27), (28), and (30). Let the τ_j be orthogonal, and assume that for every j*

$$\zeta_j|V_j| < \infty, \qquad \zeta_j V_j = 0 \ \text{a.s. on } \{Z_j(\tau_j) = 0\}, \tag{33}$$
$$E|U_j(\tau_j, \kappa_j)| < \infty, \qquad E|V_j(\tau_j, \kappa_j)|^{p_j} < \infty, \tag{34}$$

where $p_1, \ldots, p_m > 0$ with $\sum_j p_j^{-1} \leq 1$. Then

$$E \prod_{j \leq m} V_j(\tau_j, \kappa_j) = 0.$$

Proof: Let $\eta_1, \ldots, \eta_m$ denote the compensators of the pairs (τ_j, κ_j), and define the martingales $M_1, \ldots, M_m$ as in (31). Fix any $i \neq j$ in $\{1, \ldots, m\}$, and choose some predictable times $\sigma_1, \sigma_2, \ldots$ such that $\{t > 0; \Delta\bar{\eta}_i > 0\} = \bigcup_k [\sigma_k]$ a.s. (cf. FMP 25.17). By dual predictable projection (FMP 25.13) and orthogonality, we have for any $k \in \mathbb{N}$

$$E\delta_{\sigma_k}[\tau_j] = P\{\tau_j = \sigma_k\} = E\delta_{\tau_j}[\sigma_k] = E\eta_j[\sigma_k] = 0.$$

Summing over k gives $\eta_i[\tau_j] = 0$ a.s., which shows that the M_j are strongly orthogonal, in the sense that $[M_i, M_j] = 0$ a.s. for all $i \neq j$. Next integrate repeatedly by parts, to conclude that the product $M = \prod_j M_j$ is a local martingale. Writing $p = (\sum_j p_j^{-1})^{-1}$ and using Hölder's inequality, Lemma 4.27 (ii), and (33)–(34), we obtain

$$\|M^*\|_1 \le \|M^*\|_p \le \prod_j \|M_j^*\|_{p_j} \lesssim \prod_j \|V_{\tau_j, \kappa_j}\|_{p_j} < \infty.$$

Thus, M is a uniformly integrable martingale, and so by Lemma 4.27 (i)

$$E \prod_j V_j(\tau_j, \kappa_j) = E \prod_j M_j(\infty) = EM(\infty) = EM(0) = 0. \qquad \square$$

We are now ready to state the main result of this section.

Theorem 4.29 *(mapping to independence) For $1 \le j \le m$, consider an adapted pair (τ_j, κ_j) in $(0, \infty) \times K_j$ with discounted compensator ζ_j and a predictable mapping T_j of $(0, \infty) \times K_j$ into a probability space $(S_j, \mathcal{S}_j, \mu_j)$ such that $\zeta_j \circ T_j^{-1} \le \mu_j$ a.s. Let the τ_j be orthogonal. Then the random elements $\gamma_j = T_j(\tau_j, \kappa_j)$ in S_j are independent with distributions μ_j.*

Proof: Fix any sets $B_j \in \mathcal{S}_j$, $j \le m$, and introduce on $\mathbb{R}_+ \times K_j$ the bounded, predictable processes

$$V_j(t, x) = 1_{B_j} \circ T_j(t, x) - \mu_j B_j, \quad t \ge 0, \ x \in K_j, \ j \le m. \tag{35}$$

Then by definitions and hypotheses

$$\int_0^t \! \int V_j \, d\zeta_j = \int_0^t \! \int 1_{B_j}(T_j) \, d\zeta_j - \mu_j B_j(1 - Z_j(t)) \le Z_j(t) \, \mu_j B_j.$$

Replacing B_j by its complement B_j^c will only affect the sign of V, and combining the two estimates yields

$$-Z_j(t) \, \mu_j B_j^c \le \int_0^t \! \int V_j \, d\zeta_j \le Z_j(t) \, \mu_j B_j. \tag{36}$$

In particular, $|\zeta_j V_j| \le Z_j(\tau_j)$ a.s., and so $\zeta_j V_j = 0$ a.s. on $\{Z_j(\tau_j) = 0\}$. Using (30), (35), and (36), we obtain

$$-1 \le 1_{B_j} \circ T_j - 1 = V_j - \mu_j B_j^c \le U_j \le V_j + \mu_j B_j = 1_{B_j} \circ T \le 1,$$

which implies $|U_j| \le 1$. The hypotheses of Corollary 4.28 are then fulfilled with $p_1, \ldots, p_m = m$, and we conclude that

$$E \prod_j \left(1_{B_j}(\gamma_j) - \mu_j B_j \right) = E \prod_j V_j(\tau_j, \kappa_j) = 0. \tag{37}$$

We may use the last relation to show that

$$P \bigcap_{j \le m} \{\gamma_j \in B_j\} = \prod_{j \le m} \mu_j B_j. \tag{38}$$

For $m = 1$, this is obvious from (37). Proceeding by induction, we assume (38) to be true for up to $m - 1$ pairs (τ_j, κ_j), and turn to the case of m such pairs. Expanding the product on the left of (37) and applying the induction hypothesis to all lower order terms, we see that all but two of the 2^m terms will cancel out, which leaves us with the desired relation (38). □

We may extend the last result to a conditional statement, akin to Theorem 4.5, involving an additional continuous local martingale.

Corollary 4.30 (*extended mapping property*) *Let $\gamma_1, \ldots, \gamma_m$ be such as in Theorem 4.29, except that the μ_j are now allowed to be random and $\mathcal{F}_0$-measurable. Define a process X on S, as in Theorem 4.5, in terms of a continuous local martingale M in $\mathbb{R}^d$ and some predictable processes U_t, $t \in S$, such that the associated co-variation process ρ on S^2 is $\mathcal{F}_0$-measurable. Then conditionally on $\mathcal{F}_0$, the γ_j are independent with distributions μ_j, and X is independent, Gaussian with covariance function ρ.*

Proof: Let $V_1, \ldots, V_m$ be such as in the last proof, and define the associated martingales $M_1, \ldots, M_m$ as in (31). Also introduce the exponential local martingale $Z = \exp(iN + \frac{1}{2}[N])$, where N is such as in the proof of Theorem 4.5. Then $(Z - 1) \prod_j M_j$ is conditionally a uniformly integrable martingale, and we get as in (37)

$$E^{\mathcal{F}_0}(Z_\infty - 1) \prod_j \left(1_{B_j}(\gamma_j) - \mu_j B_j\right) = 0.$$

Noting that the same relation holds without the factor $Z_\infty - 1$, we may proceed recursively as before to obtain

$$E^{\mathcal{F}_0} \exp(iN_\infty) \prod_j 1_{B_j}(\gamma_j) = \exp(-\frac{1}{2}[N]_\infty) \prod_j \mu_j B_j.$$

Applying the uniqueness theorem for characteristic functions to the bounded measure $\tilde{P} = \prod_j 1_{B_j}(\gamma_j) \cdot P^{\mathcal{F}_0}$, we conclude that

$$P^{\mathcal{F}_0}\{X \in A; \ \gamma_j \in B_j, \ j \le m\} = \nu_\rho(A) \prod_j \mu_j B_j,$$

where ν_ρ denotes the centered Gaussian distribution on $\mathbb{R}^T$ with covariance function ρ. □

Our final aim is to show how a randomization can be used to resolve possible discontinuities and multiplicities. Given a point process ξ on some space S, we may form a *uniform randomization* η of ξ by adding some independent $U(0, 1)$ marks to the unit atoms of ξ, as explained in FMP 12.7. Given two filtrations $\mathcal{F}$ and $\mathcal{G}$ on a probability space $(\Omega, \mathcal{A}, P)$, we say as in FMP 18.4 that $\mathcal{G}$ is a *standard extension* of $\mathcal{F}$ if $\mathcal{F}_t \subset \mathcal{G}_t \perp\!\!\!\perp_{\mathcal{F}_t} \mathcal{F}$ for all $t \ge 0$. This is the minimal condition required to preserve all adaptedness and conditioning properties.

Lemma 4.31 *(randomization) Consider an $\mathcal{F}$-adapted point process ξ on $S \times \mathbb{R}_+$ with compensator $\hat{\xi}$ and a uniform randomization η on $S \times \mathbb{R}_+ \times [0,1]$. Let $\mathcal{G}$ be the right-continuous filtration induced by $\mathcal{F}$ and η. Then $\mathcal{G}$ is a standard extension of $\mathcal{F}$, and η has $\mathcal{G}$-compensator $\hat{\eta} = \hat{\xi} \otimes \lambda$.*

Proof: For any $t \geq 0$, let ξ_t and η_t denote the restrictions of ξ and η to $[0,t] \times S$ and $[0,t] \times S \times [0,1]$, respectively, and put $\eta'_t = \eta - \eta_t$. Then by construction

$$\eta \perp\!\!\!\perp_\xi \mathcal{F}, \qquad \eta_t \perp\!\!\!\perp_\xi \eta'_t, \qquad \eta_t \perp\!\!\!\perp_{\xi_t} \xi.$$

Using the first of these relations and then combining with the other two, invoking the chain rule for conditional independence (FMP 6.8), we obtain

$$\eta_t \perp\!\!\!\perp_{\xi, \eta'_t} \mathcal{F}, \qquad \eta_t \perp\!\!\!\perp_{\xi_t} (\eta'_t, \mathcal{F}),$$

and so

$$\eta_t \perp\!\!\!\perp_{\mathcal{F}_t} (\eta'_t, \mathcal{F}), \qquad (\eta_t, \mathcal{F}_t) \perp\!\!\!\perp_{\mathcal{F}_t} (\eta'_t, \mathcal{F}).$$

Approximating from the right in the last relation gives $\mathcal{G}_t \perp\!\!\!\perp_{\mathcal{F}_t} \mathcal{F}$, which shows that $\mathcal{G}$ is a standard extension of $\mathcal{F}$.

Using the latter property, the chain rule for conditional expectations, the relation $\eta \perp\!\!\!\perp_\xi \mathcal{F}$, Fubini's theorem, and the definitions of randomization and compensation, we get on $(t, \infty) \times S \times [0,1]$ for arbitrary $t \geq 0$

$$\begin{aligned}
E[\eta | \mathcal{G}_t] &= E[\eta | \mathcal{F}_t] = E[E[\eta | \mathcal{F}_t, \xi] | \mathcal{F}_t] = E[E[\eta | \xi] | \mathcal{F}_t] \\
&= E[\xi \otimes \lambda | \mathcal{F}_t] = E[\xi | \mathcal{F}_t] \otimes \lambda = E[\hat{\xi} | \mathcal{F}_t] \otimes \lambda \\
&= E[\hat{\xi} \otimes \lambda | \mathcal{F}_t] = E[\hat{\xi} \otimes \lambda | \mathcal{G}_t].
\end{aligned}$$

Since $\hat{\eta} = \hat{\xi} \otimes \lambda$ is $\mathcal{F}$-predictable and hence even $\mathcal{G}$-predictable, we conclude that $\hat{\eta}$ is indeed a $\mathcal{G}$-compensator of η. $\qquad \square$

For a first application of the preceding theory, we consider the transformation of a simple, unbounded point process ξ on $(0, \infty)$ to a homogeneous Poisson process. The quasi-left-continuous case, where the compensator $\hat{\xi}$ is continuous, is classical and appears in FMP 25.26.

Corollary 4.32 *(time change to Poisson) Let $\xi = \sum_j \delta_{\tau_j}$ be a simple, unbounded, $\mathcal{F}$-adapted point process on $(0, \infty)$ with compensator $\hat{\xi}$, and consider some i.i.d. $U(0,1)$ random variables $\kappa_1, \kappa_2, \ldots \perp\!\!\!\perp \mathcal{F}$. Put $\rho_t = \kappa_j$ on $\{t = \tau_j\}$ for all j, and let $\rho_t = 1$ otherwise. Then the times*

$$\sigma_j = \hat{\xi}^c(0, \tau_j] - \sum_{t \leq \tau_j} \log(1 - \rho_t \hat{\xi}\{t\}), \quad j \in \mathbb{N}, \tag{39}$$

form a unit rate Poisson process on $\mathbb{R}_+$.

When ξ is quasi-left-continuous, we note that (39) reduces to $\sigma_j = \hat{\xi}(0, \tau_j]$, in agreement with Theorem 4.5 or FMP 25.26.

Proof: We shall use Theorem 4.29 to show that the differences $\sigma_j - \sigma_{j-1}$ are independent and exponentially distributed with mean 1, where $\sigma_0 = 0$. Since the τ_j are orthogonal, it is then enough to consider σ_1. Letting $\mathcal{G}$ be the right-continuous filtration induced by $\mathcal{F}$ and the pairs (σ_j, κ_j), we see from Lemma 4.31 that (τ_1, κ_1) has $\mathcal{G}$-compensator $\eta = \hat{\xi} \otimes \lambda$ on $[0, \tau_1] \times [0, 1]$. Let ζ denote the associated discounted version with projection $\bar{\zeta}$ on $\mathbb{R}_+$, and put $Z_t = 1 - \bar{\zeta}(0, t]$. We may also introduce on $\mathbb{R}_+ \times [0, 1]$ the $\mathcal{G}$-predictable processes T and $V = e^{-T}$, given by

$$
\begin{aligned}
T(t, x) &= \bar{\eta}^c(0, t] - \sum_{s < t} \log(1 - \bar{\eta}\{s\}) - \log(1 - x\,\bar{\eta}\{t\}), \\
V(t, x) &= \exp(-\bar{\eta}^c(0, t])\,(1 - x\,\bar{\eta}\{t\}) \prod_{s < t}(1 - \bar{\eta}\{s\}) \\
&= Z_{t-}(1 - x\,\bar{\eta}\{t\}),
\end{aligned}
$$

where the last equality comes from (27). Noting that, for any random variable γ with distribution function F and for arbitrary $t \in \mathbb{R}$,

$$
\begin{aligned}
P\{F(\gamma) \le F(t)\} &= P\{\gamma \le t\} = F(t), \\
P\{F(\gamma) \le F(t-)\} &= P\{\gamma < t\} = F(t-),
\end{aligned}
$$

we obtain $\zeta \circ V^{-1} \le \lambda$ on $[0, 1]$. Hence, Theorem 4.29 shows that $V(\tau_1, \kappa_1) = e^{-\sigma_1}$ is $U(0, 1)$, which yields the desired distribution for $\sigma_1 = T(\tau_1, \kappa_1)$. □

We can also use the preceding theory to give an alternative proof of Theorem 4.7, at least for processes on $[0, 1]$.

Second proof of Theorem 4.7, for $I = [0, 1]$: Extending the original filtration $\mathcal{F}$, as explained in Lemma 2.22, we may assume that the coefficients in the representation of X are $\mathcal{F}_0$-measurable. Our first step is to truncate the sum of centered jumps in the representation of Theorem 2.18. Then let X_n denote the remainder after the first n terms, and write $X_n = M_n + \hat{X}_n$. Noting that $\mathrm{tr}[M_n]_1 = \sum_{j > n} |\beta_j|^2 \to 0$ and using the BDG-inequalities in FMP 26.12, we obtain $(M_n \circ V^{-1})_t \xrightarrow{P} 0$ for every $t \in [0, 1]$. Next we may proceed as in the proof of Theorem 2.25, which again involves the use of Theorem 2.23, to see that

$$
\begin{aligned}
E\left[\int_0^1 |d\hat{X}_n| \,\Big|\, \mathcal{F}_0\right] &\lesssim E[X_n^* | \mathcal{F}_0] \lesssim (\mathrm{tr}[X_n])_1^{1/2} \\
&= \left(\sum_{j > n} |\beta_j|^2\right)^{1/2} \to 0.
\end{aligned}
$$

This implies $(X_n \circ V^{-1})_t \xrightarrow{P} 0$ for all $t \in [0, 1]$, which reduces the proof to the case of finitely many jumps.

Here it is enough to consider a jointly exchangeable pair, consisting of a marked point process ξ on $[0, 1]$ and a Brownian bridge B in $\mathbb{R}^d$. By Lemma 4.31 we may finally reduce to the case where ξ has a.s. distinct marks. It is then equivalent to consider finitely many optional times $\tau_1, \ldots, \tau_m$ such that B and the point processes $\xi_j = \delta_{\tau_j}$ are jointly exchangeable.

To apply Corollary 4.30, we may choose the continuous martingale M to be the integral with respect to the Brownian motion in Lemma 4.20, where the predictable process V is replaced by the associated indicator functions $U_r^t = 1\{V_r \leq t\}$, $r \in [0,1]$, for arbitrary $t \in [0,1]$. Then the co-variations of the lemma become

$$
\begin{aligned}
\int_0^1 (U_r^s - \overline{U}_r^s)(U_r^t - \overline{U}_r^t)\, dr &= \int_0^1 U_r^s U_r^t\, dr - \lambda U^s \cdot \lambda U^t \\
&= \lambda U^{s \wedge t} - \lambda U^s \cdot \lambda U^t \\
&= s \wedge t - st = E B_s B_t,
\end{aligned}
$$

as required.

Turning to the random measures $\xi_j = \delta_{\tau_j}$, we recall from Theorem 2.8 that the associated compensators η_j are a.s. given by

$$
\eta_j[0,t] = \int_0^{t \wedge \tau_j} \frac{ds}{1-s} = -\log(1 - t \wedge \tau_j), \quad t \in [0,1].
$$

Since the η_j are diffuse, the corresponding discounted compensators ζ_j are obtained from (27) as

$$
\zeta_j[0,t] = 1 - e^{-\eta_j[0,t]} = 1 - (1 - t \wedge \tau_j) = t \wedge \tau_j, \quad t \in [0,1].
$$

This implies $\zeta_j = \lambda([0,\tau_j] \cap \cdot) \leq \lambda$ a.s., and so $\zeta_j \circ V^{-1} \leq \lambda \circ V^{-1} = \lambda$ a.s. We may now conclude from Corollary 4.30 that

$$
(B \circ V^{-1}, V_{\tau_1}, \ldots, V_{\tau_m}) \stackrel{d}{=} (B, \tau_1, \ldots, \tau_m),
$$

which yields the required relation $X \circ V^{-1} \stackrel{d}{=} X$. $\qquad\square$

Chapter 5

Decoupling Identities

The aim of this chapter is to establish, under suitable regularity and other conditions, the various decoupling or Wald-type identities for product moments of predictable sums and integrals with respect to exchangeable or contractable sequences or processes. The statements supplement the predictable sampling and mapping theorems of the previous chapter and yield some deeper insight into those propositions.

The main results appear in Sections 5.2 and 5.4–6. Thus, the decoupling identities for predictable sums with respect to finite or infinite exchangeable sequences are given in Section 5.2. Moment identities for stochastic integrals with respect to exchangeable processes on $[0, 1]$ are established in Section 5.4, and the corresponding formulas for processes on $\mathbb{R}_+$ are given in Section 5.5. Finally, Section 5.6 deals with some tetrahedral moment identities for contractable sums and integrals.

The proofs require some auxiliary results for predictable sums and integrals. Thus, some norm estimates are established in Section 5.1, and in Section 5.3 we present a rather subtle martingale representation of stochastic integrals with respect to exchangeable processes on $[0, 1]$, along with some further norm estimates. In the final Section 5.7, we indicate how the present identities are related to the principal results of Chapter 4.

5.1 Integrability and Norm Estimates

To prepare for the moment identities for infinite predictable sums, we need some explicit criteria for summability and existence of higher moments. In this section, we also treat the analogous problems in continuous time, where the i.i.d. sums are replaced by stochastic integrals with respect to Lévy processes. The more difficult case of exchangeable integrals on $[0, 1]$ is postponed until Section 5.3.

First we discuss infinite sums of the form $\sum_k \xi_k \eta_k$, where the sequence $\xi = (\xi_k)$ is i.i.d. and $\eta = (\eta_k)$ is predictable. Our present aim is to examine the convergence and integrability of such series.

Proposition 5.1 *(predictable i.i.d. sums) Consider some infinite random sequences $\xi = (\xi_k)$ and $\eta = (\eta_k)$ in $\mathbb{R}$, where ξ is $\mathcal{F}$-i.i.d. and η is $\mathcal{F}$-predictable. Then for any $p \geq 1$ we have*

$$E \sup_n \left| \sum_{k \leq n} \xi_k \eta_k \right|^p \lesssim |E\xi_1|^p E\left(\sum_k |\eta_k| \right)^p$$
$$+ E|\xi_1|^p E\left(\sum_k |\eta_k|^{p \wedge 2} \right)^{(p \vee 2)/2}.$$

Moreover, the series on the left converges a.s. whenever the right-hand side is finite.

We may clearly assume that $\xi_1 \in L^p$, so that $E\xi_1$ exists. When $E\xi_1 = 0$, we note that the first term on the right vanishes.

Proof: Put $S_n = \sum_{k \leq n} \xi_k \eta_k$ and $S^* = \sup_n |S_n|$. Applying a BDG-inequality (FMP 26.12) to the local martingale

$$M_n = \sum_{k \leq n} (\xi_k - E\xi_k)\, \eta_k, \quad n \in \mathbb{Z}_+,$$

we see that

$$ES^{*p} \lesssim |E\xi_1|^p E\left(\sum_k |\eta_k| \right)^p + E\left(\sum_k (\xi_k - E\xi_k)^2 \eta_k^2 \right)^{p/2}.$$

Continuing recursively in m steps, where $2^m \in (p, 2p]$, we get

$$ES^{*p} \lesssim \sum_{0 \leq r < m} \left| E\xi_1^{(r)} \right|^{p2^{-r}} E\left(\sum_k |\eta_k|^{2^r} \right)^{p2^{-r}} + E\left(\sum_k \xi_k^{(m)} \eta_k^{2^m} \right)^{p2^{-m}}. \quad (1)$$

where the random variables $\xi_k^{(r)}$ are defined recursively by

$$\xi_k^{(0)} = \xi_k; \qquad \xi_k^{(r+1)} = (\xi_k^{(r)} - E\xi_k^{(r)})^2, \quad 0 \leq r < m.$$

The argument is justified by the fact that

$$\left| E\xi_k^{(r)} \right|^{p2^{-r}} \leq E\left| \xi_k^{(r)} \right|^{p2^{-r}} \leq E|\xi_k|^p, \quad 0 \leq r < m, \quad (2)$$

where the first relation holds by Jensen's inequality and the second relation, valid even for $r = m$, follows recursively from the estimates

$$E\left| \xi_k^{(r+1)} \right|^{p2^{-r-1}} = E\left| \xi_k^{(r)} - E\xi_k^{(r)} \right|^{p2^{-r}}$$
$$\lesssim E\left| \xi_k^{(r)} \right|^{p2^{-r}} + \left| E\xi_k^{(r)} \right|^{p2^{-r}} \leq E\left| \xi_k^{(r)} \right|^{p2^{-r}}.$$

Using sub-additivity and term-wise independence, and applying the second relation in (2), we obtain

$$E\left(\sum_k \xi_k^{(m)} \eta_k^{2^m} \right)^{p2^{-m}} \leq E \sum_k \left| \xi_k^{(m)} \right|^{p2^{-m}} |\eta_k|^p$$
$$= E\left| \xi_1^{(m)} \right|^{p2^{-m}} E \sum_k |\eta_k|^p$$
$$\lesssim E|\xi_1|^p E\left(\sum_k |\eta_k|^{p \wedge 2} \right)^{(p \vee 2)/2}. \quad (3)$$

Similarly, by sub-additivity, we have

$$\left(\sum_k |\eta_k|^{2^r}\right)^{p2^{-r}} \le \left(\sum_k |\eta_k|^{p\wedge 2}\right)^{(p\vee 2)/2}, \quad r \ge 1. \tag{4}$$

The desired estimate follows by insertion of (2), (3), and (4) into (1).

To prove the last assertion, suppose that the right-hand side is finite. Then

$$E\left(\sum_k |\eta_k E\xi_k|\right)^p = |E\xi_1|^p\, E\left(\sum_k |\eta_k|\right)^p < \infty.$$

Taking differences, we conclude that $EM^{*p} < \infty$. In particular, M is an L^1-bounded martingale, and hence converges a.s. The a.s. convergence of S_n then follows by combination. $\qquad\square$

Let us now examine the convergence and integrability of stochastic integrals $V \cdot X$, where X is a Lévy process and V is predictable. Assuming the first moments of X to be finite, say $EX_t \equiv \alpha t$, we may introduce the associated characteristics (α, σ^2, ν).

Proposition 5.2 *(predictable Lévy integrals) Consider an integrable $\mathcal{F}$-Lévy process X with characteristics (α, σ^2, ν) and an $\mathcal{F}$-predictable process V, and put $\nu_p = \int |x|^p \nu(dx)$. Then for any $p \ge 1$ we have*

$$E(V \cdot X)^{*p} \underset{\sim}{<} |\alpha|^p E(\lambda|V|)^p + \sigma^p E(\lambda V^2)^{p/2}$$
$$+ \left(\nu_p + (\nu_{p\wedge 2})^{(p\vee 2)/2}\right) E\left(\lambda|V|^p + (\lambda|V|^{p\wedge 2})^{(p\vee 2)/2}\right), \tag{5}$$

in the sense that the integral process $V \cdot X$ exists and satisfies (5) whenever the bound is finite. In that case, the limit $(V \cdot X)_\infty$ also exists a.s.

When X is continuous, the stated estimate remains valid for arbitrary $p > 0$. In the other extreme case, when $\alpha = \sigma = 0$, the bound in (5) reduces to $\nu_p \lambda|V|^p$ when $p \in [1, 2]$, and for $p \ge 2$ we get

$$E(V \cdot X)^p \underset{\sim}{<} \left(\nu_p + (\nu_2)^{p/2}\right) E\left(\lambda|V|^p + (\lambda V^2)^{p/2}\right).$$

Our proof of Proposition 5.2 requires an elementary inequality.

Lemma 5.3 *(norm interpolation) For measurable functions f on an arbitrary measure space, we have*

$$\|f\|_r \le \|f\|_p \vee \|f\|_q, \quad 0 < p \le r \le q.$$

Proof: We may take $f \ge 0$ and $p < r < q$, so that $p^{-1} > r^{-1} > q^{-1}$. Setting $s = p^{-1}$ and $t = q^{-1}$, we may choose $a = 1 - b \in (0, 1)$ such that $as + bt = r^{-1}$. Then Hölder's inequality yields

$$\begin{aligned}
\|f\|_r &= \|f^{a+b}\|_{1/(as+bt)} \le \|f^a\|_{1/as}\|f^b\|_{1/bt} \\
&= \|f\|_p^a \|f\|_q^b \le \|f\|_p \vee \|f\|_q. \qquad\square
\end{aligned}$$

Proof of Proposition 5.2: We consider separately the drift, diffusion, and purely discontinuous martingale components of X. For $X_t \equiv \alpha t$ we have

$$E(V \cdot X)^{*p} = |\alpha|^p E(V \cdot \lambda)^{*p} \leq |\alpha|^p E(\lambda|V|)^p,$$

provided that the right-hand side is finite. Next suppose that $X_t \equiv \sigma B_t$ for a Brownian motion B and a constant $\sigma \geq 0$. Then a BDG-inequality for continuous martingales (FMP 17.7) yields

$$E(V \cdot X)^p = \sigma^p E(V \cdot B)^{*p} \lesssim \sigma^p E(V^2 \cdot [B])_\infty^{p/2} = \sigma^p E(\lambda V^2)^{p/2},$$

again as long as the right-hand side is finite.

Now suppose instead that X is a purely discontinuous martingale such that the last term in (5) is finite. Then define recursively the processes

$$X^{(0)} = X; \qquad X^{(r)} = [X^{(r-1)}], \quad r \in \mathbb{N},$$

and note that for any $r \geq 1$ and $t > 0$

$$X_t^{(r)} = \sum_{s \leq t} (\Delta X_s)^{2^r}, \qquad EX_t^{(r)} = t\, \nu_{2^r}. \tag{6}$$

Choosing $m \in \mathbb{N}$ to be such that $2^m \in (p, 2p]$, we see from Lemma 5.3 that $\nu_{2^r} < \infty$ for $1 \leq r < m$. Hence, the processes $X^{(r)}$ are finite and integrable for $0 \leq r < m$, and the compensated processes

$$M_t^{(r)} = X_t^{(r)} - EX_t^{(r)}, \quad t \geq 0, \;\; 0 \leq r < m,$$

are martingales. Combining (6) with a BDG-inequality for general local martingales (FMP 26.12), we get for $0 \leq r < m$

$$E(V^{2^r} \cdot X^{(r)})_\infty^{p2^{-r}} \lesssim (\nu_{2^r})^{p2^{-r}} E(\lambda V^{2^r})^{p2^{-r}} + E\left(V^{2^{r+1}} \cdot X^{(r+1)}\right)_\infty^{p2^{-r-1}},$$

where finiteness on the right guarantees that the integrals $V^{2^r} \cdot EX^{(r)}$ and $V^{2^r} \cdot M^{(r)}$ exist, for the latter according to FMP 26.13. Adding the estimates for $0 \leq r < m$ gives

$$E(V \cdot X)^{*p} \lesssim \sum_{1 \leq r < m} (\nu_{2^r})^{p2^{-r}} E\left(\lambda V^{2^r}\right)^{p2^{-r}} + E\left(V^{2^m} \cdot X^{(m)}\right)_\infty^{p2^{-m}}, \tag{7}$$

where finiteness on the right again ensures the existence of the stochastic integral $V \cdot X$ as a uniformly integrable martingale.

By subadditivity and dual predictable projection, we see from (6) that

$$\begin{aligned}
E\left(V^{2^m} \cdot X^{(m)}\right)_\infty^{p2^{-m}} &= E\left(\sum_t (V_t \Delta X_t)^{2^m}\right)^{p2^{-m}} \\
&\leq E\sum_t |V_t \Delta X_t|^p = \nu_p\, E\lambda|V|^p.
\end{aligned}$$

Furthermore, Lemma 5.3 gives for $1 \le r < m$

$$(\nu_{2^r})^{p2^{-r}} \le \nu_p + (\nu_{p \wedge 2})^{(p \vee 2)/2},$$

$$\left(\lambda V^{2^r}\right)^{p2^{-r}} \le \lambda |V|^p + \left(\lambda |V|^{p \wedge 2}\right)^{(p \vee 2)/2}.$$

Inserting all those estimates into (7) yields the required bound in (5). Since $V \cdot (X - EX)$ is a uniformly integrable martingale when this bound is finite, the last assertion follows immediately by martingale convergence. $\square$

We proceed with a simple algebraic identity. Write $\sum'_{k_1,\ldots,k_d}$ for summation over all d-tuples $(k_1,\ldots,k_d)$ with distinct components $k_1,\ldots,k_d$.

Lemma 5.4 *(diagonal decomposition) There exist some constants c_π, indexed by partitions π of $\{1,\ldots,d\}$, such that whenever $x_k^j \in \mathbb{R}$, $j \le d$, $k \in \mathbb{N}$, with $\prod_j \sum_k |x_k^j| < \infty$, we have*

$$\sum\nolimits'_{k_1,\ldots,k_d} \prod_j x_{k_j}^j = \sum\nolimits_\pi c_\pi \prod_{J \in \pi} \sum_k \prod_{j \in J} x_k^j. \tag{8}$$

Proof: The result is obvious for $d = 1$. Proceeding by induction, assume the statement to be true in dimensions $< d$, and turn to the d-dimensional case. Let S_J denote the inner sum on the right of (8). For each $i < d$, write $J_i = J$ when $i \notin J$ and $J_i = J \cup \{d\}$ when $i \in J$. Summing over distinct indices $k_1,\ldots,k_d \in \mathbb{N}$ and arbitrary partitions π of $\{1,\ldots,d-1\}$, and applying the induction hypothesis for fixed $i < d$ to the array $y_k^j = x_k^j$ for $j \ne i$ and $y_k^i = x_k^i x_k^d$, we obtain

$$
\begin{aligned}
\sum_{k_1 \ldots, k_d} \prod_{j \le d} x_{k_j}^j &= \sum_{k_1,\ldots,k_{d-1}} \sum_{k_d \ne k_1,\ldots,k_{d-1}} x_{k_d}^d \prod_{j < d} x_{k_j}^j \\
&= \sum_{k_1,\ldots,k_{d-1}} \left(S_d - \sum_{i<d} x_{k_i}^d\right) \prod_{j<d} x_{k_j}^j \\
&= S_d \sum_{k_1,\ldots,k_{d-1}} \prod_{j<d} x_{k_j}^j - \sum_{i<d} \sum_{k_1,\ldots,k_{d-1}} x_{k_i}^i x_{k_i}^d \prod_{j \ne i,d} x_{k_j}^j \\
&= \sum_\pi c_\pi S_d \prod_{J \in \pi} S_J - \sum_{i<d} \sum_\pi c_\pi \prod_{J \in \pi} S_{J_i}.
\end{aligned}
$$

This extends the result to dimension d and hence completes the induction. $\square$

In continuous time, we need a decomposition for products of general semi-martingales, extending the formula for stochastic integration by parts (FMP 26.6). For any semi-martingales $X^1,\ldots,X^d$, we introduce the general variations $[X]^J$, $J \subset \{1,\ldots,d\}$, given for $|J| = 1$ and 2 by

$$[X]_t^j = X_t^j - X_0^j, \quad [X]_t^{ij} = [X^i, X^j]_t, \qquad i \ne j \text{ in } \{1,\ldots,d\},$$

and then extended recursively to higher orders by the formula $[X]^{I \cup J} = [[X]^I, [X]^J]$, for any disjoint sets I and J. Thus, for any $J \subset \{1,\ldots,d\}$ with $|J| > 2$, we have

$$[X]_t^J = \sum\nolimits_{s \le t} \prod\nolimits_{j \in J} \Delta X_s^j, \quad t \ge 0.$$

We also need to introduce the *tetrahedral regions*

$$\Delta_m = \{(s_1, \ldots, s_m) \in [0,1]^m;\ s_1 < \cdots < s_m\}, \quad m \in \mathbb{N}.$$

Given a set J, we say that $J_1, \ldots, J_m \subset J$ form an *ordered partition* of J if the J_k are disjoint and nonempty with union J.

Lemma 5.5 *(tetrahedral decomposition) Let $X^1, \ldots, X^d$ be semi-martingales starting at 0. Then*

$$X^1_t \cdots X^d_t = \sum_{J_1, \ldots, J_m} \int \cdots \int_{t \Delta_m} d[X]^{J_1}_{s_1} \cdots d[X]^{J_m}_{s_m}, \quad t \geq 0,$$

where the summation extends over all ordered partitions $J_1, \ldots, J_m$ of the set $\{1, \ldots, d\}$.

Here the multiple stochastic integral on the right should be interpreted in the sense of repeated stochastic integration. Thus, the asserted decomposition is equivalent to

$$X^1_t \cdots X^d_t = \sum_{J_1, \ldots, J_m} \int_0^t d[X]^{J_m}_{s_m} \int_0^{s_m-} d[X]^{J_{m-1}}_{s_{m-1}} \cdots \int_0^{s_2-} d[X]^{J_1}_{s_1}.$$

In particular, the formula for $d = 2$ becomes

$$X^1_t X^2_t = \int_0^t d[X]^{ij}_s + \int_0^t X^1_{s-} dX^2_s + \int_0^t X^2_{s-} dX^1_s,$$

which agrees with the general rule for integration by parts (FMP 26.6).

Proof: First we show that

$$d(X^1 \cdots X^d)_t = \sum_J X^{J^c}_{t-} d[X]^J_t, \quad t \geq 0, \tag{9}$$

where $X^I = \prod_{i \in I} X^i$ and the summation extends over all nonempty subsets $J \subset \{1, \ldots, d\}$. This holds trivially for $d = 1$. Proceeding by induction, we assume (9) to be true for up to $d-1$ factors. Turning to the case of d factors, we may integrate by parts to get

$$\begin{aligned}
d(X^1 \cdots X^d)_t &= X^1_{t-} d(X^2 \cdots X^d)_t + (X^2 \cdots X^d)_{t-} dX^1_t \\
&\quad + d[X^1, X^2 \cdots X^d]_t.
\end{aligned} \tag{10}$$

Using the induction hypothesis and the chain rule in FMP 16.2 (ii), we obtain

$$X^1_{t-} d(X^2 \cdots X^d)_t = \sum_J X^1_{t-} X^{J^c}_{t-} d[X]^J_t = \sum_J X^{J^c \cup \{1\}}_{t-} d[X]^J_t, \tag{11}$$

where the summation extends over all nonempty subsets $J \subset \{2, \ldots, d\}$ with complements J^c. Similarly, combining the induction hypothesis with the co-variation property in FMP 26.6 (v) and the recursive property of higher order variations, we see that

$$d[X^1, X^2 \cdots X^d]_t = \sum_J X^{J^c}_{t-} d[X^1, [X]^J]_t = \sum_J X^{J^c}_{t-} d[X]^{J \cup \{1\}}_t, \tag{12}$$

where the summation is again over all nonempty subsets $J \subset \{2, \ldots, d\}$. Inserting (11) and (12) into (10), we obtain (9) for the case of d factors, which completes the induction. The required formula follows from (9) by iteration in finitely many steps, or, equivalently, by another induction argument. □

We also need the following basic projection property.

Lemma 5.6 *(optional projection, Dellacherie) Consider a measurable process X with $EX^* < \infty$ and a progressive process Y, such that for any optional time τ*

$$E[X_\tau; \tau < \infty] = E[Y_\tau; \tau < \infty]. \tag{13}$$

Then for any rcll, adapted process A of locally finite variation such that $E \int |X| \, |dA| < \infty$, we have $E \int X \, dA = E \int Y \, dA$.

Proof: By FMP 2.18 we may assume that A is non-decreasing. Replacing τ by its restrictions to the $\mathcal{F}_\tau$-measurable sets $B_\pm = \{\pm Y_\tau > 0, \tau < \infty\}$ (FMP 7.5, 25.4), we obtain

$$
\begin{aligned}
E[|Y_\tau|; \tau < \infty] &= E[Y_\tau; B_+] - E[Y_\tau; B_-] \\
&= E[X_\tau; B_+] - E[X_\tau; B_-] \\
&\leq E[|X_\tau|; \tau < \infty].
\end{aligned}
$$

Applying this to the first-passage times $\tau_s = \inf\{t \geq 0; A_t > s\}$ and using an elementary substitution (FMP 1.22) and Fubini's theorem, we obtain

$$
\begin{aligned}
E \int |Y_t| \, dA_t &= \int E[|Y_{\tau_s}|; \tau_s < \infty] \, ds \\
&\leq \int E[|X_{\tau_s}|; \tau_s < \infty] \, ds \\
&= E \int |X_t| \, dA_t < \infty.
\end{aligned}
$$

This justifies that we apply Fubini's theorem to the original integral, and we get by (13)

$$
\begin{aligned}
E \int Y_t \, dA_t &= \int E[Y_{\tau_s}; \tau_s < \infty] \, ds \\
&= \int E[X_{\tau_s}; \tau_s < \infty] \, ds = E \int X_t \, dA_t.
\end{aligned}
$$
□

Let us also quote an elementary martingale result.

Lemma 5.7 *(integration by parts) Let M and N be martingales with $\|M^*\|_p \vee \|N^*\|_q < \infty$, where $p, q > 1$ with $p^{-1} + q^{-1} \leq 1$. Then*

$$E(M_\infty N_\infty) = E(M_0 N_0) + E[M, N]_\infty.$$

Proof: Integration by parts (FMP 26.6) yields

$$M_t N_t = M_0 N_0 + (M_- \cdot N)_t + (N_- \cdot M)_t + [M, N]_t.$$

Since $\|M^* N^*\|_1 < \infty$ by Hölder's inequality, it remains to show that the integral terms $M_- \cdot N$ and $N_- \cdot M$ are uniformly integrable martingales. Now both processes are local martingales (FMP 26.13), and so the Hölder and BDG inequalities (FMP 26.12) yield

$$\begin{aligned}
E(M_- \cdot N)^* &\lesssim E[M_- \cdot N]_\infty^{1/2} = E(M_-^2 \cdot [N])_\infty^{1/2} \\
&\leq E(M^* [N]_\infty^{1/2}) \leq \|M^*\|_p \|[N]_\infty^{1/2}\|_q \\
&\lesssim \|M^*\|_p \|N^*\|_q < \infty,
\end{aligned}$$

and similarly with M and N interchanged. The required properties now follow by dominated convergence. □

We conclude with an elementary integral estimate.

Lemma 5.8 *(norms of averages) Let f be a locally integrable function on $\mathbb{R}_+$, and define $\bar{f}_t = t^{-1} \int_0^t f_s ds$. Then for any $p > 1$ and $r \geq 0$ we have*

$$\int_0^\infty |\bar{f}_t|^p \, t^{-r} \, dt \leq \left(\frac{p}{r+p-1} \right)^p \int_0^\infty |f_t|^p \, t^{-r} \, dt. \tag{14}$$

Proof: Since $|\bar{f}_t| \leq t^{-1} \int_0^t |f_s| ds$, we may assume that $f \geq 0$, and since $f = 0$ a.e. implies $\bar{f} \equiv 0$, we may further assume that $\lambda f > 0$. Finally, by monotone convergence we may reduce to the case where f is bounded with compact support in $(0, \infty)$, in which case the left-hand side of (14) is finite and strictly positive. Letting $p^{-1} + q^{-1} = 1$, writing $F_t = (f \cdot \lambda)_t$ and $c = p/(r+p-1)$, and using integration by parts and Hölder's inequality, we obtain

$$\begin{aligned}
\int_0^\infty \bar{f}_t^p \, t^{-r} \, dt &= \int_0^\infty F_t^p \, t^{-r-p} \, dt = c \int_0^\infty F_t^{p-1} f_t \, t^{1-r-p} \, dt \\
&= c \int_0^\infty \bar{f}_t^{p-1} f_t \, t^{-r} \, dt \\
&\leq c \left(\int_0^\infty \bar{f}_t^p \, t^{-r} \, dt \right)^{1/q} \left(\int_0^\infty f_t^p \, t^{-r} \, dt \right)^{1/p},
\end{aligned}$$

and (14) follows as we divide by the second factor on the right and raise both sides to the pth power. □

5.2 Exchangeable Sums

Here we begin with the moment identities for finite sums involving random n-sequences $\xi = (\xi_k^j)$ and $\eta = (\eta_k^j)$ in $\mathbb{R}^d$, where $j \in \{1, \ldots, d\}$ and $k \in \{1, \ldots, n\}$. It is often convenient to write

$$\xi^j = (\xi_1^j, \ldots, \xi_n^j), \qquad \xi_k = (\xi_k^1, \ldots, \xi_k^d).$$

Given any filtration $\mathcal{F} = (\mathcal{F}_k; \ k = 0, \ldots, n)$ and a partition $\pi = \{J_1, \ldots, J_m\}$ of $\{1, \ldots, d\}$ into disjoint, nonempty subsets, we say that the projections $\xi^J = (\xi^j; \ j \in J)$ of ξ onto the subspaces $\mathbb{R}^J$ are *separately $\mathcal{F}$-exchangeable* if they are $\mathcal{F}$-adapted and such that the shifted sequences $\theta^k \xi^{J_r}$ are separately exchangeable, conditionally on $\mathcal{F}_k$, for every $k \in [0, n)$. Recall that an exchangeable sequence $\xi = (\xi_1, \ldots, \xi_n)$ is *extreme* iff it is a.s. an urn sequence, so that the sum $\sum_k f(\xi_k)$ is a.s. non-random for every measurable function f.

For any nonempty subset $J \subset \{1, \ldots, d\}$, we introduce the sums

$$R_J = \sum_k \prod_{j \in J} \xi_k^j, \qquad S_J = \sum_k \prod_{j \in J} \eta_k^j.$$

Given a partition π of $\{1, \ldots, d\}$, we define

$$R_\pi = \prod_{J \in \pi} R_J, \qquad S_\pi = \prod_{J \in \pi} S_J.$$

A relation $\pi \prec \pi'$ or $\pi' \succ \pi$ between two partitions π and π' means by definition that π' is a refinement of π, in the sense that every set $I \in \pi'$ is contained in some $J \in \pi$.

Theorem 5.9 *(moments of finite sums) Consider some random n-sequences ξ and η in $\mathbb{R}^d$ and a partition π of $\{1, \ldots, d\}$, such that the projections ξ^J, $J \in \pi$, are extreme, separately $\mathcal{F}$-exchangeable and η is $\mathcal{F}$-predictable with $\sum_k E \prod_j |\eta_k^j| < \infty$. Suppose that S_I is a.s. non-random for any $I \subset J \in \pi$. Then*

$$E \prod_{j \le d} \sum_{k \le n} \xi_k^j \eta_k^j = \sum_{\pi_1, \pi_2} c_{\pi_1, \pi_2} R_{\pi_1} S_{\pi_2}$$

for some constants $c_{\pi_1, \pi_2} = c_{\pi_2, \pi_1}$, where the summation on the right extends over all partitions $\pi_1, \pi_2 \succ \pi$.

In particular, the moments for $d = 1, 2$ and $n \ge 2$ become

$$E_1 = \frac{R_1 S_1}{n}, \qquad E_{12} = \frac{n R_{12} S_{12} - R_{12} S_1 S_2 - R_1 R_2 S_{12} + R_1 R_2 S_1 S_2}{n(n-1)}.$$

If $R_j = S_j = 0$ for all singletons j, then for $d = 2, 3$ and $n \ge d$ we get

$$E_{12} = \frac{R_{12} S_{12}}{n-1}, \qquad E_{123} = \frac{n R_{123} S_{123}}{(n-1)(n-2)}.$$

The formulas simplify when π is nontrivial.

The remarkable (indeed surprising) fact is that the result is always the same, regardless of the dependence between ξ and η. Thus, under the stated conditions, we can evaluate the product moment *as if* the two sequences were independent. By Fubini's theorem we may even assume that η is non-random. If $\tilde{\eta}$ is another $\mathcal{F}$-predictable sequence, satisfying the same integrability condition and such that

$$\sum_k \prod_{j \in I} \eta_k^j = \sum_k \prod_{j \in I} \tilde{\eta}_k^j, \qquad I \subset J \in \pi,$$

then
$$E \prod_j \sum_k \xi_k^j \eta_k^j = E \prod_j \sum_k \xi_k^j \tilde{\eta}_k^j.$$

This holds in particular when $\tilde{\eta} \perp\!\!\!\perp \xi$ with $\tilde{\eta} \stackrel{d}{=} \eta$. Taking $\xi^1 = \cdots = \xi^d = \xi$ and $\eta^1 = \cdots = \eta^d = \eta$, we get for $p \in \mathbb{N}$ the relation

$$E\left(\sum_k \xi_k \eta_k\right)^p = E\left(\sum_k \xi_k \tilde{\eta}_k\right)^p, \tag{15}$$

valid whenever η and $\tilde{\eta}$ are predictable sequences in L^p such that the sums

$$S_h = \sum_k \eta_k^h = \sum_k \tilde{\eta}_k^h, \quad h = 1, \dots, p,$$

are a.s. non-random. The identity in (15) remains true, under appropriate conditions, when each term in the two sums is interpreted as an inner product in $\mathbb{R}^d$.

To prove Theorem 5.9, we begin with a simple special case.

Lemma 5.10 *(constant weights) The assertion of Theorem 5.9 is true when η is non-random.*

Proof: Since the ξ^J, $J \in \pi$, are independent by Lemma 2.21, the moment E_d on the left can be factored accordingly, which reduces the proof to the case where the basic partition consists of the single set $\{1, \dots, d\}$. Summing over arbitrary partitions π of $\{1, \dots, d\}$ and distinct indices $k_J \leq n$, $J \in \pi$, we get the product moment

$$\begin{aligned}
E_d &= \sum_\pi \sum_{(k_J)} E \prod_{J \in \pi} \prod_{j \in J} \xi_{k_J}^j \eta_{k_J}^j \\
&= \sum_\pi \sum_{(k_J)} \left(E \prod_{J \in \pi} \prod_{j \in J} \xi_{k_J}^j \right) \left(\prod_{J \in \pi} \prod_{j \in J} \eta_{k_J}^j \right) \\
&= \sum_\pi \frac{(n - |\pi|)!}{n!} \left(\sum_{(k_J)} \prod_J \prod_j \xi_{k_J}^j \right) \left(\sum_{(k_J)} \prod_J \prod_j \eta_{k_J}^j \right).
\end{aligned}$$

By Lemma 5.4, the inner sums on the right are linear combinations of the products $R_{\pi'}$ or $S_{\pi'}$, respectively, for arbitrary partitions $\pi' \prec \pi$, which shows that E_d has the stated form. $\square$

The assertion for general η will be proved by induction on n. It is then convenient to consider a slightly stronger conditional version. Given an $\mathcal{F}$-exchangeable sequence $\xi = (\xi_1, \dots, \xi_n)$ in S, we say that ξ is $\mathcal{F}$-*extreme* if $\sum_k f(\xi_k)$ is $\mathcal{F}_0$-measurable for every measurable function $f \colon S \to \mathbb{R}$.

Lemma 5.11 *(conditional moments) Let ξ, η, and π be such as in Theorem 5.9, except that we now allow the projections ξ^J, $J \in \pi$, to be $\mathcal{F}$-extreme and the sums S_I with $I \subset J \in \pi$ to be $\mathcal{F}_0$-measurable. Then the stated formula holds for the conditional moment $E^{\mathcal{F}_0} \prod_j \sum_k \xi_k^j \eta_k^j$.*

Proof: The statement is obvious for $n = 1$, since η is then $\mathcal{F}_0$-measurable. Now assume the result to be true for sequences of length $< n$, and proceed to the case of n-sequences ξ and η. The transfer theorem (FMP 6.10) allows us to construct another n-sequence $\tilde{\eta}$ satisfying

$$(\tilde{\eta}, \mathcal{F}_0) \stackrel{d}{=} (\eta, \mathcal{F}_0), \qquad \tilde{\eta} \perp\!\!\!\perp_{\mathcal{F}_0} \mathcal{F}_n. \tag{16}$$

Then $\tilde{\eta}$ is predictable with respect to the extended filtration $\tilde{\mathcal{F}}_k = \mathcal{F}_k \vee \sigma\{\tilde{\eta}_1, \ldots, \tilde{\eta}_{k+1}\}$, $k = 0, \ldots, n$. Furthermore, for any $k \leq n$, we see from (16) that $(\tilde{\eta}, \mathcal{F}_k) \perp\!\!\!\perp_{\mathcal{F}_k} \mathcal{F}_n$ and hence $\theta^k \xi \perp\!\!\!\perp_{\tilde{\mathcal{F}}_k} \tilde{\mathcal{F}}_k$, which shows that ξ remains separately exchangeable with respect to $\tilde{\mathcal{F}}$. Since ξ_1 is $\mathcal{F}_1$-measurable, it is also clear that the π-components of $\theta\xi$ are conditionally extreme under both $\mathcal{F}_1$ and $\tilde{\mathcal{F}}_1$. Finally, the reduced sums

$$S_J' = \sum\nolimits_{k>1} \prod\nolimits_{j \in J} \eta_k^j = S_J - \prod\nolimits_{j \in J} \eta_1^j, \quad J \subset K \in \pi,$$

are $\mathcal{F}_1$-measurable and agree a.s. with the corresponding sums for $\tilde{\eta}$.

We can now conclude from the induction hypothesis that a.s.

$$E^{\mathcal{F}_1} \prod_{j \in J} \sum_{k>1} \xi_k^j \eta_k^j = E^{\tilde{\mathcal{F}}_1} \prod_{j \in J} \sum_{k>1} \xi_k^j \tilde{\eta}_k^j, \quad J \subset \{1, \ldots, d\}.$$

Noting that $\eta_1^i = \tilde{\eta}_1^i$ a.s., we obtain

$$
\begin{aligned}
E^{\mathcal{F}_0} \prod_{j \leq d} \sum_{k \leq n} \xi_k^j \eta_k^j &= E^{\mathcal{F}_0} \sum_{J \in 2^d} \prod_{j \notin J} (\xi_1^j \eta_1^j) \, E^{\mathcal{F}_1} \prod_{j \in J} \sum_{k>1} \xi_k^j \eta_k^j \\
&= E^{\mathcal{F}_0} \sum_{J \in 2^d} \prod_{j \notin J} (\xi_1^j \tilde{\eta}_1^j) \, E^{\tilde{\mathcal{F}}_1} \prod_{j \in J} \sum_{k>1} \xi_k^j \tilde{\eta}_k^j \\
&= E^{\mathcal{F}_0} \prod_{j \leq d} \sum_{k \leq n} \xi_k^j \tilde{\eta}_k^j.
\end{aligned}
$$

Since $\xi \perp\!\!\!\perp_{\mathcal{F}_0} \tilde{\eta}$, we see from Lemma 5.10 and Fubini's theorem that the right-hand side has the stated form. This completes the induction, and the result follows. □

We turn to the basic decoupling identities for infinite sums. Though fewer sums S_J are now required to be non-random, we need to impose some more delicate integrability conditions. More specifically, we assume the existence of some constants $p_1, \ldots, p_d \geq 1$ with $\sum_j p_j^{-1} \leq 1$ such that, for every $j \in \{1, \ldots, d\}$,

$$|E\xi_1^j| \, E\left(\sum\nolimits_k |\eta_k^j|\right)^{p_j} + E|\xi_1^j|^{p_j} \, E\left(\sum\nolimits_k |\eta_k^j|^{p_j \wedge 2}\right)^{(p_j \vee 2)/2} < \infty. \tag{17}$$

Note that the first term vanishes when $E\xi_1^j = 0$.

For any nonempty subset $J \subset \{1, \ldots, d\}$, we introduce the moments and sums

$$m_J = E \prod\nolimits_{j \in J} \xi_1^j, \qquad S_J = \sum\nolimits_k \prod\nolimits_{j \in J} \eta_k^j,$$

whenever they exist. Furthermore, for any partition π of $\{1, \ldots, d\}$, we consider the products

$$m_\pi = \prod_{J \in \pi} m_J, \qquad S_\pi = \prod_{J \in \pi} S_J.$$

Theorem 5.12 *(moments of infinite series) Consider some infinite random sequences ξ and η in $\mathbb{R}^d$ and a partition π of $\{1, \ldots, d\}$, such that the projections ξ^J, $J \in \pi$, are independent $\mathcal{F}$-i.i.d. and η is $\mathcal{F}$-predictable. Suppose that (17) holds for some $p_1, \ldots, p_d > 0$ with $\sum_j p_j^{-1} \leq 1$ and that S_I is a.s. non-random for any $I \subset J \in \pi$ with $1 < |I| < d$, as well as for $I = \{j\}$ when $m_j \neq 0$ and $d \geq 2$. Then*

$$E \prod_{j \leq d} \sum_{k \geq 1} \xi_k^j \eta_k^j = \sum_{\pi_1, \pi_2} c_{\pi_1, \pi_2} m_{\pi_1} E S_{\pi_2},$$

for some constants c_{π_1, π_2}, where the summation extends over all partitions $\pi_1, \pi_2 \succ \pi$ with $\pi_1 \succ \pi_2$.

Note that the expected value $E S_{\pi_2}$ on the right can be replaced by S_{π_2} for all nontrivial partitions π_2. If $m_{\pi_1} = 0$, we regard the corresponding terms as 0. For $d = 1, 2$ we get the moments

$$E_1 = m_1 E S_1, \qquad E_{12} = (m_{12} - m_1 m_2) E S_{12} + m_1 m_2 S_1 S_2.$$

If $m_j = 0$ for all j, then for $d = 2, 3$ we have

$$E_{12} = m_{12} E S_{12}, \qquad E_{123} = m_{123} E S_{123}.$$

Again it is remarkable that, under the stated conditions, the product moment depends only on the marginal distributions of ξ and η, and hence can be evaluated as if the two sequences were independent. Indeed, if $\tilde{\eta}$ is another predictable sequence, satisfying the same integrability conditions and such that the sums S_I with $I \subset J \in \pi$ agree for η and $\tilde{\eta}$, then

$$E \prod_j \sum_k \xi_k^j \eta_k^j = E \prod_j \sum_k \xi_k^j \tilde{\eta}_k^j.$$

Specializing to the case where $\xi^j = \xi$, $\eta^j = \eta$, and $\tilde{\eta}^j = \tilde{\eta}$ for all j, we get as before an identity between ordinary moments of the form

$$E \Big(\sum_k \xi_k \eta_k \Big)^p = E \Big(\sum_k \xi_k \tilde{\eta}_k \Big)^p,$$

valid under appropriate integrability and constancy conditions. In particular, (17) reduces in this case to

$$|E\xi_1| \, E \Big(\sum_k |\eta_k| \Big)^p + E|\xi_1|^p \, E \Big(\sum_k |\eta_k|^{p \wedge 2} \Big)^{(p \vee 2)/2} < \infty. \tag{18}$$

Again we begin the proof with a special case.

Lemma 5.13 *(constant weights)* *The assertion of Theorem 5.12 is true when η is non-random with finite support.*

Proof: Again we may assume that the basic partition consists of the single set $\{1,\ldots,d\}$. Summing over arbitrary partitions π of $\{1,\ldots,d\}$ and distinct indices $k_J \in \mathbb{N}$, $J \in \pi$, we obtain the moment

$$E_d = \sum_\pi \sum_{(k_J)} E \prod_{J\in\pi} \prod_{j\in J} \xi_{k_J}^j \eta_{k_J}^j = \sum_\pi m_\pi \sum_{(k_J)} \prod_{J\in\pi} \prod_{j\in J} \eta_{k_J}^j.$$

By Lemma 5.4 the inner sum on the right is a linear combination of products $S_{\pi'}$ for partitions $\pi' \prec \pi$, which shows that E_d has the required form. □

To establish the general assertion, it is again convenient to prove a formally stronger conditional version of the statement.

Lemma 5.14 *(conditional moments)* *Let ξ, η, and π be such as in Theorem 5.12, except that S_I is now allowed to be $\mathcal{F}_0$-measurable for any $I \subset J \in \pi$ with $1 < |I| < d$, as well as for $I = \{j\}$ when $m_j \neq 0$ and $d \geq 2$. Then the stated formula holds for the $\mathcal{F}_0$-conditional moments.*

Proof: We begin with a formal argument that ignores all matters of convergence, to be followed by a detailed justification. Proceeding by induction on d, we fix a $d \in \mathbb{N}$ and assume the result to be true in any lower dimension. Define

$$S_J^n = \sum_{k>n} \prod_{j\in J} \eta_k^j = S_J - \sum_{k\leq n} \prod_{j\in J} \eta_k^j, \quad n \in \mathbb{Z}_+, \; J \subset \{1,\ldots,d\},$$

and note that the sequence (S_J^n) is predictable for any $J \subset I \in \pi$ with $1 < |J| < d$, as well as for singletons $J = \{j\}$ with $m_j \neq 0$ when $d \geq 2$. The induction hypothesis yields

$$E^{\mathcal{F}_n} \prod_{j\in J} \sum_{k>n} \xi_k^j \eta_k^j = \sum_{\pi_1,\pi_2} c_{\pi_1,\pi_2} m_{\pi_1} S_{\pi_2}^n, \quad |J| < d,$$

for some constants c_{π_1,π_2}, where the summation extends over all partitions $\pi_1 \succ \pi_2 \succ (J \cap \pi)$. Summing over nonempty subsets $J \subset \{1,\ldots,d\}$ and partitions $\pi_1 \succ \pi_2 \succ (J^c \cap \pi)$ and conditioning in the n-th term below, first on $\mathcal{F}_n$ and then on $\mathcal{F}_{n-1}$, we obtain

$$\begin{aligned}
E^{\mathcal{F}_0} \prod_j \sum_k \xi_k^j \eta_k^j &= E^{\mathcal{F}_0} \sum_{J\neq\emptyset} \sum_{n\geq 1} \left(\prod_{j\in J} \xi_n^j \eta_n^j \right) \left(\prod_{j\notin J} \sum_{k>n} \xi_k^j \eta_k^j \right) \\
&= \sum_{J\neq\emptyset} \sum_{\pi_1,\pi_2} c_{\pi_1,\pi_2} m_{\pi_1} \sum_{n\geq 1} S_{\pi_2}^n E^{\mathcal{F}_0} \prod_{j\in J} \xi_n^j \eta_n^j \\
&= \sum_{J\neq\emptyset} m_J \sum_{\pi_1,\pi_2} c_{\pi_1,\pi_2} m_{\pi_1} \sum_{n\geq 1} S_{\pi_2}^n E^{\mathcal{F}_0} \prod_{j\in J} \eta_n^j.
\end{aligned} \tag{19}$$

Here the underlying $\mathcal{F}_0$-measurability follows from the facts that $|J^c| < d$, that every $I \in \pi_2$ lies in some $I' \in \pi$, and that $\{j\} \in \pi_2$ implies $\{j\} \in \pi_1$.

Now let $\tilde{\xi} \perp\!\!\!\perp \mathcal{F}_\infty$ with $\tilde{\xi} \stackrel{d}{=} \xi$, and put $\tilde{\mathcal{F}}_n = \mathcal{F}_n \vee \sigma\{\tilde{\xi}_1, \ldots, \tilde{\xi}_n\}$. Then $\tilde{\mathcal{F}}_0 = \mathcal{F}_0$, and so by (19)

$$E^{\mathcal{F}_0} \prod_j \sum_k \xi_k^j \eta_k^j = E^{\mathcal{F}_0} \prod_j \sum_k \tilde{\xi}_k^j \eta_k^j.$$

Since $\tilde{\xi} \perp\!\!\!\perp (\eta, \mathcal{F}_0)$, we see from Fubini's theorem and Lemma 5.13 that the right-hand side has the stated form. This completes the induction.

To justify the previous argument, we may first invoke (17) and Proposition 5.1 to see that the series $\sum_k \xi_k^j \eta_k^j$ converges a.s. for every j, with partial sums satisfying

$$E \sup_n \left| \sum_{k \le n} \xi_k^j \eta_k^j \right|^{p_j} < \infty, \quad j = 1, \ldots, d. \tag{20}$$

The existence of the moments in (19) is then ensured by Hölder's inequality.

Next we may fix an arbitrary partition π' of $\{1, \ldots, d\}$ and conclude from Hölder's and Jensen's inequalities that

$$
\begin{aligned}
E \prod_{J \in \pi'} \sum_{k \ge 1} \prod_{j \in J} |\eta_k^j| &\le E \prod_j \left(\sum_k |\eta_k^j|^{|J|} \right)^{1/|J|} \\
&\le \prod_j \left\{ E \left(\sum_k |\eta_k^j|^{|J|} \right)^{p_j/|J|} \right\}^{1/p_j},
\end{aligned}
\tag{21}
$$

where the set J on the right is defined by $j \in J \in \pi'$. By subadditivity, the last expectations admit the estimates

$$E \left(\sum_k |\eta_k^j|^{|J|} \right)^{p_j/|J|} \le E \left(\sum_k |\eta_k^j|^{p_j \wedge 2} \right)^{(p_j \vee 2)/2}, \quad |J| \ge 2.$$

In view of (17), the left-hand side of (21) is then finite for all partitions π' such that $m_j \ne 0$ when $\{j\} \in \pi'$. In particular, the sum S_J converges when either $|J| \ge 2$ or $J = \{j\}$ with $m_j \ne 0$.

For any $m \in \mathbb{N}$, we have

$$
\begin{aligned}
\prod_j \sum_k \xi_k^j \eta_k^j &= \prod_j \left(\sum_{k \le m} \xi_k^j \eta_k^j + \sum_{k > m} \xi_k^j \eta_k^j \right) \\
&= \sum_{J \ne \emptyset} \sum_{n \le m} \prod_{j \in J} (\xi_n^j \eta_n^j) \prod_{i \notin J} \sum_{k > n} \xi_k^i \eta_k^i + \prod_{j \le d} \sum_{k > m} \xi_k^j \eta_k^j.
\end{aligned}
$$

To estimate the last term, we may use (20), Hölder's inequality, and dominated convergence as $m \to \infty$ to see that

$$E \prod_j \left| \sum_{k > m} \xi_k^j \eta_k^j \right| \le \prod_j \left\| \sum_{k > m} \xi_k^j \eta_k^j \right\|_{p_j} \to 0.$$

The previous formula then extends to $m = \infty$, in the sense of convergence in L^1, which justifies the first two relations in (19). $\square$

5.3 Martingale Representations

Here our aim is to prepare for the moment identities in continuous time by studying stochastic integrals of the form $V \cdot X$, where V is predictable and X is an exchangeable or contractable process on $[0, 1]$. We begin with some general norm and continuity properties, depending only on the semi-martingale properties of contractable processes in Chapter 2. Those results are needed to prove the tetrahedral decoupling identity in Theorem 5.30. Recall from Theorem 2.23 that if $\|X_t\|_p < \infty$ for some $t \in (0, 1)$ and $p \geq 1$, then $\|X^*\|_p < \infty$.

Proposition 5.15 *(contractable integrals) Consider on $[0, 1]$ an $\mathcal{F}$-contractable process X with $\|X^*\|_p < \infty$ and an $\mathcal{F}$-predictable process V with $\|V^*\|_q < \infty$, where $p, q > 1$ with $r^{-1} = p^{-1} + q^{-1} \leq 1$. Then*

(i) *$V \cdot X$ exists and satisfies $\|(V \cdot X)^*\|_r \leq \|X^*\|_p \|V^*\|_q < \infty$;*

(ii) *for any predictable processes $V_1, V_2, \ldots$ with $|V_t| \geq |V_t^n| \to 0$ for all $t \in [0, 1]$ a.s., we have $\|(V_n \cdot X)^*\|_r \to 0$.*

Proof: (i) By Proposition 2.27 we may write $X = M \cdot \lambda + N$ a.s., where M and N are L^p-martingales on $[0, 1)$ with

$$\| \lambda |M| \|_p \vee \|N^*\|_p \lesssim \|X^*\|_p < \infty.$$

Hence, $V \cdot X = VM \cdot \lambda + V \cdot N$ a.s., whenever these integrals exist. Now Hölder's inequality yields

$$
\begin{aligned}
\|(VM \cdot \lambda)^*\|_r &\leq \| \lambda |VM| \|_r \leq \|V^* \lambda |M| \|_r \\
&\leq \|V^*\|_q \| \lambda |M| \|_p \\
&\lesssim \|V^*\|_q \|X^*\|_p < \infty.
\end{aligned}
$$

Next we may use the Hölder and BDG inequalities (FMP 26.12) to see that

$$
\begin{aligned}
\|(V \cdot N)^*\|_r &\lesssim \|[V \cdot N]_1^{1/2}\|_r = \|(V^2 \cdot [N])_1^{1/2}\|_r \\
&\leq \|V^* [N]_1^{1/2}\|_r \leq \|V^*\|_q \|[N]_1^{1/2}\|_p \\
&\lesssim \|V^*\|_q \|N^*\|_p \lesssim \|V^*\|_q \|X^*\|_p < \infty,
\end{aligned}
$$

which also proves the existence of $V \cdot N$ (FMP 26.13). Finally, we may conclude from Minkowski's inequality that

$$
\begin{aligned}
\|(V \cdot X)^*\|_r &\leq \|(VM \cdot \lambda)^*\|_r + \|(V \cdot N)^*\|_r \\
&\lesssim \|V^*\|_q \|X^*\|_p < \infty.
\end{aligned}
$$

(ii) By the previous estimates and dominated convergence,

$$
\begin{aligned}
\|(V_n \cdot X)^*\|_r &\leq \|(V_n M \cdot \lambda)^*\|_r + \|(V_n \cdot N)^*\|_r \\
&\lesssim \| \lambda |V_n M| \|_r + \|(V_n^2 \cdot [N])_1^{1/2}\|_r \to 0. \qquad \square
\end{aligned}
$$

We continue with a more detailed study of stochastic integrals $V \cdot X$ on $[0, 1]$, in the special case where X is extreme, exchangeable. By Theorems 2.18 and 3.15 we may then assume that X has a representation

$$X_t = \alpha t + \sigma B_t + \sum_j \beta_j (1\{\tau_j \le t\} - t), \quad t \in [0, 1].$$

We begin with some sufficient conditions for integrability.

Theorem 5.16 *(exchangeable integrals)* *Consider on $[0, 1]$ an extreme, $\mathcal{F}$-exchangeable process X with characteristics $(\alpha, \sigma^2, \beta)$ and an $\mathcal{F}$-predictable process V. Fix a $p \in (0, 2]$ with $\sum_j |\beta_j|^p < \infty$, and suppose that $\sigma = 0$ when $p < 2$ and $\alpha = \sum_j \beta_j$ when $p \le 1$. Then*

(i) *for $p \in (0, 1]$ and $\lambda |V|^p < \infty$ a.s., we have $\int_0^1 |V_t| |dX_t| < \infty$ a.s.;*

(ii) *if $p \in (1, 2]$ and $\int_0^1 |V_t|^p (1 - t)^{-\varepsilon} dt < \infty$ a.s. for some $\varepsilon > 0$, the integral $V \cdot X$ exists on $[0, 1]$ and satisfies*

$$\int_0^1 V dX = \alpha \lambda V + \sigma \int_0^1 V dB + \sum_j \beta_j (V_{\tau_j} - \lambda V) \quad a.s. \qquad (22)$$

Proof: (i) In this case we have

$$X_t = \sum_j \beta_j 1\{\tau_j \le t\}, \quad t \in [0, 1],$$

where $\tau_1, \tau_2, \ldots$ are i.i.d. $U(0, 1)$ and $\sum_j |\beta_j| < \infty$, and we may clearly assume that $V \ge 0$ and $\beta_j \ge 0$ for all j. Since $\sum_j \beta_j^p < \infty$, we may also introduce the exchangeable process

$$Y_t = \sum_j \beta_j^p 1\{\tau_j \le t\}, \quad t \in [0, 1].$$

By Theorem 2.8 or 2.13 we note that the compensator $\hat{Y}$ of Y is absolutely continuous with the martingale density

$$N_t = (1 - t)^{-1} \sum_j \beta_j^p 1\{\tau_j > t\}, \quad t \in [0, 1).$$

Introducing the optional times

$$\sigma_n = \sup\{t \in [0, 1]; \ (V^p N \cdot \lambda)_t \le n\}, \quad n \in \mathbb{N},$$

we get by subadditivity and dual predictable projection (FMP 25.22)

$$
\begin{aligned}
E(V \cdot X)_{\sigma_n}^p &= E\Big(\sum_j \beta_j V_{\tau_j} 1\{\tau_j \le \sigma_n\} \Big)^p \\
&\le E \sum_j \beta_j^p V_{\tau_j}^p 1\{\tau_j \le \sigma_n\} \\
&= E(V^p \cdot Y)_{\sigma_n} = E(V^p N \cdot \lambda)_{\sigma_n} \le n,
\end{aligned}
$$

and so $(V \cdot X)_{\sigma_n} < \infty$ a.s. for all n. It remains to notice that $\sigma_n = 1$ for all but finitely many n, since N, as a positive martingale, is automatically L^1-bounded and hence a.s. bounded.

(ii) Here we may assume that $\alpha = 0$, and also that $V \geq 0$ and $\beta_j \geq 0$ for all j. First we consider integration with respect to the L^2-bounded martingale $X - \hat{X}$ on $[0, 1]$, where $\hat{X}$ denotes the compensator of X. If X is continuous, then $X - \hat{X} = \sigma B$ for some Brownian motion B, and $V \cdot (X - \hat{X})$ reduces to the Itô integral $\sigma V \cdot B$, which exists iff $\sigma V \in L^2(\lambda)$ a.s. (FMP 17.11).

Next let $\sigma = 0$. Defining Y, N, and σ_n as before and using Jensen's inequality, subadditivity, and dual predictable projection, we obtain

$$
\begin{aligned}
\left(E(V^2 \cdot [X])_{\sigma_n}^{1/2} \right)^p &\leq E(V^2 \cdot [X])_{\sigma_n}^{p/2} \\
&= E\left(\sum_j \beta_j^2 V_{\tau_j}^2 1\{\tau_j \leq \sigma_n\} \right)^{p/2} \\
&\leq E \sum_j \beta_j^p V_{\tau_j}^p 1\{\tau_j \leq \sigma_n\} \\
&= E(V^p \cdot Y)_{\sigma_n} = E(V^p N \cdot \lambda)_{\sigma_n} \leq n.
\end{aligned}
$$

By FMP 26.13 we conclude that $V \cdot X$ exists on $[0, \sigma_n]$, and the integrability on $[0, 1]$ follows since $\sigma_n = 1$ for all sufficiently large n.

It remains to prove the existence of the integral $V \cdot \hat{X}$. Then recall from Theorem 2.13 that $\hat{X} = -M \cdot \lambda$, where $M_t = X_t/(1-t)$ on $[0, 1)$. Hence, by Hölder's inequality,

$$
\int_0^1 V_t \, |d\hat{X}_t| = \int_0^1 \frac{V_t |X_t| \, dt}{1-t} \leq \left(\int_0^1 \frac{V_t^p \, dt}{(1-t)^\varepsilon} \right)^{1/p} \left(\int_0^1 \frac{|X_t|^q \, dt}{(1-t)^{q'}} \right)^{1/q}, \tag{23}
$$

where $p^{-1} + q^{-1} = 1$ and $q' = (1 - \varepsilon/p)q < q$. Here the first factor on the right is finite by hypothesis. Since this remains true for any smaller value of $\varepsilon > 0$, we may assume that $q' \in (1, q)$, which allows us to choose a $p' > p$ such that $(p')^{-1} + (q')^{-1} = 1$. Then by Theorem 2.32 we have $|X_t|^{p'} \lesssim (1-t)$ a.s., and therefore

$$
|X_t|^q (1-t)^{-q'} \lesssim (1-t)^{-q' + q/p'}, \quad t \in [0, 1),
$$

which is integrable on $[0, 1]$ since $q' - q/p' < q' - q'/p' = 1$. This shows that even the second factor in (23) is finite.

To prove (22), we may clearly assume that $\alpha = \sigma = 0$. Since the remaining formula is obvious for finite sums, it suffices to show that $(V \cdot X_n)_1 \xrightarrow{P} 0$, where

$$
X_n(t) = \sum_{j>n} \beta_j (1\{\tau_j \leq t\} - t), \quad t \in [0, 1]. \tag{24}
$$

This time we introduce the optional times

$$
\sigma_n = \sup\{t \in [0, 1]; \, (V^p N_n \cdot \lambda)_t \leq 1\}, \quad n \in \mathbf{Z}_+,
$$

where

$$
N_n(t) = (1-t)^{-1} \sum_{j>n} \beta_j^p 1\{\tau_j > t\}, \quad t \in [0, 1).
$$

Writing M_n for the martingale component of X_n and using a BDG-inequality (FMP 26.12), we get as before

$$E(V \cdot M_n)^{*p}_{\sigma_n} \lesssim E(V^2 \cdot [X_n])^{p/2}_{\sigma_n} \leq E(V^p N_n \cdot \lambda)_{\sigma_n},$$

and so

$$E[(V \cdot M_n)^{*p} \wedge 1] \lesssim E[(V^p N_n \cdot \lambda)_1 \wedge 1] + P\{\sigma_n < 1\}. \tag{25}$$

Since $(V^p N_n \cdot \lambda)_1 \xrightarrow{P} 0$ by dominated convergence, and hence $\sigma_n = 1$ for all but finitely many n, the right-hand side of (25) tends to 0 by dominated convergence, which shows that $(V \cdot M_n)^* \xrightarrow{P} 0$.

To prove the corresponding result for the compensator $V \cdot \hat{X}_n$, choose q, q', and p' as before and put $r = 1/p'$, so that $0 < r < p^{-1} < 1$. Then by Hölder's inequality,

$$
\begin{aligned}
\int_0^1 V_t \, |d\hat{X}^n_t| &= \int_0^1 \frac{V_t \, |X^n_t| \, dt}{1-t} \\
&\leq \int_0^1 \frac{V_t \, dt}{(1-t)^{1-r}} \sup_{t<1} \frac{|X^n_t|}{(1-t)^r} \\
&\leq \left(\int_0^1 \frac{V_t^p \, dt}{(1-t)^\varepsilon} \right)^{1/p} \left(\int_0^1 (1-t)^{qr-q'} dt \right)^{1/q} \sup_{t<1} \frac{|X^n_t|}{(1-t)^r}.
\end{aligned}
$$

Here the first two factors on the right are finite, as before, and the third one tends to 0 a.s. by Proposition 2.33. □

We turn to the special case where the integral λV is a.s. $\mathcal{F}_0$-measurable. Then introduce the predictable process

$$U_t = V_t - (1-t)^{-1} \int_t^1 V_s \, ds = V_t - \overline{V}_t, \quad t \in [0,1),$$

where $\overline{V}_t = \int_t^1 V_s \, ds / (1-t)$. The following representation generalizes the one for Brownian bridge integrals in Lemma 4.20.

Theorem 5.17 *(martingale integral) Let X be an extreme, $\mathcal{F}$-exchangeable process on $[0,1]$ with martingale component M and characteristics (α, σ, β), and let V be an $\mathcal{F}$-predictable process with $\mathcal{F}_0$-measurable integral λV, satisfying the conditions in Theorem 5.16 for some $p > 1$. Then*

$$\int_0^1 V_t \, dX_t = \alpha \lambda V + \int_0^1 U_t \, dM_t \quad a.s., \tag{26}$$

and for any $p \geq 1$ and $q > 2p$ we have

$$E(U \cdot M)^{*p} \lesssim \sigma^p \, E(\lambda V^2)^{p/2} + (E\lambda |V|^q)^{p/q} \left(\sum_j |\beta_j|^{p \wedge 2} \right)^{(p \vee 2)/2}. \tag{27}$$

If the latter bound is finite, then $U \cdot M$ is a true martingale on $[0,1]$, and the series in (22) converges in L^p.

Proof: We may clearly assume that $\alpha = 0$. Then recall from Theorem 2.13 that $X = M - N \cdot \lambda$ on $[0, 1)$, where N denotes the martingale $N_t = X_t/(1-t)$. Writing the latter relation as $X_t = (1 - t)N_t$ and integrating by parts, we obtain

$$dM_t = dX_t + N_t dt = (1 - t)dN_t. \tag{28}$$

Next we may integrate by parts and use the $\mathcal{F}_0$-measurability of λV to get, for any $t \in [0, 1)$,

$$
\begin{aligned}
\int_0^t V_s N_s ds &= N_t \int_0^t V_s ds - \int_0^t dN_s \int_0^s V_r dr \\
&= N_t \left(\lambda V - \int_t^1 V_s ds \right) - \int_0^t dN_s \left(\lambda V - \int_s^1 V_r dr \right) \\
&= \int_0^t dN_s \int_s^1 V_r dr - N_t \int_t^1 V_s ds.
\end{aligned}
$$

Using the semi-martingale decomposition of X, equation (28), and the definition of U, we conclude that

$$\int_0^t V_s dX_s = \int_0^t U_s dM_s + N_t \int_t^1 V_s ds, \quad t \in [0, 1), \tag{29}$$

where the first two integrals exist on $[0, 1]$ by Theorem 5.16 and its proof, combined with Lemma 5.8. Applying Hölder's inequality with $p^{-1} + q^{-1} = 1$ gives

$$
\begin{aligned}
\left| N_t \int_t^1 V_s ds \right| &\leq |N_t| \left(\int_t^1 (1 - s)^{\varepsilon q/p} ds \right)^{1/q} \left(\int_t^1 |V_s|^p (1 - s)^{-\varepsilon} ds \right)^{1/p} \\
&\lesssim |X_t| (1 - t)^{-(1-\varepsilon)/p},
\end{aligned}
$$

which tends a.s. to 0 as $t \to 1$ by Theorem 2.32. Equation (26) now follows from (29), as we let $t \to 1$ and use the continuity of the two stochastic integrals (FMP 26.13).

To prove (27), we note that M has quadratic variation

$$[M]_t = [X]_t = \sigma^2 t + \sum_j \beta_j^2 1\{\tau_j \leq t\}, \quad t \in [0, 1],$$

and hence, by a BDG-inequality (FMP 26.12),

$$
\begin{aligned}
E(U \cdot M)^{*p} &\lesssim E(U^2 \cdot [M])_1^{p/2} \\
&= E\left(\sigma^2 \lambda U^2 + \sum_j \beta_j^2 U_{\tau_j}^2 \right)^{p/2} \\
&\lesssim \sigma^p E(\lambda U^2)^{p/2} + E\left(\sum_j \beta_j^2 U_{\tau_j}^2 \right)^{p/2}.
\end{aligned}
$$

Since $\lambda U^2 \leq \lambda V^2$ by Lemma 4.20 (ii), it remains to estimate the second term on the right.

Then introduce for any $p > 0$ the process

$$Y_t^p = \sum_j |\beta_j|^p \, 1\{\tau_j \leq t\}, \quad t \in [0, 1],$$

and recall from Theorem 2.13 that Y^p has compensator

$$d\hat{Y}_t^p = (1 - t)^{-1} \sum_j |\beta_j|^p \, 1\{\tau_j > t\} \, dt, \quad t \in [0, 1).$$

If $p \leq 2$ and $q > 2p$, we get by subadditivity, dual predictable projection, Hölder's inequality with $r = (1 - p/q)^{-1}$, and Lemma 5.8

$$
\begin{aligned}
E\left(\sum_j \beta_j^2 U_{\tau_j}^2\right)^{p/2} &\leq E \sum_j |\beta_j \, U_{\tau_j}|^p = E \int_0^1 |U_t|^p \, dY_t^p \\
&= \sum_j |\beta_j|^p \, E \int_0^{\tau_j} |U_t|^p \, (1 - t)^{-1} dt \\
&\leq \sum_j |\beta_j|^p \, (E\lambda |U|^q)^{p/q} \left(E \int_0^{\tau_j} (1 - t)^{-r} dt\right)^{1/r} \\
&\lesssim \sum_j |\beta_j|^p \, (E\lambda |V|^q)^{p/q},
\end{aligned}
$$

where the last step relies on the fact that, by Fubini's theorem,

$$E \int_0^{\tau_j} (1 - t)^{-r} dt = \int_0^1 (1 - t)^{1-r} dt = (2 - r)^{-1} < \infty.$$

If instead $p > 2$, we may use Hölder's inequality, and then proceed as before to get

$$
\begin{aligned}
E\left(\sum_j \beta_j^2 U_{\tau_j}^2\right)^{p/2} &\leq \left(\sum_j \beta_j^2\right)^{p/2-1} E \sum_k \beta_k^2 |U_{\tau_k}|^p \\
&= \left(\sum_j \beta_j^2\right)^{p/2-1} \sum_k \beta_k^2 E \int_0^{\tau_k} |U_t|^p \, (1 - t)^{-1} dt \\
&\lesssim \left(\sum_j \beta_j^2\right)^{p/2} (E\lambda |V|^q)^{p/q}.
\end{aligned}
$$

This completes the proof of (27). When the bound is finite, we conclude from FMP 26.13 that $U \cdot M$ is an L^p-martingale on $[0, 1]$. To prove the last assertion, we may apply (22), (26), and (27) to the tail processes X_n in (24) with associated martingale components M_n to obtain

$$
\begin{aligned}
E\left|\sum_{j>n} \beta_j \, (V_{\tau_j} - \lambda V)\right|^p &= E|(V \cdot X_n)_1|^p = E|(U \cdot M_n)_1|^p \\
&\lesssim (E\lambda |V|^q)^{p/q} \left(\sum_{j>n} |\beta_j|^{p \wedge 2}\right)^{(p \vee 2)/2},
\end{aligned}
$$

which tends to 0 as $n \to \infty$ by the convergence of the series on the right. $\square$

5.4   Exchangeable Integrals

Here we establish a general moment identity for $\mathbb{R}^d$-valued processes $X = (X^j)$ and $V = (V^j)$ on $[0,1]$, where X is exchangeable and V is predictable. For any $J \subset \{1,\ldots,d\}$ we introduce the product $V_t^J = \prod_{j \in J} V_t^j$. When $|V| \in L^1$ a.s., we may consider the centered processes $\hat{V}_t^j = V_t^j - \lambda V^j$ along with their products $\hat{V}_t^J = \prod_{j \in J} \hat{V}_t^j$, which remain predictable whenever λV^j is $\mathcal{F}_0$-measurable for all j. If X has characteristics (α, ρ, β), we define

$$\beta_k^J = \prod_{j \in J} \beta_k^j, \qquad \beta_J = \sum_k \delta_{\beta_k^J}, \qquad B_J = \sum_k \beta_k^J.$$

Theorem 5.18 *(moments of integrals on $[0,1]$) Let X and V be $\mathbb{R}^d$-valued processes on $[0,1]$, where X is extreme, $\mathcal{F}$-exchangeable with directing triple (α, ρ, β) and V is $\mathcal{F}$-predictable with $E\lambda|V^j|^{p_j} < \infty$, $j \leq d$, for some $p_1,\ldots,p_d > 0$ with $\sum_j p_j^{-1} < \frac{1}{2}$. Suppose that the products $\alpha_i \lambda V^i$, $\rho_{jk} \lambda \hat{V}^{jk}$, and $\beta_J \lambda \hat{V}^J$ are a.s. non-random. Then*

$$E \prod_j \int_0^1 V^j dX^j = \sum_\pi \prod_i (\alpha_i \lambda V^i) \prod_{j,k} (\rho_{jk} \lambda \hat{V}^{ij}) \prod_J B_J P_\pi(\lambda \hat{V}^I),$$

for some polynomials $P_\pi(\lambda \hat{V}^I)$ in the integrals $\lambda \hat{V}^I$ with $I \subset J \in \pi$, where the summation extends over all partitions π of $\{1,\ldots,d\}$ into singletons $\{i\}$, pairs $\{j,k\}$, and sets J with $|J| \geq 2$.

Under the stated hypotheses, the first moment equals $E_1 = \alpha_1 \lambda V^1$. Imposing the further condition $\alpha_j \lambda V^j \equiv 0$, we have the second and third order moments

$$E_{12} = (\rho_{12} + B_{12}) \lambda \hat{V}^{12}, \qquad E_{123} = B_{123} \lambda \hat{V}^{123}.$$

If U is another predictable process satisfying the same integrability conditions and such that $\lambda U^J = \lambda V^J$ a.s. for all $J \subset \{1,\ldots,d\}$, then

$$E \prod_{j \leq d} (U^j \cdot X^j)_1 = E \prod_{j \leq d} (V^j \cdot X^j)_1.$$

In particular, we obtain a *decoupling* identity for product moments by choosing U to be independent of X with the same distribution as V. Specializing to the case of equal components, we get for real-valued processes X, U, and V the moment identity

$$E(U \cdot X)_1^n = E(V \cdot X)_1^n,$$

valid under the assumption that λU^k and λV^k agree and are $\mathcal{F}_0$-measurable for all $k \leq n$. The result remains true for $\mathbb{R}^d$-valued processes with $V \cdot X = \sum_j (V^j \cdot X^j)$, provided that the integrals $\lambda \prod_j (U^j)^{n_j}$ and $\lambda \prod_j (V^j)^{n_j}$ agree and are $\mathcal{F}_0$-measurable for all $n_1,\ldots,n_d \in \mathbb{Z}_+$ with $\sum_j n_j \leq n$.

The following result is helpful to deal with contributions from the continuous martingale components.

Lemma 5.19 *(martingale product) Let $M^1, \ldots, M^d$ be continuous $\mathcal{F}$-martingales starting at 0, such that $\rho_{ij} = [M^i, M^j]_\infty$ is a.s. non-random for $i \neq j$, and suppose that M^j is L^{p_j}-bounded for each j, where $p_1, \ldots, p_d > 0$ with $\sum_j p_j^{-1} = p^{-1} \leq 1$. Then the process*

$$M_t = E^{\mathcal{F}_t} \prod_{j \leq d} M^j_\infty, \quad t \geq 0,$$

is a continuous, L^p-bounded martingale satisfying

$$M_0 = E \prod_{j \leq d} M^j_\infty = \sum_\pi \prod_{i,j} \rho_{ij} \quad a.s.,$$

where the summation extends over all partitions π (if any) of $\{1, \ldots, d\}$ into pairs $\{i, j\}$.

Proof: By Lemma 5.5 we have for $t \geq 0$

$$M^1_t \cdots M^d_t = \sum_{J_1, \ldots, J_m} \int \cdots \int_{t\Delta_m} d[M]^{J_1} \cdots d[M]^{J_m},$$

where the summation extends over all ordered partitions of the set $\{1, \ldots, d\}$ into pairs or singletons $J_1, \ldots, J_m$. Decomposing the sum according to the last occurrence of a singleton set $\{i\}$ and applying Lemma 5.5 twice again, we obtain

$$\prod_{j \leq d} M^j_t = V_t + \sum_{I,J,i} \int_0^t dV^J_s \int_0^s M^I_r dM^i_r + \sum_{i \leq d} \int_0^t M^{\{i\}^c}_s dM^i_s,$$

where the first summation extends over all partitions of $\{1, \ldots, d\}$ into two sets I and $J \neq \emptyset$ and a singleton $\{i\}$. Here the processes M^I, V^J, and V are given by

$$M^I_t = \prod_{j \in I} M^j_t, \qquad V^J_t = \sum_{J_1, \ldots, J_m} \prod_{k \leq m} [M]^{J_k}_t, \qquad V_t = V^{1,\ldots,d}_t,$$

where the summation extends over all partitions (if any) of J into pairs J_k. In particular, $V^J = 0$ when $|J|$ is odd. Integration by parts yields

$$\prod_{j \leq d} M^j_t = V_t + \sum_{I,J,i} \left(V^J_t \int_0^t M^I_s dM^i_s - \int_0^t V^J_s M^I_s dM^i_s \right) + \sum_{i \leq d} \int_0^t M^{\{i\}^c}_s dM^i_s.$$

Letting $t \to \infty$ and noting that V^J_∞ is a.s. a constant, we get

$$\prod_{j \leq d} M^j_\infty = V_\infty + \sum_{I,J,i} \int_0^\infty (V^J_\infty - V^J_s) M^I_s dM^i_s + \sum_{i \leq d} \int_0^\infty M^{\{i\}^c}_s dM^i_s.$$

If the stochastic integrals are uniformly integrable martingales, we obtain

$$M_t = V_\infty + \sum_{I,J,i} \int_0^t (V^J_\infty - V^J_s) M^I_s dM^i_s + \sum_{i \leq d} \int_0^t M^{\{i\}^c}_s dM^i_s,$$

and the assertions follow.

To justify the formal computations, we write $p_{ij}^{-1} = p_i^{-1} + p_j^{-1}$ and use the Hölder, BDG, and Courrège inequalities (FMP 1.29, 17.7, 17.9) to get

$$
\begin{aligned}
\|[M^i, M^j]^*\|_{p_{ij}} &\leq \|[M^i]_\infty^{1/2} [M^j]_\infty^{1/2}\|_{p_{ij}} \\
&\leq \|[M^i]_\infty^{1/2}\|_{p_i} \|[M^j]_\infty^{1/2}\|_{p_j} \\
&\lesssim \|(M^i)^*\|_{p_i} \|(M^j)^*\|_{p_j}.
\end{aligned}
$$

Hence, by the same inequalities, we have for any I, J, i, and $J_1, \ldots, J_m$ as before

$$
\begin{aligned}
\Big\|\big((V_\infty^J &- V^J)M^I \cdot M^i\big)^*\Big\|_p \\
&\lesssim \Big\|\big((V_\infty^J - V^J)^2 (M^I)^2 \cdot [M^i]\big)_\infty^{1/2}\Big\|_p \\
&\lesssim \big\|(V^J)^* (M^I)^* [M^i]_\infty^{1/2}\big\|_p \\
&\leq \sum_{J_1, \ldots, J_m} \prod_{k \leq m} \|(V^{J_k})^*\|_{p_{J_k}} \prod_{h \in I} \|(M^h)^*\|_{p_h} \|(M^i)^*\|_{p_i} \\
&\lesssim \prod_j \|(M^j)^*\|_{p_j} < \infty,
\end{aligned}
$$

as required. Similar estimates yield the same bound for the norm $\|(M^{\{i\}^c} \cdot M^i)^*\|_p$. $\qquad\square$

To deal with contributions from the jumps of X, we need a result that is closely related to Corollary 4.28.

Lemma 5.20 (*predictable product*) *Let M be a continuous $\mathcal{F}$-martingale on $[0,1]$, let $V_1, \ldots, V_d$ be $\mathcal{F}$-predictable processes on $[0,1]$ with $\lambda V_1 = \cdots = \lambda V_d = 0$ a.s., and let $\tau_1, \ldots, \tau_d$ be i.i.d. $U(0,1)$ and such that the processes $X_t^j = 1\{\tau_j \leq t\}$ are $\mathcal{F}$-exchangeable. Suppose that $E|M^*|^p < \infty$ and $E\lambda|V_j|^{p_j} < \infty$ for all j, where $p, p_1, \ldots, p_d > 0$ with $p^{-1} + 2\sum_j p_j^{-1} < 1$. Then*

$$
E M_1 \prod_{j \leq d} V_j(\tau_j) = 0.
$$

Proof: Writing $V^j = V_j$ and noting that $\lambda|V^j|^{p_j} < \infty$ a.s., we see from Theorems 2.8 and 5.17 that

$$
V_{\tau_j}^j = \int_0^1 V_t^j dX_t^j = \int_0^1 U_t^j dM_t^j, \quad j = 1, \ldots, d,
$$

where for any $j \leq d$ and $t \in [0,1)$,

$$
\begin{aligned}
M_t^j &= 1\{\tau_j \leq t\} + \log(1 - t \wedge \tau_j), \\
U_t^j &= V_t^j + (1-t)^{-1} \int_0^t V_s^j ds. \tag{30}
\end{aligned}
$$

Putting $N^j = N_j = U^j \cdot M^j$ and noting that $M, N^1, \ldots, N^d$ are strongly orthogonal martingales, we get by repeated integration by parts

$$M_t \prod_j N_t^j = \int_0^t dM_s \prod_j N_s^j + \sum_i \int_0^t M_s dN_s^i \prod_{j \neq i} N_s^j.$$

It remains to show that each term on the right is a uniformly integrable martingale, since in that case

$$E \, M_1 \prod_j V_{\tau_j}^j = E \, M_1 \prod_j N_1^j = E \, M_0 \prod_j N_0^j = 0.$$

Then conclude from the estimate in Theorem 5.17 that $E(N^j)^{*q_j} < \infty$ whenever $0 < 2q_j < p_j$. Since $p^{-1} + 2\sum_j p_j^{-1} < 1$, we may choose the q_j such that $p^{-1} + \sum_j q_j^{-1} \leq 1$. Using the BDG and Hölder inequalities, we obtain

$$
\begin{aligned}
E\Big(\prod_j N_j \cdot M\Big)^* &\lesssim E\Big(\prod_j N_j^2 \cdot [M]\Big)_1^{1/2} \leq E\,[M]_1^{1/2} \prod_j N_j^* \\
&\leq \|[M]_1^{1/2}\|_p \prod_j \|N_j^*\|_{q_j} \\
&\lesssim \|M^*\|_p \prod_j \|N_j^*\|_{q_j} < \infty,
\end{aligned}
$$

as required. Similarly, for $i = 1, \ldots, d$, we have

$$
\begin{aligned}
E\Big(M \prod_{j \neq i} N_j \cdot N_i\Big)^* &\lesssim E\Big(M^2 \prod_{j \neq i} N_j^2 \cdot [N_i]\Big)_1^{1/2} \\
&\leq E\,M^* \,[N_i]_1^{1/2} \prod_{j \neq i} N_j^* \\
&\leq \|M^*\|_p \,\|[N_i]_1^{1/2}\|_{q_i} \prod_{j \neq i} \|N_j^*\|_{q_j} \\
&\lesssim \|M^*\|_p \prod_j \|N_j^*\|_{q_j} < \infty.
\end{aligned}
$$
$\square$

Proof of Theorem 5.18: Under the stated moment conditions, we see from Theorems 5.16 and 5.17 that the product moment $E \prod_j (V^j \cdot X^j)_1$ exists and can be evaluated by term-wise integration, according to (22). In particular, this allows us to assume that X has finitely many jumps. Writing $\hat{X}_t^j = X_t^j - \alpha_j t$ and noting that $(V^j \cdot \hat{X}^j)_1 = (\hat{V}^j \cdot \hat{X}^j)_1$, we obtain

$$
\begin{aligned}
E \prod_j (V^j \cdot X^j)_1 &= E \prod_j \big(\alpha_j \lambda V^j + (\hat{V}^j \cdot \hat{X}^j)_1\big) \\
&= \sum_J \prod_{j \notin J} (\alpha_j \lambda V^j)\, E \prod_{j \in J} (\hat{V}^j \cdot \hat{X}^j)_1,
\end{aligned}
$$

where the summation extends over all subsets $J \subset \{1, \ldots, d\}$. This reduces the discussion to the case where $\alpha_j = 0$ and $\lambda V^j = 0$ for all j. By Lemma 2.22 we may finally assume that the individual terms in the representation of X are jointly (indeed even separately) $\mathcal{F}$-exchangeable.

Now let M^j denote the continuous martingale component of X^j, and let U^j be given by (30). Writing $N^j = U^j \cdot M^j$, we get by Lemma 4.20

$$
\begin{aligned}
(V^j \cdot B^j)_1 &= (U^j \cdot M^j)_1 = N_1^j \\
[N^i, N^j]_1 &= (U^{ij} \cdot [M^i, M^j])_1 = \rho_{ij} \lambda U^{ij} = \rho_{ij} \lambda V^{ij}.
\end{aligned}
$$

Let us now define $p_J^{-1} = \sum_{j \in J} p_j^{-1}$ for any $J \subset \{1, \ldots, d\}$. By Lemma 5.19 there exists an L^{p_J}-bounded, continuous martingale M^J satisfying

$$M_1^J = \prod_{j \in J} N_1^j = \prod_{j \in J} (V^j \cdot B^j)_1, \tag{31}$$

$$EM_1^J = \sum_\pi \prod_{i,j} [N^i, N^j]_1 = \sum_\pi \prod_{i,j} \rho_{ij} \lambda V^{ij}, \tag{32}$$

where the summations in (32) extend over all partitions π (if any) of J into pairs $\{i, j\}$.

Next we may use the decomposition in Theorem 5.16, together with (31), to obtain

$$
\begin{aligned}
E \prod_j (V^j \cdot X^j)_1 &= E \prod_j \Big((V^j \cdot B^j)_1 + \sum_k \beta_k^j V^j(\tau_k) \Big) \\
&= E \sum_{I,\pi} M_1^I \sum_{(k_J)} \prod_{J \in \pi} \beta_{k_J}^J V^J(\tau_{k_J}) \\
&= \sum_{I,\pi} \sum_{(k_J)} \prod_{J \in \pi} \beta_{k_J}^J \, E \, M_1^I \prod_{J \in \pi} V^J(\tau_{k_J}), \tag{33}
\end{aligned}
$$

where the outer summations extend over all subsets $I \subset \{1, \ldots, d\}$ and partitions π of I^c, and the inner summations extend over all sets of distinct indices k_J, $J \in \pi$. Writing $V^J = (V^J - \lambda V^J) + \lambda V^J$ and using Lemma 5.20 and (32), we get for fixed I, π, and (k_J)

$$
E \, M_1^I \prod_{J \in \pi} V^J(\tau_{k_J}) = E \, M_1^I \prod_{J \in \pi} \lambda V^J = \prod_{J \in \pi} \lambda V^J \sum_{\pi'} \prod_{i,j} \rho_{ij} \lambda V^{ij},
$$

where the summation extends over all partitions π' of I into pairs $\{i, j\}$. Inserting this into (33) yields

$$
E \prod_j (V^j \cdot X^j)_1 = \sum_\pi \prod_{i,j} (\rho_{ij} \lambda V^{ij}) \prod_J \lambda V^J \sum_{(k_J)} \prod_J \beta_{k_J}^J,
$$

where the outer summation extends over all partitions π of $\{1, \ldots, d\}$ into pairs $\{i, j\}$ and subsets J with $|J| \geq 2$, and where the indices k_J are distinct, as before. It remains to note that, by Lemma 5.4, the inner sum on the right is a polynomial in the sums B^K, where K is a nonempty union of sets J from the partition π. $\qquad \square$

5.5 Lévy Integrals

Here we prove a general moment identity for $\mathbb{R}^d$-valued processes $X = (X^j)$ and $V = (V^j)$ on $\mathbb{R}_+$, where X is an integrable Lévy process with characteristics (α, ρ, ν) and V is predictable. In order to apply Proposition 5.2, we need to assume the existence of some constants $p_1, \ldots, p_d \geq 1$ satisfying

$$
\begin{aligned}
|\alpha_j| E(\lambda|V^j|)^{p_j} &+ \rho_{jj} E(\lambda|V^j|^2)^{p_j/2} \\
&+ \int |x_j|^{p_j} \nu(dx) \, E\big((\lambda|V^j|^{p_j \wedge 2})^{(p_j \vee 2)/2} + \lambda|V^j|^{p_j} \big) < \infty. \tag{34}
\end{aligned}
$$

For any $J \subset \{1, \ldots, d\}$, we define

$$\nu_J = \int \prod_{j \in J} x_j \, \nu(dx), \qquad V_t^J = \prod_{j \in J} V_t^j.$$

Theorem 5.21 *(moments of Lévy integrals) Let X and V be $\mathbb{R}^d$-valued processes on $\mathbb{R}_+$, where X is $\mathcal{F}$-Lévy with characteristics (α, ρ, ν) and V is an $\mathcal{F}$-predictable process satisfying (34), for some $p_1, \ldots, p_d > 0$ with $\sum_j p_j^{-1} \leq 1$. Suppose that the products $\alpha_i \lambda V^i$ (when $d \geq 2$), $\rho_{jk} \lambda V^{jk}$ (when $d \geq 3$), and $\nu_J \lambda V^J$ (for $2 \leq |J| < d$) are a.s. non-random. Then*

$$E \prod_{j \in d} \int_0^\infty V^j dX^j = E \sum_\pi \prod_i (\alpha_i \lambda V^i) \prod_{j,k} \big((\rho_{jk} + \nu_{jk}) \, \lambda V^{jk}\big) \prod_J (\nu_J \lambda V^J),$$

where the summation extends over all partitions π of $\{1, \ldots, d\}$ into singletons $\{i\}$, pairs $\{j, k\}$, and subsets J with $|J| \geq 3$.

As before, we note that the product moment depends only on the marginal distributions of X and V. Choosing $\tilde{V}$ to be independent of X with the same distribution as V, we obtain

$$E \prod_{j \leq d} (V^j \cdot X^j)_\infty = E \prod_{j \leq d} (\tilde{V}^j \cdot X^j)_\infty.$$

In the special case where $X^j = X$, $V^j = V$, and $\tilde{V}^j = \tilde{V}$ for all j, we obtain for simple moments the equality

$$E(V \cdot X)_\infty^n = E(\tilde{V} \cdot X)_\infty^n,$$

under appropriate constancy and moment conditions.

Several lemmas will be needed for the proof of Theorem 5.21. We begin with some technical estimates. Recall that $p_J^{-1} = \sum_{j \in J} p_j^{-1}$ for any nonempty subset $J \subset \{1, \ldots, d\}$.

Lemma 5.22 *(product moments) Under the hypotheses of Theorem 5.21, we have for any $J \subset \{1, \ldots, d\}$ with $|J| > 1$*

$$\int \prod_{j \in J} |x_j|^p \, \nu(dx) \, E(\lambda |V_J|^p)^{p_J/p} < \infty, \quad 1 \leq p \leq p_J.$$

Proof: Excluding the trivial case where $\int |x_j| \nu(dx) = 0$ or $\lambda |V_j| = 0$ for some $j \in J$, we have by hypothesis

$$\int |x_j|^{p_j} \nu(dx) < \infty, \quad E(\lambda |V_j|^{p_j \wedge 2})^{(p_j \vee 2)/2} + E\lambda |V_j|^{p_j} < \infty, \tag{35}$$

for any $j \in J$. Hence, by Hölder's inequality,

$$\int \prod_{j \in J} |x_j|^{p_J} \nu(dx) < \infty, \qquad E\lambda |V_J|^{p_J} < \infty. \tag{36}$$

Next we note that $x^{p \wedge 2} \leq (x^2 \wedge 1) + x^p$ for $x, p > 0$. By (35) we get for any $j \in J$

$$\int |x_j|^{p_j \wedge 2} \nu(dx) \leq \int (x_j^2 \wedge 1) \nu(dx) + \int |x_j|^{p_j} \nu(dx) < \infty,$$

and so by (35) and Lemma 5.3

$$\int |x_j|^q \nu(dx) < \infty, \qquad E(\lambda|V_j|^q)^{p_j/q} < \infty, \qquad p_j \wedge 2 \leq q \leq p_j. \tag{37}$$

Since

$$\sum_{j \in J} p_j^{-1} \leq 1 \leq \tfrac{1}{2}|J| \leq \sum_{j \in J} (p_j \wedge 2)^{-1},$$

we may choose some constants q_j in $[p_j \wedge 2, p_j]$, $j \in J$, with $\sum_{j \in J} q_j^{-1} = 1$, and conclude from Hölder's inequality and (37) that

$$\int \prod_{j \in J} |x_j|\, \nu(dx) \leq \prod_{j \in J} \left(\int |x_j|^{q_j} \nu(dx) \right)^{1/q_j} < \infty,$$

$$E(\lambda|V_J|)^{p_J} \leq E \prod_{j \in J} (\lambda|V_j|^{q_j})^{p_J/q_j}$$

$$\leq \prod_{j \in J} \left(E(\lambda|V_j|^{q_j})^{p_j/q_j} \right)^{p_J/p_j} < \infty.$$

The assertion now follows by (36) and Lemma 5.3.   □

Next we need to estimate the higher order variations $[X]^J$ of a semi-martingale X, introduced in connection with Lemma 5.5. Recall that when X is purely discontinuous,

$$[X]_t^J = \sum_{s \leq t} \prod_{j \in J} \Delta X_s^j, \qquad t \geq 0, \ |J| > 1.$$

Lemma 5.23 *(variation integrals)* If $\rho = 0$ in Theorem 5.21, then for any $J \subset \{1, \ldots, d\}$ with $|J| > 1$ we have

$$E\left(\int |V_J d[X]^J| \right)^{p_J} \lesssim \sup_{p \in [1, p_J]} \left(\int \prod_{j \in J} |x_j|^p \nu(dx) \right)^{p_J/p}$$

$$\times \ \sup_{p \in [1, p_J]} E(\lambda|V_J|^p)^{p_J/p} < \infty.$$

Proof: The total variation process $\int |d[X]^J|$ is a subordinator of pure jump type, whose Lévy measure equals the image of ν under the mapping $x \mapsto \prod_{j \in J} |x_j|$. The stated estimate then follows from Proposition 5.2, and the finiteness on the right is clear from Lemmas 5.3 and 5.22.   □

Our next aim is to construct a bounded measure on $\mathbb{R}^d$ with given product moments c_J, $J \in 2^d$, where 2^d denotes the class of subsets of $\{1, \ldots, d\}$.

Lemma 5.24 *(moment fitting)* Given any $c_J \in \mathbb{R}$, $J \in 2^d \setminus \{\emptyset\}$, there exists a measure μ on $S = \{-1, 1\}^d$ with $\mu S = \sum_J |c_J|$ such that

$$\int \mu(dx) \prod_{j \in J} x_j = c_J, \qquad J \in 2^d \setminus \{\emptyset\}.$$

Proof: We begin with the case where $c_I = \sigma 1\{I = J\}$ for some fixed subset $J \neq \emptyset$ and sign $\sigma = \pm 1$. Then fix any $k \in J$ and put $J_k = J \setminus \{k\}$. Introduce some i.i.d. random variables ξ_j, $j \neq k$, with $P\{\xi_j = \pm 1\} = \frac{1}{2}$, and define $\xi_k = \sigma \prod_{j \in J_k} \xi_j$. Let μ_J^σ denote the distribution of $(\xi_1, \ldots, \xi_d)$. If $k \notin I$ or $I \setminus J \neq \emptyset$, we may choose an $h \in I$ such that ξ_h is independent of all remaining variables ξ_i, $i \in I \setminus \{h\} \equiv I_h$, in which case

$$\int \prod_{i \in I} x_i \, \mu_J^\sigma(dx) = E \prod_{i \in I} \xi_i = E\xi_h \, E \prod_{i \in I_h} \xi_i = 0.$$

If instead $k \in I \subset J$, we have

$$\begin{aligned} \int \prod_{i \in I} x_i \, \mu_J^\sigma(dx) &= E \prod_{i \in I} \xi_i = \sigma E \prod_{i \in I} \xi_i \prod_{j \in J} \xi_j \\ &= \sigma E \prod_{j \in J \setminus I} \xi_j = \sigma \prod_{j \in J \setminus I} E\xi_j = \sigma 1\{I = J\}. \end{aligned}$$

This shows that

$$\int \prod_{i \in I} x_i \, \mu_J^\sigma(dx) = \sigma \, 1\{I = J\}, \quad I, J \in 2^d \setminus \{\emptyset\}. \tag{38}$$

For general moments c_J, we put $c_J^{\pm} = (\pm c_J) \vee 0$ and define

$$\mu = \sum_I (c_I^+ \mu_I^+ + c_I^- \mu_I^-),$$

where the summation extends over all $I \in 2^d \setminus \{\emptyset\}$. Then (38) gives

$$\begin{aligned} \mu S &= \sum_I (c_I^+ + c_I^-) = \sum_I |c_I|, \\ \int \mu(dx) \prod_{j \in J} x_i &= \sum_I (c_I^+ - c_I^-) 1\{I = J\} = c_J, \end{aligned}$$

as required. $\qquad\square$

The next result reduces the proof of Theorem 5.21 to the case where X has bounded and isolated jumps. To avoid trivial exceptions, we may henceforth assume that $E\lambda|V_j| > 0$ for all $j \leq d$ (which entails no loss of generality).

Lemma 5.25 *(truncation) In the context of Theorem 5.21, there exist some Lévy processes $X^n = (X_j^n)$ with characteristics (α, ρ, ν_n), defined with respect to standard extensions $\mathcal{F}^n$ of $\mathcal{F}$, where the ν_n are bounded with bounded supports and moments $\nu_n^J = \nu_J$, $|J| > 1$, such that*

$$\prod_j \int_0^\infty V_j \, dX_j^n \to \prod_j \int_0^\infty V_j \, dX_j \quad in \ L^1.$$

Proof: For each $n \in \mathbb{N}$, let Y^n denote the Lévy process obtained from X by omission of all centered jumps of modulus $> n$ or $< n^{-1}$. Then Y^n has characteristics $(\alpha, \rho, \hat{\nu}_n)$, where $\hat{\nu}_n$ denotes the restriction of ν to the set $B_n = \{x \in \mathbb{R}^d; \, n^{-1} \leq |x| \leq n\}$, and we may introduce the associated moment deficiencies

$$c_n^J = \int_{B_n^c} \nu(dx) \prod_{j \in J} x_j, \quad J \in 2^d, \quad |J| > 1. \tag{39}$$

For every $n \in \mathbb{N}$, there exists by Lemma 5.24 a measure μ_n on $\{-1,1\}^d$ with total mass $\|\mu_n\| = \sum_J |c_n^J| < \infty$ such that

$$\int \mu_n(dx) \prod_{j \in J} x_j = c_n^J, \quad J \in 2^d, \quad |J| > 1. \tag{40}$$

Next we may introduce, on a suitably extended probability space, a Lévy process $Z^n \perp\!\!\!\perp \mathcal{F}$ with characteristics $(0, 0, \mu_n)$. Then $X^n = Y^n + Z^n$ is again Lévy with characteristics (α, ρ, ν_n), where $\nu_n = \hat{\nu}_n + \mu_n$, and we note that X^n remains Lévy with respect to the right-continuous and complete filtration $\mathcal{F}^n$ induced by $\mathcal{F}$ and Z^n. To see that $\mathcal{F}^n$ is a standard extension of $\mathcal{F}$, we need to verify that $(Z_s^n; \, s \leq t) \perp\!\!\!\perp_{\mathcal{F}_t} \mathcal{F}$ for all $t \geq 0$, which is clear from the independence of Z and $\mathcal{F}$. We also see from (39) and (40) that ν_n and ν have the same product moments $\nu_n^J = \nu_J$, $|J| > 1$.

To prove the asserted approximation property, we may assume that, by (34) and Lemma 5.22,

$$\int \left(|x_j|^{p_j} \vee |x_j|^{p_j \wedge 2} \right) \nu(dx) < \infty, \qquad \int \nu(dx) \prod_{j \in J} |x_j| < \infty,$$

for all $j \in \{1, \ldots, d\}$ and $J \in 2^d$ with $|J| > 1$. Writing $\hat{\nu}_n' = \nu - \hat{\nu}_n$, we get by (40) and dominated convergence as $n \to \infty$

$$\int \left(|x_j|^{p_j} \vee |x_j|^{p_j \wedge 2} \right) \hat{\nu}_n'(dx) \to 0,$$

$$\int \left(|x_j|^{p_j} \vee |x_j|^{p_j \wedge 2} \right) \mu_n(dx) = \mu_n\{-1,1\}^d = \sum_J |c_n^J|$$

$$\leq \sum_J \int \hat{\nu}_n'(dx) \prod_{j \in J} |x_j| \to 0,$$

where the summations extend over all $J \in 2^d$ with $|J| > 1$. By (34) and Proposition 5.2 it follows that

$$\|(V_j \cdot (X_j - X_j^n))^*\|_{p_j} \leq \|(V_j \cdot (X_j - Y_j^n))^*\|_{p_j} + \|(V_j \cdot Z_j^n)^*\|_{p_j} \to 0.$$

Since also $\|(V_j \cdot X_j)^*\|_{p_j} < \infty$ by the same proposition, we see from Hölder's inequality that

$$E \left| \prod_j \int_0^\infty V_j \, dX_j - \prod_j \int_0^\infty V_j \, dX_j^n \right|$$

$$\leq \sum_k E \left| \prod_{j<k} (V_j \cdot X_j)^* \prod_{j>k} (V_j \cdot X_j^n)^* \, (V_k \cdot (X_k - X_k^n))^* \right|$$

$$\leq \sum_k \prod_{j<k} \|(V_j \cdot X_j)^*\|_{p_j} \prod_{j>k} \|(V_j \cdot X_j^n)^*\|_{p_j}$$
$$\times \|(V_k \cdot (X_k - X_k^n))^*\|_{p_k} \to 0,$$

where the summations extend over all $k \in \{1, \ldots, d\}$. $\qquad\qquad \square$

We are now ready to prove our key lemma. Here $\int_a^b$ denotes integration over $(a, b]$, and we will often write Lebesgue integrals $\int f(s)ds$ as $\int f$.

Lemma 5.26 *(key identity)* *In the setting of Theorem 5.21, let $\alpha = \rho = 0$, and let ν be bounded with bounded support. Consider a continuous martingale M with $\|M^*\|_p < \infty$, where $p^{-1} + \sum_j p_j^{-1} \le 1$, and suppose that $\nu_J \lambda V_J$ is a.s. non-random even for $J = \{1, \ldots, d\}$, unless M is a constant. Then for any optional time τ we have*

$$E\, M_\infty \prod_j \int_\tau^\infty V_j dX_j = E\, M_\tau \sum_\pi \prod_J \nu_J \int_\tau^\infty V_J(s)ds,$$

where the summation extends over all partitions π of $\{1, \ldots, d\}$ into sets J with $|J| > 1$.

Proof: For $d = 0$ the assertion reduces to $EM_\infty = EM_\tau$, which holds by optional sampling. Proceeding by induction, we assume the asserted formula to be valid for up to $d - 1$ factors. To extend the result to products of d integrals, we may fix any $T > 0$ and proceed through the following chain of equalities, where each step is explained in detail below:

$$E\, M_\infty \prod_{j \le d} \int_\tau^\infty V_j\, dX_j - E\, M_\infty \prod_{j \le d} \int_{\tau \vee T}^\infty V_j\, dX_j$$

$$= E\, M_\infty \sum_J \int_\tau^{\tau \vee T} V_t^J\, d[X]_t^J \prod_{j \notin J} \int_t^\infty V_j\, dX_j$$

$$= E \sum_J \int_\tau^{\tau \vee T} V_t^J\, M_t\, d[X]_t^J \sum_{\pi'} \prod_I \nu_I \int_t^\infty V_I$$

$$= E \sum_J \nu_J \int_\tau^{\tau \vee T} V_t^J\, M_t\, dt \sum_{\pi'} \prod_I \nu_I \int_t^\infty V_I$$

$$= E\, M_\infty \sum_J \nu_J \int_\tau^{\tau \vee T} V_t^J\, dt \sum_{\pi'} \prod_I \nu_I \int_t^\infty V_I$$

$$= E\, M_\infty \sum_\pi \prod_J \nu_J \int_\tau^\infty V_J - E\, M_\infty \sum_\pi \prod_J \nu_J \int_{\tau \vee T}^\infty V_J$$

$$= E\, M_\tau \sum_\pi \prod_J \nu_J \int_\tau^\infty V_J - E\, M_\infty \sum_\pi \prod_J \nu_J \int_{\tau \vee T}^\infty V_J,$$

where $\nu_J = 0$ for $|J| = 1$ and the summations extend over all nonempty subsets $J \subset \{1, \ldots, d\}$, all partitions π' of J^c into sets I with $|I| > 1$, and all partitions π of $\{1, \ldots, d\}$ into sets J with $|J| > 1$.

As promised, we proceed with a careful justification of the individual steps. First we note that the first and fifth equalities hold by the tetrahedral decomposition in Lemma 5.5, though in the equivalent form of (9), written in reverse time as

$$d(X^1 \cdots X^d)_t = \sum_J (X_\infty^{J^c} - X_t^{J^c})\, d[X]_t^J, \quad t \ge 0.$$

The time reversal is justified in the present case, since the relevant processes $V_J \cdot X_J$ and $V_J \cdot \lambda$ have locally finite variation, so that the associated integrals are of elementary Stieltjes type.

The second and fourth equalities are due to optional projection, as explicated by Lemma 5.6. The underlying sampling property is in step two

$$E\, M_\infty \prod_{j \notin J} \int_\tau^\infty V_j\, dX_j = E\, M_\tau \sum_{\pi'} \prod_{I \in \pi'} \nu_I \int_\tau^\infty V_I,$$

which holds by the induction hypothesis, whereas in step four it is simply the optional sampling formula $EM_\infty = EM_\tau$. The associated finite-variation processes of Lemma 5.6 are given, respectively, by

$$A_t = \int_{\tau \vee t}^{(\tau \vee T) \wedge t} V_J\, d[X]^J, \qquad B_t = \int_{\tau \vee t}^{(\tau \vee T) \wedge t} V_J(s)\, ds \prod_I \nu_I \int_s^\infty V_I.$$

The third step uses dual predictable projection (FMP 25.13), where the integrator $[X]_t^J$ is replaced by its compensator, equal to $\nu_J t$ when $|J| > 1$ and to 0 when $|J| = 1$. The associated integrand is

$$U_t = 1_{(\tau, \tau \vee T]}(t)\, V_J(t)\, M_t \prod_I \nu_I \int_t^\infty V_I, \quad t \geq 0,$$

where each factor is predictable, the first one by left-continuity and adaptedness, the second one by hypothesis, and the remaining ones by continuity and adaptedness. The latter property is due to the fact that $\nu_I \lambda V_J$ is a.s. non-random whenever $|I| < d$.

The sixth step is trivial when M is a constant and is otherwise due to the optional sampling formula $E[M_\infty | \mathcal{F}_\tau] = M_\tau$, combined with the fact that

$$\nu_J \int_\tau^\infty V_J = \nu_J \lambda V_J - \nu_J \int_0^\tau V_J$$

is $\mathcal{F}_\tau$-measurable for all J by FMP 7.5.

The truncation involving T is essential in steps two and three, to ensure the required integrability when $|J| = 1$. The appropriate integrability conditions are then easy to verify from (34), using Hölder's inequality and Lemmas 5.22 and 5.23. Similarly, as $T \to \infty$, we get

$$\left| E\, M_\infty \prod_j \int_{\tau \vee T}^\infty V_j dX_j \right| \leq \|M^*\|_p \prod_j \left\| \int_{\tau \vee T}^\infty V_j\, dX_j \right\|_{p_j} \to 0,$$

$$\left| E\, M_\infty \prod_J \nu_J \int_{\tau \vee T}^\infty V_J \right| \leq \|M^*\|_p \prod_J |\nu_J| \left\| \int_{\tau \vee T}^\infty |V_J| \right\|_{p_J} \to 0,$$

by the same conditions and estimates, which proves the asserted identity. □

After all these preparations, we are finally ready to prove the main result of this section.

Proof of Theorem 5.21: From Proposition 5.2 we see that $V \cdot (X - EX)$ is a uniformly integrable martingale, which proves the assertion for $d = 1$. When $d \geq 2$, we note that the products $\alpha_j \lambda V_j$ exist and are a.s. non-random. Writing $Y_t = X_t - \alpha t$ and expanding the product, we obtain

$$E \prod_j \int_0^\infty V_j \, dX_j = \sum_J \prod_{j \in J} (\alpha_j \lambda V_j) \, E \prod_{j \notin J} \int_0^\infty V_j \, dY_j,$$

where the summation extends over all subsets $J \subset \{1, \ldots, d\}$. This reduces the argument to the case of centered processes X. By Lemma 5.25 we may also assume that ν is bounded and has bounded support.

Now write B and Y for the continuous and purely discontinuous martingale components of X, and define $M_j = (V_j \cdot B_j)$, so that

$$[M_i, M_j]_\infty = (V_i V_j \cdot [B_i, B_j])_\infty = \rho_{ij} \lambda V_{ij}, \quad i \neq j. \tag{41}$$

When $d \geq 3$, the right-hand side is a.s. non-random by hypothesis, and so by Lemma 5.19 there exist some continuous martingales M_J, indexed by nonempty subsets $J \subset \{1, \ldots, d\}$, such that $\|M_J^*\|_{p_J} < \infty$ and

$$M_J(\infty) = \prod_{j \in J} M_j(\infty), \qquad EM_J = \sum_\pi \prod_{i,j} \rho_{ij} \lambda V_{ij},$$

where the summation extends over all partitions (if any) of J into pairs $\{i, j\}$. Putting $M_\emptyset \equiv 1$, we get by Lemma 5.26 with $\tau = 0$

$$
\begin{aligned}
E \prod_{j \leq d} \int_0^\infty V_j \, dX_j &= E \sum_J M_J(\infty) \prod_{j \notin J} \int_0^\infty V_j \, dY_j \\
&= E \sum_\pi \prod_{i,j} (\rho_{ij} \lambda V_{ij}) \prod_J (\nu_J \lambda V_J), \tag{42}
\end{aligned}
$$

where the summations extend over all subsets $J \subset \{1, \ldots, d\}$, respectively over all partitions π of $\{1, \ldots, d\}$ into pairs $\{i, j\}$ and subsets J with $|J| \geq 2$.

This completes the proof for $d \neq 2$. When $d = 2$, we note that the first equality in (42) remains valid with $M_{12} = M_1 M_2$. So does the second one, since by Lemma 5.7 and (41)

$$EM_1(\infty) M_2(\infty) = E[M_1, M_2]_\infty = E\rho_{ij} \lambda V_{ij}. \qquad \square$$

5.6　Contractable Sums and Integrals

Here we first consider random n-sequences $\xi = (\xi_k^j)$ and $\eta = (\eta_k^j)$ in $\mathbb{R}^d$, where $j = 1, \ldots, d$ and $k = 1, \ldots, n$. The integrability problems being elementary in this case, we may restrict our attention to bounded random variables. For typographical convenience, we shall often write binomial coefficients as $(n /\!/ k) = n! / k! (n - k)!$.

Theorem 5.27 *(tetrahedral moments of sequences)* *Let ξ and η be bounded random n-sequences in $\mathbb{R}^d$, where ξ is $\mathcal{F}$-contractable and η is $\mathcal{F}$-predictable, and suppose that the sums*

$$S_i = \sum_{k_1 < \cdots < k_d} \prod_{j \geq i} \eta_{k_j}^j, \quad i = 1, \ldots, d, \tag{43}$$

are $\mathcal{F}_0$-measurable. Then

$$E \sum_{k_1 < \cdots < k_d} \prod_{j \leq d} (\xi_{k_j}^j \eta_{k_j}^j) = (n/\!/d)^{-1} E \sum_{h_1 < \cdots < h_d} \prod_{i \leq d} \xi_{h_j}^i \sum_{k_1 < \cdots < k_d} \prod_{j \leq d} \eta_{k_j}^j.$$

Thus, if $\tilde\eta$ is another bounded and $\mathcal{F}$-predictable sequence with the same $\mathcal{F}_0$-measurable sums S_i as in (43), we have

$$E \sum_{k_1 < \cdots < k_d} \prod_{j \leq d} \xi_{k_j}^j \, \eta_{k_j}^j = E \sum_{k_1 < \cdots < k_d} \prod_{j \leq d} \xi_{k_j}^j \, \tilde\eta_{k_j}^j.$$

In particular, we obtain a *tetrahedral decoupling identity* by choosing $\tilde\eta$ to be conditionally independent of ξ with the same distribution as η. The last result leads easily to a decoupling identity for ordinary product moments.

Corollary 5.28 *(product moments)* *Let $\xi = (\xi_k^j)$ and $\eta = (\eta_k^j)$ be finite, bounded random sequences in $\mathbb{R}^d$, where ξ is $\mathcal{F}$-contractable and η is $\mathcal{F}$-predictable, and suppose that the sums*

$$S_{J_1,\ldots,J_m} = \sum_{k_1 < \cdots < k_m} \prod_{r \leq m} \prod_{j \in J_r} \eta_{k_r}^j, \quad J_1, \ldots, J_m \subset \{1, \ldots, d\} \ \text{disjoint},$$

are $\mathcal{F}_0$-measurable. Then

$$E \prod_{j \leq d} \sum_{k \leq n} \xi_k^j \, \eta_k^j = \sum_{J_1,\ldots,J_m} (n/\!/m)^{-1} E \sum_{h_1 < \cdots < h_m} \prod_{i \leq m} \xi_{h_i}^{J_i} \sum_{k_1 < \cdots < k_m} \prod_{j \leq m} \eta_{k_j}^{J_j},$$

where the outer summation on the right extends over all ordered partitions $J_1, \ldots, J_m$ of $\{1, \ldots, d\}$.

Proof: This follows from Theorem 5.27 by means of the elementary, tetrahedral decomposition

$$\prod_{j \leq d} \sum_{k \leq n} x_k^j = \sum_{J_1,\ldots,J_m} \sum_{k_1 < \cdots < k_m} \prod_{r \leq m} \prod_{j \in J_r} x_{k_r}^j,$$

where the outer summation on the right extends over all ordered partitions $J_1, \ldots, J_m$ of $\{1, \ldots, d\}$. □

In particular, we note that if $\tilde\eta$ is another bounded and $\mathcal{F}$-predictable sequence with the same $\mathcal{F}_0$-measurable sums $S_{J_1,\ldots,J_m}$, then

$$E \prod_{j \leq d} \sum_{k \leq n} \xi_k^j \, \eta_k^j = E \prod_{j \leq d} \sum_{k \leq n} \xi_k^j \, \tilde\eta_k^j.$$

Specializing to the case where $\xi_k^j = \xi_k$, $\eta_k^j = \eta_k$, and $\tilde{\eta}_k^j = \tilde{\eta}_k$ for all j and k, we get the moment identity

$$E\Big(\sum_{k\leq n}\xi_k\eta_k\Big)^r = E\Big(\sum_{k\leq n}\xi_k\tilde{\eta}_k\Big)^r,$$

valid whenever the sums

$$S_{r_1,\ldots,r_d} = \sum_{k_1<\ldots<k_d}\prod_{j\leq d}\eta_{k_j}^{r_j} = \sum_{k_1<\ldots<k_d}\prod_{j\leq d}\tilde{\eta}_{k_j}^{r_j},$$

are a.s. non-random for all $r_1,\ldots,r_d \in \mathbf{Z}_+$ with $r_1 + \cdots + r_d \leq r$.

A simple lemma is needed for the proof of Theorem 5.27.

Lemma 5.29 *(predictable sums)* *Let $\eta = (\eta_k^j)$ be an $\mathcal{F}$-predictable sequence in $\mathbb{R}^d$ of length $n \geq d$ such that the sums in (43) are $\mathcal{F}_0$-measurable. Then the predictability extends to the sequence*

$$T_h = \sum_{h<k_1<\cdots<k_d}\prod_{j\leq d}\eta_{k_j}^j, \quad 0 \leq h < n.$$

Proof: For $d = 1$ we have

$$T_h = \sum_{k>h}\eta_k = S_1 - \sum_{k\leq h}\eta_k, \quad 0 \leq h < n,$$

which is clearly $\mathcal{F}_{h-1}$-measurable, where $\mathcal{F}_{-1} = \mathcal{F}_0$ by convention. This proves the result for $d = 1$. Now assume the statement to be true in dimension $d - 1$. Turning to the d-dimensional case, we write

$$T_h = S_1 - \sum_{k\leq h}\eta_k^1 \sum_{k<k_2<\cdots<k_d}\prod_{j\geq 2}\eta_{k_j}^j, \quad 0 \leq h \leq n - d.$$

Applying the induction hypothesis to the $(d-1)$-fold inner sum, we see that the k-th term on the right is $\mathcal{F}_{k-1}$-measurable. Hence, T_h is $\mathcal{F}_{h-1}$-measurable, and so the sequence (T_h) is $\mathcal{F}$-predictable, which proves the result in d dimensions and completes the induction. $\square$

Proof of Theorem 5.27: Using repeatedly the $\mathcal{F}$-contractability of ξ, the $\mathcal{F}$-predictability of η, and Lemma 5.29, we get

$$E\sum_{k_1<\cdots<k_d}\prod_{j\leq d}(\xi_{k_j}^j\eta_{k_j}^j) = E\,\xi_n^d\sum_{k_1<\cdots<k_d}\prod_{j<d}(\xi_{k_j}^j\eta_{k_j}^j)\,\eta_{k_d}^d$$

$$= E\,\xi_n^d\sum_{k_1<\cdots<k_{d-1}}\prod_{j<d}(\xi_{k_j}^j\eta_{k_j}^j)\sum_{k_d>k_{d-1}}\eta_{k_d}^d$$

$$= E\,\xi_{n-1}^{d-1}\xi_n^d\sum_{k_1<\cdots<k_{d-1}}\prod_{j\leq d-2}(\xi_{k_j}^j\eta_{k_j}^j)\,\eta_{k_{d-1}}^{d-1}\sum_{k_d>k_{d-1}}\eta_{k_d}^d$$

$$= E\,\xi_{n-1}^{d-1}\xi_n^d\sum_{k_1<\cdots<k_{d-2}}\prod_{j\leq d-2}(\xi_{k_j}^j\eta_{k_j}^j)\sum_{k_d>k_{d-1}>k_{d-2}}\eta_{k_{d-1}}^{d-1}\eta_{k_d}^d$$

$$= \cdots = E\prod_{i\leq d}\xi_{n-d+i}^i\sum_{k_1<\cdots<k_d}\prod_{j\leq d}\eta_{k_j}^j$$

$$= (n/\!\!/d)^{-1}E\sum_{h_1<\cdots<h_d}\prod_{i\leq d}\xi_{h_i}^i\sum_{k_1<\cdots<k_d}\prod_{j\leq d}\eta_{k_j}^j. \qquad \square$$

Turning to the case of contractable processes on $[0, 1]$, we introduce the *tetrahedral regions*

$$\Delta_k = \left\{ (s_1, \dots, s_k) \in [0, 1]^k;\ s_1 < \cdots < s_k \right\}, \quad k \in \mathbb{N}.$$

Theorem 5.30 *(tetrahedral moments of processes)* *Let X and V be $\mathbb{R}^d$-valued processes on $[0, 1]$, where X is $\mathcal{F}$-contractable with $\|(X^j)^*\|_{p_j} < \infty$, $j \leq d$, for some $p_1, \dots, p_d > 0$ with $\sum_j p_j^{-1} \leq 1$ and V is bounded and $\mathcal{F}$-predictable, and suppose that the integrals*

$$\eta_k = \int \cdots \int_{\Delta_{d-k+1}} \prod_{j \geq k} V^j \, d\lambda, \quad k = 1, \dots, d, \tag{44}$$

are $\mathcal{F}_0$-measurable. Then

$$E \int \cdots \int_{\Delta_d} \prod_{j \leq d} V^j dX^j = d! \, E \int \cdots \int_{\Delta_d} \prod_{i \leq d} dX^i \int \cdots \int_{\Delta_d} \prod_{j \leq d} V^j d\lambda.$$

In particular, we note that if $\tilde{V}$ is $\mathcal{F}$-predictable with the same $\mathcal{F}_0$-measurable integrals η_k, then

$$E \int \cdots \int_{\Delta_k} \prod_{j \leq d} V^j dX^j = E \int \cdots \int_{\Delta_k} \prod_{j \leq d} \tilde{V}^j dX^j.$$

Choosing $\tilde{V}$ to be conditionally independent of X with the same distribution as V, we obtain a tetrahedral decoupling identity. As in the discrete case, we may also use the last result to derive a corresponding identity for ordinary product moments. Then write $V_s^J = \prod_{j \in J} V_s^j$, for convenience.

Corollary 5.31 *(product moments)* *Let X and V be $\mathbb{R}^d$-valued processes on $[0, 1]$, where X is $\mathcal{F}$-contractable with $\|(X^j)^*\|_{p_j} < \infty$, $j \leq d$, for some $p_1, \dots, p_d > 0$ with $\sum_j p_j^{-1} \leq 1$ and V is bounded and $\mathcal{F}$-predictable, and suppose that the integrals*

$$\eta_{J_1, \dots, J_m} = \int \cdots \int_{\Delta_m} \prod_{k \leq m} V^{J_k} d\lambda, \quad J_1, \dots, J_m \subset \{1, \dots, d\} \text{ disjoint},$$

are $\mathcal{F}_0$-measurable. Then

$$E \prod_{j \leq d} \int_0^1 V^j dX^j = \sum_{J_1, \dots, J_k} k! \, E \int \cdots \int_{\Delta_k} \prod_{i \leq k} dX^{J_i} \int \cdots \int_{\Delta_k} \prod_{j \leq k} V^{J_j} d\lambda,$$

where the summation extends over all ordered partitions $J_1, \dots, J_k$ of the set $\{1, \dots, d\}$.

In particular, we see that if $\tilde{V}$ is $\mathcal{F}$-predictable with the same integrals $\eta_{J_1, \dots, J_m}$, then

$$E \prod_{j \leq d} \int_0^1 V^j dX^j = E \prod_{j \leq d} \int_0^1 \tilde{V}^j dX^j.$$

Specializing to the case where $X_t^j = X_t$ and $V_t^j = V_t$ for all j and t, we get the moment identities

$$E(V \cdot X)_1^r = E(\tilde{V} \cdot X)_1^r,$$

valid whenever the integrals

$$\eta_{r_1,\ldots,r_m} = \int \cdots \int_{\Delta_m} \prod_{k \leq m} V_{s_k}^{r_k} ds_k$$

are $\mathcal{F}_0$-measurable for all $r_1,\ldots,r_m \in \mathbf{Z}_+$ with $r_1 + \cdots + r_m \leq r$.

The following identity involving stochastic and Lebesgue integrals will play a crucial role in our proof of Theorem 5.30. Given a random variable ρ, we say that X is $\mathcal{F}$-*contractable over* ρ, if the contractability holds with respect to the extended filtration $\mathcal{G}_t = \sigma(\mathcal{F}_t, \rho)$.

Lemma 5.32 (*integral reduction*) *Consider some real processes X and V on $[0,1]$ and a random variable ρ, where X is $\mathcal{F}$-contractable over ρ and V is $\mathcal{F}$-predictable. Suppose that $\|X^*\|_p \vee \|V^*\|_q \vee \|\rho\|_r < \infty$ for some $p,q,r > 0$ with $p^{-1} + q^{-1} + r^{-1} \leq 1$. Then for any $t,h > 0$ with $t + h \leq 1$, we have*

$$h\,E\,\rho \int_0^t V_s dX_s = E\,\rho\,(X_{t+h} - X_t) \int_0^t V_s ds.$$

Proof: Since ρV is predictable with respect to the extended filtration $\mathcal{G}_t = \mathcal{F}_t \vee \sigma(\rho)$, and since $\|\rho V^*\|_s \leq \|\rho\|_r \|V^*\|_q < \infty$ by Hölder's inequality with $s^{-1} = q^{-1} + r^{-1}$, we may take $\rho = 1$ and assume that $p^{-1} + q^{-1} \leq 1$.

We begin with a formal computation, to be justified afterwards under the stated norm conditions. Then recall from Theorem 2.13 that X is a special $\mathcal{F}$-semi-martingale, whose compensator $\hat{X}$ admits a martingale density M on $[0,1)$. Using repeatedly the definition of M, the martingale properties of $N = X - \hat{X}$ and M, and Fubini's theorem, we get for any $t, h > 0$ with $t + h \leq 1$

$$
\begin{aligned}
h\,E \int_0^t V_s dX_s &= h\,E \int_0^t V_s d\hat{X}_s = h\,E \int_0^t V_s M_s ds \\
&= h \int_0^t E(V_s M_s)\,ds = h \int_0^t E(V_s M_t)\,ds \\
&= h\,E\,M_t \int_0^t V_s ds = \int_t^{t+h} E\left(M_u \int_0^t V_s ds\right) du \\
&= E\,(\hat{X}_{t+h} - \hat{X}_t) \int_0^t V_s ds \\
&= E\,(X_{t+h} - X_t) \int_0^t V_s ds.
\end{aligned}
$$

Here the first two expressions exist by Proposition 5.15 and its proof. To justify the first equality, we recall from the same proof that

$$\|(V \cdot N)^*\|_r \lesssim \|(V^2 \cdot [N])_1^{1/2}\|_r \lesssim \|V^*\|_q \|X^*\|_p < \infty,$$

where $r^{-1} = p^{-1} + q^{-1} \leq 1$. This shows that $V \cdot N$ is a true martingale (FMP 26.13), and the relation follows.

The second equality holds by the definition of M and the elementary chain rule in FMP 1.23. To justify the use of Fubini's theorem in the third step, we see from Jensen's inequality and the proof of Proposition 5.15 that $E\lambda|VM| \leq \|V^*\|_q \|\lambda|M|\|_p < \infty$. To justify the fourth equality, we note that $E|V_s M_t| \leq \|V_s\|_q \|M_t\|_p < \infty$ by Hölder's inequality. We may then use the martingale property of M to write

$$E(V_s M_t) = E(V_s E[M_t|\mathcal{F}_s]) = E(V_s M_s).$$

The next three steps use Fubini's theorem, the martingale property of M, and Fubini's theorem again, in each case with a similar justification as before. The last step uses the martingale property of N and is justified by the fact that $E|N_{t+h}|\lambda|V| \leq \|N^*\|_p \|V^*\|_q < \infty$.   □

We also need the following elementary rule of stochastic calculus.

Lemma 5.33 (integration by parts) *Let X, V, and U be processes on $[0, 1]$, where X is an $\mathcal{F}$-semi-martingale, V is $\mathcal{F}$-predictable and X-integrable, and U is progressively $\mathcal{F}$-measurable with $\lambda|U| < \infty$ a.s. and such that λU is $\mathcal{F}_0$-measurable. Then*

$$\int_0^1 U_t dt \int_0^{t-} V_s dX_s = \int_0^1 V_t dX_t \int_t^1 U_s ds.$$

Proof: Write $A = U \cdot \lambda$ and $Y = V \cdot X$, and note that $[A, Y] = 0$ since A continuous and of bounded variation. Now integrate by parts (FMP 26.6) to obtain

$$\int_0^1 Y_{t-} dA_t = A_1 Y_1 - \int_0^1 A_t dY_t = \int_0^1 (A_1 - A_t)\, dY_t,$$

which is equivalent to the asserted formula.   □

We proceed with a continuous-time version of Lemma 5.29.

Lemma 5.34 (predictable integrals) *Let $V^1, \ldots, V^d$ be bounded, $\mathcal{F}$-predictable processes on $[0, 1]$ such that the integrals in (44) are $\mathcal{F}_0$-measurable. Then the predictability carries over to the process*

$$U_t = \int_t^1 V_{s_1}^1 ds_1 \int_{s_1}^1 V_{s_2}^2 ds_2 \int_{s_2}^1 \cdots \int_{s_{d-1}}^1 V_{s_d}^d ds_d, \quad t \in [0, 1].$$

Proof: For $d = 1$ we have

$$U_t = \eta_1 - \int_0^t V_s^1 ds, \quad t \in [0, 1],$$

which is continuous and adapted, hence predictable. Now assume the assertion to be true in dimension $d - 1$. Turning to the d-dimensional case, we may write

$$U_t = \eta_1 - \int_0^t V_{s_1}^1 ds_1 \int_{s_1}^1 V_{s_2}^2 ds_2 \int_{s_2}^1 \cdots \int_{s_{d-1}}^1 V_{s_d}^d ds_d, \quad t \in [0,1].$$

Applying the induction hypothesis to the $(d - 1)$-fold integral in the variables $s_2, \ldots, s_d$, we see that the integrand of the remaining outer integral is a predictable function of s_1. The process U is then continuous and adapted, hence predictable, which proves the statement in dimension d and completes the induction. $\qquad \square$

Finally we need a continuity property of tetrahedral integrals.

Lemma 5.35 *(approximation)* *Let X, V, and $V_1, V_2, \ldots$ be $\mathbb{R}^d$-valued processes on $[0,1]$, where X is $\mathcal{F}$-contractable and $V, V_1, V_2, \ldots$ are uniformly bounded and $\mathcal{F}$-predictable with $V_n \to V$. Suppose that $\|(X^j)^*\|_{p_j} < \infty$, $j \le d$, for some $p_1, \ldots, p_d > 0$ with $\sum_j p_j^{-1} \le 1$. Then*

$$\int \cdots \int_{\Delta_d} \prod_j V_n^j dX^j \to \int \cdots \int_{\Delta_d} \prod_j V^j dX^j \quad in \ L^1. \tag{45}$$

Proof: Changing one component at a time, we see that the difference in (45) is bounded in L^1 by

$$\sum_{k \le d} E \left| \int \cdots \int_{\Delta_d} \prod_{i<k} (V^i dX^i) (V_n^k - V^k) dX^k \prod_{j>k} V_n^j dX^j \right|$$

$$= \sum_{k \le d} E \left| \int \cdots \int_{\Delta_{d-k}} U_-^k (V_n^k - V^k) dX^k \prod_{j>k} V_n^j dX^j \right|, \tag{46}$$

where

$$U_t^k = \int \cdots \int_{t\Delta_{k-1}} \prod_{i<k} V^i dX^i, \quad t \in [0,1], \ k \le d.$$

Using Proposition 5.15 (i) repeatedly, we see that $\|(U^k)^*\|_{q_k} < \infty$ for all k, where $q_k^{-1} = \sum_{i<k} p_i^{-1}$. By part (ii) of the same proposition it follows that

$$\|(U_-^k (V_n^k - V^k) \cdot X^k)^*\|_{r_k} \to 0, \quad k \le d,$$

where $r_k^{-1} = \sum_{j \le k} p_j^{-1}$. We may finally use Proposition 5.15 (i), repeatedly as before, to see that the k-th term of (46) tends to 0 as $n \to \infty$. $\qquad \square$

We may now prove Theorem 5.30 under a simplifying assumption.

Lemma 5.36 *(case of bounded support)* *The assertion of Theorem 5.30 is true when $V^1, \ldots, V^d$ are supported by some interval $[0, b]$ with $b < 1$.*

Proof: Fix any $u \in [b, 1)$, and define

$$t_k = 1 - (1-u)\frac{d-k}{d}, \qquad \rho_k = \frac{X^k_{t_k} - X^k_{t_{k-1}}}{t_k - t_{k-1}}, \qquad k \le d.$$

For $k = 1, \ldots, d$, we introduce the processes

$$T^k_t = \int \cdots \int_{t\Delta_k} \prod_{j \le k} V^j dX^j,$$

$$U^k_t = \int_t^1 V^k_{s_k} ds_k \int_{s_k}^1 V^{k+1}_{s_{k+1}} ds_{k+1} \int_{s_{k+1}}^1 \cdots \int_{s_{d-1}}^1 V^d_{s_d} ds_d,$$

and put $T^0_t \equiv U^{d+1}_t \equiv 1$. Note that the T^k are semi-martingales and the U^k are predictable by Lemma 5.34. Using Lemma 5.33 and the definitions of T^k and U^k, we get for suitable k

$$\int_0^1 V^k_s T^{k-1}_{s-} U^{k+1}_s ds = \int_0^1 V^k_s U^{k+1}_s ds \int_0^{s-} V^{k-1}_r T^{k-2}_{r-} dX^{k-1}_r$$

$$= \int_0^1 V^{k-1}_r T^{k-2}_{r-} dX^{k-1}_r \int_r^1 V^k_s U^{k+1}_s ds$$

$$= \int_0^1 V^{k-1}_r T^{k-2}_{r-} U^k_r dX^{k-1}_r.$$

Writing $\pi_k = \rho_k \cdots \rho_d$ for $k = 1, \ldots, d+1$ and using Lemma 5.32, we obtain

$$E \pi_{k+1} \int_0^1 V^k_s T^{k-1}_{s-} U^{k+1}_s dX^k_s = E \pi_k \int_0^1 V^k_s T^{k-1}_{s-} U^{k+1}_s ds$$

$$= E \pi_k \int_0^1 V^{k-1}_s T^{k-2}_{s-} U^k_s dX^{k-1}_s,$$

where the required integrability follows by repeated use of Proposition 5.15. Iterating the latter relations yields

$$ET^d_1 = E \pi_{d+1} \int_0^1 V^d_s T^{d-1}_{s-} U^{d+1}_s dX^d_s$$

$$= E \pi_1 \int_0^1 V^1_s T^0_{s-} U^2_s ds = E \pi_1 U^1_0,$$

and since $U^1_0 = \eta_1$ by Fubini's theorem, we obtain

$$E \int \cdots \int_{\Delta_d} \prod_j V^j dX^j = E(\rho_1 \cdots \rho_d)\eta_1$$

$$= E \eta_1 E[\rho_1 \cdots \rho_d | \mathcal{F}_0]. \tag{47}$$

In particular, we may take $V^1 = \cdots = V^d = 1_A$ on $[0, u]$ for any $A \in \mathcal{F}_0$ to obtain

$$E 1_A \int \cdots \int_{u\Delta_d} dX^1 \cdots dX^d = (u^d/d!)\, E[\rho_1 \cdots \rho_d; A],$$

which implies

$$d!\, E^{\mathcal{F}_0} \int \cdots \int_{u\Delta_d} dX^1 \cdots dX^d = u^d\, E[\rho_1 \cdots \rho_d | \mathcal{F}_0].$$

Inserting this into (47) yields

$$
\begin{aligned}
u^d\, E \int \cdots \int_{\Delta_d} \prod_k V^k dX^k &= d!\, E\, \eta_1 E^{\mathcal{F}_0} \int \cdots \int_{u\Delta_d} \prod_k dX^k \\
&= d!\, E\, \eta_1 \int \cdots \int_{u\Delta_d} \prod_k dX^k,
\end{aligned}
$$

and the desired relation follows as we let $u \to 1$ and use Lemma 5.35. $\qquad\square$

Our final step is to show how the proof of Theorem 5.30 can be reduced to the special case of Lemma 5.36.

Lemma 5.37 *(truncation)* *In the setting of Theorem 5.30, suppose that $|V^k| \le c < \infty$ for all $k \le d$. Then for any $\varepsilon \in (0, \frac{1}{2}]$ there exist some predictable processes $\tilde{V}^1, \ldots, \tilde{V}^d$ on $[0,1]$ with a.s. the same values of the integrals (44), and such that for any $k \le d$*

$$
\begin{aligned}
\tilde{V}^k &= V^k &\text{on} \quad [0, 1-2\varepsilon], \\
|\tilde{V}^k| &\le 2c &\text{on} \quad (1-2\varepsilon, 1-\varepsilon], \\
\tilde{V}^k &= 0 &\text{on} \quad (1-\varepsilon, 1].
\end{aligned}
$$

Proof: On $[0,1]$ we introduce the random signed measures

$$\xi_k B = \int_B V_s^k ds, \quad B \in \mathcal{B}[0,1], \ 1 \le k \le d,$$

which are adapted by the predictability of the V^k. Equation (44) and the constraint $|V^k| \le c$ translate into the conditions

$$(\xi_k \otimes \cdots \otimes \xi_d)\Delta_{d-k+1} = \eta_k, \qquad |\xi_k[a,b]| \le (b-a)c, \qquad (48)$$

for any a, b, and k with $0 \le a \le b \le 1$ and $1 \le k \le d$, where it is enough to take $a, b \in \mathbb{Q}$. Both conditions being measurable in the random elements $\xi_1, \ldots, \xi_d$ and $\eta_1, \ldots, \eta_d$, they can be summarized in the form

$$F(\xi_1, \ldots, \xi_d; \eta_1, \ldots, \eta_d) = 0 \quad \text{a.s.},$$

for some measurable function F on the appropriate product space.

Now fix any $\varepsilon \in (0, \frac{1}{2}]$, and let $\xi_1', \ldots, \xi_d'$ denote the restrictions of $\xi_1, \ldots, \xi_d$ to $[0, 1-2\varepsilon]$. By Lemma A1.6 there exist some random signed measures $\hat{\xi}_1, \ldots, \hat{\xi}_d$, measurable with respect to $\xi_1', \ldots, \xi_d'$ and $\eta_1, \ldots, \eta_d$, such that $\hat{\xi}_k = \xi_k$ on $[0, 1-2\varepsilon]$ for every k and

$$F(\hat{\xi}_1, \ldots, \hat{\xi}_d; \eta_1, \ldots, \eta_d) = 0 \quad \text{a.s.}$$

In other words, the $\hat{\xi}_k$ are $\mathcal{F}_{1-2\varepsilon}$-measurable and satisfy (48). In particular, they are a.s. absolutely continuous (FMP 2.21) and possess densities $\hat{V}_1, \ldots, \hat{V}_d$ with values in the interval $[-c, c]$ (FMP 2.15). By FMP 7.26 we may choose the $\hat{V}_k$ to be product-measurable with respect to the σ-fields $\mathcal{B}[0, 1]$ and $\mathcal{F}_{1-2\varepsilon}$, and since $\hat{\xi}_k = \xi_k$ on $[0, 1 - 2\varepsilon]$, we may assume that $\hat{V}_k = V_k$ on the same interval. From (48) we note that the $\hat{V}_k$ satisfy (44) with the same values η_k.

Now introduce the function

$$f(t) = t - \tfrac{1}{2}(t - 1 + 2\varepsilon) \vee 0, \quad t \in [0, 1], \tag{49}$$

and note that $(f^{\otimes k})^{-1}\Delta_k = \Delta_k$ for all $k \in \mathbb{N}$ since f is strictly increasing. Hence, the first relation in (48) remains true for the random signed measures $\tilde{\xi}_k = \hat{\xi}_k \circ f^{-1}$, $k = 1, \ldots, d$. Inverting (49), we see that, a.s. on $(1 - 2\varepsilon, 1 - \varepsilon]$, the $\tilde{\xi}_k$ have densities

$$\tilde{V}_t^k = 2\hat{V}^k(2t - 1 + 2\varepsilon) \in [-2c, 2c], \quad \varepsilon \le 1 - t < 2\varepsilon, \ k \le d.$$

Since $\tilde{\xi}_k = \xi_k$ on $[0, 1 - 2\varepsilon]$ and $\tilde{\xi}_k = 0$ on $(1 - \varepsilon, 1]$, we may further choose $\tilde{V}_t^k$ to be equal to V_t^k and 0, respectively, on those intervals. Then (44) remains true for the densities $\tilde{V}^k$, and the required predictability is clear from the corresponding property for the V^k, together with the fact that $\tilde{V}^k$ is $\mathcal{B}[0, 1] \otimes \mathcal{F}_{1-2\varepsilon}$-measurable on $(1 - 2\varepsilon, 1]$. $\qquad\square$

We may now complete the proof of the main result.

Proof of Theorem 5.30: For any $\varepsilon \in (0, \tfrac{1}{2}]$, we may choose the corresponding processes $V_\varepsilon^1, \ldots, V_\varepsilon^d$ as in Lemma 5.36 and conclude from Lemma 5.37 that the stated identity holds with each V^k replaced by V_ε^k. Since the latter processes are uniformly bounded and satisfy $V_\varepsilon^k \to V^k$ on $[0, 1)$ as $\varepsilon \to 0$, the required formula follows by Lemma 5.35. $\qquad\square$

5.7 Predictable Sampling Revisited

The decoupling identities implicit in Theorems 5.9 and 5.12 can be used to give simple proofs of the predictable sampling Theorem 4.2 and its multivariate version, Proposition 4.4. Here we go directly to the latter and more general result.

Theorem 5.38 (*predictable sampling*) *Let* $\xi_j = (\xi_{jk})$, $j \le d$, *be finite or infinite, separately $\mathcal{F}$-exchangeable sequences indexed by I, and for every j consider some a.s. distinct, $\mathcal{F}$-predictable times τ_{jk} in I. Then* $(\xi_{jk}) \overset{d}{=} (\tilde{\xi}_{jk})$, *where* $\tilde{\xi}_{jk} = \xi_{j,\tau_{jk}}$.

Proof: By a monotone class argument, it is enough to consider the events $\{\tilde{\xi}_{jk} \in B\}$ for finitely many measurable sets B, and by a suitable mapping we may then reduce to the case where the state space is $[0, 1]$. Next we may extend the filtration $\mathcal{F}$, if necessary, to reduce to the case where each component ξ_j is $\mathcal{F}$-extreme. By conditioning on $\mathcal{F}_0$, we may finally assume that all components of ξ are extreme.

Given the predictable times τ_{jk}, $j \leq d$, $k \leq n$, we introduce the associated *allocation sequences*

$$\alpha_{jk} = \inf\{l \leq n; \tau_{jl} = k\}, \quad j \leq d, \ k \in I,$$

and note that $\alpha_{jk} \in \mathcal{F}_{k-1}$ for all j and k. Fix any constants $c_{jk} \in \mathbb{R}$, $j \leq d$, $k \leq n$, put $c_{j,\infty} = 0$, and define $\gamma_{jk} = c_{j,\alpha_{jk}}$. Then for fixed $j \leq d$, the finite variables α_{jk} form a permutation of $1, \ldots, n$, and so

$$\sum_k \gamma_{jk}^m = \sum_k c_{j,\alpha_{jk}}^m = \sum_k c_{jk}^m, \quad j \leq d, \ m \in \mathbb{N}.$$

Using Theorem 5.9 or 5.12 and the separate exchangeability of ξ, we get for every $m \in \mathbb{N}$

$$
\begin{aligned}
E\Big(\sum_j \sum_k c_{jk}\, \tilde{\xi}_{jk}\Big)^m &= E\Big(\sum_j \sum_k \gamma_{jk}\, \xi_{jk}\Big)^m \\
&= m! \sum_{m_1,\ldots,m_d} E \prod_j \Big(\sum_k \gamma_{jk}\, \xi_{jk}\Big)^{m_j}/m_j! \\
&= m! \sum_{m_1,\ldots,m_d} E \prod_j \Big(\sum_k c_{jk}\, \xi_{jk}\Big)^{m_j}/m_j! \\
&= E\Big(\sum_j \sum_k c_{jk}\, \xi_{jk}\Big)^m.
\end{aligned}
$$

Here the double sums in the extreme members are bounded by constants, and so their distributions are determined by the moments. Hence,

$$\sum_j \sum_k c_{jk}\, \xi_{jk} \overset{d}{=} \sum_j \sum_k c_{jk}\, \tilde{\xi}_{jk}.$$

Since the coefficients c_{jk} are arbitrary, we may use the Cramér–Wold theorem to conclude that $(\xi_{jk}) \overset{d}{=} (\tilde{\xi}_{jk})$. $\qquad\square$

Similarly, we can use the moment identities in Theorems 5.18 and 5.21 to give an alternative proof of the predictable mapping Theorem 4.7.

Theorem 5.39 *(predictable mapping) Let X be an $\mathbb{R}^d$-valued, $\mathcal{F}$-exchangeable process on $I = [0, 1]$ or $\mathbb{R}_+$, and consider an $\mathcal{F}$-predictable process V on I such that $V \circ \lambda^{-1} = \lambda$ a.s. Then $X \circ V^{-1} \overset{d}{=} X$.*

Here we need a technical lemma of independent interest.

Lemma 5.40 *(exponential moments) Let X be an extreme, exchangeable process on $[0, 1]$ or a Lévy process on $\mathbb{R}_+$ with bounded jumps. Then $Ee^{uX_t} < \infty$ for all $u \in \mathbb{R}$ and $t \in [0, 1]$ or $\mathbb{R}_+$, respectively.*

Proof: For processes X on $[0, 1]$, we have by Theorems 2.18 and 3.15 a representation

$$X_t = \alpha t + \sigma B_t + \sum_j \beta_j (1\{\tau_j \leq t\} - t), \quad t \in [0, 1],$$

where B is a Brownian bridge and $\tau_1, \tau_2, \ldots$ are independent and i.i.d. $U(0, 1)$. By Hölder's inequality we may consider each of the three components separately, and by scaling we may take $u = 1$. The first term is trivial, and for the second one we get

$$Ee^{\sigma B_t} = \exp\left(\tfrac{1}{2}\sigma^2 t(1 - t)\right) < \infty.$$

To deal with the jump component, we first assume that $\beta_j = 0$ for all but finitely many j. Then for $\alpha = \sigma = 0$ we get

$$\begin{aligned}
Ee^{X_t} &= E \exp \sum_j \beta_j (1\{\tau_j \leq t\} - t) \\
&= \prod_j E \exp(\beta_j (1\{\tau_j \leq t\} - t)) \\
&= \prod_j e^{-\beta_j t}\left(1 - t(1 - e^{\beta_j})\right) \\
&= \exp \sum_j \left\{\log\left(1 - t(1 - e^{\beta_j})\right) - \beta_j t\right\}.
\end{aligned}$$

Noting that, for fixed $t \in [0, 1]$,

$$\log(1 - t(1 - e^x)) - xt = O(x^2), \quad x \to 0,$$

and recalling that in general $\sum_j \beta_j^2 < \infty$, we see that the same computation applies to the general case, which yields a finite value for Ee^{X_t}.

For Lévy processes X we have instead

$$X_t = \alpha t + \sigma B_t + \int_0^t \int x(\eta - E\eta)(ds\, dx), \quad t \geq 0,$$

where B is a Brownian motion and η is an independent Poisson process on $\mathbb{R}_+ \times \mathbb{R}$ with intensity $\lambda \otimes \nu$ (FMP 15.4). Again it is enough to consider the jump component. Assuming first that ν is bounded, and letting $\alpha = \sigma = 0$, we get as in FMP 15.8

$$Ee^{uX_t} = \exp\left(t \int (e^{ux} - 1 - ux)\nu(dx)\right), \quad t \geq 0.$$

Noting that $e^x - 1 - x = O(x^2)$ and recalling that in general $\int(x^2 \wedge 1)\nu(dx) < \infty$, we see that the formula extends to arbitrary Lévy measures ν with bounded support, which yields the asserted finiteness of Ee^{uX_t}. $\quad\square$

Proof of Theorem 5.39: First assume that X is extreme with bounded jumps. By the Cramér–Wold theorem (FMP 5.5) it suffices to show that, for any constants $t_k \in I$ and $c_{jk} \in \mathbb{R}$,

$$\sum_{j,k} c_{jk} (X^j \circ V^{-1})_{t_k} \overset{d}{=} \sum_{j,k} c_{jk} X_{t_k}^j.$$

Since X has finite exponential moments by Lemma 5.40, it is then enough to verify the moment relations

$$E\left(\sum_{j,k}c_{jk}(X^j \circ V^{-1})_{t_k}\right)^n = E\left(\sum_{j,k}c_{jk}X^j_{t_k}\right)^n, \quad n \in \mathbb{N}.$$

Expanding each side according to the multinomial theorem, we may reduce the assertion to

$$E\prod_j\left(\sum_k c_{jk}(X^j \circ V^{-1})_{t_k}\right)^{n_j} = E\prod_j\left(\sum_k c_{jk}X^j_{t_k}\right)^{n_j},$$

where $n_1, \ldots, n_d \in \mathbb{Z}_+$ are arbitrary. This will follow from Theorem 5.18 or 5.21, respectively, if we can only show that

$$\int \prod_j\left(\sum_k c_{jk}1\{V_s \le t_k\}\right)^{n_j} ds = \int \prod_j\left(\sum_k c_{jk}1\{s \le t_k\}\right)^{n_j} ds,$$

again for arbitrary $n_1, \ldots, n_d \in \mathbb{Z}_+$. Now the last relation follows, by the substitution rule for Lebesgue–Stieltjes integrals (FMP 1.22), from the fact that $\lambda \circ V^{-1} = \lambda$ a.s.

Turning to the case of general Lévy processes, we may write $X = X^n + J^n$, where J^n denotes the compound Poisson process formed by all jumps of X greater than n. Writing Y^n for the corresponding simple Poisson process, we see from the previous case that

$$(Y^n \circ V^{-1})_t \overset{d}{=} Y^n_t \overset{P}{\to} 0, \quad t \ge 0,$$

which implies $(J^n \circ V^{-1})_t \overset{P}{\to} 0$ for all t. We may then let $n \to \infty$ in the relation $X^n \circ V^{-1} \overset{d}{=} X^n$ to obtain $X \circ V^{-1} \overset{d}{=} X$.

Now let X be a general $\mathcal{F}$-exchangeable process on $I = [0,1]$ or $\mathbb{R}_+$, directed by $\gamma = (\alpha, \rho, \beta)$ or (α, ρ, ν), respectively. Then X remains exchangeable with respect to the extended filtration $\mathcal{G}_t = \sigma(\mathcal{F}_t, \gamma)$, $t \in I$, and V clearly remains $\mathcal{G}$-predictable. It is also clear from FMP 26.4 that the process $X \circ V^{-1}$ is invariant under a change to filtration $\mathcal{G}$, and also under conditioning with respect to $\mathcal{G}_0$. The assertion now follows from the result in the extreme case, applied to the conditional distributions given $\mathcal{G}_0$. $\quad\square$

Next we show how the optional skipping theorem, Proposition 4.1, can be derived as an easy consequence of Corollary 5.28.

Proposition 5.41 *(optional skipping)* Let $\xi = (\xi_j)$ *be a finite or infinite, $\mathcal{F}$-contractable sequence indexed by I, and let $\tau_1 < \cdots < \tau_m$ be $\mathcal{F}$-predictable times in I. Then* $(\xi_{\tau_1}, \ldots, \xi_{\tau_m}) \overset{d}{=} (\xi_1, \ldots, \xi_m)$.

Proof: By a suitable transformation and truncation, we may assume that $\xi = (\xi_j)$ is a finite sequence of bounded random variables. Letting $\tau_1 < \cdots < \tau_m$ be predictable times in the index set $\{1, \ldots, n\}$ of ξ, we need to show that, for any $c_1, \ldots, c_m \in \mathbb{R}$ and $p \in \mathbb{N}$,

$$E\Big(\sum\nolimits_{k \leq m} c_k \, \xi_{\tau_k}\Big)^p = E\Big(\sum\nolimits_{k \leq m} c_k \, \xi_k\Big)^p. \tag{50}$$

Then introduce the predictable sequence

$$\alpha_j = \inf\{k; \, \tau_k = j\}, \quad j \leq n,$$

where $\alpha_j = \infty$ if $\tau_k \neq j$ for all k. Putting $c_\infty = 0$, we note that

$$\sum\nolimits_{k \leq m} c_k \, \xi_{\tau_k} = \sum\nolimits_{j \leq n} c_{\alpha_j} \xi_j.$$

Hence, (50) will follow from Corollary 5.28 if we can only show that

$$\sum_{h_1 < \cdots < h_r} \prod_{i \leq r} c_{\alpha_{h_i}}^{p_i} = \sum_{k_1 < \cdots < k_r} \prod_{j \leq r} c_{k_j}^{p_j},$$

for any $r \leq m$ and $p_1, \ldots, p_r$ in $\mathbb{Z}_+$. Here the product on the left vanishes unless $(h_1, \ldots, h_r) = (\tau_{k_1}, \ldots, \tau_{k_r})$ for some $k_1 < \ldots < k_r$, in which case $\alpha_{h_i} = k_i$ for all i. Since every sequence $k_1 < \cdots < k_r$ occurs exactly once, the two sums are indeed equal. $\qquad\square$

We may finally consider a version of Theorem 4.18, the predictable contraction property of contractable processes. Here our new proof is based on the moment identities of Corollary 5.31. Unfortunately, the present argument seems to require a strong integrability condition.

Theorem 5.42 *(predictable contraction) Let X be an $\mathcal{F}$-contractable process on $[0, 1]$ with finite exponential moments, and consider an $\mathcal{F}$-predictable set $A \subset [0, 1]$ with $\lambda A \geq h$ a.s. Then $C_A X \stackrel{d}{=} X$ on $[0, h)$.*

Proof: Arguing as before, we need to show that

$$E\Big(\int f \, d(C_A X)\Big)^n = E\Big(\int f \, dX\Big)^n, \quad n \in \mathbb{N},$$

for any step function

$$f(t) = \sum\nolimits_{j \leq m} c_j \, 1\{t \leq t_j\}, \quad t \in [0, 1],$$

where $c_1, \ldots, c_m \in \mathbb{R}$ and $t_1, \ldots, t_m \in [0, h]$ are arbitrary. Then introduce the optional times

$$\sigma_j = \inf\{s \in [0, 1]; \, \lambda(A \cap [0, s]) > t_j\}, \quad j = 1, \ldots, m,$$

and consider the predictable process

$$V_s = 1_A(s) \sum_{j \le m} c_j \, 1\{s \le \sigma_j\}, \quad s \in [0,1],$$

Since $\int f \, d(C_A X) = \int V \, dX$ by the definition of $C_A X$, it suffices by Corollary 5.31 to show that

$$\int \cdots \int_{\Delta_k} \prod_{j \le k} V_{s_j}^{p_j} \, ds_j = \int \cdots \int_{\Delta_k} \prod_{j \le k} f_{s_j}^{p_j} \, ds_j, \qquad (51)$$

for arbitrary $k \in \mathbb{N}$ and $p_1, \ldots, p_k \in \mathbb{Z}_+$.

Then introduce the right-continuous inverse $\tau \colon [0,h] \to A$ of the mapping $A_s = \lambda(A \cap [0,s])$, and note as in FMP 2.14 that $\lambda \circ \tau^{-1} = 1_A \cdot \lambda$ on $[0, \tau_h]$. Since τ is strictly increasing with $\tau_{t_j} = \sigma_j$, it is also clear that $V \circ \tau = f$. For the same reason, the k-fold tensor product $\tau^{\otimes k}$ maps Δ_k into itself, and similarly for the complement Δ_k^c. Relation (51) now follows by the substitution rule for Lebesgue–Stieltjes integrals (FMP 1.22). $\square$

Chapter 6

Homogeneity and Reflections

In this chapter, we study the invariance in distribution of a process X on $\mathbb{R}_+$ or $[0,1]$ under optional shifts or reflections, where the underlying times τ are typically restricted by the condition $X_\tau = 0$. Just as in the classical case of regenerative processes, this leads to an Itô-type representation in terms of an exchangeable point process of excursions, the associated time scale being given by a local time random measure supported by the zero set $\Xi = \{t; X_t = 0\}$ of X.

After some preliminary discussion in Section 6.1, we characterize in Section 6.2 the locally homogeneous processes on $\mathbb{R}_+$, and in Section 6.3 we consider the corresponding results for reflection invariant processes on $[0,1]$. Further propositions involving the local time random measure and its local intensity are given in Section 6.4. In Section 6.5 we show how the distribution of a sequence of independent and exponentially or uniformly distributed random variables is essentially preserved under mapping by the local time process. Next, Section 6.6 provides a comparison of the local hitting probabilities of Ξ with the corresponding probabilities for a suitably defined stationary version $\tilde{\Xi}$.

Homogeneity at several states is considered in Section 6.7, which leads to characterizations of proper and mixed Markov processes. The results suggests the developments in the final Section 6.8, where we show, under suitable regularity conditions, that the homogeneity and independence parts of the strong Markov property are essentially equivalent.

6.1 Symmetries and Dichotomies

Consider an rcll process X on $\mathbb{R}_+$ taking values in a Polish space S. We assume that X is *recurrent* at some specified state $0 \in S$, in the sense that the random set $\Xi = \{t \geq 0; X_t = 0\}$ is a.s. unbounded. Letting X be adapted to a right-continuous and complete filtration $\mathcal{F}$, it is said to be *(locally) $\mathcal{F}$-homogeneous* at 0 if $\theta_\tau X \overset{d}{=} X$ for every $\mathcal{F}$-optional time $\tau < \infty$ with $X_\tau = 0$. This holds in particular for the hitting times $\tau_r = \inf\{t \geq r; X_t = 0\}$, $r \geq 0$, which are optional since X is progressively measurable (FMP 7.7). Note that the $\mathcal{F}$-homogeneity at 0 implies $X_0 = 0$ a.s.

Our first result is a conditional form of the homogeneity.

Lemma 6.1 *(conditional homogeneity) Let X be $\mathcal{F}$-homogeneous at 0. Then for any optional times $\sigma \leq \tau < \infty$ with $X_\sigma = X_\tau = 0$, we have*

$$P[\theta_\sigma X \in \cdot | \mathcal{F}_\sigma] = P[\theta_\tau X \in \cdot | \mathcal{F}_\sigma] \quad a.s.$$

Proof: Writing σ_A for the restriction of σ to a set $A \in \mathcal{F}_\sigma$ and using the $\mathcal{F}$-homogeneity at 0, we get

$$\theta_{\sigma_A \wedge \tau} X \overset{d}{=} \theta_\tau X, \quad A \in \mathcal{F}_\sigma.$$

Noting that $\sigma_A \wedge \tau$ equals σ on A and τ on A^c, we get for any $A \in \mathcal{F}_\sigma$

$$
\begin{aligned}
E[P[\theta_\sigma X \in \cdot | \mathcal{F}_\sigma]; A] &= P[\theta_\sigma X \in \cdot; A] = P[\theta_\tau X \in \cdot; A] \\
&= E[P[\theta_\tau X \in \cdot | \mathcal{F}_\sigma]; A],
\end{aligned}
$$

and the assertion follows. □

For a more detailed description of X, we note that the complement $\overline{\Xi}^c$ is open and hence a countable union of open intervals. Since the points of $\overline{\Xi} \setminus \Xi$ are isolated from the right in Ξ, by the right-continuity of X, it follows that Ξ^c is a countable union of *excursion intervals* of the form (u, v) or $[u, v)$. The associated *excursions* of X are given by $Y_t = X_{(t+u) \wedge v}$, $t \geq 0$. Note that each Y belongs to the set D_0 of excursion paths $x \in D(\mathbb{R}_+, S)$, such that $x_t \neq 0$ for $0 < t < l(x)$ and $x_t = 0$ for all $t \geq l(x)$, where $l(x) > 0$ is referred to as the *length* of excursion x. Write $\Xi \cdot \lambda = 1_\Xi \cdot \lambda$.

The following zero-infinity laws describe the possible forms of Ξ.

Proposition 6.2 *(dichotomies on $\mathbb{R}_+$) Let X be rcll and $\mathcal{F}$-homogeneous at 0, and put $\Xi = \{X = 0\}$. Then a.s.*

(i) *for any $\varepsilon > 0$, X has 0 or ∞ many excursions of length $> \varepsilon$;*

(ii) *Ξ is either nowhere dense or a locally finite union of intervals;*

(iii) *either $\lambda \Xi = 0$, or $\Xi \cdot \lambda$ is unbounded with support $\overline{\Xi}$;*

(iv) *either Ξ is locally finite, or $\overline{\Xi}$ is perfect.*

Proof: (i) Let A_ε be the set of paths with at least one excursion of length $> \varepsilon$. Since X is $\mathcal{F}$-homogeneous at 0, we get for any $r > 0$

$$P\{X \in A_\varepsilon, \theta_{\tau_r} X \notin A_\varepsilon\} = P\{X \in A_\varepsilon\} - P\{\theta_{\tau_r} X \in A_\varepsilon\} = 0,$$

which implies

$$\{X \in A_\varepsilon\} = \bigcap_{r > 0} \{\theta_{\tau_r} X \in A_\varepsilon\} \quad a.s.$$

Thus, if X has at least one excursion longer than ε, then such excursions exist beyond every time $r > 0$, which means that their number is infinite.

(ii) Put $\sigma = \inf \Xi^c$, and note that $\{\sigma = 0\} \in \mathcal{F}_0$. By (i) and Lemma 6.1 we can then deal separately with the cases $\sigma = 0$, $\sigma \in (0, \infty)$, and $\sigma = \infty$.

When $\sigma = \infty$, we have $\Xi = \mathbb{R}_+$, and Ξ is trivially of the second type. Next suppose that $\sigma = 0$ a.s. By the $\mathcal{F}$-homogeneity at 0, we have

$$\inf(\Xi^c \cap [\tau_r, \infty)) = \tau_r \quad \text{a.s.}, \quad r \geq 0.$$

Applying this to all $r \in \mathbb{Q}_+$, we conclude that Ξ is a.s. nowhere dense.

Now let $0 < \sigma < \infty$ a.s. Applying the $\mathcal{F}$-homogeneity at 0 to the times τ_r, we see in particular that every excursion interval is followed by an interval in Ξ of positive length. In particular, X has a first excursion starting at σ and ending at some time $\gamma_1 > \sigma$. Proceeding recursively, we may identify an infinite sequence of excursions, separated by Ξ-intervals with left endpoints $\gamma_1 < \gamma_2 < \cdots$. Putting $\gamma_\infty = \lim_n \gamma_n$ and using the $\mathcal{F}$-homogeneity at 0, we see that $\gamma_\infty - \gamma_1 \overset{d}{=} \gamma_\infty$. Hence,

$$E[\exp(-\gamma_\infty + \gamma_1) - \exp(-\gamma_\infty)] = 0,$$

and since $\gamma_1 \geq 0$, we conclude that $\gamma_\infty - \gamma_1 = \gamma_\infty$ a.s. Since even $\gamma_1 > 0$ a.s., we obtain $\gamma_\infty = \infty$ a.s., which shows that Ξ is a locally finite union of intervals.

(iii) Here we put $\sigma = \inf \text{supp}(\Xi \cdot \lambda)$ and note as before that $\{\sigma = 0\} \in \mathcal{F}_0$. By Lemma 6.1 we may then consider the cases $\sigma = 0$ and $\sigma > 0$ separately. In the latter case, the $\mathcal{F}$-homogeneity at 0 yields $\sigma \wedge \tau_r \notin \text{supp}(\Xi \cdot \lambda)$ a.s. for every $r > 0$. Hence, $\sigma > \tau_r$ a.s. for all $r \geq 0$. But then $\sigma = \infty$ a.s., which means that $\lambda\Xi = 0$ a.s.

Next suppose that $\sigma = 0$. The $\mathcal{F}$-homogeneity at 0 yields $\tau_r \in \text{supp}(\Xi \cdot \lambda)$ a.s. for every $r \geq 0$, and since the times τ_r with $r \in \mathbb{Q}_+$ are dense in Ξ, we obtain

$$\overline{\Xi} \subset \text{supp}(\Xi \cdot \lambda) \subset \overline{\Xi} \quad \text{a.s.},$$

which shows that $\Xi \cdot \lambda$ has support $\overline{\Xi}$. Now define $\alpha_r = \lambda(\Xi \cap [r, \infty))$ for any $r \geq 0$. Applying the $\mathcal{F}$-homogeneity at 0 to the times τ_r, we see that $\alpha_r \overset{d}{=} \alpha_0$ for all r. Hence,

$$E(e^{-\alpha_r} - e^{-\alpha_0}) = 0, \quad r \geq 0,$$

and since $\alpha_r \leq \alpha_0$, we obtain $\alpha_r \equiv \alpha_0$ a.s. If $\alpha_0 < \infty$, then by dominated convergence

$$\lambda\Xi = \alpha_0 = \alpha_r \to 0, \quad r \to \infty,$$

which contradicts the condition $\sigma = 0$. Thus, $\lambda\Xi = \alpha_0 = \infty$ a.s.

(iv) Defining $\sigma = \inf(\Xi \cap (0, \infty))$, we have again $\{\sigma = 0\} \in \mathcal{F}_0$, which allows us to treat the cases $\sigma = 0$ and $\sigma > 0$ separately. In the former case, we may use the $\mathcal{F}$-homogeneity at 0 to see that, a.s. for fixed $r \geq 0$, the time τ_r is a limit point from the right. Applying this to all $r \in \mathbb{Q}_+$, we conclude that Ξ has a.s. no isolated points, which means that $\overline{\Xi}$ is a.s. perfect.

Next suppose that $\sigma > 0$ a.s. By the $\mathcal{F}$-homogeneity at 0, the time τ_r is then a.s. isolated from the right for every $r \geq 0$. In particular, this applies to all right endpoints of excursion intervals. Thus, σ is isolated in Ξ, and proceeding recursively, we see that Ξ begins with a sequence of isolated points $0 < \sigma_1 < \sigma_2 < \cdots$. Writing $\sigma_\infty = \lim_n \sigma_n$ and using the $\mathcal{F}$-homogeneity at time σ_1, we obtain $\sigma_\infty - \sigma_1 \overset{d}{=} \sigma_\infty$. Since $\sigma_1 \geq 0$, we may strengthen this to an a.s. equality, and since even $\sigma_1 > 0$ a.s., we conclude that $\sigma_\infty = \infty$ a.s. Thus, $\sigma_n \to \infty$ a.s., which means that Ξ is a.s. locally finite. □

Let us now consider the analogous invariance condition for rcll processes X on $[0, 1]$ with $X_0 = X_1 = 0$, taking values in a Polish space S with a specified element 0. For every random time τ in $[0, 1]$ with $X_\tau = 0$, we define the reflected process $R_\tau X$ as follows.

First we form a new process $\tilde{X}$ on $[0, 1]$ by reversing X on $[\tau, 1]$, so that

$$\tilde{X}_t = \begin{cases} X_t, & t \in [0, \tau], \\ X_{1-t+\tau}, & t \in (\tau, 1]. \end{cases}$$

Next we replace every excursion Y of $\tilde{X}$ within the interval $[\tau, 1]$ by its reversal $\tilde{Y}$. Thus, if $\tilde{X}$ has the excursion intervals $[u_j, v_j)$ or (u_j, v_j) in $[\tau, 1]$, $j \in \mathbb{N}$, we define

$$(R_\tau X)_t = \tilde{X}_{u_j + v_j - t}, \quad t \in [u_j, v_j), \ j \in \mathbb{N}.$$

We also put $(R_\tau X)_{v_j} = 0$, unless v_j is also the left endpoint of another excursion interval. For all other values of t, we put $(R_\tau X)_t = \tilde{X}_t$. The reversal $R_\tau B$ of a set $B \subset [0, 1]$ is defined by

$$R_\tau B = (B \cap [0, \tau]) \cup ((1 - B + \tau) \cap (\tau, 1]).$$

Informally, we get $R_\tau X$ from X by reversing the order of the excursions after time τ. Note, however, that the construction may cause two or more excursions of X to coalesce into a single excursion for $R_\tau X$. The next result shows that $R_\tau X$ is again rcll and that, whenever $R_\tau X \overset{d}{=} X$, the processes X and $R_\tau X$ have a.s. the same set of excursions.

Lemma 6.3 *(reflection) Let X be an rcll process on $[0, 1]$ with $X_0 = X_1 = 0$. Then $R_\tau X$ is again rcll, for any random time τ in $[0, 1]$ with $X_\tau = 0$. If $R_\tau X \overset{d}{=} X$ holds in addition, then X and $R_\tau X$ have a.s. the same set of excursions from 0, and the zero sets $\Xi = \{X = 0\}$ and $\Xi_\tau = \{R_\tau X = 0\}$ are a.s. related by $R_\tau \overline{\Xi} = \overline{\Xi}_\tau$.*

Proof: If $(u, v]$ or (u, v) is an excursion interval of $\tilde{X}$, then $R_\tau X$ is clearly rcll on $[u, v]$, by the corresponding property of X. We also note that $R_\tau X$ is trivially rcll on $[0, \tau]$. Now define $C = \{\tilde{X} = 0\} \cap [\tau, 1]$, and let t be a limit point from the right in $\overline{C}$. Then $\tilde{X}_{t+} = 0$, which implies $(R_\tau X)_{t+} = 0$. Since also $(R_\tau X)_t = 0$ in this case, we conclude that $R_\tau X$ is right-continuous at t. A similar argument shows that $(R_\tau X)_{t-} = 0$ for every limit point from the left in $\overline{C}$. Since a point of $\overline{C} \setminus \{1\}$ that is not a limit from the right must be the left endpoint of an excursion interval, and correspondingly for limit points from the left and right endpoints, the argument shows that $R_\tau X$ is rcll.

Now suppose that $R_\tau X \overset{d}{=} X$. Let β and $\tilde{\beta}$ denote the point processes of excursion lengths of X and $R_\tau X$. Since the excursions are universally measurable functions of the underlying process, we conclude that $\beta \overset{d}{=} \tilde{\beta}$. Now two or more (even infinitely many) excursions in X of lengths b_j, $j \le m$, may coalesce into a single excursion in $R_\tau X$ of length $\sum_j b_j$. By the subadditivity of $1 - e^{-x}$, we get in that case

$$\exp\left(-\sum_{j>n} b_j\right) - \exp\left(-\sum_{j \ge n} b_j\right) \le 1 - e^{-b_n}, \quad n \le m,$$

with strict inequality for $n < m$. Summing over $n \le m$ yields

$$1 - \exp\left(-\sum_j b_j\right) < \sum_j \left(1 - e^{-b_j}\right),$$

which shows that

$$\int (1 - e^{-x})\, \tilde{\beta}(dx) \le \int (1 - e^{-x})\, \beta(dx), \tag{1}$$

with equality iff $\beta = \tilde{\beta}$. Since the relation $\beta \overset{d}{=} \tilde{\beta}$ yields a.s. equality in (1), we conclude that $\beta = \tilde{\beta}$ a.s. Thus, coalescence is a.s. excluded when $R_\tau X \overset{d}{=} X$, and so, with probability 1, the two processes have then the same excursions. In particular, the excursion intervals of $\tilde{X}$ and $R_\tau X$ agree a.s. apart from endpoints, and the a.s. relation $R_\tau \overline{\Xi} = \overline{\Xi}_\tau$ follows.   □

Let us now assume that X is adapted to a right-continuous and complete filtration $\mathcal{F}$ on $[0, 1]$. We say that X has the *strong reflection property* at 0 if $R_\tau X \overset{d}{=} X$ for every $\mathcal{F}$-optional time τ in $[0, 1]$ with $X_\tau = 0$. It is often useful to consider the following conditional version.

Lemma 6.4 *(conditional reflection invariance) Let X be an rcll, $\mathcal{F}$-adapted process on $[0, 1]$ with $X_0 = X_1 = 0$, satisfying the strong reflection property at 0. Then for any optional time τ with $X_\tau = 0$, we have*

$$P[R_\tau X \in \cdot | \mathcal{F}_\tau] = P[X \in \cdot | \mathcal{F}_\tau] \quad a.s.$$

Proof: For any $A \in \mathcal{F}_\tau$, the restriction $\tau_A = \tau 1_A + 1_{A^c}$ is again an optional time in $\Xi = \{X = 0\}$. The reflection property at τ_A yields

$$\begin{aligned} E[P[R_\tau X \in \cdot | \mathcal{F}_\tau];\, A] &= P[R_\tau X \in \cdot;\, A] = P[X \in \cdot;\, A] \\ &= E[P[X \in \cdot | \mathcal{F}_\tau];\, A], \end{aligned}$$

and the desired relation follows since A was arbitrary.   □

Our next result is a finite-interval version of Proposition 6.2.

Proposition 6.5 *(dichotomies on $[0, 1]$)* *Let X be an rcll, $\mathcal{F}$-adapted process on $[0, 1]$ with $X_0 = X_1 = 0$, satisfying the strong reflection property at 0, and put $\Xi = \{X = 0\}$. Then a.s.*

(i) *Ξ is either nowhere dense or a finite union of intervals;*

(ii) *either $\lambda \Xi = 0$, or $\Xi \cdot \lambda$ has support $\overline{\Xi}$;*

(iii) *either Ξ is finite, or $\overline{\Xi}$ is perfect.*

Proof: (i) As in the proof of Proposition 6.2, we may put $\sigma = \inf \Xi^c$ and note that $\{\sigma = 0\} \in \mathcal{F}_0$. By Lemma 6.4 we can then treat the cases $\sigma = 0$ and $\sigma > 0$ separately. In the former case, $\sup \Xi^c = 1$ by the reflection property at 0 and Lemma 6.3. For any $r \in [0, 1)$, we can use the reflection property at τ_r together with the same lemma to see that τ_r is a limit point from the right of Ξ^c, a.s. on the set $\{\tau_r < 1\}$. Since this holds with a common null set for all rational r, we conclude that Ξ is a.s. nowhere dense.

Next suppose that $\sigma > 0$. Using two reflections, as before, we conclude that every optional time τ with $X_\tau = 0$ is followed by a Ξ-interval of positive length, a.s. on the set $\{\tau < 1\}$. In particular, every X-excursion ending before time 1 is followed by a non-degenerate Ξ-interval. Next consider the infimum τ of all limit points from the right of both Ξ and Ξ^c, where $\tau = 1$ when no such time exists, and note that τ is optional with $X_\tau = 0$. Hence, τ is followed by a non-degenerate Ξ-interval, a.s. on $\{\tau < 1\}$. Since this contradicts the definition of τ when $\tau < 1$, we obtain $\tau = 1$ a.s., which means that a.s. no points have the stated property. By the symmetry on $[0, 1]$, Ξ and Ξ^c have a.s. no common limit points from the left either. Hence, by combination, X has only finitely many excursions, separated by intervals of positive lengths. Finally, a reflection on $[0, 1]$ shows that even the last excursion is followed by a non-degenerate Ξ-interval.

(ii) First we note that $\Xi \cdot \lambda = \overline{\Xi} \cdot \lambda$ since $\overline{\Xi} \setminus \Xi$ is countable. Writing $\sigma = \inf \operatorname{supp}(\Xi \cdot \lambda)$, we have again $\{\sigma = 0\} \in \mathcal{F}_0$, which justifies a separate treatment of the cases $\sigma = 0$ and $\sigma > 0$. In the former case, we may apply two reflections, as before, to see that every time τ_r belongs to the support of $\Xi \cdot \lambda$, a.s. on the set $\{\tau_r < 1\}$. Since this holds simultaneously for all rational points $r \in [0, 1)$, outside a fixed null set, we conclude that $\overline{\Xi} \setminus \{1\} \subset \operatorname{supp}(\Xi \cdot \lambda)$ a.s. Finally, since 0 belongs to the support of $\Xi \cdot \lambda$, the same thing is true for the point 1, owing to the symmetry of $\overline{\Xi}$ on the whole interval.

Now assume instead that $\sigma > 0$ a.s. Reflecting twice, as before, we see that σ is followed by an interval outside of $\operatorname{supp}(\Xi \cdot \lambda)$, a.s. on $\{\sigma < 1\}$. Since this contradicts the definition of σ when $\sigma < 1$, we obtain $\sigma = 1$ a.s., which means that $\lambda \Xi = 0$ a.s. in this case.

(iii) Putting $\sigma = \inf(\Xi \cap (0, 1])$, we have again $\{\sigma = 0\} \in \mathcal{F}_0$, which allows us to treat the cases $\sigma = 0$ and $\sigma > 0$ separately. In the former case, we may apply the usual two reflections to see that every time τ_r is a limit point from the right of Ξ, a.s. on $\{\tau_r < 1\}$. Since this holds simultaneously for all rational times $r \in [0, 1)$, we conclude that Ξ has a.s. no isolated points in

$[0, 1)$. We may finally use a reflection on $[0, 1]$ to we see that, with probability one, the point 1 is not isolated in Ξ either. Hence, in this case $\overline{\overline{\Xi}}$ is a.s. perfect.

If instead $\sigma > 0$, the usual reflection argument shows that every optional time τ with $X_\tau = 0$ is a.s. isolated from the right in Ξ, a.s. on $\{\tau < 1\}$. In particular, we may define τ to be the infimum of all limit points from the right in Ξ, taking $\tau = 1$ when no such times exist. Since τ itself has the stated property on the set $\{\tau < 1\}$, we get a contradiction unless $\tau = 1$ a.s. Thus, all points of Ξ are a.s. isolated from the right. By reflection on $[0, 1]$, we see that they are also a.s. isolated from the left. Hence, $\overline{\overline{\Xi}}$ has a.s. only isolated points and therefore Ξ must be finite.   □

The previous symmetries may also be considered for random subsets Ξ of $I = \mathbb{R}_+$ or $[0, 1]$, without reference to any underlying process X. Given a right-continuous and complete filtration $\mathcal{F}$ on I, it is natural to require Ξ to be *progressively measurable* as a random subset of $I \times \Omega$, in the sense that $\Xi \cap [0, t] \in \mathcal{B} \otimes \mathcal{F}_t$ for every $t \in I$. This ensures that the times $\tau_s = \inf(\Xi \cap [s, \infty))$ will be optional for all $s \in I$ (FMP 7.7). Motivated by the case where $\Xi = \{X = 0\}$ for some rcll process X, it is often convenient to assume in addition that Ξ be *closed on the left*, in the sense that for any decreasing sequence $t_1, t_2, \ldots$ in Ξ, the limit also lies in Ξ.

It is now clear how to extend the previous symmetry conditions to random sets. Thus, letting Ξ be progressively measurable and closed on the left in $\mathbb{R}_+$, we say that Ξ is *$\mathcal{F}$-homogeneous* if $\theta_\tau \Xi \overset{d}{=} \Xi$ for every optional time τ in Ξ, where $\theta_\tau \Xi$ denotes the shifted set $(\Xi - \tau) \cap \mathbb{R}_+$, and the distribution of Ξ is specified by the hitting probabilities $P\{\Xi \cap G \neq \emptyset\}$ for any open subsets $G \subset \mathbb{R}_+$. For the *reflection property* of a random set Ξ in $[0, 1]$ with $0, 1 \in \Xi$, we require instead that $R_\tau \Xi \overset{d}{=} \Xi$ for all optional times τ in Ξ, where R_τ denotes the reversal of Ξ on the interval $[\tau, 1]$. Most results for locally homogeneous or strongly reflective processes X remain valid for homogeneous or reflective random sets Ξ, provided we replace the excursions of X by the contiguous intervals of Ξ.

6.2   Local Homogeneity

A stronger condition than local homogeneity is for X to be regenerative, where we add the independence requirement $\mathcal{F}_\tau \perp\!\!\!\perp \theta_\tau X$. Thus, X is said to be *$\mathcal{F}$-regenerative* at 0 if $P[\theta_\tau X \in \cdot | \mathcal{F}_\tau] = P_0$ a.s. for every optional time $\tau < \infty$ with $X_\tau = 0$, where $P_0 = \mathcal{L}(X)$. We shall prove that X is $\mathcal{F}$-homogeneous at 0 iff it is *conditionally $\mathcal{F}$-regenerative* at 0, in the sense that there exists a σ-field $\mathcal{I}$ satisfying

$$P[\theta_\tau X \in \cdot | \mathcal{F}_\tau, \mathcal{I}] = P[X \in \cdot | \mathcal{I}], \tag{2}$$

for every optional time τ with $X_\tau = 0$. This means that X is conditionally $\mathcal{F}$-homogeneous at 0 and satisfies $\theta_\tau X \perp\!\!\!\perp_\mathcal{I} \mathcal{F}_\tau$, whenever $\tau < \infty$ is optional with $X_\tau = 0$.

Theorem 6.6 *(homogeneity and regeneration) Let X be an $\mathcal{F}$-adapted, rcll process on $\mathbb{R}_+$, taking values in a Polish space S and recurrent at a state $0 \in S$ with $X_0 = 0$ a.s. Then these conditions are equivalent:*

(i) *X is $\mathcal{F}$-homogeneous at 0,*
(ii) *X is conditionally $\mathcal{F}$-regenerative at 0.*

Proof: Assume that X is $\mathcal{F}$-homogeneous at 0. Writing $\Xi = \{X = 0\}$ and using Proposition 6.2 (i), (ii), and (iv), we see that Ω is a.s. a disjoint union of the measurable sets

$$
\begin{aligned}
\Omega_1 &= \{\Xi = \mathbb{R}_+\}, \\
\Omega_2 &= \{\Xi \text{ is infinite but locally finite}\}, \\
\Omega_3 &= \{\Xi \text{ is an infinite but locally finite union of intervals}\}, \\
\Omega_4 &= \{\Xi \text{ is nowhere dense with a perfect closure}\}.
\end{aligned}
$$

By part (i) of the same result, we may further decompose the last of these events according to the length of the longest excursion. More precisely, for every $n \in \mathbb{N}$, we may consider the set C_n of paths of the last type with infinitely many excursions of length $\geq n^{-1}$ but none of length $\geq (n-1)^{-1}$, where 0^{-1} is interpreted as ∞. This gives an a.s. decomposition of the path space $D(\mathbb{R}_+, S)$ into countably many measurable subsets $B_1, B_2, \ldots$. Each of these sets is a.s. invariant, in the sense that a.s. $X \in B_n$ iff $\theta_\tau X \in B_n$ for every $n \in \mathbb{N}$ and for any optional time $\tau < \infty$ with $X_\tau = 0$. Thus, the $\mathcal{F}$-homogeneity at 0 remains valid under each of the conditional distributions $P[\,\cdot\,|A_n]$ with $PA_n > 0$, where $A_n = \{X \in B_n\}$.

Now suppose that the regenerative property holds conditionally when X is restricted to any one of the sets B_n. Applying this result to the conditional distributions $P_n = P[\,\cdot\,|A_n]$ with $PA_n > 0$, we see that there exist some σ-fields $\mathcal{I}_1, \mathcal{I}_2, \ldots$ satisfying

$$
P_n[\theta_\tau X \in \cdot | \mathcal{F}_\tau, \mathcal{I}_n] = P_n[X \in \cdot | \mathcal{I}_n], \quad n \in \mathbb{N}.
$$

Introducing the σ-field

$$
\mathcal{I} = \sigma\{\mathcal{I}_n \cap A_n; n \in \mathbb{N}\},
$$

we get in the general case

$$
\begin{aligned}
P[\theta_\tau X \in \cdot | \mathcal{F}_\tau, \mathcal{I}] &= \sum_n P[\theta_\tau X \in \cdot\,; A_n | \mathcal{F}_\tau, \mathcal{I}] \\
&= \sum_n 1_{A_n} P_n[\theta_\tau X \in \cdot | \mathcal{F}_\tau, \mathcal{I}_n] \\
&= \sum_n 1_{A_n} P_n[X \in \cdot | \mathcal{I}_n] \\
&= \sum_n P[X \in \cdot\,; A_n | \mathcal{I}] = P[X \in \cdot | \mathcal{I}],
\end{aligned}
$$

where the second and fourth relations are justified by the fact that, for any σ-field $\mathcal{F}$ and events $C \in \mathcal{A}$, $F \in \mathcal{F}$, and $I \in \mathcal{I}_n$,

$$
\begin{aligned}
E[1_{A_n} P_n[C|\mathcal{F}, \mathcal{I}_n]; F \cap I] &= E[P_n[C \cap F \cap I|\mathcal{F}, \mathcal{I}_n]; A_n] \\
&= P(A_n) P_n(C \cap F \cap I) \\
&= P(A_n \cap C \cap F \cap I).
\end{aligned}
$$

This reduces the discussion to the case where $X \in B_n$ a.s. for some fixed n.

First we take Ξ to be infinite but locally finite, say with points $0 = \sigma_0 < \sigma_1 < \cdots$. Let $Y_1, Y_2, \ldots$ be the associated excursions of X, and introduce the discrete filtration $\mathcal{G}_n = \mathcal{F}_{\sigma_n}$, $n \in \mathbb{Z}_+$, so that the sequence $Y = (Y_n)$ is adapted to $\mathcal{G} = (\mathcal{G}_n)$. For any $\mathcal{G}$-optional time $\kappa < \infty$ we note that $\tau = \sigma_\kappa$ is $\mathcal{F}$-optional, since for any $t \geq 0$

$$
\{\tau \leq t\} = \{\sigma_\kappa \leq t\} = \bigcup_k \{\sigma_k \leq t,\, \kappa = k\} \in \mathcal{F}_t
$$

by the definition of $\mathcal{F}_{\sigma_k}$. Hence, the $\mathcal{F}$-homogeneity of X yields $\theta_\kappa Y \overset{d}{=} Y$, and so by Proposition 2.1 the sequence Y is conditionally $\mathcal{G}$-i.i.d., given a suitable σ-field $\mathcal{I}$. Relation (2) now follows by the local property of conditional expectations (FMP 6.2).

Next we consider the case where $\lambda \Xi = \infty$ a.s. For every $n \in \mathbb{N}$, we introduce the random times

$$
\sigma_k^n = \inf\{t \geq 0;\ \lambda(\Xi \cap [0, t]) > k 2^{-n}\}, \quad k \in \mathbb{Z}_+, \tag{3}
$$

which are optional since $\mathcal{F}$ is right-continuous and the process $\lambda(\Xi \cap [0, t])$ is non-decreasing and adapted, due to Fubini's theorem (FMP 7.2, 7.6). We also note that $\sigma_k^n \in \Xi$ a.s. for all n and k, by the continuity of $\lambda(\Xi \cap [0, t])$ and the right continuity of X.

Now introduce the processes

$$
Y_k^n(t) = X((t + \sigma_{k-1}^n) \wedge \sigma_k^n), \quad t \geq 0,\ n, k \in \mathbb{N}, \tag{4}
$$

and note that, for every $n \in \mathbb{N}$, the sequence $Y^n = (Y_k^n)$ is adapted to the discrete filtration $\mathcal{G}_k^n = \mathcal{F}_{\sigma_k^n}$, $k \in \mathbb{Z}_+$. For any $\mathcal{G}^n$-optional time $\kappa < \infty$, we see as before that the time $\tau = \sigma_\kappa^n$ is $\mathcal{F}$-optional with values in Ξ. Hence, the $\mathcal{F}$-homogeneity yields $\theta_\tau X \overset{d}{=} X$, which translates into the condition $\theta_\kappa Y^n \overset{d}{=} Y^n$. By Proposition 2.1 and Corollary 1.6, the sequence Y^n is then conditionally $\mathcal{G}^n$-i.i.d., given the tail σ-field $\mathcal{T}_n$ of Y^n. Noting that $\theta_{\sigma_k^n} X$ can be measurably recovered from $\theta_k Y^n$ and that the tail σ-field $\mathcal{T}_n = \mathcal{T}$ is independent of n, we get

$$
P[\theta_\tau X \in \cdot|\mathcal{F}_\tau, \mathcal{T}] = P[X \in \cdot|\mathcal{T}] \quad \text{a.s.}, \tag{5}
$$

first for the special times $\tau = \sigma_k^n$, and then, by FMP 6.2, for any $\mathcal{F}$-optional time τ taking values in the set $\{\sigma_k^n\}$.

Now fix an arbitrary $\mathcal{F}$-optional time τ in Ξ, and introduce the approximating optional times

$$\tau_n = \inf\{\sigma_k^n \geq \tau;\ k \in \mathbb{Z}_+\}, \quad n \in \mathbb{N}. \tag{6}$$

The τ_n are non-increasing since $\sigma_k^n = \sigma_{2k}^{n+1}$ for all n and k. Recalling that $\theta_\tau X \overset{d}{=} X$ by $\mathcal{F}$-homogeneity and that 0 belongs to the support of $\Xi \cdot \lambda$ by Proposition 6.2, we see that in fact $\tau_n \downarrow \tau$ a.s. Letting $\tau = \tau_n$ in (5) and taking conditional expectations with respect to $\mathcal{F}_\tau \vee \mathcal{T}$, we obtain

$$P[\theta_{\tau_n} X \in \cdot | \mathcal{F}_\tau, \mathcal{T}] = P[X \in \cdot | \mathcal{T}] \text{ a.s.}, \quad n \in \mathbb{N},$$

and (5) follows for the general choice of τ, by the right continuity of X and dominated convergence. This completes the proof of (ii) when $\lambda \Xi = \infty$.

The last case to consider is when Ξ is nowhere dense with perfect closure and Ξ^c contains infinitely many intervals of length $> \varepsilon$. By scaling we may take $\varepsilon = 1$. For every $n \in \mathbb{N}$ we introduce the right endpoints $\sigma_1^n < \sigma_2^n < \cdots$ of the successive excursion intervals of length $> 2^{-n}$ and put $\sigma_0^n = 0$. Note that the times σ_k^n are optional and take values in Ξ. The intermediate processes Y_k^n are again defined as in (4), and the associated discrete filtrations are given by $\mathcal{G}_k^n = \mathcal{F}_{\sigma_k^n}$. Using Proposition 2.1 and Corollary 1.6, we see as before that

$$P[\theta_\tau X \in \cdot | \mathcal{F}_\tau, \mathcal{I}_n] = P[X \in \cdot | \mathcal{I}_n] \text{ a.s.}, \quad n \in \mathbb{N}, \tag{7}$$

for any optional time τ taking values in the set $\{\sigma_0^n, \sigma_1^n, \ldots\}$, where $\mathcal{I}_n$ denotes the invariant σ-field induced by the sequence $Y^n = (Y_k^n)$.

To deduce the general formula, we note that the σ-fields $\mathcal{I}_n$ are non-increasing, since the sets $\{\sigma_0^n, \sigma_1^n, \ldots\}$ are increasing in n. For the same reason, (7) remains true for the times $\tau = \sigma_j^m$ with $m \leq n$. The relation is equivalent to the a.s. conditions

$$\theta_\tau X \perp\!\!\!\perp_{\mathcal{I}_n} \mathcal{F}_\tau, \qquad P[\theta_\tau X \in \cdot | \mathcal{I}_n] = P[X \in \cdot | \mathcal{I}_n],$$

and in each formula we may let $n \to \infty$ to obtain the same result for the σ-field $\mathcal{I} = \bigcap_n \mathcal{I}_n$. Equation (2) now follows by the local property of conditional expectations (FMP 6.2), for any optional time τ that takes values in the countable set $\{\sigma_k^n\}$. For a general optional time τ in Ξ, the shifted set $\theta_\tau \Xi \overset{d}{=} \Xi$ is again nowhere dense with a perfect closure containing 0. We may then approximate, as in (6), by some optional times $\tau_n \downarrow \tau$ of the special type, and (2) follows as before by the right continuity of X. This completes the proof of the implication (i) $\Rightarrow$ (ii). The reverse assertion is obvious. $\quad\square$

When X is $\mathcal{F}$-homogeneous at 0, we may also establish a conditional form of the classical *Itô representation* (FMP 22.11). More precisely, assuming $\overline{\Xi}$ to be perfect, we shall prove the existence of a diffuse *local time* random measure ξ with support $\overline{\Xi}$, such that the excursions of X, plotted against the associated cumulative local times, form a Cox process η on $\mathbb{R}_+ \times D_0$ directed by an invariant random measure $\lambda \otimes \nu$ on $\mathbb{R}_+ \times D_0$. The alternative is for Ξ to have only isolated points, in which case the excursions of X form an exchangeable sequence in D_0.

When $\lambda\Xi = 0$ a.s. and $\mathcal{F}$ is the right-continuous and complete filtration induced by X, there is also an elementary characterization in terms of infinite sequences. Here we say that the excursions of X are *exchangeable* if, for any $\varepsilon > 0$, the excursions longer than ε form an exchangeable sequence in D_0, given that such excursions exist. Armed with these definitions, we may now state the main characterizations of locally homogeneous processes on $\mathbb{R}_+$.

We proceed to show that any locally homogeneous process X in a Polish space S has a conditional Itô representation in terms of an exchangeable point process on $\mathbb{R}_+ \times D_0$. This implies that the excursions of X are exchangeable, and the two conditions are equivalent when $\lambda\Xi = 0$.

Theorem 6.7 *(representation on $\mathbb{R}_+$) Let X be an $\mathcal{F}$-adapted, rcll process on $\mathbb{R}_+$, taking values in a Polish space S, and recurrent and $\mathcal{F}$-homogeneous at a state $0 \in S$. Then*

(i) *X has a conditional Itô representation at 0,*

(ii) *X has exchangeable excursions from 0.*

Conversely, (i) *implies that X is conditionally regenerative for the induced filtration at every optional time in $\Xi = \{X = 0\}$ that avoids the left endpoints in Ξ^c. Furthermore,* (i) *and* (ii) *are equivalent when $\lambda\Xi = 0$ a.s.*

A couple of lemmas will be needed for the proof. We begin with an elementary comparison of σ-fields.

Lemma 6.8 *(induced filtration) Let X be an rcll process inducing a right-continuous filtration $\mathcal{F}$, and consider some $\mathcal{F}$-optional times σ and τ. Then*

$$\mathcal{F}_\sigma \cap \{\sigma < \tau\} \subset \sigma(X^\tau, \tau) \subset \mathcal{F}_\tau.$$

Proof: The second inclusion is well-known (FMP 7.5). To prove the first one, we introduce the optional times

$$\sigma_n = \inf\{k/n \geq \sigma; \ k \in \mathbb{Z}_+\}, \quad n \in \mathbb{N}.$$

Then for any $A \in \mathcal{F}_\sigma \subset \mathcal{F}_{\sigma_n}$ and $s_k \downarrow s \geq 0$ we have

$$
\begin{aligned}
A \cap \{\sigma_n = s < \tau\} &\in \mathcal{F}_s \cap \{\tau > s\} = \mathcal{F}_s \cap \bigcup_k \{\tau > s_k\} \\
&\subset \bigcup_k \sigma\{X^{s_k}; \tau > s_k\} \\
&\subset \bigcup_k \sigma\{X^\tau; \tau > s_k\} \subset \sigma(X^\tau, \tau),
\end{aligned}
$$

and so

$$A \cap \{\sigma_n < \tau\} = \bigcap_s (A \cap \{\sigma_n = s < \tau\}) \in \sigma(X^\tau, \tau),$$

where the intersection extends over rational $s \geq 0$. Since $\sigma_n \downarrow \sigma$ a.s., we get

$$
\begin{aligned}
A \cap \{\sigma < \tau\} &= A \cap \bigcup_n \{\sigma_n < \tau\} \\
&= \bigcup_n (A \cap \{\sigma_n < \tau\}) \in \sigma(X^\tau, \tau). \qquad \square
\end{aligned}
$$

We proceed to show how certain conditional properties can be strengthened to relations between suitable conditional distributions. The point is that the original property is typically stated in terms of uncountably many conditions, each of which holds outside an associated null set.

Lemma 6.9 *(conditional distributions) Let ξ and η be random elements in some Borel spaces S and T, fix a σ-field $\mathcal{I}$ in Ω, and consider a regular conditional distribution $\mu = P[(\xi, \eta) \in \cdot | \mathcal{I}]$ with marginals μ_1 and μ_2. Then*

  (i) $P[\xi \in \cdot | \mathcal{I}] = P[\eta \in \cdot | \mathcal{I}]$ *iff* $\mu_1 = \mu_2$ *a.s.,*

  (ii) $\xi \perp\!\!\!\perp_\mathcal{I} \eta$ *iff* $\mu = \mu_1 \otimes \mu_2$ *a.s.*

Proof: (i) Here it is clearly understood that $S = T$. Since S is Borel, the associated σ-field contains a countable, measure-determining class $\mathcal{C}$. The first relation yields a.s. $\mu_1 B = \mu_2 B$ for all $B \in \mathcal{C}$, which implies $\mu_1 = \mu_2$ a.s. The reverse implication is obvious.

(ii) Here we may choose some countable, measure-determining classes $\mathcal{C}_1$ and $\mathcal{C}_2$ in S and T. The conditional independence yields

$$
\mu(A \times B) = \mu_1(A)\,\mu_2(B) \quad \text{a.s.,} \quad A \in \mathcal{C}_1,\ B \in \mathcal{C}_2,
$$

which extends with a fixed null set to any measurable subsets of S and T. The a.s. relation $\mu = \mu_1 \otimes \mu_2$ now follows by the uniqueness of product measures (FMP 1.27). The reverse assertion is again obvious. $\square$

We are now ready to prove the stated representation theorem.

Proof of Theorem 6.7: Suppose that X is $\mathcal{F}$-conditionally regenerative at 0, given some σ-field $\mathcal{I}$. In the discrete case, the representation in terms of exchangeable sequences follows immediately by iteration. Turning to the perfect case, we may assume that $\mathcal{F}$ is the filtration induced by X. For any optional time τ in Ξ, we have

$$
P[\theta_\tau X \in \cdot | X^\tau, \tau, \mathcal{I}] = P[X \in \cdot | \mathcal{I}] \quad \text{a.s.,} \tag{8}
$$

which is equivalent to the a.s. relations

$$
\theta_\tau X \perp\!\!\!\perp_\mathcal{I} (X^\tau, \tau), \qquad P[\theta_\tau X \in \cdot | \mathcal{I}] = P[X \in \cdot | \mathcal{I}]. \tag{9}
$$

Since τ is X-measurable, Lemma 6.9 shows that the last relations are fulfilled outside a fixed null set, in the sense that the conditions

$$
\theta_\tau X \perp\!\!\!\perp (X^\tau, \tau), \qquad \theta_\tau X \overset{d}{=} X, \tag{10}
$$

hold a.s. under $P[X \in \cdot | \mathcal{I}]$ for every fixed τ.

In particular, we note that (10) holds simultaneously, outside a fixed null set, for each of the optional times σ_k^n introduced earlier. By the local property of conditional expectations, the result remains valid, simultaneously, for all optional times τ taking values in the set $\{\sigma_k^n\}$. This is because the event $\tau = \sigma_k^n$ is measurably determined by the pair (X^τ, τ).

Now fix an arbitrary optional time τ in Ξ, and note that τ is a.s. a limit point from the right in Ξ. Defining the approximating times $\tau_n \downarrow \tau$ as in (6), though now with strict inequality, we note that (10) holds a.s. under $P[X \in \cdot | \mathcal{I}]$ with τ replaced by τ_n. Since $\mathcal{F}_\tau \subset \sigma(X^{\tau_n}, \tau_n)$ by Lemma 6.8, we get

$$\theta_{\tau_n} X \perp\!\!\!\perp \mathcal{F}_\tau, \qquad \theta_{\tau_n} X \overset{d}{=} X,$$

a.s. under the same conditional law, and (10) follows for τ itself by the right continuity of X. This shows that X is a.s. regenerative under $P[X \in \cdot | \mathcal{I}]$, and the desired representation follows from the classical result for regenerative processes (FMP 22.11) by means of Lemma A1.5.

Now assume instead condition (i), so that X has a conditional Itô representation. To show that is conditionally regenerative or to deduce condition (ii), we may consider separately the cases where Ξ is discrete, where $\lambda\Xi = \infty$, and where Ξ is nowhere dense with perfect closure and X has infinitely many excursions longer than some ε. The discrete case being elementary, we turn to the case where $\lambda\Xi = \infty$. Since the excursion point process η is Cox and directed by a product random measure $\lambda \otimes \nu$, we note that

$$P[\theta_s \eta \in \cdot | \eta^s, \nu] = P[\eta \in \cdot | \nu] \quad \text{a.s.},$$

where η^s denotes the restriction of η to $[0, s] \times D_0$. This translates immediately into the conditions in (10), a.s. under the distribution $P[X \in \cdot | \nu]$ and for any optional time $\tau = \sigma_k^n$ as in (3). That in turn implies (9) with $\mathcal{I} = \sigma(\nu)$, which is equivalent to (8). Arguing as before, we may finally extend the latter relation to (2) for the filtration $\mathcal{F}$ induced by X and for any $\mathcal{F}$-optional time τ that avoids the left endpoints in Ξ^c.

A similar argument applies when Ξ is nowhere dense with perfect closure and X has excursions longer than some $\varepsilon > 0$. Here we define $\sigma_1^n < \sigma_2^n < \cdots$ for each $n > \varepsilon^{-1}$ as the right endpoints of excursion intervals longer than n^{-1}. To obtain (10) in this case with $\tau = \sigma_k^n$ and $\mathcal{I} = \sigma(\nu)$, we need to apply the strong Markov property of η under the conditional distributions $P[\eta \in \cdot | \nu]$. The argument may then be completed as before. This proves the first part of the last statement. The same argument shows that (i) implies (ii).

It remains to show that (ii) implies (i) when $\lambda\Xi = 0$ a.s. By suitable conditioning, we may then assume that X has a.s. infinitely many excursions longer than some $\varepsilon > 0$. For any $h \in (0, \varepsilon)$, de Finetti's Theorem 1.1 shows that the excursions longer than h are conditionally i.i.d., and we may introduce their common conditional distribution ν_h, which is a random probability measure on the associated path space D_h. Then for any $h \leq k$ in $(0, \varepsilon)$, the excursions in D_k are conditionally i.i.d. with distribution $\nu_h[\,\cdot\,|D_k]$. Invoking the uniqueness part of Proposition 1.4, we conclude that the random measures ν_h are related by

$$\nu_k = \nu_h[\,\cdot\,|D_k] \quad \text{a.s.}, \quad 0 < h < k < \varepsilon.$$

Proceeding as in FMP 22.10, we may then construct a σ-finite random measure ν on D_0 such that

$$\nu_h = \nu[\,\cdot\,|D_h] \quad \text{a.s.}, \quad 0 < h < \varepsilon.$$

The normalization of ν is arbitrary and can be chosen such that $\nu = \nu_h$ a.s. on D_h for some fixed $h \in (0, \varepsilon)$.

If $\nu D_0 < \infty$, the excursions of X are conditionally i.i.d. with distribution $\nu/\nu D_0$, which implies the required representation. Assuming next that $\nu D_0 = \infty$, we may introduce a Cox process β on $\mathbb{R}_+ \times D_0$ directed by $\lambda \otimes \nu$. For every $m \in \mathbb{N}$, let $Y_1^m, Y_2^m, \ldots$ denote the X-excursions longer than m^{-1}, where $Y_k^m = 0$ for all k if no such excursions exist. Writing $\tilde{Y}_1^m, \tilde{Y}_2^m, \ldots$ for the corresponding excursion sequences obtained from β, we note that the arrays (Y_k^m) and $(\tilde{Y}_k^m)$ have the same distribution. Hence, the transfer theorem (FMP 6.10) ensures the existence of a Cox process $\eta \overset{d}{=} \beta$ on $\mathbb{R}_+ \times D_0$ generating the original array (Y_k^m). Since $\lambda \Xi = 0$, it is clear that η provides the required Itô representation of X. $\qquad\square$

The previous results remain valid with obvious changes for homogeneous random sets Ξ in $\mathbb{R}_+$. In this case, the Itô representation is expressed in terms of a Cox process η on $\mathbb{R}_+ \times (0, \infty)$, along with a random drift parameter $\alpha \geq 0$. Equivalently, the set $\overline{\Xi}$ is the closed range of a non-decreasing, exchangeable process T on $\mathbb{R}_+$, known from Theorem 1.19 to be a mixture of subordinators. In particular, the lengths of all contiguous intervals longer than an arbitrary $\varepsilon > 0$ form an exchangeable sequence in $(0, \infty)$. It is often convenient to refer to random sets Ξ of the indicated type as *exchangeable sets* in $\mathbb{R}_+$.

6.3 Reflection Invariance

We turn to the corresponding result for reflection invariant processes on $[0, 1]$, taking values in a Polish space S with a specified element 0. Recall that D_0 denotes the set of excursion paths from 0, and write $l(u)$ for the length of excursion u. Given a marked point process η on $[0, 1] \times D_0$, we say that the process X on $[0, 1]$ with $X_0 = X_1 = 0$ is *generated* by η if the excursions of X are given by the marks in η, in such a way that a point at (s, y) gives rise to an excursion with endpoints

$$T_{s\pm} = \alpha s + \int_0^{s\pm} \int l(u)\, \eta(dr\, du),$$

where $\alpha = 1 - \eta l$ is assumed to be nonnegative. To simplify the subsequent discussion, we may henceforth take $\mathcal{F}$ to be the right-continuous and complete filtration induced by X.

Theorem 6.10 (*representation on* $[0,1]$) *Let X be an rcll process on* $[0,1]$ *with induced filtration $\mathcal{F}$, taking values in a Polish space S and such that $X_0 = X_1 = 0 \in S$ a.s. Suppose that X satisfies the strong reflection property at 0. Then*

(i) *X is generated by an exchangeable point process on* $[0,1] \times D_0$,

(ii) *X has exchangeable excursions from 0.*

Conversely, (i) *implies $R_\tau X \overset{d}{=} X$ for any optional time τ in $\Xi = \{X = 0\}$ that a.s. avoids the left endpoints in Ξ^c. Furthermore,* (i) *and* (ii) *are equivalent when $\lambda \Xi = 0$ a.s.*

Proof: By Lemma 6.4 and Proposition 6.5, we can apply an elementary conditioning to reduce the proof to the special cases where Ξ is finite, where $\lambda \Xi > 0$, or where $\overline{\Xi}$ is perfect and nowhere dense with $\lambda \overline{\Xi} = 0$. First suppose that Ξ is finite, say with points

$$0 = \sigma_0 < \sigma_1 < \cdots < \sigma_\kappa = 1.$$

Defining $\sigma_k = 1$ for $k \geq \kappa$, we note that the σ_k are optional, and so the reflection property yields $R_{\sigma_k} X \overset{d}{=} X$ for all n. Since κ is invariant under reflections, we conclude that

$$P[R_{\sigma_k} X \in \cdot | \kappa = n] = P[X \in \cdot | \kappa = n], \quad 0 \leq k < n.$$

Writing $Y_1, \ldots, Y_\kappa$ for the excursions of X, we obtain

$$P[(Y_1, \ldots, Y_k, Y_n, \ldots, Y_{k+1}) \in \cdot | \kappa = n] = P[(Y_1, \ldots, Y_n) \in \cdot | \kappa = n].$$

Since any permutation of $1, \ldots, n$ is generated by reflections of type $1, \ldots, k$, $n, \ldots, k+1$, it follows that $Y_1, \ldots, Y_n$ are conditionally exchangeable, given $\kappa = n$. The same argument applies to the case of finitely many excursions separated by intervals. If instead the number of excursions is infinite, we see in the same way that, for any $\varepsilon > 0$, the excursions longer than ε are conditionally exchangeable, given their total number. This shows that the strong reflection property implies (ii), and it also yields the representation in (i) when Ξ is finite.

To prove the same representation when $\overline{\Xi}$ is perfect and nowhere dense with Lebesgue measure 0, let β be the point process of excursions of X, and introduce a uniform randomization $\tilde{\beta}$ of β. For any $m \in \mathbb{N}$, consider the X-excursions $Y_1^m, Y_2^m, \ldots$ of length $> m^{-1}$, listed in their order of occurrence, and write $\tilde{Y}_1^m, \tilde{Y}_2^m, \ldots$ for the corresponding sequence derived from $\tilde{\beta}$. By exchangeability the arrays (Y_k^m) and $(\tilde{Y}_k^m)$ have the same distribution, and so by the transfer theorem (FMP 6.10) there exists a point process $\eta \overset{d}{=} \tilde{\beta}$

generating the original array (Y_k^m). Since $\lambda \Xi = 0$, this also yields the required representation of X.

It remains to consider the case where $\lambda \Xi > 0$. Then define the optional times σ_k^n as in (3), except that the infimum of the empty set is now taken to be 1. Introduce the intervals $I_{nk} = 2^{-n}[k-1, k)$, $n, k \in \mathbb{Z}_+$, and consider for fixed $s \in [0, 1]$ the random times

$$\tau_n = \sum_k 1_{I_{nk}}(s \lambda \Xi) \sigma_k^n, \quad n \in \mathbb{N}.$$

Using the reflection property of X and noting that the random variable $\lambda \Xi$ is X-measurable and invariant under reflections, we get

$$
\begin{aligned}
P\{R_{\tau_n}X \in \cdot\} &= \sum_k P[R_{\sigma_k^n}X \in \cdot;\; s\lambda\Xi \in I_{nk}] \\
&= \sum_k P[X \in \cdot;\; s\lambda\Xi \in I_{nk}] = P\{X \in \cdot\},
\end{aligned}
$$

which shows that $R_{\tau_n}X \stackrel{d}{=} X$ for all n.

Now introduce the local time process

$$L_t = \lambda(\Xi \cap [0, t])/\lambda\Xi, \quad t \in [0, 1],$$

along with its right-continuous inverse

$$T_r = \inf\{t \in [0, 1];\; L_t > r\}, \quad r \in [0, 1].$$

Put $\tau = T_s$ for the specified number $s \in [0, 1]$, and note that $\tau_n = T_{s_n}$, where

$$s_n = \inf\{(k2^{-n}/\lambda\Xi) > s;\; k \in \mathbb{Z}_+\} \wedge 1, \quad n \in \mathbb{N}.$$

Since $s_n \downarrow s$, we obtain $\tau_n \downarrow \tau$.

To extend the reflection property to the time τ, we note that $R_{\tau_n}X = R_\tau X = X$ on $[0, \tau)$. Furthermore,

$$(R_{\tau_n}X)_t = (R_\tau X)_{t-\tau_n+\tau}, \quad t \in (\tau_n, 1),$$

and as $n \to \infty$ we get $(R_{\tau_n}X)_t \to (R_\tau X)_{t-}$ on $(\tau, 1)$. Since also $(R_{\tau_n}X)_1 = (R_\tau X)_1 = 0$ by our construction of the reflected process, we have $(R_{\tau_n}X)_t \to (R_\tau X)_t$ a.s. for every t in some dense set $B \subset [0, 1]$ containing 1. Taking limits in the relation $R_{\tau_n}X \stackrel{d}{=} X$, we conclude that $R_\tau X \stackrel{d}{=} X$ on the same set B, which extends to the entire interval $[0, 1]$ by the right-continuity on each side.

We now create a simple point process η on $[0, 1] \times D_0$ by plotting the excursions of X against their local times. Since $\Xi \cdot \lambda$ has support $\overline{\Xi}$ by Proposition 6.5 (ii), we see that η is a marked point process on $[0, 1] \times D_0$, in the sense that a.s. $\eta(\{t\} \times D_0) = 0$ or 1 for all $t \in [0, 1]$. The reflected process $R_\tau X$ has a.s. the same excursions as X by Lemma 6.3, and since $\overline{\Xi} \setminus \Xi$ and $\overline{\Xi}_\tau \setminus \Xi_\tau$ are countable, the same lemma yields a.s.

$$R_\tau(\Xi \cdot \lambda) = R_\tau(\overline{\Xi} \cdot \lambda) = \overline{\Xi}_\tau \cdot \lambda = \Xi_\tau \cdot \lambda.$$

Hence, an excursion of X with local time $t \in (s, 1)$ is reflected into one for $R_\tau X$ with local time $1 - t + s$. In other words, the point process $\tilde{\eta}$ associated with $R_\tau X$ equals the reflection R_s of η, and so the relation $R_\tau X \overset{d}{=} X$ implies $R_s \eta \overset{d}{=} \eta$. Since s was arbitrary, we conclude from Theorem 1.15 that η is exchangeable, and the desired representation follows by Lemma 1.24.

To prove the converse assertion, assume condition (i). Let $\mathcal{G}$ denote the right-continuous and complete filtration on $[0, 1]$ induced by η and its projection $\beta = \eta(\cdot \times [0, 1])$, and note that the right-continuous inverse $T = L^{-1}$ is adapted to $\mathcal{G}$. Using the continuity of L and the right-continuity of T, we get

$$\{L_t \leq s\} = \{t \leq T_s\} \in \mathcal{G}_s, \quad s, t \in [0, 1],$$

which shows that the times L_t are $\mathcal{G}$-optional. We may then introduce the time-changed filtration $\tilde{\mathcal{F}} = (\mathcal{G}_{L_t})$. Since X^t can be measurably constructed from β, along with the restriction of η to any interval $[0, L_{t+\varepsilon}]$ with $\varepsilon > 0$, and since the latter is $\tilde{\mathcal{F}}_{t+\varepsilon}$-measurable by FMP 7.5, we obtain $\mathcal{F}_t \subset \tilde{\mathcal{F}}_{t+\varepsilon} \downarrow \tilde{\mathcal{F}}_t$ as $\varepsilon \to 0$, which shows that $\mathcal{F} \subset \tilde{\mathcal{F}}$.

Now consider any $\mathcal{F}$-optional time τ in $[0, 1]$, and define $\sigma = L_\tau$. By the continuity of L we have

$$\{\sigma < s\} = \{L_\tau < s\} = \bigcup_r \{\tau < r, \, L_r < s\}, \quad s \in [0, 1], \tag{11}$$

where the union extends over all rational $r \in [0, 1]$. Noting that $\{\tau < r\} \in \mathcal{F}_r \subset \tilde{\mathcal{F}}_r = \mathcal{G}_{L_r}$ and using the right-continuity of $\mathcal{G}$, we see that the right-hand side of (11) belongs to $\mathcal{G}_s$ (FMP 7.2), which means that σ is $\mathcal{G}$-optional. Hence, Proposition 2.6 yields $R_\sigma \eta \overset{d}{=} \eta$, and it follows easily that $R_{T_\sigma} X \overset{d}{=} X$. Assuming that τ lies in Ξ and avoids the left endpoint of every excursion interval, we have $T_\sigma = T(L_\tau) = \tau$, which yields the required relation $R_\tau X \overset{d}{=} X$. The last assertion was proved implicitly already by the previous discussion. $\square$

The last theorem remains valid with obvious changes for reflective random sets Ξ in $[0, 1]$. Here the representation is in terms of an exchangeable point process η on $[0, 1] \times (0, 1]$, describing the lengths of the contiguous intervals of Ξ, along with a random drift parameter $\alpha \geq 0$. Equivalently, we may represent $\overline{\Xi}$ as the range of a non-decreasing, exchangeable process T on $[0, 1]$ with $T_1 = 1$ a.s., as described by Theorem 1.25. Note in particular that the lengths of all contiguous intervals exceeding an arbitrary $\varepsilon > 0$ now form an exchangeable sequence of random length. It is again convenient to refer to random sets Ξ of the indicated type as *exchangeable sets* in $[0, 1]$.

6.4   Local Time and Intensity

In the last two theorems we have seen that if the process X, defined in a Polish space S with a specified element 0, is locally homogeneous on $\mathbb{R}_+$ or satisfies the strong reflection property on $[0, 1]$, in either case with respect to the state 0, then it can be represented in terms of an exchangeable point process η on $\mathbb{R}_+ \times D_0$ or $[0,1] \times D_0$, where D_0 denotes the set of excursion paths from 0.

For a complete description, we also need to specify a random variable $\alpha \geq 0$, determining the amount of time spent at 0. Then any point (s, u) of η corresponds to an excursion u of X with endpoints $T_{s\pm}$, given as before by

$$T_{s\pm} = \alpha s + \int_0^{s\pm} \int l(u) \, \eta(dr \, du), \qquad (12)$$

where $l(u)$ denotes the length of excursion $u \in D_0$. For processes on $[0,1]$ we have $\alpha \perp\!\!\!\perp_\beta \eta$, where β denotes the projection of η onto D_0; for processes on $\mathbb{R}_+$ the corresponding condition is $\alpha \perp\!\!\!\perp_\nu \eta$, where $\lambda \otimes \nu$ is the directing random measure of the Cox process η. In the former case we may choose $\alpha = \lambda\Xi = 1 - \beta l$, where $\Xi = \{X_t = 0\}$, which makes the *directing* pair (α, β) uniquely and measurably determined by X.

The infinite-interval case is more subtle, since the corresponding directing pair (α, ν) is only determined up to a normalization. For most purposes, it is convenient to normalize (α, ν) by the condition

$$\alpha + \int (1 - e^{l(u)}) \, \nu(du) = 1. \qquad (13)$$

In the trivial case where $X \equiv 0$, we may choose $\alpha = 1$ and $\nu = 0$, in agreement with (13). In general, the process $T_s = T_{s+}$ given by (12) is conditionally a subordinator directed by the pair (α, ν).

The following result gives the existence and uniqueness of the corresponding representation of X, as well as a relationship between the pair (α, ν) and the σ-field $\mathcal{I}$ occurring in (2).

Proposition 6.11 *(directing elements)* *For any process X as in Theorem 6.6, there exist a σ-finite random measure ν on D_0 and a random variable $\alpha \geq 0$ such that the excursions of X are given by a Cox process $\eta \perp\!\!\!\perp_\nu \alpha$ on $\mathbb{R}_+ \times D_0$ directed by $\lambda \otimes \nu$, where an atom of η at (s, y) corresponds to an excursion with the endpoints in (12). If the pair (α, ν) is normalized by (13), it becomes a.s. unique, and (2) holds with $\mathcal{I} = \sigma(\alpha, \nu)$.*

Proof: The existence of the stated representation was proved already in Theorem 6.7, where we showed that X is regenerative for the induced filtration $\mathcal{F}$, a.s. with respect to the conditional distribution $P[X \in \cdot | \mathcal{I}]$. (Note that this property is much stronger than the conditional statement (2) established in Theorem 6.6.) The required Itô representation, valid under the conditional distribution by FMP 22.11 and 22.13, extends to the original setting by Lemma A1.5. This gives α and ν as $\mathcal{I}$-measurable random elements. Similar representations are obtainable by elementary arguments, whenever Ξ is locally finite or a locally finite union of intervals.

By the law of large numbers, the random measure ν is a.s. unique up to a random normalization, and the corresponding variable α is then uniquely determined via (12). Now introduce the scaling functions $S_c(t) = ct$, $t \geq 0$. For any random variable $\gamma \perp\!\!\!\perp_{\alpha,\nu} \eta$, the scaled point process $\eta_\gamma = \eta \circ S_\gamma^{-1}$ is again Cox with directing random measure $\lambda \otimes (\nu/\gamma)$ (FMP 12.3). Since an atom of η at (s, y) gives rise to an atom of η_γ at $(\gamma s, y)$, expression (12) for the excursion endpoints remains valid for η_γ, provided we replace α by α/γ. In particular, we may define γ by the left-hand side of (13), which is a.s. finite and strictly positive by FMP 12.13, so that (13) becomes fulfilled for the new pair $(\alpha/\gamma, \nu/\gamma)$. Changing the notation, if necessary, we may henceforth assume that the same condition holds already for the original pair (α, ν).

Now let $\mu_{a,m}$ denote the distribution of a regenerative process X, based on a directing pair (a, m) normalized as in (13). Then (2) yields

$$P[\theta_\tau X \in \cdot | \mathcal{F}_\tau, \mathcal{I}] = P[X \in \cdot | \mathcal{I}] = \mu_{\alpha,\nu}.$$

Since (α, ν) is $\mathcal{I}$-measurable, the chain rule for conditional expectations gives

$$P[\theta_\tau X \in \cdot | \mathcal{F}_\tau, \alpha, \nu] = P[X \in \cdot | \alpha, \nu] = \mu_{\alpha,\nu},$$

which shows that (2) remains true with $\mathcal{I}$ replaced by $\sigma(\alpha, \nu)$. $\qquad\square$

If ν or β is unbounded or if $\alpha > 0$, then T is strictly increasing and admits a continuous inverse

$$L_t = \inf\{s \geq 0; \, T_s > t\}, \quad t \in I,$$

called the *local time* of X at 0. Note that L is unique up to a random factor that depends, for $I = \mathbb{R}_+$, on the normalization of (α, ν). When $I = [0, 1]$ we may normalize L by the condition $L_1 = 1$, whereas for $I = \mathbb{R}_+$ the most natural normalization is given by (13). We may also introduce the *local time random measure* $\xi = \lambda \circ T^{-1}$ on I, also characterized by the condition $\xi[0, t] = L_t$ for all $t \in I$.

Lemma 6.12 *(local time)* Let X be an rcll process on $I = \mathbb{R}_+$ or $[0, 1]$, taking values in a Polish space S and exchangeable at a state $0 \in S$. Suppose that the set $\Xi = \{X = 0\}$ has a perfect closure containing 0. Then the local time random measure ξ of X at 0 is a.s. diffuse with support $\overline{\Xi}$, and ξ is a.s. unique up to a normalization.

Proof: The proof for regenerative processes, given in FMP 22.11, depends on the fact that the generating subordinator T is strictly increasing. The result extends immediately to the conditionally regenerative case. The same argument applies to the local time of an exchangeable set in $[0, 1]$, if we can only show that the generating process T is strictly increasing. This is trivially true when $\alpha > 0$, and if $\alpha = 0$ and $\beta D_0 = \infty$ it follows easily from the law of large numbers. $\qquad\square$

For the next few results we restrict our attention to exchangeable random sets Ξ in $[0, 1]$, though similar results holds for regenerative and related sets in $\mathbb{R}_+$. Our aim is to show how the local time random measure ξ of Ξ can be constructed by a simple a.s. approximation. Then introduce the symmetric neighborhoods

$$\Xi^h = (\Xi + \tfrac{1}{2}[-h, h]) \cap [0, 1], \quad h > 0,$$

so that Ξ^h consists of all points in $[0, 1]$ at a distance $\leq h/2$ from Ξ. Equivalently, $x \in \Xi^h$ iff $I_x^h \cap \Xi \neq \emptyset$, where $I_x^h = x + \tfrac{1}{2}[-h, h]$. Let $\Xi^h \cdot \lambda$ denote the restriction of Lebesgue measure λ to the set Ξ^h.

Proposition 6.13 *(approximation) Let Ξ be an a.s. perfect, exchangeable set in $[0, 1]$ with local time ξ, and put $\alpha = \lambda\Xi$. Then $\xi = \Xi \cdot \lambda / \alpha$ when $\alpha > 0$, and in general*

$$\xi_h \equiv \frac{\Xi^h \cdot \lambda}{\lambda\Xi^h} \xrightarrow{w} \xi \quad a.s., \quad h \to 0. \tag{14}$$

Proof: First suppose that $\alpha > 0$. We need to verify the relation

$$\bigcap_{s \in (0,1)} (T_{s-}, T_s)^c = \{T_s; \, s \geq 0\} \quad \text{a.e. } \lambda. \tag{15}$$

Here the inclusion $\supset$ holds since $T_t \geq T_s$ when $s \leq t$, whereas $T_t \leq T_{s-}$ when $s > t$. Conversely, let $t \in \bigcap_s [T_{s-}, T_s)^c$ be arbitrary and put $\sigma = \inf\{s; \, T_s > t\}$. Then clearly $T_{\sigma-} \leq t \leq T_\sigma$ and $t \notin [T_{\sigma-}, T_\sigma)$, which implies $t = T_\sigma$. Thus, the inclusion $\subset$ in (15) is also true, possibly apart from the countable set of left endpoints of intervals (T_{s-}, T_s).

Now fix any $t \in [0, 1]$, and conclude from (15) and the countable additivity of λ that

$$\begin{aligned}
\lambda(\Xi \cap [0, T_t]) &= \lambda\{T_s; \, s \leq t\} = \lambda \bigcap_{s \leq t} (T_{s-}, T_s)^c \cap [0, T_t] \\
&= T_t - \sum_{s \leq t} \Delta T_s = \alpha t = \alpha\lambda[0, t] \\
&= \alpha\lambda \circ T^{-1}[0, T_t] = \alpha\xi[0, T_t].
\end{aligned}$$

Hence, the distribution functions of the measures $\Xi \cdot \lambda$ and $\alpha\xi$ agree on the set (15), and it remains to note that both measures are diffuse and give no charge to the intervals (T_{s-}, T_s).

The approximation in (14) is obvious when $\alpha > 0$ and $\beta(0, \infty) < \infty$, since Ξ is then a finite union of intervals. If instead $\beta(0, \infty) = \infty$, the law of large numbers yields

$$\lim_{r \to 0} \frac{\sum_j 1\{\beta_j > r, \, \tau_j \leq s\}}{\sum_j 1\{\beta_j > r\}} = s \quad \text{a.s.}, \quad s \in [0, 1].$$

Using Fubini's theorem and letting $h \to 0$, we get a.s.

$$\xi_h[0, T_s] = \frac{\sum_j (\beta_j \wedge h) 1\{\tau_j \leq s\}}{\sum_j (\beta_j \wedge h)} = \frac{\int_0^h dr \sum_j 1\{\beta_j > r, \, \tau_j \leq s\}}{\int_0^h dr \sum_j 1\{\beta_j > r\}} \to s.$$

Furthermore, we note that a.s.

$$\xi[0, T_s] = \lambda\{r \in [0, 1]; \, T_r \leq T_s\} = \lambda[0, s] = s, \quad s \in [0, 1],$$

since T is strictly increasing. Thus, by combination,

$$\xi_h[0, T_s] \to \xi[0, T_s] = s \text{ a.s.}, \quad s \in [0, 1],$$

where the exceptional null set can be chosen, by monotonicity, to be independent of s.

Now fix any $t \in [0, 1]$, and choose $s \in [0, 1]$ such that $T_{s-} \leq t \leq T_s$. For any $\varepsilon > 0$ we get, outside a fixed null set,

$$s - \varepsilon \leftarrow \xi_h[0, T_{s-\varepsilon}] \leq \xi_h[0, t] \leq \xi_h[0, T_s] \to s,$$

$$s - \varepsilon = \xi[0, T_{s-\varepsilon}] \leq \xi[0, t] \leq \xi[0, T_s] = s.$$

Since ε was arbitrary, we conclude that a.s. $\xi_h[0, t] \to s = \xi[0, t]$ for all t, which implies $\xi_h \xrightarrow{w} \xi$ a.s. $\qquad\square$

The distribution of an exchangeable set in $\mathbb{R}_+$ or $[0, 1]$ is a mixture of *extreme* sets where the directing pair (α, ν) or (α, β) is non-random, and for many purposes it is enough to study the extreme case. If Ξ is an extreme, exchangeable set in $[0, 1]$ with $\alpha = \lambda\Xi > 0$, then Fubini's theorem yields, for any Borel set $B \subset [0, 1]$,

$$\alpha E\xi B = E\lambda(\Xi \cap B) = E\int_B 1\{t \in \Xi\} \, dt = \int_B P\{t \in \Xi\} \, dt,$$

which shows that the intensity measure $E\xi$ is absolutely continuous with density $p_t = P\{t \in \Xi\}/\alpha$. In other words, the probability for Ξ to hit a fixed point $t \in [0, 1]$ equals αp_t. Since $\lambda(\overline{\Xi} \setminus \Xi) = 0$, we may replace Ξ by $\overline{\Xi}$. By Fatou's lemma, we get for any $t_n \to t$ in $[0, 1]$

$$\limsup_{n \to \infty} P\{t_n \in \overline{\Xi}\} \leq E \limsup_{n \to \infty} 1\{t_n \in \overline{\Xi}\} \leq E1\{t \in \overline{\Xi}\} = P\{t \in \overline{\Xi}\},$$

which shows that the modified density is upper semi-continuous.

We proceed to examine the existence and continuity properties of the density of $E\xi$ when $\alpha = 0$. (The corresponding hitting probabilities will be studied in a later section.) Here it is useful to impose a regularity condition on Ξ. Given an exchangeable set Ξ in $[0, 1]$ directed by $(0, \beta)$, we may introduce the associated *indices of regularity*

$$\rho = \sup\Big\{r \geq 0; \, \lim_{u \to \infty} u^{2-r} \sum_k \beta_k^2 \, 1\{u\beta_k \leq 1\} = \infty\Big\},$$

$$\rho' = \inf\Big\{r \geq 0; \, \int_0^1 x^r \beta(dx) < \infty\Big\}.$$

We have already seen, in Theorem 2.32, Proposition 2.33, and Corollaries 3.29 and 3.30, how ρ' gives information about the local behavior of the generating process T, hence also of the local time L of Ξ. For our present purposes, however, index ρ seems to be more appropriate. The following result shows how the two indices are related.

Lemma 6.14 *(indices of regularity) The indices ρ and ρ' of an exchangeable set Ξ in $[0, 1]$ satisfy $0 \leq \rho \leq \rho' \leq 1$ a.s. For any $r \in [0, 1]$, we can choose Ξ such that $\rho = \rho' = r$ a.s.*

Proof: For any $r \in [0, 1]$ and $u > 0$, we have

$$u^2 \sum_k \beta_k^2 \, 1\{u\beta_k \leq 1\} = \int_0^{1/u} (ux)^2 \beta(dx)$$
$$\leq \int_0^{1/u} (ux)^r \beta(dx) \leq u^r \int_0^{\infty} x^r \beta(dx).$$

Since the right-hand side is finite when $\rho' < r$, so is the limit in the definition of ρ, which shows that even $\rho < r$. Since β always satisfies $\int_0^1 x\beta(dx) < \infty$, it is also clear that ρ and ρ' are both bounded by 1. This completes the proof of the first assertion.

To prove the second assertion, let us first replace β by the the measure $\nu(dx) = x^{-r-1}dx$ on $(0, 1]$ for an arbitrary $r \in [0, 1)$, so that $\rho' = r$. To evaluate ρ in this case, we note that

$$u^{2-p} \int_0^{1/u} x^2 \, \nu(dx) = (2 - r)^{-1} u^{r-p}, \quad p > 0,$$

which tends to ∞ as $u \to \infty$ iff $r > p$. Hence, $\rho = \rho' = r$ a.s. Since the same relations hold for suitable discrete approximations of ν, the a.s. equalities $\rho = \rho' = r$ may occur for every $r \in [0, 1)$. To give an example where $\rho = \rho' = 1$, we may consider suitable discrete approximations of the measure

$$\nu(dx) = |x \log x|^{-2} dx, \quad x \in (0, \tfrac{1}{2}).$$

Here clearly $\int_0^1 x^r \nu(dx) < \infty$ iff $r \geq 1$, which shows that $\rho' = 1$. Comparing with the measures $x^{-r-1}dx$ for arbitrary $r < 1$, we see that also $\rho = 1$, as required.   □

For the remainder of this section, we assume that Ξ is extreme and perfect with $\lambda\Xi = 0$ and $\rho > 0$. The distributions $\mathcal{L}(T_s)$, $s \in (0, 1)$, are then absolutely continuous with continuous densities $p_{s,t}$, given by the Fourier inversion formula

$$p_{s,t} = (2\pi)^{-1} \int_{-\infty}^{\infty} e^{-itu} \, Ee^{iuT_s} \, du, \quad s \in (0, 1), \; t \in \mathbb{R}. \tag{16}$$

The required integrability is clear from the following estimate.

Lemma 6.15 *(characteristic functions) For any $a \in (0, \rho)$, we have*

$$\int_{-\infty}^{\infty} \sup_{s \in [h, 1-h]} \left| Ee^{iuT_s} \right| du \lesssim h^{-1/a} < \infty, \quad h \in (0, \tfrac{1}{2}].$$

Proof: Let τ by $U(0,1)$. By an elementary Taylor expansion, we have for any $r \in [0,1]$, $h \in (0, \frac{1}{2}]$, and $s \in [h, 1-h]$

$$
\begin{aligned}
|E \exp(ir1\{\tau \le s\})|^2 &= |se^{ir} + 1 - s|^2 \\
&= 1 - 2s(1-s)(1 - \cos r) \\
&\le 1 - hr^2(1 - r^2/12) \\
&\le 1 - 2hr^2/3 \le \exp(-2hr^2/3),
\end{aligned}
$$

and so for any $r \ge 0$ and $s \in [h, 1-h]$,

$$
|E \exp(ir1\{\tau \le s\})| \le \exp\left(-hr^2 1\{r \le 1\}/3\right).
$$

Hence, by the independence of $\tau_1, \tau_2, \ldots$, we get for any $u \ge 0$ and $s \in [h, 1-h]$

$$
\begin{aligned}
\left|Ee^{iuT_s}\right| &= \left|E \prod_k \exp(iu\beta_k 1\{\tau_k \le s\})\right| \\
&= \prod_k |E \exp(iu\beta_k 1\{\tau_k \le s\})| \\
&\le \prod_k \exp\left(-hu^2 \beta_k^2 1\{u\beta_k \le 1\}/3\right) \\
&= \exp\left(-hu^2 \sum_k \beta_k^2 1\{u\beta_k \le 1\}/3\right).
\end{aligned}
$$

Now fix any $a \in (0, \rho)$. By the definition of ρ, there exists a constant $c > 0$ such that

$$
u^2 \sum_k \beta_k^2 1\{u\beta_k \le 1\} \ge 3cu^a, \quad u \ge 1. \tag{17}
$$

Then

$$
\begin{aligned}
\int_1^\infty \sup_{s \in [h, 1-h]} \left|Ee^{iuT_s}\right| du &\le \int_1^\infty \exp(-chu^a)\, du \\
&\le (ch)^{-1/a} \int_0^\infty \exp(-u^a)\, du \lesssim h^{-1/a}.
\end{aligned}
$$

The interval $(-\infty, -1]$ gives the same contribution, and the integral over $(-1, 1)$ is bounded by $2 \lesssim h^{-1/a}$. $\qquad\square$

When $\rho > 0$, we may integrate (16) with respect to $s \in (0,1)$ to obtain, for the intensity measure $E\xi$, the density

$$
p_t = \int_0^1 p_{s,t}\, ds, \quad t \in [0,1]. \tag{18}
$$

To describe the continuity properties of p, we may introduce the sets $S_0 \subset S_1 \subset \cdots \subset [0,1]$, where S_n consists of all sums $\sum_{k \le m} \beta_{j_k}$ with $m \le n$ and distinct indices $j_1, \ldots, j_m \in \mathbb{N}$. In particular, $S_0 = \{0\}$ and $S_1 = \{0, \beta_1, \beta_2, \ldots\}$. Write $1 - S_n = \{1 - s; \, s \in S_n\}$.

Theorem 6.16 *(local intensity) Let ξ be the local time random measure of an extreme, exchangeable set Ξ in $[0, 1]$ with index $\rho > 0$, and put $d = [\rho^{-1}] - 1$. Then $E\xi$ is absolutely continuous with a lower semi-continuous density p given by (16) and (18). The latter is also right-continuous on S_d^c and left-continuous on $(1 - S_d)^c$.*

Proof: To see that $p = (p_t)$ is a density of $E\xi$, let $B \in \mathcal{B}[0, 1]$ be arbitrary. Recalling that $\xi = \lambda \circ T^{-1}$ and using Fubini's theorem twice, we obtain

$$E\xi B = E \int_0^1 1\{T_s \in B\}\, ds = \int_0^1 P\{T_s \in B\}\, ds$$
$$= \int_0^1 ds \int_B p_{s,t}\, dt = \int_B dt \int_0^1 p_{s,t}\, ds = \int_B p_t\, dt.$$

The densities $p_{s,t}$ in (16) are jointly continuous in $s \in (0, 1)$ and $t \in [0, 1]$, by Lemma 6.15 and dominated convergence. Letting $t_n \to t$, we get by Fatou's lemma

$$\liminf_{n \to \infty} p_{t_n} \geq \int_0^1 \liminf_{n \to \infty} p_{s,t_n}\, ds = \int_0^1 p_{s,t}\, ds = p_t,$$

which shows that p is lower semi-continuous.

Now define the right neighborhoods of S_d by

$$S_d^h = (S_d + [0, h]) \cap [0, 1], \quad h > 0.$$

Fixing any $h > 0$, we may choose $n > d$ so large that $\sum_{j>n} \beta_j \leq h$. Let σ denote variable number $d+1$ in magnitude among the times $\tau_1, \ldots, \tau_n$. Then as $s \downarrow 0$ for fixed n, we have

$$P\{\sigma \leq s\} = \sum_{k \in (d,n]} (n /\!\!/ k)\, s^k (1 - s)^{n-k} \lesssim s^{d+1}, \tag{19}$$

where the $(n /\!\!/ k)$ are binomial coefficients. Writing

$$T_s = \sum_{j \leq n} \beta_j 1\{\tau_j \leq s\} + \sum_{j>n} \beta_j 1\{\tau_j \leq s\} = T_s' + T_s'',$$

we note that $T_s'' \leq h$ for all s and $T_s' \in S_d$ when $s < \sigma$. Hence, $T_s \in S_d^h$ on the set $\{\sigma > s\}$.

Now put $\mu_s = \mathcal{L}(T_s)$ and $\nu_s = \mathcal{L}(T_s'')$, and define $\mu^u = \int_0^u \mu_s ds$ for $u \in (0, 1)$. In view of the previous remarks together with formula (19), Fubini's theorem, Fourier inversion, and Lemma 6.15, we get for any Borel set $B \subset (S_d^h)^c$ and constant $a \in ((d+2)^{-1}, \rho)$

$$\mu^u B = \int_0^u P\{T_s \in B\}\, ds = \int_0^u P\{T_s \in B, \sigma \leq s\}\, ds$$
$$= \int_0^u E[P[T_s'' \in B - T_s' | T']; \sigma \leq s]\, ds$$
$$\leq (2\pi)^{-1}\lambda B \int_0^u P\{\sigma \leq s\}\, \|\hat{\nu}_s\|_1\, ds$$
$$\lesssim \lambda B \int_0^u s^{d+1-1/a}\, ds < \infty. \tag{20}$$

By the joint continuity of $p_{s,t}$, the measure $\mu^v - \mu^u$ has the continuous density $p_t^{u,v} = \int_u^v p_{s,t} ds$ on $[0,1]$ for any $u < v$ in $(0,1)$. Differentiating in (20), we get as $v \to 0$

$$\sup_{t \notin S_d^h} p_t^{u,v} \lesssim \int_0^v s^{d+1-1/a}\, ds \to 0,$$

which shows that $p^v = p^{0,v} \to 0$, uniformly on the closure $\overline{(S_d^h)^c}$. Hence $p^{1/2}$ is continuous on the same set.

Since S_d is closed and its elements are isolated from the left, we have

$$S_d^c = \bigcup_{h>0} (S_d^h)^c, \qquad S_d \subset \bigcup_{h>0} \overline{(S_d^h)^c|},$$

where the last bar denotes closure on the right. Thus, the density $p^{1/2}$ is in fact continuous on S_d^c and left-continuous on $S_d \cap (0,1]$. In other words, $p^{1/2}$ is left-continuous on $(0,1]$ and right-continuous on S_d^c. A similar argument shows that the function $p - p^{1/2} = p^{1/2,1}$ is right-continuous on $[0,1)$ and left-continuous on $(1 - S_d)^c$. The asserted continuity properties of p follows by combination of the two statements. □

6.5 Exponential and Uniform Sampling

In this section we consider some results about random sampling or truncation for exchangeable random sets Ξ in $\mathbb{R}_+$ or $[0,1]$, where the sampling or stopping times are independent of Ξ and Poisson or uniformly distributed, respectively. The results in the infinite-interval case depend on the proper normalization of the local time process $L_t = \xi[0,t]$, which is chosen in accordance with (13) unless otherwise specified.

Theorem 6.17 (*Poisson and uniform sampling*) *Let X be an rcll process on $I = \mathbb{R}_+$ or $[0,1]$, taking values in a Polish space S, and exchangeable at a state 0 with normalized local time L, directing pair (α, ν) or (α, β), and excursion point process η. For any $\tau_1, \tau_2, \ldots$ in I, let $\sigma_1, \sigma_2, \ldots$ be the distinct elements of the sequence $L_{\tau_1}, L_{\tau_2}, \ldots$, and write η_n for the restriction of η to the complement $\{\sigma_1, \ldots, \sigma_n\}^c$. Then*

(i) *if $I = \mathbb{R}_+$ and $\tau_1 < \tau_2 < \cdots$ form a unit rate Poisson process independent of X, we have*

$$(\tau_n) \overset{d}{=} (\sigma_n) \perp\!\!\!\perp (\alpha, \nu, \eta_\infty);$$

(ii) *if $I = [0,1]$ and $\tau_1, \tau_2, \ldots$ are i.i.d. $U(0,1)$ and independent of X, we have*

$$(\tau_n) \overset{d}{=} (\sigma_n); \qquad (\sigma_1, \ldots, \sigma_n) \perp\!\!\!\perp (\alpha, \beta, \eta_n), \quad n \in \mathbb{N}.$$

In either case, the sequence $\sigma_1, \sigma_2, \ldots$ is a.s. infinite. This is true for (i) since $\tau_n \to \infty$ and L has unbounded support $\overline{\Xi}$. It also holds in case (ii), due to the law of large numbers and the continuity of L. The relation $(\tau_n) \overset{d}{=} (\sigma_n)$ is easy to prove in both cases by means of Theorem 1.16.

Proof: (i) Let N denote the Poisson process formed by $\tau_1, \tau_2, \ldots$. Since T is exchangeable on $\mathbb{R}_+$ and independent of N, we see from the cited theorem that the point process $N \circ T$ is again exchangeable. The same thing is then true for the simple point process $(N \circ T)^*$, obtained from $N \circ T$ by reduction of all multiplicities to 1. Hence, the latter process is mixed Poisson. To find the rate of $(N \circ T)^*$, we note that the probability for at least one point τ_k to fall in an interval of length $h > 0$ equals $1 - e^{-h}$. Using the normalization in (13), we obtain

$$
\begin{aligned}
E[(N \circ T)_1^* | \alpha, \nu] &= \alpha + E \int_0^1 \!\!\int (1 - e^{-l(x)})\, \eta(dr\, dx) \\
&= \alpha + \int (1 - e^{-l(x)})\, \nu(dx) = 1.
\end{aligned}
$$

It remains to notice the the jumps of $N \circ T$ occur at times $\sigma_1, \sigma_2, \ldots$.

To prove the stated independence we define $\zeta = \sum_k \delta_{\sigma_k}$, so that ζ is the random measure corresponding to $(N \circ T)^*$. Next we may write $\zeta = \zeta' + \zeta''$, where ζ' arises from points τ_k hitting the excursion intervals of X and ζ'' from those falling in the set Ξ. Then the pair (η_∞, ζ') is a randomization of η, in the sense of FMP 12.2, whereas ζ'' is a Cox process directed by $\alpha\lambda$. Since $(\eta_\infty, \zeta') \perp\!\!\!\perp_\eta \nu$, we see from FMP 12.3 that (η_∞, ζ') is again a Cox process, and the relation $(\eta_\infty, \zeta') \perp\!\!\!\perp_{\eta,\alpha} \zeta''$ guarantees that the entire triple $(\eta_\infty, \zeta', \zeta'')$ is Cox. Hence, so is the pair (η_∞, ζ), and since ζ was seen to be Poisson, the independence $\eta_\infty \perp\!\!\!\perp \zeta$ follows. Applying this result to the conditional distributions, given (α, ν), we obtain the stronger statement $(\alpha, \nu, \eta_\infty) \perp\!\!\!\perp \zeta$.

(ii) Consider the measure-valued process

$$
Y_t = \sum_k \delta_k 1\{\tau_k \le t\}, \quad t \in [0, 1],
$$

which is clearly exchangeable. Then so is the composition $Y \circ T$ by Theorem 1.16. The jumps of the latter process are mutually singular point processes on $\mathbb{N}$, which we enumerate according to the lowest order term δ_k included, say as $\gamma_1, \gamma_2, \ldots$. The corresponding jump times then become $\sigma_1, \sigma_2, \ldots$, and Theorem 1.25 shows that the σ_k are again i.i.d. $U(0, 1)$.

To prove the asserted independence in this case, it is enough to show that $\eta_n \perp\!\!\!\perp (\sigma_1, \ldots, \sigma_n)$, since the general statement will then follow by conditioning on (α, β). Proceeding by induction on n, we shall prove that each η_n is a uniform randomization of some point process $\beta^{(n)}$ and satisfies $\eta_n \perp\!\!\!\perp (\sigma_1, \ldots, \sigma_n)$. Assuming this to be true for all indices less than n, we obtain

$$
\eta_{n-1} \perp\!\!\!\perp (\sigma_1, \ldots, \sigma_{n-1}), \qquad \sigma_n \perp\!\!\!\perp_{\eta_{n-1}} (\sigma_1, \ldots, \sigma_{n-1}).
$$

Hence, the chain rule for conditional independence (FMP 6.8) yields

$$(\sigma_n, \eta_{n-1}) \perp\!\!\!\perp (\sigma_1, \ldots, \sigma_{n-1}). \tag{21}$$

By the construction of σ_n and the induction hypotheses for indices 1 and $n-1$, we note that also $\sigma_n \perp\!\!\!\perp \eta_n$, and that η_n is a uniform randomization of some point process $\beta^{(n)}$. Since η_n is a measurable function of (σ_n, η_{n-1}), we conclude from (21) and FMP 3.8 that $\eta_n \perp\!\!\!\perp (\sigma_1, \ldots, \sigma_n)$. This completes the induction step from $n-1$ to n.

It remains to prove the assertion for $n=1$. Thus, we need to verify the independence $\sigma_1 \perp\!\!\!\perp \eta_1$ and show that η_1 is a uniform randomization of some point process $\beta^{(1)}$. Here we can take the original process β to be simple, since we may otherwise attach some independent and i.i.d. $U(0,1)$ marks to the excursions of X, through a uniform randomization of η. We can also reduce, via a suitable conditioning, to the case where the measure $\beta = \sum_k \delta_{\beta_k}$ is non-random. Re-labeling $\tau_1 = \tau$ and $\sigma_1 = \sigma$, we may write $\eta = \sum_k \delta_{\beta_k, \sigma_k}$, where $\sigma_1, \sigma_2, \ldots$ are i.i.d. $U(0,1)$.

Now define $\kappa = k$ when τ hits excursion interval number k, and put $\kappa = 0$ when $\tau \in \Xi$. Since τ is $U(0,1)$ and independent of X, we note that a.s.

$$P[\kappa = k \mid \sigma_1, \sigma_2, \ldots] = l(\beta_k), \quad k \in \mathbb{N},$$
$$P[\kappa = 0, \ \sigma \in \cdot \mid \sigma_1, \sigma_2, \ldots] = \alpha\lambda. \tag{22}$$

In particular, κ is independent of (σ_k), and we get

$$P[(\sigma_k) \in \cdot \mid \kappa] = P\{(\sigma_k) \in \cdot\} = \lambda^\infty. \tag{23}$$

When $\kappa > 0$, we have

$$\sigma = \sigma_\kappa, \qquad \eta_1 = \eta - \delta_{\beta_\kappa, \sigma_\kappa} = \sum_{k \neq \kappa} \delta_{\beta_k, \sigma_k},$$

which yields the required distribution of (σ, η_1), conditionally on $\kappa > 0$. For $\kappa = 0$, we get instead from (22) and (23)

$$P\{(\sigma_k) \in A, \ \sigma \in B, \ \kappa = 0\} = \alpha\,\lambda^\infty A \cdot \lambda B,$$

which implies

$$P[(\sigma, \sigma_1, \sigma_2, \ldots) \in \cdot \mid \kappa = 0] = \lambda^\infty.$$

Since $\eta_1 = \eta$ when $\kappa = 0$, the required conditional distribution follows again for the pair (σ, η_1).   □

If X is recurrent and conditionally regenerative at 0 with $X_0 = 0$, then by Lemma 6.27 the directing triple (α, ν), normalized by (13), is a.s. unique and measurably determined by X. Though the uniqueness fails when $\Xi = \{X = 0\}$ is a.s. bounded, we may still recover the distribution of (α, ν) from that of X.

To deal with the uniqueness problem for processes on $\mathbb{R}_+$, we begin with the case where the last excursion is infinite. In the present context, it may be more natural to normalize (α, ν) by the condition $\nu\{l = \infty\} = 1$.

Proposition 6.18 *(infinite excursions) Let X be an rcll process in S with $X_0 = 0$, where X is conditionally regenerative at 0 and the set $\Xi = \{X = 0\}$ is a.s. bounded. Write (α, ν) for the associated characteristics, normalized by $\nu\{l = \infty\} = 1$, and let ν' denote the restriction of ν to $\{l < \infty\}$. Then $\mathcal{L}(\alpha, \nu')$ is determined by $\mathcal{L}(X)$.*

Our proof is based on an elementary result for random variables.

Lemma 6.19 *(exponential scaling) For any random variable $\xi \geq 0$, let $\sigma \perp\!\!\!\perp \xi$ be exponentially distributed with mean 1, and put $\eta = \xi\sigma$. Then $\mathcal{L}(\xi)$ and $\mathcal{L}(\eta)$ determine each other uniquely.*

Proof: Using Fubini's theorem twice, we get for any $u > 0$

$$
\begin{aligned}
Ee^{-\eta/u} &= Ee^{-\xi\sigma/u} = E\int_0^\infty e^{-s}e^{-\xi s/u}\,ds \\
&= u\,E\int_0^\infty e^{-ut}e^{-t\xi}\,dt \\
&= u\int_0^\infty e^{-ut}\,Ee^{-t\xi}\,dt.
\end{aligned}
$$

Applying the uniqueness theorem for Laplace transforms (FMP 5.3), we see that $\mathcal{L}(\eta)$ determines the continuous function $Ee^{-t\xi}$ on $\mathbb{R}_+$. By another application of the same theorem, we conclude that $\mathcal{L}(\eta)$ determines even $\mathcal{L}(\xi)$. The reverse statement is obvious.   □

Proof of Proposition 6.18: Consider a Cox process η on $\mathbb{R}_+ \times D_0$ directed by $\lambda \otimes \nu'$, and let Y be a conditionally independent process with distribution $\nu'' = \nu - \nu'$. Next introduce an independent random variable $\sigma > 0$ with a standard exponential distribution. Then X has clearly the same distribution as the process generated by α and $\eta^\sigma = \eta \cap [0, \sigma]$, followed by the infinite excursion Y. The transfer theorem allows us to assume that the two processes agree a.s.

Since $\sigma \perp\!\!\!\perp X$, a scaling by a factor σ^{-1} yields another representation of X, in terms of a Cox process $\tilde{\eta}$ on $[0, 1] \times D_0$, directed by $\tilde{\nu} = \sigma\nu'$ along with the random drift rate $\tilde{\alpha} = \sigma\alpha$, followed by the same infinite excursion Y. Here $\tilde{\alpha}$ and $\tilde{\beta} = \tilde{\eta}([0, 1] \times \cdot)$ are measurable functions of X, which ensures $\mathcal{L}(\tilde{\alpha}, \tilde{\beta})$ to be determined by $\mathcal{L}(X)$. Hence, so is $\mathcal{L}(\tilde{\alpha}, \tilde{\eta})$ since $\tilde{\eta}$ is a uniform randomization of $\tilde{\beta}$ with $\tilde{\eta} \perp\!\!\!\perp_{\tilde{\beta}} \tilde{\alpha}$. Noting that the pair $(\tilde{\alpha}, \tilde{\eta})$ is exchangeable, we conclude as in Theorem 3.20 and Lemma 1.20 that $\mathcal{L}(X)$ determines $\mathcal{L}(\tilde{\alpha}, \tilde{\nu}) = \mathcal{L}(\sigma\alpha, \sigma\nu')$. Finally, Lemma 6.19 shows that the latter distribution determines $\mathcal{L}(\alpha, \nu')$.   □

Returning to the case of an unbounded, homogeneous set Ξ in $\mathbb{R}_+$, we may construct a bounded set of the same type by truncating Ξ at a suitable random time τ. Thus, we may form a bounded, homogeneous set $\Xi^\tau = \Xi \cap [0, \tau]$ by choosing τ to be an independent, exponentially distributed random variable. Similarly, given an exchangeable set Ξ in $[0,1]$ and an independent $U(0,1)$ random variable, we may consider the truncated random set Ξ^τ. In both cases, the truncated set has again a representation in terms of a non-decreasing, exchangeable process T on $\mathbb{R}_+$ or $[0,1]$, respectively, though the jumps of T are now allowed to be infinite. Equivalently, the directing random measure ν or β may now have an atom at ∞.

The following result shows that, in either case, the distribution of Ξ^τ determines that of the original set Ξ.

Theorem 6.20 *(exponential or uniform truncation) Let Ξ be an exchangeable set in $[0,1]$ or $\mathbb{R}_+$ directed by (α, β) or (α, ν), respectively, and let $\tau \perp\!\!\!\perp \Xi$ be $U(0,1)$ or exponentially distributed with mean 1. Then $\mathcal{L}(\alpha, \beta)$ or $\mathcal{L}(\alpha, \nu)$ is determined by $\mathcal{L}(\Xi^\tau)$.*

Proof: Beginning with the case of exchangeable sets Ξ in $\mathbb{R}_+$, we may embed the exponential time τ as the first point of a unit rate Poisson process $\xi \perp\!\!\!\perp \Xi$ on $\mathbb{R}_+$. Let η denote the Cox process on $\mathbb{R}_+ \times (0, \infty)$ representing Ξ. To construct the corresponding point process $\hat{\eta}$ representing Ξ^τ, we note that every contiguous interval hit by at least one point of ξ is replaced by an infinite interval. In addition, we need to introduce an infinite interval wherever τ hits Ξ. On the local time scale, a unit mass of η at a point (s, x) is retained with probability e^{-x} and moved to infinity with probability $1 - e^{-x}$. Additional points at infinity are created according to a mixed Poisson process with the constant rate α.

The two components of $\hat{\eta}$ are conditionally independent of ν and also mutually independent, given α and η. Hence, $\hat{\eta}$ is again a Cox process, whose restriction to $(0, \infty)$ is directed by the measure $\hat{\nu}(dx) = e^{-x}\nu(dx)$ (FMP 12.3). Furthermore, Theorem 6.17 shows that the points at infinity occur at a constant rate 1. The drift parameter α clearly remains the same after thinning, and so the stopped process is directed by a pair $(\alpha, \tilde{\nu})$, where $\tilde{\nu} = \hat{\nu}$ on $(0, \infty)$ and $\tilde{\nu}\{\infty\} = 1$. By Lemma 6.18 we see that $\mathcal{L}(\Xi^\tau)$ determines $\mathcal{L}(\alpha, \hat{\nu})$, which in turn determines $\mathcal{L}(\alpha, \nu')$, since the formula $\nu'(dx) = e^x \hat{\nu}(dx)$ exhibits ν' as a measurable function of $\hat{\nu}$ (FMP 1.41). This completes the proof for processes on $\mathbb{R}_+$.

Turning to the finite-interval case, we may form another point process β' on $(0,1)$ by letting $\beta' = \beta - \delta_x$ when τ hits a contiguous interval of length x and taking $\beta' = \beta$ if $\tau \in \Xi$. Let us also write η and η' for the associated representing point processes, with the possible interval at τ included or excluded, and recall from Theorem 6.17 that $\sigma = L_\tau$ is independent of η'. Hence, the lengths of all contiguous intervals occurring up to time τ form a σ-thinning $\hat{\beta}'$ of β', conditionally on σ. Since σ is $U(0,1)$ by Theorem 6.17, we see from FMP 12.2 that the pair $(\sigma\alpha, \hat{\beta}')$ has Laplace transform

$$E \exp\left(-u\sigma\alpha - \hat{\beta}'f\right) = \int_0^1 E \exp\left(-up\alpha + \beta' \log\left(1 - p(1 - e^{-f})\right)\right) dp,$$

where $u \geq 0$ and $f \geq 0$ is any measurable function on $(0, \infty)$. Letting $t > 0$ be arbitrary and substituting

$$u = tv, \quad 1 - e^{-f} = tg, \quad f = -\log(1 - tg),$$

we get

$$E \exp\left(-tv\sigma\alpha + \hat{\beta}' \log(1 - tg)\right)$$
$$= \int_0^1 E \exp(-tvp\alpha + \beta' \log(1 - ptg)) \, dp$$
$$= t^{-1} \int_0^t E \exp(-vs\alpha + \beta' \log(1 - sg)) \, ds.$$

Multiplying by t and differentiating at $t = 1$ for fixed g with $\|g\| < 1$, we obtain

$$E \exp(-v\alpha + \beta' \log(1 - g))$$
$$= \frac{d}{dt} \left\{ tE \exp\left(-tv\sigma\alpha + \hat{\beta}' \log(1 - tg)\right) \right\}\Big|_{t=1}.$$

The reverse substitution

$$g = 1 - e^{-h}, \qquad h = -\log(1 - g)$$

yields the equivalent formula

$$E \exp(-v\alpha - \beta' h)$$
$$= \frac{d}{dt} \left\{ tE \exp\left(-tv\sigma\alpha + \hat{\beta}' \log\left(1 - t(1 - e^{-h})\right)\right) \right\}\Big|_{t=1},$$

valid for any bounded, measurable function $h \geq 0$ on $(0, \infty)$. Invoking the uniqueness theorem for Laplace transforms, we conclude that $\mathcal{L}(\sigma\alpha, \hat{\beta}')$ determines $\mathcal{L}(\alpha, \beta')$. The assertion now follows since the pair (α, β') determines (α, β), due to the normalization $\alpha + \beta l = 1$.   □

The last result leads easily to corresponding criteria for convergence in distribution, in the spirit of Chapter 3. Given an exchangeable set Ξ in $\mathbb{R}_+$ or $[0, 1]$, we may then introduce the associated truncated set $\hat{\Xi} = \Xi^\tau = \Xi \cap [0, \tau]$, where τ is an independent random variable with a standard exponential or uniform distribution. Normalizing the directing pair (α, ν) or (α, β) of Ξ by the condition $\alpha + \nu(1 - e^{-l}) = 1$ or $\alpha + \beta l = 1$, respectively, where $l(x) = x$, we may define the associated random probability measures $\mu = \mu_{\alpha, \nu}$ or $\mu_{\alpha, \beta}$ on $\mathbb{R}_+$ by

$$\mu_{\alpha, \nu} = \alpha\delta_0 + (1 - e^{-l}) \cdot \nu, \tag{24}$$
$$\mu_{\alpha, \beta} = \alpha\delta_0 + l \cdot \beta. \tag{25}$$

In the former case, we may allow the set Ξ to be bounded, corresponding to the possibility for ν to have positive mass at infinity. Accordingly, the measure μ in (24) is regarded as defined on the compactified interval $[0, \infty]$, with a possible atom at ∞ of size $\mu\{\infty\} = \nu\{\infty\}$.

Given some random sets Ξ and $\Xi_1, \Xi_2, \ldots$ in a suitable metric space S, we write $\Xi_n \overset{d}{\to} \Xi$ for convergence in distribution with respect to the Fell topology (FMP 16.28, A2.5) of the corresponding closures $\overline{\Xi}_n$ and $\overline{\Xi}$. For the directing random measures μ of exchangeable sets Ξ in $\mathbb{R}_+$ or $[0, 1]$, the relevant mode of convergence is with respect to the weak topology on the compactified interval $[0, \infty]$ or $[0, 1]$, which allows for the possibility of some mass to escape to infinity.

Corollary 6.21 *(convergence of truncated sets)* *For every* $n \in \mathbb{N}$, *let* Ξ_n *be an exchangeable set in* $\mathbb{R}_+$ *or* $[0, 1]$ *directed by* (α_n, ν_n) *or* (α_n, β_n), *respectively, form* $\hat{\Xi}_n$ *by exponential or uniform truncation, and define* μ_n *as in* (24) *or* (25). *Then these three conditions are equivalent:*

(i) $\Xi_n \overset{d}{\to}$ *some* Ξ,

(ii) $\hat{\Xi}_n \overset{d}{\to}$ *some* $\hat{\Xi}$,

(iii) $\mu_n \overset{wd}{\longrightarrow}$ *some* μ.

When the statements are true, we may choose the limits Ξ, $\hat{\Xi}$, *and* μ *to be related in the same way as* Ξ_n, $\hat{\Xi}_n$, *and* μ_n.

Proof: Assuming (iii), we see from Theorems 3.4 and 3.5 that the associated generating processes $T_1, T_2, \ldots$ and T satisfy $T_n \overset{d}{\to} T$, with respect to the Skorohod topology on $D(\mathbb{R}_+, [0, \infty])$ or $D([0, 1], [0, 1])$. Condition (i) then follows by continuity (FMP 4.27). Since (i) implies (ii) by continuous mapping, it remains to show that (ii) implies (iii).

Then assume that $\hat{\Xi}_n \overset{d}{\to} A$. The sequence (μ_n) is uniformly bounded and hence relatively compact in distribution (FMP 16.15). If $\mu_n \overset{wd}{\longrightarrow} \mu$ along a subsequence, then the implication (iii) $\Rightarrow$ (ii) yields $\hat{\Xi}_n \overset{d}{\to} \hat{\Xi}$ for the associated random sets, where Ξ is an exchangeable set corresponding to the measure μ. This gives $A \overset{d}{=} \hat{\Xi}$, and so by Theorem 6.20 the distribution of μ is unique, and we may assume that $A = \hat{\Xi}$ a.s. The convergence $\mu_n \overset{wd}{\longrightarrow} \mu$ then remains valid along the original sequence, which proves (iii) with the indicated relationship between $\hat{\Xi}$ and μ. $\qquad\square$

6.6   Hitting Points and Intervals

Here we study the probability for an exchangeable set Ξ to hit a fixed point or short interval. We consider only the case of sets in $[0, 1]$, though similar results hold for sets in $\mathbb{R}_+$. In a previous section we have seen that, if Ξ is an extreme, exchangeable set in $[0, 1]$ with directing pair (α, β) satisfying $\alpha > 0$, then $P\{t \in \Xi\} = \alpha p_t$ for a suitable density p of $E\xi$. This suggests that, when $\alpha = 0$, the hitting probability $P\{t \in \Xi\}$ should be 0 for any

$t \in (0, 1)$. Though this is indeed true, the proof in the general case is quite complicated, and so we restrict our attention to the *regular* case, where the index of regularity ρ is strictly positive.

Theorem 6.22 *(hitting fixed points, Berbee)* *Let Ξ be a regular, exchangeable set in $[0, 1]$ with $\lambda\Xi = 0$. Then $t \notin \Xi$ a.s. for every $t \in (0, 1)$.*

Proof: We may clearly assume that Ξ is extreme. Let T denote the exchangeable process generating Ξ, with jumps β_j occurring at times τ_j, and write L for the associated local time process T^{-1}. Suppose that $P\{t \in \Xi\} > 0$ for some $t \in (0, 1)$. For a fixed $n > d = [\rho^{-1}] - 1$, construct the process T^n from T by omitting the n largest jumps, let Ξ_n be the random set generated by T^n, and write L^n for the local time of Ξ_n. Putting $\kappa_j = 1\{\tau_j \leq L_t\}$, $j \leq n$, we may choose $k_1, \ldots, k_n \in \{0, 1\}$ such that

$$
\begin{aligned}
0 &< P\{t \in \Xi,\, L_t < 1,\, \kappa_1 = k_1, \ldots, \kappa_n = k_n\} \\
&= P\{t_n \in \Xi_n,\, L^n_{t_n} < 1,\, \kappa_1 = k_1, \ldots, \kappa_n = k_n\} \\
&\leq P\{t_n \in \Xi_n,\, L^n_{t_n} < 1\},
\end{aligned}
$$

where $t_n = t - \sum_j k_j \beta_j$. This shows in particular that

$$
t_n < \sum_{j>n} \beta_j < \inf(1 - S_d).
$$

Next define $\kappa^n_j = 1\{\tau_j \leq L^n_{t_n}\}$. Using the independence of Ξ_n and $\tau_1, \ldots, \tau_n$ and noting that $P[\kappa^n_j = 0 | \Xi_n] > 0$ for all $j \leq n$, we obtain

$$
\begin{aligned}
P\{t_n \in \Xi\} &\geq P\{t_n \in \Xi,\, \kappa^n_1 = \cdots = \kappa^n_n = 0\} \\
&= P\{t_n \in \Xi_n;\, \kappa^n_1 = \cdots = \kappa^n_n = 0\} > 0.
\end{aligned}
$$

Replacing t by t_n, we may henceforth assume that $t \notin 1 - S_d$.

Reflecting Ξ at the optional time $\tau = 1 - (1 - t)1\{t \in \Xi\}$, we see from Theorem 2.32 (ii) with $f(x) = x$ that

$$
\lim_{h \to 0} h^{-1}\xi[t, t + h] = \infty \quad \text{a.s. on } \{t \in \Xi\}.
$$

Hence, by Fatou's lemma,

$$
\liminf_{h \to 0} h^{-1}E\xi[t, t + h] \geq E \liminf_{h \to 0} h^{-1}\xi[t, t + h] = \infty,
$$

which shows that $E\xi$ has no bounded density on the right of t. By symmetry there is no bounded density on the left side either. Thus, Theorem 6.16 yields $t \in S_d \cap (1 - S_d)$, which contradicts our additional hypothesis $t \notin 1 - S_d$. This shows that indeed $t \notin \Xi$ a.s. for every $t \in (0, 1)$. $\square$

The last theorem suggests that we study instead the left and right distances of Ξ to a fixed point $t \in (0, 1)$. For a striking formulation of our results, we may introduce a cyclically stationary version $\tilde{\Xi}$ of the original set, formed by a periodic continuation of Ξ, followed by a shift by an independent $U(0,1)$ random variable ϑ. Thus, for any $t \in [0,1]$, we have $t \in \tilde{\Xi}$ iff either $t - \vartheta$ or $t - \vartheta + 1$ lies in Ξ.

We begin with a two-sided estimate. Then, for $t \in [0,1]$, we define

$$\sigma_t^- = \sup(\overline{\Xi} \cap [0,t]), \quad \sigma_t^+ = \inf(\overline{\Xi} \cap [t,1]), \quad \delta_t = \sigma_t^+ - \sigma_t^-.$$

The corresponding quantities for the stationary and periodic version $\tilde{\Xi}$ will be denoted by $\tilde{\sigma}_t^\pm$ and $\tilde{\delta}_t$, respectively.

Theorem 6.23 (*endpoint distributions*) *Let Ξ be a regular, extreme, exchangeable set in $[0,1]$ with local intensity p given by Theorem 6.16. Then for any continuity point t of p, we have*

$$\frac{P\{(\sigma_s^-, \sigma_s^+) \in B, \delta_s \leq h\}}{P\{(\tilde{\sigma}_s^-, \tilde{\sigma}_s^+) \in B, \tilde{\delta}_s \leq h\}} \to p_t \quad as \ s \to t \ and \ h \to 0,$$

uniformly for sets $B \in \mathcal{B}([0,1]^2)$ such that the denominator is positive.

To appreciate this result, we note that the probability in the denominator is independent of s and can be easily computed explicitly. A similar remark applies to Theorem 6.25 below.

Our proof is based on an asymptotic relationship, clearly of interest in its own right, between the local time intensity $E\xi$ and the endpoint distributions $\mathcal{L}(T_{\tau_n\pm})$. As before, T denotes the exchangeable process generating the set Ξ, and $\tau_1, \tau_2, \ldots$ are the i.i.d. $U(0,1)$ random variables occurring in the representation of T.

Lemma 6.24 (*endpoint densities*) *For any regular, extreme, exchangeable set Ξ in $[0,1]$, the distributions of the interval endpoints $\sigma_n^\pm = T(\tau_n\pm)$ are absolutely continuous with densities $p_n^\pm$ such that, for any continuity point t of p and times $t_n \to t$, we have*

$$p_n^\pm(t_n) \to p(t).$$

Proof: By symmetry it is enough to consider the left endpoints $\sigma_n^- = T_{\tau_n-}$. Let T^n be the process obtained from T by omission of the n-th jump $\beta_n 1\{\tau_n \leq s\}$, and note that τ_n is $U(0,1)$ and independent of T^n. Writing $\mu_s^n = \mathcal{L}(T_s^n)$, we get by Fubini's theorem

$$\mathcal{L}(T_{\tau_n-}) = \mathcal{L}(T_{\tau_n}^n) = \int_0^1 \mu_s^n \, ds, \quad n \in \mathbb{N}. \tag{26}$$

Letting $a \in (0, \rho)$ be arbitrary and choosing a $c > 0$ satisfying (17), we get as in Lemma 6.15, for the associated characteristic functions $\hat{\mu}_{s,u}^n$ and for any times $s \in (0, \frac{1}{2}]$,

$$
\begin{aligned}
\int_1^\infty \sup_n |\hat{\mu}_{s,u}^n| \, du &\leq \int_1^\infty \sup_n \exp\Big(-su^2 \sum_{k \notin n} \beta_k^2 1\{u\beta_k \leq 1\}/3\Big) \, du \\
&\leq \int_1^\infty \exp(-s(cu^a - 1)) \, du \\
&\leq e^s (sc)^{-1/a} \int_0^\infty \exp(-u^a) \, du \lesssim s^{-1/a}.
\end{aligned}
$$

Combining with the corresponding estimate for $s \in (\frac{1}{2}, 1)$, we get for any $\varepsilon > 0$

$$
\int_\varepsilon^{1-\varepsilon} ds \int_\infty^\infty \sup_n |\hat{\mu}_{s,u}^n| \, du \lesssim \int_\varepsilon^{1/2} s^{-1/a} \, ds \leq \varepsilon^{-1/a} < \infty. \tag{27}
$$

For $s \in (0, 1)$, the measures μ_s^n are again absolutely continuous with continuous densities $p_{s,t}^n$ obtained by Fourier inversion, and we note that

$$
\begin{aligned}
2\pi(p_{s,t}^n - p_{s,t}) &= \int_{-\infty}^\infty e^{-itu} (\hat{\mu}_{s,u}^n - \hat{\mu}_{s,u}) \, du \\
&= s \int_{-\infty}^\infty e^{-itu} \hat{\mu}_{s,u}^n (1 - e^{iu\beta_n}) \, du.
\end{aligned}
$$

Writing

$$
p_t^\varepsilon = \int_\varepsilon^{1-\varepsilon} p_{s,t} \, ds, \qquad p_t^{n,\varepsilon} = \int_\varepsilon^{1-\varepsilon} p_{s,t}^n \, ds,
$$

and using dominated convergence based on (27), we get as $n \to \infty$ for fixed $\varepsilon \in (0, \frac{1}{2}]$

$$
\sup_t |p_t^{n,\varepsilon} - p_t^\varepsilon| \leq \pi^{-1} \int_\varepsilon^{1-\varepsilon} s \, ds \int_{-\infty}^\infty \big| \hat{\mu}_{s,u}^n \sin(u\beta_n/2) \big| \, du \to 0. \tag{28}
$$

Noting that $p_{s,t} = p_{s,t}^n = 0$ for $t \notin (0, 1)$, we get by Fubini's theorem

$$
p_{s,t} = (1-s) p_{s,t}^n + s p_{s,t-\beta_n}^n, \quad s \in (0, 1), \, t \in \mathbb{R}.
$$

Solving for the functions on the right gives

$$
p_{s,t}^n \leq 2 p_{s,t} + 2 p_{s,t+\beta_n}, \quad s, t \in (0, 1). \tag{29}
$$

Fixing a continuity point t_0 of p and putting $p_t^n = \int_0^1 p_{s,t}^n \, ds$, we obtain

$$
\begin{aligned}
|p_t^n - p_{t_0}| &\leq (p_t^n - p_t^{n,\varepsilon}) + |p_t^{n,\varepsilon} - p_t^\varepsilon| + |p_t^\varepsilon - p_{t_0}| \\
&\leq 3(p_t - p_t^\varepsilon) + 2(p_{t+\beta_n} - p_{t+\beta_n}^\varepsilon) + |p_t^{n,\varepsilon} - p_t^\varepsilon| + |p_t - p_{t_0}|.
\end{aligned}
$$

Using (28) and the continuity of p and p^ε, we get

$$
\limsup_{t \to t_0, n \to \infty} |p_t^n - p_{t_0}| \leq 5(p_{t_0} - p_{t_0}^\varepsilon),
$$

which tends to 0 as $\varepsilon \to 0$ since $p_{t_0} < \infty$. Hence, the left-hand side equals 0, which means that $p_t^n \to p_{t_0}$. It remains to note that p^n is a density of $\mathcal{L}(\sigma_n^-)$, since for any $B \in \mathcal{B}[0,1]$

$$
\begin{aligned}
P\{\sigma_n^- \in B\} &= \int_0^1 \mu_s^n B \, ds = \int_0^1 ds \int_B p_{s,t} \, dt \\
&= \int_B dt \int_0^1 p_{s,t} \, ds = \int_B p_t \, dt,
\end{aligned}
$$

in view of (26) and Fubini's theorem. $\qquad\square$

Proof of Theorem 6.23: Given any Borel set $B \subset [0,1]^2$, we define

$$
\begin{aligned}
B_s^n &= \{r \in (s, s + \beta_n);\ (r - \beta_n, r) \in B\}, & s \in [0,1],\ n \in \mathbb{N}, \\
I_h &= \{n;\ \beta_n \le h\}, & h > 0.
\end{aligned}
$$

Using Theorem 6.22 and Lemma 6.24, we get as $s \to t$

$$
\begin{aligned}
P\{(\sigma_s^-, \sigma_s^+) \in B,\ \delta_s \le h\} &= \sum\nolimits_{n \in I_h} P\{T_{\tau_n} \in B_s^n\} \\
&= \sum\nolimits_{n \in I_h} \int_0^1 1\{r \in B_s^n\} p_r^{n+} \, dr \\
&\sim p_t \sum\nolimits_{n \in I_h} \int_0^1 1\{r \in B_s^n\} \, dr \\
&= p_t \sum\nolimits_{n \in I_h} P\{\tilde{T}_{\tau_n} \in B_s^n\} \\
&= p_t \, P\{(\tilde{\sigma}_s^-, \tilde{\sigma}_s^+) \in B,\ \tilde{\delta}_s \le h\},
\end{aligned}
$$

with the obvious interpretation when $p_t = 0$. Here $\tilde{T}_{\tau_n}$ denotes the right endpoint of the n-th interval in $\tilde{\Xi}$, so that $\tilde{T}_{\tau_n} = T_{\tau_n} + \vartheta$ modulo 1. $\qquad\square$

Imposing slightly stronger conditions, we can derive similar estimates for the corresponding one-sided distributions. Equivalently, we may consider the asymptotic probability that Ξ will hit a short interval $I \subset [0,1]$. By $I \to t$ we mean that both endpoints of I tend to t.

Theorem 6.25 *(hitting fixed intervals) Let Ξ be a regular, extreme, exchangeable set in $[0,1]$ with local intensity p. Then for I restricted to subintervals of $[0,1]$, we have*

$$
\frac{P\{\Xi \cap I \ne \emptyset\}}{P\{\tilde{\Xi} \cap I \ne \emptyset\}} \to p_t \quad \text{as } I \to t, \quad t \in (0,1) \text{ a.e. } \lambda.
$$

Proof: Let p be the density in Theorem 6.16, and fix any continuity point t of p, such that p is also continuous at $t \pm \beta_n$ for every $n \in \mathbb{N}$. The exceptional t-set is clearly countable and hence has Lebesgue measure 0. Consider any interval $I = [s, s + l]$, and let $h > 0$ be arbitrary. Note that

$$
\begin{aligned}
P\{\Xi \cap I \ne \emptyset\} &= P\{\sigma_s^+ \le s + l\} \\
&= P\{\sigma_s^+ \le s + l,\ \delta_s \le h\} + P\{\sigma_s^+ \le s + l,\ \delta_s > h\},
\end{aligned}
$$

and similarly for $\tilde{\Xi}$, $\tilde{\sigma}_s^+$, and $\tilde{\delta}_s$. By (29) the densities p^{n+} are bounded near t by some constants $c_n > 0$, and so for small enough $|s - t|$ and l,

$$P\{\sigma_s^+ \le s + l, \, \delta_s > h\} = \sum_{n \notin I_h} \int_s^{s+l} p_r^{n+} dr \le l \sum_{n \notin I_h} c_n.$$

A similar relation holds trivially for $\tilde{\sigma}_s^+$ and $\tilde{\delta}_s$, with c_n replaced by 1. Furthermore, we get by monotone convergence as $l \to 0$

$$l^{-1} P\{\tilde{\sigma}_s^+ \le s + l\} = l^{-1} \sum_n (\beta_n \wedge l) = \int_0^1 (rl^{-1} \wedge 1) \, \beta(dr) \to \infty,$$

and so for fixed $h > 0$

$$\frac{P\{\sigma_s^+ \le s + l, \, \delta_s > h\}}{P\{\tilde{\sigma}_s^+ \le s + l\}} \vee \frac{P\{\tilde{\sigma}_s^+ \le s + l, \, \tilde{\delta}_s > h\}}{P\{\tilde{\sigma}_s^+ \le s + l\}} \to 0. \qquad (30)$$

Now fix any $\varepsilon > 0$. For $|s - t|$, l, and h small enough, Theorem 6.23 yields

$$(1 - \varepsilon) p_t < \frac{P\{\sigma_s^+ \le s + l, \, \delta_s \le h\}}{P\{\tilde{\sigma}_s^+ \le s + l, \, \tilde{\delta}_s \le h\}} < (1 + \varepsilon) p_t.$$

Hence, in view of (30), we get for small enough $l > 0$

$$(1 - \varepsilon) p_t < \frac{P\{\sigma_s^+ \le s + l\}}{P\{\tilde{\sigma}_s^+ \le s + l\}} < (1 + \varepsilon) p_t,$$

and the assertion follows since ε was arbitrary. $\qquad \square$

6.7 Markov Properties

Here we consider processes X that are locally $\mathcal{F}$-homogeneous at every point in some measurable subset C of the state space $(S, \mathcal{S})$. Imposing suitable regularity conditions, we show that such a process X is *conditionally Markov* on C. More precisely, there exist a σ-field $\mathcal{I}$ and some $\mathcal{I}$-measurable random probability measures μ_x, $x \in C$, on the path space of X such that

$$P[\theta_\tau X \in \cdot | \mathcal{F}_\tau, \mathcal{I}] = \mu_{X_\tau} \quad \text{a.s. on } \{X_\tau \in C\}, \qquad (31)$$

for any $\mathcal{F}$-optional time τ. Versions of this statement are true for processes in both discrete and continuous time. In the former case, we note that the paths of X and the filtration $\mathcal{F}$ are automatically continuous.

Theorem 6.26 *(local homogeneity and Markov property) Let X be an rcll process on $\mathbb{R}_+$ or $\mathbb{Z}_+$, adapted to a right-continuous, complete filtration $\mathcal{F}$, taking values in a Polish space S, and recurrent and $\mathcal{F}$-homogeneous at every state in a countable set $C \subset S$. Then X is conditionally, strongly $\mathcal{F}$-Markov on C, given some σ-field $\mathcal{I}$.*

Two lemmas are needed for the proof. We begin with a simple ergodic property, which shows how the directing random elements (α, ν) of a conditionally regenerative process X can be measurably recovered, up to a normalization, from the path of X.

Lemma 6.27 *(ratio limit laws) Let X be an rcll process on $\mathbb{R}_+$, taking values in a Polish space S, and recurrent and $\mathcal{F}$-homogeneous at a state 0 with directing pair (α, ν). Write $\eta_t B$ for the number of complete excursions in the set B up to time t, and let ζ_t be the amount of time in $[0, t]$ spent at 0. Then as $t \to \infty$, a.s. on $\{\nu B \in (0, \infty)\}$ for any measurable sets $A, B \subset D_0$, we have*

$$\frac{\eta_t A}{\eta_t B} \to \frac{\nu A}{\nu B}, \qquad \frac{\zeta_t}{\eta_t B} \to \frac{\alpha}{\nu B}.$$

Proof: Defining T as in (12) in terms of the Itô representation at 0, we see from FMP 22.11 and the law of large numbers for a stationary Poisson process that, a.s. as $s \to \infty$,

$$s^{-1} \eta_{T_s} B \to \nu B \qquad s^{-1} \zeta_{T_s} = \alpha.$$

Hence, a.s. on $\{\nu B \in (0, \infty)\}$,

$$\frac{\eta_{T_s} A}{\eta_{T_s} B} \to \frac{\nu A}{\nu B}, \qquad \frac{\zeta_{T_s}}{\eta_{T_s} B} \to \frac{\alpha}{\nu B},$$

and the assertions follow since T is a.s. non-decreasing with range $\Xi = \{X = 0\}$. $\square$

We also need the following result for conditionally regenerative processes, such as those in Theorem 6.6. Given such a process X, we put $\tau_0 = 0$, and proceed recursively to define the optional times

$$\tau_n = \inf\{t > \tau_{n-1} + 1; \ X_t = 0\}, \quad n \in \mathbb{N}. \tag{32}$$

Let us now introduce the processes

$$Y_n(t) = X((t + \tau_{n-1}) \wedge \tau_n), \quad t \geq 0, \ n \in \mathbb{N}, \tag{33}$$

so that Y_n is the path of X between τ_{n-1} and τ_n, shifted back to the origin. From the conditional regeneration property of X we see that the sequence (Y_n) is exchangeable, hence conditionally i.i.d., where the common conditional distribution μ is a random probability measure on the path space $D = D(\mathbb{R}_+, S)$.

Lemma 6.28 *(directing elements) Let X be an rcll process on $\mathbb{R}_+$, taking values in a Polish space S, and recurrent and $\mathcal{F}$-homogeneous at a state 0 with directing pair (α, ν). Write μ for the directing random measure of the sequence $Y_1, Y_2, \ldots$ given by (32) and (33). Then $\sigma(\mu) = \sigma(\alpha, \nu)$ a.s.*

Proof: Proposition 6.11 shows that X is conditionally regenerative given (α, ν). Iterating the regenerative property, we conclude that the Y_n are conditionally i.i.d. given the same pair. Now Lemma 6.27 shows that (α, ν) is a.s. measurably determined by X, which is in turn measurably determined by $Y_1, Y_2, \ldots$. Hence, Proposition 1.4 (ii) yields $\sigma(\alpha, \nu) = \sigma(\mu)$ a.s. $\square$

Proof of Theorem 6.26: Beginning with the discrete-time case, we may introduce, for any state $x \in C$, the associated recurrence times $\tau_0^x < \tau_1^x < \cdots$ and excursions $Y_0^x, Y_1^x, \ldots$, where Y_0^x is the path of X up to time τ_0^x. By a discrete-time version of Theorem 6.6 there exists a random probability measure ν_x, defined on the space D_x of excursions from x, such that the Y_k^x with $k \geq 1$ are conditionally i.i.d. ν_x. By the law of large numbers, we may recover ν_x from the Y_k^x through the formula

$$n^{-1} \sum\nolimits_{k \leq n} 1_B(Y_k^x) \to \nu_x B \quad \text{a.s.,} \quad B \in \mathcal{S}^n. \tag{34}$$

For any other state $y \in C$, the excursion sequence $Y^y = (Y_k^y)$ can be measurably recovered from $Y^x = (Y_k^x)$, which shows that the associated random measure ν_y is a.s. Y^x-measurable. Indeed, since the limit in (34) is invariant under shifts of the excursion sequence, we see that ν_y is even a shift-invariant function of Y^x. By Corollary 1.6 it follows that ν_y is a.s. ν_x-measurable, and since $x, y \in C$ were arbitrary, we conclude that the σ-field $\mathcal{I} = \sigma(\nu_x)$ is a.s. independent of x.

Now introduce the random probability measures

$$\mu_x = P[\theta_{\tau_0^x} X \in \cdot | \mathcal{I}], \quad x \in C,$$

so that μ_x is the distribution of a process X starting at x and proceeding through a sequence of i.i.d. excursions with the common distribution ν_x. In view of Proposition 2.1, we may conclude from the conditional regeneration property that

$$P[\theta_\tau X \in \cdot | \mathcal{F}_\tau, \mathcal{I}] = \mu_x = \mu_{X_\tau} \quad \text{a.s.,}$$

for any $\mathcal{F}$-optional time τ satisfying $X_\tau = x$ a.s. The general statement in (31) now follows by the local property of conditional expectations (FMP 6.2).

In continuous time and for arbitrary $x \in C$, the shifted process $\theta_{\tau_0^x} X$ can be represented as in Proposition 6.11 in terms of a random pair (α_x, ν_x), where $\alpha_x \geq 0$ is a random variable and ν_x is a σ-finite random measure on the excursion space D_x. To ensure uniqueness, we may take α_x and ν_x to be normalized by (13). Arguing as before, it remains to show that the pairs (α_x, ν_x) with $x \in C$ generate the same σ-field $\mathcal{I}$. Then fix any two states $x \neq y$ in C, put $\tau_0 = \inf\{t \geq 0; X_t = y\}$, and define the processes $Y_1, Y_2, \ldots$ as in (32) and (33), though with the state 0 replaced by y. By Lemma 6.27 the pair (α_x, ν_x) is measurably determined by X and a.s. invariant under arbitrary shifts. Hence, (α_x, ν_x) is also a shift-invariant function of the sequence $Y =$

$(Y_1, Y_2, \ldots)$. Since Y is exchangeable, say with directing random measure μ_y, we conclude from Lemmas 1.6 and 6.28 that

$$\sigma(\alpha_x, \nu_x) \subset \sigma(\mu_y) = \sigma(\alpha_y, \nu_y) \quad \text{a.s.}$$

Interchanging the roles of x and y yields the reverse relation, and so, by combination, the two σ-fields agree a.s. Thus, $\mathcal{I} = \sigma(\alpha_x, \nu_x)$ is a.s. independent of x. $\qquad\square$

We turn to a related property, slightly stronger than local homogeneity. Given an rcll process X in a Polish space S, along with a right-continuous and complete filtration $\mathcal{F}$, we say that X is *globally $\mathcal{F}$-homogeneous* on a measurable subset $C \subset S$, if it is $\mathcal{F}$-adapted and there exists a probability kernel μ_x from C to $D(\mathbb{R}_+, S)$ such that, for any $\mathcal{F}$-optional time τ,

$$P[\theta_\tau X \in \cdot | X_\tau] = \mu_{X_\tau} \quad \text{a.s. on} \quad \{X_\tau \in C\}.$$

Specializing to optional times τ satisfying $X_\tau = x$ a.s. for some $x \in C$, we recover the property of local homogeneity. Let us also say that X is *recurrent* on C if it is recurrent at every state $x \in C$.

Proposition 6.29 (*global homogeneity and Markov property*) *Let X be an rcll process on $\mathbb{R}_+$, adapted to a right-continuous, complete filtration $\mathcal{F}$, taking values in a Polish space S, and recurrent and globally $\mathcal{F}$-homogeneous on a countable subset $C \subset S$ with $|C| \geq 2$. Then X satisfies the strong Markov property on C.*

Proof: Since global homogeneity implies the corresponding local property, Theorem 6.26 shows that X satisfies the conditional strong Markov property on C, given a suitable σ-field $\mathcal{I}$. Furthermore, by Proposition 6.11, we may take $\mathcal{I} = \sigma(\alpha_x, \nu_x)$ for any $x \in C$, where α_x and ν_x denote the random characteristics associated with the excursions from x. We also recall from Lemma 6.27 that $(\alpha_x, \nu_x) = f_x(X)$ a.s. for some shift-invariant, measurable function $f_x; D \to \mathbb{R}_+ \times D_x$.

Now let $y \in C$ be arbitrary, and choose an optional time τ with $X_\tau = y$ a.s. By the invariance of f_x and the local homogeneity at y, we have

$$P \circ (\alpha_x, \nu_x)^{-1} = P[f_x(\theta_\tau X) \in \cdot | X_\tau] = \mu_y \circ f_x^{-1},$$

which means that $\mu_y \circ f_x^{-1}$ is independent of y. Using the global homogeneity on C, we conclude that, for any optional time τ and measurable subsets $A \subset \mathbb{R}_+ \times D_x$ and $B \subset C$,

$$
\begin{aligned}
P\{(\alpha_x, \nu_x) \in A, X_\tau \in B\} &= E[P[f_x(\theta_\tau X) \in A | X_\tau]; X_\tau \in B] \\
&= E[\mu_{X_\tau} \circ f_x^{-1} A; X_\tau \in B] \\
&= E[P\{(\alpha_x, \nu_x) \in A\}; X_\tau \in B] \\
&= P\{(\alpha_x, \nu_x) \in A\} P\{X_\tau \in B\},
\end{aligned}
$$

which shows that $(\alpha_x, \nu_x) \perp\!\!\!\perp X_\tau$.

Since $(\alpha_x, \nu_x) = f_x(X)$ a.s., the event $I = \{(\alpha_x, \nu_x) \in A\}$ is a.s. $\mathcal{F}_\infty$-measurable, and so by FMP 3.16 there exist some sets $I_n \in \mathcal{F}_n$, $n \in \mathbb{N}$, satisfying $P(I \Delta I_n) \to 0$. Fixing any elements $u \neq v$ in C, we may next choose some optional times $\tau_n \geq n$ such that

$$X_{\tau_n} = \begin{cases} u & \text{on } I_n, \\ v & \text{on } I_n^c, \end{cases} \qquad n \in \mathbb{N}.$$

Using the independence of (α_x, ν_x) and X_{τ_n}, we obtain

$$
\begin{aligned}
P(I \cap I_n) &= P\{(\alpha_x, \nu_x) \in A,\ X_{\tau_n} = u\} \\
&= P\{(\alpha_x, \nu_x) \in A\}\, P\{X_{\tau_n} = u\} = P(I)\, P(I_n),
\end{aligned}
$$

and as $n \to \infty$ we get $P(I) = (P(I))^2$, which implies $P\{(\alpha_x, \nu_x) \in A\} = P(I) = 0$ or 1. Since A was arbitrary, it follows that $\mathcal{I} = \sigma(\alpha_x, \nu_x)$ is a.s. trivial, and so the conditional strong Markov property reduces to the usual, unconditional version. $\qquad\square$

6.8   Homogeneity and Independence

The strong Markov property of a process X may be regarded as a combination of the global homogeneity with a condition of conditional independence. Assuming some quite restrictive conditions, we have shown in Proposition 6.29 that the global homogeneity of a process implies the strong Markov property. We will now use entirely different methods to prove, again under suitable regularity conditions, that the two components of the strong Markov property are in fact essentially equivalent. Since different conditions are needed in the two directions, we treat the two implications separately. Let us write $\nu_t = \mathcal{L}(X_t)$.

Theorem 6.30 (*from independence to homogeneity*) *Let X be an $\mathcal{F}$-adapted process in a Borel space S, such that $\mathcal{F}_\tau \perp\!\!\!\perp_{X_\tau} X_{\tau+h}$ for every simple, optional time $\tau < \infty$ and for any $h > 0$. Then each of these conditions implies that X is homogeneous $\mathcal{F}$-Markov:*

  (i) *S is countable and $\mathcal{F}_0$ is non-atomic,*

  (ii) *there exist a set $A \in \mathcal{F}_0$ with $0 < P[A|X] < 1$ a.s. and a σ-finite measure ν on S with $\nu_t \ll \nu$ for all $t \geq 0$.*

Some lemmas are needed for the proof. The following extension property will play a key role throughout our discussion. Given two σ-finite measures μ and ν on the same measurable space, we write $\mu \vee \nu$ for the smallest measure dominating μ and ν, and $\mu \wedge \nu$ for the largest measure bounded by μ and ν. (For existence, see FMP 2.9.)

Lemma 6.31 *(consistency) Consider a family of measurable functions f_i, $i \in I$, between two measurable spaces S and T, along with some σ-finite measures ν and $\nu_i \ll \nu$ on S, $i \in I$, such that $f_i = f_j$ a.e. $\nu_i \wedge \nu_j$ for all $i, j \in I$. Then there exists a measurable function $f : S \to T$ satisfying $f_i = f$ a.e. ν_i for every $i \in I$.*

Proof: First suppose that $I = \mathbb{N}$. For any $i \neq j$ in $\mathbb{N}$, let $\nu_j = \nu_{ij}^a + \nu_{ij}^s$ be the Lebesgue decomposition of ν_j with respect to ν_i (FMP 2.10), so that $\nu_{ij}^a \ll \nu_i$ and $\nu_{ij}^s \perp \nu_i$. By the latter relation, we may choose $A_{ij} \in \mathcal{S}$ such that

$$\nu_{ij}^s A_{ij} = 0, \qquad \nu_i A_{ij}^c = 0, \quad i \neq j,$$

and define recursively

$$B_j = \bigcap_{i<j} B_i^c \cap \bigcap_{k>j} A_{jk}, \quad j \in \mathbb{N}.$$

We may now take $f = f_j$ on B_j for every $j \in \mathbb{N}$, and let $f = f_1$ on $B_\infty = \bigcap_j B_j^c$.

To see that $f = f_j$ a.e. ν_j, it is enough to consider the restriction to B_i for arbitrary $i < j$, since for $k > j$ (including $k = \infty$) we have

$$B_k \subset \bigcap_{i \leq j} B_i^c = \bigcap_{i<j} B_i^c \cap \left(\bigcup_{i<j} B_i \cup \bigcup_{k>j} A_{jk}^c \right) \subset \bigcup_{k>j} A_{jk}^c,$$

and hence

$$\nu_j B_k \leq \sum_{k>j} \nu_j A_{jk}^c = 0.$$

Now $f = f_i = f_j$ a.e. $\nu_i \wedge \nu_j$ on B_i, and it remains to note that $B_i \cdot \nu_j \ll \nu_i \wedge \nu_j$ since $\nu_j \ll \nu_i$ on $A_{ij} \supset B_i$.

For a general index set I, we may assume that $\nu S < \infty$ and $\nu_i \leq \nu$ for all $i \in I$. Define $c = \sup(\nu_{i_1} \vee \cdots \vee \nu_{i_n}) S$, where the supremum extends over all finite subsets $\{i_1, \ldots, i_n\} \subset I$. Then choose a sequence of such collections $\{i_1^n, \ldots, i_n^n\}$ such that $(\nu_{i_1^n} \vee \cdots \vee \nu_{i_n^n}) S \to c$. Listing the elements i_k^n in a single sequence $i_1, i_2, \ldots$, we obtain $(\nu_{i_1} \vee \cdots \vee \nu_{i_n}) S \to c$. Now the result in the countable case yields a measurable function $f : S \to T$ such that $f = f_{i_k}$ a.e. ν_{i_k} for all k. Since $\nu_i \ll \bigvee_k \nu_{i_k}$ for every $i \in I$, the relation $f = f_i$ a.e. ν_i remains generally true. $\qquad\square$

To state the next result, we need to introduce, for suitable optional times $\tau < \infty$ and constants $h > 0$, the associated probability kernels μ_h^τ on the state space S of X satisfying

$$\mu_h^\tau(X_\tau, \cdot) = P[X_{\tau+h} \in \cdot | X_\tau], \quad h > 0.$$

Such conditional distributions exist by FMP 6.3 when S is Borel.

Lemma 6.32 *(transition kernels) Let X be an $\mathcal{F}$-progressive process in a Borel space S, and consider some optional times $\sigma, \tau < \infty$ and a constant $h > 0$ such that $\mathcal{F}_\sigma \perp\!\!\!\perp_{X_\sigma} X_{\sigma+h}$ and $\mathcal{F}_\tau \perp\!\!\!\perp_{X_\tau} X_{\tau+h}$. Then*

$$\mu_h^\sigma(X_\sigma, \cdot) = \mu_h^\tau(X_\tau, \cdot) \quad \text{a.s. on } \{\sigma = \tau\}.$$

Proof: Since $\{\sigma = \tau\} \in \mathcal{F}_\sigma \cap \mathcal{F}_\tau$ by FMP 7.1, and also $\mathcal{F}_\sigma = \mathcal{F}_\tau$ and $X_{\sigma+h} = X_{\tau+h}$ on $\{\sigma = \tau\}$, we see from FMP 6.2 and the required conditional independence that, a.s. on $\{\sigma = \tau\}$ for any measurable set $B \subset S$,

$$
\begin{aligned}
\mu_h^\sigma(X_\sigma, B) &= P[X_{\sigma+h} \in B | X_\sigma] = P[X_{\sigma+h} \in B | \mathcal{F}_\sigma] \\
&= P[X_{\tau+h} \in B | \mathcal{F}_\tau] = P[X_{\tau+h} \in B | X_\tau] = \mu_h^\tau(X_\tau, B).
\end{aligned}
$$

Since S is Borel, we can choose the exceptional null set to be independent of B, and the result follows. $\qquad\square$

Proof of Theorem 6.30: (i) Since S is countable, we have $\nu_t \ll \nu$ for all $t \geq 0$, where ν is the counting measure on S. Now fix any $s, t \geq 0$ and $x \in S$ such that $\nu_s\{x\} \wedge \nu_t\{x\} > 0$. Since $\mathcal{F}_0$ is non-atomic and $EP[X_s = x | \mathcal{F}_0] = \nu_s\{x\} > 0$, we can choose a set $A \in \mathcal{F}_0$ satisfying

$$
A \subset \{P[X_s = x | \mathcal{F}_0] > 0\}, \qquad 0 < PA < \nu_t\{x\},
$$

and we note that

$$
\begin{aligned}
P[X_s = x;\, A] &= E[P[X_s = x | \mathcal{F}_0];\, A] > 0, \\
P[X_t = x;\, A^c] &\geq \nu_t\{x\} - PA > 0.
\end{aligned}
\tag{35}
$$

The random time $\tau = s1_A + t1_{A^c}$ is $\mathcal{F}_0$-measurable and hence $\mathcal{F}$-optional, and so by Lemma 6.32

$$
\begin{aligned}
\mu_h^s(X_s, \cdot) &= \mu_h^\tau(X_\tau, \cdot) \quad \text{a.s. on } A, \\
\mu_h^t(X_t, \cdot) &= \mu_h^\tau(X_\tau, \cdot) \quad \text{a.s. on } A^c.
\end{aligned}
\tag{36}
$$

Specializing to the sets in (35) gives

$$
\mu_h^s(x, \cdot) = \mu_h^\tau(x, \cdot) = \mu_h^t(x, \cdot),
$$

which shows that $\mu_h^s = \mu_h^t$ a.s. $\nu_s \wedge \nu_t$ for any $s, t \geq 0$. Regarding the kernels μ_h^t as measurable functions from S to $\mathcal{M}_1(S)$, we conclude from Lemma 6.31 that $\mu_h^t = \mu_h$ a.e. ν_t for a single probability kernel μ_h on S. Since h was arbitrary, this proves the asserted homogeneity of X.

(ii) For A as stated and any $s, t \geq 0$, we may again put $\tau = s1_A + t1_{A^c}$ and derive (36) by means of Lemma 6.32. Writing $B = \{\mu_h^s(X_s, \cdot) = \mu_h^\tau(X_s, \cdot)\}$, we see from the first relation in (36) that

$$
E[P[A|X];\, B^c] = P(A \setminus B) = 0.
$$

Since $P[A|X] > 0$ a.s., we obtain $PB^c = 0$, which means that $\mu_h^s(X_s, \cdot) = \mu_h^\tau(X_s, \cdot)$ a.s. on the entire Ω. Thus, $\mu_h^s = \mu_h^\tau$ a.e. ν_s. A similar argument yields $\mu_h^t = \mu_h^\tau$ a.e. ν_t, and so by combination $\mu_h^s = \mu_h^t$ a.e. $\nu_s \wedge \nu_t$. Since $\nu_t \ll \nu$ for all $t \geq 0$, a reference to Lemma 6.31 again completes the proof. $\qquad\square$

To state our result in the opposite direction, we need some further notation. Given a suitably measurable process X and an optional time τ, we introduce the *absorption probabilities* a_h^τ, given a.s. by

$$a_h^\tau(X_\tau) = P[\theta_\tau X \in I_h | X_\tau], \quad h > 0,$$

where I_h denotes the set of paths x with $x_{rh} = x_0$ for all $r \in \mathbb{N}$.

Theorem 6.33 *(from homogeneity to independence, Blumenthal and Getoor, Kallenberg) Let X be an $\mathcal{F}$-adapted process in a Borel space S, such that $a_h^\sigma = a_h^\tau$ and $\mu_h^\sigma = \mu_h^\tau$ a.e. $\nu_\sigma \wedge \nu_\tau$ for some $h > 0$ and for any simple, $\mathcal{F}$-optional times $\sigma, \tau < \infty$. Then $\mathcal{F}_\tau \perp\!\!\!\perp_{X_\tau} X_{\tau+h}$ for every such time τ.*

Our proof relies on the following zero–one law for absorption, stated in terms of the notation

$$A_h^\tau = \{x \in S; \, a_h^\tau(x) = 1\}, \quad h > 0. \tag{37}$$

Lemma 6.34 *(zero–one law) Let X be an $\mathcal{F}$-adapted process in a Borel space S, such that $a_h^\sigma = a_h^\tau$ a.e. $\nu_\sigma \wedge \nu_\tau$ for some $h > 0$ and for any simple, $\mathcal{F}$-optional times $\sigma, \tau < \infty$. Then for any such time τ,*

$$a_h^\tau(X_\tau) = 1_{A_h^\tau}(X_\tau) = 1_{I_h}(\theta_\tau X) \quad a.s.$$

Proof: Since S is Borel, we may assume that $S = \mathbb{R}$. Fixing any simple, optional time $\tau < \infty$ and constants $m, n \in \mathbb{N}$, we may introduce the optional times

$$\begin{aligned}
\sigma &= \inf\{t = \tau + rh; \, r \le m, \, [nX_t] \ne [nX_\tau]\}, \\
\tau_k &= \tau + (\sigma - \tau)1\{[nX_\tau] = k\}, \quad k \in \mathbb{Z},
\end{aligned}$$

where r is restricted to $\mathbb{Z}_+$. Then clearly

$$\{[nX_{\tau_k}] = k\} \subset \{[nX_{\tau+rh}] = k, \, r \le m\}, \quad k \in \mathbb{Z}.$$

By Lemma 6.31 we may assume that $a_h^{\tau_k} = a_h^\tau = a_h$ for all $k \in \mathbb{Z}_+$. Writing $b_h = 1 - a_h$, we get

$$\begin{aligned}
E[b_h(X_\tau); \, \theta_\tau X \in I_h] &= \sum_k E[b_h(X_\tau); \, \theta_\tau X \in I_h, \, [nX_\tau] = k] \\
&\le \sum_k E[b_h(X_{\tau_k}); \, [nX_{\tau_k}] = k] \\
&= \sum_k P\{\theta_{\tau_k} X \notin I_h, \, [nX_{\tau_k}] = k\} \\
&\le P\{\theta_\tau X \notin I_h, \, [nX_{\tau+rh}] = k, \, r \le m\} \\
&\le P\{\theta_\tau X \notin I_h, \, \max_{r \le m}|X_{\tau+rh} - X_\tau| \le n^{-1}\},
\end{aligned}$$

and m being arbitrary, we obtain

$$E[b_h(X_\tau); \, \theta_\tau X \in I_h] \le P\{\theta_\tau X \notin I_h, \, \sup_r|X_{\tau+rh} - X_\tau| \le n^{-1}\},$$

which tends to 0 as $n \to \infty$. Hence, $a_h(X_\tau) = 1$ a.s. on $\{\theta_\tau X \in I_h\}$. Combining this with the definition of a_h, we see that also

$$E[a_h(X_\tau); \theta_\tau X \notin I_h] = Ea_h(X_\tau) - P\{\theta_\tau X \in I_h\} = 0,$$

and so $a_h(X_\tau) = 0$ a.s. on $\{\theta_\tau X \notin I_h\}$. This shows that $a_h(X_\tau) = 1_{I_h}(\theta_\tau X)$ a.s. In particular, we have $a_h(X_\tau) \in \{0, 1\}$ a.s., and the remaining relation $a_h(X_\tau) = 1_{A_h}(X_\tau)$ follows by the definition of $A_h = A_h^\tau$. $\qquad\square$

Proof of Theorem 6.33: Since S is Borel, we may assume that $S = \mathbb{R}$. To prove the relation $\mathcal{F}_\tau \perp\!\!\!\perp_{X_\tau} X_{\tau+h}$ for a given optional time $\tau < \infty$ and constant $h > 0$, it is enough to consider optional times σ in the set $\{\tau + rh; r \in \mathbb{Z}_+\}$. Here we have $\nu_\sigma \ll \nu$ for some σ-finite measure ν on S, and so by Lemma 6.31 we may assume that $\mu_h^\sigma = \mu_h$ and $a_h^\sigma = a_h$ a.e. ν_σ for some fixed kernel μ_h and measurable function a_h.

Define the associated set A_h by (37). For any Borel set $B \subset \mathbb{R}$ and event $F \in \mathcal{F}_\tau \cap \{X_\tau \in A_h\}$, we get by Lemma 6.34

$$P[X_{\tau+h} \in B; F] = P[X_{\tau+h} \in B, \theta_\tau X \in I_h; F] = P[X_\tau \in B; F].$$

Hence, a.s. on the set $\{X_\tau \in A_h\}$, we have

$$P[X_{\tau+h} \in B | \mathcal{F}_\tau] = 1_B(X_\tau) = P[X_{\tau+h} \in B | X_\tau] = \mu_h(X_\tau, B). \tag{38}$$

Now assume instead that $F \in \mathcal{F}_\tau \cap \{X_\tau \notin A_h\}$. Fixing any $m, n \in \mathbb{N}$, we define

$$\begin{aligned}
\sigma &= \inf\{t = \tau + rh, \, r \leq m; \, [nX_t] \neq [nX_\tau]\}, \\
\tau_k &= \tau + (\sigma - \tau)1_{F^c}1\{[nX_\tau] = k\}, \quad k \in \mathbb{Z},
\end{aligned}$$

and claim that

$$\{[nX_{\tau_k}] = k\} \cap F^c \subset \{[nX_{\tau+rh}] = k, \, r \leq m\}. \tag{39}$$

In fact, assume the conditions on the left. Then $[nX_\tau] = k$, since the opposite relation would imply $\tau_k = \tau$, which yields the contradiction $[nX_{\tau_k}] = [nX_\tau] \neq k$. It follows that $\tau_k = \sigma$, and so $[nX_\sigma] = [nX_{\tau_k}] = [nX_\tau]$, which yields $[X_{\tau+rh}] = k$ for all $r \leq m$.

Letting $B \in \mathcal{B}(\mathbb{R})$ be arbitrary, we have

$$\begin{aligned}
&|E[1_B(X_{\tau+h}) - \mu_h(X_\tau, B); F]| \\
&\quad = \left| \sum_k E[1_B(X_{\tau+h}) - \mu_h(X_\tau, B); [nX_\tau] = k, F] \right| \\
&\quad = \left| \sum_k E[1_B(X_{\tau_k+h}) - \mu_h(X_{\tau_k}, B); [nX_{\tau_k}] = k, X_{\tau_k} \notin A_h, F] \right| \\
&\quad = \left| \sum_k E[1_B(X_{\tau_k+h}) - \mu_h(X_{\tau_k}, B); [nX_{\tau_k}] = k, X_{\tau_k} \notin A_h, F^c] \right| \\
&\quad \leq \sum_k P[[nX_{\tau_k}] = k, X_{\tau_k} \notin A_h; F^c] \\
&\quad \leq \sum_k P\{[nX_{\tau+rh}] = k, \, r \leq m; \, \theta_\tau X \notin I_h\} \\
&\quad \leq P\{\max_{r \leq m} |X_{\tau+rh} - X_\tau| \leq n^{-1}, \, \theta_\tau X \notin I_h\},
\end{aligned}$$

where the second relation holds since $\tau_k = \tau$ on F, the third relation follows from the fact that

$$E[1_B(X_{\tau_k+h}) - \mu_h(X_{\tau_h}, B); [nX_{\tau_k}] = k, X_{\tau_k} \notin A_h] = 0,$$

by the definition of μ_h, the fourth relation is true since $0 \leq \mu_h \leq 1$, and the fifth relation is a consequence of the set inclusion (39), Lemma 6.34, and the definition of I_h. As $m, n \to \infty$, the right-hand side tends to $P\{\theta_\tau X \in I_h, \theta_\tau X \notin I_h\} = 0$, which shows that

$$P[X_{\tau+h} \in B|\mathcal{F}_\tau] = \mu_h(X_\tau, B) \quad \text{a.s. on } \{X_\tau \notin A_h\}. \tag{40}$$

Combining this with the same relation on $\{X_\tau \in A_h\}$, obtained in (38), we see that the equality in (40) holds a.s. on all of Ω. In particular, this shows that $\mathcal{F}_\tau \perp\!\!\!\perp_{X_\tau} X_{\tau+h}$. $\qquad\square$

Chapter 7

Symmetric Arrays

This chapter, the first of three on multivariate symmetries, is devoted to a study of contractable or exchangeable arrays of arbitrary dimension. The main results are the representation theorems for separately or jointly exchangeable or contractable arrays, established in Sections 7.4 and 7.5. Each representation is stated in terms of a measurable function on a suitable space, and in Section 7.6 we examine when two such functions f and g can be used to represent the same array. Such results are needed for some applications in later chapters.

Those core portions are preceded by three sections of preliminary material, needed for the proofs of the main results. In Section 7.1, we set the stage for the subsequent developments by introducing the necessary terminology and notation. Section 7.2 includes some basic coupling results and a characterization of arrays with independent entries, and in Section 7.3 we prove some general coding principles and a crucial inversion theorem for representations in terms of U-arrays.

We also include some material needed in subsequent chapters. Thus, Section 7.6 also provides a representation theorem for nested sequences of exchangeable arrays, required in Chapter 9, and in Section 7.7 we prove a fundamental relationship between conditional distributions, which will play a crucial role in Chapter 8. The present chapter concludes with a Section 7.8 on symmetric partitions, exhibiting a version, for general symmetries, of Kingman's celebrated paint-box representation, originally established in the exchangeable case.

7.1 Notation and Basic Symmetries

Consider a d-dimensional random array $X_{k_1,\ldots,k_d}$, $k_1,\ldots,k_d \in \mathbb{N}$, with entries in an arbitrary measurable space S. In other words, assume X to be an S-valued random process on the index set $\mathbb{N}^d$. For any mapping $p\colon \mathbb{N} \to \mathbb{N}$, we define the transformed array $X \circ p$ on the same index set by

$$(X \circ p)_{k_1,\ldots,k_d} = X(p_{k_1},\ldots,p_{k_d}), \quad k_1,\ldots,k_d \in \mathbb{N},$$

where, for typographical convenience, we write $X(k) = X_k$ for any $k = (k_1,\ldots,k_d) \in \mathbb{N}^d$. We say that X is *(jointly) exchangeable* if $X \circ p \stackrel{d}{=} X$ for all permutations $p = (p_1, p_2, \ldots)$ of $\mathbb{N}$, and *(jointly) contractable* if the same relation holds whenever p is a sub-sequence of $\mathbb{N}$, so that $p_1 < p_2 < \cdots$. The main purpose of this chapter is to characterize exchangeable and contractable arrays in terms of some general representation formulas.

An array of dimension one is simply an infinite sequence, for which the notions of exchangeability and contractability are equivalent by Theorem 1.1. To see that the stated equivalence fails for $d > 1$, let $\xi_1, \xi_2, \ldots$ be i.i.d. $U(0, 1)$, and define $X_{ij} = \xi_{i \wedge j}$ for all $i, j \in \mathbb{N}$. Then for any subsequence p of $\mathbb{N}$ and for arbitrary $i, j \in \mathbb{N}$, we have

$$(X \circ p)_{ij} = \xi_{p_i \wedge p_j} = \xi_{p_{i \wedge j}} = (\xi \circ p)_{i \wedge j},$$

which implies that X is contractable. If p is instead the transposition $(2, 1, 3, 4, \ldots)$, we get a.s.

$$(X \circ p)_{12} = X_{21} = \xi_1 \neq \xi_2 = X_{23} = (X \circ p)_{13},$$

whereas $X_{12} = X_{13} = \xi_1$ a.s. Hence, in this case $X \circ p \overset{d}{\neq} X$, and X fails to be exchangeable.

When X is exchangeable or contractable, the restrictions of X to the diagonal subsets of $\mathbb{N}^d$ are exchangeable or contractable arrays in their own right. More precisely, letting $\pi = \{I_1, \ldots, I_m\}$ be an arbitrary partition of $\{1, \ldots, d\}$ into disjoint, non-empty subsets $I_1, \ldots, I_m$, we write $\overline{\mathbb{N}}^\pi$ for the set of non-diagonal vectors $k = (k_I; I \in \pi)$. We may form an m-dimensional array X^π with index set $\overline{\mathbb{N}}^\pi$ by taking $X_h^\pi = X_{k_1, \ldots, k_d}$, where $k_i = h_I$ when $i \in I \in \pi$. An array X is clearly exchangeable or contractable iff the non-diagonal parts of the arrays X^π have the same property, in the sense that $(X^\pi \circ p) \overset{d}{=} (X^\pi)$ for all permutations or sub-sequences p of $\mathbb{N}$. Since arrays of the same dimension can be combined into a single array, it is equivalent to consider the stated properties for a sequence of non-diagonal arrays of different dimension. We define the associated index set $\overline{\mathbb{N}}$ to consist of all finite sequences $(k_1, \ldots, k_d)$ with distinct entries $k_1, \ldots, k_d \in \mathbb{N}$, where $d \in \mathbb{Z}_+$ is arbitrary.

In the contractable case, we can reduce the index set even further. Thus, for any sequence $k = (k_1, \ldots, k_d)$ with $k_1 < \cdots < k_d$, we may introduce the finite array

$$\tilde{X}_k = (X_{k \circ p}) = \{X(k_{p_1}, \ldots, k_{p_d})\},$$

where p ranges over all permutations of $\{1, \ldots, d\}$. It is easy to verify that a non-diagonal array X is contractable iff the same property holds for the tetrahedral array $\tilde{X}$. Considering arrays of different dimension, we are led to choose, for our index set in the contractable case, the family $\tilde{\mathbb{N}}$ of all finite sequences $(k_1, \ldots, k_d) \in \mathbb{N}^d$ with strictly increasing entries. Since any such vector $(k_1, \ldots, k_d)$ can be identified with the set $\{k_1, \ldots, k_d\}$, we may think of $\tilde{\mathbb{N}}$, alternatively, as the class of all finite subsets of $\mathbb{N}$. Our discussion justifies that, for exchangeable and contractable arrays X, we choose the index sets to be $\overline{\mathbb{N}}$ and $\widetilde{\mathbb{N}}$, respectively. It also makes sense to consider exchangeable arrays on $\widetilde{\mathbb{N}}$, since every permutation of $\mathbb{N}$ induces, in an obvious way, a permutation of $\widetilde{\mathbb{N}}$.

Our main representations are all in terms of families of i.i.d. $U(0,1)$ random variables, here referred to, for convenience, as *U-sequences*, *U-arrays*, or *U-processes*, depending on the choice of index set. To represent exchangeable arrays on $\overline{\mathbb{N}}$ or contractable arrays on $\widetilde{\mathbb{N}}$, we consider U-arrays $\xi = (\xi_J)$ indexed by $\widetilde{\mathbb{N}}$, where J is an arbitrary, finite subset of $\mathbb{N}$. For any sequences $k = (k_1, \dots, k_d)$ in $\overline{\mathbb{N}}$ or sets $J = \{k_1, \dots, k_d\}$ in $\widetilde{\mathbb{N}}$, we introduce the arrays $\hat{\xi}_k$ or $\hat{\xi}_J$ of variables ξ_I with $I \subset J$, enumerated consistently as follows. Letting $k = (k_1, \dots, k_d)$ and $I = \{i_1, \dots, i_m\} \in 2^d$ with $i_1 < \cdots < i_m$, where 2^d denotes the class of subsets of $\{1, \dots, d\}$, we define $k \circ I = \{k_{i_1}, \dots, k_{i_m}\}$. If instead $J = \{k_1, \dots, k_d\}$ with $k_1 < \cdots < k_d$, we put $J \circ I = k \circ I$, where $k = (k_1, \dots, k_d)$. Writing $|k|$ for the dimension of the vector k and $|J|$ for the cardinality of the set J, we may now define

$$
\begin{aligned}
\hat{\xi}_J &= \{\xi_{J\circ I};\ I \in 2^{|J|}\}, & J \in \widetilde{\mathbb{N}}, \\
\hat{\xi}_k &= \{\xi_{k\circ I};\ I \in 2^{|k|}\}, & k \in \overline{\mathbb{N}}.
\end{aligned}
$$

A representing function f is said to be *symmetric* if $f(\hat{x}_h) = f(\hat{x}_k)$ whenever $h \sim k$, in the sense that h and k are permutations of each other. When the components of $k = (k_1, \dots, k_m)$ are different, we define $\tilde{k}$ as the set $\{k_1, \dots, k_d\}$.

All representations of exchangeable and contractable arrays established in this chapter are of functional type. To illustrate the basic ideas by a simple case, we may state de Finetti's classical theorem in a functional form, obtainable from Theorem 1.1 by elementary methods.

Lemma 7.1 *(functional representation)* *An infinite, random sequence ξ in a Borel space $(S, \mathcal{S})$ is contractable iff*

$$
\xi_j = f(\alpha, \vartheta_j) \ \ a.s., \quad j \in \mathbb{N}, \tag{1}
$$

for a measurable function $f : [0,1]^2 \to S$ and some i.i.d. $U(0,1)$ random variables α and $\vartheta_1, \vartheta_2, \dots$. The directing random measure ν of ξ is then given by

$$
\nu B = \int_0^1 1_B(f(\alpha, t))\, dt, \quad B \in \mathcal{S}. \tag{2}
$$

Proof: Suppose that ξ is conditionally i.i.d. with directing random measure ν. Starting from an arbitrary collection of i.i.d. $U(0,1)$ random variables $\tilde{\alpha}$ and $\tilde{\vartheta}_1, \tilde{\vartheta}_2, \dots$, we may use repeatedly the transfer theorem (FMP 6.10) to certify the existence of some measurable functions g and h between suitable spaces such that

$$
\tilde{\nu} \equiv g(\tilde{\alpha}) \stackrel{d}{=} \nu, \qquad (\tilde{\xi}_1, \tilde{\nu}) \equiv (h(\tilde{\nu}, \tilde{\vartheta}_1), \tilde{\nu}) \stackrel{d}{=} (\xi_1, \nu). \tag{3}
$$

Writing $f(s,t) = h(g(s), t)$, we get

$$\tilde{\xi}_1 = h(\tilde{\nu}, \tilde{\vartheta}_1) = f(g(\tilde{\alpha}), \tilde{\vartheta}_1) = f(\tilde{\alpha}, \tilde{\vartheta}_1),$$

and we may define, more generally,

$$\tilde{\xi}_j = f(\tilde{\alpha}, \tilde{\vartheta}_j) = h(\tilde{\nu}, \tilde{\vartheta}_j), \quad j \in \mathbb{N}.$$

Here the last expression shows that the $\tilde{\xi}_j$ are conditionally independent given $\tilde{\nu}$, and (3) yields their common conditional distribution as $\tilde{\nu}$. Hence, the sequence $\tilde{\xi} = (\tilde{\xi}_j)$ satisfies $(\tilde{\nu}, \tilde{\xi}) \overset{d}{=} (\nu, \xi)$, and we may use the transfer theorem once again (or, more directly, FMP 6.11) to obtain a.s.

$$\nu = g(\alpha), \qquad \xi_j = h(\nu, \vartheta_j), \quad j \in \mathbb{N}. \tag{4}$$

for some i.i.d. $U(0, 1)$ variables α and $\vartheta_1, \vartheta_2, \ldots$. In particular, (1) follows.

To prove (2), let $B \in \mathcal{S}$ be arbitrary, and use (4), Fubini's theorem, and the definition of ν to write

$$\begin{aligned} \nu B &= P[\xi_1 \in B | \nu] = P[h(\nu, \vartheta_j) \in B | \nu] \\ &= \int_0^1 1_B(h(\nu, t))\, dt = \int_0^1 1_B(f(\alpha, t))\, dt. \qquad \square \end{aligned}$$

All multi-variate representations established in the sequel are extensions of this result. Thus, we show in Theorem 7.15 that a random array X on $\tilde{\mathbb{N}}$, taking values in an arbitrary Borel space S, is contractable iff it can be represented in the form

$$X_J = f(\hat{\xi}_J) \text{ a.s.}, \quad J \in \tilde{\mathbb{N}},$$

where ξ is a U-array on $\tilde{\mathbb{N}}$ and f is a measurable function on the infinite union $\bigcup_{d \geq 0} [0, 1]^{2^d}$, taking values in S. A similar result holds for exchangeable arrays. Thus, we prove in Theorem 7.22 that a random array X on $\tilde{\mathbb{N}}$, taking values in a Borel space S, is exchangeable iff it has a representation

$$X_k = f(\hat{\xi}_k) \text{ a.s.}, \quad k \in \overline{\mathbb{N}}.$$

The similarity of the two representations is remarkable. In fact, as an immediate corollary, we note that an array on $\tilde{\mathbb{N}}$ is contractable iff it admits an exchangeable extension to $\overline{\mathbb{N}}$. When X is one-dimensional, the two index sets agree, and the result reduces to Ryll-Nardzewski's theorem—the fact that, for infinite random sequences, the notions of exchangeability and contractability are equivalent. Thus, the quoted result can be regarded as a multi-dimensional version of Ryll-Nardzewski's theorem.

A zero-dimensional array is just a single random element $X = X_\emptyset$, and both representation reduce to $X = f(\xi)$ for some measurable function $f : [0, 1] \to S$ and a $U(0, 1)$ random variable ξ—a well-known result that is easy to prove (cf. FMP 6.10). In one dimension, the arrays are infinite random sequences $X = (X_1, X_2, \ldots)$, and the representation becomes $X_k = f(\xi_\emptyset, \xi_k)$ a.s. for all $k \in \mathbb{N}$, which holds in the exchangeable case by Lemma 7.1. For dimensions two and three, we obtain in both cases the less obvious representations

$$
\begin{aligned}
X_{ij} &= f(\hat{\xi}_{ij}) = f(\xi_\emptyset, \xi_i, \xi_j, \xi_{ij}), \\
X_{ijk} &= f(\hat{\xi}_{ijk}) = f(\xi_\emptyset, \xi_i, \xi_j, \xi_k, \xi_{ij}, \xi_{ik}, \xi_{jk}, \xi_{ijk}),
\end{aligned}
$$

where the indices i, j, and k are all different, and we are writing $\xi_{ij} = \xi_{\{i,j\}}$ and $\xi_{ijk} = \xi_{\{i,j,k\}}$ for convenience.

We shall consider yet another type of symmetry. Given a random array $X_{k_1,\ldots,k_d}$, $k_1, \ldots, k_d \in \mathbb{N}$, we say that X is *separately exchangeable* if $X \circ (p_1, \ldots, p_d) \overset{d}{=} X$ for any permutations $p_1, \ldots, p_d$ of $\mathbb{N}$, where

$$
(X \circ (p_1, \ldots, p_d))_{k_1,\ldots,k_d} = X(p_{1,k_1}, \ldots, p_{d,k_d}), \quad k_1, \ldots, k_d \in \mathbb{N}.
$$

In other words, X is now required to be exchangeable in the one-dimensional sense, separately in each index. For arrays ξ on $\mathbb{Z}_+^d$ and for arbitrary vectors $k = (k_1, \ldots, k_d) \in \mathbb{N}^d$, we write $\hat{\xi}_k$ for the 2^d-dimensional vector with entries ξ_{h_j}, where $h_j = k_j$ or 0 for all j. As before, the notation carries over to measurable functions on suitable domains.

As a simple consequence of the previous representations, we prove in Corollary 7.23 that a Borel-space valued array X on $\mathbb{N}^d$ is separately exchangeable iff it has a representation

$$
X_k = f(\hat{\xi}_k) \text{ a.s.}, \quad k \in \mathbb{N}^d,
$$

in terms of a U-array ξ on $\mathbb{Z}_+^d$ and a measurable function f on $[0,1]^{2^d}$. The representations in dimensions zero and one are the same as before, but in higher dimensions they are fundamentally different. Thus, in dimensions two and three, we have the representations

$$
\begin{aligned}
X_{ij} &= f(\hat{\xi}_{ij}) = f(\xi_{00}, \xi_{i0}, \xi_{0j}, \xi_{ij}), \\
X_{ijk} &= f(\hat{\xi}_{ijk}) = f(\xi_{000}, \xi_{i00}, \xi_{0j0}, \xi_{00k}, \xi_{ij0}, \xi_{i0k}, \xi_{0jk}, \xi_{ijk}),
\end{aligned}
$$

where $i, j, k \in \mathbb{N}$ are arbitrary.

The intricate proofs of the mentioned representations occupy much of the remainder of the chapter. Thus, after two more sections of preliminary material, the representation of contractable arrays is proved in Section 7.4, followed by those for separately or jointly exchangeable arrays established in Section 7.5. In Section 7.6, we examine to what extent those basic representations are unique. Section 7.7 in devoted to some conditional properties of exchangeable arrays, needed in the next chapter, and Section 7.8 deals with exchangeable and related partitions.

To simplify our statements, we assume throughout the chapter that all random arrays take values in a general Borel space, unless otherwise specified.

7.2 Coupling, Extension, and Independence

Here and in the next section, we consider some preliminary results needed for the proofs of the main theorems of the chapter. The statements in the present section include some coupling results for exchangeable or contractable arrays and a characterization of representable arrays of the mentioned type with independent entries.

Given an arbitrary index set $T \subset \mathbb{R}$, we write $\tilde{T}$ for the class of all finite subsets of T. We say that a random array X on $\tilde{T}$ is *contractable* if $X \circ p \overset{d}{=} X \circ q$ for any finite, increasing sequences $p = (p_1, p_2, \ldots)$ and $q = (q_1, q_2, \ldots)$ in T of equal length. When $T = \mathbb{N}$, this clearly reduces to the previous notion of contractability. We proceed to show how pairs of exchangeable or contractable arrays can be combined through a suitable coupling.

Lemma 7.2 *(coupling by conditional independence)*

(i) *Let X, Y, and Z be random arrays on $\mathbb{N}$ with $X \perp\!\!\!\perp_Y Z$, and suppose that the pairs (X, Y) and (Y, Z) are exchangeable. Then so is the triple (X, Y, Z).*

(ii) *Let X, Y, and Z be random arrays on $\widetilde{\mathbb{Q}}$ with $X \perp\!\!\!\perp_Y Z$, and suppose that the pairs (X, Y) and (Y, Z) are contractable. Then so is the triple (X, Y, Z).*

Proof: (i) By FMP 6.3 there exists a probability kernel μ between suitable spaces satisfying

$$\mu(Y, \cdot) = P[X \in \cdot | Y] \quad \text{a.s.}$$

Let p be a finite permutation of $\mathbb{N}$, and let A and B be measurable sets in the range spaces of X and (Y, Z). Noting that $\sigma(Y \circ p) = \sigma(Y)$ since p is invertible, and using the assumed conditional independence (twice) and the exchangeability of (X, Y) and (Y, Z), we get

$$
\begin{aligned}
P\{(X, Y, Z) \circ p \in A \times B\} \\
= \; & E[P[X \circ p \in A | (Y, Z) \circ p]; \, (Y, Z) \circ p \in B] \\
= \; & E[P[X \circ p \in A | Y \circ p]; \, (Y, Z) \circ p \in B] \\
= \; & E[\mu(Y \circ p, A); \, (Y, Z) \circ p \in B] \\
= \; & E[\mu(Y, A); \, (Y, Z) \in B] \\
= \; & E[P[X \in A | Y, Z]; \, (Y, Z) \in B] \\
= \; & P\{(X, Y, Z) \in A \times B\},
\end{aligned}
$$

which extends immediately to $(X, Y, Z) \circ p \overset{d}{=} (X, Y, Z)$. The asserted exchangeability now follows since p was arbitrary.

(ii) For any $a < b$ in $\mathbb{Q}_+$ and $t \in \mathbb{Q}$, we introduce on $\mathbb{Q}$ the functions

$$
\begin{aligned}
p_{a,t}(x) &= x + a\mathbf{1}\{x > t\}, \\
p_a^b(x) &= x + a(1 - b^{-1}x)\mathbf{1}\{0 < x < b\},
\end{aligned}
$$

and put $p_a = p_{a,0}$. Then $p_a^b \circ p_b = p_b$, and since $(X,Y) \circ p_a^b \overset{d}{=} (X,Y)$ by the contractability of (X,Y), we get

$$
\begin{aligned}
(X \circ p_b, Y) &\overset{d}{=} (X \circ p_a^b \circ p_b,\ Y \circ p_a^b) \\
&= (X \circ p_b,\ Y \circ p_a^b).
\end{aligned}
$$

Hence, Lemma 1.3 yields

$$(X \circ p_b) \underset{Y \circ p_a^b}{\perp\!\!\!\perp} Y,$$

and since $\sigma\{Y \circ p_a^b\} = \sigma\{Y \circ p_a\}$ and $\sigma\{Y \circ p_b;\ b > a\} = \sigma\{Y \circ p_a\}$, we can use a monotone-class argument to obtain the extended relation

$$(X \circ p_a) \underset{Y \circ p_a}{\perp\!\!\!\perp} Y.$$

Since also $(X \circ p_a) \perp\!\!\!\perp_Y (Z \circ p_a)$ by hypothesis, the chain rule in FMP 6.8 gives

$$(X \circ p_a) \underset{Y \circ p_a}{\perp\!\!\!\perp} (Z \circ p_a).$$

Combining this with the assumed relations

$$X \perp\!\!\!\perp_Y Z, \qquad (X,Y) \circ p_a \overset{d}{=} (X,Y), \qquad (Y,Z) \circ p_a \overset{d}{=} (Y,Z),$$

we see as before that

$$(X,Y,Z) \circ p_a \overset{d}{=} (X,Y,Z), \quad a \in \mathbb{Q}_+.$$

Re-labeling $\mathbb{Q}$, if necessary, we get the same relation for the more general mappings $p_{a,t}$. Since for any sets $I, J \in \widetilde{\mathbb{Q}}$ with $|I| = |J|$ there exist some compositions p and q of such functions satisfying $p(I) = q(J)$, we obtain the required contractability. $\qquad\square$

In order to combine two or more contractable arrays on $\widetilde{\mathbb{N}}$ into a single contractable array, using the last result, we need first to extend the original arrays to $\widetilde{\mathbb{Q}}$. The required procedure is straightforward:

Lemma 7.3 *(extension) Every contractable array X on $\widetilde{\mathbb{N}}$ can be extended to a contractable array $\overline{X}$ on $\widetilde{\mathbb{Q}}$. The distribution of $\overline{X}$ is then uniquely determined by that of X.*

Proof: To construct the finite-dimensional distributions of $\overline{X}$, fix any $J_1, \ldots, J_n \in \widetilde{\mathbb{Q}}$, put $J = \bigcup_k J_k$, and let f be the unique, increasing bijection from J to $I = \{1, \ldots, |J|\}$. Define

$$\mu_{J_1,\ldots,J_n} = \mathcal{L}(X \circ f(J_1), \ldots, X \circ f(J_n)). \tag{5}$$

To prove the consistency of these measures, put $J' = \bigcup_{k<n} J_k$, and let g and p denote the increasing bijections from J' to $I' = \{1, \ldots, |J'|\}$ and from

I' to $\bigcup_{k<n} f(J_k)$, respectively. Noting that $p \circ g = f$ on J' and using the contractability of X, we get

$$
\begin{aligned}
\mu_{J_1,\ldots,J_{n-1}} &= \mathcal{L}(X \circ g(J_1), \ldots, X \circ g(J_{n-1})) \\
&= \mathcal{L}(X \circ p \circ g(J_1), \ldots, X \circ p \circ g(J_{n-1})) \\
&= \mathcal{L}(X \circ f(J_1), \ldots, X \circ f(J_{n-1})) \\
&= \mu_{J_1,\ldots,J_n}(\cdot \times S_n),
\end{aligned}
$$

where S_n denotes the range space of X_{J_n}.

By the Daniell–Kolmogorov theorem (FMP 6.14) there exists a process Y on $\widetilde{\mathbb{Q}}$ with finite-dimensional distributions as in (5). Letting Y' denote the restriction of Y to $\widetilde{\mathbb{N}}$, we see from (5) that $Y' \overset{d}{=} X$, and so the transfer theorem (FMP 6.10) yields the existence of a process $\overline{X}$ on $\widetilde{\mathbb{Q}}$ with $(\overline{X}, X) \overset{d}{=} (Y, Y')$. In particular, $\overline{X}$ is a.s. an extension of X. The contractability of $\overline{X}$ follows from (5), and the uniqueness of $\mathcal{L}(\overline{X})$ is clear from the fact that, for any contractable extension of X, the finite-dimensional distributions must agree with those in (5). □

Combining the last two lemmas, we obtain a useful coupling of contractable arrays.

Corollary 7.4 *(coupling of contractable arrays)* Let X, $Y \overset{d}{=} Y'$, and Z' be random arrays on $\widetilde{\mathbb{N}}$, such that the pairs (X, Y) and (Y', Z') are contractable. Then there exists an array Z on $\widetilde{\mathbb{N}}$ with $(Y, Z) \overset{d}{=} (Y', Z')$, such that the triple (X, Y, Z) is again contractable.

Proof: By Lemma 7.3 we may extend (X, Y) and (Y', Z') to contractable arrays $(\overline{X}, \overline{Y})$ and $(\overline{Y}', \overline{Z}')$ on $\widetilde{\mathbb{Q}}$, and by the uniqueness part of the same lemma we note that $\overline{Y} \overset{d}{=} \overline{Y}'$. Hence, the transfer theorem (FMP 6.10 and 6.13) yields an array $\overline{Z}$ on $\widetilde{\mathbb{Q}}$ such that $(\overline{Y}, \overline{Z}) \overset{d}{=} (\overline{Y}', \overline{Z}')$ and $\overline{X} \perp\!\!\!\perp_{\overline{Y}} \overline{Z}$. The triple $(\overline{X}, \overline{Y}, \overline{Z})$ is then contractable by Lemma 7.2, and the same thing is true for the restriction (X, Y, Z) to $\widetilde{\mathbb{N}}$. □

We may next record a simple independence property of extensions.

Lemma 7.5 *(conditional independence of extension)* Let (X, ξ) be a contractable array on $\widetilde{\mathbb{Z}}$ with restriction (Y, η) to $\widetilde{\mathbb{N}}$, and assume that ξ has independent entries. Then $Y \perp\!\!\!\perp_\eta \xi$.

Proof: Writing

$$
p_n(k) = k - n1\{k \le 0\}, \quad k \in \mathbb{Z}, \, n \in \mathbb{N},
$$

and using the contractability of (X, ξ), we obtain

$$
(Y, \xi \circ p_n) \overset{d}{=} (Y, \xi), \quad n \in \mathbb{N}.
$$

Since also $\sigma\{\xi \circ p_n\} \subset \sigma\{\xi\}$, Lemma 1.3 yields

$$Y \underset{\xi \circ p_n}{\perp\!\!\!\perp} \xi, \quad n \in \mathbb{N}.$$

Next we note that $\bigcap_n \sigma\{\xi \circ p_n\} = \sigma\{\eta\}$ a.s. by the extended version of Kolmogorov's zero-one law in FMP 7.25. The assertion now follows by martingale convergence. □

To state the next result, define the *shell σ-field* $\mathcal{S}(X)$ of a random array X on $\mathbb{N}^d$ by

$$\mathcal{S}(X) = \bigcap_n \sigma\{X_k; \max_j k_j \geq n\}.$$

Let us say that an array X on $\mathbb{Z}_+^d$ is *separately exchangeable on $\mathbb{N}^d$* if $X \overset{d}{=} X \circ (p_1, \ldots, p_d)$ for any finite permutations $p_1, \ldots, p_d$ on $\mathbb{Z}_+$ that leave 0 invariant.

Lemma 7.6 *(shell σ-field, Aldous, Hoover) Let the random array X on $\mathbb{Z}_+^d$ be separately exchangeable on $\mathbb{N}^d$ with restriction X' to $\mathbb{N}^d$, and write $\mathcal{S}(X')$ for the shell σ-field of X'. Then*

$$X' \underset{\mathcal{S}(X')}{\perp\!\!\!\perp} (X \setminus X').$$

Proof: When $d = 1$, the array X' is just an infinite random sequence $\xi = (\xi_1, \xi_2, \ldots)$, which is exchangeable over $\eta = X_0$, and $\mathcal{S}(X')$ reduces to the tail σ-field $\mathcal{T}$ of ξ. Applying de Finetti's theorem to both ξ and (ξ, η), for the latter in the extended form of Proposition 2.1, we get a.s.

$$P[\xi \in \cdot | \mu] = \mu^\infty, \qquad P[\xi \in \cdot | \nu, \eta] = \nu^\infty,$$

for some random probability measures μ and ν on the state space S. By Proposition 1.4 (i) and (ii) it follows that $\mu = \nu$ a.s. and $\xi \perp\!\!\!\perp_\nu \eta$, and it remains to note that a.s. $\sigma(\nu) = \mathcal{T}$ by Corollary 1.6.

Proceeding by induction, assume the statement to be true in dimension $d - 1$, and turn to the case of a d-dimensional array X. Define an S^∞-valued array Y on $\mathbb{Z}_+^{d-1}$ by

$$Y_m = (X_{m,k}; k \in \mathbb{Z}_+), \quad m \in \mathbb{Z}_+^{d-1},$$

and note that Y is separately exchangeable on $\mathbb{N}^{d-1}$. Write Y' for the restriction of Y to $\mathbb{N}^{d-1}$ and X'' for the restriction of X to $\mathbb{N}^{d-1} \times \{0\}$. By the induction hypothesis,

$$(X', X'') = Y' \underset{\mathcal{S}(Y')}{\perp\!\!\!\perp} (Y \setminus Y') = X \setminus (X', X''),$$

and since

$$\mathcal{S}(Y') \subset \mathcal{S}(X') \vee \sigma(X'') \subset \sigma(Y'),$$

we conclude that

$$X' \underset{\mathcal{S}(X'), X''}{\perp\!\!\!\perp} (X \setminus X'). \tag{6}$$

Next we may apply the result for $d = 1$ to the S^∞-valued sequence

$$Z_k = (X_{m,k}; \ m \in \mathbb{N}^{d-1}), \quad k \in \mathbb{Z}_+,$$

to obtain

$$X' = (Z \setminus Z_0) \underset{\mathcal{S}(Z)}{\perp\!\!\!\perp} Z_0 = X''.$$

Since clearly

$$\mathcal{S}(Z) \subset \mathcal{S}(X') \subset \sigma(X'),$$

we conclude that

$$X' \underset{\mathcal{S}(X')}{\perp\!\!\!\perp} X''.$$

Combining this with (6) and using the chain rule in FMP 6.8, we obtain the desired relation. □

For the present purposes, we need only the following corollary. Given an array ξ on $\mathbb{Z}_+^d$ and a set $J \in 2^d$, we write ξ^J for the sub-array with index set $\mathbb{N}^J \times \{0\}^{J^c}$, and put $\hat{\xi}^J = (\xi^I; \ I \subset J)$. We also define $\xi^m = (\xi^J; \ |J| = m)$ and $\hat{\xi}^m = (\xi^J; \ |J| \le m)$. The conventions for arrays on $\tilde{\mathbb{N}}$ are similar. Put $\tilde{\mathbb{N}}_d = \{J \subset \mathbb{N}; \ |J| \le d\}$ and $\Delta\tilde{\mathbb{N}}_d = \{J \subset \mathbb{N}; \ |J| = d\}$, and similarly for $\mathbb{Q}$.

Proposition 7.7 *(independent entries)*

(i) *Let ξ be a random array on $\mathbb{Z}_+^d$ with independent entries, separately exchangeable on $\mathbb{N}^d$. Given a measurable function f, define*

$$\eta_k = f(\hat{\xi}_k), \quad k \in \mathbb{Z}_+^d.$$

Then η has again independent entries iff

$$\eta_k \perp\!\!\!\perp (\hat{\xi}_k \setminus \xi_k), \quad k \in \mathbb{Z}_+^d, \tag{7}$$

in which case

$$\eta^J \perp\!\!\!\perp (\xi \setminus \xi^J), \quad J \in 2^d. \tag{8}$$

(ii) *Let ξ be a contractable array on $\tilde{\mathbb{N}}$ with independent entries. Given a measurable function f, define*

$$\eta_J = f(\hat{\xi}_J), \quad J \in \tilde{\mathbb{N}}. \tag{9}$$

Then η has again independent entries iff

$$\eta_J \perp\!\!\!\perp (\hat{\xi}_J \setminus \xi_J), \quad J \in \tilde{\mathbb{N}}, \tag{10}$$

in which case

$$\eta^d \perp\!\!\!\perp (\xi \setminus \xi^d), \quad d \in \mathbb{Z}_+. \tag{11}$$

Proof: (i) First assume (7). Combining with the relation $\hat{\xi}_k \perp\!\!\!\perp (\xi \setminus \hat{\xi}_k)$ gives $\eta_k \perp\!\!\!\perp (\xi \setminus \xi_k)$, which implies

$$\eta_k \perp\!\!\!\perp (\xi \setminus \xi_k, \hat{\eta}^{|k|} \setminus \eta_k), \quad k \in \mathbf{Z}_+^d,$$

where $|k|$ denotes the number of positive k-components. As in FMP 3.8, this yields both the independence of the η-entries and relation (8).

Conversely, suppose that η has independent entries. Then for any $J \neq \emptyset$ in 2^d, the shell σ-field $\mathcal{S}(\eta^J)$ is trivial by Kolmogorov's zero-one law. Applying Lemma 7.6 to the $\mathbf{Z}_+^J$-indexed array $(\eta^J, \hat{\xi}^J \setminus \xi^J)$, which is clearly separately exchangeable on $\mathbf{N}^J$, we obtain $\eta^J \perp\!\!\!\perp (\hat{\xi}^J \setminus \xi^J)$, which implies (7).

(ii) First assume (10). Since also $\hat{\xi}_J \perp\!\!\!\perp (\xi \setminus \hat{\xi}_J)$, we obtain $\eta_J \perp\!\!\!\perp (\xi \setminus \xi_J)$, which implies

$$\eta_J \perp\!\!\!\perp (\xi \setminus \xi^d, \hat{\eta}^d \setminus \eta^J), \quad J \in \widetilde{\mathbf{N}}_d, \ d \in \mathbf{Z}_+.$$

This shows that η has independent entries and satisfies (11).

Conversely, suppose that η has independent entries. Extend ξ to a contractable array on $\widetilde{\mathbf{Q}}$, and then extend η accordingly to $\widetilde{\mathbf{Q}}$ by means of (9). Fixing any $d \in \mathbf{Z}_+$, we define

$$(\xi'_k, \eta'_k) = (\xi, \eta)_{\{i - k_i^{-1}; \, i \leq d, \, k_i > 0\}}, \quad k \in \mathbf{Z}_+^d.$$

Then ξ' is clearly separately exchangeable on $\mathbf{N}^d$ with independent entries, and (9) yields $\eta'_k = f(\hat{\xi}'_k)$ for all $k \in \mathbf{Z}_+^d$. Being part of η, the array η' has again independent entries, and so in view of (i) we have $\eta'_k \perp\!\!\!\perp (\hat{\xi}'_k \setminus \xi'_k)$ for every $k \in \mathbf{N}^d$, which is equivalent to $\eta_J \perp\!\!\!\perp (\hat{\xi}_J \setminus \xi_J)$ for suitable sets $J \in \Delta\widetilde{\mathbf{Q}}_d$. The general relation (10) now follows by the contractability of ξ. $\qquad \square$

7.3 Coding and Inversion

Here we establish some further propositions, needed to prove the main results of the chapter. The former include some coding results, for random elements invariant under finite or compact groups, as well as two crucial inversion theorems for representable arrays with independent entries. We begin with a general coding property for random arrays.

Lemma 7.8 *(coding) Let $\xi = (\xi_j)$ and $\eta = (\eta_j)$ be random arrays on an arbitrary index set I satisfying*

$$(\xi_i, \eta_i) \stackrel{d}{=} (\xi_j, \eta_j), \quad i, j \in I, \tag{12}$$

$$\xi_j \perp\!\!\!\perp_{\eta_j} (\xi \setminus \xi_j, \eta), \quad j \in I. \tag{13}$$

Then there exist some measurable functions f and g on suitable spaces, such that for any U-array $\vartheta \perp\!\!\!\perp (\xi, \eta)$ on I, the random variables

$$\zeta_j = g(\xi_j, \eta_j, \vartheta_j), \quad j \in I, \tag{14}$$

form a U-array $\zeta \perp\!\!\!\perp \eta$ satisfying

$$\xi_j = f(\eta_j, \zeta_j) \quad a.s., \quad j \in I. \tag{15}$$

Proof: Fix any $i \in I$. By the transfer theorem in FMP 6.10, we may choose a measurable function f between suitable spaces such that

$$(\xi_i, \eta_i) \overset{d}{=} (f(\eta_i, \vartheta_i), \eta_i).$$

By another application of the same result, we may next choose some measurable functions g and h, such that the random elements

$$\zeta_i = g(\xi_i, \eta_i, \vartheta_i), \qquad \tilde{\eta}_i = h(\xi_i, \eta_i, \vartheta_i)$$

satisfy

$$(\xi_i, \eta_i) = (f(\tilde{\eta}_i, \zeta_i), \tilde{\eta}_i) \text{ a.s.}, \qquad (\tilde{\eta}_i, \zeta_i) \overset{d}{=} (\eta_i, \vartheta_i).$$

In particular, we get $\tilde{\eta}_i = \eta_i$ a.s., and so (15) holds for $j = i$ with ζ_i as in (14). Furthermore, ζ_i is $U(0,1)$ and independent of η_i. The two statements extend by (12) to arbitrary $j \in I$.

Combining (13) with the relations $\vartheta \perp\!\!\!\perp (\xi, \eta)$ and $\vartheta_j \perp\!\!\!\perp (\vartheta \setminus \vartheta_j)$, we see that

$$(\xi_j, \eta_j, \vartheta_j) \perp\!\!\!\perp_{\eta_j} (\xi \setminus \xi_j, \eta, \vartheta \setminus \vartheta_j), \quad j \in I,$$

which implies

$$\zeta_j \perp\!\!\!\perp_{\eta_j} (\zeta \setminus \zeta_j, \eta), \quad j \in I.$$

Combining this with the relation $\zeta_j \perp\!\!\!\perp \eta_j$ and using the chain rule in FMP 6.8, we conclude that $\zeta_j \perp\!\!\!\perp (\zeta \setminus \zeta_j, \eta)$. By iteration (FMP 3.8) it follows that the array η and the elements ζ_j, $j \in I$, are all independent. Thus, ζ is indeed a U-array independent of η. $\qquad\square$

We turn to a general statement about symmetric representations, where the invariance is defined in terms of a finite group G, acting measurably on a Borel space S. Introduce the associated stabilizers (symmetry groups)

$$G_s = \{g \in G;\ gs = s\}, \quad s \in S.$$

Proposition 7.9 *(symmetric coding) Let G be a finite group, acting measurably on a Borel space S, and consider some random elements ξ and η in S such that $g(\xi, \eta) \overset{d}{=} (\xi, \eta)$ for all $g \in G$ and $G_\eta \subset G_\xi$ a.s. Then there exist some measurable functions $f: S \times [0,1] \to S$ and $b: S^2 \times [0,1] \to [0,1]$ with*

$$b(gx, gy, t) = b(x, y, t), \quad x, y \in S,\ g \in G,\ t \in [0,1], \tag{16}$$

such that whenever $\vartheta \perp\!\!\!\perp (\xi, \eta)$ is $U(0,1)$, the random variable $\zeta = b(\xi, \eta, \vartheta)$ is $U(0,1)$ with $\zeta \perp\!\!\!\perp \eta$ and satisfies

$$g\xi = f(g\eta, \zeta) \text{ a.s.}, \quad g \in G. \tag{17}$$

Our construction proceeds in steps. First we need to construct an invariant function $h: S \to G$. Here we write $h_s = h(s)$ and let h_s^{-1} denote the inverse of h_s as an element of G.

Lemma 7.10 *(invariant function) Let G be a finite group, acting measurably on a Borel space S. Then there exists a measurable function $h\colon S \to G$ such that*

$$h_{gs}gs = h_s s, \quad g \in G, \ s \in S. \tag{18}$$

For any mapping $b\colon S \to S$, the function

$$f(s) = h_s^{-1}b(h_s s), \quad s \in S, \tag{19}$$

satisfies

$$G_s \subset G_{f(s)} \quad \Rightarrow \quad f(gs) = gf(s), \quad g \in G. \tag{20}$$

Proof: We may assume that $S \in \mathcal{B}[0,1]$. For any $s \in S$, let M_s denote the set of elements $g \in G$ maximizing gs. Then for any $g \in G$ and $s \in S$,

$$M_{gs}g = \{hg;\ h \in M_{gs}\} = \{hg \in M_s;\ h \in G\} = \{h \in M_s\} = M_s. \tag{21}$$

We also note that M_s is a left coset of G_s, in the sense that

$$M_s = hG_s = \{hg;\ g \in G_s\}, \quad s \in S, \ h \in M_s. \tag{22}$$

In fact, for fixed s and h we have $g \in M_s$ iff $gs = hs$, which is equivalent to $h^{-1}g \in G_s$ and hence to $g \in hG_s$.

Fixing an enumeration $g_1, \ldots, g_m$ of G, we define h_s to be the first element g_k belonging to M_s. This clearly defines a measurable function $h\colon S \to G$. Using (21) and (22), we get for any s and g

$$h_{gs}g \in M_{gs}g = M_s = h_s G_s,$$

and so

$$h_s^{-1}h_{gs}g \in G_s, \quad s \in S, \ g \in G,$$

which implies (18).

Now consider an arbitrary mapping $b\colon S \to S$, and define the associated function f by (19). To prove (20), fix any $g \in G$ and an $s \in S$ with $G_s \subset G_{f(s)}$. By (18) we have

$$g^{-1}h_{gs}^{-1}h_s \in G_s \subset G_{f(s)},$$

and so by (18) and (19),

$$\begin{aligned}
f(gs) &= h_{gs}^{-1}b(h_{gs}gs) = h_{gs}^{-1}b(h_s s) \\
&= h_{gs}^{-1}h_s f(s) = gf(s).
\end{aligned} \qquad \square$$

Using the function h of the last lemma along with a Haar measure λ on G, we can now construct the desired function f satisfying (17).

Lemma 7.11 *(representing function) Let G be a compact metric group, acting measurably on a Borel space S. Let ξ and η be random elements in S such that $G_\eta \subset G_\xi$ a.s. and $g(\xi, \eta) \stackrel{d}{=} (\xi, \eta)$ for all $g \in G$, and consider a measurable function $h\colon S \to G$ satisfying (18). Then there exist a measurable function $f\colon S \times [0,1] \to S$ and a $U(0,1)$ random variable $\zeta \perp\!\!\!\perp \eta$ such that (17) holds outside a fixed P-null set. Here ζ can be replaced by any random variable ζ' with $(\xi, \eta, \zeta) \stackrel{d}{=} (\xi, \eta, \zeta')$.*

Proof: By (18) we have

$$h_\eta^{-1} h_{g\eta} g \in G_\eta \subset G_\xi, \quad g \in G,$$

outside a fixed null set. Hence, the random elements $\tilde{\xi} = h_\eta \xi$ and $\tilde{\eta} = h_\eta \eta$ satisfy

$$(\tilde{\xi}, \tilde{\eta}) = h_\eta(\xi, \eta) = h_{g\eta} g(\xi, \eta), \quad g \in G. \tag{23}$$

Writing λ for the normalized Haar measure (FMP 2.27) on G, we introduce a random element $\gamma \perp\!\!\!\perp (\xi, \eta)$ in G with distribution λ. Using Fubini's theorem and the invariance of $\mathcal{L}(\xi, \eta)$ and $\mathcal{L}(\gamma)$, we get

$$\gamma(\xi, \eta) \stackrel{d}{=} (\xi, \eta), \qquad (\gamma h_\eta, \xi, \eta) \stackrel{d}{=} (\gamma, \xi, \eta). \tag{24}$$

By (23) and (24) we have

$$
\begin{aligned}
(\eta, \tilde{\xi}, \tilde{\eta}) &= (\eta, h_\eta \xi, h_\eta \eta) \\
&\stackrel{d}{=} (\gamma\eta, h_{\gamma\eta} \gamma \xi, h_{\gamma\eta} \gamma \eta) \\
&= (\gamma\eta, h_\eta \xi, h_\eta \eta) \\
&\stackrel{d}{=} (\gamma h_\eta \eta, h_\eta \xi, h_\eta \eta) = (\gamma\tilde{\eta}, \tilde{\xi}, \tilde{\eta}).
\end{aligned}
$$

Since also $\gamma\tilde{\eta} \perp\!\!\!\perp_{\tilde{\eta}} \tilde{\xi}$ by the independence $\gamma \perp\!\!\!\perp (\xi, \eta)$, we conclude that $\eta \perp\!\!\!\perp_{\tilde{\eta}} \tilde{\xi}$. Hence, by FMP 6.13 there exist a measurable function $b\colon S \times [0,1] \to S$ and a $U(0,1)$ random variable $\zeta \perp\!\!\!\perp \eta$ such that

$$h_\eta \xi = \tilde{\xi} = b(\tilde{\eta}, \zeta) = b(h_\eta \eta, \zeta) \quad \text{a.s.}$$

Putting

$$f(s, t) = h_s^{-1} b(h_s s, t), \quad s \in S,\ t \in [0,1],$$

we obtain $\xi = f(\eta, \zeta)$ a.s. Since also $G_\eta \subset G_\xi$ a.s. by hypothesis, Lemma 7.10 shows that (17) holds outside a fixed null set. If $(\xi, \eta, \zeta) \stackrel{d}{=} (\xi, \eta, \zeta')$, then even $\xi = f(\eta, \zeta')$ a.s. since the diagonal in S^2 is measurable, and (17) remains valid with ζ replaced by ζ'. $\qquad\square$

Proof of Proposition 7.9: By Lemmas 7.10 and 7.11 there exist a measurable function $f\colon S \times [0,1] \to S$ and a $U(0,1)$ random variable $\zeta \perp\!\!\!\perp \eta$ satisfying (17) a.s. Writing $g\zeta = \zeta$, we conclude that even

$$g(\xi, \eta, \zeta) \stackrel{d}{=} (\xi, \eta, \zeta), \quad g \in G,$$

and trivially $G_{\xi,\eta} \subset G = G_\zeta$. Applying the same lemmas to the triple (ξ, η, ζ), we may next choose a measurable function $b: S^2 \times [0,1] \to [0,1]$ and a $U(0,1)$ random variable $\vartheta \perp\!\!\!\perp (\xi, \eta)$ such that a.s.

$$\zeta = g\zeta = b(g\xi, g\eta, \vartheta), \quad g \in G.$$

The same relation being a.s. true for the symmetrized version

$$\tilde{b}(x, y, t) = |G|^{-1} \sum\nolimits_{g \in G} b(gx, gy, t), \quad x, y \in S, \ t \in [0,1],$$

we may assume that b satisfies (16). Finally, (17) remains true with ζ replaced by $\zeta' = b(\xi, \eta, \vartheta')$ for an arbitrary $U(0,1)$ random variable $\vartheta' \perp\!\!\!\perp (\xi, \eta)$, since in this case $(\xi, \eta, \zeta) \overset{d}{=} (\xi, \eta, \zeta')$.   □

We continue with some elementary properties of symmetric functions. Recall that, if two arrays ξ and η on $\widetilde{\mathbb{N}}$ are related by $\eta_J = f(\hat{\xi}_J)$, then the function $\hat{f}$ is given by $\hat{\eta}_J = \hat{f}(\hat{\xi}_J)$. Thus,

$$\hat{f}(x_I; I \in 2^d) = \Big(f(x_{J \circ I}; I \in 2^{|J|}); J \in 2^d\Big), \quad d \in \mathbb{Z}_+. \tag{25}$$

Lemma 7.12 *(symmetric functions) Let ξ be a random array on $\widetilde{\mathbb{N}}$, and let f and g be measurable functions between suitable spaces.*

(i) *If $\eta_J = f(\hat{\xi}_J)$ on $\widetilde{\mathbb{N}}$, then even $\hat{\eta}_J = \hat{f}(\hat{\xi}_J)$ on $\widetilde{\mathbb{N}}$.*
(ii) *If $\eta_k = f(\hat{\xi}_k)$ on $\overline{\mathbb{N}}$ and f is symmetric, then $\hat{\eta}_k = \hat{f}(\hat{\xi}_k)$ on $\overline{\mathbb{N}}$.*
(iii) *If f and g are symmetric, then so is $f \circ \hat{g}$.*
(iv) *If ξ is exchangeable and f is symmetric, then the array $\eta_J = f(\hat{\xi}_J)$ on $\widetilde{\mathbb{N}}$ is again exchangeable.*

Proof: (i) This is just the definition of $\hat{f}$, restated here for comparison.

(ii) Letting $k \in \overline{\mathbb{N}}$ and $J \in 2^{|k|}$, and using the symmetry of f and the definition of $\hat{\xi}$, we get

$$f(\xi_{k \circ (J \circ I)}; I \in 2^{|J|}) = f(\xi_{(k \circ J) \circ I}; I \in 2^{|J|}) = f(\hat{\xi}_{k \circ J}).$$

Hence, by the definitions of $\hat{\xi}$, $\hat{f}$, η, and $\hat{\eta}$,

$$\begin{aligned}
\hat{f}(\hat{\xi}_k) &= \hat{f}(\xi_{k \circ I}; I \in 2^{|k|}) \\
&= (f(\xi_{k \circ (J \circ I)}; I \in 2^{|J|}); J \in 2^{|k|}) \\
&= (f(\hat{\xi}_{k \circ J}); J \in 2^{|k|}) \\
&= (\eta_{k \circ J}; J \in 2^{|k|}) = \hat{\eta}_k.
\end{aligned}$$

(iii) Put $\eta_k = g(\hat{\xi}_k)$ and $h = f \circ \hat{g}$. Applying (ii) to g and using the symmetry of f, we get for any $k \in \overline{\mathbb{N}}$

$$\begin{aligned}
h(\hat{\xi}_k) &= f \circ \hat{g}(\hat{\xi}_k) = f(\hat{\eta}_k) \\
&= f(\hat{\eta}_{\tilde{k}}) = f \circ \hat{g}(\hat{\xi}_{\tilde{k}}) = h(\hat{\xi}_{\tilde{k}}).
\end{aligned}$$

(iv) Fix any set $J \in \tilde{\mathbb{N}}$ and a permutation p on $\mathbb{N}$. Using the symmetry of f, we get as in (ii)

$$
\begin{aligned}
(\eta \circ p)_J &= \eta_{p \circ J} = f(\hat{\xi}_{p \circ J}) \\
&= f(\xi_{(p \circ J) \circ I};\ I \in 2^{|J|}) \\
&= f(\xi_{p \circ (J \circ I)};\ I \in 2^{|J|}) \\
&= f((\xi \circ p)_{J \circ I};\ I \in 2^{|J|}) = f((\xi \circ p)_J^{\wedge}).
\end{aligned}
$$

This shows that $\eta \circ p = g(\xi \circ p)$ for a suitable function g, and so the exchangeability of ξ carries over to η. $\qquad\square$

We proceed with another key result, which allows us to solve for ξ in the representation $\eta_J = f(\hat{\xi}_J)$, provided that even η has independent entries.

Proposition 7.13 *(inversion on $\tilde{\mathbb{N}}$) Let ξ be a contractable array on $\tilde{\mathbb{N}}$ with independent entries, and fix a measurable function f such that the array*

$$
\eta_J = f(\hat{\xi}_J), \quad J \in \tilde{\mathbb{N}}, \tag{26}
$$

has again independent entries. Then there exist some measurable functions g and h such that, for any U-array $\vartheta \perp\!\!\!\perp \xi$ on $\tilde{\mathbb{N}}$, the variables

$$
\zeta_J = g(\hat{\xi}_J, \vartheta_J), \quad J \in \tilde{\mathbb{N}}, \tag{27}
$$

form a U-array $\zeta \perp\!\!\!\perp \eta$ satisfying

$$
\xi_J = h(\hat{\eta}_J, \hat{\zeta}_J) \ a.s., \quad J \in \tilde{\mathbb{N}}. \tag{28}
$$

If f is symmetric, we can choose g and h to have the same property.

Proof: For convenience, we may take ξ and η to have real entries. Write ξ^d and $\Delta \xi^d$ for the restrictions of ξ to $\tilde{\mathbb{N}}_d$ and $\Delta\tilde{\mathbb{N}}_d$, respectively, and similarly for η, ζ, and ϑ. Our construction of g and h proceeds recursively with respect to the dimension d. For $d = 0$, we may apply Lemma 7.8 to the single random pair $(\xi_\emptyset, \eta_\emptyset)$, to ensure the existence of some functions g and h on suitable sub-spaces such that, whenever $\vartheta_\emptyset \perp\!\!\!\perp \xi$ is $U(0,1)$, the random variable $\zeta_\emptyset = g(\xi_\emptyset, \vartheta_\emptyset)$ becomes $U(0,1)$ with $\zeta_\emptyset \perp\!\!\!\perp \eta_\emptyset$ and satisfies $\xi_\emptyset = h(\eta_\emptyset, \zeta_\emptyset)$ a.s.

Now assume that g and h have already been constructed on $\bigcup_{k<d} \mathbb{R}^{2^k+1}$ and $\bigcup_{k<d} \mathbb{R}^{2^{k+1}}$, respectively, such that the array ζ^{d-1}, given by (27) in terms of an arbitrary U-array $\vartheta \perp\!\!\!\perp \xi$, is again a U-array independent of η^{d-1} and satisfying (28) on $\tilde{\mathbb{N}}_d$. To extend the construction to dimension d, we note that

$$
\xi_J \perp\!\!\!\perp_{\hat{\xi}'_J} (\xi^d \setminus \xi_J), \quad J \in \Delta\tilde{\mathbb{N}}_d,
$$

since ξ has independent entries. By (26) it follows that

$$
\xi_J \perp\!\!\!\perp_{\hat{\xi}'_J, \eta_J} (\xi^d \setminus \xi_J, \eta^d), \quad J \in \Delta\tilde{\mathbb{N}}_d.
$$

We also note that the pairs $(\hat{\xi}_J, \eta_J) = (\xi_J, \hat{\xi}'_J, \eta_J)$ have the same distribution for all $J \in \Delta\tilde{\mathbb{N}}_d$, due to the contractability of ξ. Hence, Lemma 7.8 ensures the existence of some measurable functions g_d on $\mathbb{R}^{2^d+1}$ and h_d on $\mathbb{R}^{2^{d+1}}$, such that the random variables

$$\zeta_J = g_d(\hat{\xi}_J, \eta_J, \vartheta_J), \quad J \in \Delta\tilde{\mathbb{N}}_d, \tag{29}$$

form a U-array $\Delta\zeta^d$ satisfying

$$\Delta\zeta^d \perp\!\!\!\perp (\xi^{d-1}, \eta^d), \tag{30}$$

$$\xi_J = h_d(\hat{\xi}'_J, \eta_J, \zeta_J) \text{ a.s.}, \quad J \in \Delta\tilde{\mathbb{N}}_d. \tag{31}$$

Inserting (26) into (29) and (28)—in the version for $\tilde{\mathbb{N}}_{d-1}$—into (31), we obtain

$$\zeta_J = g_d\big(\hat{\xi}_J, f(\hat{\xi}_J), \vartheta_J\big),$$

$$\xi_J = h_d\big(\hat{h}'(\hat{\eta}'_J, \hat{\zeta}'_J), \eta_J, \zeta_J\big) \text{ a.s.}, \quad J \in \Delta\tilde{\mathbb{N}}_d,$$

for some measurable function $\hat{h}'$. This shows that (27) and (28) remain valid on $\tilde{\mathbb{N}}_d$ for suitable extensions of g and h.

To complete the recursion, we need to prove that the arrays ζ^{d-1}, $\Delta\zeta^d$, and η^d are independent. Then note that $\vartheta^{d-1} \perp\!\!\!\perp (\xi^d, \Delta\vartheta^d)$, since ϑ is a U-array independent of ξ. By (26) and (27) it follows that

$$\vartheta^{d-1} \perp\!\!\!\perp (\xi^{d-1}, \Delta\eta^d, \Delta\zeta^d).$$

Combining this with (30) gives

$$\Delta\zeta^d \perp\!\!\!\perp (\xi^{d-1}, \Delta\eta^d, \vartheta^{d-1}),$$

and by (26) and (27) it follows that $\Delta\zeta^d \perp\!\!\!\perp (\eta^d, \zeta^{d-1})$.

Now only $\eta^d \perp\!\!\!\perp \zeta^{d-1}$ remains to be proved. Since ξ and η have independent entries and are related by (26), we may infer from Proposition 7.7 (ii) that $\Delta\eta^d \perp\!\!\!\perp \xi^{d-1}$. Since $\vartheta \perp\!\!\!\perp (\xi, \eta)$, the last relation extends to $\Delta\eta^d \perp\!\!\!\perp (\xi^{d-1}, \vartheta^{d-1})$, which implies $\Delta\eta^d \perp\!\!\!\perp (\eta^{d-1}, \zeta^{d-1})$ by (26) and (27). Since also $\eta^{d-1} \perp\!\!\!\perp \zeta^{d-1}$ by the induction hypothesis, we conclude that indeed $\eta^d \perp\!\!\!\perp \zeta^{d-1}$.

We turn to a proof of the last assertion. Here the construction for $d = 0$ is the same as before. Now fix any $d \in \mathbb{N}$, and assume that some measurable and symmetric functions g and h have already been constructed on appropriate subspaces, such that (27) defines a U-array $\zeta^{d-1} \perp\!\!\!\perp \eta^{d-1}$ on $\tilde{\mathbb{N}}_{d-1}$ satisfying (28) on $\tilde{\mathbb{N}}_{d-1}$ for any U-array $\vartheta \perp\!\!\!\perp \xi$.

To extend g to $\mathbb{R}^{2^d+1}$ and h to $\mathbb{R}^{2^{d+1}}$, we note that the pair (ξ, η) is exchangeable, due to the symmetry of f and Lemma 7.12 (iv). By Proposition 7.9, applied to $(\hat{\xi}_J, \eta_J)$ for fixed $J \in \Delta\tilde{\mathbb{N}}_d$, there exist some measurable functions g_d and h_d, the former even symmetric, such that the random variables ζ_J in (29) are $U(0, 1)$ and satisfy

$$\zeta_J \perp\!\!\!\perp (\hat{\xi}'_J, \eta_J), \qquad J \in \Delta\tilde{\mathbb{N}}_d, \tag{32}$$

$$\xi_{\bar{k}} = h_d(\hat{\xi}'_{\bar{k}}, \eta_{\bar{k}}, \zeta_{\bar{k}}) \text{ a.s.}, \qquad k \in \Delta\overline{\mathbb{N}}_d, \tag{33}$$

where $\hat{\xi}'_J = \hat{\xi}_J \setminus \xi_J$ and $\Delta\overline{\mathbb{N}}_d = \{k \in \overline{\mathbb{N}}; |k| = d\}$. Replacing h_d by its symmetrized version, which again satisfies (33), we may assume that even h_d is symmetric. Inserting (26) into (29) and (28)—in the version for $\widetilde{\mathbb{N}}_{d-1}$—into (33), we obtain the desired symmetric extensions of g and h.

To prove the required independence relations, we note that

$$(\xi_J, \vartheta_J) \perp\!\!\!\perp (\xi^d \setminus \xi_J, \vartheta^d \setminus \vartheta_J), \quad J \in \Delta\widetilde{\mathbb{N}}_d,$$

since (ξ, ϑ) has independent entries. Hence, by (26)

$$(\hat{\xi}_J, \vartheta_J) \underset{\hat{\xi}'_J, \eta_J}{\perp\!\!\!\perp} (\eta^d, \xi^d \setminus \xi_J, \vartheta^d \setminus \vartheta_J), \quad J \in \Delta\widetilde{\mathbb{N}}_d,$$

and so by (27)

$$\zeta_J \underset{\hat{\xi}'_J, \eta_J}{\perp\!\!\!\perp} (\eta^d, \zeta^d \setminus \zeta_J), \quad J \in \Delta\widetilde{\mathbb{N}}_d.$$

Combining with (32) and using the chain rule in FMP 6.8, we obtain

$$\zeta_J \perp\!\!\!\perp (\eta^d, \zeta^d \setminus \zeta_J), \quad J \in \Delta\widetilde{\mathbb{N}}_d,$$

which shows that the ζ_J are mutually independent and independent of the pair (ζ^{d-1}, η^d). It remains to show that $\zeta^{d-1} \perp\!\!\!\perp \eta^d$. Here the previous proof applies without changes. $\qquad\square$

We also need a version of the last result for separately exchangeable arrays.

Proposition 7.14 (*inversion on* $\mathbb{Z}_+^d$) *Let the U-arrays ξ and η on $\mathbb{Z}_+^d$ be related by*

$$\eta_k = f(\hat{\xi}_k), \quad k \in \mathbb{Z}_+^d, \tag{34}$$

for some measurable function f. Then there exist a third U-array $\zeta \perp\!\!\!\perp \eta$ on $\mathbb{Z}_+^d$ and a measurable function g such that

$$\xi_k = g(\hat{\eta}_k, \hat{\zeta}_k) \ a.s., \quad k \in \mathbb{Z}_+^d. \tag{35}$$

Proof: For any $J \in 2^d$, we may identify the set $\mathbb{N}^J \times \{0\}^{J^c}$ with $\mathbb{N}^J$. Write ξ^J for the restriction of ξ to $\mathbb{N}^J$, and put $\hat{\xi}^J = (\xi^I; I \subset J)$ and $\hat{\xi}'_k = \hat{\xi}_k \setminus \xi_k$. Since $\xi_k \perp\!\!\!\perp (\xi^J \setminus \xi_k)$ for any $k \in \mathbb{N}^J$, we see from (34) and FMP 6.7-8 that

$$\xi_k \underset{\hat{\xi}'_k, \eta_k}{\perp\!\!\!\perp} (\hat{\xi}^J \setminus \xi_k, \eta^J), \quad k \in \mathbb{N}^J, \ J \in 2^d.$$

Furthermore, we note that the distribution of $(\hat{\xi}_k, \eta_k)$ is the same for all $k \in \mathbb{N}^J$ with fixed $J \in 2^d$. Hence, by Lemma 7.8, there exist some measurable functions G_J and some U-arrays ζ^J on $\mathbb{N}^J$ with

$$\zeta^J \perp\!\!\!\perp (\hat{\xi}^J \setminus \xi^J, \eta^J), \quad J \in 2^d, \tag{36}$$

such that
$$\xi_k = G_J(\hat{\xi}'_k, \eta_k, \zeta_k) \text{ a.s.}, \quad k \in \mathbb{N}^J, \ J \in 2^d.$$
Iterating this relation yields (35) for a suitable function g.

As a further consequence of Lemma 7.8, we can choose
$$\zeta_k = h(\hat{\xi}_k, \vartheta_k), \quad k \in \mathbb{Z}_+^d, \tag{37}$$
for some measurable function h between suitable spaces and an arbitrary U-array $\vartheta \perp\!\!\!\perp \xi$ on $\mathbb{Z}_+^d$. Noting that
$$(\hat{\xi}^J, \vartheta^J) \perp\!\!\!\perp (\xi \setminus \hat{\xi}^J, \ \vartheta \setminus \vartheta^J), \quad J \in 2^d,$$
we see from (34) and (37) that
$$(\hat{\xi}^J, \eta^J, \zeta^J) \perp\!\!\!\perp (\xi \setminus \hat{\xi}^J, \ \vartheta \setminus \vartheta^J), \quad J \in 2^d,$$
which together with (36) yields
$$\zeta^J \perp\!\!\!\perp (\xi \setminus \xi^J, \ \eta^J, \ \vartheta \setminus \vartheta^J), \quad J \in 2^d.$$
Hence, by (34) and (37)
$$\zeta^J \perp\!\!\!\perp (\hat{\eta}^{|J|}, \ \hat{\zeta}^{|J|} \setminus \zeta^J), \quad J \in 2^d, \tag{38}$$
where $\eta^m = (\eta^J; |J| = m)$. In particular, the arrays ζ^J are independent and hence form a U-array ζ on $\mathbb{Z}_+^d$.

Since both ξ and η have independent entries, Proposition 7.7 (i) yields $\eta^J \perp\!\!\!\perp (\xi \setminus \xi^J, \vartheta)$ for every $J \in 2^J$, and so by (34) and (37)
$$\eta^J \perp\!\!\!\perp (\hat{\eta}^{|J|} \setminus \eta^J, \ \hat{\zeta}^{|J|} \setminus \zeta^J), \quad J \in 2^d.$$
From this relation and (38) we get
$$\zeta^m \perp\!\!\!\perp (\hat{\eta}^m, \hat{\zeta}^{m-1}), \quad \eta^m \perp\!\!\!\perp (\hat{\eta}^{m-1}, \hat{\zeta}^{m-1}), \qquad m \le d.$$
Proceeding by induction on m, we conclude that the arrays $\zeta^0, \ldots, \zeta^m$ and $\eta^0, \ldots, \eta^m$ are all independent. In particular, we have $\zeta \perp\!\!\!\perp \eta$. $\qquad\square$

7.4 Contractable Arrays

The purpose of this section is to prove the following basic representation theorem for contractable arrays on $\widetilde{\mathbb{N}}$.

Theorem 7.15 (*jointly contractable arrays*) *Let X be a random array on $\widetilde{\mathbb{N}}$ with values in a Borel space S. Then X is contractable iff there exist a U-array ξ on $\widetilde{\mathbb{N}}$ and a measurable function $f \colon \bigcup_{d \ge 0} [0,1]^{2^d} \to S$ such that*
$$X_J = f(\hat{\xi}_J) \text{ a.s.}, \quad J \in \widetilde{\mathbb{N}}.$$
Here we can choose f to be symmetric iff X is exchangeable.

The following extension property is an immediate consequence of the main result. Identifying the elements of $\tilde{\mathbb{N}}$ with vectors $k = (k_1, \ldots, k_d)$ with $k_1 < \cdots < k_d$, we may regard the index set $\tilde{\mathbb{N}}$ as a subset of $\overline{\mathbb{N}}$, and it makes sense to consider extensions of a random array X on $\tilde{\mathbb{N}}$ to the larger index set $\overline{\mathbb{N}}$.

Corollary 7.16 *(extension criterion) Let X be a random array on $\tilde{\mathbb{N}}$ with values in a Borel space. Then X is contractable iff it can be extended to an exchangeable array on $\overline{\mathbb{N}}$.*

Proof: Suppose that X is contractable. Then by Theorem 7.15 it has a representation $X_J = f(\hat{\xi}_J)$, $J \in \tilde{\mathbb{N}}$, in terms of a U-array ξ and a measurable function f, and we may define an extension Y to $\overline{\mathbb{N}}$ by $Y_k = f(\hat{\xi}_k)$, $k \in \overline{\mathbb{N}}$. Since ξ is trivially exchangeable, we note that Y has the same property. This proves the necessity of our condition. The sufficiency is obvious, since every exchangeable array is also contractable.   □

For convenience, we divide the proof of Theorem 7.15 into a sequence of lemmas. To state our first lemma, we say that a random array is *representable* if can be represented as in Theorem 7.15. Recall that all random arrays are assumed to take values in an arbitrary Borel space.

Lemma 7.17 *(augmentation on $\tilde{\mathbb{N}}_d$) For a fixed $d \in \mathbb{Z}_+$, assume that all contractable arrays on $\tilde{\mathbb{N}}_d$ are representable. Consider a contractable array (X, ξ) on $\tilde{\mathbb{N}}_d$, where ξ has independent entries. Then there exist a U-array $\eta \perp\!\!\!\perp \xi$ on $\tilde{\mathbb{N}}_d$ and a measurable function f such that*

$$X_J = f(\hat{\xi}_J, \hat{\eta}_J) \quad a.s., \quad J \in \tilde{\mathbb{N}}_d.$$

Proof: Since the pair (X, ξ) is representable, there exist a U-array ζ and some measurable functions g and h between suitable spaces such that a.s.

$$X_J = g(\hat{\zeta}_J), \quad \xi_J = h(\hat{\zeta}_J), \quad J \in \tilde{\mathbb{N}}_d. \tag{39}$$

By Proposition 7.13, applied to the second of the two equations, there exist a U-array $\eta \perp\!\!\!\perp \xi$ and a measurable function k such that

$$\zeta_J = k(\hat{\xi}_J, \hat{\eta}_J) \quad a.s., \quad J \in \tilde{\mathbb{N}}_d.$$

Combining this with the first relation in (39), we obtain the required representation of X with $f = g \circ \hat{k}$.   □

The following construction is crucial for the proof of our main result. Let $\tilde{\mathbb{Z}}_-$ denote the class of finite subsets of $\mathbb{Z}_- = -\mathbb{Z}_+$.

Lemma 7.18 *(key construction) Let $\overline{X}$ be a contractable array on $\widetilde{\mathbb{Z}}$ with restriction X to $\widetilde{\mathbb{N}}$, and consider the arrays*

$$
\begin{aligned}
Y_J &= \{\overline{X}_{I \cup J};\ \emptyset \neq I \in \widetilde{\mathbb{Z}}_-\}, & J &\in \widetilde{\mathbb{N}}, \\
X_J^n &= X_{\{n\} \cup (J+n)}, & & \\
Y_J^n &= (Y_{J+n}, Y_{\{n\} \cup (J+n)}), & J &\in \widetilde{\mathbb{N}},\ n \in \mathbb{N}.
\end{aligned}
$$

Then the pairs (X, Y) and (X^n, Y^n) are again contractable, and the latter are equally distributed with

$$
X^n \perp\!\!\!\perp_{Y^n} (\overline{X} \setminus X^n), \quad n \in \mathbb{N}.
$$

Proof: Noting that, for $J \in \widetilde{\mathbb{N}}$ and $n \in \mathbb{N}$,

$$
\begin{aligned}
(X_J, Y_J) &= \{\overline{X}_{I \cup J};\ I \in \widetilde{\mathbb{Z}}_-\}, \\
(X_J^n, Y_J^n) &= \{\overline{X}_{I \cup (J+n)};\ \emptyset \neq I \in (\mathbb{Z}_- \cup \{n\})^\sim\},
\end{aligned}
$$

and using the contractability of $\overline{X}$, we see that (X, Y) and (X^n, Y^n) are again contractable and that the latter pairs are equally distributed. It remains only to prove the asserted conditional independence.

Then write $\overline{X} = (Q^n, R^n)$ for each $n \in \mathbb{N}$, where Q^n denotes the restriction of X to $(\mathbb{N} + n - 1)^\sim$ and $R^n = \overline{X} \setminus Q^n$. Introduce the mappings

$$
p_n(k) = k - (n - 1)1\{k < n\}, \quad k \in \mathbb{Z},\ n \in \mathbb{N},
$$

and conclude from the contractability of $\overline{X}$ that

$$
(Q^n, R^n) = \overline{X} \overset{d}{=} \overline{X} \circ p_n = (Q^n, R^n \circ p_n).
$$

By Lemma 1.3 it follows that $R^n \perp\!\!\!\perp_{R^n \circ p_n} Q^n$, which is equivalent to

$$
(Y, X^1, \ldots, X^{n-1}) \perp\!\!\!\perp_{Y^n} (X^n, X^{n+1}, \ldots\ ; X_\emptyset).
$$

Replacing n by $n + 1$ and noting that Y^{n+1} is a sub-array of Y^n, we see that also

$$
(Y, X^1, \ldots, X^n) \perp\!\!\!\perp_{Y^n} (X^{n+1}, X^{n+2}, \ldots\ ; X_\emptyset).
$$

Combining the two formulas, we get by FMP 3.8

$$
X^n \perp\!\!\!\perp_{Y^n} (Y, X^1, \ldots, X^{n-1}, X^{n+1}, \ldots\ ; X_\emptyset) = \overline{X} \setminus X^n,
$$

as required. $\qquad\square$

The representation theorem will first be proved for contractable arrays of bounded dimension. Here we need a recursive construction based on the previous lemma.

Lemma 7.19 *(recursion) For a fixed $d \in \mathbb{N}$, assume that all contractable arrays on $\widetilde{\mathbb{N}}_{d-1}$ are representable, and let X be a contractable array on $\widetilde{\mathbb{N}}_d$. Then there exists a U-array η on $\widetilde{\mathbb{N}}_{d-1}$ such that the pair (X, η) is contractable, the element $X_\emptyset$ is $\eta_\emptyset$-measurable, and the arrays*

$$
\begin{aligned}
X_J^n &= X_{\{n\} \cup (J+n)}, \\
\eta_J^n &= \eta_{J+n-1}, \qquad J \in \widetilde{\mathbb{N}}_{d-1}, \ n \in \mathbb{N},
\end{aligned}
$$

satisfy

$$
X^n \perp\!\!\!\perp_{\eta^n} (X \setminus X^n, \eta), \quad n \in \mathbb{N}.
$$

For any contractable array X on $\widetilde{\mathbb{N}}_d$, we can construct a contractable extension to $\widetilde{\mathbb{N}}$ by taking $X_J = 0$ when $|J| > d$. By this device, we may henceforth regard any array on $\widetilde{\mathbb{N}}_d$ as defined on the larger index set $\widetilde{\mathbb{N}}$.

Proof: Defining Y as in Lemma 7.18 in terms of a contractable extension of X to $\widetilde{\mathbb{Z}}$, we note that $(X_\emptyset, Y)$ is a contractable array on $\widetilde{\mathbb{N}}_{d-1}$. Hence, by hypothesis, it has an a.s. representation

$$
X_\emptyset = f(\xi_\emptyset), \qquad Y_J = g(\hat{\xi}_J), \quad J \in \widetilde{\mathbb{N}}_{d-1}, \tag{40}
$$

in terms of some measurable functions f and g between suitable spaces and a U-array ξ on $\widetilde{\mathbb{N}}_{d-1}$. The pairs (X, Y) and (Y, ξ), regarded as contractable arrays on $\widetilde{\mathbb{N}}$, admit extensions as in Lemma 7.3 to contractable arrays $(\overline{X}, \overline{Y})$ and $(\overline{Y}', \bar{\xi})$ on $\widetilde{\mathbb{Q}}$. Here clearly $(X_\emptyset, \overline{Y}) \overset{d}{=} (X_\emptyset, \overline{Y}')$, and so the transfer theorem (FMP 6.10, 6.13) yields the existence of a U-array $\bar{\eta}$ on $\widetilde{\mathbb{Q}}_{d-1}$ satisfying

$$
(X_\emptyset, \overline{Y}, \bar{\eta}) \overset{d}{=} (X_\emptyset, \overline{Y}', \bar{\xi}), \qquad \bar{\eta} \perp\!\!\!\perp_{X_\emptyset, \overline{Y}} \overline{X}. \tag{41}
$$

The a.s. representations in (40) now extend to $\widetilde{\mathbb{Q}}_{d-1}$, though with (Y, ξ) replaced by $(\overline{Y}, \bar{\eta})$, and Lemma 7.2 shows that $(\overline{X}, \overline{Y}, \bar{\eta})$ remains contractable on $\widetilde{\mathbb{Q}}$.

By Lemma 7.18 we have

$$
X^n \perp\!\!\!\perp_Y (X \setminus X^n), \quad n \in \mathbb{N},
$$

and by contractability and martingale convergence we may replace Y by the pair $(X_\emptyset, \overline{Y})$. Combining this with the second relation in (41), using a conditional version of FMP 3.8, we obtain

$$
X^n \perp\!\!\!\perp_{X_\emptyset, \overline{Y}} (X \setminus X^n, \bar{\eta}), \quad n \in \mathbb{N}.
$$

Since $X_\emptyset$ and $\overline{Y}$ are measurable functions of $\bar{\eta}$, due to the extended representation in (40), we conclude that

$$
X^n \perp\!\!\!\perp_{\bar{\eta}} (X \setminus X^n), \quad n \in \mathbb{N}. \tag{42}
$$

By the contractability of $(\overline{X}, \bar{\eta})$, we may replace $\bar{\eta}$ by its restriction $\bar{\eta}_\varepsilon$ to the index set $\widetilde{T}_\varepsilon$, where $T_\varepsilon = \bigcup_{k>0}(k - \varepsilon, k]$ with $\varepsilon > 0$ rational. Since $\sigma\{\bar{\eta}_\varepsilon\} \downarrow \sigma\{\eta\}$ by the extended version of Kolmogorov's zero-one law in FMP 7.25, we conclude that, by reverse martingale convergence, (42) remains true with $\bar{\eta}$ replaced by η. Furthermore, we see from Lemma 7.5 that

$$X^n \perp\!\!\!\perp_{\eta^n} \eta, \quad n \in \mathbb{N},$$

and so by combination, using the chain rule in FMP 6.8, we obtain the required conditional independence. $\qquad \square$

We are now ready to establish the representation of contractable arrays, beginning with the case of bounded dimensions.

Lemma 7.20 *(bounded dimensions) Contractable arrays on $\widetilde{\mathbb{N}}_d$ are representable for every $d \in \mathbb{Z}_+$.*

Proof: The result for $d = 0$ is elementary and follows from Lemma 7.8 with $\eta = 0$. Proceeding by induction, we fix a $d \in \mathbb{N}$ and, assuming all contractable arrays on $\widetilde{\mathbb{N}}_{d-1}$ to be representable, consider a contractable array X on $\widetilde{\mathbb{N}}_d$. Let the U-array η on $\widetilde{\mathbb{N}}_{d-1}$ be such as in Lemma 7.19, and extend η to a U-array on $\widetilde{\mathbb{N}}_d$ by choosing the variables η_J, $J \in \Delta\widetilde{\mathbb{N}}_d$, to be i.i.d. $U(0,1)$, independently of X and all previously defined η-variables. Then (X, η) remains contractable, the random element $X_\emptyset$ is $\eta_\emptyset$-measurable, and the arrays

$$\begin{aligned} X_J^n &= X_{\{n\} \cup (J+n)}, \\ \eta_J^n &= (\eta_{J+n}, \eta_{\{n\} \cup (J+n)}), \quad J \in \widetilde{\mathbb{N}}_{d-1}, \ n \in \mathbb{N}, \end{aligned}$$

(differing only slightly from those in Lemma 7.19) satisfy

$$X^n \perp\!\!\!\perp_{\eta^n} (X \setminus X^n, \eta), \quad n \in \mathbb{N}. \tag{43}$$

The pairs (X^n, η^n) are equally distributed and inherit the contractability from (X, η). Furthermore, the elements of η^n are mutually independent for fixed n and uniformly distributed over the unit square $[0,1]^2$. Hence, Lemma 7.17 ensures the existence of some U-arrays $\zeta^n \perp\!\!\!\perp \eta^n$ on $\widetilde{\mathbb{N}}_{d-1}$ and a measurable function G between suitable spaces such that

$$X_J^n = G(\hat{\eta}_J^n, \hat{\zeta}_J^n) \text{ a.s.}, \quad J \in \widetilde{\mathbb{N}}_{d-1}, \ n \in \mathbb{N}. \tag{44}$$

In fact, by transfer (FMP 6.10) we may choose

$$\zeta^n = h(X^n, \eta^n, \vartheta_n), \quad n \in \mathbb{N}, \tag{45}$$

for some U-sequence $\vartheta = (\vartheta_n) \perp\!\!\!\perp (X, \eta)$ and a measurable function h, so that by FMP 6.13

$$\zeta^n \perp\!\!\!\perp_{X^n, \eta^n} (\eta, \zeta \setminus \zeta^n), \quad n \in \mathbb{N}.$$

On the other hand, by (43) and (45) together with the independence of the ϑ_n, we have

$$X^n \perp\!\!\!\perp_{\eta^n} (\eta, \zeta \setminus \zeta^n), \quad n \in \mathbb{N}.$$

Using the chain rule in FMP 6.8, we may combine the last two relations into the single formula

$$\zeta^n \perp\!\!\!\perp_{\eta^n} (\eta, \zeta \setminus \zeta^n), \quad n \in \mathbb{N}.$$

Since also $\zeta^n \perp\!\!\!\perp \eta^n$, the same result yields

$$\zeta^n \perp\!\!\!\perp (\eta, \zeta \setminus \zeta^n), \quad n \in \mathbb{N}.$$

This shows (FMP 3.8) that the ζ^n are mutually independent and independent of η.

The arrays ζ^n may be combined into a single U-array $\zeta \perp\!\!\!\perp \eta$ on $\widetilde{\mathbb{N}}_d \setminus \{\emptyset\}$, given by

$$\zeta_{\{n\} \cup (J+n)} = \zeta_J^n, \quad J \in \widetilde{\mathbb{N}}_{d-1}, \ n \in \mathbb{N},$$

and we may extend ζ to $\widetilde{\mathbb{N}}_d$ by choosing $\zeta_\emptyset$ to be $U(0,1)$ and independent of η and all previous ζ-variables. Then (44) becomes

$$X_J = F(\hat{\eta}_J, \hat{\zeta}_J) \text{ a.s.}, \quad \emptyset \neq J \in \widetilde{\mathbb{N}}_d, \tag{46}$$

where F is given, for $k = 1, \ldots, d$, by

$$F\big\{(y,z)_I; \ I \in 2^k\big\} = G\big\{y_{I+1}, (y,z)_{\{1\} \cup (I+1)}; \ I \in 2^{k-1}\big\}.$$

Since $X_\emptyset$ is $\eta_\emptyset$-measurable, (46) can be extended to $J = \emptyset$ through a suitable extension of F to $[0,1]^2$, e.g. by reference to FMP 1.13. By Lemma 7.8 with $\eta = 0$, we may finally choose a U-array ξ on $\widetilde{\mathbb{N}}_d$ and a measurable function $b: [0,1] \to [0,1]^2$ such that

$$(\eta_J, \zeta_J) = b(\xi_J) \text{ a.s.}, \quad J \in \widetilde{\mathbb{N}}_d.$$

Then $(\hat{\eta}_J, \hat{\zeta}_J) = \hat{b}(\hat{\xi}_J)$ a.s. for all $J \in \widetilde{\mathbb{N}}_d$, and substituting this into (46) yields the desired representation $X_J = f(\hat{\xi}_J)$ a.s. with $f = F \circ \hat{b}$. $\qquad \square$

Our next aim is to extend the representation to arrays of unbounded dimensions.

Lemma 7.21 *(unbounded dimensions) If the contractable arrays on $\widetilde{\mathbb{N}}_d$ are representable for every $d \in \mathbb{Z}_+$, then so are all contractable arrays on $\widetilde{\mathbb{N}}$.*

Proof: Let X be a contractable array on $\widetilde{\mathbb{N}}$. Our first step is to construct some independent U-arrays ξ^k on $\widetilde{\mathbb{N}}_k$, $k \in \mathbb{Z}_+$, and some measurable functions $f_0, f_1, \ldots$ on suitable spaces such that

$$X_J = f_k(\hat{\xi}_J^0, \ldots, \hat{\xi}_J^k) \text{ a.s.}, \quad J \in \Delta \widetilde{\mathbb{N}}_k, \ k \in \mathbb{Z}_+, \tag{47}$$

where $\xi_J^k = 0$ for $|J| > k$. For $k = 0$, equation (47) reduces to $X_\emptyset = f_0(\xi_\emptyset^0)$, which holds for suitable f_0 and $\xi_\emptyset^0$ by Lemma 7.20 with $d = 0$. By Corollary 7.4 we may redefine $\xi_\emptyset^0$, if necessary, such that the augmented array $(X, \xi_\emptyset^0)$ becomes contractable.

Proceeding recursively, fix a $d \in \mathbb{N}$, and assume that $\xi^0, \ldots, \xi^{d-1}$ and $f_0, \ldots, f_{d-1}$ have already been chosen to satisfy (47) for all $k < d$, where the ξ^k are independent U-arrays such that the combined array $(X, \xi^0, \ldots, \xi^{d-1})$ is contractable on $\widehat{\mathbb{N}}$. Since contractable arrays on $\widetilde{\mathbb{N}}_d$ are assumed to be representable, Lemma 7.17 ensures the existence of a U-array $\xi^d \perp\!\!\!\perp (\xi^0, \ldots, \xi^{d-1})$ on $\widetilde{\mathbb{N}}_d$ and a measurable function f_d between suitable spaces such that (47) remains true for $k = d$.

Since $(\xi^0, \ldots, \xi^d)$ is again a U-array and hence contractable, (47) yields the same property for the combined array $(X^d, \xi^0, \ldots, \xi^d)$, where X^d denotes the restriction of X to $\widetilde{\mathbb{N}}_d$. Applying Corollary 7.4 to the arrays $(X, \xi^0, \ldots, \xi^{d-1})$ and $(X^d, \xi^0, \ldots, \xi^d)$, we see that ξ^d can be redefined, if necessary, such that $(X, \xi^0, \ldots, \xi^d)$ becomes contractable. This completes the recursion and proves the existence of arrays $\xi^0, \xi^1, \ldots$ and functions $f_0, f_1, \ldots$ with the desired properties.

We now extend the sequence $\xi^0, \xi^1, \ldots$ to a U-array on $\mathbb{Z}_+ \times \widetilde{\mathbb{N}}$, by choosing the variables ξ_J^k with $|J| > k$ to be i.i.d. $U(0, 1)$ and independent of all previously constructed variables. (This may require an obvious adjustment of functions f_k, to ensure that (47) remains fulfilled.) Putting

$$F(\hat{x}_J^0, \hat{x}_J^1, \ldots) = f_k(\hat{x}_J^0, \ldots, \hat{x}_J^k), \quad J \in 2^k, \ k \in \mathbb{Z}_+,$$

we may write (47) in the form

$$X_J = F(\hat{\xi}_J^0, \hat{\xi}_J^1, \ldots) \text{ a.s.}, \quad J \in \widetilde{\mathbb{N}}.$$

By Lemma 7.8 with $\eta = 0$, we may next choose a U-array ξ on $\widetilde{\mathbb{N}}$ and a measurable function $g \colon [0, 1] \to [0, 1]^\infty$ such that

$$(\xi_J^0, \xi_J^1, \ldots) = g(\xi_J) \text{ a.s.}, \quad J \in \widetilde{\mathbb{N}}.$$

The desired representation $X_J = f(\hat{\xi}_J)$ a.s., $J \in \widetilde{\mathbb{N}}$, now follows with $f = F \circ \hat{g}$. $\qquad\square$

Proof of Theorem 7.15: In view of the last two lemmas, it remains only to prove the last statement. Anticipating Theorem 7.22 (whose proof is independent of the present result), we get for exchangeable arrays X the more general representation

$$X_{\tilde{k}} = f(\hat{\xi}_k) \text{ a.s.}, \quad k \in \overline{\mathbb{N}}.$$

This gives $2^{|J|}$ different representations of each element X_J, and averaging over those yields a representation in terms of a symmetric function $\tilde{f}$. Conversely, if an array X is representable in terms of a symmetric function f, it must be exchangeable by Lemma 7.12 (iv). $\qquad\square$

7.5   Exchangeable Arrays

The aim of this section is to prove the following closely related representation theorem for exchangeable arrays on $\overline{\mathbb{N}}$.

Theorem 7.22 *(jointly exchangeable arrays, Hoover)* *Let X be a random array on $\overline{\mathbb{N}}$ with values in a Borel space S. Then X is exchangeable iff there exist a U-array ξ on $\tilde{\mathbb{N}}$ and a measurable function $f \colon \bigcup_{d \geq 0} [0,1]^{2^d} \to S$ such that*

$$X_k = f(\hat{\xi}_k) \ \ a.s., \quad k \in \overline{\mathbb{N}}.$$

As an easy consequence, we obtain the following representation in the separately exchangeable case.

Corollary 7.23 *(separately exchangeable arrays, Aldous, Hoover)* *Let X be a random array on $\mathbb{N}^d$ with values in a Borel space S. Then X is separately exchangeable iff there exist a U-array ξ on $\mathbb{Z}_+^d$ and a measurable function $f \colon [0,1]^{2^d} \to S$ such that*

$$X_k = f(\hat{\xi}_k) \ \ a.s., \quad k \in \mathbb{N}^d.$$

Proof: Let X be separately exchangeable. It is then exchangeable even in the joint sense, and so by Theorem 7.22 the non-diagonal part of X has a representation $X_k = f(\hat{\zeta}_k)$ in terms of a U-array ζ on $\tilde{\mathbb{N}}$. Fixing any disjoint, infinite subsets $N_1, \dots, N_d$ of $\mathbb{N}$, we conclude that X has the required representation on the index set $A = N_1 \times \cdots \times N_d$, apart from an appropriate re-labeling of ζ. Now the restriction to A has the same distribution as X itself, due to the separate exchangeability of X. Using the transfer theorem in FMP 6.10, we conclude that the representation on A extends to $\mathbb{N}^d$. $\square$

Our proof of the main result, Theorem 7.22, is again divided into a sequence of lemmas. We begin with an exchangeable counterpart of Lemma 7.19. For any $k \in \overline{\mathbb{N}}$, write $h \sim k$ if h is a permutation of k, and define $\tilde{k}$ as the set of permutations of k. Let $\overline{\mathbb{N}}_d$ and $\Delta \overline{\mathbb{N}}_d$ consist of all sequences $k \in \overline{\mathbb{N}}$ with $|k| \leq d$ or $|k| = d$, respectively, and similarly for $\mathbb{Z}$. Say that an exchangeable array is *representable* if it can be expressed as in Theorem 7.22.

Lemma 7.24 *(recursion)* *For a fixed $d \in \mathbb{N}$, assume that all exchangeable arrays on $\overline{\mathbb{N}}_{d-1}$ are representable, and let X be an exchangeable array on $\overline{\mathbb{N}}_d$. Then X can be represented on $\overline{\mathbb{N}}_{d-1}$ in terms of a U-array ξ on $\tilde{\mathbb{N}}_{d-1}$, such that the pair (X, ξ) is exchangeable and the arrays $\tilde{X}_k = \{X_h; h \sim k\}$ satisfy*

$$\tilde{X}_k \perp\!\!\!\perp_{\hat{\xi}_k} (X \setminus \tilde{X}_k, \xi), \quad k \in \Delta \overline{\mathbb{N}}_d. \tag{48}$$

Proof: Let $\overline{X}$ denote a stationary extension of X to the set $\overline{\mathbb{Z}}_d$, consisting of all sequences r of at most d distinct integers $r_1, \dots, r_m$. Let r_+ denote the sub-sequence of elements $r_j > 0$. If $k \in \overline{\mathbb{N}}_{d-1}$ and $r = (r_1, \ldots, r_m) \in \overline{\mathbb{Z}}_d$ with $r_+ \sim (1, \ldots, |k|)$, we write $k \circ r = (k_{r_1}, \ldots, k_{r_m})$, where $k_{r_j} = r_j$ when $r_j \leq 0$. Introduce the array Y on $\overline{\mathbb{N}}_{d-1}$ with elements

$$Y_k = \left(\overline{X}_{k \circ r}; \ r \in \overline{\mathbb{Z}}_d, \ r_+ \sim (1, \ldots, |k|) \right), \quad k \in \overline{\mathbb{N}}_{d-1}.$$

Informally, we may think of Y_k as the set $\{\overline{X}_h; \ h_+ \sim k\}$, provided with a consistently chosen order of enumeration. The exchangeability of $\overline{X}$ implies that the pair (X, Y) is again exchangeable.

For any $k = (k_1, \ldots, k_d) \in \Delta \overline{\mathbb{N}}_d$, let $\hat{Y}_k$ denote the restriction of Y to sequences from $\tilde{k}$ of length $< d$, and note that

$$\left(\tilde{X}_k, \hat{Y}_k \right) \overset{d}{=} \left(\tilde{X}_k, X \setminus \tilde{X}_k, Y \right), \quad k \in \Delta \overline{\mathbb{N}}_d,$$

since the two arrays are restrictions of the same exchangeable array $\overline{X}$ to sequences in $\mathbb{Z}_- \cup \{\tilde{k}\}$ and $\mathbb{Z}$, respectively. Since also $\sigma(\hat{Y}_k) \subset \sigma(X \setminus \tilde{X}_k, Y)$, we see from Lemma 1.3 that

$$\tilde{X}_k \perp\!\!\!\perp_{\hat{Y}_k} (X \setminus \tilde{X}_k, Y), \quad k \in \Delta \overline{\mathbb{N}}_d. \tag{49}$$

Since exchangeable arrays on $\overline{\mathbb{N}}_{d-1}$ are representable, there exists a U-array ξ on $\tilde{\mathbb{N}}_{d-1}$ and a measurable function f between suitable spaces such that

$$(X_k, Y_k) = f(\hat{\xi}_k), \quad k \in \overline{\mathbb{N}}_{d-1}. \tag{50}$$

By FMP 6.10 and 6.13 we may assume that $\xi \perp\!\!\!\perp_{X', Y} X$, where X' denotes the restriction of X to $\overline{\mathbb{N}}_{d-1}$, in which case the pair (X, ξ) is again exchangeable by Lemma 7.2. The stated conditional independence also implies that

$$\tilde{X}_k \underset{Y, X \setminus \tilde{X}_k}{\perp\!\!\!\perp} \xi, \quad k \in \Delta \overline{\mathbb{N}}_d. \tag{51}$$

Using the chain rule in FMP 6.8, we may combine (49) and (51) into the single formula

$$\tilde{X}_k \perp\!\!\!\perp_{\hat{Y}_k} (X \setminus \tilde{X}_k, \xi), \quad k \in \Delta \overline{\mathbb{N}}_d,$$

which yields the asserted relation, since $\hat{Y}_k$ is $\hat{\xi}_k$-measurable in view of (50). $\square$

We are now ready to derive the representation of Theorem 7.22, in the special case of bounded dimensions.

Lemma 7.25 *(bounded dimensions) Exchangeable arrays on $\overline{\mathbb{N}}_d$ are representable for every $d \in \mathbb{Z}_+$.*

Proof: The result for $d = 0$ is again elementary and follows from Lemma 7.8 with $\eta = 0$. Proceeding by induction, fix a $d \in \mathbb{N}$ such that all exchangeable arrays on $\overline{\mathbb{N}}_{d-1}$ are representable, and consider an exchangeable array X on $\overline{\mathbb{N}}_d$. By Lemma 7.24 we may choose a U-array ξ on $\overline{\mathbb{N}}_{d-1}$ and a measurable function f between suitable spaces satisfying

$$X_k = f(\hat{\xi}_k) \text{ a.s.}, \quad k \in \overline{\mathbb{N}}_{d-1}, \tag{52}$$

and such that the pair (X,ξ) is exchangeable and the arrays $\tilde{X}_k = \{X_h;$ $h \sim k\}$ with $k \in \Delta\overline{\mathbb{N}}_d$ satisfy (48).

To handle symmetries involving the sets $\tilde{X}_k$, we need to fix an order of enumeration. Letting G denote the group of permutations on $1, \ldots, d$, we write

$$\tilde{X}_k = (X_{k \circ g}; g \in G), \quad \hat{\xi}_k = (\xi_{k \circ I}; I \in 2^d), \quad k \in \Delta\overline{\mathbb{N}}_d,$$

where $k \circ I = \{k_i; i \in I\}$. The exchangeability of (X, ξ) implies that the pairs $(\tilde{X}_k, \hat{\xi}_k)$ are equally distributed, and in particular

$$(\tilde{X}_{k \circ g}, \hat{\xi}_{k \circ g}) \overset{d}{=} (\tilde{X}_k, \hat{\xi}_k), \quad g \in G, \ k \in \Delta\overline{\mathbb{N}}_d.$$

Since ξ is a U-array, the arrays $\hat{\xi}_{k \circ g}$ are a.s. different for fixed $k \in \Delta\overline{\mathbb{N}}_d$, which implies that the associated symmetry groups are a.s. trivial. Hence, by Proposition 7.9, there exist some $U(0, 1)$ random variables $\zeta_k \perp\!\!\!\perp \hat{\xi}_k$ and a measurable function h such that

$$X_{k \circ g} = h(\hat{\xi}_{k \circ g}, \zeta_k) \text{ a.s.}, \quad g \in G, \ k \in \Delta\overline{\mathbb{N}}_d. \tag{53}$$

Now introduce a U-array $\vartheta \perp\!\!\!\perp \xi$ on $\Delta\tilde{\mathbb{N}}_d$, and define

$$Y_k = h(\hat{\xi}_k, \vartheta_{\tilde{k}}), \quad \tilde{Y}_k = (Y_{k \circ g}; g \in G), \quad k \in \Delta\overline{\mathbb{N}}_d. \tag{54}$$

Comparing with (53) and using the independence properties of (ξ, ϑ), we see that

$$(\tilde{Y}_k, \hat{\xi}_k) \overset{d}{=} (\tilde{X}_k, \hat{\xi}_k), \quad \tilde{Y}_k \perp\!\!\!\perp_{\hat{\xi}_k} (Y \setminus \tilde{Y}_k, \xi), \quad k \in \Delta\overline{\mathbb{N}}_d.$$

In view of (48) it follows that $(\tilde{Y}_k, \xi) \overset{d}{=} (\tilde{X}_k, \xi)$ for every $k \in \Delta\overline{\mathbb{N}}_d$, and moreover

$$\tilde{X}_k \perp\!\!\!\perp_\xi (X \setminus \tilde{X}_k), \quad \tilde{Y}_k \perp\!\!\!\perp_\xi (Y \setminus \tilde{Y}_k), \quad k \in \Delta\overline{\mathbb{N}}_d.$$

Thus, the arrays $\tilde{X}_k$ associated with different sets $\tilde{k}$ are conditionally independent given ξ, and similarly for the arrays $\tilde{Y}_k$. Writing X' for the restriction of X to $\overline{\mathbb{N}}_{d-1}$, which is ξ-measurable by (52), we obtain $(X, \xi) \overset{d}{=} (X', Y, \xi)$. Using (54) and applying the transfer theorem in the form of FMP 6.11, we conclude that

$$X_k = h(\hat{\xi}_k, \eta_k) \text{ a.s.}, \quad k \in \Delta\overline{\mathbb{N}}_d,$$

for some U-array $\eta \perp\!\!\!\perp \xi$ on $\Delta\tilde{\mathbb{N}}_d$. This, together with (52), yields the required representation of X. $\square$

Yet another lemma is needed, before we are ready to prove Theorem 7.22 in full generality. The following result is an exchangeable version of Lemma 7.17.

Lemma 7.26 *(augmentation on $\overline{\mathbb{N}}_d$)* *For a fixed $d \in \mathbb{Z}_+$, assume that all exchangeable arrays on $\overline{\mathbb{N}}_d$ are representable. Consider some arrays X on $\overline{\mathbb{N}}_d$ and ξ on $\widetilde{\mathbb{N}}_d$, where ξ has independent entries and the pair (X, ξ) is exchangeable on $\overline{\mathbb{N}}_d$. Then there exist a U-array $\eta \perp\!\!\!\perp \xi$ on $\widetilde{\mathbb{N}}_d$ and a measurable function f between suitable spaces such that*

$$X_k = f(\hat{\xi}_k, \hat{\eta}_k) \quad a.s., \quad k \in \overline{\mathbb{N}}_d.$$

Proof: Since the pair (X, ξ) is exchangeable and hence representable on $\overline{\mathbb{N}}_d$, we may choose a U-array ζ on $\widetilde{\mathbb{N}}_d$ and some measurable functions g and h between suitable spaces such that a.s.

$$X_k = g(\hat{\zeta}_k), \quad \xi_{\tilde{k}} = h(\hat{\zeta}_k), \qquad k \in \overline{\mathbb{N}}_d. \tag{55}$$

Here the latter relation remains true for the symmetrized version of h, which allows us to assume from the outset that h is symmetric. According to the last statement of Proposition 7.13, we may then choose a U-array $\eta \perp\!\!\!\perp \xi$ on $\widetilde{\mathbb{N}}_d$ and a symmetric, measurable function b such that

$$\zeta_J = b(\hat{\xi}_J, \hat{\eta}_J) \quad \text{a.s.,} \quad J \in \widetilde{\mathbb{N}}_d.$$

By the symmetry of b and Lemma 7.12 (ii), we obtain

$$\hat{\zeta}_k = \hat{b}(\hat{\xi}_k, \hat{\eta}_k) \quad \text{a.s.,} \quad k \in \overline{\mathbb{N}}_d,$$

and inserting this into (55) yields the desired formula with $f = g \circ \hat{b}$. $\qquad\square$

We are now ready to complete the main proof.

Proof of Theorem 7.22: In Lemma 7.25 we saw that the exchangeable processes on $\overline{\mathbb{N}}_d$ are representable for every $d \in \mathbb{Z}_+$. To extend the result to $\overline{\mathbb{N}}$, we may proceed as in the proof of Lemma 7.21, except that now we need to use Lemma 7.26 instead of Lemma 7.17 and Lemma 7.2 (i) in place of Lemma 7.4. $\qquad\square$

7.6   Equivalence Criteria

The previous representations are far from unique, in general. Our present aim is to characterize pairs of functions f and f' that can be used to represent the same array, when composed with suitable U-arrays ξ and ξ'. In the contractable case, it is clearly equivalent that the arrays $X_J = f(\hat{\xi}_J)$ and $X'_J = f'(\hat{\xi}_J)$ have the same distribution for a given U-array ξ. The case of exchangeable arrays is similar.

Writing λ for Lebesgue measure on $[0, 1]$, we say that a measurable function $f(\hat{x}_J)$, $J \in \widetilde{\mathbb{N}}$, *preserves λ in the highest order arguments*, if for fixed $J \in \widetilde{\mathbb{N}}$ and $\hat{x}'_J = \hat{x}_J \setminus x_J$ the mapping $x_J \mapsto f(\hat{x}_J)$ preserves λ on $[0, 1]$, where

$\hat{x}_J \setminus x_J$ denotes the array $\hat{x}_J$ with the element x_J omitted. Similarly, we say that the function $g(\hat{x}_J, \hat{y}_J)$ *maps λ^2 into λ in the highest order arguments*, if for fixed $J \in \widetilde{\mathbb{N}}$ and $(\hat{x}'_J, \hat{y}'_J)$ the function $(x_J, y_J) \mapsto g(\hat{x}_J, \hat{y}_J)$ maps λ^2 into λ. When $\eta_J = f(\hat{\xi}_J)$, we define $\hat{f}$ by the formula $\hat{\eta}_J = \hat{f}(\hat{\xi}_J)$.

Consider an random array η on $\widetilde{\mathbb{N}}$, given by

$$\eta_J = f(\hat{\xi}_J), \quad J \in \widetilde{\mathbb{N}},$$

in terms of a U-array ξ on $\widetilde{\mathbb{N}}$ and a measurable function f between suitable spaces. We shall often use the fact that f has a version that preserves λ in the highest order arguments iff η_J is $U(0,1)$ and independent of $\hat{\xi}'_J = \hat{\xi}_J \setminus \xi_J$ for every $J \in \widetilde{\mathbb{N}}$. Similar statements hold for representations of separately or jointly exchangeable arrays.

We begin with the equivalence criteria for representations of contractable arrays on $\widetilde{\mathbb{N}}$.

Theorem 7.27 *(contractable arrays) Let ξ and η be independent U-arrays on $\widetilde{\mathbb{N}}$, and fix some measurable functions $f, f' : \bigcup_{d \geq 0} [0,1]^{2^d} \to S$, where S is Borel. Then these conditions are equivalent:*

(i) $(f(\hat{\xi}_J); J \in \widetilde{\mathbb{N}}) \overset{d}{=} (f'(\hat{\xi}_J); J \in \widetilde{\mathbb{N}})$;

(ii) $f \circ \hat{g}(\hat{\xi}_J) = f' \circ \hat{g}'(\hat{\xi}_J)$ *a.s., $J \in \widetilde{\mathbb{N}}$, for some measurable functions g, g': $\bigcup_{d \geq 0} [0,1]^{2^d} \to [0,1]$ that preserve λ in the highest order arguments;*

(iii) $f(\hat{\xi}_J) = f' \circ \hat{h}(\hat{\xi}_J, \hat{\eta}_J)$ *a.s., $J \in \widetilde{\mathbb{N}}$, for a measurable function h: $\bigcup_{d \geq 0} [0,1]^{2^{d+1}} \to [0,1]$ that maps λ^2 into λ in the highest order arguments.*

Proof, (i) $\Rightarrow$ (ii): Assume (i). Then by Corollary 7.4 we may choose a U-array ζ on $\widetilde{\mathbb{N}}$, such that the pair (ξ, ζ) is contractable and

$$f(\hat{\xi}_J) = f'(\hat{\zeta}_J) \text{ a.s., } \quad J \in \widetilde{\mathbb{N}}. \tag{56}$$

Next, Theorem 7.15 yields the existence of a U-array χ on $\widetilde{\mathbb{N}}$ and some measurable functions g and g' between suitable spaces such that a.s.

$$\xi_J = g(\hat{\chi}_J), \quad \zeta_J = g'(\hat{\chi}_J), \qquad J \in \widetilde{\mathbb{N}}. \tag{57}$$

By Proposition 7.7 (ii) we can modify g and g' to become λ-preserving in the highest order arguments. Substituting (57) into (56) yields the relation in (ii) with ξ replaced by χ.

(ii) $\Rightarrow$ (iii): Assuming (ii), we define

$$\chi_J = g(\hat{\xi}_J), \quad \zeta_J = g'(\hat{\xi}_J), \qquad J \in \widetilde{\mathbb{N}}. \tag{58}$$

Since g and g' preserve λ in the highest order arguments, Proposition 7.7 (ii) shows that χ and ζ are again U-arrays on $\widetilde{\mathbb{N}}$. By Proposition 7.13 we may then choose a measurable function b and a U-array $\gamma \perp\!\!\!\perp \chi$ on $\widetilde{\mathbb{N}}$ such that

$$\xi_J = b(\hat{\chi}_J, \hat{\gamma}_J) \text{ a.s., } \quad J \in \widetilde{\mathbb{N}}.$$

Substituting this into (58) gives

$$\zeta_J = g' \circ \hat{b}(\hat{\chi}_J, \hat{\gamma}_J) \text{ a.s.}, \quad J \in \tilde{\mathbb{N}},$$

which, when inserted into (ii), yields the relation in (iii) with $h = g' \circ \hat{b}$ and with (ξ, η) replaced by (χ, γ). By Proposition 7.7 (ii) we can finally modify h into a function that maps λ^2 into λ in the highest order arguments.

(iii) $\Rightarrow$ (i): Assuming (iii), we may define

$$\zeta_J = h(\hat{\xi}_J, \hat{\eta}_J), \quad J \in \tilde{\mathbb{N}}. \tag{59}$$

Since h maps λ^2 into λ in the highest order arguments, Proposition 7.7 (ii) shows that ζ is again a U-array on $\tilde{\mathbb{N}}$. Hence,

$$(f(\hat{\xi}_J); \ J \in \tilde{\mathbb{N}}) = (f'(\hat{\zeta}_J); \ J \in \tilde{\mathbb{N}}) \stackrel{d}{=} (f'(\hat{\xi}_J); \ J \in \tilde{\mathbb{N}}),$$

which proves (i). $\qquad\qquad\qquad\qquad\qquad\qquad\qquad\qquad\qquad\qquad\qquad\square$

The result for jointly exchangeable arrays on $\overline{\mathbb{N}}$ or $\tilde{\mathbb{N}}$ is similar:

Theorem 7.28 *(jointly exchangeable arrays, Hoover)* *For any ξ, η, f, and f' as in Theorem 7.27, these conditions are equivalent:*

(i) $(f(\hat{\xi}_k); \ k \in \overline{\mathbb{N}}) \stackrel{d}{=} (f'(\hat{\xi}_k); \ k \in \overline{\mathbb{N}})$;

(ii) $f \circ \hat{g}(\hat{\xi}_J) = f' \circ \hat{g}'(\hat{\xi}_J)$ *a.s.*, $J \in \tilde{\mathbb{N}}$, *for some symmetric, measurable functions $g, g' : \bigcup_{d \geq 0} [0,1]^{2^d} \to [0,1]$ that preserve λ in the highest order arguments;*

(iii) $f(\hat{\xi}_J) = f' \circ \hat{h}(\hat{\xi}_J, \hat{\eta}_J)$ *a.s.*, $J \in \tilde{\mathbb{N}}$, *for a symmetric, measurable function $h : \bigcup_{d \geq 0} [0,1]^{2^{d+1}} \to [0,1]$ that maps λ^2 into λ in the highest order arguments.*

The last result also covers the case of exchangeable arrays on $\tilde{\mathbb{N}}$, since by Theorem 7.15 we may then choose the representing functions f and f' to be symmetric, in which case (i) reduces to the condition

$$(f(\hat{\xi}_J); \ J \in \tilde{\mathbb{N}}) \stackrel{d}{=} (f'(\hat{\xi}_J); \ J \in \tilde{\mathbb{N}}).$$

Proof. (i) $\Rightarrow$ (ii): Assume (i). Then by Lemma 7.2 (i) there exists a U-array ζ on $\tilde{\mathbb{N}}$, such that (ξ, ζ) is exchangeable and

$$f(\hat{\xi}_k) = f'(\hat{\zeta}_k) \text{ a.s.}, \quad k \in \overline{\mathbb{N}}.$$

By Theorem 7.22 we may next choose a U-array χ on $\tilde{\mathbb{N}}$ and some measurable functions g and g' between suitable spaces such that a.s.

$$\xi_{\bar{k}} = g(\hat{\chi}_k), \quad \zeta_{\bar{k}} = g'(\hat{\chi}_k), \quad k \in \overline{\mathbb{N}}.$$

This clearly remains true with g and g' replaced by their symmetrized versions, and by Proposition 7.7 (ii) we may further assume that both functions preserve λ in the highest order arguments. Condition (ii) now follows by substitution, as before.

(ii) $\Rightarrow$ (iii): Here we may proceed as in the proof of Theorem 7.27, with only some minor changes: The last statement of Proposition 7.13 now enables us to choose a symmetric version of b. Then Lemma 7.12 (iii) shows that even $h = g' \circ \hat{b}$ is symmetric.

(iii) $\Rightarrow$ (i): Assuming (iii), we may define ζ by (59), and we note as before that ζ is a U-array on $\overline{\mathbb{N}}$. Since h is symmetric, (59) extends by Lemma 7.12 (ii) to the form

$$\hat{\zeta}_k = \hat{h}(\hat{\xi}_k, \hat{\eta}_k), \quad k \in \overline{\mathbb{N}}.$$

Furthermore, we may use the exchangeability of (ξ, η) to extend the relation in (iii) to

$$f(\hat{\xi}_k) = f' \circ \hat{h}(\hat{\xi}_k, \hat{\eta}_k), \quad k \in \overline{\mathbb{N}}.$$

Hence, by combination

$$(f(\hat{\xi}_k);\ k \in \overline{\mathbb{N}}) = (f'(\hat{\zeta}_k);\ k \in \overline{\mathbb{N}}) \stackrel{d}{=} (f'(\hat{\xi}_k);\ k \in \overline{\mathbb{N}}),$$

which proves (i). $\qquad\qquad\qquad\qquad\qquad\qquad\qquad\qquad\qquad\square$

The corresponding equivalence criteria for separately exchangeable arrays may be less obvious. Assuming $\eta_k = f(\hat{\xi}_k)$ for all $k \in \mathbb{Z}_+^d$, we now define the function $\hat{f}$ on $\mathbb{Z}_+^d$ by $\hat{\eta}_k = \hat{f}(\hat{\xi}_k)$.

Theorem 7.29 (*separately exchangeable arrays, Hoover*) *Let ξ and η be independent U-arrays on $\mathbb{Z}_+^d$, and fix some measurable functions $f, f' : [0,1]^{2^d} \to S$, where S is Borel. Then these conditions are equivalent:*

(i) $(f(\hat{\xi}_k);\ k \in \mathbb{N}^d) \stackrel{d}{=} (f'(\hat{\xi}_k);\ k \in \mathbb{N}^d)$;

(ii) $f \circ \hat{g}(\hat{\xi}_k) = f' \circ \hat{g}'(\hat{\xi}_k)$ *a.s.*, $k \in \mathbb{N}^d$, *for some measurable functions* $g, g' : \bigcup_{J \in 2^d} [0,1]^{2^J} \to [0,1]$ *that preserve λ in the highest order arguments;*

(iii) $f(\hat{\xi}_k) = f' \circ \hat{h}(\hat{\xi}_k, \hat{\eta}_k)$ *a.s.*, $k \in \mathbb{N}^d$, *for a measurable function* $h : \bigcup_{J \in 2^d} [0,1]^{2 \cdot 2^J} \to [0,1]$ *that maps λ^2 into λ in the highest order arguments.*

Proof. (i) $\Rightarrow$ (ii): Assuming (i), there exists by Lemma 7.2 (i) a U-array ζ on $\mathbb{Z}_+^d$, such that the pair (ξ, ζ) is separately exchangeable and

$$f(\hat{\xi}_k) = f'(\hat{\zeta}_k) \text{ a.s.}, \quad k \in \mathbb{Z}_+^d.$$

By an obvious extension of Corollary 7.23, there exist a U-array χ on $\mathbb{Z}_+^d$ and some measurable functions g and g' between suitable spaces such that a.s.

$$\xi_k = g(\hat{\chi}_k), \quad \zeta_k = g'(\hat{\chi}_k), \qquad k \in \mathbb{Z}_+^d.$$

Finally, Proposition 7.7 (i) allows us to construct modifications of g and g' that preserve λ in the highest order arguments.

(ii) $\Rightarrow$ (iii): Assuming (ii), we may define

$$\chi_k = g(\hat{\xi}_k), \quad \zeta_k = g'(\hat{\xi}_k), \qquad k \in \mathbf{Z}_+^d,$$

and we note that χ and ζ are U-arrays by Proposition 7.7 (i). By Proposition 7.14 there exist a U-array $\gamma \perp\!\!\!\perp \chi$ on $\mathbf{Z}_+^d$ and a measurable function b satisfying

$$\xi_k = b(\hat{\chi}_k, \hat{\gamma}_k) \text{ a.s.}, \quad k \in \mathbf{Z}_+^d.$$

Hence, by combination

$$\zeta_k = g' \circ \hat{b}(\hat{\chi}_k, \hat{\gamma}_k) \text{ a.s.}, \quad k \in \mathbf{Z}_+^d.$$

Finally, Proposition 7.7 (i) shows that the function $h = g' \circ \hat{b}$ has an equivalent version that maps λ into λ^2 in the highest order arguments.

(iii) $\Rightarrow$ (i): Assuming (iii), we see from Proposition 7.7 (i) that

$$\zeta_k = h(\hat{\xi}_k, \hat{\eta}_k), \quad k \in \mathbf{Z}_+^d,$$

defines a U-array ζ on $\mathbf{Z}_+^d$. $\square$

For an interesting and useful application of the previous results, we show how the representation in Theorem 7.22 can be extended to any nested sequence of exchangeable arrays. The statement will be needed to prove a main result in Chapter 9. To explain the terminology, consider two random arrays $X = (X_{ij})$ and $Y = (Y_{ij})$ on $\mathbf{N}^2$ satisfying $X \subset Y$, in the sense that X is the sub-array of Y obtained by omitting the jth row and column whenever $Y_{jj} \neq B$ for some measurable set B. If Y is jointly exchangeable and ergodic, then the sequence (Y_{jj}) is i.i.d. by de Finetti's theorem, and the set of indices j with $Y_{jj} \in B$ has density $p = P\{Y_{11} \in B\}$. It is clear that X is again ergodic, exchangeable.

Now Theorem 7.22 shows that Y has an a.s. representation

$$Y_{ij} = g(\xi_i, \xi_j, \zeta_{ij}), \quad i, j \in \mathbf{N},$$

in terms of some i.i.d. $U(0,1)$ random variables ξ_i and $\zeta_{ij} = \zeta_{ji}$, $i \leq j$, and a measurable function g on $[0,1]^3$, where we may assume that $g(x, x, z) = g_1(x)$ for some function g_1 on $[0,1]$. Writing $A = g_1^{-1}B$ and letting $h \colon [0,1] \to A$ be measurable with $\lambda \circ h^{-1} = 1_A \lambda / p$, we note that X is representable in terms of the function

$$f(x, y, z) = g(h(x), h(y), z), \quad x, y, z \in [0,1].$$

Choosing g_1 such that $A = [0, p]$, we get in particular

$$f(x, y, z) = g(px, py, z), \quad x, y, z \in [0, 1]. \tag{60}$$

This suggests that, for any nested sequence $X^1 \subset X^2 \subset \ldots$ of jointly ergodic, exchangeable arrays on $\mathbb{N}^2$, the combined array (X^n) should be representable in terms of a single function f on $\mathbb{R}_+^2 \times [0,1]$. If we could reverse the previous construction and recover a version of the function g in (60) in terms of f, then by continuing recursively we would obtain a representing function for the whole sequence. Unfortunately, the indicated construction seems to be impossible in general, and a more sophisticated approach is required.

Proposition 7.30 *(nested arrays) Let $X^1 \subset X^2 \subset \ldots$ be a nested sequence of jointly ergodic, exchangeable arrays on $\mathbb{N}^2$, where X^1 has density $p_n = r_n^{-1}$ in X^n for each n. Then there exists a measurable function f on $\mathbb{R}_+^2 \times [0,1]$ such that the X^n have representing functions*

$$f_n(x,y,z) = f(r_n x, r_n y, z), \quad x,y,z \in [0,1], \quad n \in \mathbb{N}.$$

Proof: Let us first consider only two jointly ergodic, exchangeable arrays $X \subset Y$, where X has density $p = r^{-1}$ in Y. Given a representing function f of X, we need to construct a suitable extension of f, serving as a representing function of Y. Then consider any representing function g of Y, chosen such that X is representable in terms of the function

$$\hat{g}(x,y,z) = g(px, py, z), \quad x,y,z \in [0,1].$$

Then Theorem 7.28 shows that the functions f, g, and $\hat{g}$ are related on $[0,1]^6$, a.e. with respect to λ^6, by

$$
\begin{aligned}
f(x,y,z) &= \hat{g}(h_1(x,x'), h_1(y,y'), h_2(x,x',y,y',z,z')) \\
&= g(ph_1(x,x'), ph_1(y,y'), h_2(x,x',y,y',z,z')), \quad (61)
\end{aligned}
$$

for some measurable functions $h_1 \colon [0,1]^2 \to [0,1]$ and $h_2 \colon [0,1]^6 \to [0,1]$ that map λ^2 into λ in the last two arguments, where h_2 is also symmetric in the pairs (x,x') and (y,y').

We now extend h_1 to a measurable function $\hat{h}_1$ from $[0,r] \times [0,1]$ to $[0,r]$, still mapping λ^2 into λ. Similarly, we extend h_2 to a symmetric, measurable function $\hat{h}_2$ from $[0,r]^2 \times [0,1]^4$ to $[0,1]$, where the extension is in the x- and y-coordinates, such that λ^2 continues to be mapped into λ in the last two variables. Next define a function $\hat{f}$ on $[0,r]^2 \times [0,1]^4$ by

$$\hat{f}(x,x',y,y',z,z') = g\big(p\hat{h}_1(x,x'), p\hat{h}_1(y,y'), \hat{h}_2(x,x',y,y',z,z')\big).$$

Comparing with (61), we see that $\hat{f}$ is an extension of f, in the sense that

$$\hat{f}(x,x',y,y',z,z') = f(x,y,z) \quad \text{on } [0,1]^6 \text{ a.e. } \lambda^6.$$

Furthermore, if ξ_i, ξ_i' and $\zeta_{ij}, \zeta_{ij}', i \leq j$, are i.i.d. $U(0,1)$, then the same thing is true for the variables

$$\eta_i = p\hat{h}_1(r\xi_i, \xi_i'), \qquad \eta_{ij} = \hat{h}_2(r\xi_i, \xi_i', r\xi_j, \xi_j', \zeta_{ij}, \zeta_{ij}').$$

Hence, the array

$$\hat{Y}_{ij} = \hat{f}(r\xi_i, \xi_i', r\xi_j, \xi_j', \zeta_{ij}, \zeta_{ij}'), \quad i, j \in \mathbb{N},$$

has the same distribution as Y, and it follows that Y itself can be represented by the function $f(rx, x', ry, y', z, z')$ on $[0,1]^6$.

We return to our nested sequence of jointly ergodic, exchangeable arrays $X^1 \subset X^2 \subset \cdots$. Proceeding recursively, as before, we may construct a sequence of measurable functions f_n on $[0, r_n]^2 \times [0,1]^{3n-2}$, each providing a representation

$$X_{ij}^n = f_n(r_n \xi_i^n, \tilde{\xi}_i^n, r_n \xi_j^n, \tilde{\xi}_j^n, \zeta_{ij}^n) \text{ a.s.}, \quad i, j, n \in \mathbb{N},$$

for some independent and uniformly distributed random vectors ξ_i^n in $[0,1]$, $\tilde{\xi}_i^n$ in $[0,1]^{n-1}$, and $\zeta_{ij}^n = \zeta_{ji}^n$ in $[0,1]^n$, where the f_n are successive extensions of each other in the sense that, for any $m \le n$,

$$f_m(x, y, z) = f_n(x, x', y, y', z, z') \text{ on } [0, r_m]^2 \times [0,1]^{3n-2} \text{ a.e. } \lambda^{3n}.$$

Regarding the functions f_n as defined on the infinite-dimensional product spaces

$$S_n = ([0, r_n] \times [0,1]^\infty)^2 \times [0,1]^\infty, \quad n \in \mathbb{N},$$

though each f_n depends only on the first n variables in each group, we may introduce their common extension $\hat{f}$ to the space

$$S_\infty = (\mathbb{R}_+ \times [0,1]^\infty)^2 \times [0,1]^\infty.$$

This enables us to represent the arrays X^n in the form

$$X_{ij}^n = \hat{f}(r_n \xi_i^n, \tilde{\xi}_i^n, r_n \xi_j^n, \tilde{\xi}_j^n, \zeta_{ij}^n) \text{ a.s.}, \quad i, j, n \in \mathbb{N},$$

for some independent and uniformly distributed random elements ξ_i^n in $[0,1]$ and $\tilde{\xi}_i^n$, $\zeta_{ij}^n = \zeta_{ji}^n$ in $[0,1]^\infty$. Here the latter may be regarded as infinite sequences of i.i.d. $U(0,1)$ random variables.

Let us finally introduce a measurable function h from $\mathbb{R}_+ \times [0,1]^\infty$ to $\mathbb{R}_+$, mapping $[0, r_n] \times [0,1]^\infty$ into $[0, r_n]$ for every n and such that $\lambda^\infty \circ h^{-1} = \lambda$. Putting

$$f(x, y, z) = \hat{f}(h(x), h(y), h(z)), \quad x, y \in \mathbb{R}_+, \ z \in [0,1],$$

and letting ξ_i and $\zeta_{ij} = \zeta_{ji}$ be i.i.d. $U(0,1)$ random variables, we see that the arrays

$$Y_{ij}^n = f(r_n \xi_i, r_n \xi_j, \zeta_{ij}), \quad i, j, n \in \mathbb{N},$$

have the same distributions as $X^1, X^2, \ldots$. Hence, the transfer theorem (FMP 6.10) yields

$$X_{ij}^n = f(r_n \eta_i^n, r_n \eta_j^n, \chi_{ij}^n) \text{ a.s.}, \quad i, j, n \in \mathbb{N},$$

for suitable arrays of i.i.d. $U(0,1)$ random variables η_i^n and $\chi_{ij}^n = \chi_{ji}^n$, $n \in \mathbb{N}$. This shows that f has indeed the required property. $\square$

7.7 Conditional Distributions

In this section we prove some a.s. relations between conditional distributions of exchangeable arrays, which allow us to move freely between different representations. Recall that 2^d denotes the collection of subsets of $\{1, \ldots, d\}$. We say that a class $\mathcal{J} \subset 2^d$ is *ideal* if $I \subset J \in \mathcal{J}$ implies $I \in \mathcal{J}$. This holds in particular for the classes $2^J = \{I; I \subset J\}$, $J \in 2^d$; in general $\mathcal{J}$ is ideal iff $\mathcal{J} = \bigcup_{J \in \mathcal{J}} 2^J$. For any array Y indexed by $\mathbb{Z}^d$ or $\mathbb{Z}_+^d$, we introduce the sub-arrays $Y^J = (Y_k; k_J > 0, k_{J^c} \leq 0)$, $J \in 2^d$, where $k_J > 0$ means that $k_j > 0$ for all $j \in J$, and similarly for $k_{J^c} \leq 0$. For subsets $\mathcal{J} \subset 2^d$ we write $Y^{\mathcal{J}} = \{Y^J; J \in \mathcal{J}\}$, and similarly for $\xi^{\mathcal{J}}$.

We are now ready to state our main result about conditioning in separately exchangeable arrays.

Proposition 7.31 *(separately exchangeable arrays) Let X be a separately exchangeable array on $\mathbb{N}^d$, representable in terms of a U-array ξ on $\mathbb{Z}_+^d$, and let Y be a stationary extension of X to $\mathbb{Z}^d$. Then for any ideal set $\mathcal{J} \subset 2^d$, we have*

$$P[X \in \cdot | \xi^{\mathcal{J}}] = P[X \in \cdot | Y^{\mathcal{J}}] \quad a.s.$$

Our proof is based on the following lemma. For convenience, we may take the representing U-array ξ to be indexed by $\hat{\mathbb{N}}_d = (\mathbb{N}^J; J \in 2^d)$ rather than $\mathbb{Z}_+^d$, which allows us to construct an extension Y of X to $\mathbb{Z}^d$ by extending ξ to a U-array η on $\hat{\mathbb{Z}}_d = (\mathbb{Z}^J; J \in 2^d)$. We may then write ξ^J for the restriction of ξ to $\mathbb{N}^J$, and similarly for η.

Lemma 7.32 *(conditional independence) Let X be a separately exchangeable array on $\mathbb{N}^d$, representable in terms of a U-array ξ on $\hat{\mathbb{N}}_d$, and let (Y, η) be a stationary extension of (X, ξ) to $\mathbb{Z}^d$. Then*

(i) $Y \perp\!\!\!\perp_{Y^{\mathcal{J}}} \eta^{\mathcal{J}}$ *for every ideal set $\mathcal{J} \subset 2^d$,*

(ii) $X \perp\!\!\!\perp_{\xi^{\mathcal{J}}} Y^{\mathcal{J}}$ *for every ideal set $\mathcal{J} \subset 2^d$.*

Proof: (i) For any $j \in J \in 2^d$ we consider the sequence

$$\zeta_h = (\hat{Y}_k^J; k_j = h), \quad h \in \mathbb{Z}.$$

Putting $I = J \setminus \{j\}$ and noting that $\hat{\eta}^I$ is invariant in the j-th index, we see from the exchangeability of (Y, η) that the sequence $(\zeta_h, \hat{\eta}^I)$, $h \in \mathbb{Z}$, is exchangeable. Hence, Lemma 1.3 yields $\zeta \perp\!\!\!\perp_{\zeta_-} \hat{\eta}^I$, where ζ_- denotes the restriction of ζ to $\mathbb{Z}_-$, and so

$$\hat{Y}^J \perp\!\!\!\perp_{\hat{Y}^I} \hat{\eta}^I, \quad I \subset J \in 2^d, \ |J \setminus I| = 1.$$

Fixing any $J \in 2^d$ with $|J| = m$ and choosing $J = J_m \subset \cdots \subset J_d$ with $|J_k| = k$ for all k, we obtain

$$\hat{Y}^{J_{k+1}} \perp\!\!\!\perp_{\hat{Y}^{J_k}} \hat{\eta}^J, \quad k = m, \ldots, d-1.$$

Noting that
$$\hat{Y}^J = \hat{Y}^{J_m} \subset \cdots \subset \hat{Y}^{J_d} = Y,$$

in the sense of inclusion of the corresponding index sets, we may use the chain rule in FMP 6.8 to see that

$$Y \perp\!\!\!\perp_{\hat{Y}^J} \hat{\eta}^J, \quad J \in 2^d.$$

Applying the latter relation for any $\mathcal{I} \subset 2^d$ to the combined array $Z = (Y, \eta^{\mathcal{I}})$, which is again separately exchangeable with representing U-array η, we obtain

$$(Y, \eta^{\mathcal{I}}) \perp\!\!\!\perp_{\hat{Z}^J} \hat{\eta}^J, \quad J \in 2^d.$$

If the set $\mathcal{J} \subset 2^d$ is ideal and contains $\mathcal{I}$, then

$$\hat{Z}^J \subset Z^J = (Y^J, \eta^{\mathcal{I}}) \subset (Y, \eta^{\mathcal{I}}), \quad J \in \mathcal{J},$$

and it follows that

$$Y \underset{Y^J, \eta^{\mathcal{I}}}{\perp\!\!\!\perp} \hat{\eta}^J, \quad J \in \mathcal{J}.$$

Fixing an enumeration $\mathcal{J} = \{J_1, \ldots, J_m\}$ and writing $\mathcal{J}_k = \{J_1, \ldots, J_k\}$ for $k = 0, \ldots, m$, we get in particular

$$Y \underset{Y^{\mathcal{J}}, \eta^{\mathcal{J}_{k-1}}}{\perp\!\!\!\perp} \hat{\eta}^{J_k}, \quad k = 1, \ldots, m,$$

where $\eta^{\mathcal{J}_0} = \emptyset$. The asserted relation now follows by the chain rule for conditional independence in FMP 6.8.

(ii) Since $\xi \perp\!\!\!\perp (\eta \setminus \xi)$, we have

$$X \perp\!\!\!\perp_{\xi^{\mathcal{J}}} (\xi^{\mathcal{J}}, \eta \setminus \xi), \quad \mathcal{J} \subset 2^d.$$

If $\mathcal{J}$ is ideal, then the representation of $Y^{\mathcal{J}}$ involves only elements from $(\xi^{\mathcal{J}}, \eta \setminus \xi)$, and the assertion follows. $\qquad\square$

Proof of Proposition 7.31: If Y is representable in terms of a stationary extension of ξ to $\hat{\mathbb{Z}}_d$, then by Lemma 7.32 (i) and (ii) we have a.s.

$$P[X \in \cdot |Y^{\mathcal{J}}] = P[X \in \cdot |Y^{\mathcal{J}}, \xi^{\mathcal{J}}] = P[X \in \cdot |\xi^{\mathcal{J}}]. \tag{62}$$

In the general case, let $\bar{\eta}$ be a U-array on $\hat{\mathbb{Z}}_d$ representing Y, and let η denote the restriction of $\bar{\eta}$ to $\hat{\mathbb{N}}_d$. Introduce a third U-array $\zeta \perp\!\!\!\perp_X (\xi, \eta)$ on $\hat{\mathbb{N}}_d$ satisfying $(\zeta, X) \overset{d}{=} (\xi, X)$, and note that the triples (X, ξ, ζ) and (X, η, ζ) are separately exchangeable on $\mathbb{N}^d$ by Lemma 7.2. In particular, they admit stationary extensions $(U, \bar{\xi}, \bar{\zeta})$ and $(V, \bar{\eta}', \bar{\zeta}')$ to $\mathbb{Z}^d$, and we note that U is representable in terms of both $\bar{\xi}$ and $\bar{\zeta}$, whereas V is representable in terms of both $\bar{\eta}'$ and $\bar{\zeta}'$. Hence, (62) applies to all five triples

$$(X, Y, \eta), \quad (X, V, \eta), \quad (X, V, \zeta), \quad (X, U, \zeta), \quad (X, U, \xi),$$

and we get

$$P[X \in \cdot | Y^{\mathcal{J}}] = P[X \in \cdot | \eta^{\mathcal{J}}] = P[X \in \cdot | V^{\mathcal{J}}]$$
$$= P[X \in \cdot | \zeta^{\mathcal{J}}] = P[X \in \cdot | U^{\mathcal{J}}] = P[X \in \cdot | \xi^{\mathcal{J}}]. \qquad \square$$

To state the corresponding result in the jointly exchangeable case, recall that the representing array ξ is then indexed by $\tilde{\mathbb{N}}_d = \{J \subset \mathbb{N}; |J| \le d\}$. For any $k \le d$, let ξ_k and $\hat{\xi}_k$ denote the restrictions of ξ to $\Delta \tilde{\mathbb{N}}_k$ and $\tilde{\mathbb{N}}_k$, respectively. Write $Y_\emptyset = Y^\emptyset$ for convenience.

Proposition 7.33 *(jointly exchangeable arrays)* *Let X be a jointly exchangeable array on $\mathbb{N}^d$, representable in terms of each of the U-arrays ξ and η on $\tilde{\mathbb{N}}_d$, and let Y be a stationary extension of X to $\mathbb{Z}^d$. Then*
  (i) $P[X \in \cdot | \hat{\xi}_k] = P[X \in \cdot | \hat{\eta}_k]$ *a.s. for all $k \le d$,*
  (ii) $P[X \in \cdot | \xi_\emptyset] = P[X \in \cdot | Y_\emptyset]$ *a.s.*

Proof: (i) We may assume that the pair (ξ, η) is exchangeable, since we can otherwise introduce a third representing U-array $\zeta \perp\!\!\!\perp_X (\xi, \eta)$, so that (ξ, ζ) and (η, ζ) become exchangeable by Lemma 7.2. We may then use the result in the special case to conclude that

$$P[X \in \cdot | \hat{\xi}_k] = P[X \in \cdot | \hat{\zeta}_k] = P[X \in \cdot | \hat{\eta}_k] \text{ a.s.,} \quad k \le d.$$

If (ξ, η) is exchangeable, then both ξ and η can be represented as in Theorem 7.15 in terms of a common U-array ζ on $\tilde{\mathbb{N}}_d$. Assuming the assertion to be true for each of the pairs (ξ, ζ) and (η, ζ), we obtain

$$P[X \in \cdot | \hat{\xi}_k] = P[X \in \cdot | \hat{\zeta}_k] = P[X \in \cdot | \hat{\eta}_k] \text{ a.s.,} \quad k \le d.$$

This reduces the argument to the case where η is representable in terms of ξ, so that $\eta_J = g(\hat{\xi}_J)$ for some measurable function g. Since η has independent entries, we see from Proposition 7.7 (ii) that $\eta_{m+1} \perp\!\!\!\perp \hat{\xi}_m$ for every $m < d$. Noting that $\hat{\eta}_m$ is representable in terms of $\hat{\xi}_m$, we get

$$(\hat{\xi}_k, \hat{\eta}_m) \perp\!\!\!\perp \eta_{m+1}, \quad 0 \le k \le m < d,$$

and so by iteration

$$\hat{\xi}_k \perp\!\!\!\perp (\eta \setminus \hat{\eta}_k), \quad 0 \le k < d. \tag{63}$$

Now assume that X has a representation $f(\hat{\eta})$ as in Theorem 7.22. Combining with the formula $\eta = g(\hat{\xi})$, we obtain

$$X = f(\hat{\eta}) = f(\hat{\eta}_k, \eta \setminus \hat{\eta}_k) = f(\hat{g}(\hat{\xi}_k), \eta \setminus \hat{\eta}_k) \text{ a.s.,} \quad k \le d. \tag{64}$$

For any bounded, measurable function h on the range space of X, write $F = h \circ f$, and use Fubini's theorem in the version of FMP 3.11 or 6.4 to conclude from (63), (64), and the relation $\hat{\eta}_k \perp\!\!\!\perp (\eta \setminus \hat{\eta}_k)$ that

$$E[h(X)|\hat{\xi}_k] = EF(\hat{g}(\hat{x}_k), \eta \setminus \hat{\eta}_k)|_{x=\xi}$$
$$= EF(\hat{y}_k, \eta \setminus \hat{\eta}_k)|_{y=\eta} = E[h(X)|\hat{\eta}_k],$$

which implies the asserted equation.

(ii) Here we may assume that Y is represented by an extension of ξ. In fact, we may otherwise introduce a U-array $\bar{\eta}$ on $\widetilde{\mathbb{Z}}_d$ representing Y and write η for the restriction of $\bar{\eta}$ to $\widetilde{\mathbb{N}}_d$. Assuming the result to be true in the special case and using part (i), we obtain

$$P[X \in \cdot | Y_\emptyset] = P[X \in \cdot | \eta_\emptyset] = P[X \in \cdot | \xi_\emptyset] \quad \text{a.s.}$$

Under this extra hypothesis, the pair $(Y, \xi_\emptyset)$ is again exchangeable. Hence, for any bijection $p \colon \mathbb{Z}_- \to \mathbb{Z}$ we get $(Y \circ p, \xi_\emptyset) \overset{d}{=} (Y_\emptyset, \xi_\emptyset)$, and Lemma 1.3 yields $X \perp\!\!\!\perp_{Y_\emptyset} \xi_\emptyset$. On the other hand we note that $X \perp\!\!\!\perp_{\xi_\emptyset} Y_\emptyset$, since $\xi_\emptyset$ is the only ξ-element in common for the representations of X and $Y_\emptyset$. Combining the two relations, we get

$$P[X \in \cdot | Y_\emptyset] = P[X \in \cdot | Y_\emptyset, \xi_\emptyset] = P[X \in \cdot | \xi_\emptyset] \quad \text{a.s.} \qquad \square$$

Next we show how the exchangeability of a random array is preserved, in two different ways, under conditioning as in Propositions 7.31 or 7.33. To avoid repetitions, here and in the subsequent lemma, we include even the rotatable case. This requires only the basic definitions (see Chapter 8), and the proofs are essentially the same as in the exchangeable case.

Recall that a set I is said to be μ-*invariant* with respect to a group G of measurable transformations if $\mu(I \bigtriangleup g^{-1}I) = 0$ for every $g \in G$. We define a distribution μ to be *ergodic* if every μ-invariant set has measure 0 or 1.

Lemma 7.34 *(preservation laws) Let X be a separately or jointly exchangeable or rotatable array on $\mathbb{N}^d$ with stationary extension Y to $\mathbb{Z}^d$. Then*

(i) X *is conditionally ergodic, separately or jointly exchangeable or rotatable, given $Y_\emptyset$;*

(ii) *the family $(E[X|Y^{\mathcal{J}}]; \mathcal{J} \subset 2^d)$, assuming it exists, is again separately or jointly exchangeable or rotatable.*

In other words, (ii) holds for any real-valued array X with integrable entries. Here we prove only the invariance part of (i), the ergodicity assertion being established after the next lemma.

Partial proof: (i) In the separately exchangeable case, let $p = (p_1, \ldots, p_d)$, where $p_1, \ldots, p_d$ are finite permutations on $\mathbb{N}$. The exchangeability of Y yields $(X \circ p, Y_\emptyset) \overset{d}{=} (X, Y_\emptyset)$, and so $P[X \circ p \in \cdot | Y_\emptyset] = P[X \in \cdot | Y_\emptyset]$ a.s. Since the set of permutations p is countable, we may choose the exceptional null set to be independent of p, and the conditional exchangeability follows. In the jointly exchangeable case, the same proof applies with $p_1 = \cdots = p_d$.

The rotatable case involves the additional complication of an uncountable class of transformations. However, every rotation T can be approximated by rotations from a countable set $T_1, T_2, \ldots$, and the previous argument applies with obvious changes to the countable family $\{T_n\}$.

(ii) By FMP 1.13 we have $E[X|Y^{\mathcal{J}}] = f_{\mathcal{J}}(Y^{\mathcal{J}})$ for some measurable functions $f_{\mathcal{J}}$, taking values in the set of all real-valued arrays on $\mathbb{N}^d$. Now let p be as before, except that each p_i should now be regarded as a permutation on $\mathbb{Z}$ that leaves $\mathbb{Z}_-$ invariant. Then $(Y \circ p)^{\mathcal{J}}$ is a permutation of $Y^{\mathcal{J}}$ generating the same σ-field, and so

$$E[X|Y^{\mathcal{J}}] \circ p = E[X \circ p|Y^{\mathcal{J}}] = E[X \circ p|(Y \circ p)^{\mathcal{J}}] = f_{\mathcal{J}} \circ (Y \circ p)^{\mathcal{J}},$$

where the last relation follows from the exchangeability of Y. Hence, by the same property,

$$
\begin{aligned}
(E[X|Y^{\mathcal{J}}]; \mathcal{J} \subset 2^d) \circ p &= (f_{\mathcal{J}} \circ (Y \circ p)^{\mathcal{J}}; \mathcal{J} \subset 2^d) \\
&\stackrel{d}{=} (f_{\mathcal{J}} \circ Y^{\mathcal{J}}; \mathcal{J} \subset 2^d) \\
&= (E[X|Y^{\mathcal{J}}]; \mathcal{J} \subset 2^d).
\end{aligned}
$$

A similar argument applies to the rotatable case. $\qquad\square$

We proceed to characterize ergodicity for exchangeable arrays on $\mathbb{N}^d$. Here it is again convenient to include the rotatable case. An array X on $\mathbb{N}^d$ is said to be *dissociated* if, for any disjoint sets $I_1, \dots, I_m \in \mathbb{N}$, the restrictions of X to the index sets $I_1^d, \dots, I_m^d$ are independent. It is clearly enough in this condition to take $m = 2$. Let us also say that X is representable in terms of a U-array $\xi \setminus \xi_\emptyset$, if it has a representation $X_k = f(\hat{\xi}_k)$ that does not involve the variable $\xi_\emptyset$.

Lemma 7.35 *(ergodicity, Aldous) Let X be a separately or jointly exchangeable or rotatable array on $\mathbb{N}^d$. Then these conditions are equivalent:*

(i) *X is ergodic,*

(ii) *X is dissociated,*

(iii) *X is representable in terms of a U-array $\xi \setminus \xi_\emptyset$.*

Proof, (i) $\Rightarrow$ (ii): Assume (i), and let Y be a stationary extension of X to $\mathbb{Z}^d$. Then the random measure $P[X \in \cdot|Y_\emptyset]$ is a.s. invariant by Lemma 7.34 (i), and so by (i) and Lemma A1.2 it is a.s. non-random and equal to $\mathcal{L}(X)$. Hence, $X \perp\!\!\!\perp Y_\emptyset$, and (ii) follows by the exchangeability of Y.

(ii) $\Rightarrow$ (iii): Assuming (ii), let X have a representation $f(\hat{\xi})$ as in Theorem 7.22 or Corollary 7.23, and let Y be a stationary extension of X to $\mathbb{Z}^d$. By Proposition 7.31 or 7.33 and Fubini's theorem we get a.s.

$$
\begin{aligned}
P\{X \in \cdot\} &= P[X \in \cdot|Y_\emptyset] = P[X \in \cdot|\xi_\emptyset] \\
&= P[f(\hat{\xi}) \in \cdot|\xi_\emptyset] = P\{f(\hat{\xi} \setminus \xi_\emptyset, x) \in \cdot\}|_{x=\xi_\emptyset},
\end{aligned}
$$

which implies

$$X \stackrel{d}{=} f(\hat{\xi} \setminus \xi_\emptyset, x), \quad x \in [0,1] \text{ a.e. } \lambda.$$

Fixing any $x \in [0,1]$ satisfying the last relation and using the transfer theorem in FMP 6.10, we obtain $X = f(\tilde{\xi} \setminus \tilde{\xi}_\emptyset, x)$ a.s. for some $\tilde{\xi} \overset{d}{=} \xi$.

(iii) $\Rightarrow$ (ii): This is clear since $\xi_\emptyset$ is the only common element of ξ in the representations of X and $Y^\emptyset$.

(ii) $\Rightarrow$ (i): We may proceed as in the classical proof of the Hewitt–Savage zero–one law (FMP 3.14–15). Thus, writing $\mu = \mathcal{L}(X)$, we consider any μ-invariant, measurable set I of arrays on $\mathbb{N}^d$. Choose some approximating sets A_n, depending only on the entries in $\{1,\ldots,n\}^d$, and let B_n be the shifted sets, described in terms of the elements in $\{n+1,\ldots,2n\}^d$. Using the invariance of μ and I and the μ-independence of A_n and B_n, we get

$$\begin{aligned} |\mu I - (\mu I)^2| &\leq |\mu I - \mu(A_n \cap B_n)| + |(\mu I)^2 - (\mu A_n)(\mu B_n)| \\ &\leq 2\mu(I \Delta A_n) + 2\mu(I \Delta B_n) = 4\mu(I \Delta A_n) \to 0, \end{aligned}$$

which shows that $\mu I = (\mu I)^2$. Hence, $\mu I = 0$ or 1. $\qquad\square$

End of proof of Lemma 7.34: Considering an a.s. representation $X = f(\hat{\xi})$ and using Proposition 7.31 or 7.33 and Fubini's theorem, we get a.s.

$$P[X \in \cdot|Y_\emptyset] = P[X \in \cdot|\xi_\emptyset] = P\{f(\hat{\xi} \setminus \xi_\emptyset, x) \in \cdot\}|_{x=\xi_\emptyset}.$$

Here the array $f(\hat{\xi} \setminus \xi_\emptyset, x)$ is again exchangeable or rotatable, and it is also ergodic by Lemma 7.35. Hence, the conditional distribution $P[X \in \cdot|Y_\emptyset]$ is a.s. ergodic. $\qquad\square$

We proceed with a rather technical extension result that will be needed in the next chapter. Here we write $\mathbb{N}'_d$ for the non-diagonal part of $\mathbb{N}^d$ and define $\hat{\mathbb{N}}'_d = \bigcup_J \mathbb{N}'_J$, where $\mathbb{N}'_J$ denotes the non-diagonal part of $\mathbb{N}^J$ for arbitrary $J \in 2^d$. Similarly, if $R = N_1 \times \cdots \times N_d$ for some disjoint sets $N_1,\ldots,N_d \subset \mathbb{N}$, we write $\hat{R} = \bigcup_J N_J$, where N_J denotes the product of the sets N_j with $j \in J$. Note that $\hat{\mathbb{N}}'_d$ can be regarded as a subset of $\mathbb{Z}^d_+$ and that $\hat{R}$ can be identified with $\bar{N}_1 \times \cdots \times \bar{N}_d$, where $\bar{N}_j = N_j \cup \{0\}$ for all j. However, the joint exchangeability on $\hat{\mathbb{N}}'_d$ is always defined with respect to permutations on $\mathbb{N}$.

Lemma 7.36 *(extension from rectangular set) Let X be a jointly exchangeable array on $\mathbb{N}'_d$ such that*

$$X_k = f(\hat{\xi}_k) \quad a.s., \quad k \in N_1 \times \cdots \times N_d \equiv R, \tag{65}$$

for some disjoint, infinite sets $N_1,\ldots,N_d \subset \mathbb{N}$, a measurable function f on $[0,1]^{2^d}$, and a U-array ξ on $\hat{R}$. Then there exists a jointly exchangeable array η on $\hat{\mathbb{N}}'_d$ with $\eta \overset{d}{=} \xi$ on $\hat{R}$ such that

$$X_k = f(\hat{\eta}_k) \quad a.s., \quad k \in \mathbb{N}'_d. \tag{66}$$

If X is ergodic and the function $f(\hat{x}_k)$ is independent of $x_\emptyset$, then even $\eta \setminus \eta_\emptyset$ can be chosen to be ergodic.

Proof: Since X is jointly exchangeable, Theorem 7.22 yields a representation

$$X_k = g(\hat{\zeta}_k) \text{ a.s.,} \quad k \in \mathbb{N}'_d, \tag{67}$$

in terms of of a U-array ζ on $\tilde{\mathbb{N}}_d$ and a measurable function g on $[0,1]^{2^d}$. Comparing with (65) and using Theorem 7.29, we obtain

$$g(\hat{\zeta}_k) = f \circ \hat{h}(\hat{\zeta}_k, \hat{\vartheta}_k) \text{ a.s.,} \quad k \in R, \tag{68}$$

for some measurable function h on $\bigcup_{J \in 2^d} [0,1]^{2 \cdot 2^J}$ that maps λ^2 into λ in the highest order arguments, where ϑ is an arbitrary U-array on $\tilde{\mathbb{N}}_d$ independent of ζ. Since the combined U-array (ζ, ϑ) is trivially exchangeable, equation (68) extends to $\mathbb{N}'_d$, and we may substitute into (67) to obtain (66) with

$$\eta_k^J = h^J(\hat{\zeta}_k, \hat{\vartheta}_k), \quad k \in \mathbb{N}'_J, \ J \in 2^d. \tag{69}$$

The joint exchangeability of η follows from that of (ζ, ϑ), and Proposition 7.7 (i) gives $\eta \overset{d}{=} \xi$ on $\hat{R}$.

In the ergodic case, let $c \in [0,1]^2$ be arbitrary, and define an array η^c as in (69), except that we now replace $(\zeta_\emptyset, \vartheta_\emptyset)$ by c. Next replace the element $\eta_\emptyset^c$ in η^c by a $U(0,1)$ random variable $\chi_\emptyset \perp\!\!\!\perp (\zeta, \vartheta)$ to form an array χ^c. Note that χ^c is again jointly exchangeable. Since η is a U-array on $\hat{R}$, Proposition 7.31 shows that the $\hat{R}$-restriction of $\eta \setminus \eta_\emptyset$ is independent of $(\zeta_\emptyset, \vartheta_\emptyset)$, and so Fubini's theorem yields $\chi^c \overset{d}{=} \eta \overset{d}{=} \xi$ on $\hat{R}$ for λ^2-almost every c. Next we see from Lemma 7.35 that X is dissociated, which implies $X \perp\!\!\!\perp (\zeta_\emptyset, \vartheta_\emptyset)$ by Proposition 7.33. Defining X^c as in (66), but now with η replaced by η^c or χ^c, we get $X^c \overset{d}{=} X$ for almost every c by Fubini's theorem. For any non-exceptional c, the transfer theorem in FMP 6.10 ensures that (66) remains true with η replaced by a suitable array $\tilde{\eta} \overset{d}{=} \chi^c$. Finally, we note that $\tilde{\eta} \overset{d}{=} \xi$ on $\hat{R}$ and that $\tilde{\eta} \setminus \tilde{\eta}_\emptyset \overset{d}{=} \eta^c \setminus \eta_\emptyset^c$ is ergodic by Lemma 7.35. $\qquad\square$

We conclude with a simple algebraic fact concerning the ideal classes in 2^d. Let $\mathcal{P}_d$ denote the class of partitions of the set $\{1, \ldots, d\}$ into disjoint, non-empty subsets. We say that a set S is *separated* by a class of subsets $\mathcal{C} \subset 2^S$, if for any $x \neq y$ in S there exists a set $A \in \mathcal{C}$ that contains exactly one of the points x and y.

Lemma 7.37 *(separation)* *The class $\mathcal{P}_d$ is separated by the family*

$$\mathcal{P}_J = \{\pi \in \mathcal{P}_d; \ \pi \subset J\}, \quad J \subset 2^d \text{ ideal.}$$

Proof: Fix any partition $\pi = \{J_1, \ldots, J_m\}$ of $\{1, \ldots, d\}$, and note that the classes $\mathcal{J} = \bigcup_k 2^{J_k}$ and $\mathcal{J}_k = \mathcal{J} \setminus J_k$, $k \leq m$, are ideal. We claim that

$$\{\pi\} = \mathcal{P}_{\mathcal{J}} \cap \mathcal{P}_{\mathcal{J}_1}^c \cap \cdots \cap \mathcal{P}_{\mathcal{J}_m}^c. \tag{70}$$

Here the inclusion $\subset$ is obvious, since $\pi \subset \mathcal{J}$ and $\pi \not\subset \mathcal{J}_k$ for all $k \leq m$. Conversely, suppose that $\pi' \in \mathcal{P}_d$ belongs to the right-hand side of (70), so that $\pi' \subset \mathcal{J}$ and $\pi' \not\subset \mathcal{J}_k$ for all k. Since $\mathcal{J} \setminus \mathcal{J}_k = \{J_k\}$, we obtain $J_k \in \pi'$ for all k, which means that $\pi \subset \pi'$. Since $\pi, \pi' \in \mathcal{P}_d$, it follows that $\pi = \pi'$, as required.

Now consider any partition $\pi' \in \mathcal{P}_d$ with $\pi' \neq \pi$. Then by (70) we have either $\pi' \notin \mathcal{P}_\mathcal{J}$ or $\pi' \in \mathcal{P}_{\mathcal{J}_k}$ for some $k \leq m$. In the former case, π and π' are clearly separated by $\mathcal{P}_\mathcal{J}$; in the latter case they are separated by $\mathcal{P}_{\mathcal{J}_k}$. $\quad\square$

7.8 Symmetric Partitions

Here we consider partitions R of $\mathbb{N}$ into finitely or infinitely many disjoint, non-empty subsets $A_1, A_2, \ldots$. Writing $i \sim j$ if i and j belong to the same set A_n, we may define the associated *indicator array* r on $\mathbb{N}^2$ by

$$r_{ij} = 1\{i \sim j\}, \quad i, j \in \mathbb{N}.$$

The indicator arrays may be characterized intrinsically by the relations

$$r_{ii} = 1, \qquad r_{ij} = r_{ji}, \qquad r_{ij}r_{jk}(1 - r_{ik}) = 0, \tag{71}$$

for arbitrary $i, j, k \in \mathbb{N}$. For any mapping $p \colon \mathbb{N} \to \mathbb{N}$, we note that the class of non-empty, inverse sets $p^{-1}A_n$ is again a partition of $\mathbb{N}$, here denoted by $p^{-1}R$. Note that $p^{-1}R$ has the indicator array $r \circ p$, given by

$$(r \circ p)_{ij} = r_{p_i, p_j}, \quad i, j \in \mathbb{N}.$$

Any $\{0, 1\}$-valued random array X on $\mathbb{N}^2$ satisfying the restrictions in (71) defines a *random partition* R of $\mathbb{N}$. It is often convenient to identify the two objects and refer to the process X as a random partition of $\mathbb{N}$. We say that R is *exchangeable* if $p^{-1}R \overset{d}{=} R$ for all permutations p of $\mathbb{N}$ and *contractable* if the same relation holds for all sub-sequences p of $\mathbb{N}$. In terms of the associated indicator array X, the condition becomes $X \circ p \overset{d}{=} X$ for all permutations or sub-sequences p of $\mathbb{N}$. In other words, R is exchangeable or contractable iff the array X is jointly exchangeable or contractable, respectively.

More generally, we may consider an arbitrary collection T of injective maps $p \colon \mathbb{N} \to \mathbb{N}$ and say that the random partition R is T-invariant in distribution or simply T-*symmetric* if $p^{-1}R \overset{d}{=} R$ for all $p \in T$. In particular, this covers the cases of separately or jointly exchangeable or contractable partitions of $\mathbb{N}^d$ for arbitrary d. Our basic result is the fact that X is T-symmetric iff can be represented in the form

$$X_{ij} = 1\{\xi_i = \xi_j\}, \quad i, j \in \mathbb{N},$$

for some T-symmetric sequence of random variables $\xi = (\xi_j)$.

We may consider the more general case where a random mark in some measurable space S is attached to each subset of R. Letting $\kappa_1, \kappa_2, \ldots$ be the sequence of marks associated with the elements of $\mathbb{N}$, and writing $i \sim j$ whenever i and j belong to the same class of R, we define the associated *indicator array* X on $\mathbb{N}^2$ by

$$X_{ij} = \kappa_i 1\{i \sim j\}, \quad i, j \in \mathbb{N}. \tag{72}$$

Then R is said to be *T-symmetric* if X is jointly T-symmetric, in the sense that $X \circ p \stackrel{d}{=} X$ for every $p \in T$. We may now state the basic representation theorem for symmetrically distributed partitions of $\mathbb{N}$.

Theorem 7.38 *(paint-box representation, Kingman, Kallenberg) Let R be a random partition of $\mathbb{N}$ with marks in a Borel space S, and consider a family T of injective maps on $\mathbb{N}$. Then R is T-symmetric iff there exist a T-symmetric sequence of random variables $\xi_1, \xi_2, \ldots$ and a measurable function $b \colon \mathbb{R} \to S$, such that R has the indicator array*

$$X_{ij} = b(\xi_i) 1\{\xi_i = \xi_j\} \text{ a.s.}, \quad i, j \in \mathbb{N}. \tag{73}$$

Our proof is based on an elementary algebraic fact. For any partition R on $\mathbb{N}$ with associated indicator array r, we define the mappings $m(r) \colon \mathbb{N} \to \mathbb{N}$ and $k(r) \colon \mathbb{N} \to S$ by

$$
\begin{aligned}
m_j(r) &= \min\{i \in \mathbb{N}; \ i \sim j\}, \\
k_j(r) &= r_{jj},
\end{aligned}
\qquad j \in \mathbb{N}. \tag{74}
$$

Lemma 7.39 *(mapping of lead elements) For any indicator array r on $\mathbb{N}^2$ and injection p on $\mathbb{N}$, there exists an injection q on $\mathbb{N}$, depending measurably on r and p, such that*

$$q \circ m(r \circ p) = m(r) \circ p. \tag{75}$$

Proof: Let $A_1, A_2, \ldots$ be the partition classes associated with r, listed in their order of first appearance, and define for all k

$$
\begin{aligned}
B_k &= p^{-1} A_k, \quad a_k = \inf A_k, \quad b_k = \inf B_k, \\
K &= \{k \in \mathbb{N}; \ B_k \neq \emptyset\}, \\
I &= \{a_k; \ k \in K\}, \quad J = \{b_k; \ k \in K\}.
\end{aligned}
$$

Since $|B_k| \leq |A_k|$ for all k by the injectivity of p, we see that the $\mathbb{N}$-complements I^c and J^c satisfy

$$
\begin{aligned}
|J^c| &= \bigcup_{k \in K} |B_k \setminus \{b_k\}| = \sum_{k \in K} (|B_k| - 1) \\
&\leq \sum_{k \in K} (|A_k| - 1) = \bigcup_{k \in K} |A_k \setminus \{a_k\}| \leq |I^c|.
\end{aligned}
$$

Introducing the increasing enumerations

$$I^c = \{i_1, i_2, \ldots\}, \qquad J^c = \{j_1, j_2, \ldots\},$$

we may define an (r,p)-measurable injection q on $\mathbb{N}$ by

$$q(b_k) = a_k, \qquad k \in K, \tag{76}$$
$$q(j_n) = i_n, \qquad n \le |J^c|.$$

To verify (75), fix any $j \in \mathbb{N}$, and note that $j \in B_k$ for some $k \in K$. Then $p_j \in A_k$, and so by (76)

$$q(m_j(r \circ p)) = q(b_k) = a_k = m_{p_j}(r). \qquad \qquad \square$$

Proof of Theorem 7.38: The sufficiency is immediate from (82), since

$$(X \circ p)_{ij} = X_{p_i,p_j} = b(\xi_{p_i})1\{\xi_{p_i} = \xi_{p_j}\}, \quad i,j \in \mathbb{N}.$$

Conversely, suppose that X is T-symmetric. Letting $\vartheta_1, \vartheta_2, \ldots$ be i.i.d. $U(0,1)$ and independent of X, we define

$$\eta_j = \vartheta \circ m_j(X), \quad \kappa_j = k_j(X), \qquad j \in \mathbb{N}, \tag{77}$$

where m and k are given by (74). Since the ϑ_j are a.s. distinct, we see from (72) that

$$X_{ij} = \kappa_i 1\{\eta_i = \eta_j\}, \quad i,j \in \mathbb{N}. \tag{78}$$

Now fix any $p \in T$. By Lemma 7.39 we may choose a random injection $q(X)$ on $\mathbb{N}$ satisfying

$$q(X) \circ m_j(X \circ p) = m(X) \circ p_j, \quad j \in \mathbb{N}. \tag{79}$$

Using (77) and (79), the T-symmetry of X and exchangeability of ϑ, the independence of X and ϑ, and Fubini's theorem, we get for any measurable function $f \ge 0$ on $[0,1]^\infty \times S^\infty$

$$
\begin{aligned}
Ef(\eta \circ p, \kappa \circ p) &= Ef(\vartheta \circ m(X) \circ p, k(X) \circ p) \\
&= Ef(\vartheta \circ q(X) \circ m(X \circ p), k(X \circ p)) \\
&= E[Ef(\vartheta \circ q(r) \circ m(r \circ p), k(r \circ p))]_{r=X} \\
&= E[Ef(\vartheta \circ m(r \circ p), k(r \circ p))]_{r=X} \\
&= Ef(\vartheta \circ m(X \circ p), k(X \circ p)) \\
&= E[Ef(t \circ m(X \circ p), k(X \circ p))]_{t=\vartheta} \\
&= E[Ef(t \circ m(X), k(X))]_{t=\vartheta} \\
&= Ef(\vartheta \circ m(X), k(X)) = E(\eta, \kappa),
\end{aligned}
$$

which shows that $(\eta, \kappa) \circ p \overset{d}{=} (\eta, \kappa)$. Since $p \in T$ was arbitrary, the pair (η, κ) is then T-symmetric.

Since S is Borel, so is $[0,1] \times S$, and there exists a Borel isomorphism g from $[0,1] \times S$ onto a Borel set $B \subset [0,1]$. The inverse function g^{-1} can be extended to a measurable mapping $h: [0,1] \to [0,1] \times S$, which is clearly one-to-one on B with inverse g. Define $\xi_j = g(\eta_j, \kappa_j)$, and note that the sequence $\xi = (\xi_j)$ is again T-symmetric. It is also clear that

$$\eta_i = \eta_j \quad \Leftrightarrow \quad (\eta_i, \kappa_i) = (\eta_j, \kappa_j) \quad \Leftrightarrow \quad \xi_i = \xi_j. \tag{80}$$

Letting π denote the natural projection $[0,1] \times S \to S$ and putting $b = \pi \circ h$, we obtain

$$b(\xi_j) = \pi(\eta_j, \kappa_j) = \kappa_j, \quad j \in \mathbb{N}. \tag{81}$$

We may finally combine (78), (80), and (81) to get

$$X_{ij} = \kappa_i 1\{\eta_i = \eta_j\} = b(\xi_i) 1\{\xi_i = \xi_j\}, \quad i, j \in \mathbb{N}. \qquad \square$$

A similar proof yields the following more general multi-variate result, where we consider a sequence of random partitions $R_1, R_2, \ldots$ of $\mathbb{N}$, exchangeable with respect to a class T of injections on $\mathbb{N}$.

Corollary 7.40 *(sequence of partitions)* *Let $R_1, R_2, \ldots$ be random partitions of $\mathbb{N}$ with marks in a Borel space S, and consider a family T of injections on $\mathbb{N}$. Then the sequence $R = (R_k)$ is T-symmetric iff there exist a random array $\xi = (\xi_j^n)$ on $\mathbb{N}^2$ which is T-symmetric in index j, along with a measurable function $b \colon \mathbb{R} \to S$, such that the R_n have indicator arrays*

$$X_{ij}^n = b(\xi_i^n) 1\{\xi_i^n = \xi_j^n\} \quad a.s., \quad i, j, n \in \mathbb{N}. \tag{82}$$

Though contractable arrays X on $\mathbb{N}^2$ may not be exchangeable in general, the two notions are equivalent when X is the indicator array of a random partition of $\mathbb{N}$:

Corollary 7.41 *(contractable and exchangeable partitions)* *A marked partition of $\mathbb{N}$ is exchangeable iff it is contractable.*

Proof: If the partition R is contractable, then by Theorem 7.38 the associated indicator process X can be represented as in (73) in terms of a contractable sequence $\xi = (\xi_j)$. Since ξ is even exchangeable by Theorem 1.1, (73) shows that the same thing is true for R. $\qquad \square$

The equivalence of exchangeability and contractability fails for partitions of $\mathbb{N}^d$ with $d > 1$. For example, the non-random partition $A_k = \{(i,j) \in \mathbb{N}^2; \, i \wedge j = k\}$ of $\mathbb{N}^2$ is clearly contractable but not exchangeable. For a more interesting example, let $\eta_{ij}, \, i, j \in \mathbb{N}$, be i.i.d. Bernoulli random variables with $E\eta_{ij} = \frac{1}{2}$, and consider the partition R of $\mathbb{N}^2$ generated by the array

$$\xi_{ij} = \eta_{ij} + \eta_{i \wedge j, \, i \wedge j}, \quad i, j \in \mathbb{N}.$$

Then R is contractable but not exchangeable, since for $\tilde{\xi} = \xi \circ p$ with $p = (2, 1, 3, 4, \ldots)$ we have

$$P\{\xi_{12} = \xi_{13}\} = 1/2, \qquad P\{\tilde{\xi}_{12} = \tilde{\xi}_{13}\} = 3/8.$$

Let us now specialize to the case of exchangeable partitions R of $\mathbb{N}$. Here Theorem 7.38 shows that each class A_k of R is either a singleton or an infinite set with *density* $\beta_k > 0$, in the sense that $n^{-1}|A_k \cap [0, n]| \to \beta_k$ a.s. In fact, writing μ for the directing random measure of the generating sequence $\xi = (\xi_j)$, we see that the β_k are simply the atom sizes of μ. We denote the associated marks in S by κ_k. The mark distribution of the singleton sets is given by $\alpha = \mu_d \circ b^{-1}$, where μ_d denotes the diffuse component of μ, and we note that $\alpha S + \sum_j \beta_j = 1$. We define the *directing random measure* of R as the random probability measure

$$\nu = \delta_0 \otimes \alpha + \sum_j \beta_j \, \delta_{\beta_j, \kappa_j} \tag{83}$$

on $[0, 1] \times S$. Our terminology is justified by the following result.

Proposition 7.42 *(continuity, Kingman, Kallenberg) Let* $X^1, X^2, \ldots$ *be indicator arrays of some exchangeable partitions of* $\mathbb{N}$ *with marks in a Polish space* S, *and let* $\nu_1, \nu_2, \ldots$ *denote the associated directing random measures. Then* $X^n \xrightarrow{d}$ *some* X *in* S^∞ *iff* $\nu_n \xrightarrow{wd}$ *some* ν *on* $[0, 1] \times S$, *in which case* X *can be chosen to be the indicator array of an exchangeable partition directed by* ν.

Note that the statement contains the corresponding uniqueness assertion, the fact that the distributions of X and ν determine each other uniquely. Some lemmas are needed for the proof. First we construct an exchangeable partition associated with a random probability measure ν as in (83).

Lemma 7.43 *(construction) Given a random measure* ν *as in* (83) *and an independent* U-*sequence* $\vartheta_1, \vartheta_2, \ldots$, *define a random probability measure* $\tilde{\nu}$ *on* $[0, 1]^2 \times S$ *by*

$$\tilde{\nu} = \delta_0 \otimes \lambda \otimes \alpha + \sum_j \beta_j \, \delta_{\beta_j, \vartheta_j, \kappa_j}. \tag{84}$$

Let $\xi_1, \xi_2, \ldots$ *be an exchangeable sequence in* $[0, 1]^2 \times S$ *directed by* $\tilde{\nu}$, *and form an array* X *on* $\mathbb{N}^2$ *as in* (73), *where* b *is the natural projection of* $[0, 1]^2 \times S$ *onto* S. *Then* X *generates an exchangeable, marked partition of* $\mathbb{N}$ *with directing random measure* ν.

Proof: The array X clearly generates an exchangeable partition R of $\mathbb{N}$ with marks in S. Each atom $\beta_j \delta_{\beta_j, \vartheta_j, \kappa_j}$ of $\tilde{\nu}$ gives rise to a partition class of R with density β_j and mark κ_j, and the mark distribution of the singleton sets in R is given by α. Thus, R has directing random measure ν. $\qquad \square$

The purpose of the randomization in (84) was to make sure that the new marks $\tilde{\kappa}_j = (\vartheta_j, \kappa_j)$ will be distinct and the new measure $\tilde{\alpha} = \lambda \otimes \alpha$ diffuse. If these conditions are already fulfilled for the original marks κ_j and measure α, we can use the same construction based on the random measure ν. We can now prove the sufficiency part of Theorem 7.42 in a special case.

Lemma 7.44 *(simple, ergodic case) Let X and $X^1, X^2, \ldots$ be indicator arrays of some exchangeable, marked partitions of $\mathbb{N}$ with directing measures ν and $\nu_1, \nu_2, \ldots$, where ν is given by (83) and*

$$\nu_n = \delta_0 \otimes \alpha_n + \sum_j \beta_{nj}\, \delta_{\beta_{nj}, \kappa_{nj}}, \quad n \in \mathbb{N}.$$

Assume that α is a.s. diffuse and the κ_j are a.s. distinct, and similarly for the measures α_n and marks κ_{nj}. Then $\nu_n \xrightarrow{wd} \nu$ implies $X^n \xrightarrow{d} X$.

Proof: By Lemmas 7.43 and A2.4 we may assume that the measures ν and $\nu_1, \nu_2, \ldots$ are non-random. Introducing an i.i.d. sequence $\xi = (\xi_j)$ in $[0,1] \times S$ based on the distribution ν, we see as in Lemma 7.43 that (73) defines an indicator array $\tilde{X} \overset{d}{=} X$. Similarly, for every $n \in \mathbb{N}$, we can generate an array $\tilde{X}^n \overset{d}{=} X^n$ by means of an i.i.d. sequence $\xi^n = (\xi_j^n)$ with distribution ν_n. Since $\nu_n \xrightarrow{w} \nu$ on $[0,1] \times S$, we see from FMP 4.29 that $\xi^n \xrightarrow{d} \xi$ in $[0,1]^\infty \times S^\infty$. To show that $X^n \xrightarrow{d} X$ in S^∞, we may use Lemma A2.4 again to reduce to the case where ξ and the ξ^n are non-random with $\xi_j^n \to \xi_j$ for every j. Since the components of ξ lie a.s. in the support of ν and repetition may only occur in the set $(0,1] \times S$, we may assume the same properties for the fixed sequence ξ. A corresponding assumption can be made for every sequence ξ^n.

By the continuity of the projection b, it is clear that $\xi_i^n \to \xi_i$ implies $b(\xi_i^n) \to b(\xi_i)$ for all $i \in \mathbb{N}$. Thus, it remains to show that $\xi^n \to \xi$ implies

$$1\{\xi_i^n = \xi_j^n\} \to 1\{\xi_i = \xi_j\}, \quad i, j \in \mathbb{N}.$$

This is obvious when $\xi_i \neq \xi_j$, since in that case even $\xi_i^n \neq \xi_j^n$ for large enough n. It is also obvious when $i = j$. It remains to assume that $\xi_i = \xi_j = (\beta_k, \kappa_k)$ for some $i \neq j$ and k. Since the atoms of ν are isolated, we may choose a neighborhood G of (β_k, κ_k) such that ν has no other supporting point in the closure $\bar{G}$. Then by weak convergence

$$\beta_k = \nu G \leq \liminf_n \nu_n G \leq \limsup_n \nu_n \bar{G} \leq \nu \bar{G} = \beta_k,$$

and so $\nu_n G \to \beta_k$. If G is small enough, then from the special form of the measures ν_n it is clear that even the latter have eventually only a single atom in G. Since also $\xi_i^n \to \xi_i$ and $\xi_j^n \to \xi_j$ a.s., we conclude that $\xi_i^n = \xi_j^n$ for all sufficiently large n, as required. $\qquad\square$

Before proceeding to the general case, we need to show that the randomization in Lemma 7.43 is continuous in distribution with respect to the weak topologies on $[0,1] \times S$ and $[0,1]^2 \times S$.

Lemma 7.45 *(randomization) Let ν and $\nu_1, \nu_2, \ldots$ be random probability measures on $[0,1] \times S$ of the form (83), and define the associated random measures $\tilde{\nu}$ and $\tilde{\nu}_1, \tilde{\nu}_2, \ldots$ on $[0,1]^2 \times S$ as in (84). Then $\nu_n \xrightarrow{wd} \nu$ implies $\tilde{\nu}_n \xrightarrow{wd} \tilde{\nu}$.*

Proof: For any bounded, continuous function $f \geq 0$ on $[0, 1]^2 \times S$, we may define an associated bounded, measurable function $\tilde{f} \geq 0$ on $[0, 1] \times S$ by

$$\tilde{f}(r, x) = \begin{cases} -r^{-1} \log \int_0^1 e^{-rf(r,t,x)} \, dt, & r > 0, \\ \int_0^1 f(0, t, x) \, dt, & r = 0. \end{cases}$$

Then FMP 12.2 (iii) yields

$$E \exp(-\tilde{\nu}f) = E \exp(-\nu \tilde{f}),$$

and similarly for the measures ν_n and $\tilde{\nu}_n$. If even $\tilde{f}$ can be shown to be continuous, then the convergence $\nu_n \xrightarrow{wd} \nu$ implies

$$Ee^{-\tilde{\nu}_n f} = Ee^{-\nu_n \tilde{f}} \to Ee^{-\nu \tilde{f}} = Ee^{-\tilde{\nu} f},$$

and so $\tilde{\nu}_n f \xrightarrow{d} \tilde{\nu} f$ for all f. By Theorem A2.3 it follows that $\tilde{\nu}_n \xrightarrow{wd} \tilde{\nu}$.

To prove the required continuity, we need to show that $\tilde{f}(r_n, x_n) \to \tilde{f}(r, x)$ as $r_n \to r$ in $[0, 1]$ and $x_n \to x$ in S. This is obvious by dominated convergence when $r > 0$, and also when $r_n \equiv r = 0$. In the remaining case where $0 < r_n \to r = 0$, we may use Taylor's formula and dominated convergence to obtain

$$
\begin{aligned}
\tilde{f}(r_n, x_n) &= -r_n^{-1} \log \int_0^1 \exp(-r_n f(r_n, t, x_n)) \, dt \\
&= -r_n^{-1} \log \int_0^1 (1 - r_n f(r_n, t, x_n) + O(r_n^2)) \, dt \\
&= -r_n^{-1} \log \left(1 - r_n \int_0^1 f(r_n, t, x_n) \, dt + O(r_n^2) \right) \\
&= \int_0^1 f(r_n, t, x_n) \, dt + O(r_n) \\
&\to \int_0^1 f(0, t, x) \, dt = \tilde{f}(0, x) = \tilde{f}(r, x). \qquad \square
\end{aligned}
$$

Proof of Proposition 7.42: Suppose that $\nu_n \xrightarrow{wd} \nu$. If ν and the ν_n are non-random and satisfy the conditions of Lemma 7.44, then $X^n \xrightarrow{d} X$ by the same lemma. By Lemma A2.4 the last statement extends to any random measures ν and ν_n satisfying the same condition. In the general case, define $\tilde{\nu}$ and $\tilde{\nu}_n$ as in (84), and note that $\tilde{\nu}_n \xrightarrow{wd} \tilde{\nu}$ by Lemma 7.45. Using the result in the special case, we conclude that $\tilde{X}^n \xrightarrow{d} \tilde{X}$, where $\tilde{X}$ and the $\tilde{X}^n$ are indicator arrays corresponding to $\tilde{\nu}$ and $\tilde{\nu}_n$, taking values in $[0, 1] \times S$. Then by continuity

$$X^n \stackrel{d}{=} b \circ \tilde{X}^n \xrightarrow{d} b \circ \tilde{X} \stackrel{d}{=} X,$$

which implies $X^n \xrightarrow{d} X$.

Conversely, suppose that $X^n \xrightarrow{d} X$. Noting that

$$E\nu_n \circ b^{-1} = \mathcal{L}(\kappa_i^n) = \mathcal{L}(X_{ii}^n) \xrightarrow{w} \mathcal{L}(X_{ii}),$$

we see from Prohorov's theorem that the sequence of measures on the left is tight in $\mathcal{M}_1(S)$. Then the sequence $(E\nu_n)$ is tight in $\mathcal{M}_1([0,1] \times S)$, and so by Lemma A2.2 it follows that the sequence (ν_n) is weakly tight on the same space. Using Prohorov's theorem in the other direction, we conclude that any sub-sequence $N' \subset \mathbb{N}$ contains a further sub-sequence N'', such that $\nu_n \xrightarrow{wd} \nu$ along N'' for some random probability measure ν on $[0,1] \times S$. Note that ν must then be of the form (83). Now the direct assertion yields $X^n \xrightarrow{d} Y$ along N'', where Y is an indicator array corresponding to ν. But then $Y \overset{d}{=} X$, and since ν is measurably determined by Y, the distribution of ν is unique. Hence, the convergence $\nu_n \xrightarrow{wd} \nu$ remains valid along the original sequence, and we may choose ν to be the directing random measure of X. $\qquad\qquad\square$

We conclude with a criterion for extremality.

Lemma 7.46 *(extremality) If the array X in Theorem 7.38 is extreme, then the sequence ξ can be chosen to have the same property. The converse statement is also true when T is a group.*

Proof: Suppose that the T-invariant array X is extreme and represented as in (73) in terms of a T-invariant sequence ξ. Letting μ denote the distribution of ξ and writing (73) in the form $X = f(\xi)$, we see that $\mathcal{L}(X) = \mu \circ f^{-1}$. Now Theorem A1.3 yields $\mu = \int m\nu(dm)$, where ν is a probability measure on the set of extreme, T-invariant distributions m, and so $\mu \circ f^{-1} = \int (m \circ f^{-1})\nu(dm)$. Here the measures $m \circ f^{-1}$ are again distributions of T-invariant indicator arrays on $\mathbb{N}^2$, and by the extremality of $\mu \circ f^{-1}$ we have $m \circ f^{-1} = \mu \circ f^{-1}$ a.s. ν. Fixing an extreme measure m satisfying this relation and choosing a sequence η with distribution m, we obtain $X \overset{d}{=} f(\eta)$. Finally, the transfer theorem (FMP 6.10) yields a random sequence $\tilde{\eta} \overset{d}{=} \eta$ with $X = f(\tilde{\eta})$ a.s.

Now let T be a group, and suppose that $X = f(\xi)$ for an extreme, T-invariant sequence ξ. From Lemma A1.2 we see that ξ is ergodic, and since $X \circ p = f(\xi \circ p)$ for any $p \in T$, even X is ergodic by Lemma A1.1. Since T is a group, we may use Lemma A1.2 in the other direction to conclude that X is extreme. $\qquad\qquad\square$

Chapter 8

Multi-variate Rotations

In this chapter we continue our discussion of multi-variate symmetries with a study of higher-dimensional rotations. Here the basic representations are stated, most naturally, in terms of iso-normal Gaussian processes and their tensor products—the associated multiple Wiener–Itô integrals. Our analysis also leads to some representations of exchangeable or contractable processes in higher dimensions.

Several sections of preliminary material are required before we are ready to establish our main results. Thus, it is not until Section 8.5 that we can prove the first general representation theorem, for the case of separately rotatable arrays and functionals. In the jointly rotatable case, we need to master the symmetric functionals in Section 8.6, before we are able in Section 8.7 to deal with the more difficult case of random arrays. Those representations, in turn, provide the tools for analyzing, in the final Sections 8.8 and 8.9, the structure of separately or jointly exchangeable or contractable random sheets on a Euclidean space.

The basic representations are discussed, most conveniently, in an abstract Hilbert space setting, the basic notation being explained in Section 8.1. Section 8.2 contains some auxiliary results for Gaussian processes, and in Section 8.3 we consider some basic propositions for continuous, linear, random functionals (CLRFs) on suitable product spaces. Finally, Section 8.4 provides some key lemmas needed to prove our main results. Some further discussion of multiple stochastic integrals is provided by Appendix A3.

8.1 Rotational Symmetries

Throughout this chapter, we let H denote a real, infinite-dimensional, separable Hilbert space, and we write $H^{\otimes n}$ for the n-fold tensor product $H \otimes \cdots \otimes H$. Without loss of generality, we may assume that $H = L^2(S, \mathcal{S}, \mu)$ for some σ-finite measure μ with infinite support, in which case $H^{\otimes n}$ can be identified with the space $L^2(S^n, \mathcal{S}^{\otimes n}, \mu^{\otimes n})$ and $\bigotimes_{k \leq n} f_k = f_1 \otimes \cdots \otimes f_n$ with the function $f_1(t_1) \cdots f_n(t_n)$ on S^n. Given an ortho-normal basis (ONB) $h_1, h_2, \ldots$ in H, we recall that the tensor products $\bigotimes_{j \leq n} h_{k_j}$ with arbitrary $k_1, \ldots, k_n \in \mathbb{N}$ form an ONB in $H^{\otimes n}$.

By a *continuous, linear, random functional (CLRF)* X on H we mean a real-valued process Xf, $f \in H$, enjoying the linearity and continuity properties

$$X(af + bg) = aXf + bXg \text{ a.s.}, \quad f, g \in H, \ a, b \in \mathbb{R},$$

$$Xf_n \xrightarrow{P} 0, \quad f_1, f_2, \ldots \in H, \ \|f_n\| \to 0.$$

Equivalently, we may think of X as a continuous, linear operator from H to $L^0(P)$, the space of real-valued random variables on some fixed probability space $(\Omega, \mathcal{A}, P)$, endowed with the topology of convergence in probability.

A basic example is given by an *iso-normal Gaussian process (G-process)* on H, defined as a centered Gaussian process η on H satisfying $E(\eta f \, \eta g) = \langle f, g \rangle$ for all $f, g \in H$, where $\langle \cdot, \cdot \rangle$ denotes the inner product in H. The latter process is clearly *rotatable*, in the sense that $\eta \circ U \overset{d}{=} \eta$ or $\eta(Uf) \overset{d}{=} \eta f$, $f \in H$, for every unitary operator U on H. In fact, we saw in Proposition 1.31 that a linear random functional ξ on l^2 is rotatable iff $\xi = \sigma \eta$ a.s. for some G-process η on l^2 and an independent random variable $\sigma \geq 0$.

To motivate the subsequent developments, we may state the elementary Proposition 1.31 in the following abstract form.

Lemma 8.1 *(rotatable functionals) A CLRF ξ on H is rotatable iff $\xi = \sigma \eta$ a.s. for some G-process η and an independent random variable $\sigma \geq 0$. The latter is then a.s. unique.*

Proof: The result is a special case of Theorem 1.37. For a direct proof, consider an ONB $h_1, h_2, \ldots$ in H, and define $\xi_j = \xi h_j$ for all j. Then the sequence (ξ_j) is again rotatable, and so by Proposition 1.31 we have $\xi_j = \sigma \eta_j$ a.s. for all j, where the η_j are i.i.d. $N(0, 1)$ and $\sigma \geq 0$ is an independent random variable. Now define a G-process η on H by $\eta f = \sum_j \langle f, h_j \rangle \eta_j$ for all $f \in H$, where the sum converges in L^2, and note that a.s.

$$\xi h_j = \xi_j = \sigma \eta_j = \sigma \eta h_j, \quad j \in \mathbb{N}.$$

Using the linearity and continuity of both ξ and η, we conclude that $\xi = \sigma \eta$ a.s. $\qquad \square$

An advantage with the latter formulation is that it contains not only the discrete version in Proposition 1.31, but also the corresponding continuous-time statement, characterizing processes on $[0, 1]$ or $\mathbb{R}_+$ with rotatable increments in terms of a Brownian motion on the same interval.

To introduce the multi-variate rotations of CLRFs on product spaces, we define the tensor product of some unitary operators $U_1, \ldots, U_d$ on H as the unique, unitary operator $\bigotimes_k U_k$ on $H^{\otimes d}$ satisfying

$$\left(\bigotimes_k U_k \right)\left(\bigotimes_k f_k \right) = \bigotimes_k (U_k f_k), \quad f_1, \ldots, f_d \in H.$$

In particular, we may write $U^{\otimes d}$ for the d-fold tensor product $U \otimes \cdots \otimes U$. Given a CLRF X on $H^{\otimes d}$, we say that X is *jointly rotatable* if $X \circ U^{\otimes d} \overset{d}{=} X$ for every unitary operator U on H and *separately rotatable* if $X \circ \bigotimes_k U_k \overset{d}{=} X$ for any such operators $U_1, \ldots, U_d$. Our primary aim is to characterize CLRFs with such symmetries through suitable representations in terms of G-processes and their tensor products. These results will then be used to derive similar representations of jointly rotatable arrays and exchangeable or contractable random sheets.

Our basic building blocks are the multiple stochastic integrals formed by G-processes on H. To define those, we may begin with the case where $\eta_1, \ldots, \eta_n$ are independent G-processes on some infinite-dimensional, separable Hilbert spaces $H_1, \ldots, H_n$. Then $\bigotimes_k \eta_k = \eta_1 \otimes \cdots \otimes \eta_n$ is defined as the a.s. unique CLRF on the tensor product $\bigotimes_k H_k = H_1 \otimes \cdots \otimes H_n$ satisfying

$$\left(\bigotimes_k \eta_k\right)\left(\bigotimes_k f_k\right) = \prod_k \eta_k f_k \quad \text{a.s.,} \quad f_k \in H_k, \ k \le n.$$

If η is instead a single G-process on H, we define $\eta^{\otimes n}$ as the a.s. unique CLRF on $H^{\otimes n}$ satisfying

$$\eta^{\otimes n} \bigotimes_k f_k = \prod_k \eta_k f_k \quad \text{a.s.,} \quad f_1, \ldots, f_n \in H \text{ orthogonal.} \tag{1}$$

In this case, $\eta^{\otimes n} f$ is called the n-th order *multiple integral* of f with respect to η (FMP 13.21). (We avoid the usual notation $I_n f$, since it is essential in the present context to exhibit the choice of underlying G-process η.) Note that the product property in (1) may fail if $f_1, \ldots, f_n$ are not orthogonal. More generally, for independent G-processes $\eta_1, \ldots, \eta_n$ as above, we need to consider the CLRFs $\bigotimes_k \eta_k^{\otimes r_k}$ on $\bigotimes_k H_k^{\otimes r_k}$, characterized by the relations

$$\left(\bigotimes_k \eta_k^{\otimes r_k}\right)\left(\bigotimes_k \bigotimes_j f_{k,j}\right) = \prod_k \prod_j \eta_k f_{k,j} \quad \text{a.s.,}$$

where the elements $f_{k,1}, \ldots, f_{k,r_k} \in H_k$ are orthogonal for fixed k, but otherwise arbitrary.

These multiple integrals are all rotatable in different ways. Thus, $\eta^{\otimes n}$ is the basic example of a jointly rotatable CLRF on $H^{\otimes n}$. Furthermore, the CLRF $X = \bigotimes_k \eta_k$ is separately rotatable on $\bigotimes_k H_k$, in the sense that $X \circ \bigotimes_k U_k \overset{d}{=} X$ for any unitary operators U_k on H_k, $k \le n$. More generally, the CLRF $X = \bigotimes_k \eta_k^{\otimes r_k}$ satisfies the rotation invariance $X \circ \bigotimes_k U_k^{\otimes r_k} \overset{d}{=} X$ for any $U_1, \ldots, U_n$ as before. The general separately or jointly rotatable CLRFs on $H^{\otimes d}$ may essentially be written as linear combinations of multiple stochastic integrals of the indicated types.

For a more precise description, we begin with the case of separately rotatable CLRFs on $H^{\otimes d}$. Let $\mathcal{P}_d$ denote the class of partitions of the set $\{1, \ldots, d\}$ into non-empty subsets J. For every $J \in 2^d$ we consider a G-process η_J on $H \otimes H^{\otimes J} = H \otimes \bigotimes_{j \in J} H$. Furthermore, we introduce for every $\pi \in \mathcal{P}_d$ an element $\alpha_\pi \in H^{\otimes \pi} = \bigotimes_{J \in \pi} H$. Then we may show that a CLRF X on $H^{\otimes d}$ is separately rotatable iff it is a mixture of CLRFs of the form

$$Xf = \sum_{\pi \in \mathcal{P}_d} \left(\bigotimes_{J \in \pi} \eta_J \right) (\alpha_\pi \otimes f), \quad f \in H^{\otimes d}, \tag{2}$$

where the various G-processes η_J are assumed to be independent.

The jointly rotatable case is more complicated. Here the building blocks are multiple integrals formed by some independent G-processes η_r on $H^{\otimes(1+r)}$, $r = 1, \ldots, d$. It is also important in this case to keep track of the order of the component spaces. For this purpose, we introduce the class $\mathcal{O}_d$ of partitions π of the set $\{1, \ldots, d\}$ into sequences $k = (k_1, \ldots, k_r)$ of positive length $|k| = r$ and with distinct elements $k_j \leq d$. The associated sets $\tilde{k} = \{k_1, \ldots, k_r\}$ will then form a partition of $\{1, \ldots, d\}$ in the usual sense, but for each partition class we also need to specify an *order* of the elements. The general jointly rotatable CLRF X on $H^{\otimes d}$ may now be written as a mixture of CLRFs of the form

$$Xf = \sum_{\pi \in \mathcal{O}_d} \left(\bigotimes_{k \in \pi} \eta_{|k|} \right) (\alpha_\pi \otimes f), \quad f \in H^{\otimes d}. \tag{3}$$

Here we define the multiple integral in the π-th term, for any tensor products $\alpha = \bigotimes_{k \in \pi} \alpha_k$ in $H^{\otimes \pi}$ and $f = \bigotimes_{j \leq d} f_j$ in $H^{\otimes d}$ with orthogonal factors α_k and f_j in H, by the formula

$$\left(\bigotimes_{k \in \pi} \eta_{|k|} \right) (\alpha \otimes f) = \prod_{k \in \pi} \eta_{|k|} \left(\alpha_k \otimes \bigotimes_{j \leq |k|} f_{k_j} \right).$$

It is often suggestive and convenient to write the representations (2) and (3) symbolically as

$$X = \sum_{\pi \in \mathcal{P}_d} \alpha_\pi^* \bigotimes_{J \in \pi} \eta_J, \qquad X = \sum_{\pi \in \mathcal{O}_d} \alpha_\pi^* \bigotimes_{k \in \pi} \eta_{|k|}, \tag{4}$$

where we may think of α_π^* as a formal adjoint of the random operator $\alpha_\pi \otimes (\cdot)$ from $H^{\otimes d}$ to $H^{\otimes \pi} \otimes H^{\otimes d}$.

To proceed from here to the general representation formulas, we need only consider the α_π as *random* elements in the appropriate Hilbert spaces, independent of the G-processes η_J or η_r. This requires us to make sense of integrals like $X\varphi$, where X is a CLRF on H and φ is an independent random element in H. Here we need φ to be *measurable*, in the sense that $\langle \varphi, f \rangle$ is a random variable for every $f \in H$. Fixing an arbitrary ONB $h_1, h_2, \ldots$ in H, we may then define

$$X\varphi = \sum_i \langle \varphi, h_i \rangle X h_i,$$

where the sum converges in probability and the limit is a.s. independent of the choice of basis. When interpreted in this way, the integrals in (2) and (3) are well-defined even for random α_π, and the two formulas yield the most general representations of separately or jointly rotatable CLRFs X on $H^{\otimes d}$. For convenience, we may still write these formulas as in (4).

The preceding equations can also be written in coordinate form, for any choice of ONB $h_1, h_2, \ldots$ in H. Let us then introduce the d-dimensional random array

$$X_k = X_{k_1, \ldots, k_d} = X \bigotimes_{j \leq d} h_{k_j}, \quad k = (k_1, \ldots, k_d) \in \mathbb{N}^d,$$

and note that (X_k) is separately or jointly rotatable in the obvious sense iff the corresponding property holds for the underlying CLRF X. Using the Itô expansion of multiple stochastic integrals in terms of Hermite polynomials (FMP 13.25), we may then write (2) and (3) in an elementary form, involving suitable G-arrays of random variables. (A *G-array* is defined as an array of independent $N(0, 1)$ random variables.) In particular, for two-dimensional, separately rotatable arrays, we get the a.s. representation

$$X_{ij} = \alpha_0 \, \eta_{ij}^0 + \sum_{h,k} \alpha_{hk} \, \eta_{hi}^1 \, \eta_{kj}^2, \quad i, j \in \mathbb{N},$$

where the variables η_{ij}^0, η_{hi}^1 and η_{kj}^2 form a G-array, independent of the set of coefficient variables α_0 and α_{hk}. By a suitable diagonalization, we can reduce this to the simpler formula

$$X_{ij} = \alpha_0 \, \eta_{ij}^0 + \sum_k \alpha_k \, \eta_{ki}^1 \, \eta_{kj}^2, \quad i, j \in \mathbb{N}, \tag{5}$$

involving a possibly different G-array (η_{ij}^k).

In the jointly rotatable case, we get instead

$$X_{ij} = \alpha_0 \, \eta_{ij}^0 + \alpha_0' \, \eta_{ji}^0 + \sum_{h,k} \alpha_{hk} \, (\eta_{hi} \, \eta_{kj} - \delta_{ij} \, \delta_{hk}), \quad i, j \in \mathbb{N}. \tag{6}$$

When X is symmetric, in the sense that $X_{ij} = X_{ji}$, we can simplify the latter representation to

$$X_{ij} = \alpha_0 \, (\eta_{ij}^0 + \eta_{ji}^0) + \sum_k \alpha_k \, (\eta_{ki} \, \eta_{kj} - \delta_{ij}), \quad i, j \in \mathbb{N}. \tag{7}$$

Though (5) gives the most general representation of a separately rotatable array on $\mathbb{N}^2$, the expressions in (6) and (7) require an additional diagonal term $\rho \delta_{ij}$, to allow for an arbitrary jointly rotatable array X to be representable. The situation in higher dimensions is similar, except that now the diagonal terms become more complicated and may even involve some Gaussian random variables. Those terms did not show up in the original representations, because the associated operators turn out to be discontinuous.

Our proofs of the mentioned representations rely in a crucial way on the results for exchangeable arrays in the preceding chapter. More surprisingly, we may also proceed in the opposite direction and derive representations of exchangeable or contractable random sheets, using the present representations of rotatable arrays and CLRFs. Here a *random sheet* on a product space $\mathbb{R}_+^d \times [0, 1]^{d'}$ is defined as a continuous random field X, such that $X = 0$ on all $d + d'$ coordinate hyper-planes. We say that X is separately or jointly exchangeable or contractable if, for any regular, cubic grid with a vertex at 0, the associated array of increments has the corresponding invariance property.

For the latter representations, the proofs are based on the simple observation that any continuous process on $\mathbb{R}_+$ or $[0, 1]$ with $X_0 = 0$, taking values in a Euclidean space $\mathbb{R}^m$, is exchangeable iff it can be expressed in the form $X_t = \alpha t + \sigma B_t$ a.s. for some d-dimensional Brownian motion or bridge B, respectively, and an independent pair of random vectors α and σ. Subtracting the drift term αt, we are left with a process σB_t that is essentially rotatable. Exploiting a similar connection in higher dimensions, where the drift coefficients are again rotatable but of lower order, and proceeding recursively, we obtain a finite decomposition of X, where each term can in turn be described in terms of rotatable processes.

The argument suggests the representation of an arbitrary separately exchangeable random sheet X on $\mathbb{R}_+^d$ in the form

$$X_t = \sum_{J \in 2^d} \sum_{\pi \in \mathcal{P}_J} \left(\lambda^{J^c} \otimes \bigotimes_{I \in \pi} \eta_I \right) (\alpha_\pi \otimes [0, t]), \quad t \in \mathbb{R}_+^d,$$

where the η_I are independent G-processes on appropriate Hilbert spaces, and the summations extend over all sets $J \in 2^d$ and partitions π of J into disjoint, non-empty subsets I. The same representation holds for separately exchangeable random sheets on $[0, 1]^d$, except that now the processes η_I need to be replaced by suitably reduced versions $\hat{\eta}_I$. The representation of jointly exchangeable random sheets on $\mathbb{R}_+^d$ is similar but more complicated. In particular, some additional diagonal terms are needed in this case, similar to those required for the jointly exchangeable random arrays.

To gain a better understanding of the indicated formulas, we may consider some explicit coordinate versions, valid in the relatively simple two-dimensional case. Beginning with the separately exchangeable random sheets on $\mathbb{R}_+^2$, we get a representation

$$X_{s,t} = \rho s t + \sigma A_{s,t} + \sum_j \left(\alpha_j B_s^j C_t^j + \beta_j t B_s^j + \gamma_j s C_t^j \right), \quad s, t \geq 0,$$

for some independent Brownian motions $B^1, B^2, \ldots$ and $C^1, C^2, \ldots$ and an independent Brownian sheet A. The same formula holds for separately exchangeable sheets on $[0, 1]^2$, except that the processes B^j and C^j are now Brownian bridges, and A is a correspondingly tied-down or pinned Brownian sheet. For jointly exchangeable processes on $\mathbb{R}_+^2$, we get instead a representation of the form

$$\begin{aligned} X_{s,t} = {} & \rho s t + \sigma A_{s,t} + \sigma' A_{t,s} + \vartheta \, (s \wedge t) \\ & + \sum_{i,j} \alpha_{ij} \left(B_s^i B_t^j - (s \wedge t) \, \delta_{ij} \right) \\ & + \sum_j \left(\beta_j \, t B_s^j + \beta_j' \, s B_t^j + \gamma_j \, B_{s \wedge t}^j \right), \quad s, t \geq 0, \end{aligned}$$

where $B^1, B^2, \ldots$ are independent Brownian motions and A is an independent Brownian sheet.

In all the quoted formulas, the coefficients are themselves random and independent of the Brownian processes A, B^j and C^j. They also need to be suitably square summable, to ensure that the series occurring in the various representations will converge in probability.

8.2 Gaussian and Rotatable Processes

Here we begin with an infinite-dimensional version of Proposition 1.31, which provides the basic connection between rotatable and Gaussian processes.

Lemma 8.2 *(rotatable sequences in $\mathbb{R}^\infty$, Dawid)* *Let $X = (X_k)$ be an infinite random sequence in $\mathbb{R}^\infty$. Then X is rotatable iff the X_k are conditionally i.i.d. centered Gaussian. The conditional covariance array ρ is then a.s. unique and X-measurable, and for any stationary extension X_- of X to $\mathbb{Z}_-$ we have*

$$P[X \in \cdot|\rho] = P[X \in \cdot|X_-] \quad a.s.$$

Proof: Let X be rotatable with a stationary extension X_- to $\mathbb{Z}_-$. In particular, X is exchangeable and therefore conditionally i.i.d. ν, where ν is a random probability measure on $\mathbb{R}^\infty$. By Proposition 7.31, in the elementary case where $d = 1$, it is equivalent to condition on X_-, which shows that X is also a.s. conditionally rotatable. By Maxwell's theorem 1.32 it follows that any finite linear combination $\sum_j a_j X_1^j$ is a.s. conditionally centered Gaussian. Restricting the coefficients to $\mathbb{Q}$ and noting that the set of centered Gaussian distributions is closed under weak convergence, we conclude that, outside a fixed P-null set, all finite-dimensional projections of ν are Gaussian. The uniqueness and X-measurability of the conditional covariance array ρ are clear from Proposition 1.4, and the last assertion follows by another application of Proposition 7.31. □

The following result allows us to move back and forth between continuous, exchangeable processes on $\mathbb{R}_+$ and $[0, 1]$.

Lemma 8.3 *(scaling) For any compact, metric space S, let X and Y be continuous, $C(S)$-valued processes on $\mathbb{R}_+$ and $[0, 1)$, respectively, related by the reciprocal relations*

$$Y_s = (1 - s)X_{s/(1-s)}, \quad s \in [0, 1), \tag{8}$$
$$X_t = (1 + t)Y_{t/(1+t)}, \quad t \in \mathbb{R}_+. \tag{9}$$

Then X and Y are simultaneously exchangeable. In that case, Y can be extended to an a.s. continuous process on $[0, 1]$, and X is rotatable iff $Y_1 = 0$ a.s.

Proof: Employing a monotone-class argument, we may first reduce to the case where S is finite, so that X and Y take values in a Euclidean space $\mathbb{R}^d$. In that case, Theorem 3.15 shows that a continuous process X on $\mathbb{R}_+$ or $[0,1]$ is exchangeable iff it has an a.s. representation $X_t = \alpha t + \sigma B_t$, where B is a Brownian motion or bridge, respectively, independent of the pair (α, σ). Letting X be such a process on $\mathbb{R}_+$ and defining Y by (8), we obtain

$$Y_s = \alpha s + (1-s)\sigma B_{s/(1-s)} = \alpha s + \sigma B_s^\circ, \quad s \in [0,1),$$

where B° extends by FMP 13.6 to a Brownian bridge on $[0,1]$.

Conversely, suppose that Y is continuous and exchangeable on $[0,1)$. Then the processes Y and $Y_t' = Y(\frac{1}{2}+t) - Y(\frac{1}{2})$ have the same distribution on $[0,\frac{1}{2})$, and so by FMP 6.10 we may extend Y' to a continuous process on $[0,\frac{1}{2}]$. Putting $Y_1 = Y_{1/2} + Y_{1/2}'$, we obtain an a.s. continuous extension of Y to $[0,1]$. The exchangeability of Y extends by continuity to the entire interval $[0,1]$, and so we have an a.s. representation $Y_s = \alpha t + \sigma B_t^\circ$ involving a Brownian bridge B°. Defining X by (9), we get

$$X_t = \alpha t + (1+t)\sigma B_{t/(1+t)}^\circ = \alpha t + \sigma B_t, \quad t \geq 0,$$

where B is a Brownian motion on $\mathbb{R}_+$ by FMP 13.6.

To prove the last assertion, we note that $Y_1 = 0$ iff $\alpha = 0$, which holds by Lemma 8.2 iff X is rotatable. $\qquad\square$

The next lemma gives an elementary relationship between two Gaussian processes, corresponding to some standard representations of a Brownian bridge. Given a CLRF ξ on $H \otimes L^2(I)$ with $I = \mathbb{R}_+$ or $[0,1]$, we write $\xi_t = \xi(\cdot \otimes [0,t])$ for all $t \in I$. For any CLRFs ξ on H_1 and η on H_2, we say that the tensor product $\xi \otimes \eta$ exists, if the mapping $(f,g) \mapsto \xi f \eta g$ on $H_1 \times H_2$ can be extended to a CLRF $\xi \otimes \eta$ on $H_1 \otimes H_2$.

Lemma 8.4 *(scaling and centering)* *For any G-process ξ on $H \otimes L^2([0,1])$, there exists a G-process $\tilde{\xi}$ on $H \otimes L^2([0,1])$, such that whenever $\xi \otimes \eta$ exists for some CLRF $\eta \perp\!\!\!\perp \xi$ on H, we have*

$$(1-s)(\xi \otimes \eta)_{s/(1-s)} = \left((\tilde{\xi} - \tilde{\xi}_1 \otimes \lambda) \otimes \eta\right)_s \quad a.s., \quad s \in [0,1). \tag{10}$$

Proof: It suffices to prove the existence of a G-process $\tilde{\xi}$ satisfying

$$(1-s)\xi_{s/(1-s)} = (\tilde{\xi} - \tilde{\xi}_1 \otimes \lambda)_s \quad a.s., \quad s \in [0,1), \tag{11}$$

since (10) will then follow by the linearity and continuity of both sides. By FMP 6.10 it is then enough to show that the processes in (11) have the same distribution whenever ξ and $\tilde{\xi}$ are both G-processes. Expanding the latter processes in terms of an ONB in H, we note that the corresponding terms in (11) are independent on each side. It is then enough to consider each term separately, which reduces the discussion to the case of real-valued processes ξ and $\tilde{\xi}$. But in that case, each side is clearly a Brownian bridge on $[0, 1]$, by Lemma 8.3 or FMP 13.6. $\square$

We continue with an elementary result for Gaussian processes.

Lemma 8.5 *(orthogonality and equivalence) Let X and Y be centered, Gaussian processes on an arbitrary index set T. Then*

$$X \perp Y, \quad (X + Y) \perp (X - Y) \quad \Rightarrow \quad X \overset{d}{=} Y.$$

Proof: For any $s, t \in T$, we have

$$
\begin{aligned}
EX_sX_t - EY_sY_t &= EX_sX_t - EX_sY_t + EY_sX_t - EY_sY_t \\
&= E(X_s + Y_s)(X_t - Y_t) = 0.
\end{aligned}
$$

Hence, X and Y have the same mean and covariance functions, and the relation $X \overset{d}{=} Y$ follows by FMP 13.1. $\square$

To state the next result, recall that a linear mapping I between two Hilbert spaces H and K is called an *isometry* if $\langle If, Ig \rangle = \langle f, g \rangle$ for all $f, g \in H$. A *G-process* on H is defined as a linear isometry ξ from H to $L^2(P)$ such that ξf is centered Gaussian for every $f \in H$. The G-processes $\xi_1, \xi_2, \dots$ are said to be *jointly Gaussian* if $\xi_1 f_1 + \cdots + \xi_n f_n$ is Gaussian for all $n \in \mathbb{N}$ and $f_1, \dots, f_n \in H$.

Lemma 8.6 *(Gaussian representation) For any jointly Gaussian G-processes $\xi_1, \xi_2, \dots$ on H, there exist a G-process ξ and some linear isometries $I_1, I_2, \dots$ on H such that $\xi_k = \xi \circ I_k$ a.s. for all k.*

Proof: Let K be the closed linear subspace in $L^2(P)$ spanned by $\xi_1, \xi_2, \dots$. Fix a G-process η on H and a linear isometry $I \colon K \to H$, and define

$$I_k = I \circ \xi_k, \quad \eta_k = \eta \circ I_k, \qquad k \in \mathbb{N},$$

so that

$$\eta_k f = \eta \circ I \circ \xi_k f = \eta(I(\xi_k f)), \quad k \in \mathbb{N}, \ f \in H.$$

The η_k are again jointly centered Gaussian, and for any $j, k \in \mathbb{N}$ and $f, g \in H$ we get

$$
\begin{aligned}
E(\eta_j f)(\eta_k g) &= E(\eta I \xi_j f)(\eta I \xi_k g) \\
&= \langle I \xi_j f, I \xi_k g \rangle = E(\xi_j f)(\xi_k g).
\end{aligned}
$$

Hence, $(\eta_k) \overset{d}{=} (\xi_k)$ by FMP 13.1, and so by FMP 6.10 we may choose some $\xi \overset{d}{=} \eta$ such that $\xi_k = \xi \circ I_k$ a.s. for every k. $\square$

To state the next result, we say that a process X on a group S is *left-stationary* if the shifted process $\theta_r X$, defined by $(\theta_r X)_s = X_{rs}$, satisfies $\theta_r X \overset{d}{=} X$ for every $r \in S$.

Lemma 8.7 *(moving-average representation) For a finite group S, let $X = (X_n^s)$ be a centered, Gaussian process on $S \times \mathbb{N}$. Then X_n^s is left-stationary in $s \in S$ iff it has an a.s. representation*

$$X_n^s = \sum_{r \in S} \sum_{k \in \mathbb{N}} c_{n,k}^r \, \eta_k^{sr}, \quad s \in S, \; n \in \mathbb{N}, \tag{12}$$

in terms of a G-array $\eta = (\eta_n^s)$ on $S \times \mathbb{N}$ and some constants c_{nk}^s indexed by $S \times \mathbb{N}^2$ such that $\sum_k (c_{nk}^s)^2 < \infty$ for all s and n.

Proof: The sufficiency is clear from the stationarity of $\eta = (\eta_k^s)$, since the shifted process $\theta_s X$ has the same representation in terms of $\theta_s \eta$. Conversely, suppose that X is left-stationary. Choose an ONB $\xi_1, \xi_2, \ldots$ in the L^2-space spanned by X, put $p = |S|$, and consider the expansion

$$p^{-1/2} X_n^{s^{-1}} = \sum_k c_{n,k}^s \, \xi_k, \quad s \in S, \; n \in \mathbb{N}. \tag{13}$$

Letting $\zeta = (\zeta_k^s)$ be a G-array on $S \times \mathbb{N}$, we may define a process Y on $S \times \mathbb{N}$ by

$$Y_n^s = \sum_r \sum_k c_{n,k}^r \, \zeta_k^{sr}, \quad s \in S, \; n \in \mathbb{N}, \tag{14}$$

where the inner sum converges since $\sum_k (c_{n,k}^s)^2 < \infty$ for all s and n. Using (13) and (14), the ortho-normality of ξ and ζ, the group property of S, the left stationarity of X, and the definition of p, we get for any $s, t \in G$ and $m, n \in \mathbb{N}$

$$\begin{aligned}
E(Y_m^s \, Y_n^t) &= \sum_{u,v} \sum_{h,k} c_{m,h}^u \, c_{n,k}^v \, \delta_{su,tv} \, \delta_{h,k} \\
&= \sum_r \sum_k c_{m,k}^{(rs)^{-1}} \, c_{n,k}^{(rt)^{-1}} \\
&= p^{-1} \sum_r E(X_m^{rs} \, X_n^{rt}) = E(X_m^s \, X_n^t).
\end{aligned}$$

Since X and Y are both centered Gaussian, it follows that $X \overset{d}{=} Y$ (FMP 13.1). We may finally use the transfer theorem, in the version of FMP 6.11, to obtain a G-array $\eta = (\eta_k^s)$ satisfying (12). □

8.3 Functionals on a Product Space

By a *random element* in a Hilbert space H we mean a mapping $\varphi \colon \Omega \to H$ such that $\langle \varphi, h \rangle$ is measurable, hence a random variable, for every $h \in H$. Equivalently (FMP 1.4), φ may be regarded as a random element in H, endowed with the σ-field $\mathcal{H}$ generated by all projections $\pi_h \colon f \mapsto \langle f, h \rangle$ on H, $h \in H$. A CLRF X on H is said to be *measurable* if $Xf \colon (\omega, f) \mapsto X(\omega)f$ is a product measurable function on $\Omega \times H$. In this case, the composition $X\varphi$ is clearly a random variable for every random element φ in H (FMP 1.7 and 1.8).

Lemma 8.8 *(measurability)* *Every CLRF X on H has a measurable version $\tilde{X}$, and for any random element $\varphi \perp\!\!\!\perp X$ and ONB $h_1, h_2, \ldots$ in H, we have*

$$\sum_{k \leq n} \langle \varphi, h_k \rangle X h_k \xrightarrow{P} \tilde{X}\varphi.$$

Proof: By the continuity of X, we may choose some positive constants $\delta_n \downarrow 0$ such that

$$E(|Xf| \wedge 1) \leq 2^{-n}, \quad f \in H \text{ with } \|f\| \leq \delta_n, \quad n \in \mathbb{N}.$$

Fixing an ONB $h_1, h_2, \ldots$ in H, we define for any $f \in H$ and $n \in \mathbb{N}$

$$\begin{aligned}
f_n &= \sum_{k \leq n} \langle f, h_k \rangle h_k, \\
m_n(f) &= \inf\{k \in \mathbb{N};\ \|f - f_k\| \leq \delta_n\}, \\
F_{f,n} &= f_{m_n(f)}.
\end{aligned}$$

Then

$$\begin{aligned}
E \sum_n (|Xf - XF_{f,n}| \wedge 1) &= \sum_n E(|Xf - XF_{f,n}| \wedge 1) \\
&\leq \sum_n 2^{-n} < \infty,
\end{aligned}$$

and so the sum on the left converges a.s., which implies $XF_{f,n} \to Xf$ a.s. Constructing the variables $XF_{f,n}$ by linearity from any fixed versions of $Xh_1, Xh_2, \ldots$, we may now define $\tilde{X}f = \lim_n XF_{f,n}$ when the limit exists, and put $\tilde{X}f = 0$ otherwise. It is easy to verify that $\tilde{X}$ is a measurable version of X.

Now assume that X is measurable. Then the homogeneity, additivity, and continuity properties of X extend to

$$\begin{aligned}
X(\alpha f) &= \alpha X f \quad \text{a.s.,} \\
X(\varphi + \psi) &= X\varphi + X\psi \ \text{a.s.,} \\
\|\varphi_n\| \to 0 \ \text{a.s.} \ &\Rightarrow \ X\varphi_n \xrightarrow{P} 0,
\end{aligned}$$

whenever the random variable α and elements φ, ψ, and φ_n are independent of X. In fact, by Fubini's theorem and the linearity of X, we have

$$\begin{aligned}
P\{X(\alpha f) = \alpha X f\} &= EP\{X(af) = aXf\}|_{a=\alpha} = 1, \\
P\{X(\varphi + \psi) = X\varphi + X\psi\} &= EP\{X(f + g) = Xf + Xg\}|_{(f,g)=(\varphi,\psi)} = 1.
\end{aligned}$$

Similarly, assuming $\|\varphi_n\| \to 0$ a.s. and using Fubini's theorem, the continuity of X, and dominated convergence, we get

$$E(|X\varphi_n| \wedge 1) = EE(|Xf| \wedge 1)|_{f=\varphi_n} \to 0,$$

which means that $X\varphi_n \xrightarrow{P} 0$.

Now fix any ONB $h_1, h_2, \ldots$ of H and a random element $\varphi \perp\!\!\!\perp X$ in H. Putting $\varphi_n = \sum_{k \le n} \langle \varphi, h_k \rangle h_n$ and using the extended homogeneity and additivity properties of X, we obtain

$$X\varphi_n = \sum_{k \le n} \langle \varphi, h_k \rangle X h_k \quad \text{a.s.}$$

Since $\|\varphi - \varphi_n\| \to 0$ a.s., we conclude from the extended additivity and continuity properties that

$$X\varphi_n = X\varphi - X(\varphi - \varphi_n) \xrightarrow{P} X\varphi.$$

The final assertion follows by combination of the last two relations. $\qquad\square$

By the last result, we may henceforth assume that any CLRF X on H is measurable. We say that X is L^p-*bounded* if $\|Xf\|_p \lesssim \|f\|$. Let us also say that a CLRF X on $H^{\otimes d}$ has the form $X_1 \otimes \cdots \otimes X_d$ for some independent CLRFs $X_1, \ldots, X_d$ on H, if

$$X \bigotimes_k f_k = \prod_k X_k f_k \quad \text{a.s.}, \quad f_1, \ldots, f_d \in H.$$

Lemma 8.9 *(conditional expectation) Let X be an L^1-bounded CLRF on $H^{\otimes d}$ of the form $X_1 \otimes \cdots \otimes X_d$, where $X_1, \ldots, X_d$ are independent CLRFs on H with mean 0, and consider a random element $\varphi \perp\!\!\!\perp X$ in $H^{\otimes d}$. Then*

$$E[X\varphi | X_1, \ldots, X_k, \varphi] = 0, \quad 0 \le k < d.$$

Proof: Fix any ONB $h_1, h_2, \ldots$ in H, and let φ_n be the projection of φ onto $H_n^{\otimes d}$, where H_n denotes the linear span of $h_1, \ldots, h_n$. By Lemma 8.8,

$$X\varphi_n = \sum_{k_1, \ldots, k_d \le n} \left\langle \varphi, \bigotimes_j h_{k_j} \right\rangle \prod_j X_j h_{k_j} \quad \text{a.s.}, \quad n \in \mathbb{N},$$

and so, by Fubini's theorem,

$$E[X\varphi_n | X_1, \ldots, X_k, \varphi] = 0 \quad \text{a.s.}, \quad n \in \mathbb{N}, \ k < d.$$

Writing $Y_k = (X_1, \ldots, X_k)$ and using Jensen's inequality, Fubini's theorem, and the L^1-boundedness of X, we obtain a.s.

$$
\begin{aligned}
E[|E[X\varphi | Y_k, \varphi]| \, | \varphi] &= E[|E[X(\varphi - \varphi_n) | Y_k, \varphi]| \, | \varphi] \\
&\le E[|X(\varphi - \varphi_n)| \, | \varphi] \\
&\lesssim \|\varphi - \varphi_n\| \to 0,
\end{aligned}
$$

which implies $E[X\varphi | Y_k, \varphi] = 0$ a.s. $\qquad\square$

Our next aim is to construct suitably reduced versions of G-processes on $H \otimes L^2([0, d]^d)$, which will be needed to represent exchangeable random sheets on $[0, 1]^d$. For any $f \in H \otimes L^2([0, 1]^d)$ and $J \in 2^d$, we define the

J-average $\bar{f}_J \in H \otimes L^2([0,1]^{J^c})$ by

$$\langle \bar{f}_J, h \rangle = \langle f, h \otimes [0,1]^J \rangle, \quad h \in H \otimes L^2([0,1]^{J^c}),$$

where the existence and uniqueness follow from Riesz' theorem. Similarly, given a CLRF ξ on $H \otimes L^2([0,1]^d)$, we define a CLRF $\bar{\xi}_J$ on $H \otimes L^2([0,1]^{J^c})$ by

$$\bar{\xi}_J f = \xi(f \otimes [0,1]^J), \quad f \in H \otimes L^2([0,1]^{J^c}).$$

Writing P_J for the operator $f \mapsto \bar{f}_J \otimes [0,1]^J$ on $H \otimes L^2([0,1]^d)$, let N_j be the null space of $P_j = P_{\{j\}}$ and put $N_J = \bigcap_{j \in J} N_j$. Let A_J denote the orthogonal projection onto N_J. For any CLRF ξ on $H \otimes L^2([0,1]^d)$, we introduce the dual projection $\hat{\xi}_J = A_J^* \xi$ given by $\hat{\xi}_J f = \xi(A_J f)$, which is again a CLRF on $H \otimes L^2([0,1]^d)$. If ξ is a G-process, then $\hat{\xi}_J$ is clearly centered Gaussian with covariance function $E(\hat{\xi}_J f)(\hat{\xi}_J g) = \langle A_J f, A_J g \rangle$.

The following result gives some basic relations between the various averages and projections.

Lemma 8.10 *(projections and averages)*

(i) *The operators A_J commute and are given by*

$$A_J f = \sum_{I \subset J} (-1)^{|I|} (\bar{f}_I \otimes [0,1]^I), \quad f \in H \otimes L^2([0,1]^d), \quad J \in 2^d.$$

(ii) *For any CLRF ξ on $H \otimes L^2([0,1]^d)$, we have*

$$\hat{\xi}_J = A_J^* \xi = \sum_{I \subset J} (-1)^{|I|} (\bar{\xi}_I \otimes \lambda^I), \quad J \in 2^d.$$

Proof: (i) For any $f \in H \otimes [0,1]^d$ and $h \in H \otimes [0,1]^{J^c}$, we have

$$\begin{aligned}
\langle P_J f, h \otimes [0,1]^J \rangle &= \langle \bar{f}_J \otimes [0,1]^J, h \otimes [0,1]^J \rangle \\
&= \langle \bar{f}_J, h \rangle = \langle f, h \otimes [0,1]^J \rangle,
\end{aligned}$$

which shows that P_J is the orthogonal projection onto the subspace of elements $h \otimes [0,1]^J$. In particular, $A_j = I - P_j$ for all $j \le d$.

For any disjoint sets $I, J \in 2^d$ and elements

$$f \in H \otimes L^2([0,1]^{I^c}), \qquad h \in H \otimes L^2([0,1]^{J^c}),$$

where $h = h' \otimes h''$ with $h'' \in L^2([0,1]^I)$, we have

$$\begin{aligned}
\langle P_J(f \otimes [0,1]^I), h \otimes [0,1]^J \rangle &= \langle f \otimes [0,1]^I, h \otimes [0,1]^J \rangle \\
&= \langle f, h' \otimes [0,1]^J \rangle \langle [0,1]^I, h'' \rangle \\
&= \langle P_J f, h' \otimes [0,1]^J \rangle \langle [0,1]^I, h'' \rangle \\
&= \langle P_J f \otimes [0,1]^I, h \otimes [0,1]^J \rangle,
\end{aligned}$$

which extends by linearity and continuity to arbitrary h. Hence,

$$P_J(f \otimes [0,1]^I) = P_J f \otimes [0,1]^I, \quad I \cap J = \emptyset.$$

Since for any $I, J \in 2^d$,

$$\langle P_I P_J f, h \otimes [0,1]^{I \cup J} \rangle = \langle f, h \otimes [0,1]^{I \cup J} \rangle$$
$$= \langle P_{I \cup J} f, h \otimes [0,1]^{I \cup J} \rangle,$$

we conclude that $P_I P_J = P_{I \cup J}$. In particular, the operators P_J commute and satisfy $P_J = \prod_{j \in J} P_j$ for all J.

The commutative property carries over to the projections $A_j = I - P_j$ and their products, and for any $J \in 2^d$ we get

$$A_J' \equiv \prod_{j \in J} A_j = \prod_{j \in J} (I - P_j) = \sum_{I \subset J} (-1)^{|I|} P_I.$$

The latter operator is again a projection, and its range is clearly N_J, which means that $A_J' = A_J$.

(ii) For any tensor product $f = f' \otimes f''$ with $f' \in H \otimes L^2([0,1]^{J^c})$ and $f'' \in L^2([0,1]^J)$, we have

$$\langle \bar{f}_J, h \rangle = \langle f, h \otimes [0,1]^J \rangle$$
$$= \langle f', h \rangle \langle f'', [0,1]^J \rangle = \langle f', h \rangle \lambda^J f'',$$

which implies $\bar{f}_J = (\lambda^J f'') f'$. Hence,

$$\xi P_J f = \xi(\bar{f}_J \otimes [0,1]^J)$$
$$= \xi(f' \otimes [0,1]^J)(\lambda^J f'')$$
$$= (\bar{\xi}_J f')(\lambda^J f'') = (\bar{\xi}_J \otimes \lambda^J) f.$$

This extends by linearity and continuity to arbitrary f, and we get $\xi \circ P_J = \bar{\xi}_J \otimes \lambda^J$. Using (i), we conclude that

$$\hat{\xi}_J = \xi \circ A_J = \sum_{I \subset J} (-1)^{|I|} (\xi \circ P_I) = \sum_{I \subset J} (-1)^{|I|} (\bar{\xi}_I \otimes \lambda^J). \qquad \square$$

8.4 Preliminaries for Rotatable Arrays

The next result shows that every ergodic, separately rotatable array has finite moments of all orders.

Lemma 8.11 *(moments of rotatable arrays)* *For any ergodic, separately rotatable array X on $\mathbb{N}^d$, we have*

$$E|X_k|^p < \infty, \quad k \in \mathbb{N}^d, \ p > 0.$$

Proof: Let Y be a stationary extension of X to $\mathbb{Z}^d$. We claim that, for any $p > 0$ and $m \in \{0, \ldots, d\}$,

$$E[|Y_k|^p | Y_\emptyset] < \infty \text{ a.s.}, \quad k_1, \ldots, k_m > 0 \geq k_{m+1}, \ldots, k_d. \tag{15}$$

The statement is obvious for $m = 0$, since Y_k is $Y_\emptyset$-measurable for $k_1, \ldots, k_d \leq 0$. Now assume (15) to be true for some $m < d$. Proceeding by induction, we define $\xi_j = Y_{1,\ldots,1,j,0,\ldots,0}$, where index j occurs in position $m + 1$. The sequence (ξ_j) is rotatable, and so by Lemma 8.2 and FMP 6.10 it has an a.s. representation $\xi_j = \sigma \eta_j$, $j \in \mathbb{Z}$, in terms of a G-sequence (η_j) and an independent random variable $\sigma \geq 0$. By a conditional version of the Cauchy–Buniakovsky inequality, we get for any $j, n \in \mathbb{N}$

$$
\begin{aligned}
E^{Y_\emptyset} |\xi_j|^p &= E^{Y_\emptyset} \left(\eta_j^2 \sigma^2 \right)^{p/2} = E^{Y_\emptyset} \left(\frac{\eta_j^2 \left(\xi_{-1}^2 + \cdots + \xi_{-n}^2 \right)}{\eta_{-1}^2 + \cdots + \eta_{-n}^2} \right)^{p/2} \\
&\leq \left(E^{Y_\emptyset} \left(\frac{\eta_j^2}{\eta_{-1}^2 + \cdots + \eta_{-n}^2} \right)^p E^{Y_\emptyset} \left(\xi_{-1}^2 + \cdots + \xi_{-n}^2 \right)^p \right)^{1/2}.
\end{aligned}
$$

Using Minkowski's inequality and the induction hypothesis, we get for the second factor on the right

$$
E^{Y_\emptyset} (\xi_{-1}^2 + \cdots + \xi_{-n}^2)^p \leq n^p \, E^{Y_\emptyset} |\xi_{-1}|^{2p} < \infty \text{ a.s.}
$$

Even the first factor is a.s. finite when $n > 2p$, since it has expected value

$$
\begin{aligned}
E \left(\frac{\eta_j^2}{\eta_{-1}^2 + \cdots + \eta_{-n}^2} \right)^p &= E|\eta_j|^{2p} \, E \left(\eta_{-1}^2 + \cdots + \eta_{-n}^2 \right)^{-p} \\
&\lesssim \int_0^\infty r^{-2p} \, e^{-r^2/2} \, r^{n-1} \, dr < \infty.
\end{aligned}
$$

This shows that (15) remains true for $m + 1$, which completes the induction. In particular, we get for $m = d$

$$
E|X_k|^p = E^{Y_\emptyset} |Y_k|^p < \infty, \quad k \in \mathbb{N}^d,
$$

since Y is dissociated by Lemma 7.35.   □

The following lemma plays a key role in the proofs of our main results.

Lemma 8.12 *(separation of variables, Aldous) Consider an array X on $\mathbb{N}^3$, given by*

$$
X_{ij}^n = \sum_k c_k^n \, \varphi_{ik}^n \, \psi_{jk}^n, \quad i, j, n \in \mathbb{N},
$$

in terms of some constants c_k^n with $\sum_k (c_k^n)^2 < \infty$ for all n and some independent arrays (φ_{jk}^n) and (ψ_{jk}^n), each of which is i.i.d. in index j and orthonormal in k for fixed j and n. Further suppose that $X = (X_{ij}^n)$ is rotatable in index i. Then the array

$$
\zeta_{ik}^n = c_k^n \, \varphi_{ik}^n, \quad i, k, n \in \mathbb{N}, \tag{16}
$$

is centered, Gaussian and rotatable in index i.

Proof: By FMP 3.22 we may choose some measurable functions f_k^n and g_k^n on $[0,1]$, $k, n \in \mathbb{N}$, such that

$$\lambda \circ (f_k^n)^{-1} = P \circ (\varphi_{1k}^n)^{-1}, \quad \lambda \circ (g_k^n)^{-1} = P \circ (\psi_{1k}^n)^{-1}.$$

In particular, the sequences $f_1^n, f_2^n, \ldots$ and $g_1^n, g_2^n, \ldots$ are ortho-normal in $L^2(\lambda)$ for each n. Letting $\xi_1, \xi_2, \ldots$ and $\eta_1, \eta_2, \ldots$ be independent U-sequences on the basic probability space, we conclude from the i.i.d. assumption on (φ_{jk}^n) and (ψ_{jk}^n) that

$$\{(f_k^n(\xi_i), g_k^n(\eta_j)); \; i, j, k, n \in \mathbb{N}\} \overset{d}{=} \{(\varphi_{ik}^n, \psi_{jk}^n); \; i, j, k, n \in \mathbb{N}\}.$$

By FMP 6.10 we may then assume that a.s.

$$\varphi_{jk}^n = f_k^n(\xi_j), \quad \psi_{jk}^n = g_k^n(\eta_j), \qquad j, k, n \in \mathbb{N},$$

where the exceptional P-null set can be eliminated by a modification of the variables φ_{jk}^n and ψ_{jk}^n, which affects neither hypotheses nor assertion.

The tensor products $f_k^n \otimes g_k^n$ are again ortho-normal in k for fixed n, and the condition on the c_k^n allows us to define

$$h_n = \sum_k c_k^n (f_k^n \otimes g_k^n) \text{ a.e. } \lambda^2, \quad n \in \mathbb{N},$$

in the sense of convergence in $L^2(\lambda^2)$. Then the functions

$$\hat{h}_n(y) = h_n(\cdot, y), \quad y \in [0,1], \; n \in \mathbb{N},$$

satisfy $\hat{h}_n(y) \in L^2(\lambda)$ for $y \in [0,1]$ a.e. λ, and we may choose versions of the h_n such that $\hat{h}_n(y) \in L^2(\lambda)$ holds identically. It is easy to check that $\hat{h}_1, \hat{h}_2, \ldots$ are then measurable functions from $[0,1]$ to $L^2(\lambda)$. Introducing the closed supports B_n of the probability measures $\mu_n = \lambda \circ \hat{h}_n^{-1}$ on $L^2(\lambda)$, we note that $\hat{h}_n(y) \in B_n$ for $y \in [0,1]$ a.e. λ, and again we may modify the functions h_n on the exceptional null sets such that $\hat{h}_n(y) \in B_n$ holds identically.

For any versions of the functions h_n, we get

$$X_{ij}^n = h_n(\xi_i, \eta_j) \text{ a.s.}, \quad i, j, n \in \mathbb{N}.$$

By Proposition 7.31 and Lemma 7.34, the array $X = (X_{ij}^n)$ remains a.s. rotatable in i, conditionally on $\eta = (\eta_j)$. By Lemma 8.2, the functions $\hat{h}_{n_1}(y_1), \ldots, \hat{h}_{n_m}(y_m)$ are then jointly centered, Gaussian for any m and $n_1, \ldots, n_m \in \mathbb{N}$ and for $(y_1, \ldots, y_m) \in [0,1]^m$ a.e. λ^m. Since the product measure $\mu_{n_1} \otimes \cdots \otimes \mu_{n_m}$ on $(L^2(\lambda))^m$ has support $B_{n_1} \times \cdots \times B_{n_m}$ and the set of centered, Gaussian distributions is closed under weak convergence, the whole collection $\{\hat{h}_n(y)\}$ is then jointly centered, Gaussian.

For each $n \in \mathbb{N}$, let A_n denote the set of all $y \in [0,1]$ such that

$$\sum_k c_k^n g_k^n(y) f_k^n = \hat{h}_n(y) \text{ in } L^2(\lambda).$$

By the ortho-normality of $g_1^n, g_2^n, \ldots$ and Fubini's theorem, we have

$$\int_0^1 \sum_k (c_k^n \, g_k^n(y))^2 \, dy = \sum_k (c_k^n)^2 \, \|g_k^n\|^2 = \sum_k (c_k^n)^2 < \infty,$$

and so the sum on the left converges a.e., which implies $\lambda A_n = 1$. Next let H_n denote the closed subspace of l^2 spanned by the sequences $c_k^n g_k^n(y)$, $k \in \mathbb{N}$, with $y \in A_n$, and note that the sums $\sum_k r_k f_k^n$ with $(r_k) \in H_n$, $n \in \mathbb{N}$, are jointly centered, Gaussian. To identify H_n, let $(r_k) \perp H_n$ in l^2, so that

$$\sum_k r_k \, c_k^n \, g_k^n(y) = 0, \quad y \in A_n.$$

Using the ortho-normality of the g_k^n and the fact that $\lambda A_n = 1$, we conclude that

$$\sum_k r_k \, c_k^n \, g_k^n = 0 \quad \text{in } L^2(\lambda),$$

which implies $r_k c_k^n = 0$ for all k. Thus, the orthogonal complement of H_n consists of all sequences $(r_k) \in l^2$ such that $r_k = 0$ when $c_k^n \neq 0$, and so the space H_n itself is spanned by the sequences $(0, \ldots, 0, c_k^n, 0, \ldots)$ with c_k^n in the kth position. This shows that the functions $c_k^n f_k^n$ are jointly centered, Gaussian. The same thing is then true for the array (ζ_{ik}^n) in (16), and the assertion follows by the assumed i.i.d. property in index i. $\qquad \Box$

We may now establish a Gaussian representation for separately rotatable arrays of a special kind.

Lemma 8.13 *(arrays of product type, Aldous, Kallenberg) Consider a separately rotatable array on $\mathbb{N}^d$ of the form*

$$X_k = f(\xi_{1,k_1}, \ldots, \xi_{d,k_d}), \quad k = (k_1, \ldots, k_d) \in \mathbb{N}^d, \tag{17}$$

where (ξ_{jk}) is a U-array on $d \times \mathbb{N}$ and f is a measurable function on $[0,1]^d$. Then

$$X_k = \sum_{p \in \mathbb{N}^d} c_p \prod_{i \leq d} \eta_{k_i, p_i}^i \quad a.s., \quad k \in \mathbb{N}^d, \tag{18}$$

for a G-array $(\eta_{k,p}^i)$ on $d \times \mathbb{N}^2$ and some real constants c_p, $p \in \mathbb{N}^d$, with $\sum_p c_p^2 < \infty$.

Proof: By Lemma 7.35 we note that X is ergodic, and so by Lemma 8.11 it has finite moments of all orders, and in particular $f \in L^2(\lambda^d)$. We shall prove that $f \in H_1 \otimes \cdots \otimes H_d$, where each H_j is a Hilbert space of centered, Gaussian functions on the Lebesgue unit interval. Then assume that, for fixed $m < d$,

$$f \in H_1 \otimes \cdots \otimes H_m \otimes L^2(\lambda^{d-m}), \tag{19}$$

with $H_1, \ldots, H_m$ such as stated. To extend the result to index $m+1$, we may use Lemma A3.1 to express f in the form $\sum_j b_j (g_j \otimes h_j)$, for some orthonormal sequences

$$g_j \in H_1 \otimes \cdots H_m \otimes L^2(\lambda^{d-m-1}), \qquad h_j \in L^2(\lambda),$$

where the factor h_j occurs in coordinate $m+1$. Writing H_{m+1} for the Hilbert space in $L^2(\lambda)$ spanned by the functions $b_j h_j$, we see that (19) remains true for index $m + 1$. To complete the induction, it remains to note that the functions $b_j h_j$ are jointly centered, Gaussian by Lemma 8.12.

For every $i \leq d$, we may now choose an ONB $h_{i1}, h_{i2}, \ldots$ in H_i, so that the random variables

$$\eta^i_{k,p} = h_{i,p}(\xi_{i,k}), \quad i, k, p \in \mathbb{N}, \tag{20}$$

become i.i.d. $N(0,1)$. Since the tensor products $h_{1,p_1} \otimes \cdots \otimes h_{d,p_d}$ form an ONB in $H_1 \otimes \cdots \otimes H_d$, we have an expansion

$$f = \sum_{p \in \mathbb{N}^d} c_p \bigotimes_{i \leq d} h_{i,p_i}$$

with $\sum_p c_p^2 < \infty$, and (18) follows by means of (17) and (20). $\qquad \square$

For any $k \in \mathbb{N}^d$ and $J \in 2^d$, we write $k_J = (k_j; \ j \in J)$ and $X^J_h = (X_k; \ k_J = h)$, $h \in \mathbb{N}^J$, so that the X^J_h are sub-arrays of X corresponding to fixed values of the indices k_j with $j \in J$. We say that X is *J-rotatable* if its distribution is invariant under *arbitrary* rotations in the indices $k_J \in \mathbb{N}^J$. This holds by Lemma 8.2 iff the arrays X^J_h, $h \in \mathbb{N}^J$, are conditionally i.i.d. centered, Gaussian. An array X on $\mathbb{N}^d$ that is *J*-rotatable for $J = \{1, \ldots, d\}$ is said to be *totally rotatable*. The following result gives a condition for a separately rotatable array to be *J*-rotatable.

Proposition 8.14 *(J-rotatable arrays, Aldous, Kallenberg) Let X be an $\mathbb{R}^\infty$-valued, separately rotatable array on $\mathbb{N}^d$, representable in terms of a U-array ξ on $\mathcal{J} \cap \{J\}$, where $\mathcal{J} \subset 2^d$ is ideal and $J \notin \mathcal{J}$, and suppose that $E[X|\xi^{\mathcal{J}}] = 0$ a.s. Then X is J-rotatable.*

Proof: We may clearly assume that $|J| \geq 2$. For fixed $j \in J$, we see from Lemma 8.2 that the arrays X^j_k, $k \in \mathbb{N}$, on $\mathbb{N}^{d-1}$ are conditionally i.i.d., centered, Gaussian, given the associated random covariance function ρ on $(\mathbb{N}^{d-1})^2$. By the same lemma, it is equivalent to condition on $\hat{Y}^{j^c}$ for any stationary extension Y of X to $\mathbb{Z}^d$, where $j^c = \{j\}^c = \{1, \ldots, d\} \setminus \{j\}$. Writing $\mathcal{J}_j = \{I \in \mathcal{J}; \ j \notin I\}$, we see from Proposition 7.31 that even

$$P[X \in \cdot|\rho] = P[X \in \cdot|\xi^{\mathcal{J}_j}] = P[X \in \cdot|Y^{\mathcal{J}_j}]. \tag{21}$$

Now consider any $h, k \in \mathbb{N}^d$ with $h_j = k_j$ but $h_J \neq k_J$, and note that $X_h \perp\!\!\!\perp_{\xi^{\mathcal{J}}} X_k$ since $\xi^J_{h_J} \perp\!\!\!\perp_{\xi^{\mathcal{J}}} \xi^J_{k_J}$. Writing $h' = h_{j^c}$ and $k' = k_{j^c}$, and using (21), along with the chain rule for conditional expectations and the hypothetical relation $E[X|\xi^{\mathcal{J}}] = 0$, we obtain

$$\begin{aligned}
\rho_{h',k'} &= E[X_h X_k | \xi^{\mathcal{J}_j}] \\
&= E\left[E[X_h X_k | \xi^{\mathcal{J}}] \big| \xi^{\mathcal{J}_j} \right] \\
&= E\left[E[X_h | \xi^{\mathcal{J}}] \, E[X_k | \xi^{\mathcal{J}}] \big| \xi^{\mathcal{J}_j} \right] = 0. \tag{22}
\end{aligned}$$

Next let T be a rotation in an arbitrary index $i \in J \setminus \{j\}$, and use (21) and (22), together with the relations $TY^{\mathcal{J}_j} = (TY)^{\mathcal{J}_j}$ and $TY \stackrel{d}{=} Y$, to obtain

$$
\begin{aligned}
E[(TX)_h\,(TX)_k\,|\,\rho] &= E[(TX)_h\,(TX)_k\,|\,Y^{\mathcal{J}_j}] \\
&= E[(TX)_h\,(TX)_k\,|\,(TY)^{\mathcal{J}_j}] \\
&\stackrel{d}{=} E[X_h X_k\,|\,Y^{\mathcal{J}_j}] = \rho_{h',k'} = 0.
\end{aligned}
\tag{23}
$$

In particular, we may fix any $p \neq q$ in $\mathbb{N}$ and choose T such that

$$
(TX)_p^i = \frac{X_p^i + X_q^i}{\sqrt{2}}, \qquad (TX)_q^i = \frac{X_p^i - X_q^i}{\sqrt{2}}.
\tag{24}
$$

Using (22) and (23), the latter with T as in (24), we see from Lemma 8.5 that

$$
P[X_p^i \in \cdot\,|\rho] = P[X_q^i \in \cdot\,|\rho], \quad p, q \in \mathbb{N}, \ i \in J,
$$

and iterating this for different values of $i \in J$ yields

$$
P[X_h^J \in \cdot\,|\rho] = P[X_k^J \in \cdot\,|\rho], \quad h, k \in \mathbb{N}^J.
$$

Combining with (22), we see that the arrays X_k^J, $k \in \mathbb{N}^J$, are conditionally i.i.d. centered Gaussian, which implies the asserted J-rotatability of X. $\quad\square$

We proceed to show that any separately rotatable collection of totally rotatable arrays is again totally rotatable.

Corollary 8.15 *(combined arrays) Let $X = (X^n)$ be a separately rotatable family of totally rotatable arrays $X^1, X^2, \dots$ on $\mathbb{N}^d$. Then X is again totally rotatable.*

Proof: Since X is separately exchangeable, it can be represented in terms of a U-array ξ on $\bigcup_J \mathbb{N}^J$. By Lemma 8.2 we have also a representation of each sub-array X^n in the form $X_k^n = \sigma_n \eta_k^n$, $k \in \mathbb{N}^d$, for a G-array $\eta^n = (\eta_k^n)$ on $\mathbb{N}^d$ and an independent random variable σ_n. Applying Proposition 7.31 with $\mathcal{J} = 2^d \setminus \{1, \dots, d\}$ gives

$$
E[X^n|\xi^{\mathcal{J}}] = E[X^n|\sigma_n] = 0, \quad n \in \mathbb{N},
$$

and so X is totally rotatable by Proposition 8.14. $\quad\square$

Next we show that any J-rotatable arrays X_J on $\mathbb{N}^J$, $J \in 2^d$, are conditionally independent whenever the whole family is separately exchangeable. Here the exchangeability is with respect to permutations of the indices in $\mathbb{N}^J$ for every J; the sets J are not affected.

Proposition 8.16 *(conditional independence) Let X be a separately exchangeable family of $\mathbb{R}^\infty$-valued arrays X_J on $\mathbb{N}^J$, $J \in \tilde{\mathbb{N}}$, such that X_J is J-rotatable for every J. Then the X_J are conditionally independent, given $X_\emptyset$ and the set of conditional covariance arrays ρ_J.*

Proof: By martingale convergence (FMP 7.23), it is enough to consider the arrays X_J with $J \subset \{1, \dots, d\}$ for a fixed $d \in \mathbb{N}$. Let $\bar{X}_J$ denote the J^c-invariant extension of X_J to $\mathbb{N}^d$, and note that the combined array $\bar{X} = (\bar{X}_J)$ remains separately exchangeable. Introduce a stationary extension Y of $\bar{X}$ to $\mathbb{Z}^d$, and write Y^J for the restriction of Y to $\mathbb{N}^J \times \mathbb{Z}_-^{J^c}$. Note that X_J is trivially Y^J-measurable by construction, and that ρ_J is $Y^\emptyset$-measurable by the law of large numbers.

Now fix an ideal class $\mathcal{J} \subset 2^d$ different from $\emptyset$ and 2^d, along with a set $J \in 2^d \setminus \mathcal{J}$. Letting ξ be a representing U-array of X, we get by Proposition 7.31

$$P[X_J \in \cdot | \rho_J] = P[X_J \in \cdot | \xi^J] = P[X_J \in \cdot | Y^J] \quad \text{a.s.},$$

and since ρ_J is Y^J-measurable, we obtain $X_J \perp\!\!\!\perp_{\rho_J} Y^J$ by FMP 6.6. Since also $\alpha = X_\emptyset$, $\rho = (\rho_J)$, and all the X_I, $I \in \mathcal{J}$, are Y^J-measurable, we conclude that

$$X_J \perp\!\!\!\perp_{\alpha,\rho} (X_I; I \in \mathcal{J}), \quad J \notin \mathcal{J}, \quad \mathcal{J} \subset 2^d \text{ ideal.} \tag{25}$$

Choosing an enumeration $2^d = \{J_1, \dots, J_{2^d}\}$ with $|J_1| \le \cdots \le |J_{2^d}|$ and noting that the sets $\mathcal{J}_k = \{J_1, \dots, J_k\}$ are ideal, we get by (25)

$$(X_{J_1}, \dots, X_{J_k}) \perp\!\!\!\perp_{\alpha,\rho} X_{J_{k+1}}, \quad k < 2^d,$$

and the asserted conditional independence follows by FMP 3.8. $\square$

We also need the following coupling of ergodic, rotatable arrays.

Lemma 8.17 *(coupling)* *Let X and $Y_1, Y_2, \dots$ be random arrays on $\mathbb{N}^d$ such that the pairs (X, Y_k) are separately or jointly rotatable. Then there exist some arrays $\tilde{Y}_k$ with $(X, \tilde{Y}_k) \stackrel{d}{=} (X, Y_k)$ such that the whole family $(X, \tilde{Y}) = (X, \tilde{Y}_1, \tilde{Y}_2, \dots)$ is separately or jointly rotatable. If the pairs (X, Y_k) are ergodic, we can arrange for $(X, \tilde{Y})$ to have the same property.*

Proof: By FMP 6.10 and 6.13, we may assume that $Y_1, Y_2, \dots$ are conditionally independent given X. Arguing as in the proof of Lemma 7.2, we see that the combined array $U = (X, Y_1, Y_2, \dots)$ is again separately or jointly rotatable. Letting $U_\emptyset$ be a stationary extension of U to $\mathbb{Z}_-^d$ and using Lemma 7.34, we conclude that U is ergodic, separately or jointly rotatable, conditionally on $U_\emptyset$. Since the pairs (X, Y_k) are already ergodic, hence extreme by Lemma A1.2, it follows that $P[(X, Y_k) \in \cdot | U_\emptyset] = \mathcal{L}(X, Y_k)$ a.s. for every k. Now fix any $\omega \in \Omega$ outside the exceptional P-null sets, and define $\mu = P[U \in \cdot | U_\emptyset](\omega)$. By another application of FMP 6.10, we may choose arrays $\tilde{Y}_1, \tilde{Y}_2, \dots$ such that $\tilde{U} = (X, \tilde{Y}_1, \tilde{Y}_2, \dots)$ has distribution μ. Then $\tilde{U}$ is ergodic, separately or jointly rotatable, and $(X, \tilde{Y}_k) \stackrel{d}{=} (X, Y_k)$ for all k. $\square$

8.5   Separately Rotatable Arrays and Functionals

We are now ready to prove the basic representation theorem for separately rotatable random functionals on a tensor product $H^{\otimes d}$, stated in terms of independent G-processes and the associated multiple stochastic integrals. Recall that $\mathcal{P}_d$ denotes the class of partitions π of the set $\{1, \dots, d\}$ into non-empty subsets J.

Theorem 8.18 *(separately rotatable functionals) A CLRF X on $H^{\otimes d}$ is separately rotatable iff it has an a.s. representation*

$$Xf = \sum_{\pi \in \mathcal{P}_d} \Big(\bigotimes_{J \in \pi} \eta_J\Big)(\alpha_\pi \otimes f), \quad f \in H^{\otimes d}, \tag{26}$$

in terms of some independent G-processes η_J on $H \otimes H^{\otimes J}$, $J \in 2^d \setminus \{\emptyset\}$, and an independent collection of random elements α_π in $H^{\otimes \pi}$, $\pi \in \mathcal{P}_d$. The latter can then be chosen to be non-random iff X is ergodic.

The result is essentially equivalent to the following representation of separately rotatable arrays on $\mathbb{N}^d$, which will be proved first.

Lemma 8.19 *(separately rotatable arrays, Aldous, Kallenberg) A random array X on $\mathbb{N}^d$ is separately rotatable iff it has an a.s. representation*

$$X_k = \sum_{\pi \in \mathcal{P}_d} \sum_{l \in \mathbf{N}^\pi} \alpha_l^\pi \prod_{J \in \pi} \eta_{k_J, l_J}^J, \quad k \in \mathbb{N}^d, \tag{27}$$

in terms of a G-array η_{kl}^J, $k \in \mathbb{N}^J$, $l \in \mathbb{N}$, $J \in 2^d \setminus \{\emptyset\}$, and an independent collection of random variables α_l^π, $l \in \mathbb{N}^\pi$, $\pi \in \mathcal{P}_d$, with $\sum_l (\alpha_l^\pi)^2 < \infty$ a.s. The latter can then be chosen to be non-random iff X is ergodic.

Proof: First suppose that X has the stated representation. Note that the inner sum in (27) converges a.s. in L^2, conditionally on the coefficients α_l^π, since the products of η-variables are orthonormal. In particular, the sum is then a.s. independent of the order of terms. The separate rotatability of X is clear from the sufficiency part of Lemma 8.2. When the coefficients are non-random, we see from Lemma 7.35 that X is ergodic.

Now consider any separately rotatable array X on $\mathbb{N}^d$. Since X is separately exchangeable, it can be represented as in Corollary 7.23 in terms of a U-array ξ on $\mathbf{Z}_+^d$. By Proposition 7.31, Lemma 7.34, and Lemma A1.5 it is enough to show that, whenever X is representable in terms of $\xi \setminus \xi_\emptyset$, it can also be represented as in (27) with constant coefficients.

This holds for $d = 1$ by Lemma 8.2. Proceeding by induction, we assume the statement to be true for all separately rotatable arrays of dimension $< d$. Turning to the d-dimensional case, we may fix an enumeration $2^d = \{J_1, \dots, J_{2^d}\}$ with $|J_1| \le \dots \le |J_{2^d}|$, and note that the sets $\mathcal{J}_k = \{J_1, \dots, J_k\}$ are ideal. Now define

$$\begin{aligned}
X^{d+1} &= E[X | \xi^{\mathcal{J}_{d+1}}] = E[X | \xi^1, \dots, \xi^d], \\
X^k &= E[X | \xi^{\mathcal{J}_k}] - E[X | \xi^{\mathcal{J}_{k-1}}], \quad d + 2 \le k \le 2^d, \tag{28}
\end{aligned}$$

where the conditional expectations exist by Lemma 8.11, and note that $X = X^{d+1} + \cdots + X^{2^d}$. The combined array (X^k) is again separately rotatable by Proposition 7.31 and Lemma 7.34, and X^k is representable in terms of $\xi^{\mathcal{J}_k} \setminus \xi_\emptyset$ for every k. In particular, the combined array (X^k) is then ergodic by Lemma 7.35.

For $k \geq d + 2$ we see from (28) that $E[X^k | \xi^{\mathcal{J}_k}] = 0$ a.s., and so the array X^k is J_k-rotatable by Lemma 8.14. We also note that X^k remains ergodic with respect to the larger class of J_k-rotations. Since $|J_k| \geq 2$, the induction hypothesis yields a representation of X^k as in (27), though with non-random coefficients, in terms of a G-array γ^k. A similar representation holds for X^{d+1} by Lemma 8.13.

From (27) we see that the pairs (X^k, γ^k) are separately rotatable, and they are further ergodic by Lemma 7.35. Hence, Lemma 8.17 allows us to choose versions of the arrays γ^k, such that the combined array $\gamma = (\gamma^k)$ becomes ergodic, separately rotatable. For each $J \in 2^d$, the sub-array γ^J of $\mathbb{N}^J$-indexed random variables is then J-rotatable by Corollary 8.15. Since it is also inherits the ergodicity from γ, it is centered Gaussian by Lemma 8.2. Finally, we see from Proposition 8.16 that the γ^J are independent.

For every sub-array γ_i^J, we may now introduce an ONB (η_{ij}^J) in the generated Hilbert space to obtain a representation

$$\gamma_{ik}^J = \sum_j c_{kj}^J \eta_{ij}^J, \quad i \in \mathbb{N}^J, \ k \in \mathbb{N}^{J^c}, \ J \in 2^d, \tag{29}$$

where we can choose the coefficients c_{kj}^J to be independent of i since the γ_i^J have the same distribution for fixed J. The η_i^J are clearly independent G-arrays, and so the whole collection $\eta = (\eta_{ij}^J)$ is again a G-array. To deduce (27), it remains to substitute (29) into the earlier representations of the X^k in terms of γ^k. $\qquad \square$

Now consider any separately rotatable CLRF X on $H^{\otimes d}$. Given any ONB $h_1, h_2, \ldots$ in H, we may introduce the associated array

$$X_k = X \bigotimes_{j \leq d} h_{k_j}, \quad k = (k_1, \ldots, k_d) \in \mathbb{N}^d, \tag{30}$$

which is again separately rotatable. Conversely, we proceed to show that any separately rotatable array (X_k) can be extended to a separately rotatable CLRF X on $(l^2)^{\otimes d}$, in the sense that (30) holds for the coordinate sequences $h_j = (0, \ldots, 0, 1, 0, \ldots)$ with 1 occurring in the jth position. To appreciate this result, we note that the corresponding statement in the jointly rotatable case is false.

For an explicit construction, suppose that (X_k) is given by (27). We will show that the associated CLRF X can then be represented as in (26), where the G-processes η_J on $H \otimes H^{\otimes J}$ and the random elements α_π in $H^{\otimes \pi}$ are determined by the equations

$$\eta_{lk}^J \;=\; \eta_J\Big(h_k \otimes \bigotimes\nolimits_{j\in J} h_{l_j}\Big), \qquad k\in\mathbb{N},\ l\in\mathbb{N}^J,\ J\in 2^d\setminus\{\emptyset\}, \qquad (31)$$

$$\alpha_k^\pi \;=\; \big\langle \alpha_\pi,\ \bigotimes\nolimits_{J\in\pi} h_{k_j}\big\rangle, \qquad k\in\mathbb{N}^\pi,\ \pi\in\mathcal{P}_d. \qquad (32)$$

Proposition 8.20 *(extension)* *Any separately rotatable array* (X_k) *on* $\mathbb{N}^d$ *admits an a.s. unique extension to a separately rotatable CLRF* X *on* $(l^2)^{\otimes d}$. *Specifically, if* (X_k) *is given by* (27) *in terms of a G-array* (η_{kl}^J) *and an independent set of coefficients* α_l^π, *we may define* X *by* (26)*, where the G-processes* η_J *and random elements* α_π *are given by* (31) *and* (32)*. Furthermore,* (X_k) *and* X *are simultaneously ergodic.*

Proof: The random element α_π is a.s. determined by (32) since $\sum_k(\alpha_k^\pi)^2 < \infty$ a.s. As for η_J, it extends by linearity to an isonormal, centered, Gaussian process on the linear span of the basis elements in $H\otimes H^J$, and then by continuity to a G-process on the entire space. The multiple stochastic integrals $\bigotimes_{J\in\pi}\eta_J$ are again CLRFs on the associated tensor products $H^{\otimes\pi}\otimes H^{\otimes d}$, and the expression in (26) is well defined by Lemma 8.8. The separate rotatability of X is clear from the corresponding property of the G-processes η_J. The extension property (30) is an easy consequence of formulas (31) and (32), and the a.s. uniqueness follows from (30) together with the linearity and continuity of X.

To prove the last assertion, we note that the distributions $\mu = \mathcal{L}(\{X_k\})$ and $\tilde\mu = \mathcal{L}(X)$ determine each other uniquely via (30). If $\mu = a\mu_1 + b\mu_2$ for some $a,b>0$ with $a+b=1$ and some separately rotatable distributions μ_1 and μ_2, then $\tilde\mu$ has the corresponding decomposition $a\tilde\mu_1 + b\tilde\mu_2$, and conversely. Since $\mu_1 = \mu_2$ iff $\tilde\mu_1 = \tilde\mu_2$, we see that μ and $\tilde\mu$ are simultaneously extreme. It remains to note that, in either case, the notions of extremality and ergodicity are equivalent by Lemma A1.2. $\square$

It is now easy to complete the proof of the main result.

Proof of Theorem 8.18: Suppose that X is separately rotatable. For any ONB $h_1, h_2, \ldots$ in H, we may define an associated separately rotatable array (X_k) on $\mathbb{N}^d$. By Lemma 8.19, the latter can be represented as in (27) in terms of a G-array (η_{kl}^J) and an independent set of coefficients α_l^π. Defining the G-processes η_J and random elements α_π by (31) and (32), we see as in Proposition 8.20 that (30) holds for the CLRF $\tilde X$ on $H^{\otimes d}$ given by (26). By linearity and continuity we have $X = \tilde X$ a.s., which shows that X itself has the a.s. representation (26). Conversely, any CLRF X of the form (26) is clearly separately rotatable.

From Proposition 8.20 we see that X and (X_k) are simultaneously ergodic, and Lemma 8.19 shows that (X_k) is ergodic iff the coefficients α_l^π can be chosen to be non-random. By (32) it is equivalent that the random elements α_π be non-random, and the last assertion follows. $\square$

We now examine to what extent the representation in Theorem 8.18 is unique. By a *random isometry* on H we mean a random, linear operator I on H such that $\langle If, Ig \rangle = \langle f, g \rangle$ a.s. for any $f, g \in H$. Here If is required to be measurable, hence a random variable, for every $f \in H$. By Lemma 8.8 we can then choose I to be product measurable on $\Omega \times H$, which ensures $I\varphi$ to be well defined, even for random elements φ of H. For any random isometries $I_1, \ldots, I_d$ on H, the tensor product $\bigotimes_k I_k$ is clearly a random isometry on $H^{\otimes d}$.

Theorem 8.21 *(uniqueness) Let X and Y be separately rotatable CLRFs on $H^{\otimes d}$, representable as in Theorem 8.18 in terms of some random arrays $\alpha = (\alpha_\pi)$ and $\beta = (\beta_\pi)$, respectively, on $\mathcal{P}_d$. Then*

(i) $X = 0$ *a.s. iff $\alpha = 0$ a.s.;*

(ii) $X = Y$ *a.s. iff the πth terms agree a.s. for every $\pi \in \mathcal{P}_d$;*

(iii) *α and β may represent the same separately rotatable CLRF, iff there exist some random isometries I_J and I'_J on H, $J \in 2^d \setminus \{\emptyset\}$, such that a.s.*

$$\left(\bigotimes_{J \in \pi} I_J \right) \alpha_\pi = \left(\bigotimes_{J \in \pi} I'_J \right) \beta_\pi, \quad \pi \in \mathcal{P}_d; \tag{33}$$

(iv) *$X \overset{d}{=} Y$ iff there exist some random isometries I_J and I'_J on H, $J \in 2^d \setminus \{\emptyset\}$, such that (33) holds in distribution.*

The first property in (iii) means that there exist some G-processes $\eta = (\eta_J) \perp\!\!\!\perp \alpha$ and $\zeta = (\zeta_J) \perp\!\!\!\perp \beta$ as in Theorem 8.18 such that

$$\sum_{\pi \in \mathcal{P}_d} \left(\bigotimes_{J \in \pi} \eta_J \right) (\alpha_\pi \otimes f) = \sum_{\pi \in \mathcal{P}_d} \left(\bigotimes_{J \in \pi} \zeta_J \right) (\beta_\pi \otimes f), \tag{34}$$

a.s. for every $f \in H^{\otimes d}$. For convenience, we may often write the tensor products in (33) as I_π and I'_π, and those in (34) as η_π and ζ_π.

Proof: (ii) We shall prove the assertion when both summations in (34) are restricted to some set $S \subset \mathcal{P}_d$. The statement is trivially true when $|S| = 1$. Proceeding by induction, we assume the statement to be true whenever $|S| < m$ for some $m > 1$. Now let $|S| = m$. Then by Lemma 7.37 we may choose an ideal class $\mathcal{J} \subset 2^d$, such that the associated family $\mathcal{P}_\mathcal{J}$ of partitions $\pi \subset \mathcal{J}$ satisfies $0 < |S \cap \mathcal{P}_\mathcal{J}| < m$. By Proposition 7.31 we have

$$E[X \,|\, \alpha, \eta^\mathcal{J}] = E[X \,|\, \beta, \zeta^\mathcal{J}] \quad \text{a.s.},$$

which is equivalent, by Lemma 8.9, to equation (34) with both summations restricted to $S \cap \mathcal{P}_\mathcal{J}$. Subtracting this from the original relation (34), we obtain the same formula with summations over $S \cap \mathcal{P}_\mathcal{J}^c$. In either case, the desired term-wise equality follows by the induction hypothesis. This completes the induction and proves the assertion for every S. It remains to take $S = \mathcal{P}_d$.

(i) If $X = 0$, then by (ii)

$$\xi_\pi(\alpha_\pi \otimes f) = 0 \text{ a.s.}, \quad \pi \in \mathcal{P}_d, \ f \in H^{\otimes d}.$$

This remains a.s. true under conditioning on α, and so by Lemma 8.8 and Fubini's theorem we may assume that α is non-random. Since ξ_π is an isometry from $H^{\otimes d}$ to $L^2(P)$, we get

$$\|\alpha_\pi\|^2 \|f\|^2 = \|\alpha_\pi \otimes f\|^2 = E|\xi_\pi(\alpha_\pi \otimes f)|^2 = 0.$$

Choosing $f \neq 0$, we conclude that $\alpha_\pi = 0$.

(iii) First assume (33). Introduce some independent G-processes ξ_k^J on H, $k \in \mathbb{N}^J$, $J \in 2^d \setminus \{\emptyset\}$, and define

$$(\eta_k^J, \zeta_k^J) = \xi_k^J \circ (I_J, I_J'), \quad k \in \mathbb{N}^J, \ J \in 2^d \setminus \{\emptyset\}. \tag{35}$$

Since the I_J and I_J' are isometries, we see from Fubini's theorem that the arrays $\eta = (\eta_k^J)$ and $\zeta = (\zeta_k^J)$ are again G-processes satisfying $\eta \perp\!\!\!\perp \alpha$ and $\zeta \perp\!\!\!\perp \beta$. Fixing any ONB $h_1, h_2, \ldots$ in H, we define the G-processes ξ_J, η_J, and ζ_J on $H \otimes H^{\otimes J}$ by

$$(\xi_J, \eta_J, \zeta_J) \left(\cdot \otimes \bigotimes\nolimits_{j \in J} h_{k_j} \right) = (\xi_k^J, \eta_k^J, \zeta_k^J), \quad k \in \mathbb{N}^J, \ J \in 2^d \setminus \{\emptyset\}. \tag{36}$$

From (33) and (35) we may conclude that

$$
\begin{aligned}
\sum\nolimits_{\pi \in \mathcal{P}_d} \eta_\pi(\alpha_\pi \otimes f) &= \sum\nolimits_{\pi \in \mathcal{P}_d} \xi_\pi(I_\pi \alpha_\pi \otimes f) \\
&= \sum\nolimits_{\pi \in \mathcal{P}_d} \xi_\pi(I_\pi' \beta_\pi \otimes f) \\
&= \sum\nolimits_{\pi \in \mathcal{P}_d} \zeta_\pi(\beta_\pi \otimes f),
\end{aligned}
$$

a.s. for every $f \in H^{\otimes d}$, which proves (34).

To prove the reverse implication, suppose that X is represented by both expressions in (34), for some non-random α and β. Since the pairs (X, η) and (X, ζ) are ergodic, separately rotatable, there exist by Lemma 8.17 some G-processes $\tilde{\eta}$ and $\tilde{\zeta}$ with $(X, \tilde{\eta}) \overset{d}{=} (X, \eta)$ and $(X, \tilde{\zeta}) \overset{d}{=} (X, \zeta)$, such that the pair $(\tilde{\eta}, \tilde{\zeta})$ is again ergodic, separately rotatable. To simplify the notation, we may assume the same property already for (η, ζ). Then Corollary 8.15 shows that the pair (η_J, ζ_J) is ergodic, J-rotatable for every J, and so by Lemma 8.2 the associated coordinate processes (η_k^J, ζ_k^J) in (36) are centered Gaussian and i.i.d. in $k \in \mathbb{N}^J$ for fixed J. We also note that the pairs (η_J, ζ_J) are independent by Proposition 8.16.

By Lemma 8.6 we may choose some isometries I_J and I_J' and some independent G-processes ξ_k^J on H, $k \in \mathbb{N}^J$, $J \in 2^d \setminus \{\emptyset\}$, satisfying (35). Defining the G-processes ξ_J by (36), we see from (34) and (35) that

$$\sum_{\pi \in \mathcal{P}_d} \xi_\pi (I_\pi \alpha_\pi \otimes f) = \sum_{\pi \in \mathcal{P}_d} \xi_\pi (I'_\pi \beta_\pi \otimes f) \text{ a.s.,} \quad f \in H^{\otimes d}.$$

Relation (33) now follows by means of (i).

For general, possibly random α and β, we see from Proposition 7.31, Lemma 8.8, and Fubini's theorem that

$$Q_\alpha = P[X \in \cdot | \alpha] = P[X \in \cdot | \beta] = Q_\beta \text{ a.s.,} \tag{37}$$

where Q_α denotes the distribution of X in Theorem 8.18 when the underlying array $\alpha = (\alpha_\pi)$ is non-random. Assuming first that $Q_\alpha = Q_\beta$ for some non-random α and β, we see from FMP 6.10 that (34) holds for suitable G-processes η and ζ. By the non-random version of (iii), we conclude that even (33) is true for some isometries I_J and I'_J. For general α and β, we may fix any $\omega \in \Omega$ outside the exceptional null set in (37), and conclude from the previous case that (33) holds a.s. for suitable isometries $I_J(\omega)$ and $I'_J(\omega)$.

To choose measurable versions of the functions I_J and I'_J, we may write condition (33), together with the isometry properties of $I = (I_J)$ and $I' = (I'_J)$, in the form $g(\alpha, \beta, I, I') = 0$ for some measurable function g on a suitable space. Then the set

$$A = \{(\omega, I, I'); \, g(\alpha(\omega), \beta(\omega), I, I') = 0\}$$

is product measurable, and the previous argument shows that the projection $\bar{A}$ onto Ω satisfies $P(\bar{A}) = 1$. By the general section theorem in FMP A1.4, we may then choose some (α, β)-measurable functions I and I' such that $(\omega, I(\omega), I'(\omega)) \in A$ a.s. The associated components I_J and I'_J are random isometries satisfying (33) a.s.

(iv) First assume that (33) holds in distribution. By FMP 6.10 we may then choose a pair $(\tilde{I}', \tilde{\beta}) \overset{d}{=} (I', \beta)$ satisfying (33) a.s. By part (iii) we conclude that (34) holds with β replaced by $\tilde{\beta}$, for suitable choices of η and ζ. Letting $\tilde{\zeta} \overset{d}{=} \zeta$ be independent of β, we obtain $(\beta, \tilde{\zeta}) \overset{d}{=} (\tilde{\beta}, \zeta)$, and so the CLRF Y based on β and $\tilde{\zeta}$ satisfies $Y \overset{d}{=} X$.

Conversely, suppose that $X \overset{d}{=} Y$. By FMP 6.10 we may choose a $\tilde{\beta} \overset{d}{=} \beta$ such that both α and $\tilde{\beta}$ represent X. Then part (iii) yields the existence of some random isometries I_J and I'_J satisfying (33), though with β replaced by $\tilde{\beta}$. By FMP 6.10 we may next choose a set of random isometries $\tilde{I}' = (\tilde{I}'_J)$ such that $(\beta, \tilde{I}') \overset{d}{=} (\tilde{\beta}, I')$. Then (33) remains true in distribution for the original β, except that now the isometries I'_J need to be replaced by $\tilde{I}'_J$. $\quad \square$

We may next extend Theorem 8.18 to sets of random functionals of different dimension. As before, we use $\tilde{\mathbb{N}}$ to denote the class of finite subsets of $\mathbb{N}$. For any $J \in 2^d$, let $\mathcal{P}_J$ be the class of partitions π of J into non-empty subsets I.

Corollary 8.22 *(combined random functionals)* *For any CLRFs* X_J *on* $H^{\otimes J}$, $J \in \tilde{\mathbb{N}}$, *the combined array* $X = (X_J)$ *is separately rotatable iff it has an a.s. representation*

$$X_J f = \sum_{\pi \in \mathcal{P}_J} \Big(\bigotimes_{I \in \pi} \eta_I \Big) (\alpha_\pi \otimes f), \quad f \in H^{\otimes J}, \ J \in \tilde{\mathbb{N}}, \tag{38}$$

in terms of some independent G-processes η_J *on* $H \otimes H^{\otimes J}$, $J \in \tilde{\mathbb{N}} \setminus \{\emptyset\}$, *and an independent collection of random elements* α_π *in* $H^{\otimes \pi}$, $\pi \in \mathcal{P}_J$, $J \in \tilde{\mathbb{N}}$. *The latter can then be chosen to be non-random iff* X *is ergodic.*

Proof: Suppose that X is separately rotatable. To prove (38), it suffices by Lemma A1.5 to assume that X is ergodic, and to prove that X has then the desired representation for some non-random elements α_π. Now Theorem 8.18 shows that each process X_J has such a representation in terms of some elements $\alpha_\pi \in H$, $\pi \in \mathcal{P}_J$, and some independent G-processes γ_I^J on $H \otimes H^{\otimes I}$, $\emptyset \neq I \subset J$. By Lemma 8.17 we may choose the combined array $\gamma = (\gamma_I^J; I \subset J \in \tilde{\mathbb{N}})$ to be ergodic, separately rotatable. The array $\gamma_I = (\gamma_I^J; J \supset I)$ is then I-rotatable for every I by Corollary 8.15, and the processes γ_I are independent by Proposition 8.16. By Lemma 8.6 we may then represent the processes γ_I^J as isometric images of some independent G-processes η_I on $H \otimes H^{\otimes I}$, $I \neq \emptyset$, and (38) follows by substitution into the original formulas. $\qquad \Box$

The next result requires a minimality condition on the representation in Lemma 8.19, akin to the notion of minimality for ortho-normal expansions on tensor products of Hilbert spaces, discussed in connection with Lemma A3.1. Here we fix a separately rotatable array X on $\mathbb{N}^d$, admitting a representation as in (27) with constant coefficients α_l^π. For any $J \in 2^d \setminus \{\emptyset\}$ and $k \in \mathbb{N}^J$, consider the Gaussian Hilbert spaces $H_k^J \subset L^2(P)$ spanned by all variables η_{kl}^J, such that $\alpha_l^\pi \neq 0$ for some $\pi \in \mathcal{P}_d$ with $J \in \pi$. It is sometimes possible for some J to reduce H_k^J to a proper sub-space, by applying a unitary operator to the associated G-array η^J. By Zorn's lemma there exists a minimal choice of spaces H_k^J such that no further reduction is possible. The associated representations are also said to be *minimal*.

The following lemma deals with the subtle problem of extending the separate rotatability of a sequence of arrays $X^1, X^2, \dots$ on $\mathbb{N}^d$ to a given family of representing G-arrays $\eta^{n,J}$. Such a result will be needed in the next section, to prove the representation theorem for jointly rotatable CLRFs.

Lemma 8.23 *(rotatability of representing arrays)* *Let* X *be a separately rotatable family of arrays* X^n *on* $\mathbb{N}^d$, *each having a minimal representation as in Lemma 8.19 with constant coefficients* $a_l^{n,\pi}$ *in terms of a G-array* η^n *on* $\bigcup_J (\mathbb{N}^J \times \mathbb{N})$, *such that the whole family* $\eta = (\eta^n)$ *is separately exchangeable. Then* η *is again separately rotatable.*

Proof: Writing the representations of Lemma 8.19 in the form $X^n = \sum_\pi X^{n,\pi}$, we note that the array $(X^{n,\pi}; \pi \in \mathcal{P}_d)$ is again ergodic, separately rotatable for every $n \in \mathbb{N}$. By Lemma 8.17 we may choose some decompositions $X^n = \sum_\pi \tilde{X}^{n,\pi}$ with

$$(\tilde{X}^{n,\pi}; \pi \in \mathcal{P}_d) \overset{d}{=} (X^{n,\pi}; \pi \in \mathcal{P}_d), \quad n \in \mathbb{N},$$

such that the combined array $(\tilde{X}^{n,\pi}; \pi \in \mathcal{P}_d, n \in \mathbb{N})$ is separately rotatable. But then $\tilde{X}^{n,\pi} = X^{n,\pi}$ a.s. by Theorem 8.21 (ii), and so the original array $(X^{n,\pi}; \pi \in \mathcal{P}_d, n \in \mathbb{N})$ has the same property. For each n and π, we may reduce η^n to a minimal array $\eta^{n,\pi}$ representing $X^{n,\pi}$, and it suffices to show that the combined array $(\eta^{n,\pi}; \pi \in \mathcal{P}_d, n \in \mathbb{N})$ is separately rotatable. This reduces the argument to the case where each X^n consists of a single component $X^{n,\pi}$ with associated partition $\pi = \pi_n$.

By symmetry it is enough to show that η is rotatable in the first index, or 1-*rotatable*. By the minimality of η and Lemma A3.1, we may then write the representation of each X^n in diagonal form as

$$X^n_{ij} = \sum_k c^n_k \, \zeta^n_{ik} \, Z^n_{jk}, \quad i \in \mathbb{N}^{I_n}, \, j \in \mathbb{N}^{I^c_n}, \tag{39}$$

where $c^n_k > 0$ for all n and k, and the summation is allowed to be finite. Here I_n denotes the component of π_n containing 1, $\zeta^n = (\zeta^n_{ik})$ is the associated G-array of variables η^{n,I_n}_{ik}, and $Z^n = (Z^n_{jk})$ is the corresponding array on $\mathbb{N}^{I^c_n} \times \mathbb{N}$, which can be chosen to be ortho-normal in k for fixed j and separately rotatable in j with a representation of the form

$$Z^n_{jk} = \sum_{h \in \mathbb{N}^{\pi'_n}} a^n_{hk} \prod_{J \in \pi'_n} \eta^{n,J}_{j_J, h_J}, \quad j \in \mathbb{N}^{I^c_n}, \, k \in \mathbb{N}, \tag{40}$$

where $\pi'_n = \pi_n \setminus \{I_n\}$. To prove the asserted rotatability of η, it remains to show that the combined array $\zeta = (\zeta^n_{ik})$ is 1-rotatable.

Since η is separately exchangeable, it can be represented as in Corollary 7.23 in terms of a U-array ξ on $\hat{\mathbb{N}}_d = \bigcup_{J \in 2^d} \mathbb{N}^J$. (For our present purposes, the latter index set is more convenient than $\mathbb{Z}^d_+$.) From Proposition 7.31 and Lemma 7.34 we see that all hypotheses remain conditionally true given $\xi_\emptyset$, and so it suffices to prove the asserted 1-rotatability of η under the same conditioning. By Fubini's theorem it is equivalent to replace $\xi_\emptyset$ by a suitable constant, so that η becomes ergodic and represented by $\xi \setminus \xi_\emptyset$.

Now introduce the ideal class $\mathcal{J} = 2^{1^c}$, consisting of all subsets of $1^c = \{2, \ldots, d\}$. Using Proposition 7.31, along with the one-dimensional version of Lemma 7.34 (i), we see that X remains conditionally 1-rotatable given $\xi^\mathcal{J}$. By an obvious re-labeling of the elements, we obtain the corresponding statement for any stationary extension $\bar{X}$ of X to $\mathbb{Z}^d$, together with a consistent extension $\bar{\xi}$ of ξ to $\hat{\mathbb{Z}}_d$. By the chain rule for conditional expectations we conclude that even $\bar{X}$ is conditionally 1-rotatable given $\xi^\mathcal{J}$. The same argument shows that each array ζ^n remains conditionally 1-rotatable given $\xi^\mathcal{J}$.

Now the ζ^n are G-arrays, hence already ergodic, and so their distributions are unaffected by this conditioning. In other words, $\zeta^n \perp\!\!\!\perp \xi^{\mathcal{J}}$ for every n.

Consistently with $\bar{\xi}$, we may now introduce for every $n \in \mathbb{N}$ a stationary extension $\bar{Z}^n$ of Z^n to $\mathbb{Z}^{I_n^c} \times \mathbb{N}$. Writing $\bar{q} = (q, \ldots, q)$ in all dimensions, we may next define the arrays Y^n on $\mathbb{N}^{d+1}$ by

$$Y_{jq}^n = \bar{X}_{jI_n, -\bar{q}}^n = \sum_k c_k^n \zeta_{jI_n, k}^n \bar{Z}_{-\bar{q}, k}^n, \quad j \in \mathbb{N}^d, \ q, n \in \mathbb{N},$$

where the second equality follows from the extended version of (39). The combined array $Y = (Y^n)$ is again conditionally 1-rotatable given $\xi^{\mathcal{J}}$, since its construction from $\bar{X}$ commutes with 1-rotations. It is also clear from (40), together with the representation of η in terms of ξ, that the array $(Z_{-\bar{q}, k}^n)$ is i.i.d. in $q \in \mathbb{N}$ and independent of ξ. Both properties clearly remain conditionally true given $\xi^{\mathcal{J}}$. Finally, we see from the representation in terms of ξ that the combined array $\zeta = (\zeta_{jI_n, k}^n)$ is i.i.d. in j_1 under the same conditioning. We may now apply Lemma 8.12 to the array

$$U_{p,q}^n = (Y_{p,i,q}^n; \ i \in \mathbb{N}^{d-1}), \quad p, q, n \in \mathbb{N},$$

subject to the $\xi^{\mathcal{J}}$-conditional distributions, to see that ζ is conditionally 1-rotatable given $\xi^{\mathcal{J}}$. This yields immediately the corresponding unconditional property. $\square$

8.6   Jointly Rotatable Functionals

Here we extend the representation theorem of the previous section to the jointly rotatable case. Recall that $\mathcal{O}_d$ denotes the class of partitions π of $\{1, \ldots, d\}$ into *ordered* subsets k of size $|k| \geq 1$.

Theorem 8.24 *(jointly rotatable functionals) A CLRF X on $H^{\otimes d}$ is jointly rotatable iff it has an a.s. representation*

$$Xf = \sum_{\pi \in \mathcal{O}_d} \Big(\bigotimes_{k \in \pi} \eta_{|k|} \Big)(\alpha_\pi \otimes f), \quad f \in H^{\otimes d}, \tag{41}$$

in terms of some independent G-processes η_m on $H^{\otimes(1+m)}$, $1 \leq m \leq d$, and an independent collection of random elements α_π in $H^{\otimes \pi}$, $\pi \in \mathcal{O}_d$. The latter can then be chosen to be non-random iff X is ergodic.

Introducing an ONB $h_1, h_2, \ldots$ in H, we may define the components α_l^π and η_{kl}^m through the formulas

$$\alpha_\pi = \sum_{l \in \mathbb{N}^\pi} \alpha_l^\pi \bigotimes_{s \in \pi} h_{l_s}, \quad \pi \in \mathcal{O}_d, \tag{42}$$

$$\eta_{kl}^m = \eta_m \Big(h_l \otimes \bigotimes_{r \leq m} h_{k_r} \Big), \quad l \in \mathbb{N}, \ k \in \mathbb{N}^m, \ m \leq d, \tag{43}$$

and express the off-diagonal part of the array $X_k = X \bigotimes_j h_{k_j}$, $k \in \mathbb{N}^d$, in the form

$$X_k = \sum_{\pi \in \mathcal{O}_d} \sum_{l \in \mathbb{N}^\pi} \alpha_l^\pi \prod_{s \in \pi} \eta_{k \circ s, l_s}^{|s|} \quad \text{a.s.,} \quad k \in \mathbb{N}_d', \tag{44}$$

where $\mathbb{N}_d'$ denotes the non-diagonal part of $\mathbb{N}^d$. We show that the two representations are equivalent.

Lemma 8.25 *(off-diagonal uniqueness criteria) Let X and Y be jointly rotatable CLRFs on $H^{\otimes d}$. Then*

(i) $X = Y$ *a.s. iff*

$$X \bigotimes_{j \le d} h_j = Y \bigotimes_{j \le d} h_j \quad \text{a.s.,} \quad h_1, \ldots, h_d \in H \quad \text{orthogonal;}$$

(ii) $X \overset{d}{=} Y$ *if there exists an ONB $h_1, h_2, \ldots$ in H such that*

$$X \bigotimes_{j \le d} h_{k_j} = Y \bigotimes_{j \le d} h_{k_j} \quad \text{a.s.,} \quad k_1, \ldots, k_d \in \mathbb{N} \quad \text{distinct.} \tag{45}$$

Proof: (i) We may clearly assume that $H = L^2([0,1], \lambda)$, and identify $H^{\otimes d}$ with $L^2([0,1]^d, \lambda^d)$. Introduce the dyadic intervals

$$I_{nk} = 2^{-n}(k-1, k], \quad k \le 2^n, \quad n \in \mathbb{N},$$

and note that the associated indicator functions f_{nk} are orthogonal for fixed n. Letting $\mathcal{L}_n$ denote the set of all linear combinations of tensor products $f_{n,k_1} \otimes \cdots \otimes f_{n,k_d}$ with distinct $k_1, \ldots, k_d \le 2^n$, we obtain $X = Y$ a.s. on $\bigcup_n \mathcal{L}_n$, and it is enough to show the latter class is dense in $L^2(\lambda^d)$. Since the continuous functions on $[0,1]^d$ are dense in $L^2(\lambda^d)$ by FMP 1.35, it remains to show that every continuous function can be approximated in L^2 by functions in $\mathcal{L}_n$. But this is clear by uniform continuity, together with the fact that the diagonal parts D_n of the 2^{-n}-grids in $[0,1]^d$ satisfy $\lambda^d D_n \le 2^{-n} \to 0$.

(ii) Let $\mathcal{L}$ denote the linear sub-space of $H^{\otimes d}$ spanned by the tensor products $h_{k_1} \otimes \cdots \otimes h_{k_d}$ with distinct $k_1, \ldots, k_d \in \mathbb{N}$, and note that (45) extends by linearity to $Xf = Yf$, a.s. for every $f \in \mathcal{L}$. By the joint rotatability of X and Y, we obtain $X \circ U^{\otimes d} \overset{d}{=} Y \circ U^{\otimes d}$ on $\mathcal{L}$ for every unitary operator U on H, which means that $X \overset{d}{=} Y$ on $U^{\otimes d} \mathcal{L}$. As before, it remains to note that the latter class is dense in $H^{\otimes d}$. □

We proceed to examine how the representation in (44) are affected by a change of basis. Writing S_m for the group of permutations of $\{1, \ldots, m\}$, we define S_π for any $\pi \in \mathcal{O}_d$ to be the class of functions p on π such that $p_s \in S_{|s|}$ for every $s \in \pi$.

Lemma 8.26 *(change of basis) Let the array X on $\mathbb{N}'_d$ be given by (44), in terms of a G-array $\eta = (\eta^m_{i,k})$ and an independent array of coefficients $\alpha = (\alpha^\pi_k)$, and suppose that η is given in terms of another G-array $\xi = (\xi^m_{i,k})$ through a linear isometry of the form*

$$\eta^m_{i,n} = \sum_{p \in S_m} \sum_{k \in \mathbb{N}} c^{m,p}_{n,k}\, \xi^m_{iop,k}, \quad i \in \mathbb{N}^m, \ m \le d, \ n \in \mathbb{N}. \tag{46}$$

Then (44) remains valid with η replaced by ξ and with α replaced by the array

$$\beta^\pi_k = \sum_{p \in S_\pi} \sum_{n \in \mathbb{N}^\pi} \alpha^{\pi o p^{-1}}_n \prod_{s \in \pi} c^{|s|,p_s}_{n_s,k_s}, \quad k \in \mathbb{N}^\pi, \ \pi \in \mathcal{O}_d. \tag{47}$$

Proof: Substituting (46) into (44), we get for any $i \in \mathbb{N}'_d$

$$
\begin{aligned}
X_i &= \sum_{\pi \in \mathcal{O}_d} \sum_{n \in \mathbb{N}^\pi} \alpha^\pi_n \prod_{s \in \pi} \sum_{p \in S_{|s|}} \sum_{k \in \mathbb{N}} c^{|s|,p}_{n_s,k}\, \xi^{|s|}_{iosop,k} \\
&= \sum_{\pi \in \mathcal{O}_d} \sum_{p \in S_\pi} \sum_{k,n \in \mathbb{N}^\pi} \alpha^\pi_n \prod_{s \in \pi} c^{|s|,p_s}_{n_s,k_s}\, \xi^{|s|}_{iosops,k_s} \\
&= \sum_{\pi' \in \mathcal{O}_d} \sum_{p \in S_{\pi'}} \sum_{k,n \in \mathbb{N}^{\pi'}} \alpha^{\pi'op^{-1}}_n \prod_{t \in \pi'} c^{|t|,p_t}_{n_t,k_t}\, \xi^{|t|}_{iot,k_t} \\
&= \sum_{\pi \in \mathcal{O}_d} \sum_{k \in \mathbb{N}^\pi} \beta^\pi_k \prod_{t \in \pi} \xi^{|t|}_{iot,k_t},
\end{aligned}
$$

as required. Here the second equality is obtained by a term-wise multiplication of infinite series, followed by a change in the order of summation, the third equality results from the substitutions $\pi \circ p = \pi'$ and $s \circ p = t$, and the fourth equality arises from another change of summation order. $\square$

The transformations (46) and (47) may be written in coordinate-free notation as

$$\eta_m = \xi_m \circ \sum_{p \in S_m} (C^m_p \otimes T_p), \qquad m \le d, \tag{48}$$

$$\beta_\pi = \sum_{p \in S_\pi} \Big(\bigotimes_{s \in \pi} C^{|s|}_{p_s} \Big) \alpha_{\pi o p^{-1}}, \qquad \pi \in \mathcal{O}_d, \tag{49}$$

in terms of some bounded linear operators C^m_p on H, $p \in S_m$, $m \le d$, where T_p denotes the permutation operator on $H^{\otimes d}$ induced by $p \in S_m$. We can now translate the last result into a similar statement for the representations of CLRFs in (41).

Corollary 8.27 *(isometric substitution) Let the CLRF X on $H^{\otimes d}$ be given by (41), in terms of some independent G-processes η_m and an independent set of random elements α_π, and suppose that η_m is given by (48) in terms of some independent G-processes ξ_m. Then (41) remains true with (η_m) replaced by (ξ_m) and with (α_π) replaced by the array (β_π) in (49).*

Proof: Define the CLRF Y as in (41), though with η replaced by ξ and with α replaced by β. Since the permutation operator T_p in (48) commutes with any unitary operator on $H^{\otimes d}$ of the form $U^{\otimes d}$, we see that the pair (ξ, η) is jointly rotatable, and so the same thing is true for the pair (X, Y). Furthermore, (45) holds by Lemma 8.26 for any ONB $h_1, h_2, \ldots$ in H. Hence, Lemma 8.25 (i) yields $X = Y$ a.s. $\qquad\square$

We are now ready to prove our crucial lemma, which gives a representation outside the diagonals of any jointly rotatable array on $\mathbb{N}^d$. The diagonal terms are more complicated in general, and the complete representation will not be derived until the next section.

Lemma 8.28 *(off-diagonal representation) Let X be a jointly rotatable array on $\mathbb{N}^d$. Then (44) holds a.s. for a G-array η^m_{kl}, $k \in \mathbb{N}^m$, $l \in \mathbb{N}$, $m \leq d$, and an independent collection of random variables α^π_l, $l \in \mathbb{N}^\pi$, $\pi \in \mathcal{O}_d$, with $\sum_l (\alpha^\pi_l)^2 < \infty$ a.s. for all π.*

Proof: In view of Proposition 7.33 and Lemmas 7.34 (i), 7.35, and A1.5, we may assume that X is ergodic, in which case we need to show that X has an off-diagonal representation as in (44) with non-random coefficients α^π_l. Then fix any disjoint, infinite sets $N_1, \ldots, N_d \subset \mathbb{N}$, and note that X is separately rotatable on $R = N_1 \times \cdots \times N_d$. Applying Lemma 7.35 to both X itself and its restriction to R, we see that even the latter array is ergodic. Hence, it can be represented as in Lemma 8.19 in terms of a G-array ξ^J_{ik}, $i \in N_J$, $k \in \mathbb{N}$, $J \in 2^d \setminus \{\emptyset\}$, and a non-random collection of coefficients. We may clearly assume that the representation is minimal, in the sense of the preceding section. By Lemma 7.36 we can modify ξ, if necessary, such that the same representation holds on the non-diagonal index set $\mathbb{N}'_d$, for some ergodic, jointly exchangeable extension of ξ to $\hat{\mathbb{N}}'_d \setminus \{0\}$.

The array of reflections $\tilde{X} = (X_{k \circ p}; k \in \mathbb{N}^d, p \in S_d)$ is again ergodic, separately rotatable on R by Lemma 7.35, and it is representable on the same set by the combined array $\tilde{\xi} = (\tilde{\xi}^J)$, where $\tilde{\xi}^J$ consists of all reflections of the processes ξ^I with $|I| = |J|$, so that $\tilde{\xi}^J_k = (\xi^I_{k \circ p})$ for arbitrary sets $I \in 2^d$ with $|I| = |J|$ and associated bijections $p \colon I \to J$. Since $\tilde{\xi}$ is again separately exchangeable on R, it is even separately rotatable on the same index set by Lemma 8.23. Hence, Corollary 8.15 shows that $\tilde{\xi}^J$ is J-rotatable on N_J for every $J \in 2^d \setminus \{\emptyset\}$, and so by Lemma 8.2 and the ergodicity of ξ it is i.i.d. centered Gaussian. Furthermore, the arrays $\tilde{\xi}^J$ are independent on R by Proposition 8.16. Applying Proposition 7.7 twice, first part (i) and then part (ii), we conclude that the jointly contractable arrays $\tilde{\xi}^m = \tilde{\xi}^{\{1,\ldots,m\}}$ on $\Delta \hat{\mathbb{N}}_m = \{k \in \mathbb{N}^m; k_1 < \cdots < k_m\}$, $1 \leq m \leq d$, are mutually independent with i.i.d. entries.

The joint exchangeability of ξ implies that the array $\tilde{\xi}^m_{k \circ q} = (\xi^I_{k \circ q \circ p_I, n})$ is left-stationary in $q \in S_m$ for fixed m, where p_I denotes the unique, increasing bijection $|I| \to \{1, \ldots, m\}$. Hence, for any $m \leq d$ and $k \in \tilde{\mathbb{N}}_m$, Lemma 8.7 yields an a.s. representation

$$\xi^I_{k \circ q \circ p_I, n} = \sum_{r \in S_m} \sum_{h \in \mathbb{N}} c^{I, r}_{n, h} \, \eta_{k \circ qr, h}, \quad |I| = m, \; q \in S_m, \; n \in \mathbb{N}, \qquad (50)$$

where the variables $\eta_{l, h}$ form a G-array on $(k \circ S_m) \times \mathbb{N}$. By the independence properties of $\tilde{\xi}$ and FMP 6.10, we may choose a G-array η on the combined space $\bigcup_m (\mathbb{N}'_m \times \mathbb{N})$ satisfying (50) for all m and k. Taking q to be the identity permutation on S_m and inserting the simplified version of (50) into the earlier representation for X, we get as in Lemma 8.26 the desired representation (44). $\qquad \square$

Proof of Theorem 8.24: The CLRF X in (41) is clearly jointly rotatable for any choice of random elements α_π. Conversely, suppose that X is a jointly rotatable CLRF on $H^{\otimes d}$. Fix any ONB $h_1, h_2, \ldots$ in H, and define $X_k = X \bigotimes_j h_{k_j}$, $k \in \mathbb{N}^d$. Then the array (X_k) is jointly rotatable on $\mathbb{N}^d$, and so by Lemma 8.28 it has an off-diagonal representation (44), in terms of a G-array (η^m_{kl}) and an independent set of coefficients α^π_l such that $\sum_l (\alpha^\pi_l)^2 < \infty$ a.s. for all π. Now define the random elements α_π and G-processes η_m by (42) and (43), and let Y be the associated CLRF in (41). Then Y is again jointly rotatable, and (45) holds by construction. Hence, $X \overset{d}{=} Y$ by Lemma 8.25 (ii), and the desired representation follows by FMP 6.10 and Lemma 8.8.

Now let μ_a denote the distribution of X when $\alpha = a$ is non-random, so that in general $\mathcal{L}(X) = E\mu_\alpha$. If X is ergodic, then Lemma A1.2 yields $\mathcal{L}(X) = \mu_\alpha$ a.s., and we may choose a fixed array $a = (a_\pi)$ such that $\mathcal{L}(X) = \mu_a$. By FMP 6.10 we may finally choose a G-array $\eta = (\eta_m)$ such that X satisfies (41) with the α_π replaced by a_π.

Conversely, suppose that the α_π are non-random. Then the associated array (X_k) is ergodic by Lemma 7.35. If $\mathcal{L}(X) = c\mathcal{L}(X') + (1 - c)\mathcal{L}(X'')$ for some jointly rotatable CLRFs X' and X'' and a constant $c \in (0, 1)$, then a similar relation holds for the associated arrays (X_k), (X'_k), and (X''_k). The ergodicity of (X_k) yields $(X'_k) \overset{d}{=} (X''_k)$, which implies $X' \overset{d}{=} X''$. This shows that X is extreme, and the required ergodicity follows by Lemma A1.2. $\qquad \square$

Turning to the matter of uniqueness, we say that two partitions in $\mathcal{O}_d$ are *equivalent* if the corresponding unordered partitions agree, and write $[\pi]$ for the equivalence class containing π. The following result gives some a.s. and distributional uniqueness criteria, corresponding to those for separately rotatable functionals in Theorem 8.21. Due to the added subtleties in the jointly rotatable case, the present conditions are inevitably more complicated.

Theorem 8.29 *(uniqueness)* *Let X and Y be jointly rotatable CLRFs on $H^{\otimes d}$, representable as in Theorem 8.24 in terms of some random arrays $\alpha = (\alpha_\pi)$ and $\beta = (\beta_\pi)$, respectively, on $\mathcal{O}_d$. Then*

(i) $X = 0$ *a.s. iff* $\alpha = 0$ *a.s.;*

(ii) $X = Y$ *a.s. iff the sums over* $[\pi]$ *agree a.s. for every* $\pi \in \mathcal{O}_d$;

(iii) α *and* β *may represent the same jointly rotatable CLRF, iff there exist some random isometries on* $H^{\otimes(1+m)}$, $1 \le m \le d$, *of the form*

$$I_m = \sum_{p \in S_m} (A_p^m \otimes T_p), \qquad J_m = \sum_{p \in S_m} (B_p^m \otimes T_p), \qquad (51)$$

such that, a.s. for every $\pi \in \mathcal{O}_d$,

$$\sum_{p \in S_\pi} \Big(\bigotimes_{s \in \pi} A_{p_s}^{|s|} \Big) \alpha_{\pi \circ p^{-1}} = \sum_{p \in S_\pi} \Big(\bigotimes_{s \in \pi} B_{p_s}^{|s|} \Big) \beta_{\pi \circ p^{-1}}; \qquad (52)$$

(iv) $X \overset{d}{=} Y$ *iff there exist some random isometries* I_m *and* J_m *as in* (51) *such that* (52) *holds in distribution.*

Proof: (ii) For any $\pi \in \mathcal{O}_d$, let X^π and Y^π denote the partial sums over $[\pi]$ for the two representations. Then X^π and Y^π are both jointly rotatable, and so by Lemma 8.25 (i) it is enough to show that $X^\pi \otimes_j h_j = Y^\pi \otimes_j h_j$, a.s. for any ONB $h_1, h_2, \ldots$ in H. Then fix any disjoint, countable subsets $N_1, \ldots, N_d \subset \mathbb{N}$ with $j \in N_j$ for all j, and note that the array $X_n = X \otimes_j h_{n_j}$ is separately rotatable on the index set $R = N_1 \times \cdots \times N_d$. Furthermore, any representation (44) on $\mathbb{N}_d'$ yields an associated representation on R, similar to the one in Lemma 8.19. In particular, the πth term of the latter representation arises from the partial sum in (44) over the associated equivalence class $[\pi]$. Now Theorem 8.21 (ii) shows that the πth terms agree for the representations based on α and β. The corresponding statement is then true for the partial sums over $[\pi]$ in (44), and the desired relation follows as we take $n_j = j$ for all j.

(i) By Lemma 8.8 and Fubini's theorem we may assume that α is non-random, and from part (ii) we see that $X = 0$ implies

$$\sum_{p \in S_\pi} \xi_\pi (\alpha_{\pi \circ p} \otimes T_p f) = 0 \text{ a.s.}, \quad f \in H^{\otimes d}, \qquad (53)$$

where $\xi_\pi = \otimes_{k \in \pi} \xi_{|k|}$. Now let P_π denote the class of mappings $q \in S_d$ that only permutes components in π of equal length, and let T_q' denote the associated permutation operators on $H^{\otimes \pi}$. Choose $f = \otimes_j h_j$ for some orthonormal elements $h_1, \ldots, h_d \in H$, and note that the ortho-normality carries over to the elements $T_p f$ in $H^{\otimes d}$ for arbitrary permutations $p \in S_d$. Using the symmetry and isometry properties of the multiple stochastic integrals ξ_π (FMP 13.22), we get from (53)

$$\begin{aligned}
\|\alpha_\pi\|^2 &\leq \sum_{p\in S_\pi}\sum_{q\in P_\pi}\left\|T_q'\alpha_{\pi\circ p}\otimes T_{q\circ p}f\right\|^2 \\
&= \left\|\sum_{p\in S_\pi}\sum_{q\in P_\pi}(T_q'\alpha_{\pi\circ p}\otimes T_{q\circ p}f)\right\|^2 \\
&\lesssim E\left|\xi_\pi\circ\sum_{p\in S_\pi}\sum_{q\in P_\pi}(T_q'\alpha_{\pi\circ p}\otimes T_{q\circ p}f)\right|^2 \\
&\lesssim E\left|\sum_{p\in S_\pi}\xi_p(\alpha_{\pi\circ p}\otimes T_p f)\right|^2 = 0,
\end{aligned}$$

which implies $\alpha_\pi = 0$.

(iii) First suppose that α and β are related by (52), for some random isometries I_m and J_m as in (51). Introducing some independent G-processes ξ_m on $H^{\otimes(1+m)}$, $1 \leq m \leq d$, independent even of α and β, we consider for every m the processes $\eta_m = \xi_m \circ I_m$ and $\zeta_m = \xi_m \circ J_m$. By the isometric property of I_m and J_m together with Fubini's theorem, we see that $\eta^1, \ldots, \eta^d$ are again mutually independent G-processes, independent even of α, and similarly for $\zeta^1, \ldots, \zeta^d$. Letting γ_π denote the common expression in (52) and using Corollary 8.27, we get for any $f \in H^{\otimes d}$

$$\sum_{\pi\in\mathcal{O}_d}\eta_\pi(\alpha_\pi\otimes f) = \sum_{\pi\in\mathcal{O}_d}\xi_\pi(\gamma_\pi\otimes f) = \sum_{\pi\in\mathcal{O}_d}\zeta_\pi(\beta_\pi\otimes f),$$

and similarly for η and ζ. This shows that the pairs (α, η) and (β, ζ) represent the same jointly rotatable CLRF.

Conversely, suppose that the jointly rotatable CLRF X can be represented in terms of both α and β, along with the associated G-processes $\eta = (\eta_m)$ and $\zeta = (\zeta_m)$. Arguing as in the proof of Theorem 8.21, we may assume that α and β are non-random and the pair (η, ζ) is ergodic, jointly rotatable. By Lemma 8.7 we may then represent η and ζ as in (48), in terms of some independent G-processes ξ_m on $H^{\otimes(1+m)}$, $m \leq d$, and some isometries I_m and J_m as in (51). By Corollary 8.27 we obtain the new representations

$$Xf = \sum_{\pi\in\mathcal{O}_d}\xi_\pi(\tilde{\alpha}_\pi\otimes f) = \sum_{\pi\in\mathcal{O}_d}\xi_\pi(\tilde{\beta}_\pi\otimes f),$$

where $\tilde{\alpha}_\pi$ and $\tilde{\beta}_\pi$ denote the left- and right-hand sides of (52). Then (i) yields $\tilde{\alpha}_\pi = \tilde{\beta}_\pi$ a.s. for all $\pi \in \mathcal{O}_d$, as required.

(iv) Here we may argue as in the proof of Theorem 8.21. $\square$

8.7 Jointly Rotatable Arrays

In Proposition 8.20 we saw that every separately rotatable array on $\mathbb{N}^d$ can be extended to a separately rotatable CLRF on $(l^2)^{\otimes d}$. The corresponding statement in the jointly rotatable case is false in general, since a jointly rotatable array may include diagonal terms, whose associated random functionals fail to be continuous.

For any partition $\pi \in \mathcal{P}_d$, the associated diagonal space D_π consists of all $t \in \mathbb{R}_+^d$ such that $t_i = t_j$ whenever i and j belong to the same set $J \in \pi$. By λ^π we denote the natural projection of Lebesgue measure on $\mathbb{R}_+^\pi$ onto D_π. More precisely, letting $e_\pi \colon \mathbb{R}_+^\pi \to D_\pi$ be given by

$$e_{\pi,j}(t) = t_J, \quad j \in J \in \pi, \ t \in \mathbb{R}_+^\pi,$$

we define λ^π as the image of Lebesgue measure on $\mathbb{R}_+^\pi$ under the mapping e_π.

To construct a CLRF extending a given array X, we need to consider an underlying space $L_c^2(\mu)$, consisting of all functions $f \in L^2(\mu)$ with compact support. Continuity is then defined in the local sense, in terms of functions with uniformly bounded support. We also write $\mathcal{P}_d^2$ for the class of collections π of disjoint sets $J \in 2^d$ with $|J| = 2$, and put $\pi^c = \bigcap_{J \in \pi} J^c$. Finally, $\mathcal{O}_J$ denotes the class of partitions κ of J into ordered sets k of size $|k| \geq 1$.

Theorem 8.30 *(jointly rotatable arrays)* *An array X on $\mathbb{N}^d$ is jointly rotatable iff it can be extended to a CLRF on $L_c^2(\sum_{\pi \in \mathcal{P}_d^2} \lambda^\pi)$ of the form*

$$Xf = \sum_{\pi \in \mathcal{P}_d^2} \sum_{\kappa \in \mathcal{O}_{\pi^c}} \left(\lambda^\pi \otimes \bigotimes_{k \in \kappa} \eta_{|k|} \right) (\alpha_{\pi,\kappa} \otimes f),$$

for some independent G-processes η_m on $H \otimes L^2(\lambda^m)$, $1 \leq m \leq d$, and an independent set of random elements $\alpha_{\pi,\kappa}$ in $H^{\otimes \kappa}$, $\pi \in \mathcal{P}_d^2$, $\kappa \in \mathcal{O}_{\pi^c}$. The latter can then be chosen to be non-random iff X is ergodic.

Here the random functional Y is said to *extend* the array X if $X_k = Y(I_k)$ for all $k \in \mathbb{N}^d$, where I_k denotes the unit cube $(k - 1, k]$ in $\mathbb{R}_+^d$, 1 being the vector $(1, \ldots, 1)$. Several lemmas are needed for the proof, beginning with a simple self-similarity property, where we write $mI_k = (m(k - 1), mk]$.

Lemma 8.31 *(scaling)* *Let X be a jointly rotatable family of arrays X_d on $\mathbb{N}^d$, $d \in \mathbb{Z}_+$, and define*

$$X_d^m(k) = X_d(mI_k) = \sum_{l \in mI_k} X_d(l), \quad k \in \mathbb{N}^d, \ m \in \mathbb{N}, \ d \in \mathbb{Z}_+.$$

Then

$$(X_d^m; d \in \mathbb{Z}_+) \overset{d}{=} (m^{d/2} X_d; d \in \mathbb{Z}_+), \quad m \in \mathbb{N}. \tag{54}$$

Proof: Given any $m, n \in \mathbb{N}$, we may choose an orthogonal matrix $u = (u_{ij})$ of dimension $mn \times mn$ such that

$$u_{ij} = m^{-1/2} 1\{i \in mI_j\}, \quad i \leq mn, \ j \leq n.$$

Letting U denote the corresponding unitary operator on l^2 that affects only the first mn coordinates, we obtain

$$(X_d \circ U^{\otimes d})_k = m^{-d/2} X_d(mI_k), \quad k_1, \ldots, k_d \leq n.$$

Since X is jointly rotatable, (54) holds on the set $\bigcup_d \{1, \ldots, n\}^d$, and the assertion follows since n was arbitrary. $\qquad\square$

Our next aim is to characterize certain jointly rotatable diagonal arrays.

Lemma 8.32 *(binary arrays) Let X be a random array on $\{0,1\}^d$ such that $X_k = 0$ unless $k_1 = \cdots = k_d$. Then X is jointly rotatable iff $X_{00} = X_{11}$ a.s. when $d = 2$, and iff $X = 0$ a.s. when $d > 2$.*

Proof: Introduce the orthogonal matrix

$$U = \frac{1}{\sqrt{2}} \begin{pmatrix} 1 & 1 \\ -1 & 1 \end{pmatrix}.$$

For $d = 2$, we may use the joint rotatability of X to get

$$0 = X_{01} \overset{d}{=} (X \circ U^{\otimes 2})_{01} = (U^T X U)_{01} = \tfrac{1}{2}(X_{11} - X_{00}),$$

which implies $X_{00} = X_{11}$ a.s. Conversely, diagonal arrays with this property are trivially jointly rotatable, since $I \circ U^{\otimes 2} = U^T I U = U^T U = I$.

For $d > 2$, we define

$$U_m = U^{\otimes m} \otimes I^{\otimes(d-m)}, \quad m \leq d.$$

Using the joint rotatability of X and noting that rotations preserve the Euclidean norm, we get

$$\begin{aligned}
0 &= \sum_k X_{01k}^2 \overset{d}{=} \sum_k (X \circ U_d)_{01k}^2 \\
&= \sum_k (X \circ U_{d-1})_{01k}^2 = \cdots = \sum_k (X \circ U_2)_{01k}^2 \\
&= (X \circ U_2)_{010}^2 + (X \circ U_2)_{011}^2 \\
&= (X_0^2 + X_1^2)/4,
\end{aligned}$$

where $\mathbf{0} = (0,\ldots,0)$ and $\mathbf{1} = (1,\ldots,1)$ in any dimension, and all summations extend over $k \in \{0,1\}^{d-2}$. This implies $X_{\mathbf{0}}^2 = X_{\mathbf{1}}^2 = 0$ a.s. $\qquad\square$

Let us now write $\mathcal{I}^d$ for the class of bounded rectangles in $\mathbb{R}_+^d$ determined by binary rational coordinates, and put $\mathcal{I}_{01}^d = \mathcal{I}^d \cap (0,1]^d$. The next result shows how a jointly rotatable process on $\mathcal{I}^d$ can be viewed as a jointly rotatable array on $\mathbb{N}^d$, whose entries are themselves processes on $\mathcal{I}_{01}^d$.

Lemma 8.33 *(array of grid processes) Let X be a finitely additive, jointly rotatable process on $\mathcal{I}^d$, fix an $n \in \mathbb{N}$, and define the processes Y_j on $\mathcal{I}_{01}^d$ by*

$$Y_j(B) = X(2^{-n}(B + j)), \quad B \in \mathcal{I}_{01}^d, \ j \in \mathbb{Z}_+^d.$$

Then the array $Y = (Y_j)$ is again jointly rotatable.

Proof: We need to show that $Y \circ U^{\otimes d} \overset{d}{=} Y$ for any rotation U on l^2 that affects only finitely many coordinates. Then fix any $m \in \mathbb{N}$, and define for $j \in \mathbb{Z}_+^d$ and $k \in (\mathbf{0}, 2^m \mathbf{1}]$

$$
\begin{aligned}
Z_{jk} &= Y_j(2^{-m} I_k) = X(2^{-n}(2^{-m} I_k + j)) \\
&= X(2^{-m-n}(I_k + j2^m)) \\
&= X(2^{-m-n} I_{k+j2^m}).
\end{aligned}
$$

The array $Z = (Z_{jk})$ is clearly jointly rotatable in the pair (j, k). We need to show that Z_{jk} is also jointly rotatable in index j alone. This will follow if we can show that every rotation in j is also a rotation in the pair (j, k). The last statement follows from the fact that, for any orthogonal $r \times r$ matrix $U = (U_{ij})$, the matrix $V_{(i,h),(j,k)} = U_{ij}\delta_{hk}$ is again orthogonal of dimension $r2^m \times r2^m$. $\qquad\square$

The next result concerns extensions of processes on the non-diagonal part $\mathcal{I}_d'$ of $\mathcal{I}^d$. Given such a process X, we introduce the associated *family of reflections* $\tilde{X} = (X^p)$, where p is an arbitrary permutation of $1, \ldots, d$ and

$$
X^p(2^{-n} I_k) = X(2^{-n} I_{k \circ p}), \quad n \in \mathbb{N}, \ k \in \mathbb{N}_d'.
$$

Lemma 8.34 *(off-diagonal component) Let X be a finitely additive, jointly exchangeable process on $\mathcal{I}_d'$, such that the associated family of reflections is separately rotatable on every rectangle in $\mathcal{I}_d'$. Then X can be extended to a jointly rotatable CLRF on $L^2(\lambda^d)$.*

Proof: For every $n \in \mathbb{Z}$, let X_n denote the restriction of X to the non-diagonal, cubic grid of mesh size 2^n, regarded as an array on $\mathbb{N}_d'$. We claim that X_n can be represented as in (44), in terms of a G-array $\eta = (\eta_{kl}^m)$ and an independent array of random coefficients α_l^π. Then note that X_n is again jointly exchangeable, and is therefore representable in terms of a U-array ξ. By Proposition 7.33 and Lemma 7.34 we see that, conditionally on $\xi_\emptyset$, the array X_n is ergodic, jointly exchangeable, whereas the associated array of reflections $\tilde{X}_n = (X_n^p)$ is separately rotatable on every non-diagonal rectangle. We may now derive the desired representation of X_n, first in the ergodic case, as in the proof of Lemma 8.28, and then in general by means of Lemma A1.5.

Now fix any ONB $h_1, h_2, \ldots$ in $L^2(\lambda)$, and define the random elements α_π in $L^2(\lambda)$ as in (42), based on the coefficients α_k^π in the representation of X_0. Introducing some independent G-processes η_m on $L^2(\lambda^{m+1})$, $m = 1, \ldots, d$, we may next define a jointly rotatable CLRF Y on $L^2(\lambda^d)$ by the right-hand side of (41). Let Y_n denote the restriction of Y to the non-diagonal cubic grid in $\mathbb{R}_+^d$ of mesh size 2^n, and note that $Y_0 \overset{d}{=} X_0$ by construction. Since every X_n and Y_n can be extended to a jointly rotatable array on $\mathbb{N}^d$, we get by Lemma 8.31

$$X_n \overset{d}{=} 2^{nd/2} X_0 \overset{d}{=} 2^{nd/2} Y_0 \overset{d}{=} Y_n, \quad n \in \mathbb{Z},$$

which implies $X \overset{d}{=} Y$ on $\mathcal{I}'_d$. Finally, by FMP 6.10, there exists a CLRF $\bar{X} \overset{d}{=} Y$ on $L^2(\lambda^d)$ such that $X = \bar{X}$ a.s. on $\mathcal{I}'_d$. $\qquad\square$

Given a sub-class $\mathcal{C} \subset \mathcal{P}_d$, we say that a partition $\pi \in \mathcal{C}$ is *maximal* in $\mathcal{C}$ if $\pi \prec \pi' \in \mathcal{C}$ implies $\pi = \pi'$, where $\pi \prec \pi'$ means that π' is a refinement of π. Defining the *restriction* of a process X on $\mathcal{I}^d$ to a set $B \subset \mathbb{R}^d_+$ as the restriction to the class of dyadic subsets $A \in \mathcal{I}_d \cap B$, we say that X is *supported* by B if its restriction to B^c vanishes a.s. For any $\pi \in \mathcal{P}_d$, we write X_π for the restriction of X to the class of rectangles $A_1 \times \cdots \times A_d \in \mathcal{I}_d$, such that $A_i \cap A_j = \emptyset$ whenever i and j belong to different sets in π. If X is supported by $\bigcup_{\pi \in \mathcal{C}} D_\pi$ and π is maximal in $\mathcal{C}$, then X_π is clearly supported by D_π. We may refer to X_π as the restriction of X to the non-diagonal part $D'_\pi = D_\pi \cap \bigcap_{\pi'} D^c_{\pi'}$, where the last intersection extends over all partitions $\pi' \prec \pi$ with $\pi' \neq \pi$. For any $\pi \in \mathcal{P}_d$ and $k \leq d$, we define $\pi_k = \{I \in \pi;\ |I| = k\}$.

Lemma 8.35 *(maximal components) Let X be a finitely additive, jointly rotatable process on $\mathcal{I}^d$, supported by $\bigcup_{\pi \in \mathcal{C}} D_\pi$ for some $\mathcal{C} \subset \mathcal{P}_d$. For any maximal partition $\pi \in \mathcal{C}$, let X_π be the restriction of X to D'_π. Then $X_\pi = 0$ a.s. unless $\pi = \pi_1 \cup \pi_2$, in which case X_π can be extended to a CLRF on $L^2_c(\lambda^\pi)$ of the form $\lambda^{\pi_2} \otimes Y_\pi$, where Y_π is a jointly rotatable CLRF on $L^2(\lambda^{\pi_1})$.*

Proof: For every $n \in \mathbb{N}$, let X^n be the array of contributions of X to the regular cubic grid of mesh size 2^n, and note that X^n is again jointly rotatable. Fix any maximal partition $\pi \in \mathcal{C}$ and a set $J \in \pi$ with $|J| \geq 2$, and put $\pi' = \pi \setminus \{J\}$. Let $m \in D_{\pi'} \cap \mathbb{N}^{J^c}$ be based on distinct numbers $m_I \in \mathbb{N}$, $I \in \pi'$, and choose $h \neq k$ in $\mathbb{N} \setminus \{m_I;\ I \in \pi'\}$. Since π is maximal in $\mathcal{C}$, the array X^n vanishes a.s. outside the main diagonal of the cubic array $\{h, k\}^J \times \{m\}$. Considering joint rotations involving only indices h and k, we see from Lemma 8.32 that $X^n_{k,\dots,k,m}$ is a.s. independent of k and equals 0 when $|J| > 2$. Since J was arbitrary, it follows that $X^n_\pi = 0$ a.s. unless $\pi = \pi_1 \cup \pi_2$, in which case X^n_π depends only on the indices in π_1. In the latter case, we denote the corresponding array on $\mathbb{N}'_{\pi_1}$ by Y^n.

To see how the arrays Y^n are related for different n, fix any $m < n$ in $\mathbb{N}$, and choose $r \in \mathbb{N}^{\pi_1}$ and $k \in \mathbb{N}^{\pi_2}$ such that the sets $I_k \times I_r$ and $I_k \times 2^{n-m} I_r$ are both non-diagonal in $\mathbb{R}^\pi_+$. Using the additivity and invariance properties of the arrays X^n, we get

$$\begin{aligned}
Y^m(2^{n-m} I_r) &= X^m(I^2_k \times 2^{n-m} I_r) \\
&= 2^{-(n-m)|\pi_2|} X^m(2^{n-m}(I^2_k \times I_r)) \\
&= 2^{-(n-m)|\pi_2|} X^n(I^2_k \times I_r) \\
&= 2^{-(n-m)|\pi_2|} Y^n(I_r).
\end{aligned}$$

Without ambiguity, we may then define a finitely additive, random set function Y_π on $\mathcal{I}'_{\pi_1}$ through the formula

$$Y_\pi(2^n I_r) = 2^{-n|\pi_2|} Y^n(I_r), \quad r \in \mathbb{N}'_{\pi_1}, \; n \in \mathbb{Z}.$$

The joint rotatability of X^n implies that Y^n is jointly exchangeable, and also that the corresponding array of reflections $\tilde{Y}^n$ is separately rotatable on every non-diagonal rectangle in $\mathbb{N}^{\pi_1}$. The process Y_π has then the same properties, and so by Lemma 8.34 it extends to a jointly rotatable CLRF on $L^2(\lambda^{\pi_1})$. To see that the CLRF $\lambda^{\pi_2} \otimes Y_\pi$ on $L^2_c(\lambda^\pi)$ is an extension of X_π, we may consider any k and r as before and conclude from the definitions of X^n, Y^n, Y_π, λ^{π_2}, and $\lambda^{\pi_2} \otimes Y_\pi$ that

$$\begin{aligned}
X_\pi(2^n(I_k^2 \times I_r)) &= X^n(I_k^2 \times I_r) = Y^n(I_r) \\
&= 2^{n|\pi_2|} Y_\pi(2^n I_r) \\
&= \lambda^{\pi_2}(2^n I_k^2) Y_\pi(2^n I_r) \\
&= (\lambda^{\pi_2} \otimes Y_\pi)(2^n(I_k^2 \times I_r)). \qquad \square
\end{aligned}$$

The next result ensures that the various diagonal components combine into a jointly rotatable process.

Lemma 8.36 (*combined grid processes*) *Let X be a finitely additive, jointly rotatable process on $\mathcal{I}^d$. Fix a linear sub-space $D \subset \mathbb{R}^d$ with invariant measure λ_D, and let A be a finite union of linear sub-spaces of $\mathbb{R}^d$ such that $\lambda_D A = 0$. Write X' for the restriction of X to A^c, and assume that X' can be extended to a CLRF Y on $L^2_c(\lambda_D)$. Then the pair (X, Y) is again jointly rotatable on $\mathcal{I}^d$.*

Proof: Introduce the arrays X^m and Y^m, recording the contributions of X and Y to the regular cubic grid of mesh size 2^{-m}. We need to show that the pair (X^m, Y^m) is jointly rotatable for every $m \in \mathbb{N}$. Then for every $n \geq m$, consider the union A_n of all cubes in the regular 2^{-n}-grid intersecting A, and form the periodic continuations

$$A^m = \bigcup_k (A + k2^{-m}), \qquad A_n^m = \bigcup_k (A_n + k2^{-m}).$$

Write Y_n^m for the array of contributions of Y to the complementary sets $2^{-m} I_k \setminus A_n^m$. Since X and Y agree on $(A_n^m)^c$, Lemma 8.33 shows that the pair (X^m, Y_n^m) is jointly rotatable for fixed n. Now by dominated convergence as $n \to \infty$, for any bounded Borel set $B \subset \mathbb{R}^d_+$, we have

$$\lambda_D(A_n^m \cap B) \to \lambda_D(A^m \cap B) = 0.$$

Hence, the continuity of Y yields $Y_n^m \overset{P}{\to} Y^m$, and so the joint rotatability of (X^m, Y_n^m) carries over to the pair (X^m, Y^m). $\qquad \square$

We may now derive the required diagonal decomposition, using a recursive argument based on the last two lemmas.

Lemma 8.37 *(diagonal decomposition) Every finitely additive, jointly rotatable process X on $\mathcal{I}^d$ can be extended to a CLRF on $L_c^2(\sum_{\pi\in\mathcal{P}_d^2}\lambda^\pi)$ of the form*

$$X = \sum_{\pi\in\mathcal{P}_d^2} (\lambda^{\pi_2}\otimes Y_\pi) \quad a.s., \tag{55}$$

for some jointly rotatable family of CLRFs Y_π on $L^2(\lambda^{\pi_1})$, $\pi\in\mathcal{P}_d^2$. This decomposition is a.s. unique.

Proof: Assume that X is supported by $\bigcup_{\pi\in\mathcal{C}} D_\pi$ for some $\mathcal{C}\subset\mathcal{P}_d$. We claim that X is then representable as in (55), with the summation restricted to $\mathcal{C}$. The desired decomposition will then follow as we take $\mathcal{C}=\mathcal{P}_d$.

The statement is vacuously true for $\mathcal{C}=\emptyset$. Next, assuming the result to be true for $|\mathcal{C}|<m$, we proceed to the case where $|\mathcal{C}|=m$. Letting π be a maximal partition in $\mathcal{C}$, we may introduce the restriction X_π of X to D_π', the non-diagonal part of D_π. By Lemma 8.35 we have $X_\pi=0$ a.s. when $\pi\neq\pi_1\cup\pi_2$, in which case $\mathcal{C}$ can be replaced by the set $\mathcal{C}\setminus\{\pi\}$ with cardinality $m-1$. The assertion then follows by the induction hypothesis.

If instead $\pi=\pi_1\cup\pi_2$, then X_π extends by Lemma 8.35 to a CLRF on $L^2(\lambda^\pi)$ of the form $\lambda^{\pi_2}\otimes Y_\pi$, where Y_π is a jointly rotatable CLRF on $L^2(\lambda^{\pi_1})$. Furthermore, we see from Lemma 8.36 that the pair (X,X_π) is again jointly rotatable. But then $X'=X-X_\pi$ is jointly rotatable and supported by $\bigcup_{\pi'\in\mathcal{C}'} D_{\pi'}$, where $\mathcal{C}'=\mathcal{C}\setminus\{\pi\}$, and so the induction hypothesis yields an a.s. representation $X'=\sum_{\pi'\in\mathcal{C}'}(\lambda^{\pi_2'}\otimes Y_{\pi'})$, for some CLRFs $Y_{\pi'}$ on $L^2(\lambda^{\pi_1'})$ such that the entire family $Y'=(Y_{\pi'})$ is jointly rotatable. Here the pairs (X',X_π), (X',Y'), and (X_π,Y_π) are jointly rotatable, and since Y_π and Y' are measurable functions of X_π and X', respectively, the joint rotatability extends by Lemma 8.17 to the whole collection (Y',Y_π). This completes the induction and establishes the desired decomposition.

The uniqueness can be proved recursively by a similar argument. Thus, assuming that X has two decompositions as in (55), both with summation restricted to $\mathcal{C}\subset\mathcal{P}_d$, we note that X is supported by $\bigcup_{\pi\in\mathcal{C}} D_\pi$. For any maximal element $\pi\in\mathcal{C}$, we may introduce the restriction X_π of X to D_π', which agrees a.s. with the corresponding terms in (55) when $\pi=\pi_1\cup\pi_2$, and equals 0 otherwise. In the former case, the factorization $X_\pi=\lambda^{\pi_2}\otimes Y_\pi$ is clearly a.s. unique. Subtracting the common πth term from the two decompositions, we may reduce the discussion to the case of summation over $\mathcal{C}'=\mathcal{C}\setminus\{\pi\}$. $\square$

Proof of Theorem 8.30: For every $n\in\mathbb{N}$, we may define a finitely additive set function X^n on the cubic grid $\mathcal{I}_n^d=(2^{-n}I_k;\ k\in\mathbb{N}^d)$ by

$$X^n(2^{-n}I_k)=2^{-nd}X_k, \quad k\in\mathbb{N}^d,\ n\in\mathbb{N}.$$

Then Lemma 8.31 yields $X^m\overset{d}{=}X^n$ on $\mathcal{I}_m^d$ for all $m<n$, and so by FMP 6.14 there exists a finitely additive process $\overline{X}$ on $\mathcal{I}^d$ satisfying $\overline{X}(I_k)=X_k$

a.s. for all $k \in \mathbb{N}^d$, and such that moreover

$$\left(\overline{X}(2^{-n} I_k); \; k \in \mathbb{N}^d\right) \overset{d}{=} \left(2^{-nd} X_k; \; k \in \mathbb{N}^d\right), \quad n \in \mathbb{Z}_+.$$

In particular, we note that $\overline{X}$ is again jointly rotatable.

By Lemma 8.37 we may extend the process $\overline{X}$ to a CLRF on the space $L_c^2(\sum_{\pi \in \mathcal{P}_d^2} \lambda^\pi)$ and obtain a decomposition as in (55), in terms of a jointly rotatable array of CLRFs Y_π on $L^2(\lambda^{\pi_1})$, $\pi \in \mathcal{P}_d^2$. Proceeding as in the proof of Theorem 8.24, we may finally construct the required representation of the functionals Y_π, in terms of some independent G-processes η_m on $H \otimes L^2(\lambda^m)$, $1 \leq m \leq d$, along with an independent collection of random elements $\alpha_{\pi,\kappa}$ in suitable Hilbert spaces. The last statement can be proved by a similar argument as before. $\qquad\square$

8.8 Separately Exchangeable Sheets

We may now use the methods and results of the previous sections to derive representations of exchangeable random sheets. By a *random sheet* on a rectangular space $T = \mathbb{R}_+^{d'} \times [0,1]^{d''}$ we mean a continuous random process X on T such that $X_t = 0$ whenever $\prod_j t_j = 0$. Separate or joint exchangeability of X is defined in terms of the d-dimensional increments of X over an arbitrary cubic grid in T, where $d = d' + d''$.

Let $\hat{\mathcal{P}}_d$ denote the class of families π of disjoint subsets $J \in 2^d \setminus \{\emptyset\}$, so that $\hat{\mathcal{P}}_d = \bigcup_{J \in 2^d} \mathcal{P}_J$, where $\mathcal{P}_J$ is the class of partitions of J into non-empty subsets I. Put $\pi^c = \bigcap_{J \in \pi} J^c$. Given a random functional η and a measurable set B, we often write ηB instead of $\eta 1_B$.

Theorem 8.38 (*separately exchangeable sheets on $\mathbb{R}_+^d$*) *A process X on $\mathbb{R}_+^d$ is separately exchangeable with a continuous version iff it has an a.s. representation*

$$X_t = \sum_{\pi \in \hat{\mathcal{P}}_d} \left(\lambda^{\pi^c} \otimes \bigotimes\nolimits_{J \in \pi} \eta_J\right)(\alpha_\pi \otimes [0,t]), \quad t \in \mathbb{R}_+^d, \qquad (56)$$

in terms of some independent G-processes η_J on $H \otimes L^2(\lambda^J)$, $J \in 2^d \setminus \{\emptyset\}$, and an independent collection of random elements α_π in $H^{\otimes \pi}$, $\pi \in \hat{\mathcal{P}}_d$. The latter can then be chosen to be non-random iff X is ergodic.

Our proof is based on a sequence of lemmas. Our first aim is to show that the processes in (56) are continuous.

Lemma 8.39 (*Hölder continuity*) *The process X in (56) has a version that is a.s. locally Hölder continuous with exponent p for every $p \in (0, \frac{1}{2})$. This remains true if $\bigotimes_J \eta_J$ is replaced by $\bigotimes_J \eta_{|J|}$ for some independent G-processes η_k on $H \otimes L^2(\lambda^k)$, $k \leq d$.*

Proof: By Fubini's theorem and Lemma 8.8 we may assume that the elements α_π are non-random. It is also enough to consider a single term

$$X_t^\pi = \left(\prod_{j\in\pi^c} t_j\right)\left(\bigotimes_{J\in\pi}\eta_J\right)(\alpha_\pi \otimes [0,t']), \quad t \in \mathbb{R}_+^d,$$

for an arbitrary $\pi \in \hat{\mathcal{P}}_d$, where $t' = (t_j;\ j \notin \pi^c)$. Write Y^π for the second factor on the right and put $m = \sum_{J\in\pi}|J|$. Using Lemma A3.3, we get for fixed $q > 0$ and arbitrary $s, t \in \mathbb{R}_+^m$

$$
\begin{aligned}
\|Y_s^\pi - Y_t^\pi\|_q &\lesssim \left\|\alpha_\pi \otimes (1_{[0,s]} - 1_{[0,t]})\right\| \\
&= \|\alpha_\pi\|\,\|1_{[0,s]} - 1_{[0,t]}\|_2 \\
&\lesssim \|\alpha_\pi\|\left(|s-t|(|s| \vee |t|)^{m-1}\right)^{1/2}.
\end{aligned}
$$

Hence, for fixed $q > 0$ and α and for bounded $s, t \in \mathbb{R}_+^m$,

$$E|Y_s^\pi - Y_t^\pi|^q \lesssim |s-t|^{q/2}. \tag{57}$$

Fixing any $p \in (0, \frac{1}{2})$, we may now choose q so large that $q > 2(m + pq)$. Then the bound in (57) can be replaced by $|s-t|^{m+pq}$, and we may conclude from the Kolmogorov–Chentsov criterion in FMP 3.23 that Y^π has a locally Hölder continuous version with exponent p. The same thing is then true for the original process X^π on $\mathbb{R}_+^d$, and the assertion follows. $\qquad\square$

The main idea behind the proof of Theorem 8.38 is to decompose X into rotatable processes, each of which can be represented as in Theorem 8.18. We begin with an elementary decomposition of processes on $[0,1]^d$.

Lemma 8.40 *(centering decomposition) Given a set $T \subset [0,1]$ containing 0 and 1, let X be a process on T^d such that $X_t = 0$ whenever $\prod_j t_j = 0$. Then X has a unique decomposition*

$$X_t = \sum_{J\in 2^d} X_{t_J}^J \prod_{i\notin J} t_i, \quad t = (t_1,\ldots,t_d) \in T^d, \tag{58}$$

where each X^J is a process on T^J such that $X_t^J = 0$ whenever $\prod_j t_j(1-t_j) = 0$. When $T = [0,1]$, the process X is [ergodic] separately exchangeable iff the family (X^J) has the same properties.

Proof: Assuming the X^J to have the stated properties, we get by (58) for any $J \in 2^d$

$$X_{t_J}^J = X_t - \sum_{I\in\hat{J}'} X_{t_I}^I \prod_{i\in J\setminus I} t_i, \quad t \in T^J \times \{1\}^{J^c}, \tag{59}$$

where $\hat{J}'$ denotes the class of proper subsets of J. The asserted uniqueness now follows by induction on $|J|$. Conversely, we may construct the X^J recursively by means of (59), and for $d = \{1,\ldots,1\}$ we note that (58) and (59) are equivalent. We also see from (59) that $X_t^J = 0$ when $\prod_j t_j = 0$.

To see that $X_t^J = 0$ even when $\max_{j \in J} t_j = 1$, assume this to be true for all processes X^I with $|I| < |J|$. Fix any $j \in J$ and put $K = J \setminus \{j\}$. Applying (59) to both J and K and using the induction hypothesis, we get for any $t \in T^K \times \{1\}^{K^c}$

$$X_{t_J}^J = X_t - \sum_{I \in \tilde{J}'} X_{t_I}^I \prod_{i \in I^c} t_i = X_t - \sum_{I \subset K} X_{t_I}^I \prod_{i \in I^c} t_i = 0.$$

This completes the induction and proves the claim. The last assertion follows easily by means of Lemma A1.1. □

Combining the last result with Lemma 8.3, we obtain a similar decomposition of a separately rotatable random sheet X on $\mathbb{R}_+^d$ into separately rotatable components X^J.

Lemma 8.41 *(rotatable decomposition)* *Any separately exchangeable random sheet X on $\mathbb{R}_+^d$ has an a.s. unique decomposition (58) on $\mathbb{R}_+^d$, in terms of a separately rotatable family of random sheets X^J on $\mathbb{R}_+^J$, $J \in 2^d$.*

Proof: Applying the first transformation of Lemma 8.3 in each coordinate, we obtain a separately exchangeable random sheet Y on $[0,1]^d$. Next we may decompose Y as in Lemma 8.40, in terms of some continuous processes Y^J on $[0,1]^J$, $J \in 2^d$, satisfying $Y_t^J = 0$ whenever $\prod_j t_j(1-t_j) = 0$, and such that the whole array (Y^J) is separately exchangeable. Applying the second transformation in Lemma 8.3 to each component Y^J, we obtain a separately rotatable array of random sheets X^J on $\mathbb{R}_+^J$, $J \in 2^d$, and we note that (58) remains valid on $\mathbb{R}_+^d$ for the processes X and X^J.

To prove the asserted uniqueness, suppose that X has also a decomposition (58) in terms of a separately rotatable array of random sheets $\tilde{X}^J$ on $\mathbb{R}_+^J$. Applying the first transformation in Lemma 8.3 to the processes X and $\tilde{X}^J$, we obtain a representation of Y in terms of some processes $\tilde{Y}^J$ with $\tilde{Y}_t^J = 0$ whenever $\prod_j t_j(1-t_j) = 0$. The uniqueness in Lemma 8.40 yields $\tilde{Y}^J = Y^J$ a.s. for every J, and we may finally employ the second transformation in Lemma 8.3 to get $\tilde{X}^J = X^J$ a.s. for all J. □

We also need a criterion for ergodicity. Let us say that a random sheet X is *dissociated* if the corresponding increment process Y has this property. If X is separately or jointly exchangeable, then Y can be extended to a process on $\mathbb{R}^d$ with the same property, and we write $Y_\emptyset$ for the restriction of the latter process to $\mathbb{R}_-^d$.

Lemma 8.42 *(ergodicity)* *Let X be a separately or jointly exchangeable random sheet on $\mathbb{R}_+^d$ with extension Y to $\mathbb{R}^d$. Then X is conditionally ergodic exchangeable given $Y_\emptyset$, and X is ergodic iff it is dissociated.*

Proof: The last assertion can be proved in the same way as the equivalence of (i) and (ii) in Lemma 7.35. To prove the first assertion, we note as in Lemma 7.34 that X remains separately or jointly exchangeable, conditionally on $Y_\emptyset$. Applying the mentioned results to the increments over an arbitrary cubic grid, we see that X is also conditionally dissociated, and the asserted ergodicity follows as before. □

Proof of Theorem 8.38: A process X with representation (56) is clearly separately exchangeable, and by Lemma 8.39 it has a continuous version. Conversely, let X be a separately exchangeable random sheet on $\mathbb{R}^d_+$. Then Lemma 8.41 yields a decomposition as in (58), in terms of some random sheets X^J on $\mathbb{R}^J_+$, $J \in 2^d$, such that the whole array (X^J) is separately rotatable. By Lemma 8.22, supplemented by FMP 6.10, there exist some independent G-processes η_J on $H \otimes L^2(\mathbb{R}^J_+)$, $J \in \tilde{\mathbb{N}} \setminus \{\emptyset\}$, and an independent collection of random elements α_π in $H^{\otimes \pi}$, $\pi \in \mathcal{P}_J$, $J \in \tilde{\mathbb{N}}$, such that a.s.

$$X^J_t = \sum_{\pi \in \mathcal{P}_J} \left(\bigotimes_{I \in \pi} \eta_I \right) (\alpha_\pi \otimes [0, t]), \quad t \in \mathbb{N}^J, \ J \in 2^d. \tag{60}$$

To extend the representation to arbitrary $t \in \mathbb{R}^J_+$, $J \in 2^d$, we define the processes Y^J on $\mathbb{R}^J_+$ by the right-hand side of (60). For every $m \in \mathbb{N}$, we may next introduce the arrays

$$X^{J,m}_k = X^J_{k/m}, \quad Y^{J,m}_k = Y^J_{k/m}, \qquad k \in \mathbb{N}^J, \ J \in 2^d.$$

Then for fixed m, Lemma 8.31 yields

$$(X^{J,m}) \stackrel{d}{=} (m^{-|J|/2} X^{J,1}) = (m^{-|J|/2} Y^{J,1}) \stackrel{d}{=} (Y^{J,m}),$$

which implies $(X^J) \stackrel{d}{=} (Y^J)$ on $\mathbb{Q}^d_+$. By FMP 6.10 we may then assume that (60) holds a.s. on $\mathbb{Q}^J_+$, $J \in 2^d$. The representations extend to the index sets $\mathbb{R}^J_+$ by the continuity of each side, and (56) follows.

To prove the last assertion, let Q_α be the distribution of X when $\alpha = (\alpha_\pi)$ is non-random, and note that $P[X \in \cdot | \alpha] = Q_\alpha$ a.s. in general by Lemma 8.8 and Fubini's theorem. If X is ergodic, it is also extreme by Lemma A1.2, and so the measure Q_α is a.s. non-random. But then $\mathcal{L}(X) = Q_a$ for some fixed array $a = (a_\pi)$, and so by FMP 6.10 we have an a.s. representation (56) with $\alpha = a$. Conversely, if (56) holds for some non-random α, then X is clearly dissociated, and the ergodicity follows by Lemma 8.42. □

Given a CLRF ξ on $H \otimes L^2(\mathbb{R}^{J^c}_+ \times [0, 1]^J)$, we refer to the associated process $\hat{\xi} = \hat{\xi}_J$ in Lemma 8.10 as the *reduced* version of ξ. Writing $X_t = \xi(\cdot \otimes [0, t])$ and $\hat{X}_t = \hat{\xi}(\cdot \otimes [0, t])$ for any $t \in [0, 1]^J$, we note that $\hat{X}_t$ vanishes on the boundary of $[0, 1]^J$. In fact, assuming $t_j = 1$ and putting $J' = J \setminus \{j\}$, we get

$$\begin{aligned} A_J(f \otimes [0, t]) &= A_{J'} A_j(f \otimes [0, t]) \\ &= A_{J'}(I - P_j)(f \otimes [0, t]) = 0, \end{aligned}$$

by Lemma 8.10 (i) and the definition of P_j.

In the special case where ξ is a G-process on $L^2(\mathbb{R}_+^{J^c} \times [0,1]^J)$, we recall from Lemma 8.39 that X has a continuous version, which is of course a Brownian sheet on $\mathbb{R}_+^{J^c} \times [0,1]^J$. The corresponding process $\hat{X}$ is often referred to, suggestively, as a *pinned Brownian sheet* or a *Brownian sail* on the same space. The existence of the associated multiple integrals is clear from Lemmas A3.4 and 8.10.

After all these preparations, we are ready to state the representation theorem for separately exchangeable random sheets on $\mathbb{R}_+^{J^c} \times [0,1]^J$. Note that Theorem 8.38 gives the special case where $J = \emptyset$.

Theorem 8.43 (*separately exchangeable sheets on $[0,1]^J \times \mathbb{R}_+^{J^c}$*) *For any* $J \in 2^d$, *a process* X *on* $[0,1]^J \times \mathbb{R}^{J^c}$ *is separately exchangeable with a continuous version iff it has an a.s. representation as in* (56), *though with each* η_I *replaced by its reduced version* $\hat{\eta}_I$ *on* $H \otimes L^2([0,1]^{I \cap J} \times \mathbb{R}_+^{I \setminus J})$.

Proof: Applying the second transformation in Lemma 8.3 to each coordinate $j \in J$, we obtain a separately exchangeable random sheet on $\mathbb{R}_+^d$, which can be represented as in Theorem 8.38 in terms of some independent G-processes η_I and an independent set of random elements α_π. By Lemma 8.4 the reverse transformations are equivalent to the dual projections A_j^*, applied successively, in each coordinate $j \in I \cap J$, to some independent G-processes $\tilde{\eta}_I$ on $H \otimes L^2([0,1]^{I \cap J} \times \mathbb{R}_+^{I \setminus J})$. According to Lemma 8.10, the latter construction amounts to replacing each process $\tilde{\eta}_I$ by its reduced version $A_{I \cap J}^* \tilde{\eta}_I = \hat{\eta}_I$. $\qquad\square$

8.9 Jointly Exchangeable or Contractable Sheets

The aim of this section is to derive some representations of jointly exchangeable or contractable random sheets on $\mathbb{R}_+^d$. Beginning with the exchangeable case, we introduce the class $\hat{\mathcal{O}}_d$ of disjoint collections of non-empty, ordered subsets of $\{1, \ldots, d\}$, so that $\hat{\mathcal{O}}_d = \bigcup_{J \in 2^d} \mathcal{O}_J$, where $\mathcal{O}_J$ denotes the class of partitions κ of J into ordered subsets k of length $|k| \geq 1$. For any $t \in \mathbb{R}_+^d$ and $\pi \in \mathcal{P}_d$, we define the vector $\hat{t}_\pi = (\hat{t}_{\pi,J}) \in \mathbb{R}_+^\pi$ by $\hat{t}_{\pi,J} = \min_{j \in J} t_j$, $J \in \pi$.

Theorem 8.44 (*jointly exchangeable sheets*) *A process* X *on* $\mathbb{R}_+^d$ *is jointly exchangeable with a continuous version iff it has an a.s. representation*

$$X_t = \sum_{\pi \in \mathcal{P}_d} \sum_{\kappa \in \hat{\mathcal{O}}_\pi} \left(\lambda^{\kappa^c} \otimes \bigotimes\nolimits_{k \in \kappa} \eta_{|k|}\right)(\alpha_{\pi,\kappa} \otimes [0, \hat{t}_\pi]), \quad t \in \mathbb{R}_+^d,$$

in terms of some independent G-processes η_m *on* $H \otimes L^2(\lambda^m)$, $1 \leq m \leq d$, *and an independent collection of random elements* $\alpha_{\pi,\kappa}$ *in* $H^{\otimes\kappa}$, $\kappa \in \hat{\mathcal{O}}_\pi$, $\pi \in \mathcal{P}_d$. *The latter can then be chosen to be non-random iff* X *is ergodic.*

Some lemmas are needed for the proof, beginning with the following uniqueness assertion. Recall that $\mathbb{N}_d'$ denotes the non-diagonal part of $\mathbb{N}^d$.

Lemma 8.45 *(centering decomposition) Let X be a random array on $\mathbb{N}'_d$ of the form*

$$X_k = \sum_{J \in 2^d} X^J_{k_J}, \quad k \in \mathbb{N}'_d,$$

for some jointly rotatable arrays X^J on $\mathbb{N}^J$, $J \in 2^d$. Then this decomposition is a.s. unique on $\mathbb{N}'_d$.

Proof: To prove the uniqueness for a given $k = (k_1, \ldots, k_d) \in \mathbb{N}'_d$, we may choose some disjoint, infinite sets $N_1, \ldots, N_d \subset \mathbb{N}$ containing $k_1, \ldots, k_d$, respectively, and prove the uniqueness on $R = N_1 \times \cdots \times N_d$. Since the X^J are separately rotatable on R, we may transform the index set and assume instead that X^J is separately rotatable on $\mathbb{N}^J$ for every $J \in 2^d$.

Then define the arrays S on $\mathbb{Z}^d_+$ and S^J on $\mathbb{Z}^J_+$, $J \in 2^d$, by

$$S_n = \sum_{k \leq n} X_k, \qquad S^J_n = \sum_{k \leq n} X^J_k,$$

and note that S and the S^J satisfy the relation in Lemma 8.40 on the index set $\mathbb{Z}^d_+$. Applying the first transformation of Lemma 8.3 in each coordinate, we see that the resulting processes Y and Y^J satisfy the same relation on the index set T^d, where $T = \{n/(1+n); n \in \mathbb{Z}_+\}$. By the separate rotatability of the S^J and Lemma 8.2, we have $n_j^{-1} S^J_n \to 0$ a.s. as $n_j \to \infty$ for fixed n_{j^c}, $j \in J$, which implies $Y^J_{t_j} \to 0$ a.s. as $t_j \to 1$ along T. This suggests that we define $Y^J_t = 0$ when $\max_{j \in J} t_j = 1$. From (58) we see that even Y extends by continuity in each variable, so that the formula remains a.s. true on the closure $\bar{T}^d$. Then Lemma 8.40 shows that the processes Y^J are a.s. unique. The a.s. uniqueness carries over to the summation processes S^J and their underlying arrays of increments X^J, via the second transformation in Lemma 8.3. □

To state the next result, recall that $\mathcal{I}'_d$ denotes the set of non-diagonal, dyadic rectangles in $\mathbb{R}^d_+$. Given an additive process X on $\mathcal{I}'_d$, we say that X generates a continuous process on the rectangle $I = (a, b] \in \mathcal{I}'_d$, if the process $X^a(t) = X(a, t]$, defined for dyadic $t \in [a, b]$, has a continuous extension to the whole rectangle.

Lemma 8.46 *(off-diagonal representation) Let X be a finitely additive, jointly exchangeable process on $\mathcal{I}'_d$, generating a continuous process on every non-diagonal rectangle. Then X can be extended to a CLRF on $L^2_c(\lambda^d)$ of the form*

$$Xf = \sum_{\pi \in \tilde{\mathcal{O}}_d} \left(\lambda^{\pi^c} \otimes \bigotimes\nolimits_{k \in \pi} \eta_{|k|} \right) (\alpha_\pi \otimes f), \quad f \in L^2_c(\lambda^d), \tag{61}$$

for some independent G-processes η_m on $H \otimes L^2(\lambda^m)$, $1 \leq m \leq d$, and an independent set of random elements α_π in $H^{\otimes \pi}$, $\pi \in \tilde{\mathcal{O}}_d$.

Proof: First we may reduce to the ergodic case by means of Lemmas 8.8, 8.42, and A1.5. Fixing any disjoint, infinite sets $N_1, \ldots, N_d \subset \mathbb{N}$, we note that the process $\tilde{X}$ of reflections is separately exchangeable on the corresponding union of unit cubes. By the second statement in Lemma 8.42, applied in both directions, we see that $\tilde{X}$ is dissociated and hence ergodic. It is also clear that the stated continuity property of X carries over to $\tilde{X}$. Applying a vector-valued version of Theorem 8.38 to the latter process, we obtain an a.s. representation as in (56), in terms of some independent G-processes η_J on $H \otimes L^2(\lambda^J)$ and some non-random elements $\alpha_{\pi,p} \in H^{\otimes \pi}$, $\pi \in \hat{\mathcal{P}}_d$, $p \in S_d$. In particular, this gives an a.s. decomposition of $\tilde{X}$, as in Lemma 8.41, in terms of some processes X^J on $\mathcal{I}'_J$, such that the associated family of reflections $\tilde{X}^J$ is separately rotatable on the index set $R = N_1 \times \cdots \times N_d$.

Now let $X^{J,n}$ denote the restriction of X^J to the non-diagonal cubic grid of mesh size 2^{-n}. Proceeding as in Lemma 8.28 for fixed $n \in \mathbb{Z}_+$, we may derive a representation of the array $(X^{J,n})$ consistent with (61). The expression for $n = 0$ extends by FMP 6.10 to a CLRF Y on $L_c^2(\lambda^d)$ as in (61), and we may write Y^n and $Y^{J,n}$ for the associated restrictions to the regular cubic grid of size 2^{-n}. The latter arrays clearly satisfy the scaling relations in Lemma 8.31, and by Lemma 8.45 the same relations hold for the arrays $X^{J,n}$. Arguing as in the proof of Theorem 8.38, we conclude that $(X^{J,n}) \overset{d}{=} (Y^{J,n})$ for every n, which implies $X \overset{d}{=} Y$ on $\mathcal{I}'_d$. By FMP 6.10 we may then redefine Y, along with the underlying G-processes η_J, such that $X = Y$ a.s. on $\mathcal{I}'_d$. This gives the desired extension of X. $\qquad\square$

We also need the following decomposition of a jointly exchangeable random sheet X into its diagonal components.

Lemma 8.47 *(diagonal decomposition) Any jointly exchangeable random sheet X on $\mathbb{R}_+^d$ has an a.s. unique decomposition*

$$X_t = \sum_{\pi \in \mathcal{P}_d} X^\pi(\hat{t}_\pi), \quad t \in \mathbb{R}_+^d, \tag{62}$$

in terms of a jointly exchangeable family of random sheets X^π on $\mathbb{R}_+^\pi$, generated by some CLRFs Y^π on $L_c^2(\lambda^\pi)$, $\pi \in \mathcal{P}_d$.

Proof: Let Y be the process of increments associated with X, and suppose that Y is supported by $\bigcup_{\pi \in \mathcal{C}} D_\pi$ for some $\mathcal{C} \subset \mathcal{P}_d$. We shall prove by induction that Y has a decomposition as in (62) with summation over $\pi \in \mathcal{C}$. This is trivially true when $\mathcal{C} = \emptyset$. Assuming the statement to hold when $|\mathcal{C}| < m$, we proceed to the case where $|\mathcal{C}| = m$. Then choose a maximal partition $\pi \in \mathcal{C}$, and define an associated process Y^π on $\mathcal{I}'_\pi$ by

$$Y^\pi \bigotimes_{J \in \pi} B_J = Y \bigotimes_{J \in \pi} B_J^J, \tag{63}$$

for any disjoint, dyadic intervals B_J, $J \in \pi$. The maximality of π implies that Y^π is finitely additive, and Y^π inherits the joint exchangeability from Y. We also note that Y^π generates a continuous process on every non-diagonal rectangle. Hence, Y^π extends by Lemma 8.46 to a CLRF on $L^2_c(\lambda^\pi)$.

Now introduce the associated process $X^\pi_t = Y^\pi[0, t]$, $t \in \mathbb{R}^\pi_+$, and conclude from Lemmas 8.39 and 8.46 that X^π is again jointly exchangeable and admits a continuous version. Define $X'_t = X_t - X^\pi(\hat{t}_\pi)$, $t \in \mathbb{R}^d_+$, and let Y' denote the corresponding increment process. Then (63) shows that Y' is supported by $\bigcup_{\pi' \in \mathcal{C}'} D_{\pi'}$, where $\mathcal{C}' = \mathcal{C} \backslash \{\pi\}$. Furthermore, the pair (X', X^π) is again jointly exchangeable by Lemma A1.1. In particular, the induction hypothesis yields a decomposition of X' as in (62), in terms of some processes $X^{\pi'}$, $\pi' \in \mathcal{C}'$, with the required properties.

To complete the induction, we need to show that the entire family $\{(X^{\pi'}),$ $X^\pi\}$ is jointly exchangeable. Then recall that the pairs $\{X', (X^{\pi'})\}$ and (X', X^π) are both jointly exchangeable. By Lemma 7.2 (i) we can then choose some processes $\tilde{X}^{\pi'}$, $\pi' \in \mathcal{C}$, with $\{X', (\tilde{X}^{\pi'})\} \stackrel{d}{=} \{X', (X^{\pi'})\}$, such that the whole triple $\{X', (\tilde{X}^{\pi'}), X^\pi\}$ is jointly exchangeable. By the obvious uniqueness of the decomposition, the original triple $\{X', (X^{\pi'}), X^\pi\}$ has the same property, and the assertion follows. $\square$

Proof of Theorem 8.44: By the usual argument based on Lemma A1.5, it is enough to consider the ergodic case. First we may decompose X as in Lemma 8.47 in terms of some random sheets X^π on $\mathbb{R}^\pi_+$, $\pi \in \mathcal{P}_d$, such that each process X^π is generated by a CLRF Y^π on $L^2_c(\lambda^\pi)$. Next we may decompose each Y^π as in Lemma 8.46, in terms of some jointly rotatable processes $Y^{\pi,J}$, $J \subset \pi$. The combined array is then jointly exchangeable, by the uniqueness of the decompositions together with Lemma 7.2 (i). Proceeding as in the proof of Lemma 8.28, we may finally represent the processes $Y^{\pi,J}$ as in Theorem 8.24, in terms of a common collection of independent G-processes η_m on $H \otimes L^2(\lambda^m)$, $1 \leq m \leq d$. The desired representation of X arises as the integrated form of the resulting representation for Y. $\square$

Turning to the contractable case, let $\widetilde{\mathcal{P}}_d$ denote the class of partitions of $\{1, \ldots, d\}$ into *sequences* $\pi = (J_1, \ldots, J_k)$ of disjoint, non-empty subsets. For every $m \in \mathbb{N}$, we consider the tetrahedral region $\Delta_m = \{t \in \mathbb{R}^m_+; \, t_1 \leq \cdots \leq t_m\}$ and define $\Delta(t) = \Delta_m \cap [0, t]$, $t \in \mathbb{R}^m_+$.

Theorem 8.48 (*jointly contractable sheets*) *A process X on $\mathbb{R}^d_+$ is jointly contractable with a continuous version iff it has an a.s. representation*

$$X_t = \sum_{\pi \in \widetilde{\mathcal{P}}_d} \sum_{\kappa \in \hat{\mathcal{P}}_\pi} \left(\lambda^{\kappa^c} \otimes \bigotimes_{J \in \kappa} \eta_{|J|} \right) \left(\alpha_{\pi,\kappa} \otimes \Delta(\hat{t}_\pi) \right), \quad t \in \mathbb{R}^d_+, \qquad (64)$$

in terms of some independent G-processes η_m on $H \otimes L^2(\Delta_m)$, $1 \leq m \leq d$, and an independent collection of random elements $\alpha_{\pi,\kappa}$ in $H^{\otimes \kappa}$, $\kappa \in \hat{\mathcal{P}}_\pi$, $\pi \in \widetilde{\mathcal{P}}_d$.

Again we need a couple of lemmas. The basic relationship between jointly exchangeable and contractable random sheets is given by the following continuous-time version of the fundamental Corollary 7.16. Here we define $\Delta \mathcal{I}'_d = \Delta_d \cap \mathcal{I}'_d$.

Lemma 8.49 *(exchangeable extension) Let X be a finitely additive, jointly contractable process on $\Delta \mathcal{I}'_d$, generating a continuous process on every non-diagonal rectangle. Then X can be extended to a jointly exchangeable process on $\mathcal{I}'_d$ with the same properties.*

Proof: Let X_n denote the restriction of X to a regular cubic grid in $\Delta \mathcal{I}'_d$ with mesh size 2^{-n}. By Corollary 7.16 we may extend each X_n to a jointly exchangeable array Y_n on the corresponding cubic grid in $\mathcal{I}'_d$. Using the same notation Y_n for the generated additive process, we see that Y_n remains jointly exchangeable on the 2^{-m}-grid in $\mathcal{I}'_d$ for every $m \leq n$. In particular, the distribution of $Y_n(B)$ is independent of $n \geq m$ for every non-diagonal 2^{-m}-cube B. Defining $Y_n = 0$ on the 2^{-m}-grids with $m > n$, we conclude that the sequence (Y_n) is tight on the countable index set $\mathcal{I}'_d$, and hence converges in distribution along a sub-sequence toward a process Y on $\mathcal{I}'_d$.

In particular, we have $Y \overset{d}{=} X$ on $\Delta \mathcal{I}'_d$, and so by FMP 6.10 we may assume that $Y = X$ a.s. on $\Delta \mathcal{I}'_d$. The finite additivity and joint exchangeability of the arrays Y_n clearly extend to the limit Y. In particular, the finite-dimensional distributions of Y within a fixed rectangle $B \in \mathcal{I}'_d$ agree with those of X within the reflected rectangle $\tilde{B} \in \Delta \mathcal{I}'_d$, apart from the order of coordinates. Since X generates a continuous process on $\tilde{B}$, the same thing is a.s. true for the process Y on B, again by virtue of FMP 6.10. $\quad\square$

The last result enables us to use the previously established representations of jointly exchangeable processes to derive similar formulas in the contractable case. This will first be accomplished in a non-diagonal setting.

Lemma 8.50 *(off-diagonal representation) Any process X as in Lemma 8.49 can be extended to a CLRF on $L_c^2(\Delta_d)$ of the form*

$$Xf = \sum_{\pi \in \hat{\mathcal{P}}_d} \left(\lambda^{\pi^c} \otimes \bigotimes_{J \in \pi} \eta_{|J|} \right) (\alpha_\pi \otimes f), \quad f \in L_c^2(\Delta_d), \tag{65}$$

for some independent G-processes η_m on $H \otimes L^2(\Delta_m)$, $1 \leq m \leq d$, and an independent collection of random elements α_π in $H^{\otimes \pi}$, $\pi \in \hat{\mathcal{P}}_d$.

Proof: By Lemma 8.49 we may extend X to a finitely additive, jointly exchangeable process on $\mathcal{I}'_d$. The latter may in turn be represented as in Lemma 8.46, in terms of some independent G-processes η_m on $H \otimes L^2(\lambda^m)$, $1 \leq m \leq d$. The representation may be written in the form (65), except that $\eta_1, \ldots, \eta_d$ are now replaced the corresponding reflected processes $\tilde{\eta}_m$ on $H_m \otimes L^2(\lambda^m)$, $1 \leq m \leq d$, where H_m is the direct sum of $m!$ copies of H.

Now the representation on Δ_d involves only the restrictions of the processes $\tilde{\eta}_m$ to the sub-spaces $H_m \otimes L^2(\Delta_m)$, which are again independent G-processes, since their components are obtained from the η_m by reflection from disjoint tetrahedral regions. Since the Hilbert spaces H_m are isomorphic to H, we may finally derive the desired representation by a suitable set of isometric transformations.   □

Proof of Theorem 8.48: Using Lemma 8.50 recursively as in the proof of Lemma 8.47, we may decompose X into diagonal components X^π on $\mathbb{R}^\pi_+$, $\pi \in \mathcal{P}_d$, where each X^π is generated by a CLRF Y^π on $L^2_c(\lambda^\pi)$. Next we may represent each process Y^π in terms of the corresponding reflected array $\tilde{Y}^\pi$ on Δ_π, whose components are CLRFs on $L^2_c(\Delta_{|\pi|})$. By a semi-group version of Lemma A1.1, the whole collection $(\tilde{Y}^\pi)$ is jointly contractable.

Proceeding as in Lemma 8.49, we may next extend the array $(\tilde{Y}^\pi)$ to a jointly exchangeable family of processes Z^π, $\pi \in \widehat{\mathcal{P}}_d$, where each Z^π is a CLRF on $L^2_c(\lambda^\pi)$. The array (Z^π) may be represented as in Lemma 8.46, in terms of some independent G-processes η_m on $H \otimes L^2(\mathbb{R}^m_+)$, $1 \leq m \leq d$, along with an independent set of random elements $\alpha_{\pi,\kappa}$. The resulting representation of the original, tetrahedral array $(\tilde{Y}^\pi)$ may be simplified as in Lemma 8.50, and (64) follows as we rewrite the latter expression in integrated form. □

In conclusion, we conjecture that the representation in Theorem 8.44 remains valid for jointly exchangeable random sheets on $[0,1]^d$, provided we replace the G-processes η_m by their reduced versions, as defined in the previous sections. The statement is clearly equivalent to a jointly exchangeable, multi-variate version of Lemma 8.3.

Chapter 9

Symmetric Measures in the Plane

In this final chapter, we consider yet another case of multi-variate symmetries, now for exchangeable random measures on a finite or infinite rectangle, which leads to representations in terms of Poisson processes and sequences of independent $U(0,1)$ random variables. In two dimensions there are only five different cases to consider, depending on whether the symmetry is separate or joint and the rectangle is a square, a quadrant, or an infinite strip. Here the representations on a square are derived in Section 9.3, the one on a strip in Section 9.4, and those on a quadrant in Section 9.6.

The remaining sections deal with various introductory material, needed for the proofs of the main results. Thus, the basic symmetries are considered in Section 9.1, along with a brief discussion of the contractable case. Some auxiliary propositions for product symmetries and ergodic decompositions appear in Section 9.2. Finally, Section 9.5 contains some technical lemmas needed to establish the representations in the quadrant case. Some proofs in this chapter also rely on results from the general theory of ergodic decompositions, as summarized in Appendix A1.

9.1 Notions of Invariance

A random measure ξ on a rectangle $R = I_1 \times \cdots \times I_d$ in $\mathbb{R}^d$ is said to be *separately exchangeable*, if for any measure-preserving transformations $f_1, \ldots, f_d$ on $I_1, \ldots, I_d$, respectively, we have

$$\xi \circ \left(\bigotimes_j f_j \right)^{-1} \overset{d}{=} \xi. \qquad (1)$$

When $I_1 = \cdots = I_d = I$, we also say that ξ is *jointly exchangeable* if (1) holds for any measure-preserving transformation $f_1 = \cdots = f_d = f$ on I. Our aim is to derive general representations of separately or jointly exchangeable random measures on a two-dimensional rectangle $I_1 \times I_2$. Here there are clearly five cases to consider, depending on whether I_1 and I_2 are finite or infinite. Though similar representations can be expected to hold in higher dimensions, the technical and notational complications when $d > 2$ seem bewildering.

For technical reasons, it is often convenient to restrict $f_1, \ldots, f_d$ in (1) to the set of finite permutations of dyadic intervals, in which case we need to consider only transpositions of such intervals, in the sense of Chapter 1. This approach has the advantage that each symmetry is defined in terms of a countable group of transformations, which allows us to apply some powerful results from the general theory of ergodic decompositions. An obvious disadvantage is the loss of mathematical elegance and flexibility.

An even more elementary approach would be to consider an arbitrary rectangular grid, and require the corresponding discrete symmetry condition to be satisfied for the associated array of increments. In the case of joint exchangeability, we need the grid to be cubic and invariant under permutations of the coordinates. This approach has the benefits of a reduction to the more elementary case of exchangeable arrays, discussed extensively already in the previous chapter. A disadvantage is that the underlying transformations are no longer applied to ξ itself, as required by the general theory.

Fortunately no distinctions are necessary, since all three approaches to multi-variate exchangeability for random measures turn out to be equivalent. This allows us to move freely between the various interpretations of symmetry. A similar result holds in the contractable case, to be addressed in Proposition 9.3.

Proposition 9.1 *(equivalent symmetries) For random measures ξ on a rectangle $I_1 \times \cdots \times I_d$ in $\mathbb{R}^d$, it is equivalent to base the definitions of separate or joint exchangeability on the following classes of transformations:*

(i) *the measure-preserving transformations on $I_1, \ldots, I_d$,*

(ii) *the transpositions of dyadic intervals in $I_1, \ldots, I_d$,*

(iii) *the permutations of the increments over any dyadic grid.*

The result implies the equivalence of the corresponding notions of extremality, and by Lemma A1.2 it is also equivalent that ξ be ergodic for the transformations in (ii). If nothing else is said, ergodicity will henceforth be understood in the sense of (ii).

A simple lemma will be helpful for the proof. Here an interval of the form $I_k^n = 2^{-n}(k - 1, k]$, $k \in \mathbb{N}$, is said to be *n-dyadic*.

Lemma 9.2 *(approximation) For any λ-preserving transformation f on $[0, 1]$, there exist some λ-preserving transformations $f_n \to f$ a.e. λ, such that each f_n permutes the n-dyadic intervals in $[0, 1]$.*

Proof: Define $A_k^n = f^{-1}I_k^n$ for $n \in \mathbb{N}$ and $k \leq 2^n$, and note that the sets A_k^n are disjoint for fixed n with $\lambda A_k^n = 2^{-n}$. Arguing as in the proof of Theorem 1.18, we may next choose some disjoint, dyadic sets $B_k^n \subset [0, 1]$ such that

$$\lambda B_k^n = 2^{-n}, \qquad \sum_k \lambda(A_k^n \,\Delta\, B_k^n) < 2^{-n}.$$

Define f_n to be the unique function on $[0,1]$ such that, for every $k \le 2^n$, the restriction of f_n to B_k^n is increasing and λ-preserving with range I_k^n. Then each f_n is clearly λ-preserving on $[0,1]$ and satisfies

$$\lambda\{|f_n - f| > 2^{-n}\} < 2^{-n}, \quad n \in \mathbb{N}.$$

Hence, by the Borel–Cantelli lemma,

$$\lambda\{f_n \not\to f\} \le \lambda\{\|f_n - f\| > 2^{-n} \text{ i.o.}\} = 0,$$

which means that $f_n \to f$ a.e. λ. Changing the numbering, if necessary, we can finally ensure that each f_n permutes only n-dyadic intervals. $\square$

Proof of Proposition 9.1: The three notions of exchangeability are clearly related by (i) $\Rightarrow$ (ii) $\Rightarrow$ (iii), and by a monotone class argument we have even (iii) $\Rightarrow$ (ii). Thus, it remains to prove that (ii) $\Rightarrow$ (i). Here we consider only the case of joint exchangeability, the separately exchangeable case being similar. By a simple truncation argument, we may further reduce to the case of random measures on $[0,1]^d$.

Then let the random measure ξ on $[0,1]^d$ be jointly exchangeable in the sense of (ii). Fix any λ-preserving transformation f on $[0,1]$. By Lemma 9.2 we may choose some transformations $f_1, f_2, \ldots$ of type (ii) such that $f_n \to f$ a.e. λ. In other words, we have $f_n \to f$ on some measurable set $A \subset [0,1]$ with $\lambda A = 1$. This clearly implies $f_n^{\otimes d} \to f^{\otimes d}$ on A^d.

The one-dimensional projections $\xi_1, \ldots, \xi_d$ on $[0,1]$ are again exchangeable in the sense of (ii), and this remains true in the sense of (i) by Theorem 1.18. In particular, $\xi_i A^c = 0$ a.s. for $1 \le i \le d$, and therefore $\xi(A^d)^c = 0$ a.s. But then $\xi\{f_n^{\otimes d} \not\to f^{\otimes d}\} = 0$ a.s., and so by FMP 4.7

$$\xi \overset{d}{=} \xi \circ (f_n^{\otimes d})^{-1} \overset{w}{\to} \xi \circ (f^{\otimes d})^{-1},$$

which implies $\xi \circ (f^{\otimes d})^{-1} \overset{d}{=} \xi$. Since f was arbitrary, we conclude that ξ is also jointly exchangeable in the sense of (i). $\square$

Contractability of random measures is defined, most naturally, on the *tetrahedral* regions

$$\Delta_d = \{(s_1, \ldots, s_d) \in \mathbb{R}_+^d;\ s_1 < \cdots < s_d\}, \quad d \in \mathbb{N}.$$

Again we may identify three different versions of the definition, corresponding to the notions of exchangeability considered in Proposition 9.1. The following version of Corollary 7.16 and Lemma 8.49 characterizes multi-variate contractability for random measures in terms of the corresponding exchangeable notion.

Proposition 9.3 *(contractable random measures, Casukhela) A random measure on Δ_d is contractable, in either sense of Proposition 9.1, iff it can be extended to an exchangeable random measure on $\mathbb{R}_+^d$.*

Proof: The sufficiency is clear, since for any exchangeable random measure ξ on $\mathbb{R}_+^d$, the restriction of ξ to Δ_d is contractable in the sense of either definition. Conversely, suppose that ξ is contractable on Δ_d in the sense of (iii). By a monotone class argument, the contractability remains true in the sense of (ii). For every $n \in \mathbb{N}$, let η_n denote the array of restrictions of ξ to the n-dyadic cubes $2^{-n}[k-1, k)$, $k \in \mathbb{N}^d \cap \Delta_d$. Then ξ_n is again contractable, and so by Corollary 7.16 it has an exchangeable extension $\tilde{\eta}_n$ to the non-diagonal part of $\mathbb{N}^d$. Define $\tilde{\xi}_n$ as the corresponding extension of ξ to $\mathbb{R}_+^d$.

Let $\mathcal{U}_n$ denote the set of finite unions of n-dyadic cubes in $\mathbb{R}_+^d$. For any $B \in \mathcal{B}(\mathbb{R}_+^d)$, write $\hat{B}$ for the reflection of B into $\bar{\Delta}_d$, and note that

$$P\{\tilde{\xi}_n B > r\} \le 2^d P\{\xi \hat{B} > r2^{-d}\}, \quad B \in \mathcal{U}_m, \ m \le n, \ r > 0. \tag{2}$$

In particular, the sequence $(\tilde{\xi}_n)$ is tight, and so we have convergence $\tilde{\xi}_n \overset{d}{\to} \tilde{\xi}$ along a sub-sequence $N' \subset \mathbb{N}$ for some random measure $\tilde{\xi}$ on $\mathbb{R}_+^d$.

To see that $\tilde{\xi}$ is exchangeable, fix any $B \in \mathcal{U} = \bigcup_n \mathcal{U}_n$, and choose some sets $C_1, C_2, \ldots \in \mathcal{U}$ such that

$$\partial B \subset C_k^\circ \subset C_k \downarrow \partial B.$$

Fixing any $r > 0$ and $k \in \mathbb{N}$ and using FMP 4.25 and (2), we get for n restricted to N'

$$
\begin{aligned}
P\{\tilde{\xi}\partial B > r\} &\le P\{\tilde{\xi}C_k^\circ > r\} \le \liminf_n P\{\tilde{\xi}_n C_k^\circ > r\} \\
&\le \liminf_n P\{\tilde{\xi}_n C_k > r\} \\
&\le 2^d P\{\xi \hat{C}_k > r2^{-d}\}.
\end{aligned}
$$

Letting $k \to \infty$ and noting that $\xi\widehat{\partial B} = 0$ a.s. by contractability, we get

$$P\{\tilde{\xi}\partial B > r\} \le 2^d P\{\xi\widehat{\partial B} \ge r2^{-d}\} = 0, \quad r > 0,$$

which shows that $\tilde{\xi}\partial B = 0$ a.s. for every $B \in \mathcal{U}$.

For any $B_1, \ldots, B_k \in \mathcal{U}$, we may now apply FMP 16.16 to obtain

$$(\tilde{\xi}_n B_1, \ldots, \tilde{\xi}_n B_k) \overset{d}{\to} (\tilde{\xi}B_1, \ldots, \tilde{\xi}B_k).$$

Assuming $B_1, \ldots, B_k$ to be disjoint, non-diagonal, and m-dydic cubes, and using the exchangeability of $\tilde{\eta}_n$ for $n \ge m$, we conclude that $\tilde{\xi}$ is again exchangeable on $\mathcal{U}_m$. Since m was arbitrary, it follows that $\tilde{\xi}$ is exchangeable in the sense of (ii), and so by Proposition 9.1 it remains exchangeable in the sense of (i). Noting that $\tilde{\xi}_n = \xi$ on Δ_d for every n, we have also $\tilde{\xi} \overset{d}{=} \xi$ on Δ_d, and so by FMP 6.10 we may assume that $\tilde{\xi} = \xi$ a.s. on Δ_d. Then $\tilde{\xi}$ is indeed an exchangeable extension of ξ to $\mathbb{R}_+^d$. $\qquad\square$

9.2   General Prerequisites

In this section we prove some general results, needed to prove the multi-variate representations in subsequent sections. We begin with a common situation where the symmetry of a pair of random elements is preserved by a measurable function. Given a family $\mathcal{T}$ of measurable transformations of a space S, we say that a random element ξ in S is $\mathcal{T}$-*symmetric* if $T\xi \overset{d}{=} \xi$ for every $T \in \mathcal{T}$. Exchangeability is clearly a special case.

Lemma 9.4 *(composite symmetries) Let S, S', and S'' be measurable spaces equipped with families of measurable transformations $\mathcal{T}$, $\mathcal{T}'$, and $\mathcal{T}''$, respectively. Fix a measurable function $f: S' \times S'' \to S$, and suppose that for every $T \in \mathcal{T}$ there exist some $T' \in \mathcal{T}'$ and $T''_x \in \mathcal{T}''$, $x \in S'$, such that $T''_x y$ is product measurable on $S' \times S''$ and*

$$T \circ f(x, y) = f(T'x, T''_x y), \quad x \in S', \ y \in S''.$$

Then for any independent, $\mathcal{T}'$- and $\mathcal{T}''$-symmetric random elements ξ in S' and η in S'', their image $\zeta = f(\xi, \eta)$ in S is $\mathcal{T}$-symmetric.

Proof: For any $T \in \mathcal{T}$, choose $T' \in \mathcal{T}'$ and $T''_x \in \mathcal{T}''$, $x \in S'$, with the stated properties, and consider any measurable function $g \geq 0$ on $S' \times S''$. By Fubini's theorem and the independence and symmetry of ξ and η, we have

$$
\begin{aligned}
Eg(T'\xi, T''_\xi \eta) &= (Eg(T'x, T''_x \eta))_{x=\xi} \\
&= (Eg(T'x, \eta))_{x=\xi} = Eg(T'\xi, \eta) \\
&= (Eg(T'\xi, y))_{y=\eta} \\
&= (Eg(\xi, y))_{y=\eta} = Eg(\xi, \eta).
\end{aligned}
$$

In particular, we may take $g = 1_A \circ f$ for any measurable set $A \subset S$ to obtain

$$
\begin{aligned}
P\{T\zeta \in A\} &= P\{f(T'\xi, T''_\xi \eta) \in A\} \\
&= P\{f(\xi, \eta) \in A\} = P\{\zeta \in A\},
\end{aligned}
$$

which shows that ζ is $\mathcal{T}$-symmetric. $\square$

We continue with some results for invariance on product spaces. Given some measurable spaces S_k equipped with families of measurable transformations $\mathcal{T}_k$, we say that a random element $\xi = (\xi_1, \dots, \xi_n)$ in the product space $S = S_1 \times \cdots \times S_n$ is *separately symmetric* if its distribution is invariant under arbitrary transformations $T_1 \otimes \cdots \otimes T_n$ in $\mathcal{T}_1 \times \cdots \times \mathcal{T}_n$, in the sense that

$$(T_1\xi_1, \dots, T_n\xi_n) \overset{d}{=} (\xi_1, \dots, \xi_n), \quad T_k \in \mathcal{T}_k, \ k \leq n.$$

We may also consider the notion of *separate ergodicity* with respect to the same class of transformations on $S_1 \times \cdots \times S_n$.

Lemma 9.5 *(separate product symmetries) For every $k \leq n$, let S_k be a Borel space endowed with a countable group of measurable transformations $\mathcal{T}_k$ and associated invariant σ-field $\mathcal{I}_k$. Consider a random element $\xi = (\xi_1, \ldots, \xi_n)$ in $S_1 \times \cdots \times S_n$, and put $\mathcal{J}_k = \xi_k^{-1}\mathcal{I}_k$. Then these conditions are equivalent:*

(i) *ξ is separately symmetric,*

(ii) *ξ_k is conditionally symmetric given $(\xi_j;\, j \neq k)$ for every $k \leq n$,*

(iii) *the ξ_k are conditionally independent and symmetric given $\mathcal{J}_1 \vee \cdots \vee \mathcal{J}_n$.*

When those statements are true, we have $\xi_k \perp\!\!\!\perp_{\mathcal{J}_k} (\xi_j;\, j \neq k)$ for all k, and ξ is ergodic iff the same property holds for $\xi_1, \ldots, \xi_n$.

Proof: Assuming (i), we get for any $k \leq n$ and $T \in \mathcal{T}_k$

$$P[T\xi_k \in \cdot \,|\, \xi_j;\, j \neq k] = P[\xi_k \in \cdot \,|\, \xi_j;\, j \neq k] \quad \text{a.s.,}$$

and (ii) follows since $\mathcal{T}_k$ is countable. Since (ii) trivially implies (i), the two conditions are equivalent. It is also clear that (iii) implies (i). Conversely, we see from (i) and the definition of $\mathcal{J}_k$ that, for any $k \leq n$ and $T \in \mathcal{T}_k$,

$$(T\xi_k, \mathcal{J}_1, \ldots, \mathcal{J}_n) \stackrel{d}{=} (\xi_k, \mathcal{J}_1, \ldots, \mathcal{J}_n),$$

which shows that $\xi_1, \ldots, \xi_n$ are conditionally symmetric, given $\mathcal{J}_1 \vee \cdots \vee \mathcal{J}_n$. Next we note that

$$P[\xi_k \in \cdot \,|\, \mathcal{J}_k] = E[P[\xi_k \in \cdot \,|\, \mathcal{J}_k;\, \xi_j,\, j \neq k]|\, \mathcal{J}_k] \quad \text{a.s.}$$

Writing this in terms of regular conditional distributions, and noting that the left-hand side is a.s. ergodic and hence extreme by Lemmas A1.2 and A1.4, we obtain

$$P[\xi_k \in \cdot \,|\, \mathcal{J}_k;\, \xi_j,\, j \neq k] = P[\xi_k \in \cdot \,|\, \mathcal{J}_k] \quad \text{a.s.,}$$

which means that

$$\xi_k \perp\!\!\!\perp_{\mathcal{J}_k} (\xi_j,\, j \neq k), \quad k \leq n.$$

This implies

$$\xi_k \perp\!\!\!\perp_{\mathcal{J}_1 \vee \cdots \vee \mathcal{J}_n} (\xi_j,\, j \neq k), \quad k \leq n,$$

and so $\xi_1, \ldots, \xi_n$ are conditionally independent given $\mathcal{J}_1 \vee \cdots \vee \mathcal{J}_n$. This completes the proof of the implication (i) $\Rightarrow$ (iii), and shows that all three conditions (i)–(iii) are equivalent.

To prove the last assertion, we may assume that ξ is defined on the canonical probability space, so that

$$\mathcal{J}_k = S_1 \times \cdots \times S_{k-1} \times \mathcal{I}_k \times S_{k+1} \times \cdots \times S_n, \quad 1 \leq k \leq n.$$

Letting ξ be ergodic and noting that $\mathcal{J}_1, \ldots, \mathcal{J}_n$ are invariant under $\mathcal{T}_1 \otimes \cdots \otimes \mathcal{T}_n$, we see that the latter σ-fields are trivial, which means that even $\xi_1, \ldots, \xi_n$

are ergodic. Conversely, assuming $\xi_1, \ldots, \xi_n$ to be ergodic, we see from (iii) that they are independent, and so the distribution $\mu = \mathcal{L}(\xi)$ can be written as $\mu_1 \otimes \cdots \otimes \mu_n$. If $\mu = c\mu' + (1-c)\mu''$ for some invariant distributions μ' and μ'' and a constant $c \in (0, 1)$, then $\mathcal{J}_1, \ldots, \mathcal{J}_n$ remain trivial under μ' and μ'', and we conclude as before that

$$\mu' = \mu_1' \otimes \cdots \otimes \mu_n', \qquad \mu'' = \mu_1'' \otimes \cdots \otimes \mu_n'',$$

for some invariant distributions μ_k' and μ_k''. But then

$$\mu_k = c\mu_k' + (1-c)\mu_k'', \quad k = 1, \ldots, n,$$

and since the μ_k are ergodic, hence even extreme by Lemma A1.2, we conclude that $\mu_k' = \mu_k'' = \mu_k$ for all k. Then also $\mu' = \mu'' = \mu$, which shows that μ is extreme and hence ergodic by the same lemma. $\qquad\square$

For the corresponding notions of joint symmetry or ergodicity to make sense, we need the component spaces S_k with associated classes of transformations $\mathcal{T}_k$ to be the same for all k. Given a class $\mathcal{T}$ of measurable transformations on a measurable space S, we say that $\xi = (\xi_1, \ldots, \xi_n)$ is *jointly symmetric* if its distribution is invariant under any mapping $T^{\otimes n}$ with $T \in \mathcal{T}$, so that

$$(T\xi_1, \ldots, T\xi_n) \overset{d}{=} (\xi_1, \ldots, \xi_n), \quad T \in \mathcal{T}.$$

Joint ergodicity is defined with respect to the same class of transformations $T^{\otimes n}$. Here we need only a simple result in the two-dimensional case.

Lemma 9.6 (*joint product symmetries*) *Let $\mathcal{T}$ be a group of measurable transformations on a measurable space S. Consider some random elements ξ and η in S, where ξ is $\mathcal{T}$-symmetric, η is conditionally $\mathcal{T}$-symmetric given ξ, and the pair (ξ, η) is jointly $\mathcal{T}$-ergodic. Then ξ and η are independent and ergodic, $\mathcal{T}$-symmetric.*

Proof: The symmetry of η is clear from the corresponding conditional property. If $I \subset S$ is measurable and $\mathcal{T}$-invariant a.s. $\mathcal{L}(\xi)$, then $I \times S$ is $\mathcal{T}$-invariant in S^2, a.s. with respect to (ξ, η), and so

$$P\{\xi \in I\} = P\{(\xi, \eta) \in I \times S\} = 0 \text{ or } 1.$$

This shows that ξ is $\mathcal{T}$-ergodic. The same argument shows that even η is $\mathcal{T}$-ergodic. Since $\mathcal{T}$ is a group, we conclude from Lemma A1.2 that $\mathcal{L}(\eta)$ is extreme as a $\mathcal{T}$-invariant distribution. In particular, the decomposition $P\{\eta \in \cdot\} = EP[\eta \in \cdot | \xi]$ yields $P[\eta \in \cdot | \xi] = P\{\eta \in \cdot\}$ a.s., and the asserted independence $\xi \perp\!\!\!\perp \eta$ follows. $\qquad\square$

We proceed with some useful conditions for extremality.

Lemma 9.7 *(extremality) Consider some Borel spaces S, T, U, V, a measurable mapping $f: T \times U \to S$, and a random element γ in U. Let $\mathcal{C}$ be the convex hull of the measures $m_t = \mathcal{L}(\xi_t)$, $t \in T$, where $\xi_t = f(t, \gamma)$. Suppose that the functions $g: S \to V$ and $h: T \to V$ are measurable and satisfy*

(i) $g(\xi_t) = h(t)$ a.s., $t \in T$,

(ii) $h(s) = h(t)$ implies $m_s = m_t$, $s, t \in T$.

Then the m_t are extreme in $\mathcal{C}$. This holds in particular if there exists a measurable mapping $F: S \times [0, 1] \to S$ such that, for any $U(0, 1)$ random variable $\vartheta \perp\!\!\!\perp \gamma$, the random elements $\tilde{\xi}_t = F(\xi_t, \vartheta)$ satisfy $\xi_t \overset{d}{=} \tilde{\xi}_t \perp\!\!\!\perp \xi_t$.

Proof: Suppose that $m_t = \int m_s \mu(ds)$ for some $t \in T$ and $\mu \in \mathcal{M}_1(T)$. Then Fubini's theorem yields $\xi_t \overset{d}{=} \xi_\tau$, where $\tau \perp\!\!\!\perp \gamma$ with distribution μ. By (i) and Fubini's theorem we get $h(t) = h(\tau)$ a.s., and so by (ii) we have $m_t = m_\tau$ a.s., which means that $m_s = m_t$ for $s \in T$ a.e. μ. This shows that m_t is extreme in $\mathcal{C}$.

Now let F be such as stated. Define

$$g(s) = \mathcal{L}(F(s, \vartheta)), \quad s \in S; \qquad h(t) = m_t, \quad t \in T,$$

and note that (ii) is trivially true. To prove (i), we may use the definitions of g, h, $\tilde{\xi}_t$, and m_t, the independence relations $\gamma \perp\!\!\!\perp \vartheta$ and $\xi \perp\!\!\!\perp \tilde{\xi}_t$, and Fubini's theorem to write

$$\begin{aligned} g(\xi_t) &= P[F(\xi_t, \vartheta) \in \cdot | \xi_t] \\ &= P[\tilde{\xi}_t \in \cdot | \xi_t] = P\{\tilde{\xi}_t \in \cdot\} \\ &= P\{\xi_t \in \cdot\} = m_t = h(t), \end{aligned}$$

a.s. for any $t \in T$. The asserted extremality now follows from the previous statement. $\qquad\square$

The last lemma leads to an easy construction of a *directing element*, associated with a symmetric random element ξ. By this we mean an a.s. ξ-measurable and invariant random element ρ, such that the distributions of ξ and ρ determine each other uniquely.

Lemma 9.8 *(directing element) In the context of Lemma 9.7, let $\mathcal{T}$ be a family of measurable transformations on S, and suppose that $\mathcal{C}$ consists of all $\mathcal{T}$-invariant distributions on S. Then every $\mathcal{T}$-symmetric random element ξ in S has an a.s. representation $f(\tau, \tilde{\gamma})$ for some $\tilde{\gamma} \overset{d}{=} \gamma$ and $\tau \perp\!\!\!\perp \tilde{\gamma}$, and $\rho = h(\tau)$ is a directing element of ξ.*

Proof: The first assertion follows from Lemma A1.5. Letting $\pi \in \mathcal{T}$ be arbitrary and using Fubini's theorem, condition (i) of Lemma 9.7, and the $\mathcal{T}$-invariance of the measures m_t, we get

$$
\begin{aligned}
P\{g(\pi\xi) = \rho\} &= P\{g \circ \pi f(\tau, \tilde{\gamma}) = h(\tau)\} \\
&= E(P\{g \circ \pi f(t, \tilde{\gamma}) = h(t)\})_{t=\tau} \\
&= E(P\{g(\pi\xi_t) = h(t)\})_{t=\tau} \\
&= E(P\{g(\xi_t) = h(t)\})_{t=\tau} = 1,
\end{aligned}
$$

which shows that $g(\pi\xi) = g(\xi) = \rho$ a.s. Hence, ρ is a.s. a measurable and $\mathcal{T}$-invariant function of ξ. In particular, $\mathcal{L}(\rho)$ is uniquely determined by $\mathcal{L}(\xi)$.

To prove the converse statement, consider another $\mathcal{T}$-symmetric random element $\xi' = f(\tau', \gamma')$ such that $h(\tau') \overset{d}{=} h(\tau) = \rho$. The transfer theorem allows us to choose $\tilde{\tau} \overset{d}{=} \tau'$ such that $h(\tilde{\tau}) = h(\tau)$ a.s. Then (ii) yields $m_\tau = m_{\tilde{\tau}} \overset{d}{=} m_{\tau'}$ a.s., and so

$$
\mathcal{L}(\xi) = Em_\tau = Em_{\tau'} = \mathcal{L}(\xi'),
$$

which shows that $\mathcal{L}(\xi)$ is also determined by $\mathcal{L}(\rho)$. $\qquad\square$

The following result is often useful to verify the extremality criterion in Lemma 9.7.

Lemma 9.9 *(kernel criterion) Fix three Borel spaces S, U, V, an index set T, and some measurable mappings $f_t \colon U \times V \to S$, $t \in T$. Consider some independent random elements α in U and η in V, let $\vartheta \perp\!\!\!\perp (\alpha, \eta)$ be $U(0, 1)$, and define*

$$
\xi_t = f_t(\alpha, \eta), \quad t \in T. \tag{3}
$$

Suppose that

$$
P[\xi_t \in \cdot \,|\alpha] = \nu(\xi_t; \cdot) \quad a.s., \quad t \in T, \tag{4}
$$

for some kernel ν on S. Then there exist a measurable mapping $g \colon S \times [0, 1] \to S$ and some random elements $\eta_t \overset{d}{=} \eta$ with $\eta_t \perp\!\!\!\perp (\alpha, \xi_t)$, $t \in T$, such that

$$
\xi_t' \equiv g(\xi_t, \vartheta) = f_t(\alpha, \eta_t) \quad a.s., \quad t \in T. \tag{5}
$$

Proof: We may clearly assume that $S = \mathbb{R}$. Letting H_t and G be the distribution functions associated with the kernels in (4), we obtain

$$
H_t(\alpha, x) = P[\xi_t \le x | \alpha] = \nu(\xi_t, (-\infty, x]) = G(\xi_t, x), \quad x \in \mathbb{R}, \tag{6}
$$

a.s. for each $t \in T$. By right continuity we note that H_t and G are product measurable, and the latter property clearly carries over to the right-continuous inverses h_t and g. Hence, by (6) and Fubini's theorem,

$$
\xi_t' \equiv g(\xi_t, \vartheta) = h_t(\alpha, \vartheta) \quad a.s., \quad t \in T. \tag{7}
$$

Since ϑ is $U(0, 1)$ and independent of α, we see from (6) and (7) that

$$
P[\xi_t' \in \cdot \,|\alpha] = P[\xi_t \in \cdot |\alpha] \quad a.s., \quad t \in T,
$$

which implies
$$(\xi'_t, \alpha) \stackrel{d}{=} (\xi_t, \alpha), \quad t \in T. \tag{8}$$

Now choose $\tilde{\eta} \stackrel{d}{=} \eta$ to be independent of $(\alpha, \eta, \vartheta)$, and note that (8) remains true with ξ_t replaced by $\tilde{\xi}_t = f_t(\alpha, \tilde{\eta})$. Since $\eta \perp\!\!\!\perp (\alpha, \tilde{\eta}, \vartheta)$, we conclude that

$$(\xi'_t, \alpha, \eta) \stackrel{d}{=} (\tilde{\xi}_t, \alpha, \eta) = (f_t(\alpha, \tilde{\eta}), \alpha, \eta), \quad t \in T.$$

Hence, the transfer theorem (FMP 6.10) ensures the existence of some random triples
$$(\alpha_t, \eta'_t, \eta_t) \stackrel{d}{=} (\alpha, \eta, \tilde{\eta}), \quad t \in T,$$
satisfying
$$(\xi'_t, \alpha, \eta) = (f_t(\alpha_t, \eta_t), \alpha_t, \eta'_t) \text{ a.s.}, \quad t \in T.$$

In particular, $\alpha_t = \alpha$ and $\eta'_t = \eta$ a.s., and so (5) holds with $\eta_t \stackrel{d}{=} \tilde{\eta} \stackrel{d}{=} \eta$ and $\eta_t \perp\!\!\!\perp (\alpha, \eta)$. Finally, $\eta_t \perp\!\!\!\perp (\alpha, \xi_t)$ follows by virtue of (3). $\qquad\square$

The next result often allows us to extend results involving parametric representations to a functional setting.

Lemma 9.10 *(parametric representation) Let $\mathcal{C}$ be the class of measurable functions from a σ-finite measure space (S, μ) to a Borel space T. Then there exist a measurable function $F: [0, 1] \times S \to T$ and a functional $p: \mathcal{C} \to [0, 1]$ such that*
$$f = F(p(f), \cdot) \text{ a.e. } \mu, \quad f \in \mathcal{C}.$$

Proof: We may clearly assume that μ is bounded and $T \in \mathcal{B}[0, 1]$, so that $\mathcal{C} \subset L^2(S, \mu) \equiv H$. Fixing any ortho-normal basis $h_1, h_2, \ldots$ in H, we define a function $h: l^2 \to L^2(S)$ by

$$h(a, \cdot) = \sum_j a_j h_j, \quad a = (a_j) \in l^2.$$

By FMP 4.32 we may choose a product measurable version of h on $l^2 \times S$. Next we define a mapping $g: \mathcal{C} \to l^2$ by

$$g_j(f) = \langle f, h_j \rangle, \quad j \in \mathbb{N}, \ f \in \mathcal{C},$$

so that $f = \sum_j g_j(f) h_j$ in $L^2(S)$, and hence

$$f = h(g(f), \cdot) \text{ a.e. } \mu, \quad f \in \mathcal{C}.$$

Introducing a Borel isomorphism $r: l^2 \to B \in \mathcal{B}[0, 1]$ and writing $p = r \circ g$, we obtain $f = F(p(f), \cdot)$ a.e. μ, where

$$F(x, \cdot) = h(r^{-1}(x), \cdot), \quad x \in B.$$

We may finally modify F to satisfy $F(x, s) \in T$ for all $x \in B$ and $s \in S$, and choose a constant value $F(x, s) = t_0 \in T$ when $x \notin B$. $\qquad\square$

In particular, we can use the last result to extend Lemma A1.5 to suitable functional representations. To be precise, suppose that the distribution $\mathcal{L}(\eta)$ is a mixture of measures belonging to a class of the form

$$M = \{\mathcal{L}(g(\xi, f(\zeta))); \ f \in \mathcal{C}\},$$

defined in terms of a measurable function g and some random elements ξ and ζ on suitable spaces. As before, we may then conclude that

$$\eta = g(\tilde{\xi}, F(\tau, \tilde{\zeta})) \ \text{a.s.},$$

for a fixed, product-measurable function F and some random elements $\tilde{\xi}$, $\tilde{\zeta}$, and τ such that $\tau \perp\!\!\!\perp (\tilde{\xi}, \tilde{\zeta}) \overset{d}{=} (\xi, \zeta)$. Applications of this type will often be made in the sequel, without further comments.

9.3 Symmetries on a Square

Recall that a random measure ξ on $[0,1]^d$ or $\mathbb{R}_+^d$ is said to be *separately* or *jointly exchangeable* if, for any regular cubic grid $(I_{k_1,\dots,k_d})$ in $\mathbb{R}_+^d$, the associated array

$$X_{k_1,\dots,k_d} = \xi I_{k_1,\dots,k_d}, \quad k_1,\dots,k_d \in \mathbb{N},$$

is separately or jointly exchangeable, respectively, in the sense of Chapter 7. In the separately exchangeable case, it is equivalent by Proposition 9.1 to require $\xi \circ (f_1 \otimes \cdots \otimes f_d)^{-1} \overset{d}{=} \xi$ for any λ-preserving transformations $f_1, \dots, f_d$ on $\mathbb{R}_+$. To define joint exchangeability, we may instead impose the weaker condition $\xi \circ (f^{\otimes d})^{-1} \overset{d}{=} \xi$, in terms of a common, λ-preserving transformation f on $\mathbb{R}_+$. The definition of contractable random measures is similar. Separate (but not joint) exchangeability can also be defined for random measures on $\mathbb{R}_+ \times [0,1]$.

In this section, we characterize the classes of separately or jointly exchangeable random measures on $[0,1]^2$, through explicit representations involving U-sequences. Similar, though more complicated, representations for exchangeable random measures on $\mathbb{R}_+ \times [0,1]$ and $\mathbb{R}_+^2$ will be derived in subsequent sections.

Theorem 9.11 *(separate exchangeability on a square)* *A random measure ξ on $[0,1]^2$ is separately exchangeable iff a.s.*

$$\xi = \sum_{i,j} \alpha_{ij}\, \delta_{\tau_i, \tau_j'} + \sum_i \beta_i (\delta_{\tau_i} \otimes \lambda) + \sum_j \beta_j' (\lambda \otimes \delta_{\tau_j'}) + \gamma \lambda^2,$$

for some independent U-sequences $\tau_1, \tau_2, \dots$ and $\tau_1', \tau_2', \dots$ and an independent set of random variables $\alpha_{ij}, \beta_i, \beta_j', \gamma \geq 0$, $i, j \in \mathbb{N}$. The latter can then be chosen to be non-random iff ξ is extreme.

To state the corresponding result in the jointly exchangeable case, let λ_D denote Lebesgue measure along the main diagonal $D = \{(s,t) \in [0,1]^2;\ s=t\}$, normalized to have total mass one.

Theorem 9.12 *(joint exchangeability on a square)* *A random measure ξ on $[0,1]^2$ is jointly exchangeable iff a.s.*

$$\xi = \sum_{i,j} \alpha_{ij}\, \delta_{\tau_i,\tau_j} + \sum_j \Big(\beta_j(\delta_{\tau_j} \otimes \lambda) + \beta'_j(\lambda \otimes \delta_{\tau_j})\Big) + \gamma\lambda^2 + \vartheta\lambda_D,$$

for a U-sequence $\tau_1, \tau_2, \ldots$ and an independent set of random variables α_{ij}, $\beta_i, \beta'_j, \gamma, \vartheta \geq 0$, $i,j \in \mathbb{N}$. The latter can then be chosen to be non-random iff ξ is extreme.

Throughout the subsequent proofs, we shall often write $A \cdot \mu$ for the restriction of a measure μ to the measurable set A.

Proof of Theorem 9.11: Any random measure of the stated form is clearly separately exchangeable. To prove that, conversely, every separately exchangeable random measure ξ on $[0,1]^2$ has the stated representation, it is enough, e.g. by Theorem A1.4 and Lemma A1.5, to assume that ξ is ergodic, and to prove that ξ has then the required representation with non-random coefficients. Then introduce the random sets

$$\begin{aligned} M_1 &= \{s \in [0,1];\ \xi(\{s\} \times [0,1]) > 0\}, \\ M_2 &= \{t \in [0,1];\ \xi([0,1] \times \{t\}) > 0\}, \end{aligned}$$

and conclude from Lemma A1.1 and Theorem 1.25 that, a.s.,

$$(M_1^c \times [0,1]) \cdot \xi = \lambda \otimes \eta_2, \qquad ([0,1] \times M_2^c) \cdot \xi = \eta_1 \otimes \lambda, \tag{9}$$

for some exchangeable random measures η_1 and η_2 on $[0,1]$. In particular, we have a.s.

$$(M_1^c \times M_2^c) \cdot \xi = (\eta_1 M_1^c)\,\lambda^2 = (\eta_2 M_2^c)\,\lambda^2,$$

where $c = \eta_1 M_1^c = \eta_2 M_2^c$ is a.s. non-random since ξ is ergodic. By the invariance of the measure $c\lambda^2$, we note that $\xi - c\lambda^2$ is again separately ergodic exchangeable. Thus, we may henceforth assume that $\xi(M_1^c \times M_2^c) = 0$.

In that case, (9) shows that ξ has an a.s. representation

$$\xi = \sum_{i,j} \alpha_{ij}\delta_{\sigma_i,\sigma'_j} + \sum_i \beta_i(\delta_{\sigma_i} \otimes \lambda) + \sum_j \beta'_j(\lambda \otimes \delta_{\sigma'_j}), \tag{10}$$

in terms of some ξ-measurable random variables $\alpha_{ij}, \beta_j, \beta'_j \geq 0$ and $\sigma_i, \sigma'_j \in [0,1]$, $i,j \in \mathbb{N}$, where the latter are a.s. distinct. We may enumerate those variables such that the sequences

$$\begin{aligned} r_i &= \xi(\{\sigma_i\} \times [0,1]) = \beta_i + \textstyle\sum_j \alpha_{ij}, & i \in \mathbb{N}, \\ r'_j &= \xi([0,1] \times \{\sigma'_j\}) = \beta'_j + \textstyle\sum_i \alpha_{ij}, & j \in \mathbb{N}, \end{aligned}$$

become non-increasing, in which case the r_i and r_j will be invariant and hence a.s. non-random, due to the ergodicity of ξ. Then so are the quantities

$$i_0 = \sup\{i \in \mathbb{N};\ r_i > 0\}, \qquad j_0 = \sup\{j \in \mathbb{N};\ r'_j > 0\}.$$

For definiteness, we may further assume that the σ_i are increasing for fixed $r_i > 0$, and similarly for the σ'_j when $r'_j > 0$ is fixed.

Independently of ξ, we next introduce some i.i.d. $U(0,1)$ random variables $\vartheta_1, \vartheta_2, \ldots$ and $\vartheta'_1, \vartheta'_2, \ldots$, and consider on $[0,1]^4$ the random measure

$$\hat{\xi} = \sum_{i,j} \alpha_{ij} \delta_{\sigma_i, \vartheta_i, \sigma'_j, \vartheta'_j} + \sum_i \beta_i (\delta_{\sigma_i, \vartheta_i} \otimes \lambda^2) + \sum_j \beta'_j (\lambda^2 \otimes \delta_{\sigma'_j, \vartheta'_j}), \qquad (11)$$

with projection ξ onto the first and third coordinates. From Lemma 9.4 we see that $\hat{\xi}$ remains separately exchangeable under transformations in the corresponding variables. Decomposing the distribution of $\hat{\xi}$ into ergodic measures Q, we note that, due to the ergodicity of ξ, the projection ξ has the same distribution under each Q, as originally under $\mathcal{L}(\hat{\xi})$. Invoking the transfer theorem, we may then reduce to the case where the measure $\hat{\xi}$ in (11) is ergodic, separately exchangeable. Though in the ergodic case the variables ϑ_i and ϑ'_j may no longer be i.i.d. $U(0,1)$, they remain a.s. distinct.

Let us now change the order of enumeration, so that the quantities r_i and r'_j remain non-increasing, but the marks ϑ_i rather than the times σ_i become increasing for fixed $r_i > 0$, and similarly for the ϑ'_j when $r'_j > 0$ is fixed. This leads to a possibly different representation of $\hat{\xi}$, which we may write as

$$\hat{\xi} = \sum_{i,j} a_{ij} \delta_{\tau_i, u_i, \tau'_j, u'_j} + \sum_i b_i (\delta_{\tau_i, u_i} \otimes \lambda^2) + \sum_j b'_j (\lambda^2 \otimes \delta_{\tau'_j, u'_j}). \qquad (12)$$

The purpose of this random permutation of indices was to ensure that the new coefficients a_{ij}, b_i, and b'_j will be invariant, hence a.s. non-random. To verify the invariance, note that the u_i and u'_j are trivially invariant for $i \leq i_0$ and $j \leq j_0$. Since clearly

$$a_{ij} = \hat{\xi}([0,1] \times \{u_i\} \times [0,1] \times \{u'_j\}), \quad i \leq i_0,\ j \leq j_0,$$

we note that even the a_{ij} are invariant functions of $\hat{\xi}$. To deal with the b_i and b'_j, we may first subtract the first sum in (12), which is obviously invariant. The remaining expression $\hat{\xi}'$ satisfies

$$
\begin{aligned}
b_i &= \hat{\xi}'([0,1] \times \{u_i\} \times [0,1]^2), & i \leq i_0, \\
b'_j &= \hat{\xi}'([0,1]^3 \times \{u'_j\}), & j \leq j_0,
\end{aligned}
$$

which shows that even b_i and b'_j are invariant functions of $\hat{\xi}$.

Now introduce on $[0,1] \times \mathbb{N}$ the marked point processes

$$\eta = \sum_{i \leq i_0} \delta_{\tau_i, i}, \qquad \eta' = \sum_{j \leq j_0} \delta_{\tau'_j, j},$$

and conclude from Lemma A1.1 that the pair (η, η') is ergodic, separately exchangeable in the sense of symmetries on product spaces, where the underlying symmetries may be defined in terms of measure-preserving transformations on $[0, 1]$. Next we see from Lemma 9.5 that η and η' are independent and ergodic, exchangeable. Finally, Lemma 1.24 shows that η and η' are uniform randomizations of their projections onto $\mathbb{N}$, which means that the sequences $\{\tau_i\}$ and $\{\tau_j'\}$ are i.i.d. $U(0,1)$. Projecting the representation (12) onto the sub-space spanned by the first and third coordinates, we obtain the required representation of ξ, with coefficients that are all non-random.

It remains to show that ξ is ergodic whenever it admits a representation with non-random coefficients a_{ij}, b_i, b_j', and c. By Lemma 9.7 it is then enough to construct a *fixed*, measurable function $F : \mathcal{M}([0,1]^2) \times [0,1] \to \mathcal{M}([0,1]^2)$ such that, for any $U(0,1)$ random variable $\vartheta \perp\!\!\!\perp \xi$, the random measure $\tilde{\xi} = F(\xi, \vartheta)$ satisfies $\xi \overset{d}{=} \tilde{\xi} \perp\!\!\!\perp \xi$. To simplify the writing, we may then omit the term $c\lambda^2$, which is clearly ξ-measurable and non-random. Next we may return to the representation of ξ in (10), in terms of the previously chosen, ξ-measurable random variables α_{ij}, β_i, β_j', σ_i, and σ_j'. Comparing this with the representation

$$\xi = \sum_{i,j} a_{ij}\delta_{\tau_i,\tau_j'} + \sum_i b_i(\delta_{\tau_i} \otimes \lambda) + \sum_j b_j'(\lambda \otimes \delta_{\tau_j'}),$$

we note that the variables σ_i, σ_j', τ_i, and τ_j' are related by

$$\tau_i = \sigma \circ \pi_i, \qquad \tau_j' = \sigma' \circ \pi_j', \tag{13}$$

for some random permutations (π_i) and (π_j') of the indices $i \leq i_0$ and $j \leq j_0$, respectively.

Now introduce some i.i.d. $U(0,1)$ random variables $\tilde{\sigma}_1, \tilde{\sigma}_2, \ldots$ and $\tilde{\sigma}_1', \tilde{\sigma}_2', \ldots$, independent of ξ and all the variables τ_i and τ_j', and define

$$\tilde{\xi} = \sum_{i,j} \alpha_{ij}\delta_{\tilde{\sigma}_i,\tilde{\sigma}_j'} + \sum_i \beta_i(\delta_{\tilde{\sigma}_i} \otimes \lambda) + \sum_j \beta_j'(\lambda \otimes \delta_{\tilde{\sigma}_j'}).$$

Comparing with (13), we see that the variables

$$\tilde{\tau}_i = \tilde{\sigma} \circ \pi_i, \qquad \tilde{\tau}_j' = \tilde{\sigma}' \circ \pi_j'$$

satisfy

$$\tilde{\xi} = \sum_{i,j} a_{ij}\delta_{\tilde{\tau}_i,\tilde{\tau}_j'} + \sum_i b_i(\delta_{\tilde{\tau}_i} \otimes \lambda) + \sum_j b_j'(\lambda \otimes \delta_{\tilde{\tau}_j'}).$$

Since the permutations (π_i) and (π_j') are independent of the sequences $(\tilde{\sigma}_i)$ and $(\tilde{\sigma}_j')$, the variables $\tilde{\tau}_i$ and $\tilde{\tau}_j'$ are again i.i.d. $U(0,1)$ and independent of ξ. Hence, $\tilde{\xi}$ is indeed independent of ξ with the same distribution. $\qquad \square$

Proof of Theorem 9.12: Consider any ergodic, jointly exchangeable random measure ξ on $[0, 1]^2$. From Lemma A1.1 we see that the random measure $\xi_D = (D \cdot \xi)(\cdot \times [0, 1])$ is ergodic, exchangeable on $[0, 1]$. Hence, by Theorem

1.25, the diffuse component of ξ_D equals $c_D \lambda$ a.s. for some constant $c_D \geq 0$, which means that $D \cdot \xi$ has a.s. the diffuse component $c_D \lambda_D$, as required. Subtracting this term from ξ, if necessary, we may henceforth assume that $c_D = 0$.

Now introduce the random set

$$M = \{s \in [0,1]; \ \xi(\{s\} \times [0,1]) \vee \xi([0,1] \times \{s\}) > 0\},$$

and note that

$$(D \cdot \xi)(M^c \times [0,1]) = (D \cdot \xi)([0,1] \times M^c) = 0$$

since $D \cdot \xi$ is purely atomic. Using the fact that ξ is separately exchangeable on $U \times V$ for any disjoint, finite, dyadic interval unions U and V, we see from Theorem 9.11 that

$$(M^c \times [0,1]) \cdot \xi = \lambda \otimes \eta_2, \qquad ([0,1] \times M^c) \cdot \xi = \eta_1 \otimes \lambda,$$

for some random measures η_1 and η_2 on $[0,1]$. In particular,

$$(M^c \times M^c) \cdot \xi = (\eta_1 M^c)\lambda^2 = (\eta_2 M^c)\lambda^2,$$

where $c = \eta_1 M^c = \eta_2 M^c$ is a.s. non-random. Subtracting the term $c\lambda^2$ from ξ, we may henceforth assume that even $c = 0$.

The measure ξ may now be written in the form (10), except that it is now preferable to enumerate the variables σ_i and σ'_j as a single sequence $\sigma_1, \sigma_2, \ldots$, arranged such that the sums

$$r_j = \beta_j + \beta'_j + \sum_i (\alpha_{ij} + \alpha_{ji}), \quad j \in \mathbb{N},$$

become non-increasing and the σ_j increasing for fixed $r_j > 0$. The sequence (r_j) will again be a.s. non-random, and hence so is the number $j_0 = \sup\{j \in \mathbb{N}; \ r_j > 0\}$. Now introduce an independent sequence of i.i.d. $U(0,1)$ random variables $\vartheta_1, \vartheta_2, \ldots$, and define $\hat{\xi}$ as in (11), though with $\sigma'_j = \sigma_j$ and $\vartheta'_j = \vartheta_j$. Re-numbering the terms, such that the marks ϑ_j rather than the times σ_j become increasing for fixed $r_j > 0$, we obtain a representation as in (12), though with $\tau'_j = \tau_j$ and $u'_j = u_j$, and we may show as before that both the coefficients a_{ij}, b_i, b'_j and the marks u_j are a.s. non-random. By Lemma A1.1, the marked point process $\eta = \sum_j \delta_{\tau_j, j}$ is exchangeable in the first coordinate, and so by Lemma 1.24 the variables τ_j are i.i.d. $U(0,1)$.

This proves the desired representation in the ergodic case. The remaining parts of the argument follow closely the preceding proof, and may therefore be omitted. $\qquad \square$

9.4   Symmetry on a Strip

For random measures on the infinite strip $\mathbb{R}_+ \times [0,1]$, only the notion of separate exchangeability makes sense. Our aim is to characterize random measures with this property through a general representation formula involving U-arrays and Poisson processes.

Theorem 9.13 *(separate exchangeability on a strip) A random measure ξ on $\mathbb{R}_+ \times [0,1]$ is separately exchangeable iff a.s.*

$$\xi = \sum_{i,j} f_j(\alpha, \vartheta_i)\, \delta_{\sigma_i, \tau_j} + \sum_{i,k} g_k(\alpha, \vartheta_i)\, \delta_{\sigma_i, \rho_{ik}}$$
$$+ \sum_i h(\alpha, \vartheta_i)(\delta_{\sigma_i} \otimes \lambda) + \sum_j \beta_j(\lambda \otimes \delta_{\tau_j}) + \gamma\lambda^2, \qquad (14)$$

for some measurable functions $f_j, g_k, h \geq 0$ on $\mathbb{R}_+^2$, a unit rate Poisson process $\{(\sigma_i, \vartheta_i)\}$ on $\mathbb{R}_+^2$, some independent U-arrays (τ_j) and (ρ_{ik}), and an independent set of random variables $\alpha, \beta_j, \gamma \geq 0$. The latter can then be chosen to be non-random iff ξ is extreme.

For the proof we need some consequences of the one-dimensional representation in Proposition 1.21. Recall that $|A|$ denotes the cardinality of the set A.

Lemma 9.14 *(supporting lines) Let the random measure ξ on $\mathbb{R}_+ \times [0,1]$ be exchangeable along $\mathbb{R}_+$ and directed by (α, ν), and define*

$$\psi_t = \alpha\{t\} + \int (1 - e^{-\mu\{t\}})\, \nu(d\mu), \quad t \in [0,1].$$

Then the set $M = \{t \in [0,1];\, \psi_t > 0\}$ is a.s. covered by some ξ-measurable random variables $\tau_1, \tau_2, \ldots$ in $[0,1]$ satisfying $\psi_{\tau_n} \downarrow 0$ a.s., and we have a.s.

$$|\operatorname{supp} \xi(\cdot \times \{t\})| \leq 1, \quad t \notin M. \qquad (15)$$

Proof: Assuming first that the pair (α, ν) is non-random and writing $\xi_r = \xi([0,r] \times \cdot)$, we note that

$$\psi_t = -\log E e^{-\xi_1\{t\}}, \quad t \in [0,1]. \qquad (16)$$

Since for any distinct numbers $t_1, t_2, \ldots \in [0,1]$,

$$\sum_n \xi_1\{t_n\} \leq \xi_1[0,1] < \infty \text{ a.s.},$$

we have $\xi_1\{t_n\} \to 0$ a.s., and so by (16) and dominated convergence we get $\psi_{t_n} \to 0$. Hence, the set $M = \{\psi_t > 0\}$ is countable and may be covered by a sequence $t_1, t_2, \ldots$ such that $\psi_{t_n} \downarrow 0$.

In the general exchangeable case, we show that for fixed $r > 0$, the atoms of the random measure ξ_r may be covered by some distinct, ξ_r-measurable random variables $\sigma_1^r, \sigma_2^r, \ldots$. Then for each n, divide $[0, 1)$ into dyadic intervals $I_{nk} = 2^{-n}[k - 1, k)$, and let $\gamma_1^n, \gamma_2^n, \ldots$ denote the right endpoints of the intervals I_{nk} with $\xi_r I_{nk} \geq \varepsilon$. When fewer than k such intervals exist, we put $\gamma_k^n = 1$. As $n \to \infty$, the γ_k^n converge toward the sites $\gamma_1, \gamma_2, \ldots$ of the atoms $\geq \varepsilon$, with the same convention when there are fewer than k such atoms. Combining the lists for different $\varepsilon > 0$, we obtain a cover of all atoms of ξ_r by a single sequence σ_k^r. The required measurability is clear by construction, and we can easily re-arrange the list, so as to make the σ_k^r distinct. Finally, we may combine the lists for different $r > 0$ into a single sequence of ξ-measurable random variables $\sigma_1, \sigma_2, \ldots$, covering the atoms of ξ_r for all r.

Next we show that the process ψ is product measurable. Then let $I_n(t)$ denote the n-dyadic interval I_{nj} containing t, and note that the step process

$$\psi_t^n = \alpha I_n(t) + \int (1 - e^{-\mu I_n(t)}) \, \nu(d\mu), \quad t \in [0, 1),$$

is trivially product measurable for each n. Since $\psi_t^n \to \psi_t$ by dominated convergence as $n \to \infty$, the measurability carries over to ψ. In particular, the random variables $\beta_k = \psi_{\sigma_k}$ are ξ-measurable, and we may order the σ_k according to the associated β-values to obtain a ξ-measurable sequence $\tau_1, \tau_2, \ldots$ such that $\psi_{\tau_j} \downarrow 0$.

By a similar argument based on the approximation $\psi_t^n \downarrow \psi_t$, we can also construct directly a sequence of (α, ν)-measurable random variables $\tau_1', \tau_2', \ldots$ covering M. Using the disintegration theorem, we get as $r \to \infty$ for fixed j

$$E[\exp(-\xi_r\{\tau_j'\}) | \alpha, \nu] = \exp(-r\psi_{\tau_j'}) \to 0 \text{ a.s. on } \{\psi_{\tau_j'} > 0\},$$

which implies $\xi_r\{\tau_j'\} \to \infty$, a.s. on $\{\psi_{\tau_j'} > 0\}$. This shows that, outside a fixed P-null set,

$$\xi(\mathbb{R}_+ \times \{t\}) = \infty, \quad t \in M.$$

In particular, M was already covered a.s. by the variables σ_j, and hence also by the original sequence $\tau_1, \tau_2, \ldots$.

To prove the last assertion, fix any $k \in \mathbb{N}$ and $s > r > 0$ in $\mathbb{Q}$, let $a \in [0, 1]$ with $a \notin M$ a.s., and define

$$\sigma = \sigma_k^r + (a - \sigma_k^r) 1_M(\sigma_k^r).$$

Noting that σ is (α, ν, ξ_r)-measurable and $(\xi_s - \xi_r) \perp\!\!\!\perp_{(\alpha,\nu)} \xi_r$, we see from the disintegration theorem that

$$\begin{aligned} E \, \xi_r\{\sigma\} \, (\xi_s - \xi_r)\{\sigma\} &= E \, \xi_r\{\sigma\} \, E[(\xi_s - \xi_r)\{\sigma\} | \alpha, \nu, \xi_r] \\ &= E \, \xi_r\{\sigma\} \, (E[\xi_{s-r}\{t\} | \alpha, \nu])_{t=\sigma} = 0. \end{aligned}$$

Since k was arbitrary and $\xi_r\{t\} = 0$ when $t \notin \{\sigma_1^r, \sigma_2^r, \ldots\}$, we obtain

$$\xi_r\{t\} \, (\xi_s - \xi_r)\{t\} = 0, \quad t \notin M, \tag{17}$$

outside a fixed P-null set, which can also be chosen to be independent of r and s. If $|\operatorname{supp}\xi(\cdot \times \{t\})| > 1$ for some $t \notin M$, there exist some rational numbers $r < s$ such that the support of $\xi(\cdot \times \{t\})$ intersects both $[0, r)$ and (r, s). This contradicts (17), and (15) follows.   □

We are now ready to prove the main result of this section.

Proof of Theorem 9.13: A random measure with the stated representation is clearly separately exchangeable. To prove the converse assertion, we consider first the case where ξ is an ergodic, separately exchangeable random measure on $\mathbb{R}_+ \times [0, 1]$. Then introduce the countable random sets

$$
\begin{aligned}
M_1 &= \{s \geq 0; \xi(\{s\} \times [0, 1]) > 0\}, \\
M_2 &= \{t \in [0, 1]; \xi(\mathbb{R}_+ \times \{t\}) > 0\},
\end{aligned}
$$

and note as before that a.s.

$$
(M_1^c \times [0, 1]) \cdot \xi = \lambda \otimes \eta_2, \qquad (\mathbb{R}_+ \times M_2^c) \cdot \xi = \eta_1 \otimes \lambda, \tag{18}
$$

for some random measures η_1 on $\mathbb{R}_+$ and η_2 on $[0, 1]$. In particular, we get $(M_1^c \times M_2^c) \cdot \xi = c\lambda^2$ a.s., where $c = \eta_2 M_2^c$ is invariant and hence a.s. non-random. Subtracting the term $c\lambda^2$ from ξ, we may henceforth assume that $\xi(M_1^c \times M_2^c) = 0$ a.s.

Now define the process ψ on $[0, 1]$ as in Lemma 9.14. The same result yields the existence of some ξ-measurable random variables $\tau_1, \tau_2, \ldots$ in $[0, 1]$ such that $r_j = \psi_{\tau_j} \downarrow 0$ a.s. Since the directing pair (α, ν) in Lemma 9.14 is a.s. determined by ξ, via the law of large numbers, it is clear that the r_j are invariant functions of ξ, and so by ergodicity they are a.s. non-random. The same thing is then true for the number $j_0 = \sup\{j \in \mathbb{N}; r_j > 0\}$ of non-zero elements r_j. Until further notice, we assume the τ_j to be increasing for fixed $r_j > 0$. Define

$$
M = \{t \in [0, 1]; \psi_t > 0\} = \{t \in [0, 1]; \xi(\mathbb{R}_+ \times \{t\}) = \infty\}.
$$

Next we choose a measurable enumeration $\sigma_1, \sigma_2, \ldots$ of M_1, which is possible since the cardinality $|M_1|$ is invariant and hence a.s. constant. For every index i, we may finally enumerate the atom sites of $\xi(\{\sigma_i\} \times \cdot)$ outside M as $\rho_{i1}, \rho_{i2}, \ldots$, where we may take $\rho_{ik} = 0$ if there are fewer than k such atoms. Then from (18) we see that ξ has an a.s. representation of the form

$$
\begin{aligned}
\xi &= \sum_{i,j} \alpha_{ij} \delta_{\sigma_i, \tau_j} + \sum_{i,k} \gamma_{ik} \delta_{\sigma_i, \rho_{ik}} \\
&\quad + \sum_i \beta_i'(\delta_{\sigma_i} \otimes \lambda) + \sum_j \beta_j(\lambda \otimes \delta_{\tau_j}),
\end{aligned} \tag{19}
$$

where we may take the γ_{ik} to be non-increasing in k for fixed i.

We now introduce an independent sequence of i.i.d. $U(0,1)$ random variables $\kappa_1, \kappa_2, \ldots$, and consider on $\mathbb{R}_+ \times [0,1]^2$ the random measure

$$\hat{\xi} = \sum_{i,j} \alpha_{ij} \delta_{\sigma_i, \tau_j, \kappa_j} + \sum_{i,k} \gamma_{ik}(\delta_{\sigma_i, \rho_{ik}} \otimes \lambda)$$
$$+ \sum_i \beta_i'(\delta_{\sigma_i} \otimes \lambda^2) + \sum_j \beta_j (\lambda \otimes \delta_{\tau_j, \kappa_j})$$

with projection ξ on $\mathbb{R}_+ \times [0,1]$. Note that $\hat{\xi}$ remains separately exchangeable in the first two coordinates by Lemma 9.4. Arguing as in the proof of Theorem 9.11, we may reduce to the case where $\hat{\xi}$ is ergodic under the same transformations, and then re-order the variables τ_j such that the κ_j become increasing for fixed $r_j > 0$. The κ_j and β_j will then be invariant and hence a.s. non-random.

To every variable σ_i we now attach a mark ϑ_i, composed of all coefficients in (19) associated with the time σ_i. More precisely, we define

$$\vartheta_i = (\Sigma_i, \beta_i', \{\alpha_{ij}\}_j, \{\gamma_{ik}\}_k), \quad i \geq 1,$$

where the redundant marks

$$\Sigma_i = \xi(\{\sigma_i\} \times [0,1]) = \beta_i' + \sum_j \alpha_{ij} + \sum_k \gamma_{ik}, \quad i \geq 1,$$

have been added to ensure that the set of pairs (σ_i, ϑ_i) will be locally finite in the infinite product space $\mathbb{R}_+ \times K$ with $K = (0, \infty) \times \mathbb{R}_+^\infty$. The coefficients α_{ij}, γ_{ik}, and β_i' may be recovered from ϑ_i through suitable projections f_j, g_k, and h, respectively, so that

$$\alpha_{ij} = f_j(\vartheta_i), \qquad \gamma_{ik} = g_k(\vartheta_i), \qquad \beta_i' = h(\vartheta_i). \tag{20}$$

Inserting these expressions into (19) yields the desired representation formula (14), except for the expected lack of any global parameter α.

To identify the joint distribution of the various variables in (14), we introduce the marked point processes

$$\eta = \sum_i \delta_{\sigma_i, \vartheta_i}, \qquad \zeta = \sum_j \delta_{\tau_j, j},$$

defined on $\mathbb{R}_+ \times K$ and $[0,1] \times \mathbb{N}$, respectively. Using the separate exchangeability and ergodicity of $\hat{\xi}$, we see from Lemma A1.1 that the pair (η, ζ) has the same properties, but now in the sense of symmetries on product spaces. The processes η and ζ are then independent by Lemma 9.5, and the variables $\tau_1, \tau_2, \ldots$ are i.i.d. $U(0,1)$ by Lemma 1.24, whereas ζ is Poisson by Lemma 1.20 with an intensity measure of the form $\lambda \otimes \nu$.

Here ν is a σ-finite measure on the Borel space K, and so by FMP 3.22 it can be generated as the image of Lebesgue measure λ on $\mathbb{R}_+$ under some measurable mapping $T \colon \mathbb{R}_+ \to \mathbb{R}_+^\infty$. Letting η' be a Poisson process on $\mathbb{R}_+^2$ with intensity measure λ^2, we see from FMP 12.3 that $\eta' \circ (I \otimes T)^{-1} \overset{d}{=} \eta$ on $\mathbb{R}_+ \times K$, where I denotes the identity mapping on $\mathbb{R}_+$. Hence, by FMP

6.10, we may assume that $\eta = \eta' \circ (I \otimes T)^{-1}$ a.s., where η' is Poisson λ^2 with $\eta' \perp\!\!\!\perp_\eta \hat{\xi}$. Writing $\eta' = \sum_i \delta_{\sigma_i, \vartheta_i'}$, we get from (20)

$$\alpha_{ij} = f_j \circ T(\vartheta_i'), \qquad \gamma_{ik} = g_k \circ T(\vartheta_i'), \qquad \beta_i' = h \circ T(\vartheta_i').$$

Here we may clearly assume that $\{(\sigma_i, \vartheta_i')\} \perp\!\!\!\perp_{\eta'} \hat{\xi}$. For convenience, we may henceforth drop the primes and take the process $\eta = \sum_i \delta_{\sigma_i, \vartheta_i}$ to be Poisson on $\mathbb{R}_+^2$ with intensity measure λ^2. This allows us to replace the awkward compositions $f_j \circ T$, $g_k \circ T$, and $h \circ T$ by their simplified versions f_j, g_k, and h, respectively.

The induced point process $\tilde{\eta} = \eta \circ (I \otimes T)^{-1}$, our previous η, is clearly invariant under measure-preserving transformations of $\hat{\xi}$ along $[0, 1]$ of the special permutation type, which shows that $\hat{\xi}$ is conditionally exchangeable given $\tilde{\eta}$. By conditional independence, this remains true under conditioning on the new process η, and even on the associated sequence of pairs (σ_i, ϑ_i). By Lemma A1.1 the conditional exchangeability carries over to the pair of marked point processes

$$\zeta = \sum_j \delta_{\tau_j, j}, \qquad \zeta' = \sum_{i,k} \delta_{\rho_{ik}, \gamma_{ik}, \sigma_i}.$$

Recalling that the times τ_j and ρ_{ik} are a.s. distinct by Lemma 9.14, we see from Lemma 1.24 that the pair (ζ, ζ') is conditionally a uniform randomization of its projection onto the associated mark space. In other words, the conditional distribution of (ζ, ζ') is the same as if the τ_j and ρ_{ik} were replaced by some independent i.i.d. $U(0, 1)$ random variables. The same thing is then true for the unconditional distribution of $\hat{\xi}$, and by FMP 6.10 we may assume that the original variables τ_j and ρ_{ik} have the stated distribution.

This proves the representation (14) in the ergodic case, for some non-random parameters $\beta_1, \beta_2, \ldots$ and γ, with no need for any additional variable α. To extend the result to the non-ergodic case, we may use Lemma 9.10 to write the functions f_j, g_k, and h in parametric form as

$$f_j(t) = \Phi(\alpha_j, t), \qquad g_k(t) = \Phi(\alpha_k', t), \qquad h(t) = \Phi(\alpha_0, t),$$

for a universal, measurable function Φ and some constants α_j, α_k', and α_0 in $[0, 1]$. In view of Lemmas A1.4 and A1.5, it is now sufficient to replace the parameters α_j, α_k', α_0, β_j, and γ by suitable random quantities, independent of all remaining variables. Invoking FMP 3.22 and 6.10, we may finally express the quantities α_j, α_k', and α_0 as measurable functions of a single random variable α, which yields the desired representation.

It remains to show that ξ is ergodic, whenever it admits a representation (14) with non-random α, γ, and $\beta_1, \beta_2, \ldots$. By Lemma 9.7 it is then enough to construct a measurable function F, independent of all functions f_j, g_k, h and parameters α, β_j, and γ, such that the random measure $\tilde{\xi} = F(\xi, \vartheta)$ satisfies $\xi \overset{d}{=} \tilde{\xi} \perp\!\!\!\perp \xi$, where ϑ is an independent $U(0, 1)$ random variable. We proceed in two steps.

First we note that, for fixed α, β_j, γ and conditionally on the variables $\tau_1, \tau_2, \ldots$ in (14), the random measure ξ has stationary, independent increments along $\mathbb{R}_+$. Hence, by the law of large numbers, there exists a kernel ν, independent of all parameters and functions in (14), such that

$$P[\xi \in \cdot \,|\, \tau_1, \tau_2, \ldots] = \nu(\xi; \,\cdot\,) \text{ a.s.}$$

Lemma 9.9 then ensures the existence of a measurable function G and some random variables $\tilde{\sigma}_i$, $\tilde{\vartheta}_i$, $\tilde{\rho}_{ik}$, $i, k \in \mathbb{N}$, independent of ξ and $\tau_1, \tau_2, \ldots$ and with the same joint distribution as σ_i, ϑ_i, ρ_{ik}, $i, k \in \mathbb{N}$, such that $\xi' = G(\xi, \vartheta)$ has the same representation as ξ, apart from a change from (σ_i, ϑ_i) to $(\tilde{\sigma}_i, \tilde{\vartheta}_i)$ and from ρ_{ik} to $\tilde{\rho}_{ik}$.

For the second step in our construction, we use a method already employed in the proof of Theorem 9.11. Omitting the trivial term $\gamma\lambda^2$, we may represent ξ' in the form (19), except that the original random variables $\tau_1, \tau_2, \ldots$ are now denoted by $\tau'_1, \tau'_2, \ldots$, so that $\tau = \tau' \circ \pi$ for some random permutation π. Next we introduce some i.i.d. $U(0,1)$ random variables $\tilde{\tau}'_1, \tilde{\tau}'_2, \ldots$, independent of all previously considered random elements, and use the same permutation π to define $\tilde{\tau} = \tilde{\tau}' \circ \pi$. Replacing the variables $\tau'_1, \tau'_2, \ldots$ by $\tilde{\tau}'_1, \tilde{\tau}'_2, \ldots$ in the representation (19) of ξ', we obtain a new random measure $\tilde{\xi}$ on $\mathbb{R}_+ \times [0,1]$. It is clearly equivalent to replace $\tau_1, \tau_2, \ldots$ by $\tilde{\tau}_1, \tilde{\tau}_2, \ldots$ in the representation (14) of ξ'.

Since $\tilde{\tau}'_1, \tilde{\tau}'_2, \ldots$ are i.i.d. $U(0,1)$ and independent of ξ, π, and all the $\tilde{\sigma}_i$, $\tilde{\vartheta}_i$, and $\tilde{\rho}_{ik}$, we note that $\tilde{\tau}_1, \tilde{\tau}_2, \ldots$ have the same properties. Hence, the variables $\tilde{\tau}_j$, $\tilde{\sigma}_i$, $\tilde{\vartheta}_i$, and $\tilde{\rho}_{ik}$ are i.i.d. $U(0,1)$ and independent of ξ, which shows that $\tilde{\xi}$ is independent of ξ with the same distribution. Note that $\tilde{\xi}$ was measurably obtained from ξ, ϑ, and the $\tilde{\tau}'_j$ only, independently of all functions and parameters occurring in (14). Using a single randomization variable $\tilde{\vartheta}$ to construct ϑ and all the $\tilde{\tau}'_j$, we may write $\tilde{\xi}$ in the form $F(\xi, \tilde{\vartheta})$ for a suitable choice of measurable function F. $\qquad\square$

It remains to examine when the expression for ξ in Theorem 9.13 converges, in the sense that the right-hand side of (14) represents an a.s. locally finite random measure on $\mathbb{R}_+ \times [0,1]$. It is then enough to consider the extreme case, where α, γ, and the β_j are all constants. In that case, we may omit α from our notation and write $\lambda f = \lambda f(\alpha, \cdot)$ for any function of the form $f(\alpha, \cdot)$.

Proposition 9.15 *(local summability)* *For fixed α, γ, and β_j, the random measure ξ in Theorem 9.13 is a.s. locally finite iff*

$$\lambda\left(1 \wedge \left(\sum_j (f_j + g_j) + h\right)\right) + \sum_j \beta_j < \infty.$$

Proof: By exchangeability it is enough to consider the contribution of ξ to the square $[0,1]^2$. Omitting α from our notation, we get

$$\xi[0,1]^2 = \sum_{i,j} f_j(\vartheta_i)1\{\sigma_i \le 1\} + \sum_{i,k} g_k(\vartheta_i)1\{\sigma_i \le 1\}$$
$$+ \sum_i h(\vartheta_i)1\{\sigma_i \le 1\} + \sum_j \beta_j + \gamma$$
$$= \eta\Big(\sum_j (f_j + g_j) + h\Big) + \sum_j \beta_j + \gamma,$$

where $\eta = \sum_i \delta_{\vartheta_i}1\{\sigma_i \le 1\}$. Since η is a unit rate Poisson process on $\mathbb{R}_+$, the stated criterion now follows from FMP 12.13 or Theorem A3.5. $\square$

9.5   Technical Preparation

Before proceeding to the representation theorems on a quadrant, we need to prove some preliminary results that follow from previously established theorems. Let us then put

$$D_1 = \{(s,t) \in [0,1]^2; \, s = t\}, \qquad W_1 = \{(s,t) \in [0,1]^2; \, s \le t\}.$$

For random measures ξ on $[0,1]^2$ and $s,t \in [0,1]$, we define

$$\tilde{\xi}(ds\,dt) = \xi(dt\,ds), \quad \xi_s(dt) = \xi(\{s\} \times dt), \quad \tilde{\xi}_t(ds) = \xi(ds \times \{t\}).$$

Let us also write

$$\mathcal{A}_1 = \{r\delta_t; \, r \ge 0, \, t \in [0,1]\} = \{\mu \in \mathcal{M}([0,1]); \, |\mathrm{supp}\,\mu| \le 1\}.$$

We begin with some criteria for *complete exchangeability*, defined as exchangeability in the one-dimensional sense with respect to the measure λ^2.

Lemma 9.16 *(complete exchangeability) For any random measure ξ on $[0,1]^2$, the following three conditions are equivalent:*

(i) ξ *is [ergodic] completely exchangeable;*

(ii) ξ *is [ergodic] separately exchangeable, and $\xi_s, \tilde{\xi}_s \in \mathcal{A}_1$ for all $s \in [0,1]$ a.s.;*

(iii) ξ *is [ergodic] jointly exchangeable, $\xi D_1 = 0$ a.s., and $\xi_s + \tilde{\xi}_s \in \mathcal{A}_1$ and $\xi_s \wedge \tilde{\xi}_s = 0$ for all $s \in [0,1]$ a.s.*

Furthermore, the next two conditions are related by (iv) $\Rightarrow$ (v):

(iv) ξ *is jointly exchangeable, $\xi D_1 = 0$ a.s., and $\xi_s + \tilde{\xi}_s \in \mathcal{A}_1$ for all $s \in [0,1]$ a.s.;*

(v) $(\xi, \tilde{\xi})$ *is completely exchangeable on W_1, and $\xi \overset{d}{=} \tilde{\xi}$.*

Proof: The implications (i) $\Rightarrow$ (ii) $\Rightarrow$ (iii) being obvious, it suffices to show that (iii) $\Rightarrow$ (i) and (iv) $\Rightarrow$ (v). Since the two proofs are very similar, we consider only the latter. Thus, assume that ξ satisfies (iv). Since the conditions in (v) are preserved by convex combinations, we may add the hypothesis that ξ be ergodic. Then Theorem 9.12 yields the representation

$$\xi = \sum_{i,j} a_{ij}\, \delta_{\tau_i,\tau_j} + c\lambda^2 \text{ a.s.,}$$

for some constants $a_{ij} \geq 0$ and $c \geq 0$ and some i.i.d. $U(0,1)$ random variables $\tau_1, \tau_2, \ldots$. The condition $\xi D_1 = 0$ a.s. implies $a_{ii} = 0$ for all i, and since a.s. $\xi_s + \tilde{\xi}_s \in \mathcal{A}_1$ for all s, the array $(a_{ij} + a_{ji})$ has at most one positive entry in every row or column. Hence, the variables τ_j come in disjoint pairs $(\sigma_k, \sigma_k') = (\tau_{i_k}, \tau_{j_k})$, so that

$$\xi = \sum_k \left(a_{i_k j_k} \delta_{\sigma_k, \sigma_k'} + a_{i_k j_k} \delta_{\sigma_k', \sigma_k} \right) + c\lambda^2 \text{ a.s.} \tag{21}$$

By independence it is enough to show that each term in (21) satisfies the conditions in (v).

The case $\xi = c\lambda^2$ being trivial, it remains to take

$$\xi = a\, \delta_{\sigma,\tau} + b\, \delta_{\tau,\sigma}$$

for some constants $a, b \geq 0$ and some independent $U(0,1)$ variables σ and τ. Then

$$(\xi, \tilde{\xi}) = (a, b)\, \delta_{\sigma,\tau} + (b, a)\, \delta_{\tau,\sigma},$$

and the symmetry $\xi \overset{d}{=} \tilde{\xi}$ follows from the relation $(\sigma, \tau) \overset{d}{=} (\tau, \sigma)$. Next we note that on W_1

$$(\xi, \tilde{\xi}) = \begin{cases} (a, b)\, \delta_{\sigma,\tau} & \text{when } \sigma < \tau, \\ (b, a)\, \delta_{\tau,\sigma} & \text{when } \tau < \sigma. \end{cases}$$

The asserted complete exchangeability now follows from the fact that

$$P[(\sigma, \tau) \in \cdot \mid \sigma < \tau] = P[(\tau, \sigma) \in \cdot \mid \tau < \sigma] = 2\lambda^2 \text{ on } W. \qquad \square$$

Since each of the following results has versions for both the separately and the jointly exchangeable case, it is convenient first to discuss the former case in detail, and then indicate how the argument may be modified to cover even the latter case. Our first aim is to characterize the component of ξ with independent increments, using the characterizations of completely exchangeable random measures on $[0,1]^2$ in Lemma 9.16. Then define

$$D = \{(s,t) \in \mathbb{R}_+^2;\ s = t\}, \qquad W = \{(s,t) \in \mathbb{R}_+^2;\ s \leq t\}.$$

The relevant representations of ξ are now of the form

$$\xi = \sum_j f(\alpha_j)\, \delta_{\sigma_j,\tau_j} + c\lambda^2, \tag{22}$$

$$\xi = \sum_j \left(f(\alpha_j)\, \delta_{\sigma_j,\tau_j} + g(\alpha_j)\, \delta_{\tau_j,\sigma_j} \right) + c\lambda^2. \tag{23}$$

Lemma 9.17 *(independent increments)* *For any random measure ξ on $\mathbb{R}_+^2$, the following three conditions are equivalent:*

(i) *ξ is ergodic, separately exchangeable, and $(\xi_s + \tilde{\xi}_s)\mathbb{R}_+ < \infty$ for all $s \geq 0$ a.s.;*

(ii) *ξ has stationary, independent increments;*

(iii) *ξ has an a.s. representation (22), for some measurable function $f \geq 0$ on $\mathbb{R}_+$, a unit rate Poisson process $\{(\sigma_i, \tau_i, \alpha_i)\}$ on $\mathbb{R}_+^3$, and a constant $c \geq 0$.*

So are the next three conditions:

(iv) *ξ is ergodic, jointly exchangeable, $\xi D = 0$ a.s., and $(\xi_s + \tilde{\xi}_s)\mathbb{R}_+ < \infty$ for all $s \geq 0$ a.s.;*

(v) *$(\xi, \tilde{\xi})$ has stationary, independent increments on W, and $\tilde{\xi} \overset{d}{=} \xi$;*

(vi) *ξ has an a.s. representation (23), for some measurable functions $f, g \geq 0$ on $\mathbb{R}_+$, a unit rate Poisson process $\{(\sigma_i, \tau_i, \alpha_i)\}$ on $\mathbb{R}_+^3$, and a constant $c \geq 0$.*

Proof: First assume (i). Then Lemma 9.14 yields a.s. $\xi_s, \tilde{\xi}_s \in \mathcal{A}_1$ for all $s \geq 0$, and so ξ is completely exchangeable by Lemma 9.16. Since, conversely, every completely exchangeable random measure is trivially separately exchangeable, ξ remains extreme in the sense of complete exchangeability, by the extremality in (i). Hence, (ii) follows from the obvious two-dimensional extension of Proposition 1.21.

Using the same result or FMP 15.4, we infer from (ii) that ξ has a representation

$$\xi = c\lambda^2 + \sum_j \beta_j \, \delta_{\sigma_j, \tau_j} \quad \text{a.s.,}$$

for some constant $c \geq 0$ and a Poisson process $\{(\sigma_j, \tau_j, \beta_j)\}$ on $\mathbb{R}_+^3$ with intensity measure of the form $\lambda^2 \otimes \nu$. Choosing $f \geq 0$ to be a measurable function on $\mathbb{R}_+$ such that $\lambda \circ f^{-1} = \nu$ on $(0, \infty)$, we see from FMP 12.3 that ξ has the same distribution as the process in (iii), and the corresponding a.s. representation follows by FMP 6.10.

Next (iii) implies that ξ is separately exchangeable with $(\xi_s + \tilde{\xi}_s)\mathbb{R}_+ < \infty$ for all $s \geq 0$ a.s. To show that ξ is also ergodic in the sense of separate exchangeability, let η be an invariant function of ξ, and note that η remains invariant under measure-preserving transformations in one coordinate. Considering transformations of permutation type and using the Hewitt–Savage zero-one law, combined with independence properties of the Poisson process, we conclude that η is a.s. a constant. Thus, ξ is ergodic and (i) follows. This proves that (i)–(iii) are equivalent.

Now assume instead condition (iv), and note that $\xi + \tilde{\xi}$ is again jointly exchangeable by Lemma A1.1. In particular, it is separately exchangeable on every set of the form $(a, b) \times (\mathbb{R}_+ \setminus (a, b))$ with $a < b$. By Lemma 9.14 it follows that, with probability one, the restriction of $\xi_s + \tilde{\xi}_s$ to $\mathbb{R}_+ \setminus (a, b)$ belongs to $\mathcal{A}_1$ for every $s \in (a, b)$. Since this holds simultaneously for all rational $a < b$ and since $\xi D = 0$ a.s. by hypothesis, we conclude that a.s. $\xi_s + \tilde{\xi}_s \in \mathcal{A}_1$ for all $s \geq 0$. Hence, Lemma 9.16 shows that the pair $(\xi, \tilde{\xi})$ is completely exchangeable on W with $\xi \overset{d}{=} \tilde{\xi}$. Then by Proposition 1.21 the random measures ξ and $\tilde{\xi}$ have diffuse components $\gamma \lambda^2$ and $\tilde{\gamma} \lambda^2$ on W for some random variables γ and $\tilde{\gamma}$, and their atomic parts on W are given by a Cox process on $W \times \mathbb{R}_+^2$ with directing random measure of the form $\lambda^2 \otimes \nu$. Here γ and $\tilde{\gamma}$ are clearly jointly invariant functions of ξ, and so is ν by the law of large numbers. By the joint ergodicity of ξ it follows that γ, $\tilde{\gamma}$, and ν are a.s. non-random, which means that $(\xi, \tilde{\xi})$ has stationary, independent increments on W. This shows that (iv) implies (v).

Next assume (v), and note that the pair $(\xi, \tilde{\xi})$ has a representation on W as before, in terms of some constants $c, \tilde{c} \geq 0$ and a measure ν. The condition $\xi \overset{d}{=} \tilde{\xi}$ on $\mathbb{R}_+^2$ yields $(\xi, \tilde{\xi}) \overset{d}{=} (\tilde{\xi}, \xi)$, which shows that $c = \tilde{c}$, whereas ν is symmetric under reflection in the diagonal D. In particular, ξ has a.s. the diffuse component $c\lambda^2$, and we may henceforth assume that ξ is purely atomic. Now choose a measurable function $(f, g) : \mathbb{R}_+ \to \mathbb{R}_+^2$ such that $\lambda \circ (f, g)^{-1} = \nu/2$ on $\mathbb{R}_+^2 \setminus \{0\}$. Let η be a Poisson process on $\mathbb{R}_+^3$ with intensity λ^3, write $\tilde{\eta}(ds\, dt \times \cdot) = \eta(dt\, ds \times \cdot)$, and define a random measure ξ' on $\mathbb{R}_+^2$ by

$$\xi'B = \int f(u)\, \eta(B \times du) + \int g(u)\, \tilde{\eta}(B \times du), \quad B \in \mathcal{B}(\mathbb{R}_+^2). \quad (24)$$

Letting I be the identity mapping on $\mathbb{R}_+^2$ and putting

$$\zeta = \eta \circ (I \otimes (f, g))^{-1}, \qquad \tilde{\zeta} = \tilde{\eta} \circ (I \otimes (f, g))^{-1},$$

we obtain

$$\begin{aligned}
(\xi', \tilde{\xi}') &= \int (f, g)(u)\, \eta(\cdot \times du) + \int (g, f)(u)\, \tilde{\eta}(\cdot \times du) \\
&= \int (x, y)\, (\zeta + \tilde{\zeta})(\cdot \times dx\, dy).
\end{aligned}$$

Since ζ and $\tilde{\zeta}$ are independent Poisson processes on $W \times (\mathbb{R}_+^2 \setminus \{0\})$ with intensity measure $\lambda^2 \otimes \nu/2$, their sum is again Poisson on the same set with intensity $\lambda^2 \otimes \nu$. Hence, $(\xi', \tilde{\xi}') \overset{d}{=} (\xi, \tilde{\xi})$ on W, which implies $\xi' \overset{d}{=} \xi$ on $\mathbb{R}_+^2$. By FMP 6.10 we infer that even ξ can be represented as in (24), which is equivalent to the representation in (vi).

Finally, assume (vi). Then the last two conditions in (iv) are obvious, whereas the joint exchangeability of ξ follows from the corresponding property of the underlying Poisson process $\eta = \sum_j \delta_{\sigma_j, \tau_j, \alpha_j}$. To prove the ergodicity of ξ, we note that if $\mu = \mathcal{L}(\xi)$ is a convex combination of some jointly exchangeable distributions μ_1 and μ_2, then even the latter satisfy the last two conditions in (iv). It is then enough to prove the extremality of μ within the set of mixtures of type (vi) distributions. But this is clear from Lemma 9.7, since the parameters γ and ν, hence the entire distribution μ, can be measurably recovered from ξ by the law of large numbers. Thus, (vi) implies (iv), which shows that even (iv)–(vi) are equivalent. $\square$

Next we consider some conditional symmetry properties of separately or jointly exchangeable random measures. For convenience, we may regard a quadruple $(\xi^1, \xi^2, \xi^3, \xi^4)$ of random measures on $[0,1]^2$ as a single $\mathbb{R}_+^4$-valued random measure ξ on the same space. The notions of separate or joint exchangeability of ξ then refer to the coordinates in $[0,1]^2$. We begin with the separately exchangeable case.

Lemma 9.18 *(conditional, separate exchangeability)* *Consider a separately exchangeable, $\mathbb{R}_+^4$-valued random measure $\xi = (\xi^1, \xi^2, \xi^3, \xi^4)$ on $[0,1]^2$, such that a.s.*

(i) *ξ^1, ξ^2, ξ^3, and ξ^4 are mutually singular;*

(ii) *$\xi_s \in \mathcal{A}_1$ for every $s \in [0,1]$ with $\xi_s^1 + \xi_s^4 \neq 0$, and similarly for $\tilde{\xi}_s$ when $\tilde{\xi}_s^2 + \tilde{\xi}_s^4 \neq 0$.*

Then

(iii) *ξ^4 is separately exchangeable, conditionally on (ξ^1, ξ^2, ξ^3);*

(iv) *ξ^1 is exchangeable in the first coordinate, conditionally on (ξ^2, ξ^3), and similarly for $\tilde{\xi}^2$, given (ξ^1, ξ^3).*

Proof: Since any a.s. property of ξ is a.s. shared by its extremal components, and since the conditional properties in (iii) and (iv) are preserved by convex combinations of the ξ-distributions, we may assume that ξ is ergodic. Then ξ can be represented as in Theorem 9.11, though with non-random coefficients in $\mathbb{R}_+^4$. From (ii) we see that, if $\xi_{\tau_i}^4 \neq 0$ for some i, then $\xi_{\tau_i} \in \mathcal{A}_1$, and so by (i) we have $(\xi^1, \xi^2, \xi^3)_{\tau_i} = 0$. Similarly, $\tilde{\xi}_{\tau_j'}^4 \neq 0$ implies $(\tilde{\xi}^1, \tilde{\xi}^2, \tilde{\xi}^3)_{\tau_j'} = 0$. Hence, (ξ^1, ξ^2, ξ^3) and ξ^4 are represented in terms of disjoint subsets of the variables τ_j and τ_j', and so $(\xi^1, \xi^2, \xi^3) \perp\!\!\!\perp \xi^4$. This implies $P[\xi^4 \in \cdot | \xi^1, \xi^2, \xi^3] = P\{\xi^4 \in \cdot\}$ a.s., and (iii) follows.

Next we see from (i) and (ii) that ξ^1 and (ξ^2, ξ^3) are represented in terms of disjoint sets of variables τ_i, and that $\xi_{\tau_i}^1 \neq 0$ implies $\xi_{\tau_i}^1 \in \mathcal{A}_1$. By suitable re-labeling we obtain an a.s. representation

$$\xi^1 = \sum_i a_i \delta_{\tau_i, \sigma_i} + \sum_i b_i (\lambda \otimes \delta_{\sigma_i}) + c\lambda^2,$$

where the τ_i are i.i.d. $U(0,1)$ and independent of (ξ^2, ξ^3) and all the σ_j. Since the joint distribution of $\tau_1, \tau_2, \ldots$ is invariant under measure-preserving transformations of $[0,1]$, it follows that ξ^1 is conditionally exchangeable in the first coordinate, given (ξ^2, ξ^3) and $\sigma_1, \sigma_2, \ldots$, which implies the first part of (iv). A similar argument proves the second part. $\square$

We turn to the corresponding result for jointly exchangeable random measures. Here we need the further notation

$$\bar{\xi} = \xi + \tilde{\xi}; \qquad \bar{\xi}_s = \xi_s + \tilde{\xi}_s, \quad s \in [0,1].$$

Lemma 9.19 *(conditional, joint exchangeability)   Consider a jointly exchangeable, $\mathbb{R}_+^4$-valued random measure $\xi = (\xi^1, \xi^2, \xi^3, \xi^4)$ on $[0,1]^2$, such that a.s.*

(i) *$\xi^1 + \tilde{\xi}^2$, $\tilde{\xi}^1 + \xi^2$, $\bar{\xi}^3$, and $\bar{\xi}^4$ are mutually singular;*

(ii) *$\bar{\xi}_s^1 + \bar{\xi}_s^2 + \bar{\xi}_s^3 + \bar{\xi}_s^4 \in \mathcal{A}_1$ for every $s \in [0,1]$ with $\xi_s^1 + \tilde{\xi}_s^2 + \tilde{\xi}_s^4 \neq 0$.*

Then

(iii) *ξ^4 is jointly exchangeable, conditionally on (ξ^1, ξ^2, ξ^3);*

(iv) *$(\xi^1, \tilde{\xi}^2)$ is exchangeable in the first coordinate, conditionally on ξ^3.*

Proof: As in the preceding proof, we may assume that ξ is ergodic, and hence can be represented as in Theorem 9.12, though with non-random, $\mathbb{R}_+^4$-valued coefficients. If $\bar{\xi}_{\tau_i}^4 \neq 0$ for some i, then (ii) yields $\bar{\xi}_{\tau_i} \in \mathcal{A}_1$, and so by (i) we have $(\bar{\xi}^1 + \bar{\xi}^2 + \bar{\xi}^3)_{\tau_i} = 0$. This shows that $\xi^4 \perp\!\!\!\perp (\xi^1, \xi^2, \xi^3)$, and (iii) follows.

Next suppose that $(\xi^1, \tilde{\xi}^2)_{\tau_i} \neq 0$ for some i. Then by (i) and (ii) we have $(\xi^1, \tilde{\xi}^2)_{\tau_i} \in \mathcal{A}_1$ and $(\tilde{\xi}^1, \xi^2, \bar{\xi}^3)_{\tau_i} = 0$. We also see from (i) that $(\xi^1, \tilde{\xi}^2)D = 0$. Thus, the representation of $(\xi^1, \tilde{\xi}^2)$ can be simplified to

$$(\xi^1, \tilde{\xi}^2) = \sum_i (a_i, a_i') \delta_{\tau_i, \sigma_i} + \sum_i (b_i, b_i')(\lambda \otimes \delta_{\sigma_i}) + (c, c')\lambda^2,$$

where the τ_i are i.i.d. $U(0,1)$ and independent of ξ^3 and all the σ_j. Condition (iv) now follows as before from the distributional invariance of the sequence $\tau_1, \tau_2, \ldots$ under measure-preserving transformations of $[0,1]$.   □

The next result gives a decomposition of ξ into conditionally independent components, defined by their relations to the horizontal and vertical supporting lines, given by Lemma 9.14. More precisely, we introduce the random sets

$$\begin{aligned} M_1 &= \{s \geq 0; \, \xi(\{s\} \times \mathbb{R}_+) = \infty\}, \\ M_2 &= \{t \geq 0; \, \xi(\mathbb{R}_+ \times \{t\}) = \infty\}, \\ M &= M_1 \cup M_2 \cup \{s \geq 0; \, \xi\{(s,s) > 0\}, \end{aligned}$$

and define

$$\begin{aligned} \xi_1 &= 1_{M_1^c \times M_2} \cdot \xi, & \xi_2 &= 1_{M_1 \times M_2^c} \cdot \xi, \\ \xi_3 &= 1_{M_1 \times M_2} \cdot \xi, & \xi_4 &= 1_{M_1^c \times M_2^c} \cdot \xi. \end{aligned} \tag{25}$$

Since M_1, M_2, and M are ξ-measurable by the quoted lemma, it follows easily that the ξ_i are measurable functions of ξ.

Lemma 9.20 *(support-line decomposition)*

(i) *Let ξ be an ergodic, separately exchangeable random measure on $\mathbb{R}_+^2$, and define $\xi_1, \dots, \xi_4$ by (25). Then there exist some random measures α_1, α_2 on $\mathbb{R}_+$ and ν_1, ν_2 on $\mathbb{R}_+ \times (0, \infty)$, such that ξ_1 and ξ_2 have a.s. diffuse components $\lambda \otimes \alpha_1$ and $\alpha_2 \otimes \lambda$, respectively, and their atom positions and sizes are given by Cox processes η_1, η_2 directed by $\lambda \otimes \nu_1$ and $\nu_2 \otimes \lambda$. Furthermore, ξ_4 and $(\alpha_1, \alpha_2, \nu_1, \nu_2, \xi_3)$ are ergodic, separately exchangeable, and*

$$\xi_4 \perp\!\!\!\perp (\xi_1, \xi_2, \xi_3), \qquad \eta_1 \perp\!\!\!\perp_{\nu_1} (\alpha_1, \xi_2, \xi_3), \qquad \eta_2 \perp\!\!\!\perp_{\nu_2} (\alpha_2, \xi_1, \xi_3).$$

(ii) *Let ξ be an ergodic, jointly exchangeable random measure on $\mathbb{R}_+^2$, and define $\xi_1, \dots, \xi_4$ as in (25), though with M_1 and M_2 replaced by M. Then there exist some random measures α_1, α_2 on $\mathbb{R}_+$ and ν on $\mathbb{R}_+ \times (\mathbb{R}_+^2 \setminus \{0\})$, such that ξ_1 and ξ_2 have a.s. diffuse components $\lambda \otimes \alpha_1$ and $\alpha_2 \otimes \lambda$, respectively, and the atom positions and sizes of $(\xi_1, \tilde{\xi}_2)$ are given by a Cox process η directed by $\lambda \otimes \nu$. Furthermore, ξ_4 and $(\alpha_1, \alpha_2, \nu, \xi_3)$ are ergodic, jointly exchangeable, and*

$$\xi_4 \perp\!\!\!\perp (\xi_1, \xi_2, \xi_3), \qquad \eta \perp\!\!\!\perp_{\nu} (\alpha_1, \alpha_2, \xi_3).$$

Here the separate and joint exchangeability or ergodicity of the vectors $(\alpha_1, \alpha_2, \nu_1, \nu_2, \xi_3)$ and $(\alpha_1, \alpha_2, \nu, \xi_3)$, respectively, are defined by the corresponding properties for

$$(\lambda \otimes \alpha_1, \; \alpha_2 \otimes \lambda, \; \lambda \otimes \nu_1, \; \lambda \otimes \nu_2, \; \xi_3), \qquad (\lambda \otimes \alpha_1, \; \lambda \otimes \alpha_2, \; \lambda \otimes \nu, \; \xi_3).$$

Proof: (i) By Lemma A1.1 the ergodic, separate exchangeability of ξ carries over to $\xi_1, \dots, \xi_4$, and even to the associated diffuse and atomic components. By Lemma 9.14 it follows that a.s. $\xi_1(\{s\} \times \cdot) \in \mathcal{A}_1$ and $\xi_2(\cdot \times \{t\}) \in \mathcal{A}_1$ for all $s, t \geq 0$. Thus, Proposition 1.21 yields the stated forms of the diffuse and atomic components of ξ_1 and ξ_2, in terms of some random measures α_1, α_2 on $\mathbb{R}_+$ and ν_1, ν_2 on $\mathbb{R}_+ \times (0, \infty)$. Noting that the α_i and ν_i are measurably determined by ξ and using Lemma A1.1 again, we see that even $(\alpha_1, \alpha_2, \nu_1, \nu_2, \xi_3)$ is ergodic, separately exchangeable.

From Lemma 9.14 we see that the hypotheses of Lemma 9.18, hence also the two conclusions, are fulfilled on every square $[0, a]^2$. By martingale convergence, the two results extend to the entire quadrant. In particular, ξ_4 is conditionally, separately exchangeable, given (ξ_1, ξ_2, ξ_3). Since ξ_4 is ergodic, the conditional distribution is a.s. constant, which means that ξ_4 is independent of (ξ_1, ξ_2, ξ_3). Next we see that ξ_1 is conditionally exchangeable in the first coordinate, given (ξ_2, ξ_3). Since α_1 and ν_1 are measurable functions of ξ_1, a.s. invariant under measure-preserving transformations of ξ_1 in the first coordinate, we conclude that ξ_1 remains conditionally exchangeable, given $(\alpha_1, \nu_1, \xi_2, \xi_3)$. The same thing is then true for the random measure η_1. Since η_1 is stationary Poisson, hence ergodic exchangeable, already under conditioning on ν_1, it follows that $\eta_1 \perp\!\!\!\perp_{\nu_1} (\alpha_1, \xi_2, \xi_3)$. A similar argument yields $\eta_2 \perp\!\!\!\perp_{\nu_2} (\alpha_2, \xi_1, \xi_3)$.

(ii) From Lemma A1.1 we see that ξ_4 and $(\xi_1, \tilde{\xi}_2)$, along with their diffuse and atomic parts, are again ergodic, jointly exchangeable. Proceeding as in the proof of the implication (iv) $\Rightarrow$ (v) in Lemma 9.17, we can easily verify the hypotheses of Lemma 9.19 for the restrictions of $\xi_1, \ldots, \xi_4$ to an arbitrary square $[0, a]^2$. Both hypotheses and conclusions are then fulfilled even on $\mathbb{R}_+^2$. In particular, $(\xi_1, \tilde{\xi}_2)$ is exchangeable in the first coordinate, and $(\xi_1 + \tilde{\xi}_2)(\{s\} \times \cdot) \in \mathcal{A}_1$ for all $s \geq 0$ a.s. Hence, Proposition 1.21 shows that ξ_1 and ξ_2 have a.s. diffuse components of the form $\lambda \otimes \alpha_1$ and $\alpha_2 \otimes \lambda$, respectively, and η is Cox with directing random measure of the form $\lambda \otimes \nu$. Furthermore, it is clear from Lemma A1.1 and the law of large numbers that even $(\alpha_1, \alpha_2, \nu, \xi_3)$ is ergodic, jointly exchangeable. The independence and conditional independence assertions may be proved as in case (i), except that Lemma 9.19 should now be used instead of Lemma 9.18. $\square$

Next we examine the structure of separately or jointly exchangeable point processes of the form

$$\xi = \sum_{i,j} \delta_{\sigma_i, \tau_j, \alpha_{ij}}, \tag{26}$$

where the sequences (σ_i) and (τ_j) are strictly increasing, and the marks α_{ij} are random elements in an arbitrary Borel space. The following result is a two-dimensional version of Theorem 1.23. Our characterizations are in terms of the array A and simple point processes η and ζ, given by

$$A = (\alpha_{ij}; \, i, j \in \mathbb{N}), \qquad \eta = \sum_i \delta_{\sigma_i}, \qquad \zeta = \sum_j \delta_{\tau_j}. \tag{27}$$

Lemma 9.21 *(separation of sites and marks) Let ξ be a simple point process on $\mathbb{R}_+^2 \times S$ of the form* (26)*, where S is Borel and the sequences (σ_i) and (τ_j) are increasing, and define the array A and the point processes η and ζ by* (27)*.*

(i) If ξ is ergodic, separately exchangeable, then A, η, and ζ are independent, and A is ergodic, separately exchangeable, whereas η and ζ are homogeneous Poisson.

(ii) If ξ is ergodic, jointly exchangeable and $\eta = \zeta$, then $A \perp\!\!\!\perp \eta$, and A is ergodic, jointly exchangeable, whereas η is homogeneous Poisson.

Proof: (i) Lemma A1.1 shows that η and ζ are ergodic, exchangeable, and so by Lemma 1.20 they are homogeneous Poisson. Regarding ξ as a point process on $\mathbb{R}_+$ with sites σ_i and associated marks $\kappa_i = (\tau_j, \alpha_{ij}, j \in \mathbb{N})$ in $(\mathbb{R}_+^2)^\infty$, $i \in \mathbb{N}$, we may infer from the latter lemma that ξ is Cox with directing measure of the form $\lambda \otimes \nu$. Thus, by FMP 12.3 the point process η and the sequence (κ_i) are conditionally independent and exchangeable, given ν. Since η is already ergodic and hence extreme, it must then be independent of (κ_i) and hence of the pair (ζ, A). Interchanging the roles of η and ζ, we conclude that η, ζ, and A are all independent.

The exchangeability of the sequence (κ_i) yields the corresponding property of A in the first index, and the symmetric argument shows that A is exchangeable in the second index as well. To see that A is ergodic, suppose that the distribution $\mu = \mathcal{L}(A)$ is a mixture of some separately exchangeable distributions μ_1 and μ_2. Then Lemma 9.4 shows that the corresponding distributions of ξ are again separately exchangeable. Since the original distribution of ξ was ergodic and hence extreme, it follows that $\mu_1 = \mu_2$, which means that even A is extreme and hence ergodic.

(ii) Here Lemmas A1.1 and 1.20 show that η is homogeneous Poisson. Now define for any $t > 0$

$$\eta^t = 1_{[0,t]} \cdot \eta, \qquad \eta_t = \eta[0,t], \qquad A_t = (\alpha_{ij};\ i,j \leq \eta_t),$$

and note that ξ remains separately exchangeable on $[0,t]^2$, conditionally on η_t. Hence, by Theorem 9.12, there exists for every $n \in \mathbb{N}$ a random permutation $\pi = (\pi_1, \ldots, \pi_n)$ of $(1, \ldots, n)$ such that, conditionally on $\eta_t = n$, the random variables $\sigma'_j = \sigma_{\pi_j}$, $j \leq n$, are i.i.d. $U(0,1)$ and independent of $A_t \circ \pi = (\alpha_{\pi_i, \pi_j};\ i,j \leq n)$. Writing π' for the inverse permutation π^{-1} and noting that the marked point process

$$\hat{\eta}^t = \sum\nolimits_{j \leq n} \delta_{\sigma_j, \pi'_j} = \sum\nolimits_{k \leq n} \delta_{\sigma'_k, k}$$

is conditionally ergodic, exchangeable given $\eta_t = n$, we see from Theorem 1.23 that η^t and π' are conditionally independent and exchangeable. Combining this with the previously noted independence, we conclude that $A_t \circ \pi \circ \pi' = A_t$ is jointly exchangeable and independent of η^t, conditionally on $\eta_t = n$. In particular, the matrix $A^n = (\alpha_{ij};\ i,j \leq n)$ is jointly exchangeable, conditionally on $\eta_t \geq n$, and the joint exchangeability of A follows as we let $t \to \infty$ and then $n \to \infty$.

From the previous discussion it is also clear that η^t is independent of A_t and exchangeable on $[0,t]$, conditionally on η_t. Hence, for any $s < t$, the process η^s is conditionally exchangeable given A_t and η_t, which implies that η^s is conditionally exchangeable, given the matrix A^n and the event $\eta_s \leq n$. Letting $n \to \infty$ and then $s \to \infty$, we conclude that η is conditionally exchangeable given A. Since η was shown to be ergodic and hence extreme, the conditional distribution is a.s. independent of A, which means that η and A are independent. The ergodicity of A can now be proved as in case (i). $\square$

We may finally combine the last result with Proposition 7.30 to obtain Poisson and functional representations for the lattice component of a separately or jointly exchangeable, marked point process ξ on $\mathbb{R}_+^2$. Since the projection of ξ onto $\mathbb{R}_+^2$ may no longer be locally finite, we need to introduce an extra mark in each coordinate. Thus, we consider marked point processes of the form

$$\xi = \sum\nolimits_{i,j} \delta_{\sigma_i, \tau_j, \alpha_i, \beta_j, \gamma_{ij}}, \tag{28}$$

where the marks α_i and β_j can be chosen to be $\mathbb{R}_+$-valued, and the γ_{ij} take values in an arbitrary Borel space S. We assume the variables in (28) to be such that the random measures

$$\eta_n = \sum_i \delta_{\sigma_i} 1\{\alpha_i \le n\}, \quad \zeta_n = \sum_j \delta_{\tau_j} 1\{\beta_j \le n\}, \qquad n \in \mathbb{N}, \qquad (29)$$

are locally finite, simple point processes on $\mathbb{R}_+$. The notions of separate or joint exchangeability or ergodicity are now defined in terms of transformations of the variables σ_i and τ_j.

Lemma 9.22 *(lattice component)* *Let ξ be a point process on $\mathbb{R}_+^4 \times S$ of the form (28), where S is Borel and the projections η_n and ζ_n in (29) are simple and locally finite.*

(i) *If ξ is ergodic, separately exchangeable, there exist some measurable functions $f, g \colon \mathbb{R}_+ \to \overline{\mathbb{R}}_+$ and $h \colon \mathbb{R}_+^2 \times [0,1] \to K$, a U-array (γ'_{ij}) on $\mathbb{N}^2$, and some independent, unit rate Poisson processes $\{(\sigma'_i, \alpha'_i)\}$ and $\{(\tau'_j, \beta'_j)\}$ on $\mathbb{R}_+^2$, such that (28) holds a.s. on $\mathbb{R}_+^4 \times K$ with $\sigma_i, \tau_j, \alpha_i, \beta_j, \gamma_{ij}$ replaced by the variables*

$$\sigma'_i, \ \tau'_j, \ f(\alpha'_i), \ g(\beta'_j), \ h(\alpha'_i, \beta'_j, \gamma'_{ij}), \quad i, j \in \mathbb{N}. \qquad (30)$$

(ii) *If ξ is ergodic, jointly exchangeable with $\sigma_i = \tau_i$ and $\alpha_i = \beta_i$, there exist some measurable functions $f \colon \mathbb{R}_+ \to \overline{\mathbb{R}}_+$ and $h \colon \mathbb{R}_+^2 \times [0,1] \to K$, a U-array $(\gamma'_{\{i,j\}})$ on $\tilde{\mathbb{N}}_2$, and an independent, unit rate Poisson process $\{(\sigma'_i, \tau'_i)\}$ on $\mathbb{R}_+^2$, such that (28) holds a.s. on $\mathbb{R}_+^4 \times K$ with $\sigma_i = \tau_i$, $\alpha_i = \beta_i$, and γ_{ij} replaced by the variables*

$$\sigma'_i, \ f(\alpha'_i), \ h(\alpha'_i, \alpha'_j, \gamma'_{\{i,j\}}), \quad i, j \in \mathbb{N}.$$

Proof: (i) By Lemma 9.21 the point processes η_n and ζ_n are homogeneous Poisson, say with densities a_n and b_n, respectively. Choosing n_0 large enough that $a_{n_0} \wedge b_{n_0} > 0$, we may write

$$\eta_n = \sum_i \delta_{\sigma_{ni}}, \quad \zeta_n = \sum_j \delta_{\tau_{nj}}, \qquad n \ge n_0,$$

where the σ_{ni} and τ_{nj} are increasing for fixed n. Comparing with (29), we see that

$$\sigma_{ni} = \sigma_{\kappa_{ni}}, \quad \tau_{nj} = \tau_{\kappa'_{nj}}, \qquad i, j \in \mathbb{N}, \ n \ge n_0,$$

for some random indices κ_{ni} and κ'_{nj}. We may introduce the corresponding quantities

$$\alpha_{ni} = \alpha_{\kappa_{ni}}, \quad \beta_{nj} = \beta_{\kappa_{nj}}, \quad \gamma_{nij} = \gamma_{\kappa_{ni}, \kappa_{nj}}, \qquad i, j \in \mathbb{N}, \ n \ge n_0,$$

and form the arrays

$$X_{ij}^n = (\alpha_{ni}, \beta_{nj}, \gamma_{nij}), \qquad i, j \in \mathbb{N}, \ n \ge n_0.$$

The X^n are separately exchangeable by Lemmas A1.1 and 9.21, and they are further nested in the sense of Chapter 7. Hence, Proposition 7.30 ensures the existence of a measurable function $F \colon \mathbb{R}_+^2 \times [0,1] \to \mathbb{R}_+^2 \times S$ and some random variables $\alpha'_{ni}, \beta'_{nj}, \gamma'_{nij}, i, j \in \mathbb{N}, n \ge n_0$, where the latter are independent for fixed n and uniformly distributed on the intervals $[0, a_n], [0, b_n]$, and $[0, 1]$, respectively, such that a.s.

$$X_{ij}^n = F(\alpha'_{ni}, \beta'_{nj}, \gamma'_{nij}), \qquad i, j \in \mathbb{N}, \ n \geq n_0.$$

Since α_{ni} is independent of j and β_{nj} of i, Fubini's theorem yields the simplified representation

$$\alpha_{ni} = f(\alpha'_{ni}), \ \ \beta_{nj} = g(\beta'_{nj}), \ \ \gamma_{nij} = h(\alpha'_{ni}, \beta'_{nj}, \gamma'_{nij}), \ \ i, j \in \mathbb{N}, \ n \geq n_0,$$

for some measurable functions $f, g \colon \mathbb{R}_+ \to \mathbb{R}_+$ and $h \colon \mathbb{R}_+^2 \times [0, 1] \to S$.

Since X^n, η_n, and ζ_n are independent for fixed $n \geq n_0$ by Lemma 9.21, the sequences (σ_{ni}) and (τ_{nj}) are independent. By FMP 6.10 we may even choose the array $(\alpha'_{ni}, \beta'_{nj}, \gamma'_{nij})$, $i, j \in \mathbb{N}$, to be independent of the two sequences. Recalling that the σ_{ni} form a Poisson process with constant rate a_n, whereas the α'_{ni} are i.i.d. $U(0, a_n)$, we see from FMP 12.3 that the pairs $(\sigma_{ni}, \alpha'_{ni})$, $i \in \mathbb{N}$, form a Poisson process on $\mathbb{R}_+ \times [0, a_n]$ with intensity measure λ^2. Similarly, the pairs (τ_{nj}, β'_{nj}), $j \in \mathbb{N}$, form a Poisson process on $\mathbb{R}_+ \times [0, b_n]$ with intensity λ^2, and the two processes are mutually independent and independent of the array γ'_{nij}, $i, j \in \mathbb{N}$.

Now consider any set of random variables σ'_i, τ'_j, α'_i, β'_j, γ'_{ij}, $i, j \in \mathbb{N}$, with the stated joint distribution, and define ξ' as in (28), but based on the variables in (30). Then clearly $\xi' \overset{d}{=} \xi$ on $\mathbb{R}_+^2 \times [0, n]^2 \times S$ for every n, and so the two distributions agree on $\mathbb{R}_+^4 \times S$ by FMP 1.1. Hence, FMP 6.10 allows us to re-define the σ'_i, τ'_j, α'_i, β'_j, and γ'_{ij} such that $\xi' = \xi$ a.s. This proves the required representation of ξ.

(ii) Here the previous proof applies with obvious changes. $\qquad\square$

9.6   Symmetries on a Quadrant

Here our aim is to characterize the classes of separately and jointly exchangeable random measures on the quadrant $\mathbb{R}_+^2$. Let us first state the representations in the two cases.

Theorem 9.23 *(separate exchangeability on a quadrant) A random measure ξ on $\mathbb{R}_+^2$ is separately exchangeable iff a.s.*

$$\begin{aligned}
\xi &= \sum_{i,j} f(\alpha, \vartheta_i, \vartheta'_j, \zeta_{ij}) \delta_{\tau_i, \tau'_j} + \sum_k l(\alpha, \eta_k) \delta_{\rho_k, \rho'_k} + \gamma \lambda^2 \\
&\quad + \sum_{i,k} g(\alpha, \vartheta_i, \chi_{ik}) \delta_{\tau_i, \sigma_{ik}} + \sum_{j,k} g'(\alpha, \vartheta'_j, \chi'_{jk}) \delta_{\sigma'_{jk}, \tau'_j} \\
&\quad + \sum_i h(\alpha, \vartheta_i)(\delta_{\tau_i} \otimes \lambda) + \sum_j h'(\alpha, \vartheta'_j)(\lambda \otimes \delta_{\tau'_j}),
\end{aligned} \qquad (31)$$

for some measurable functions $f \geq 0$ on $\mathbb{R}_+^4$, $g, g' \geq 0$ on $\mathbb{R}_+^3$, and $h, h', l \geq 0$ on $\mathbb{R}_+^2$, a U-array (ζ_{ij}) on $\mathbb{N}^2$, some independent, unit rate Poisson processes $\{(\tau_j, \vartheta_j)\}$, $\{(\tau'_j, \vartheta'_j)\}$, and $\{(\sigma_{ij}, \chi_{ij})\}_j, \{(\sigma'_{ij}, \chi'_{ij})\}_j$, $i \in \mathbb{N}$, on $\mathbb{R}_+^2$ and $\{(\rho_j, \rho'_j, \eta_j)\}$ on $\mathbb{R}_+^3$, and an independent pair of random variables $\alpha, \gamma \geq 0$. The latter can then be chosen to be non-random iff ξ is extreme.

Here it is understood that the Poisson processes are not only mutually independent, but also independent of all previously mentioned random elements. We may think of the σ_i and τ_j as the random times s and t where $\xi(\{s\} \times \mathbb{R}_+) = \infty$ and $\xi(\mathbb{R}_+ \times \{t\}) = \infty$. The sets $\{\sigma_i\} \times \mathbb{R}_+$ and $\mathbb{R}_+ \times \{\tau_j\}$ form Poisson processes of vertical and horizontal lines, and the various sums represent the contributions of ξ to the intersection points (σ_i, τ_j), the remaining parts of the lines, and the area between the lines.

Theorem 9.24 (*joint exchangeability on a quadrant*) *A random measure ξ on $\mathbb{R}_+^2$ is jointly exchangeable iff a.s.*

$$
\begin{aligned}
\xi \;=\; & \sum_{i,j} f(\alpha, \vartheta_i, \vartheta_j, \zeta_{\{i,j\}}) \delta_{\tau_i, \tau_j} + \beta \lambda_D + \gamma \lambda^2 \\
& + \sum_{j,k} \Big(g(\alpha, \vartheta_j, \chi_{jk}) \delta_{\tau_j, \sigma_{jk}} + g'(\alpha, \vartheta_j, \chi_{jk}) \delta_{\sigma_{jk}, \tau_j} \Big) \\
& + \sum_j \Big(h(\alpha, \vartheta_j)(\delta_{\tau_j} \otimes \lambda) + h'(\alpha, \vartheta_j)(\lambda \otimes \delta_{\tau_j}) \Big) \\
& + \sum_k \Big(l(\alpha, \eta_k) \delta_{\rho_k, \rho_k'} + l'(\alpha, \eta_k) \delta_{\rho_k', \rho_k} \Big),
\end{aligned}
\tag{32}
$$

for some measurable functions $f \geq 0$ on $\mathbb{R}_+^4$, $g, g' \geq 0$ on $\mathbb{R}_+^3$, and $h, h', l, l' \geq 0$ on $\mathbb{R}_+^2$, a U-array $(\zeta_{\{i,j\}})$ on $\tilde{\mathbb{N}}_2$, some independent, unit rate Poisson processes $\{(\tau_j, \vartheta_j)\}$ and $\{(\sigma_{ij}, \chi_{ij})\}_j$, $i \in \mathbb{N}$, on $\mathbb{R}_+^2$ and $\{(\rho_j, \rho_j', \eta_j)\}$ on $\mathbb{R}_+^3$, and an independent set of random variables $\alpha, \beta, \gamma \geq 0$. The latter can then be chosen to be non-random iff ξ is extreme.

Before proving these results, we show that the representation in Theorem 9.23 is an easy consequence of the one in Theorem 9.24.

Proof of Theorem 9.23 from Theorem 9.24: A separately exchangeable random measure ξ is also jointly exchangeable and can therefore be represented as in Theorem 9.24. Now define

$$
F(x, y) = (x + [x], y + [y] + 1), \quad x, y \geq 0,
$$

and note that F has range $R \times (R + 1)$, where $R = \bigcup_{k \in \mathbb{N}} [2k - 1, 2k)$. The separate exchangeability of ξ yields $\xi \circ F \overset{d}{=} \xi$, where $(\xi \circ F)B = \xi(F(B))$ for any Borel set $B \subset \mathbb{R}_+^2$. The desired representation now follows from the fact that the various Poisson processes in Theorem 9.24 have independent increments and therefore give independent contributions to the disjoint sets R and $R + 1$. $\qquad \square$

We now prove the main results of this section. Though Theorem 9.23 was shown to be a simple consequence of Theorem 9.24, it may be helpful to prove the former result first, and then indicate how the proof can be modified to cover the more difficult case of Theorem 9.24.

Proof of Theorem 9.23: To prove that any random measure ξ with a representation as in (31) is separately exchangeable, it is clearly enough to show that the marked point process

$$\eta = \sum_{i,j} \delta_{\tau_i, \tau_j', \vartheta_i, \vartheta_j', \zeta_{ij}}$$

is separately exchangeable under measure-preserving transformations in the first two coordinates. Now η is a uniform randomization of the product measure

$$\eta_1 \otimes \eta_2 = \sum_i \delta_{\tau_i, \vartheta_i} \otimes \sum_j \delta_{\tau_j', \vartheta_j'} = \sum_{i,j} \delta_{\tau_i, \vartheta_i, \tau_j', \vartheta_j'},$$

where η_1 and η_2 are independent, unit rate Poisson processes on $\mathbb{R}_+^2$. By FMP 12.3 (ii) it is then enough to prove the separate exchangeability of $\eta_1 \otimes \eta_2$, which holds by part (i) of the same result.

Now suppose that ξ is ergodic, separately exchangeable, and define $\xi_1, \ldots, \xi_4$ by (25). Then Lemma 9.20 (i) shows that ξ_4 is again ergodic, separately exchangeable and independent of (ξ_1, ξ_2, ξ_3). By the implication (i) $\Rightarrow$ (iii) of Lemma 9.17 it can be represented as in (22), which yields the second and third terms of (31), with constant α and γ. To simplify the writing, we may henceforth assume that $\xi_4 = 0$.

By Lemma 9.14 we can next choose some random variables $\sigma_1, \sigma_2, \ldots$ and $\sigma_1', \sigma_2', \ldots$ such that $M_1 \subset \{\sigma_i\}$ and $M_2 \subset \{\sigma_j'\}$ a.s. The cardinalities $|M_1|$ and $|M_2|$ are invariant and hence a.s. constant, and by FMP 11.1 we have either $M_k = \emptyset$ a.s. or $|M_k| = \infty$ a.s. for each k. In the latter case we may clearly assume that $M_1 = \{\sigma_i\}$ or $M_2 = \{\sigma_j'\}$, respectively. We also recall that the subsets $\{s \geq 0;\ \psi_s > \varepsilon\}$ and $\{s \geq 0;\ \psi_s' > \varepsilon\}$ are locally finite for every $\varepsilon > 0$, where the ψ_s and ψ_s' are defined as in Lemma 9.14.

Changing the notation, we now let α, α' and ν, ν' denote the random measures α_2, α_1 and ν_2, ν_1 in Lemma 9.20 (i). Since ξ_1 is supported by $\mathbb{R}_+ \times M_2$ and ξ_2 by $M_1 \times \mathbb{R}_+$, we see from the law of large numbers that α and α' are a.s. supported by M_1 and M_2, respectively, whereas ν and ν' are a.s. supported by $M_1 \times (0, \infty)$ and $M_2 \times (0, \infty)$. For $i, j \in \mathbb{N}$, define

$$\alpha_i = \alpha\{\sigma_i\}, \quad \alpha_j' = \alpha'\{\sigma_j'\}, \quad \gamma_{ij} = \xi_3\{(\sigma_i, \sigma_j')\},$$
$$\nu_i = \nu(\{\sigma_i\} \times \cdot), \quad \nu_j' = \nu'(\{\sigma_j'\} \times \cdot), \tag{33}$$

and introduce the marked point process

$$\zeta = \sum_{i,j} \delta_{\sigma_i, \sigma_j', \alpha_i, \alpha_j', \nu_i, \nu_j', \gamma_{ij}}. \tag{34}$$

If $|M_1| = |M_2| = \infty$ a.s., then Lemma A1.1 shows that the process ζ is ergodic, separately exchangeable in the first two coordinates. Hence, by Lemma 9.22 (i), there exist some measurable functions $f \geq 0$ on $\mathbb{R}_+^2 \times [0, 1]$ and $h, h' \geq 0$ on $\mathbb{R}_+$, some kernels G, G' from $\mathbb{R}_+$ to $(0, \infty)$, a U-array (ζ_{ij}) on $\mathbb{N}^2$, and some independent, unit rate Poisson processes $\{(\tau_i, \vartheta_i)\}$ and

$\{(\tau'_j, \vartheta'_j)\}$ on $\mathbb{R}^2_+$, such that (34) remains true with the random variables or measures $\sigma_i, \sigma'_j, \alpha_i, \alpha'_j, \nu_i, \nu'_j$, and γ_{ij} replaced by

$$\tau_i, \ \tau'_j, \ h(\vartheta_i), \ h'(\vartheta'_j), \ G(\vartheta_i), \ G'(\vartheta'_j), \ f(\vartheta_i, \vartheta'_j, \zeta_{ij}), \quad i, j \in \mathbb{N}.$$

Comparing with (33), we note that for all $i, j \in \mathbb{N}$,

$$\begin{aligned}
\alpha\{\tau_i\} &= h(\vartheta_i), \quad \alpha'\{\tau'_j\} = h'(\vartheta'_j), \\
\nu(\{\tau_i\} \times \cdot) &= G(\vartheta_i), \quad \nu'(\{\tau'_j\} \times \cdot) = G'(\vartheta'_j), \\
\xi_3\{(\tau_i, \tau'_j)\} &= f(\vartheta_i, \vartheta'_j, \zeta_{ij}).
\end{aligned} \tag{35}$$

If instead $|M_1| = \infty$ and $M_2 = \emptyset$ a.s., then $\xi_1 = \xi_3 = 0$ a.s., and we may consider the marked point process

$$\zeta_1 = \sum_i \delta_{\sigma_i, \alpha_i, \nu_i}, \tag{36}$$

which is ergodic, exchangeable in the first coordinate by Lemma A1.1. Hence, Lemma 1.20 shows that ζ_1 is homogeneous Poisson, and so there exist a measurable function $h \geq 0$ on $\mathbb{R}_+$, a kernel G from $\mathbb{R}_+$ to $(0, \infty)$, and a unit rate Poisson process $\{(\tau_i, \vartheta_i)\}$ on $\mathbb{R}^2_+$, such that (36) remains fulfilled with the random variables σ_i, α_i and measures ν_i replaced by $\tau_i, h(\vartheta_i)$, and $G(\vartheta_i)$, respectively. Note that (35) remains valid in this case, for τ'_j, ϑ'_j, and ζ_{ij} as before and with $f = h' = G' = 0$. The same argument applies to the case where $M_1 = \emptyset$ and $|M_2| = \infty$ a.s. Finally, we may take $f = h = h' = G = G' = 0$ when $M_1 = M_2 = \emptyset$ a.s.

By FMP 3.22 we may next choose some measurable functions $g, g' \geq 0$ on $\mathbb{R}^2_+$, such that on $(0, \infty)$

$$\begin{aligned}
\lambda \circ (g(x, \cdot))^{-1} &= G(x), \\
\lambda \circ (g'(x, \cdot))^{-1} &= G'(x), \quad x \geq 0.
\end{aligned} \tag{37}$$

We also introduce some independent, unit rate Poisson processes $\{(\sigma_{jk}, \chi_{jk})\}_k$ and $\{(\sigma'_{jk}, \chi'_{jk})\}_k$ on $\mathbb{R}^2_+$, $j \in \mathbb{N}$, and define

$$\begin{aligned}
\xi'_1 &= \sum_{j,k} g'(\vartheta'_j, \chi'_{jk}) \, \delta_{\sigma'_{jk}, \tau'_j} + \sum_j h'(\vartheta'_j) \, (\lambda \otimes \delta_{\tau'_j}), \\
\xi'_2 &= \sum_{i,k} g(\vartheta_i, \chi_{ik}) \, \delta_{\tau_i, \sigma_{ik}} + \sum_i h(\vartheta_i) \, (\delta_{\tau_i} \otimes \lambda).
\end{aligned} \tag{38}$$

In view of (35), we have

$$\begin{aligned}
\sum_j h'(\vartheta'_j) \, (\lambda \otimes \delta_{\tau'_j}) &= \lambda \otimes \sum_j \alpha'\{\tau'_j\} \delta_{\tau'_j} = \lambda \otimes \alpha_1, \\
\sum_i h(\vartheta_i) \, (\delta_{\tau_i} \otimes \lambda) &= \sum_i \alpha\{\tau_i\} \delta_{\tau_i} \otimes \lambda = \alpha_2 \otimes \lambda,
\end{aligned}$$

which shows that the diffuse parts of ξ'_1 and ξ'_2 agree with those of ξ_1 and ξ_2.

Next let η'_1 and η'_2 denote the point processes of atom sizes and positions of ξ'_1 and ξ'_2. Using (35), (37), and FMP 12.3, we see that, conditionally on the pairs (τ_j', ϑ_j') or (τ_i, ϑ_i), respectively, the processes η_1' and η_2' are Poisson with intensities

$$\sum_j \lambda \otimes \delta_{\tau_j'} \otimes \nu'(\{\tau_j'\} \times \cdot) = \lambda \otimes \nu_1,$$
$$\sum_i \delta_{\tau_i} \otimes \nu(\{\tau_j\} \times \cdot) \otimes \lambda = \nu_2 \otimes \lambda.$$

Under the same conditioning, it is also clear that η_1' is independent of the triple $(\alpha_1, \xi_2', \xi_3)$ while η_2' is independent of $(\alpha_2, \xi_1', \xi_3)$, which implies

$$\eta_1' \perp\!\!\!\perp_{\nu_1} (\alpha_1, \xi_2', \xi_3), \qquad \eta_2' \perp\!\!\!\perp_{\nu_2} (\alpha_2, \xi_1', \xi_3).$$

Hence, ξ_1' and ξ_2' are conditionally independent given $(\alpha_1, \alpha_2, \nu_1, \nu_2, \xi_3)$, with the same conditional distributions as ξ_1 and ξ_2, and it follows that $(\xi_1', \xi_2, \xi_3) \overset{d}{=} (\xi_1, \xi_2, \xi_3)$. By FMP 6.10 we can then re-define the sequences $\{(\sigma_{ik}', \chi_{ik}')\}$ and $\{(\sigma_{jk}, \chi_{jk})\}$, with the same properties as before, such that $\xi_1 = \xi_1'$ and $\xi_2 = \xi_2'$ a.s. This proves the representation (31) in the ergodic case, with constant values of α and γ.

To extend the representation to the general case, we may first use Lemma 9.10 to express the functions f, g, g', and h, h', l in terms of some universal functions Φ_4, Φ_3, and Φ_2 of different dimensions, along with some associated parameters α_4, α_3, α_3' and $\alpha_2, \alpha_2', \alpha_2''$. From Theorem A1.4 we see that the general distribution $\mathcal{L}(\xi)$ is a mixture of ergodic ones, and by Lemma A1.5 we get a representation in the general case, simply by substituting random variables for all parameters in the ergodic formula. To convert the resulting representation into the required form, it remains to express all randomization variables $\alpha_4, \alpha_3, \ldots$ in terms of a single random variable α.

Now suppose that ξ is representable as in (31), with α and γ non-random. Then ξ is clearly dissociated in the sense of Chapter 7, and so the distribution $\mathcal{L}(\xi)$ can be a.s. recovered from ξ by the law of large numbers. The required ergodicity of ξ now follows from Lemma 9.7 with $h(t) = m_t$. □

Proof of Theorem 9.24: First suppose that ξ is represented as in (32). Letting (ζ_{ij}) be a U-array on $\mathbb{N}^2$, independent of all representing variables, and arguing as in the preceding proof, we see that the marked point process

$$\eta = \sum_{i,j} \delta_{\tau_i, \tau_j, \vartheta_i, \vartheta_j, \zeta_{ij}}$$

is jointly exchangeable under measure-preserving transformations in the first two coordinates. This implies the same property for the symmetrized version of η, where ζ_{ij} is replaced by

$$\tilde\zeta_{\{i,j\}} = \zeta_{ij} + \zeta_{ji}, \quad i, j \in \mathbb{N}.$$

By a simple transformation in the last coordinate, we may extend the joint exchangeability to the version with $\tilde\zeta_{ij}$ replaced by $\zeta_{\{i,j\}}$, which shows that ξ is jointly exchangeable.

Now let ξ be jointly exchangeable. By Lemma A1.1 the diffuse mass along the diagonal D is ergodic, exchangeable, and so by Proposition 1.21 it is proportional to λ_D. The remainder of ξ_4 satisfies condition (iv) of Lemma 9.17, and hence is representable as in condition (vi) of the same result. Thus, ξ_4 can be represented by the second, third, and last terms of (32), for some constants α, β, γ, some measurable functions l, l', and a unit rate Poisson process $\{(\rho_j, \rho'_j, \eta_j)\}$ on $\mathbb{R}^3$. Since ξ_4 is independent of the triple (ξ_1, ξ_2, ξ_3) by Lemma 9.20 (ii), it remains to derive the representation of the latter.

The case where $M = \emptyset$ a.s. being trivial, we may assume that $|M| = \infty$ a.s. By Lemma 9.14, applied to any non-diagonal sets of the form $I \times I^c$ or $I^c \times I$, we can choose some random variables $\sigma_1, \sigma_2, \ldots$ such that $M = \{\sigma_j\}$ a.s. The random measures $(\alpha, \alpha') = (\alpha_2, \alpha_1)$ and ν of Lemma 9.20 (ii) are a.s. supported by M and $M \times (\mathbb{R}^2_+ \setminus \{0\})$, respectively, and for any $i, j \in \mathbb{N}$ we may put

$$\alpha_j = \alpha\{\sigma_j\}, \quad \alpha'_j = \alpha'\{\sigma_j\},$$
$$\nu_j = \nu(\{\sigma_j\} \times \cdot), \quad \gamma_{ij} = \xi_3\{(\sigma_i, \sigma_j)\}. \tag{39}$$

Now introduce the marked point process

$$\zeta = \sum_{i,j} \delta_{\sigma_i, \sigma_j, \alpha_i, \alpha'_j, \nu_i, \gamma_{ij}}. \tag{40}$$

From Lemma A1.1 we see that ζ is ergodic, jointly exchangeable in the first two coordinates. Hence, Lemma 28 (ii) yields the existence of some measurable functions $f \geq 0$ on $\mathbb{R}^2_+ \times [0, 1]$ and $h, h' \geq 0$ on $\mathbb{R}_+$, a kernel G from $\mathbb{R}_+$ to $\mathbb{R}^2_+ \setminus \{0\}$, a U-array $(\zeta_{\{i,j\}})$ on $\tilde{\mathbb{N}}_2$, and an independent, unit rate Poisson process $\{(\tau_j, \vartheta_j)\}$ on $\mathbb{R}^2_+$, such that (40) remains true with $\sigma_i, \alpha_i, \alpha'_j$, ν_i, and γ_{ij} replaced by

$$\tau_i, \ h(\vartheta_i), \ h'(\vartheta_j), \ G(\vartheta_i), \ f(\vartheta_i, \vartheta_j, \zeta_{\{i,j\}}), \quad i, j \in \mathbb{N}.$$

Comparing with (39), we get the relations

$$\alpha\{\tau_i\} = h(\vartheta_i), \quad \alpha'\{\tau_i\} = h'(\vartheta_i),$$
$$\nu(\{\tau_i\} \times \cdot) = G(\vartheta_i),$$
$$\xi_3\{(\tau_i, \tau_j)\} = f(\vartheta_i, \vartheta_j, \zeta_{\{i,j\}}). \tag{41}$$

By FMP 3.22 we may choose some measurable functions $g, g' \geq 0$ on $\mathbb{R}_+$ such that

$$\lambda \circ (g'(x, \cdot), g(x, \cdot))^{-1} = G(x) \text{ on } \mathbb{R}^2_+ \setminus \{0\}, \quad x \geq 0. \tag{42}$$

We may further introduce some independent, unit rate Poisson processes $\{(\sigma_{ik}, \chi_{ik})\}_k$ on $\mathbb{R}^2_+$, $i \in \mathbb{N}$. Define ξ'_1 and ξ'_2 as in (38), except that τ'_j, ϑ'_j, σ'_{jk}, and χ'_{jk} are now replaced by $\tau_j, \vartheta_j, \sigma_{jk}$, and χ_{jk}, respectively. Then ξ'_1 and ξ'_2 have a.s. diffuse components $\lambda \otimes \alpha_1$ and $\alpha_2 \otimes \lambda$. Next we see from (41), (42), and FMP 12.3 that the point process η' of atom sizes and positions of the pair $(\xi_1', \tilde{\xi}_2')$ is conditionally Poisson $\lambda \otimes \nu$ and independent of $(\alpha_1, \alpha_2, \xi_3)$, given the points (τ_i, ϑ_i), hence also given ν. Comparing with the properties of ξ_1 and ξ_2 obtained in Lemma 9.20 (ii), we conclude that $(\xi_1', \xi_2', \xi_3) \overset{d}{=} (\xi_1, \xi_2, \xi_3)$. By FMP 6.10 we can then re-define the random variables σ_{ik} and χ_{ik} such that equality holds a.s. The proof may now be completed as in case of Theorem 9.23. $\qquad\square$

It remains to derive convergence criteria for the series in Theorems 9.23 and 9.24. Again it is enough to consider the extreme case, where the parameters α, β, and γ are constants. As before, we may then simplify the notation by omitting α from our notation. It is also convenient to write $\hat{f} = f \wedge 1$ and

$$f_1 = \lambda_{23}^2 \hat{f}, \qquad f_2 = \lambda_{13}^2 \hat{f}, \qquad g_1 = \lambda_2 \hat{g},$$

where λ_{23}^2 denotes two-dimensional Lebesgue measure in the second and third coordinates, and similarly for λ_{13}^2 and λ_2.

Proposition 9.25 (*local summability*) *For fixed α, the random measure ξ in Theorem 9.23 is a.s. locally finite iff these four conditions are fulfilled:*

(i) $\lambda(\hat{l} + \hat{h} + \hat{h}') < \infty$,

(ii) $\lambda(\hat{g}_1 + \hat{g}_1') < \infty$,

(iii) $\lambda\{f_i = \infty\} = 0$ *and* $\lambda\{f_i > 1\} < \infty$ *for* $i = 1, 2$,

(iv) $\lambda[\hat{f}; f_1 \vee f_2 \leq 1] < \infty$.

In Theorem 9.24, local finiteness is equivalent to (i)–(iv), *together with the condition*

(v) $\lambda \hat{l}' + \lambda_D \lambda \hat{f} < \infty$.

Proof: It is enough to consider the convergence on $[0, 1]^2$. Omitting the constant α from our notation, we get

$$
\begin{aligned}
\xi[0,1]^2 \;=\; & \sum_{i,j} f(\vartheta_i, \vartheta_j', \zeta_{ij}) 1\{\tau_i \vee \tau_j' \leq 1\} \\
& + \sum_k l(\eta_k) 1\{\rho_k \vee \rho_k' \leq 1\} + \gamma \\
& + \sum_{i,k} g(\vartheta_i, \chi_{ik}) 1\{\tau_i \vee \sigma_{ik} \leq 1\} \\
& + \sum_{j,k} g'(\vartheta_j', \chi_{jk}') 1\{\tau_j' \vee \sigma_{jk}' \leq 1\} \\
& + \sum_i h(\vartheta_i) 1\{\tau_i \leq 1\} + \sum_j h'(\vartheta_j') 1\{\tau_j' \leq 1\}. \qquad (43)
\end{aligned}
$$

Introducing the unit rate Poisson processes on $\mathbb{R}_+$,

$$
\begin{aligned}
\eta &= \sum_k \delta_{\eta_k} 1\{\rho_k \vee \rho_k' \leq 1\}, \\
\vartheta &= \sum_i \delta_{\vartheta_i} 1\{\tau_i \leq 1\}, \qquad \vartheta' = \sum_j \delta_{\vartheta_j'} 1\{\tau_j' \leq 1\},
\end{aligned}
$$

we may write the second and last two terms of (43) as $\eta l + \vartheta h + \vartheta' h'$, which converges by FMP 12.13 or Theorem A3.5 iff our present condition (i) is fulfilled.

Next recall from FMP 4.14 that, for any independent random variables $\gamma_1, \gamma_2, \ldots \geq 0$, the series $\sum_j \gamma_j$ converges a.s. iff $\sum_j E(\gamma_j \wedge 1) < \infty$. Hence, by Fubini's theorem, the first sum in (43) converges a.s. iff

$$\sum_{i,j} \lambda \hat{f}(\vartheta_i, \vartheta'_j, \cdot) 1\{\tau_i \vee \tau'_j \leq 1\} = \vartheta \vartheta' \hat{f} < \infty \text{ a.s.,}$$

which occurs iff conditions (i)–(iii) in Theorem A3.5 are fulfilled for the function $F(s,t) = \lambda \hat{f}(s,t,\cdot)$. Noting that $F \leq 1$ and hence $\hat{F} = F$, we see that the mentioned requirements are equivalent to our present conditions (iii) and (iv).

By the cited criterion from FMP 4.14 together with Fubini's theorem, the third and fourth sums in (43) converge a.s. iff

$$\sum_i E\left[1 \wedge \sum_k g(\vartheta_i, \chi_{ik}) 1\{\sigma_{ik} \leq 1\} \Big| \vartheta_i, \tau_i\right] 1\{\tau_i \leq 1\} < \infty,$$
$$\sum_j E\left[1 \wedge \sum_k g'(\vartheta'_j, \chi'_{jk}) 1\{\sigma'_{jk} \leq 1\} \Big| \vartheta'_j, \tau'_j\right] 1\{\tau'_j \leq 1\} < \infty. \quad (44)$$

Noting that $1 \wedge x \asymp 1 - e^{-x} = \psi(x)$ and applying Lemma A3.6 (i) to the unit rate Poisson processes

$$\chi_i = \sum_k \delta_{\chi_{ik}} 1\{\sigma_{ik} \leq 1\}, \qquad \chi'_j = \sum_k \delta_{\chi'_{jk}} 1\{\sigma'_{jk} \leq 1\},$$

we see that the conditions in (44) are equivalent to

$$\sum_i \psi(\lambda \psi(g(\vartheta_i, \cdot))) 1\{\tau_i \leq 1\} \asymp \vartheta(\psi \circ \hat{g}_1) < \infty,$$
$$\sum_j \psi(\lambda \psi(g'(\vartheta'_j, \cdot))) 1\{\tau'_j \leq 1\} \asymp \vartheta'(\psi \circ \hat{g}'_1) < \infty.$$

Finally, by FMP 12.13 or Theorem A3.5, the latter conditions are equivalent to (ii).

This completes the proof for separately exchangeable random measures. The proof in the jointly exchangeable case is similar, except that the last assertion of Theorem A3.5 is then needed to characterize convergence of the first term in the representation of Theorem 9.24. Noting that the sum

$$\zeta = \sum_i \delta_{\vartheta_i, \varsigma_i} 1\{\tau_i \leq 1\}$$

represents a unit rate Poisson process on $\mathbb{R}_+ \times [0,1]$, we get in this case the extra condition $\lambda_D \lambda \hat{f} < \infty$. Finally, the term $\sum_k l'(\alpha, \eta_k) \delta_{\rho'_k, \rho_k}$ converges a.s. iff $\lambda \hat{l}' < \infty$, by the same argument as for the accompanying sum involving l. $\square$

Appendices

Most of the material included here is needed, in one way or another, for the development in the main text. Results have been deferred to an appendix for various reasons: They may not fit naturally into the basic exposition, or their proofs may be too technical or complicated to include in the regular text. In fact, some of the longer proofs are omitted altogether and replaced by references to the standard literature.

We begin, in Appendix A1, with a review of some basic results about extremal or ergodic decompositions. Though referred to explicitly only in the later chapters, the subject is clearly of fundamental importance for the entire book. Unfortunately, some key results in the area are quite deep, and their proofs often require methods outside the scope of the present exposition. Some more specialized results in this area appear in Section 9.2.

Appendix A2 contains some technical results about convergence in distribution, especially for random measures, needed in Chapter 3. In Appendix A3 we review some results about multiple stochastic integrals, required in Chapters 8 and 9, where the underlying processes may be either Gaussian or Poisson. In particular, we give a short proof of Nelson's hypercontraction theorem for multiple Wiener–Itô integrals. Next, to fill some needs in Sections 1.6 and 1.7, we list in Appendix A4 some classical results about completely monotone and positive definite functions, including the Hausdorff–Bernstein characterizations and the celebrated theorems of Bochner and Schoenberg. Finally, Appendix A5 reviews the basic theory of Palm measures and Papangelou kernels, required for our discussion in Section 2.7.

A1. Decomposition and Selection

Given a family $\mathcal{T}$ of measurable transformations on a probability space $(S, \mathcal{S}, \mu)$, we say that a set $I \in \mathcal{S}$ is *(strictly)* $\mathcal{T}$-*invariant* if $T^{-1}I = I$ for every $T \in \mathcal{T}$ and *a.s.* $\mathcal{T}$-*invariant* if $\mu(I \,\Delta\, T^{-1}I) = 0$ for every $T \in \mathcal{T}$, where $A \,\Delta\, B$ denotes the symmetric difference of A and B. Furthermore, we say that μ is *ergodic* if $\mu I = 0$ or 1 for every a.s. $\mathcal{T}$-invariant set $I \in \mathcal{S}$, and *weakly ergodic* if the same condition holds for every strictly invariant set I. A random element ξ of S is said to be $\mathcal{T}$-*symmetric* if its distribution μ is $\mathcal{T}$-*invariant*, in the sense that $\mu \circ T^{-1} = \mu$ for every $T \in \mathcal{T}$, and we say that ξ is $\mathcal{T}$-*ergodic* if the corresponding property holds for μ.

Let us begin with a common situation where the symmetry or ergodicity of a random element is preserved by a measurable mapping.

Lemma A1.1 *(preservation laws) Let $\mathcal{T}$ and $\mathcal{T}'$ be families of measurable transformations on some measurable spaces S and S', and fix a measurable mapping $f : S \to S'$ such that*

$$\{f \circ T; \ T \in \mathcal{T}\} = \{T' \circ f; \ T' \in \mathcal{T}'\}.$$

Then for any $\mathcal{T}$-symmetric or $\mathcal{T}$-ergodic random element ξ in S, the random element $\eta = f(\xi)$ in S is $\mathcal{T}'$-symmetric or $\mathcal{T}'$-ergodic, respectively.

Proof: Suppose that ξ is $\mathcal{T}$-symmetric. Letting $T' \in \mathcal{T}'$ be arbitrary and choosing $T \in \mathcal{T}$ with $f \circ T = T' \circ f$, we get

$$T' \circ \eta = T' \circ f \circ \xi = f \circ T \circ \xi \overset{d}{=} f \circ \xi = \eta,$$

which shows that η is $\mathcal{T}'$-symmetric.

Next we note that the invariant σ-fields $\mathcal{I}$ in S and $\mathcal{J}$ in S' are related by $f^{-1}\mathcal{J} \subset \mathcal{I}$. In fact, letting $J \in \mathcal{J}$ and $T \in \mathcal{T}$ be arbitrary and choosing $T' \in \mathcal{T}'$ with $T' \circ f = f \circ T$, we get a.s. $\mathcal{L}(\xi)$

$$T^{-1}f^{-1}J = f^{-1}T'^{-1}J = f^{-1}J.$$

Hence, if ξ is $\mathcal{T}$-ergodic,

$$P\{\eta \in J\} = P\{\xi \in f^{-1}J\} = 0 \text{ or } 1, \quad J \in \mathcal{J},$$

which means that η is $\mathcal{T}'$-ergodic. $\qquad\square$

For any class $\mathcal{T}$ of measurable transformations on a measurable space S, the set of $\mathcal{T}$-invariant probability measures μ on S is clearly convex. A $\mathcal{T}$-invariant distribution μ is said to be *extreme* if it has no non-trivial representation as a convex combination of invariant measures. We examine the relationship between the notions of ergodicity and extremality.

Lemma A1.2 *(ergodicity and extremality) Let $\mathcal{T}$ be a family of measurable transformations on a measurable space S, and consider a $\mathcal{T}$-invariant distribution μ on S. If μ is extreme, it is even ergodic, and the two notions are equivalent when $\mathcal{T}$ is a group. For countable groups $\mathcal{T}$, it is also equivalent that μ be weakly ergodic.*

Proof: Suppose that μ is not ergodic. Then there exists an a.s. $\mathcal{T}$-invariant set $I \in \mathcal{S}$ such that $0 < \mu I < 1$, and we get a decomposition

$$\mu = \mu(I) \, \mu[\,\cdot\,|I] + \mu(I^c) \, \mu[\,\cdot\,|I^c].$$

Here $\mu[\cdot|I] \neq \mu[\cdot|I^c]$, since $\mu[I|I] = 1$ and $\mu[I|I^c] = 0$. Furthermore, the invariance of I and μ implies that, for any $T \in \mathcal{T}$ and $B \in \mathcal{S}$,

$$
\begin{aligned}
\mu(I)\,\mu[T^{-1}B|I] &= \mu(T^{-1}B \cap I) = \mu \circ T^{-1}(B \cap I) \\
&= \mu(B \cap I) = \mu(I)\,\mu[B|I],
\end{aligned}
$$

which shows that $\mu[\cdot|I]$ is again invariant. Hence, μ is not extreme either.

Now let $\mathcal{T}$ be a group, and assume that μ is ergodic. Consider any convex combination

$$
\mu = c\mu_1 + (1 - c)\mu_2,
$$

where μ_1 and μ_2 are $\mathcal{T}$-invariant distributions on S and $c \in (0, 1)$. Introduce the Radon–Nikodým density $f = d\mu_1/d\mu$. Letting $T \in \mathcal{T}$ and $B \in \mathcal{S}$ be arbitrary and using the invariance of μ_1 and μ, we get

$$
\mu_1 B = \mu_1(TB) = \int_{TB} f\,d\mu = \int_{TB} f\,d(\mu \circ T^{-1}) = \int_B (f \circ T)\,d\mu.
$$

The uniqueness of f yields $f \circ T = f$ a.s., and so for any $a \geq 0$

$$
T^{-1}\{f \geq a\} = \{f \circ T \geq a\} = \{f \geq a\} \text{ a.s. } \mu,
$$

which means that the set $\{f \geq a\}$ is a.s. invariant. Since μ is ergodic, we get $\mu\{f \geq a\} = 0$ or 1 for every a, and it follows that f is a.s. a constant. But then $\mu_1 = \mu_2 = \mu$, and the stated decomposition is trivial. This shows that μ is extreme.

If $\mathcal{T}$ is a countable group and I is a.s. invariant, then the set $I' = \bigcap_{T \in \mathcal{T}} T^{-1}I$ is strictly invariant with $\mu(I \Delta I') = 0$. Assuming μ to be weakly ergodic, we get $\mu I = \mu I' = 0$ or 1, which shows that μ is even ergodic. $\quad\Box$

The following result identifies two cases where every invariant distribution has an integral representation in terms of extreme points. This decomposition may not be unique in general.

Theorem A1.3 *(extremal decomposition, Choquet, Kallenberg)* *Let $\mathcal{T}$ be a class of measurable transformations on a measurable space S, and assume one of these conditions:*

  (i) *$S = B^\infty$ for some Borel space B, and $\mathcal{T}$ is induced by a class of transformations on $\mathbb{N}$;*

  (ii) *S is Polish, and the set of $\mathcal{T}$-invariant distributions on S is weakly closed.*

Then every $\mathcal{T}$-invariant probability measure on S is a mixture of extreme, $\mathcal{T}$-invariant distributions.

Proof: (i) Embedding B as a Borel set in $[0, 1]$, we may regard μ as an invariant distribution on the compact space $J = [0, 1]^\infty$. The space $\mathcal{M}_1(J)$ is again compact and metrizable (cf. Rogers and Williams (1994), Theorem

81.3), and the sub-set M of $\mathcal{T}$-invariant distributions on J is convex and closed, hence compact. By a standard form of Choquet's theorem (cf. Alfsen (1971)), any measure $\mu \in M$ has then an integral representation

$$\mu A = \int m(A)\, \nu(dm), \quad A \in \mathcal{B}(J), \tag{1}$$

where ν is a probability measure on the set $\mathrm{ex}(M)$ of extreme elements in M. In particular, we note that $\nu\{m;\, mB^\infty = 1\} = 1$. Writing $R(m)$ for the restriction of m to B^∞ and putting $\pi_A m = m(A)$, we get for any Borel set $A \subset B^\infty$

$$\begin{aligned}
\mu A &= \mu(A \cap B^\infty) = \int m(A \cap B^\infty)\, \nu(dm) \\
&= \int (\pi_A \circ R)\, d\nu = \int \pi_A\, d(\nu \circ R^{-1}) \\
&= \int m(A)\, (\nu \circ R^{-1})(dm).
\end{aligned} \tag{2}$$

It remains to note that if $m \in \mathrm{ex}(M)$ with $mB^\infty = 1$, then m is an extreme, $\mathcal{T}$-invariant distribution on B^∞.

(ii) Here we may embed S as a Borel sub-set of a compact metric space J (cf. Rogers and Williams (1994), Theorem 82.5). The space $\mathcal{M}_1(J)$ is again compact and metrizable, and $\mathcal{M}_1(S)$ can be identified with the sub-set $\{\mu \in \mathcal{M}_1(J);\, \mu S = 1\}$ (op. cit., Theorem 83.7). Now the set M of all $\mathcal{T}$-invariant distributions on S remains convex as a sub-set of $\mathcal{M}_1(J)$, and its closure $\overline{M}$ in $\mathcal{M}_1(J)$ is both convex and compact. Thus, Choquet's theorem yields an integral representation as in (1), in terms of a probability measure ν on $\mathrm{ex}(\overline{M})$.

Since μ is restricted to S, we may proceed as before to obtain a representation of type (2), where $R(m)$ now denotes the restriction of m to S. It remains to show that $\mathrm{ex}(\overline{M}) \cap \mathcal{M}_1(S) \subset \mathrm{ex}(M)$. But this follows easily from the relation $\overline{M} \cap \mathcal{M}_1(S) = M$, which holds since M is closed in $\mathcal{M}_1(S)$. $\quad\square$

Under suitable regularity conditions, the ergodic decomposition is unique and can be obtained by conditioning on the invariant σ-field.

Theorem A1.4 *(decomposition by conditioning, Farrel, Varadarajan) Let $\mathcal{T}$ be a countable group of measurable transformations on a Borel space S, and consider a $\mathcal{T}$-symmetric random element ξ in S with distribution μ and invariant σ-field $\mathcal{I}_\xi$. Then the conditional distributions $P[\xi \in \cdot | \mathcal{I}_\xi]$ are a.s. ergodic and $\mathcal{T}$-invariant, and μ has the unique ergodic decomposition*

$$\mu = \int m\, \nu(dm), \qquad \nu = \mathcal{L}(P[\xi \in \cdot | \mathcal{I}_\xi]).$$

Proof: See Dynkin (1978) or Maitra (1977). $\quad\square$

The following result allows us to extend a representation from the ergodic to the general case through a suitable randomization.

Lemma A1.5 *(randomization) Given some Borel spaces S, T, and U, a measurable mapping $f \colon S \times T \to U$, and some random elements ξ in S and η in U, define $m_t = \mathcal{L}(f(\xi, t))$, $t \in T$. Then $\mathcal{L}(\eta)$ is a mixture over the class $\mathcal{C} = \{m_t\}$ iff $\eta = f(\tilde{\xi}, \tau)$ a.s. for some random elements $\tilde{\xi} \stackrel{d}{=} \xi$ in S and $\tau \perp\!\!\!\perp \tilde{\xi}$ in T. If even $P[\eta \in \cdot | \mathcal{I}] \in \mathcal{C}$ a.s. for some σ-field $\mathcal{I}$, we can choose τ to be $\mathcal{I}$-measurable.*

Proof: Put $M = \mathcal{M}_1(U)$, and note that M is again Borel. In $M \times T$ we may introduce the product measurable set

$$A = \{(\mu, t) \in M \times T; \; \mu = \mathcal{L}(f(\xi, t))\}.$$

Then, by hypothesis, $\mathcal{L}(\eta) = \int \mu \, \nu(d\mu)$ for some probability measure ν on M satisfying $\nu(\pi A) = \nu(\mathcal{C}) = 1$, where πA denotes the projection of A onto M. Hence, the general section theorem (FMP A1.4) yields a measurable mapping $g \colon M \to T$ satisfying

$$\nu\{\mu = \mathcal{L}(f(\xi, g(\mu)))\} = \nu\{(\mu, g(\mu)) \in A\} = 1.$$

Letting $\alpha \perp\!\!\!\perp \xi$ be a random element in M with distribution ν, we get $\eta \stackrel{d}{=} f(\xi, g(\alpha))$ by Fubini's theorem. Finally, the transfer theorem (FMP 6.10) yields a random pair $(\tilde{\xi}, \tau) \stackrel{d}{=} (\xi, g(\alpha))$ in $S \times T$ satisfying $\eta = f(\tilde{\xi}, \tau)$ a.s.

If $P[\eta \in \cdot | \mathcal{I}] \in \mathcal{C}$ a.s., we may instead apply the section theorem to the product-measurable set

$$A' = \{(\omega, t) \in \Omega \times T; \; P[\eta \in \cdot | \mathcal{I}](\omega) = \mathcal{L}(f(\xi, t))\},$$

to obtain an $\mathcal{I}$-measurable random element τ in T satisfying

$$P[\eta \in \cdot | \mathcal{I}] = \mathcal{L}(f(\xi, t))|_{t=\tau} \quad \text{a.s.}$$

Letting $\zeta \stackrel{d}{=} \xi$ in S with $\zeta \perp\!\!\!\perp \tau$ and using Fubini's theorem, we get a.s.

$$P[f(\zeta, \tau) \in \cdot | \tau] = P\{f(\zeta, t) \in \cdot\}|_{t=\tau} = \mu_\tau = P[\eta \in \cdot | \tau],$$

which implies $(f(\zeta, \tau), \tau) \stackrel{d}{=} (\eta, \tau)$. By FMP 6.11 we may then choose a random pair $(\tilde{\xi}, \tilde{\tau}) \stackrel{d}{=} (\zeta, \tau)$ in $S \times T$ such that

$$(f(\tilde{\xi}, \tilde{\tau}), \tilde{\tau}) = (\eta, \tau) \quad \text{a.s.}$$

In particular, we get $\tilde{\tau} = \tau$ a.s., and so $\eta = f(\tilde{\xi}, \tau)$ a.s. Finally, we note that $\tilde{\xi} \stackrel{d}{=} \zeta \stackrel{d}{=} \xi$, and also that $\tilde{\xi} \perp\!\!\!\perp \tau$ since $\zeta \perp\!\!\!\perp \tau$ by construction. $\square$

We proceed with a result on measurable selections.

Lemma A1.6 *(measurable selection) Let ξ and η be random elements in some measurable spaces S and T, where T is Borel, and let f be a measurable function on $S \times T$ such that $f(\xi, \eta) = 0$ a.s. Then there exists a ξ-measurable random element $\hat{\eta}$ in T satisfying $f(\xi, \hat{\eta}) = 0$ a.s.*

Proof: Define a measurable set $A \subset S \times T$ and its S-projection πA by

$$A = \{(s,t) \in S \times T;\ f(s,t) = 0\},$$
$$\pi A = \bigcup_{t \in T}\{s \in S;\ f(s,t) = 0\}.$$

By the general section theorem (FMP A1.4), there exists a measurable function $g\colon S \to T$ satisfying

$$(\xi, g(\xi)) \in A \text{ a.s. on } \{\xi \in \pi A\}.$$

Since $(\xi, \eta) \in A$ implies $\xi \in \pi A$, we also have

$$P\{\xi \in \pi A\} \geq P\{(\xi, \eta) \in A\} = P\{f(\xi, \eta) = 0\} = 1.$$

This proves the assertion with $\hat{\eta} = g(\xi)$. $\qquad\qquad\qquad\qquad\square$

A2. Weak Convergence

For any metric or metrizable space S, let $\hat{\mathcal{M}}(S)$ denote the space of bounded measures on S, write $\hat{\mathcal{M}}_c(S)$ for the sub-set of measures bounded by c, and let $\mathcal{M}_c(S)$ be the further sub-class of measures μ with $\mu S = c$. For any $\mu \in \hat{\mathcal{M}}(S)$, we define $\hat{\mu} = \mu/(\mu S \vee 1)$. On $\hat{\mathcal{M}}(S)$ we introduce the *weak topology*, induced by the mappings $\mu \mapsto \mu f = \int f d\mu$ for any f belonging to the space $\hat{C}_+(S)$ of bounded, continuous, non-negative functions on S. Then the weak convergence $\mu_n \xrightarrow{w} \mu$ means that $\mu_n f \to \mu f$ for every $f \in \hat{C}_+(S)$. A set $M \subset \hat{\mathcal{M}}(S)$ is said to be *tight* if

$$\inf_{K \in \mathcal{K}} \sup_{\mu \in M} \mu K^c = 0,$$

where $\mathcal{K} = \mathcal{K}(S)$ denotes the class of compact sets in S. We begin with a simple extension of Prohorov's theorem (FMP 16.3).

Lemma A2.1 *(weak compactness, Prohorov)* *For any Polish space S, a set $M \subset \hat{\mathcal{M}}(S)$ is weakly, relatively compact iff*

- (i) $\sup_{\mu \in M} \mu S < \infty$,
- (ii) M *is tight.*

Proof: First suppose that M is weakly, relatively compact. Since the mapping $\mu \mapsto \mu S$ is weakly continuous, we conclude that the set $\{\mu S;\ \mu \in M\}$ is relatively compact in $\mathbb{R}_+$, and (i) follows by the Heine–Borel theorem. To prove (ii), we note that the set $\hat{M} = \{\hat{\mu};\ \mu \in M\}$ is again weakly, relatively compact by the weak continuity of the mapping $\mu \mapsto \hat{\mu}$. By (i) we may then assume that $M \subset \hat{\mathcal{M}}_1(S)$. For fixed $s \in S$, define $\mu' = \mu + (1 - \mu S)\delta_s$. The mapping $\mu \mapsto \mu'$ is weakly continuous on $\hat{\mathcal{M}}_1(S)$, and so the set $M' = \{\mu';\ \mu \in M\}$ in $\mathcal{M}_1(S)$ is again relatively compact. But then Prohorov's theorem yields (ii) for the set M', and the same condition follows for M.

Conversely, assume conditions (i) and (ii), and fix any $\mu_1, \mu_2, \ldots \in M$. If $\liminf_n \mu_n S = 0$, then $\mu_n \xrightarrow{w} 0$ along a sub-sequence. If instead $\liminf_n \mu_n S > 0$, then by (i) we have $\mu_n S \to c \in (0, \infty)$ along a sub-sequence $N' \subset \mathbb{N}$. But then by (ii) the sequence $\tilde{\mu}_n = \mu_n / \mu_n S$, $n \in N'$, is tight, and so by Prohorov's theorem we have $\tilde{\mu}_n \xrightarrow{w} \mu$ along a further sub-sequence N''. Hence, $\mu_n = (\mu_n S) \tilde{\mu}_n \xrightarrow{w} c\mu$ along N'', which shows that M is weakly, relatively compact. $\qquad \square$

Next we prove a tightness criterion for a.s. bounded random measures.

Lemma A2.2 *(weak tightness, Prohorov, Aldous)* Let $\xi_1, \xi_2, \ldots$ be a.s. bounded random measures on a Polish space S. Then (ξ_n) is tight for the weak topology on $\hat{\mathcal{M}}(S)$ iff

(i) $(\xi_n S)$ is tight in $\mathbb{R}_+$,

(ii) $(E\hat{\xi}_n)$ is weakly tight in $\hat{\mathcal{M}}(S)$.

Proof: First suppose that (ξ_n) is weakly tight. Since the mapping $\mu \mapsto \mu S$ is weakly continuous on $\hat{\mathcal{M}}(S)$, condition (i) follows by FMP 16.4. Using the continuity of the mapping $\mu \mapsto \hat{\mu}$, we see from the same result that $(\hat{\xi}_n)$ is weakly tight. Hence, there exist some weakly compact sets $M_k \subset \hat{\mathcal{M}}_1(S)$ such that
$$P\{\hat{\xi}_n \notin M_k\} \leq 2^{-k-1}, \quad k, n \in \mathbb{N}.$$
By Lemma A2.1 we may next choose some compact sets $K_k \subset S$ such that
$$M_k \subset \{\mu; \, \mu K_k^c \leq 2^{-k-1}\}, \quad k \in \mathbb{N}.$$
Then for any n and k
$$E\hat{\xi}_n K_k^c \leq P\{\hat{\xi}_n \notin M_k\} + E[\hat{\xi}_n K_k^c; \, \xi_n \in M_k] \leq 2^{-k},$$
and (ii) follows.

Conversely, assume conditions (i) and (ii). We may then choose some constants $c_k > 0$ and some compact sets $K_k \subset S$ such that
$$P\{\xi_n S > c_k\} \leq 2^{-k}, \quad E\hat{\xi}_n K_k^c \leq 2^{-2k}, \quad k, n \in \mathbb{N}.$$
Now introduce in $\hat{\mathcal{M}}(S)$ the sets
$$M_m = \{\mu; \, \mu S \leq c_m\} \cap \bigcap_{k > m} \{\mu; \, \hat{\mu} K_k^c \leq 2^{-k}\}, \quad m \in \mathbb{N},$$
which are weakly, relatively compact by Lemma A2.1. Using the countable sub-additivity of P and Chebyshev's inequality, we get
$$
\begin{aligned}
P\{\xi_n \notin M_m\} &= P\Big(\{\xi_n S > c_m\} \cup \bigcup\nolimits_{k>m} \{\hat{\xi}_n K_k^c > 2^{-k}\}\Big) \\
&\leq P\{\xi_n S > c_m\} + \sum\nolimits_{k>m} P\{\hat{\xi}_n K_k^c > 2^{-k}\} \\
&\leq 2^{-m} + \sum\nolimits_{k>m} 2^k E\hat{\xi}_n K_k^c \\
&\leq 2^{-m} + \sum\nolimits_{k>m} 2^{-k} = 2^{-m+1} \to 0,
\end{aligned}
$$
which shows that (ξ_n) is weakly tight in $\hat{\mathcal{M}}(S)$. $\qquad \square$

The last result leads to some useful criteria for convergence in distribution of a.s. bounded random measures on a metrizable space S. When S is Polish, we write $\xi_n \xrightarrow{wd} \xi$ for such convergence with respect to the weak topology on $\hat{\mathcal{M}}(S)$. In the special case where S is locally compact, we may also consider the corresponding convergence with respect to the vague topology, here denoted by $\xi_n \xrightarrow{vd} \xi$. In the latter case, write $\overline{S}$ for the one-point compactification of S. With a slight abuse of notation, we say that $(\hat{E}\xi_n)$ is *tight* if

$$\inf_{K \in \mathcal{K}} \limsup_{n \to \infty} E[\xi_n K^c \wedge 1] = 0.$$

Theorem A2.3 *(convergence in distribution) Let $\xi, \xi_1, \xi_2, \ldots$ be a.s. bounded random measures on a Polish space S. Then these two conditions are equivalent:*

(i) $\xi_n \xrightarrow{wd} \xi$,

(ii) $\xi_n f \xrightarrow{d} \xi f$ for all $f \in \hat{C}_+(S)$.

If S is locally compact, then (i) is also equivalent to each of the following conditions:

(iii) $\xi_n \xrightarrow{vd} \xi$ and $\xi_n S \xrightarrow{d} \xi S$,

(iv) $\xi_n \xrightarrow{vd} \xi$ and $(\hat{E}\xi_n)$ is tight,

(v) $\xi_n \xrightarrow{vd} \xi$ on $\overline{S}$.

Proof: If $\xi_n \xrightarrow{wd} \xi$, then $\xi_n f \xrightarrow{d} \xi f$ for all $f \in \hat{C}_+(S)$ by continuous mapping, and $(\hat{E}\xi_n)$ is tight by Prohorov's theorem and Lemma A2.2. Hence, (i) implies conditions (ii)–(v).

Now assume instead condition (ii). Then the Cramér–Wold theorem (FMP 5.5) yields $(\xi_n f, \xi_n S) \xrightarrow{d} (\xi f, \xi S)$ for all $f \in \hat{C}_+$, and so $\hat{\xi}_n f \xrightarrow{d} \hat{\xi} f$ for the same functions f, where $\hat{\xi} = \xi/(\xi S \vee 1)$ and $\hat{\xi}_n = \xi_n/(\xi_n S \vee 1)$, as before. Since $\hat{\xi} S \leq 1$ and $\hat{\xi}_n S \leq 1$ for all n, it follows that $E\hat{\xi}_n \xrightarrow{w} E\hat{\xi}$, and so by Lemma A2.1 the sequence $(E\hat{\xi}_n)$ is weakly tight. Since also $\xi_n S \xrightarrow{d} \xi S$, Lemma A2.2 shows that the random sequence (ξ_n) is tight with respect to the weak topology. Then Prohorov's theorem (FMP 16.3) shows that (ξ_n) is also relatively compact in distribution. In other words, any sub-sequence $N' \subset \mathbb{N}$ has a further sub-sequence N'', such that $\xi_n \xrightarrow{wd} \eta$ along N'' for some random measure η on S with $\eta S < \infty$ a.s. Thus, for any $f \in \hat{C}_+$, we have both $\xi_n f \xrightarrow{d} \xi f$ and $\xi_n f \xrightarrow{d} \eta f$, and then also $\xi f \overset{d}{=} \eta f$. Applying the Cramér–Wold theorem once again, we obtain

$$(\xi f_1, \ldots, \xi f_n) \overset{d}{=} (\eta f_1, \ldots, \eta f_n), \quad f_1, \ldots, f_n \in \hat{C}_+.$$

By a monotone-class argument we conclude that $\xi \overset{d}{=} \eta$, on the σ-field $\mathcal{C}$ generated by the mappings $\mu \mapsto \mu f$ for arbitrary $f \in \hat{C}_+$. Noting that $\mathcal{C}$ agrees with the Borel σ-field on $\hat{\mathcal{M}}(S)$ since the latter space is separable, we obtain $\xi \overset{d}{=} \eta$. This shows that $\xi_n \overset{wd}{\longrightarrow} \xi$ along N'', and (i) follows since N' was arbitrary.

Now suppose that S is locally compact. In each of the cases (iii)–(v) it suffices to show that (ξ_n) is weakly tight, since the required weak convergence will then follow as before, by means of Prohorov's theorem. First we assume condition (iii). For any compact set $K \subset S$, continuous function f on S with $0 \le f \le 1_K$, and measure $\mu \in \hat{\mathcal{M}}(S)$, we note that

$$1 - e^{-\mu K^c} \le 1 - e^{-\mu(1-f)} \le e^{\mu S}(e^{-\mu f} - e^{-\mu S}).$$

Letting $c > 0$ with $\xi S \ne c$ a.s., we obtain

$$
\begin{aligned}
E(1 - e^{-\xi_n K^c}) &\le\; E[1 - e^{-\xi_n K^c}; \xi_n S \le c] + P\{\xi_n S > c\} \\
&\le\; e^c E(e^{-\xi_n f} - e^{-\xi_n S}) + P\{\xi_n S > c\} \\
&\to\; e^c E(e^{-\xi f} - e^{-\xi S}) + P\{\xi S > c\}.
\end{aligned}
$$

Here the right-hand side tends to 0 as $f \uparrow 1$ and then $c \to \infty$, which shows that $(\hat{E}\xi_n)$ is tight. The required tightness of (ξ_n) now follows by Lemma A2.2.

Now assume condition (iv). By the tightness of $(\hat{E}\xi_n)$, we may choose some compact sets $K_k \subset S$ such that

$$E[\xi_n K_k^c \wedge 1] \le 2^{-k}, \quad k, n \in \mathbb{N}.$$

Next, for every $k \in \mathbb{N}$, we may choose some $f_k \in C_K(S)$ with $f_k \ge 1_{K_k}$. Since $\xi_n K_k \le \xi_n f_k \overset{d}{\to} \xi f_k$, the sequences $(\xi_n K_k)$ are tight in $\mathbb{R}_+$, and we may choose some constants $c_k > 0$ such that

$$P\{\xi_n K_k > c_k\} \le 2^{-k}, \quad k, n \in \mathbb{N}.$$

Letting $r_k = c_k + 1$, we get by combination

$$
\begin{aligned}
P\{\xi_n S > r_k\} &\le\; P\{\xi_n K_k > c_k\} + P\{\xi_n K_k^c > 1\} \\
&\le\; 2^{-k} + 2^{-k} = 2^{-k+1},
\end{aligned}
$$

which shows that the sequence $(\xi_n S)$ is tight in $\mathbb{R}_+$. By Lemma A2.2 it follows that (ξ_n) is weakly tight in $\hat{\mathcal{M}}(S)$.

Finally, we assume condition (v). If S is compact, then $\overline{S} = S$, and the weak and vague topologies coincide. Otherwise, we have $\infty \in K^c$ in $\overline{S}$ for $K \subset \mathcal{K}(S)$, and every $f \in C_K(S)$ may be extended to a continuous function $\bar{f}$ on $\overline{S}$ satisfying $\bar{f}(\infty) = 0$. By continuous mapping,

$$\xi_n f = \xi_n \bar{f} \overset{d}{\to} \xi \bar{f} = \xi f, \quad f \in C_K(S),$$

and so $\xi_n \overset{vd}{\longrightarrow} \xi$ on S by FMP 16.16. Since also $1 \in C_K(\overline{S})$, we get

$$\xi_n S = \xi_n \overline{S} \overset{d}{\to} \xi \overline{S} = \xi S.$$

This reduces the proof to the case of (iii). $\square$

The following lemma is often useful to extend a convergence criterion from the ergodic to the general case.

Lemma A2.4 *(randomization) Let* $\mu, \mu_1, \mu_2, \ldots$ *be probability kernels between two metric spaces* S *and* T *such that* $s_n \to s$ *in* S *implies* $\mu_n(s_n, \cdot) \xrightarrow{w} \mu(s, \cdot)$ *in* $\mathcal{M}_1(T)$. *Then for any random elements* $\xi, \xi_1, \xi_2, \ldots$ *in* S *with* $\xi_n \xrightarrow{d} \xi$, *we have* $E\mu_n(\xi_n, \cdot) \xrightarrow{w} E\mu(\xi, \cdot)$. *This remains true when the constants* $s, s_1, s_2, \ldots$ *and random elements* $\xi, \xi_1, \xi_2, \ldots$ *are restricted to some measurable subsets* $S_0, S_1, \ldots \subset S$.

Proof: For any bounded, continuous function f on T, the integrals μf and $\mu_n f$ are bounded, measurable functions on S such that $s_n \to s$ implies $\mu_n f(s_n) \to \mu f(s)$. Hence, the continuous-mapping theorem in FMP 4.27 yields $\mu_n f(\xi_n) \xrightarrow{d} \mu f(\xi)$, and so $E\mu_n f(\xi_n) \to E\mu f(\xi)$. The assertion follows since f was arbitrary. The extended version follows by the same argument from the corresponding extension of FMP 4.27. $\square$

We also need the following elementary tightness criterion.

Lemma A2.5 *(hyper-contraction and tightness) Consider some random variables* $\xi_1, \xi_2, \ldots \geq 0$, *$\sigma$-fields* $\mathcal{F}_1, \mathcal{F}_2, \ldots$, *and constant* $c \in (0, \infty)$ *such that a.s.*

$$E[\xi_n^2 | \mathcal{F}_n] \leq c(E[\xi_n | \mathcal{F}_n])^2 < \infty, \quad n \in \mathbb{N}.$$

Then, assuming (ξ_n) *to be tight, so is the sequence* $\eta_n = E[\xi_n | \mathcal{F}_n]$, $n \in \mathbb{N}$.

Proof: Fix any r, ε, and $p_1, p_2, \ldots \in (0, 1)$ with $p_n \to 0$. Using the Paley–Zygmund inequality in FMP 4.1, we have a.s. on $\{\eta_n > 0\}$

$$
\begin{aligned}
0 \;<\; \frac{(1-r)^2}{c} &\leq (1-r)^2 \frac{(E[\xi_n | \mathcal{F}_n])^2}{E[\xi_n^2 | \mathcal{F}_n]} \\
&\leq P[\xi_n > r\eta_n | \mathcal{F}_n] \\
&\leq P[p_n \xi_n > r\varepsilon | \mathcal{F}_n] + 1\{p_n \eta_n < \varepsilon\}, \quad (1)
\end{aligned}
$$

which is also trivially true when $\eta_n = 0$. The tightness of (ξ_n) yields $p_n \xi_n \xrightarrow{P} 0$ by the criterion in FMP 4.9, and so the first term on the right of (1) tends to 0 in L^1, hence also in probability. Since the sum is bounded from below, we obtain $1\{p_n \eta_n < \varepsilon\} \xrightarrow{P} 1$, which shows that $p_n \eta_n \xrightarrow{P} 0$. Using FMP 4.9 in the opposite direction, we conclude that even (η_n) is tight. $\square$

A3. Multiple Stochastic Integrals

Multiple Gaussian and Poisson integrals are needed to represent processes with higher-dimensional symmetries. The former are defined, most naturally, on tensor products $\bigotimes_i H_i = H_1 \otimes \cdots \otimes H_d$ of Hilbert spaces, which are understood to be infinite-dimensional and separable, unless otherwise specified. Given an ortho-normal basis (ONB) $h_{i1}, h_{i2}, \ldots$ in each H_i, we recall that the tensor products $\bigotimes_i h_{i,j_i} = h_{1,j_1} \otimes \cdots \otimes h_{d,j_d}$ form an ONB in $\bigotimes_i H_i$. Since the H_i are isomorphic, it is often convenient to take $H_i = H$ for all i, in which case we may write $\bigotimes_i H_i = H^{\otimes d}$.

For any $f \in H_1 \otimes \cdots \otimes H_d$, we define the *supporting spaces* $M_1, \ldots, M_d$ of f to be the smallest closed, linear sub-spaces $M_i \subset H_i$, $i \leq d$, such that $f \in M_1 \otimes \cdots \otimes M_d$. More precisely, the M_i are the smallest closed, linear sub-spaces of H_i such that f has an orthogonal expansion in terms of tensor products $h_1 \otimes \cdots \otimes h_d$ with $h_i \in M_i$ for all i. (Such an expansion is said to be *minimal* if it is based on the supporting spaces M_i.) For a basis-free description of the M_i, write $H_i' = \bigotimes_{j \neq i} H_j$, and define some bounded, linear operators $A_i \colon H_i' \to H_i$ by

$$\langle g, A_i h \rangle = \langle f, g \otimes h \rangle, \quad g \in H_i,\ h \in H_i',\ i \leq d.$$

Then M_i equals $\overline{R(A_i)}$, the closed range of A_i.

The orthogonal representation of an element $f \in H_1 \otimes \cdots \otimes H_d$ clearly depends on the choice of ortho-normal bases in the d spaces. In the two-dimensional case, however, there is a simple diagonal version, which is essentially unique.

Lemma A3.1 *(diagonalization) For any $f \in H_1 \otimes H_2$, there exist a finite or infinite sequence $\lambda_1, \lambda_2, \ldots > 0$ with $\sum_j \lambda_j^2 < \infty$ and some ortho-normal sequences $\varphi_1, \varphi_2, \ldots$ in H_1 and $\psi_1, \psi_2, \ldots$ in H_2, such that*

$$f = \sum_j \lambda_j \left(\varphi_j \otimes \psi_j \right). \tag{1}$$

This representation is unique, apart from joint, orthogonal transformations of the elements φ_j and ψ_j, within sets of indices j with a common value $\lambda_j > 0$. Furthermore, the expansion in (1) is minimal.

Proof: Define a bounded linear operator $A \colon H_2 \to H_1$ and its adjoint $A^* \colon H_1 \to H_2$ by

$$\langle g, A h \rangle = \langle A^* g, h \rangle = \langle f, g \otimes h \rangle, \quad g \in H_1,\ h \in H_2. \tag{2}$$

Then A^*A is a positive, self-adjoint, and compact operator on H_2, and so it has a finite or infinite sequence of eigen-values $\lambda_1^2 \geq \lambda_2^2 \geq \cdots > 0$ with associated ortho-normal eigen-vectors $\psi_1, \psi_2, \ldots \in H_2$. For definiteness, we may choose $\lambda_j > 0$ for all j. It is easy to check that the elements $\varphi_j = \lambda_j^{-1} A \psi_j \in H_1$ are ortho-normal eigen-vectors of the operator AA^* on H_1 with the same eigen-values, and that

$$A \psi_j = \lambda_j \varphi_j, \qquad A^* \varphi_j = \lambda_j \psi_j. \tag{3}$$

This remains true for any extension of the sequences (φ_i) and (ψ_i) to ONBs of H_1 and H_2. Since the tensor products $\varphi_i \otimes \psi_j$ form an ONB in $H_1 \otimes H_2$, we have a representation

$$f = \sum_{i,j} c_{ij} \, (\varphi_i \otimes \psi_j),$$

for some constants c_{ij} with $\sum_{i,j} c_{ij}^2 < \infty$. Using (2) and (3), we obtain

$$c_{ij} = \langle f, \varphi_i \otimes \psi_j \rangle = \langle A\psi_j, \varphi_i \rangle = \lambda_j \langle \varphi_j, \varphi_i \rangle = \lambda_j \delta_{ij},$$

and (1) follows.

Conversely, suppose that (1) holds for some constants $\lambda_j > 0$ and ortho-normal elements $\varphi_j \in H_1$ and $\psi_j \in H_2$. Combining this with (2) yields (3), and so the λ_j^2 are eigen-values of A^*A with associated ortho-normal eigen-vectors ψ_j. This implies the asserted uniqueness of the λ_j and ψ_j. Since any other sets of eigen-vectors ψ_j' and φ_j' must satisfy (3), the two sets are related by a common set of orthogonal transformations. Finally, the minimality of the representation (1) is clear from (3). □

A process η on a Hilbert space H is said to be *iso-normal* if $\eta h \in L^2$ for every $h \in H$ and the mapping η preserves inner products, so that $E(\eta h \, \eta k) = \langle h, k \rangle$. By a *G-process* on H we mean an iso-normal, centered Gaussian process on H. More generally, we define a *continuous, linear, random functional (CLRF)* on H as a process η on H such that

$$\eta(ah + bk) = a\eta h + b\eta k \quad \text{a.s.}, \quad h, k \in H, \ a, b \in \mathbb{R},$$
$$\eta h_n \xrightarrow{P} 0, \qquad \|h_n\| \to 0.$$

We need the following basic existence and uniqueness result, valid for multiple stochastic integrals based on independent or identical G-processes.

Theorem A3.2 (*multiple Gaussian integrals, Wiener, Itô*) *Let $\eta_1, \ldots, \eta_d$ be independent G-processes on some Hilbert spaces $H_1, \ldots, H_d$, and fix any $k_1, \ldots, k_d \in \mathbb{N}$. Then there exists an a.s. unique CLRF $\eta = \bigotimes_i \eta^{\otimes k_i}$ on $H = \bigotimes_i H_i^{\otimes k_i}$ such that*

$$\left(\bigotimes_i \eta_i^{\otimes k_i} \right) \left(\bigotimes_{i,j} f_{ij} \right) = \prod_{i,j} \eta_i f_{ij},$$

whenever the elements $f_{ij} \in H_i$ are orthogonal in j for fixed i. The functional η is L^2-bounded with mean 0.

Proof: The result for $d = 1$ is classical (cf. FMP 13.21). In general, we may introduce the G-process $\zeta = \bigoplus_i \eta_i$ on the Hilbert space $H = \bigoplus_i H_i$ and put $\chi = \zeta^{\otimes k}$, where $k = \sum_i k_i$. The restriction η of χ to $\bigotimes_i H_i^{\otimes k_i}$ has clearly the desired properties. □

We also need some norm estimates for multiple Wiener–Itô integrals.

Lemma A3.3 *(hyper-contraction, Nelson) Let $\eta_1, \ldots, \eta_d$ be independent G-processes on some Hilbert spaces $H_1, \ldots, H_d$, and fix any $p > 0$ and $k_1, \ldots, k_d \in \mathbb{N}$. Then we have, uniformly in f,*

$$\left\| \left(\bigotimes_i \eta_i^{\otimes k_i} \right) f \right\|_p \lesssim \|f\|, \quad f \in \bigotimes_i H_i^{\otimes k_i}.$$

Proof: Considering the G-process $\eta = \eta_1 + \cdots + \eta_d$ in $H = \bigoplus_i H_i$, we may write the multiple integral $(\bigotimes_i \eta_i^{\otimes k_i}) f$ in the form $\eta^{\otimes k} g$ for a suitable element $g \in H^{\otimes k}$ with $\|g\| = \|f\|$, where $k = \sum_i k_i$. It is then enough to prove that, for any $p > 0$ and $n \in \mathbb{N}$,

$$\|\eta^{\otimes n} f\|_p \lesssim \|f\|, \quad f \in H^{\otimes n}. \tag{4}$$

Here we may take $H = L^2(\mathbb{R}_+, \lambda)$ and let η be generated by a standard Brownian motion B on $\mathbb{R}_+$, in which case f may be regarded as an element of $L^2(\lambda^n)$. Considering separately each of the $n!$ tetrahedral regions in $\mathbb{R}_+^n$, we may further reduce to the case where f is supported by the set

$$\Delta_n = \{(t_1, \ldots, t_n) \in \mathbb{R}_+^n; \ t_1 < \cdots < t_n\}.$$

Then $\eta^{\otimes n} f$ can be written as an iterated Itô integral (FMP 18.13)

$$\begin{aligned}
\eta^{\otimes n} f &= \int dB_{t_n} \int dB_{t_{n-1}} \cdots \int f(t_1, \ldots, t_n) \, dB_{t_1} \\
&= \int (\eta^{\otimes(n-1)} \hat{f}_t) \, dB_t,
\end{aligned} \tag{5}$$

where $\hat{f}_{t_n}(t_1, \ldots, t_{n-1}) = f(t_1, \ldots, t_n)$.

For $n = 1$ we note that ηf is $N(0, \|f\|_2^2)$, and therefore

$$\|\eta f\|_p = \|\eta(f/\|f\|_2)\|_p \|f\|_2 \lesssim \|f\|_2, \quad f \in L^2(\lambda),$$

as required. Now suppose that (4) holds for all multiple integrals up to order $n - 1$. Using the representation (5), a BDG-inequality from FMP 17.7, the extended Minkowski inequality in FMP 1.30, the induction hypothesis, and Fubini's theorem, we get for any $p \geq 2$

$$\begin{aligned}
\|\eta^{\otimes n} f\|_p &= \left\| \int (\eta^{\otimes(n-1)} \hat{f}_t) \, dB_t \right\|_p \\
&\lesssim \left\| \int (\eta^{\otimes(n-1)} \hat{f}_t)^2 \, dt \right\|_{p/2}^{1/2} \\
&\leq \left(\int \|\eta^{\otimes(n-1)} \hat{f}_t\|_p^2 \, dt \right)^{1/2} \\
&\lesssim \left(\int \|\hat{f}_t\|_2^2 \, dt \right)^{1/2} = \|f\|_2.
\end{aligned}$$

Taking $p = 2$ and using Jensen's inequality, we get for any $p \leq 2$

$$\|\eta^{\otimes n} f\|_p \leq \|\eta^{\otimes n} f\|_2 \lesssim \|f\|_2, \quad f \in L^2(\lambda^n).$$

This completes the induction and proves (4).   □

Next we construct the tensor product $X \otimes h$ of a CLRF X on H with a fixed element $h \in H$.

Lemma A3.4 *(mixed multiple integrals) For any CLRF X on H and element $h \in H$, there exists an a.s. unique CLRF $X \otimes h$ on $H^{\otimes 2}$ such that*

$$(X \otimes h)(f \otimes g) = (Xf)\langle h, g \rangle \quad a.s., \quad f, g \in H. \tag{6}$$

Proof: For any $h \in H$, define a linear operator A_h from H to $H^{\otimes 2}$ by $A_h f = f \otimes h$, $f \in H$, and note that A_h is bounded since

$$\|A_h f\| = \|f \otimes h\| = \|f\|\,\|h\|, \quad f \in H.$$

The adjoint A_h^* is a bounded, linear operator from $H^{\otimes 2}$ to H, and we may define a CLRF $X \otimes h$ on $H^{\otimes 2}$ by

$$(X \otimes h)f = X A_h^* f, \quad f \in H^{\otimes 2}. \tag{7}$$

For any $f, g, k \in H$, we have

$$\langle k, A_h^*(f \otimes g) \rangle = \langle A_h k, \, f \otimes g \rangle = \langle k \otimes h, \, f \otimes g \rangle = \langle k, f \rangle \langle h, g \rangle,$$

which implies

$$A_h^*(f \otimes g) = \langle h, g \rangle f, \quad f, g \in H.$$

Combining with (7) and using the linearity of X, we obtain (6). To prove the asserted uniqueness, fix any ONB $h_1, h_2, \ldots$ of H and apply (6) to the the tensor products $h_i \otimes h_j$, $i, j \in \mathbb{N}$, which form an ONB in $H^{\otimes 2}$. The required uniqueness now follows by the linearity and continuity of X. $\qquad \square$

We turn our attention to double Poisson integrals of the form $\xi \eta f$ or $\xi^2 f$, where ξ and η are independent Poisson processes on some measurable spaces S and T. We assume the underlying intensity measures $E\xi$ and $E\eta$ to be σ-finite, and to simplify the writing we may take $S = T = \mathbb{R}_+$ and $E\xi = E\eta = \lambda$. The existence poses no problem, since the mentioned integrals can be defined as path-wise, Lebesgue-type integrals with respect to the product measures $\xi \otimes \eta$ and $\xi^2 = \xi \otimes \xi$, respectively. It is less obvious when these integrals converge.

The following result gives necessary and sufficient conditions for the a.s. convergence of the stochastic integrals ξf, $\xi^2 f$, or $\xi \eta f$, where ξ and η are independent, unit rate Poisson processes on $\mathbb{R}_+$ and $f \geq 0$ is a measurable function on $\mathbb{R}_+$ or $\mathbb{R}_+^2$, respectively. Given a measurable function $f \geq 0$ on $\mathbb{R}_+^2$, we define $f_1 = \lambda_2 \hat{f}$ and $f_2 = \lambda_1 \hat{f}$, where $\hat{f} = f \wedge 1$, and $\lambda_i f$ denotes the Lebesgue integral of f in the i-th coordinate. Also put $f^* = \sup_s |f(s)|$, and write λ_D for normalized Lebesgue measure along the diagonal $D = \{(x, y) \in \mathbb{R}_+^2; \, x = y\}$.

Theorem A3.5 *(convergence of Poisson integrals, Kallenberg and Szulga)*
*Let ξ and η be independent, unit rate Poisson processes on $\mathbb{R}_+$. Then for
any measurable function $f \geq 0$ on $\mathbb{R}_+$, we have $\xi f < \infty$ a.s. iff $\lambda \hat{f} < \infty$. If
$f \geq 0$ is instead measurable on $\mathbb{R}_+^2$ and $f_i = \lambda_j \hat{f}$ for $i \neq j$, we have $\xi \eta f < \infty$
a.s. iff these three conditions are fulfilled:*

  (i) $\lambda\{f_i = \infty\} = 0$ *for* $i = 1, 2$,
  (ii) $\lambda\{f_i > 1\} < \infty$ *for* $i = 1, 2$,
  (iii) $\lambda^2[\hat{f}; f_1 \vee f_2 \leq 1] < \infty$.

Finally, $\xi^2 f < \infty$ a.s. iff (i)–(iii) hold and $\lambda_D \hat{f} < \infty$.

Our proof is based on a sequence of lemmas. We begin with some elementary moment formulas, where we write $\psi(t) = 1 - e^{-t}$ for $t \geq 0$.

Lemma A3.6 *(moment identities)* *Let ξ and η be independent, unit rate
Poisson processes on $\mathbb{R}_+$. Then for any measurable set $B \subset \mathbb{R}_+$ or function
$f \geq 0$ on $\mathbb{R}_+$ or $\mathbb{R}_+^2$, we have*

  (i) $E\psi(\xi f) = \psi(\lambda(\psi \circ f))$,
  (ii) $P\{\xi B > 0\} = \psi(\lambda B)$,
  (iii) $E(\xi f)^2 = \lambda f^2 + (\lambda f)^2$, $E\xi\eta f = \lambda^2 f$,
  (iv) $E(\xi\eta f)^2 = \lambda^2 f^2 + \lambda(\lambda_1 f)^2 + \lambda(\lambda_2 f)^2 + (\lambda^2 f)^2$.

Proof: Statements (i) and (ii) appear in FMP 12.2. For claim (iii), note
that $E(\xi B)^2 = \lambda B + (\lambda B)^2$, and extend by linearity, independence, and
monotone convergence. To prove (iv), conclude from (iii) and Fubini's theorem that

$$
\begin{aligned}
E(\xi\eta f)^2 &= E(\xi(\eta f))^2 = E\lambda_1(\eta f)^2 + E(\lambda_1\eta f)^2 \\
&= \lambda_1 E(\eta f)^2 + E(\eta(\lambda_1 f))^2 \\
&= \lambda^2 f^2 + \lambda_1(\lambda_2 f)^2 + \lambda_2(\lambda_1 f)^2 + (\lambda^2 f)^2. \qquad \square
\end{aligned}
$$

We proceed to estimate the tails in the distribution of $\xi\eta f$. For simplicity
we let $f \leq 1$, so that $f_i = \lambda_j f$ for $i \neq j$.

Lemma A3.7 *(tail estimate)* *Let ξ be a unit rate Poisson process on $\mathbb{R}_+$,
and consider a measurable function $f : \mathbb{R}_+^2 \to [0, 1]$ with $\lambda^2 f < \infty$. Then*

$$
P\{\xi\eta f > \tfrac{1}{2}\lambda^2 f\} \gtrsim \psi\left(\frac{\lambda^2 f}{1 + f_1^* \vee f_2^*}\right).
$$

Proof: We may clearly assume that $\lambda^2 f > 0$. By Lemma A3.6 we have
$E\xi\eta f = \lambda^2 f$ and

$$
\begin{aligned}
E(\xi\eta f)^2 &\leq (\lambda^2 f)^2 + \lambda^2 f + f_1^* \lambda f_1 + f_2^* \lambda f_2 \\
&= (\lambda^2 f)^2 + (1 + f_1^* + f_2^*)\lambda^2 f,
\end{aligned}
$$

and so the Paley–Zygmund inequality in FMP 4.1 yields

$$
\begin{aligned}
P\{\xi\eta f > \tfrac{1}{2}\lambda^2 f\} \;&\geq\; \frac{(1 - \tfrac{1}{2})^2(\lambda^2 f)^2}{(\lambda^2 f)^2 + (1 + f_1^* \vee f_2^*)\lambda^2 f} \\
&\gtrsim\; \left(1 + \frac{1 + f_1^* \vee f_2^*}{\lambda^2 f}\right)^{-1} \\
&\gtrsim\; \psi\left(\frac{\lambda^2 f}{1 + f_1^* \vee f_2^*}\right).
\end{aligned}
\qquad\square
$$

Lemma A3.8 *(decoupling) Let ξ and η be independent, unit rate Poisson processes on $\mathbb{R}_+$. Then for any measurable functions $f \geq 0$ on $\mathbb{R}_+^2$ with $f = 0$ on D, we have*

$$
E\psi(\xi^2 f) \asymp E\psi(\xi\eta f).
$$

Proof: We may clearly assume that f is supported by the wedge $W = \{(s,t);\ 0 \leq s < t\}$. It is equivalent to show that

$$
E[(V \cdot \xi)_\infty \wedge 1] \asymp E[(V \cdot \eta)_\infty \wedge 1],
$$

where $V_t = \xi f(\cdot, t) \wedge 1$. Here V is clearly predictable with respect to the right-continuous and complete filtration induced by ξ and η (FMP 25.23). The random time

$$
\tau = \inf\{t \geq 0;\ (V \cdot \eta)_t > 1\}
$$

is then optional (FMP 7.6), and so the process $1\{\tau \geq t\}$ is again predictable (FMP 25.1). Noting that ξ and η are both compensated by λ (FMP 25.25), we get (FMP 25.22)

$$
\begin{aligned}
E[(V \cdot \xi)_\infty;\ \tau = \infty] \;&\leq\; E(V \cdot \xi)_\tau = E(V \cdot \lambda)_\tau = E(V \cdot \eta)_\tau \\
&\leq\; E[(V \cdot \eta)_\infty \wedge 1] + 2P\{\tau < \infty\},
\end{aligned}
$$

and therefore

$$
\begin{aligned}
E[(V \cdot \xi)_\infty \wedge 1] \;&\leq\; E[(V \cdot \xi)_\infty;\ \tau = \infty] + P\{\tau < \infty\} \\
&\leq\; E[(V \cdot \eta)_\infty \wedge 1] + 3P\{\tau < \infty\} \\
&\leq\; 4E[(V \cdot \eta)_\infty \wedge 1].
\end{aligned}
$$

The same argument applies with the roles of ξ and η interchanged. $\square$

Proof of Theorem A3.5: To prove the first assertion, let $f \geq 0$ on $\mathbb{R}_+$. Then Lemma A3.6 (i) shows that $\xi f = \infty$ a.s. iff $\lambda \hat{f} \geq \lambda(\psi \circ f) = \infty$, and so by Kolmogorov's zero-one law we have $\xi f < \infty$ a.s. iff $\lambda \hat{f} < \infty$.

Turning to the second assertion, let $f \geq 0$ on $\mathbb{R}_+^2$. Since $\xi \otimes \eta$ is a.s. simple, we have $\xi\eta f < \infty$ iff $\xi\eta \hat{f} < \infty$ a.s., which allows us to take $f \leq 1$. First assume conditions (i)–(iii). Here (i) yields

$$
E\xi\eta[f;\ f_1 \vee f_2 = \infty] \leq \sum_i \lambda^2[f;\ f_i = \infty] = 0,
$$

and so we may assume that $f_1, f_2 < \infty$. Then the first assertion yields $\eta f(s, \cdot) < \infty$ and $\xi f(\cdot, s) < \infty$ a.s. for every $s \geq 0$. Furthermore, (ii) implies $\xi\{f_1 > 1\} < \infty$ and $\eta\{f_2 > 1\} < \infty$ a.s. By Fubini's theorem we get a.s.

$$\xi\eta[f; \, f_1 \vee f_2 > 1] \leq \xi[\eta f; \, f_1 > 1] + \eta[\xi f; \, f_2 > 1] < \infty,$$

which allows us to assume that even $f_1, f_2 \leq 1$. Then (iii) yields $E\xi\eta f = \lambda^2 f < \infty$, which implies $\xi\eta f < \infty$ a.s.

Conversely, suppose that $\xi\eta f < \infty$ a.s. for some function f into $[0,1]$. By Lemma A3.6 (i) and Fubini's theorem we have

$$E\psi(\lambda\psi(t\eta f)) = E\psi(t\xi\eta f) \to 0, \quad t \downarrow 0,$$

which implies $\lambda\psi(t\eta f) \to 0$ a.s., and hence $\eta f < \infty$ a.e. $\lambda \otimes P$. By the first assertion and Fubini's theorem we get $f_1 = \lambda_2 f < \infty$ a.e. λ, and the symmetric argument yields $f_2 = \lambda_1 f < \infty$ a.e. This proves (i).

Next, Lemma A3.6 (i) yields, on the set $\{f_1 > 1\}$,

$$\begin{aligned} E\psi(\eta f) &= \psi(\lambda_2(\psi \circ f)) \geq \psi((1 - e^{-1})f_1) \\ &\geq \psi(1 - e^{-1}) \equiv c > 0. \end{aligned}$$

Hence, for any measurable set $B \subset \{f_1 > 1\}$,

$$E\lambda[1 - \psi(\eta f); B] \leq (1 - c)\lambda B,$$

and so, by Chebyshev's inequality,

$$\begin{aligned} P\{\lambda\psi(\eta f) < \tfrac{1}{2}c\lambda B\} &\leq P\{\lambda[1 - \psi(\eta f); B] > (1 - \tfrac{1}{2}c)\lambda B\} \\ &\leq \frac{E\lambda[1 - \psi(\eta f); B]}{(1 - \tfrac{1}{2}c)\lambda B} \leq \frac{1 - c}{1 - \tfrac{1}{2}c}. \end{aligned}$$

Since B was arbitrary, we conclude that

$$P\{\lambda\psi(\eta f) \geq \tfrac{1}{2}c\lambda\{f_1 > 1\}\} \geq 1 - \frac{1 - c}{1 - \tfrac{1}{2}c} = \frac{c}{2 - c} > 0.$$

Noting that $\lambda\psi(\eta f) < \infty$ a.s. by the one-dimensional result and Fubini's theorem, we obtain $\lambda\{f_1 > 1\} < \infty$. This, together with the corresponding result for f_2, proves (ii).

Finally, we may apply Lemma A3.7 to the function $f1\{f_1 \vee f_2 \leq 1\}$ to obtain

$$P\{\xi\eta f > \tfrac{1}{2}\lambda^2[f; \, f_1 \vee f_2 \leq 1]\} \geq \psi\left(\tfrac{1}{2}\lambda^2[f; \, f_1 \vee f_2 \leq 1]\right).$$

This implies (iii), since the opposite statement would yield the contradiction $P\{\xi\eta f = \infty\} > 0$.

Now turn to the last assertion. Since $1_D \cdot \xi^2$ has coordinate projections ξ, the one-dimensional result shows that $\xi^2 f 1_D < \infty$ a.s. iff $\lambda_D \hat{f} < \infty$. Thus, we may henceforth assume that $f = 0$ on D. Then Lemma A3.8 yields

$$E\psi(t\xi^2 f) \asymp E\psi(t\xi\eta f), \quad t > 0,$$

and so, as $t \to 0$,

$$P\{\xi^2 f = \infty\} \asymp P\{\xi\eta f = \infty\},$$

which shows that $\xi^2 f < \infty$ a.s. iff $\xi\eta f < \infty$ a.s.   □

A4. Complete Monotonicity

For any infinite sequence $c_0, c_1, \ldots \in \mathbb{R}$, put $\Delta c_k = c_{k+1} - c_k$, and define recursively the higher order differences by

$$\Delta^0 c_k = c_k, \quad \Delta^{n+1} c_k = \Delta(\Delta^n c_k), \qquad k, n \geq 0,$$

where all differences are with respect to k. We say that (c_k) is *completely monotone* if

$$(-1)^n \Delta^n c_k \geq 0, \quad k, n \geq 0. \tag{1}$$

The definition for finite sequences is the same, apart from the obvious restrictions on the parameters k and n.

Next we say that a function $f \colon \mathbb{R}_+ \to \mathbb{R}$ is *completely monotone*, if the sequence $f(nh)$, $n \in \mathbb{Z}_+$, is completely monotone in the discrete-time sense for every $h > 0$. Thus, we require

$$(-1)^n \Delta_h^n f(t) \geq 0, \quad t, h, n \geq 0, \tag{2}$$

where $\Delta_h f(t) = f(t + h) - f(t)$, and the higher order differences are defined recursively by

$$\Delta_h^0 f(t) = f(t), \quad \Delta_h^{n+1} f(t) = \Delta_h(\Delta_h^n f(t)), \qquad t, h, n \geq 0,$$

where all differences are now with respect to t. For functions on $[0, 1]$ the definitions are the same, apart from the obvious restrictions on t, h, and n. When $f \in C^k(\mathbb{R}_+)$ or $C^k[0, 1]$, it is easy to verify by induction that

$$\Delta_h^n f(t) = \int_0^h \cdots \int_0^h f^{(n)}(t + s_1 + \cdots + s_n) \, ds_1 \cdots ds_n,$$

for appropriate values of t, h, and n, where $f^{(n)}$ denotes the nth derivative of f. In this case, (2) is clearly equivalent to the condition

$$(-1)^n f^{(n)}(t) \geq 0, \quad t \in I^\circ, \ n \geq 0. \tag{3}$$

Sometimes it is more convenient to consider the related notion of *absolute monotonicity*, defined as in (1), (2), or (3), respectively, except that the factor $(-1)^n$ is now omitted. Note that a sequence (c_k) or function $f(t)$ is absolutely monotone on a discrete or continuous interval I iff (c_{-k}) or $f(-t)$ is completely monotone on the reflected interval $-I = \{t; \ -t \in I\}$.

We may now state the basic characterizations of completely monotone sequences and functions. For completeness, we include the elementary case of finite sequences, using the notation

$$n^{(k)} = \frac{n!}{(n-k)!} = n(n-1)\cdots(n-k+1), \quad 0 \le k \le n.$$

Theorem A4.1 *(complete monotonicity, Hausdorff, Bernstein)*

(i) *A finite sequence $c_0, \ldots, c_n \in \mathbb{R}$ is completely monotone with $c_0 = 1$ iff there exists a random variable κ in $\{0, \ldots, n\}$ such that*

$$c_k = E\kappa^{(k)}/n^{(k)}, \quad k = 0, \ldots, n.$$

(ii) *An infinite sequence $c_0, c_1, \ldots \in \mathbb{R}$ is completely monotone with $c_0 = 1$ iff there exists a random variable α in $[0, 1]$ such that*

$$c_k = E\alpha^k, \quad k \in \mathbb{Z}_+.$$

(iii) *A function $f: [0, 1] \to \mathbb{R}$ is completely monotone with $f(0) = f(0+) = 1$ iff there exists a random variable κ in $\mathbb{Z}_+$ such that*

$$f(t) = E(1-t)^\kappa, \quad t \in [0, 1].$$

(iv) *A function $f: \mathbb{R}_+ \to \mathbb{R}$ is completely monotone with $f(0) = f(0+) = 1$ iff there exists a random variable $\rho \ge 0$ such that*

$$f(t) = Ee^{-\rho t}, \quad t \ge 0.$$

In each case, the associated distribution is unique.

Proof: See Feller (1971), pp. 223, 225, 439.   □

Next we say that a function $f: \mathbb{R}^d \to \mathbb{C}$ is *non-negative definite* if

$$\sum\nolimits_{h,k} c_h \bar{c}_k f(x_h - x_k) \ge 0, \quad c_1, \ldots, c_d \in \mathbb{C}, \quad x_1, \ldots, x_d \in \mathbb{R}^d.$$

The following result characterizes non-negative definite functions in terms of characteristic functions.

Theorem A4.2 *(non-negative definite functions, Bochner) A function $f: \mathbb{R}^d \to \mathbb{C}$ is continuous and non-negative definite with $f(0) = 1$ iff there exists a random vector ξ in $\mathbb{R}^d$ such that*

$$f(t) = Ee^{it\xi}, \quad t \in \mathbb{R}^d.$$

Proof: See Feller (1971), p. 622, for the case $d = 1$. The proof for $d > 1$ is similar. □

The next result gives a remarkable connection between non-negative definite and completely monotone functions. Given a function f on $\mathbb{R}_+$, we define the functions f_n on $\mathbb{R}^n$ by

$$f_n(x_1, \ldots, x_n) = f(x_1^2 + \cdots + x_n^2), \quad x_1, \ldots, x_n \in \mathbb{R}. \tag{4}$$

Theorem A4.3 *(Fourier and Laplace transforms, Schoenberg) A continuous function $f : \mathbb{R}_+ \to \mathbb{R}$ with $f(0) = 1$ is completely monotone iff the function f_n in (4) is non-negative definite on $\mathbb{R}^n$ for every $n \in \mathbb{N}$.*

Proof: See Schoenberg (1938a), Theorem 2, Donoghue (1969), pp. 201–206, or Berg et al. (1984), pp. 144–148. □

A5. Palm and Papangelou Kernels

Given a random measure ξ on a measurable space $(S, \mathcal{S})$, we define the associated *Campbell measure* C on $S \times \mathcal{M}(S)$ by

$$Cf = E \int f(s, \xi)\, \xi(ds), \quad f \in (S \times \mathcal{M}(S))_+,$$

where $(S \times \mathcal{M}(S))_+$ denotes the class of measurable functions $f \geq 0$ on $S \times \mathcal{M}(S)$. Since ξ is assumed to be a.s. σ-finite, the same thing is true for C, and we may choose f to be strictly positive with $Cf < \infty$, in which case the projection $\nu = (f \cdot C)(\cdot \times \mathcal{M})$ is bounded and satisfies $\nu B = 0$ iff $\xi B = 0$ a.s. for every $B \in \mathcal{S}$. Any σ-finite measure ν with the latter property is called a *supporting measure* for ξ, and we note that ν is unique up to an equivalence, in the sense of mutual absolute continuity. In particular, we may choose $\nu = E\xi$ when the latter measure is σ-finite.

If S is Borel, then so is $\mathcal{M}(S)$, and there exists a kernel $Q = (Q_s)$ from S to $\mathcal{M}(S)$ satisfying the disintegration formula

$$Cf = \int \nu(ds) \int f(s, \mu)\, Q_s(d\mu), \quad f \in (S \times \mathcal{M}(S))_+, \tag{1}$$

or simply $C = \nu \otimes Q$. This can be proved in the same way as the existence of regular conditional distributions (FMP 6.3–4). The *Palm measures* Q_s of ξ are a.e. unique up to a normalization, and they can be chosen to be probability measures—the *Palm distributions* of ξ—iff $\nu = E\xi$ is σ-finite.

When ξ is a point process (i.e. integer-valued), we can also introduce the *reduced Campbell measure* C' on $S \times \mathcal{N}(S)$, given by

$$C'f = E \int f(s, \xi - \delta_s)\, \xi(ds), \quad f \in (S \times \mathcal{N}(S))_+,$$

and consider the disintegration of C' into *reduced Palm measures,*

$$C'f = \int \nu(ds) \int f(s, \mu) Q'_s(d\mu), \quad f \in (S \times \mathcal{N}(S))_+,$$

or $C' = \nu \otimes Q'$. Comparing with (1), we see that

$$\int_B \nu(ds) \int f(\mu) Q'_s(d\mu) = E \int_B f(\xi - \delta_s) \xi(ds)$$
$$= \int_B \nu(ds) \int f(\mu - \delta_s) Q_s(d\mu),$$

which implies $Q'_s = Q_s \circ (\mu - \delta_s)^{-1}$ a.e. ν, in the sense that

$$\int f(\mu) Q'_s(d\mu) = \int f(\mu - \delta_s) Q_s(d\mu), \quad s \in S \text{ a.e. } \nu.$$

If $E\xi$ is σ-finite, then again we may choose $\nu = E\xi$, in which case Q'_s may be thought of as the conditional distribution of $\xi - \delta_s$, given that $\xi\{s\} > 0$.

We may also consider a disintegration of C' in the other variable. This is especially useful when $C'(S \times \cdot) \ll \mathcal{L}(\xi)$, in the sense that $P\{\xi \in A\} = 0$ implies $C'(S \times A) = 0$, since we can then choose the supporting measure on $\mathcal{N}(S)$ to be equal to $\mathcal{L}(\xi)$. Unfortunately, the stated absolute continuity may fail in general, which makes the present theory more complicated. In the following we shall often write $1_B \cdot \xi = 1_B \xi$, for convenience.

Theorem A5.1 *(disintegration kernel, Papangelou, Kallenberg) Let C' be the reduced Campbell measure of a point process ξ on a Borel space S. Then there exists a maximal kernel R from $\mathcal{N}(S)$ to S such that*

$$E \int f(s, \xi) R(\xi, ds) \le C'f, \quad f \in (S \times \mathcal{N}(S))_+, \tag{2}$$

where the random measure $\eta = R(\xi, \cdot)$ is given on $(\text{supp } \xi)^c$ by

$$1_B \eta = \frac{E[1_B \xi; \xi B = 1|1_{B^c}\xi]}{P[\xi B = 0|1_{B^c}\xi]} \text{ a.s. on } \{\xi B = 0\}, \quad B \in \mathcal{S}. \tag{3}$$

If ξ is simple, we have $\eta = 0$ a.s. on $\text{supp } \xi$, and equality holds in (2) iff

$$P[\xi B = 0|1_{B^c}\xi] > 0 \text{ a.s. on } \{\xi B = 1\}, \quad B \in \mathcal{S}, \tag{4}$$

which then remains true on $\{\xi B < \infty\}$.

By the *maximality* of R we mean that, if R' is any other kernel satisfying (2), then $R'(\xi, \cdot) \le R(\xi, \cdot)$ a.s. The maximal solution R is called the *Papangelou kernel* of ξ, and we may refer to the associated random measure $\eta = R(\xi, \cdot)$ on S as the *Papangelou measure*. The requirement (4) for absolute continuity is often referred to as condition (Σ).

Proof: Since C' is σ-finite, we may fix a measurable function $g > 0$ on $S \times \mathcal{N}(S)$ with $C'g < \infty$ and introduce the projection $\nu = (g \cdot C')(S \times \cdot)$ on $\mathcal{N}(S)$. Then ν is a supporting measure of C' in the second variable, in the sense that $C'(S \times M) = 0$ iff $\nu M = 0$ for any measurable set $M \subset \mathcal{N}(S)$. Now consider the Lebesgue decomposition $\nu = \nu_a + \nu_s$ with respect to $\mathcal{L}(\xi)$ (FMP 2.10), and fix a measurable subset $A \subset \mathcal{N}(S)$ with $\nu_s A = 0$ and $P\{\xi \in A^c\} = 0$, so that $\nu_a = 1_A \cdot \nu$. Then $C'(S \times (A \cap \cdot)) \ll \mathcal{L}(\xi)$, in the sense that the left-hand side vanishes for any set M with $P\{\xi \in M\} = 0$, and we may introduce the associated disintegration

$$E \int f(s, \xi)\, R(\xi, ds) = (1_{S \times A} \cdot C')f \le C'f, \quad f \in (S \times \mathcal{N}(S))_+.$$

If R' is any other kernel satisfying (2), then for any f as above,

$$
\begin{aligned}
E \int f(s, \xi)\, R'(\xi, ds) &= E\!\left[\int f(s, \xi)\, R'(\xi, ds);\ \xi \in A\right] \\
&\le C'(1_{S \times A} f) = (1_{S \times A} \cdot C')f \\
&= E \int f(s, \xi)\, R(\xi, ds).
\end{aligned}
$$

Choosing $g > 0$ as before with $C'g < \infty$, we get a similar relationship between the kernels

$$\tilde{R}(\xi, ds) = g(s, \xi) R(\xi, ds), \qquad \tilde{R}'(\xi, ds) = g(s, \xi) R'(\xi, ds).$$

In particular, for any $B \in \mathcal{S}$, we may take

$$f(s, \xi) = 1_B(s)\, 1\{\tilde{R}(\xi, B) < \tilde{R}'(\xi, B)\}$$

to obtain

$$E[\tilde{R}(\xi, B) - \tilde{R}'(\xi, B);\ \tilde{R}(\xi, B) < \tilde{R}'(\xi, B)] \ge 0,$$

which implies $\tilde{R}(\xi, B) \ge \tilde{R}'(\xi, B)$ a.s. Starting from countably many such relations and extending by a monotone-class argument, we conclude that $\tilde{R}(\xi, \cdot) \ge \tilde{R}'(\xi, \cdot)$ and hence $R(\xi, \cdot) \ge R'(\xi, \cdot)$, outside a fixed null set, which establishes the required maximality.

Next we show that

$$1\{\xi B = 0\} \ll P[\xi B = 0 | 1_{B^c}\xi] \quad \text{a.s.,} \quad B \in \mathcal{S}, \tag{5}$$

where a relation $a \ll b$ between two quantities $a, b \ge 0$ means that $b = 0$ implies $a = 0$. Formula (5) follows from the fact that

$$
\begin{aligned}
P\{P[\xi B = 0 | 1_{B^c}\xi] &= 0,\ \xi B = 0\} \\
&= E[P[\xi B = 0 | 1_{B^c}\xi];\ P[\xi B = 0 | 1_{B^c}\xi] = 0] = 0.
\end{aligned}
$$

Fixing any $B \in \mathcal{S}$ and letting $M \subset \mathcal{N}(S)$ be measurable with

$$M \subset M_0 \equiv \{\mu;\ \mu B = 0,\ P[\xi B = 0 | 1_{B^c}\xi \in d\mu] > 0\},$$

we obtain

$$C'(B \times M) = P\{1_{B^c}\xi \in M, \xi B = 1, P[\xi B = 0|1_{B^c}\xi] > 0\}$$
$$\ll E[P[\xi B = 0|1_{B^c}\xi]; 1_{B^c}\xi \in M]$$
$$= P\{\xi B = 0, 1_{B^c}\xi \in M\} \le P\{\xi \in M\},$$

which shows that $C'(B \times \cdot) \ll \mathcal{L}(\xi)$ on M_0. Combining (5) with the maximality of R, we get for any $A \in \mathcal{S} \cap B$

$$E[E[\xi A; \xi B = 1|1_{B^c}\xi]; 1_{B^c}\xi \in M, P[\xi B = 0|1_{B^c}\xi] > 0]$$
$$= E[\xi A; 1_{B^c}\xi \in M, \xi B = 1, P[\xi B = 0|1_{B^c}\xi] > 0]$$
$$= C'(A \times (M \cap M_0)) = E[\eta A; \xi \in (M \cap M_0)]$$
$$= E[\eta A; 1_{B^c}\xi \in M, \xi B = 0, P[\xi B = 0|1_{B^c}\xi] > 0]$$
$$= E[R(1_{B^c}\xi, A); 1_{B^c}\xi \in M, \xi B = 0]$$
$$= E[R(1_{B^c}\xi, A) P[\xi B = 0|1_{B^c}\xi]; 1_{B^c}\xi \in M].$$

Since M was arbitrary, it follows that

$$E[\xi A; \xi B = 1|1_{B^c}\xi] 1\{P[\xi B = 0|1_{B^c}\xi] > 0\}$$
$$= R(1_{B^c}\xi, A) P[\xi B = 0|1_{B^c}\xi],$$

and so by (5) we have

$$\eta A = \frac{E[\xi A; \xi B = 1|1_{B^c}\xi]}{P[\xi B = 0|1_{B^c}\xi]} \quad \text{a.s. on } \{\xi B = 0\},$$

which extends to (3) since both sides are random measures on B.

From this point on, we assume that ξ is simple. Applying (2) to the product-measurable function

$$f(s, \mu) = \mu\{s\}, \quad s \in S, \quad \mu \in \mathcal{N}(S),$$

we obtain

$$E \int \eta\{s\} \xi(ds) = E \int \xi\{s\} \eta(ds) \le C'f$$
$$= E \int (\xi - \delta_s)\{s\} \xi(ds)$$
$$= E \int (\xi\{s\} - 1) \xi(ds) = 0.$$

Hence, $\int \eta\{s\}\xi(ds) = 0$ a.s., which implies $\eta = 0$ a.s. on $\operatorname{supp} \xi$.

Now suppose that (4) is fulfilled. Fix any $B \in \mathcal{S}$ with $\xi B < \infty$ a.s., and consider any measurable subset $M \subset \mathcal{N}(S)$ with $P\{\xi \in M\} = 0$. Assuming first that $M \subset \{\mu B = 0\}$, we get by (4)

$$C'(B \times M) = E \int_B 1\{(\xi - \delta_s)B = 0, 1_{B^c}\xi \in M\} \xi(ds)$$
$$= P\{\xi B = 1, 1_{B^c}\xi \in M\}$$
$$\ll E[P[\xi B = 0|1_{B^c}\xi]; 1_{B^c}\xi \in M]$$
$$= P\{\xi B = 0, 1_{B^c}\xi \in M\}$$
$$\le P\{\xi \in M\} = 0,$$

which shows that $C'(B \times M) = 0$.

Next let $M \subset \{\mu B < m\}$ for some $m \in \mathbb{N}$. Fix a nested sequence of countable partitions of B into measurable subsets B_{nj}, such that any two points $s \neq t$ of B eventually lie in different sets B_{nj}. By the result for $m = 1$ and dominated convergence, we obtain

$$
\begin{aligned}
C'(B \times M) &= \sum_j C'(B_{nj} \times M) \\
&= \sum_j C'(B_{nj} \times (M \cap \{\mu B_{nj} > 0\})) \\
&= \sum_j E \int_{B_{nj}} 1\{\xi - \delta_s \in M,\ \xi B_{nj} > 1\}\, \xi(ds) \\
&\leq m \sum_j P\{\xi B \leq m,\ \xi B_{nj} > 1\} \\
&= mE\Big[\sum_j 1\{\xi B_{nj} > 1\};\ \xi B \leq m\Big] \to 0,
\end{aligned}
$$

which shows that again $C'(B \times M) = 0$. The result extends by monotone convergence, first to the case $M \subset \{\mu B < \infty\}$, and then to $B = S$ for general M. Thus, (4) implies $C'(S \times \cdot) \ll \mathcal{L}(\xi)$.

Conversely, suppose that $C'(S \times \cdot) \ll \mathcal{L}(\xi)$. Then for any $n \in \mathbb{N}$, $B \in \mathcal{S}$, and measurable $M \subset \mathcal{N}(S)$, we have

$$
\begin{aligned}
nP\{\xi B = n,\ 1_{B^c}\xi \in M\} &= E[\xi B;\ \xi B = n,\ 1_{B^c}\xi \in M] \\
&= C'(B \times \{\mu B = n - 1,\ 1_{B^c}\mu \in M\}) \\
&\ll P\{\xi B = n - 1,\ 1_{B^c}\xi \in M\}.
\end{aligned}
$$

Iterating this relation and then summing over n, we obtain

$$
P\{\xi B < \infty,\ 1_{B^c}\xi \in M\} \ll P\{\xi B = 0,\ 1_{B^c}\xi \in M\},
$$

which together with (5) yields

$$
\begin{aligned}
P\{\xi B < \infty,\ &P[\xi B = 0 | 1_{B^c}\xi] = 0\} \\
&\ll P\{\xi B = 0,\ P[\xi B = 0 | 1_{B^c}\xi] = 0\} = 0.
\end{aligned}
$$

This shows that $P[\xi B = 0 | 1_{B^c}\xi] > 0$, a.s. on $\{\xi B < \infty\}$. $\qquad\square$

Historical and Bibliographical Notes

Only publications closely related to topics in the main text are mentioned. No completeness is claimed, and I apologize in advance for inevitable errors and omissions. References to my own papers are indicated by K(·).

1. The Basic Symmetries

The notion of exchangeability was introduced by HAAG (1928), who derived some formulas for finite sequences of exchangeable events, some of which are implicit already in the work of DE MOIVRE (1718–56). Further information about the early development of the subject appears in DALE (1985).

The characterization of infinite, exchangeable sequences as mixtures of i.i.d. sequences was established by DE FINETTI, first (1929, 1930) for random events, and then (1937) for general random variables. The result was extended by HEWITT and SAVAGE (1955) to random elements in a compact Hausdorff space. DUBINS and FREEDMAN (1979) and FREEDMAN (1980) showed by examples that de Finetti's theorem fails, even in its weaker mixing form, without some regularity conditions on the underlying space.

De Finetti's study of exchangeable sequences was continued by many people, including KHINCHIN (1932, 1952), DE FINETTI (1933a,b), DYNKIN (1953), LOÈVE (1960–63), ALDOUS (1982a,b), and RESSEL (1985). OLSHEN (1971, 1973) noted the equivalence of the various σ-fields occurring in the conditional form of the result. The paper by HEWITT and SAVAGE (1955) also contains the celebrated zero-one law named after these authors, extensions of which are given by ALDOUS and PITMAN (1979).

The connection between finite, exchangeable sequences and sampling from a finite population has been noted by many authors. A vast literature deals with comparative results for sampling with or without replacement, translating into comparisons of finite and infinite, exchangeable sequences. A notable result in this area is the inequality of HOEFFDING (1963), which has been generalized by many authors, including ROSÉN (1967) and PATHAK (1974). Our continuous-time version in Proposition 3.19 may be new. Error estimates in the approximation of finite, exchangeable sequences by infinite ones are given by DIACONIS and FREEDMAN (1980b).

For infinite sequences of random variables, RYLL-NARDZEWSKI (1957) noted that the properties of exchangeability and contractability are equivalent. The result fails for finite sequences, as noted by KINGMAN (1978a).

In that case, the relationship between the two notions was investigated in K(2000). Part (i) of Theorem 1.13 was conjectured by IVANOFF and WEBER (personal communication) and proved by the author.

Processes with exchangeable increments were first studied by BÜHLMANN (1960), who proved a version of Theorem 1.19. Alternative approaches appear in ACCARDI and LU (1993) and in FREEDMAN (1996). Exchangeable random measures on $[0, 1]$ were first characterized in K(1973b, 1975a); the corresponding characterizations on the product spaces $S \times \mathbb{R}_+$ and $S \times [0, 1]$ appeared in K(1990a). The relationship between contractable sequences and processes, stated in Theorem 1.23, is quoted from K(2000). General symmetry properties, such as those in Theorems 1.17 and 1.18, were first noted for random measures in K(1975–86). The fact that exchangeability is preserved under composition of independent processes was noted in K(1982).

FELLER (1966–71) noted the connection between exchangeable sequences and HAUSDORFF's (1921) characterization of absolutely monotone sequences. Further discussion on the subject appears in KIMBERLING (1973). The corresponding relationship between exchangeable processes and BERNSTEIN's (1928) characterization of completely monotone functions was explored in K(1972, 1975–86), and independently by DABONI (1975, 1982). Related remarks appear in FREEDMAN (1996). The fact, from K(1973a), that any exchangeable, simple point process on $[0, 1]$ is a mixed binomial process was also noted by both DAVIDSON (1974) and MATTHES, KERSTAN, and MECKE (1974–82). The characterization of mixed Poisson processes by the order-statistics property in Corollary 1.28 (iii) goes back to NAWROTZKI (1962) (see also FEIGIN (1979)). The result in Exercise 12 was explored by RÉNYI (1953), in the context of order statistics. The description of the linear birth (or YULE) process in Exercise 13 is due to KENDALL (1966), and alternative proofs appear in WAUGH (1970), NEUTS and RESNICK (1971), and ATHREYA and NEY (1972).

MAXWELL (1875, 1878) derived the normal distribution for the velocities of the molecules in a gas, assuming spherical symmetry and independence in orthogonal directions. The Gaussian approximation of spherically symmetric distributions on a high-dimensional sphere seems to be due to MAXWELL (op. cit.) and BOREL (1914), though the result is often attributed to POINCARÉ (1912). (See the historical remarks in EVERITT (1974), p. 134, and DIACONIS and FREEDMAN (1987).) Related discussions and error estimates appear in MCKEAN (1973), STAM (1982), GALLARDO (1983), YOR (1985), and DIACONIS and FREEDMAN (1987).

Random sequences and processes with more general symmetries than exchangeability have been studied by many authors, beginning with DE FINETTI himself (1938, 1959). In particular, FREEDMAN (1962–63) obtained the discrete- and continuous-time versions of Theorem 1.31. Alternative proofs and further discussion appear in papers by KELKER (1970), KINGMAN (1972b), EATON (1981), LETAC (1981), and SMITH (1981). The result was later recognized as equivalent to the celebrated theorem of SCHOENBERG

(1938a,b) in classical analysis.

Related symmetries, leading to mixtures of stable distributions, have been studied by many authors, including BRETAGNOLLE, DACUNHA-CASTELLE, and KRIVINE (1966), DACUNHA-CASTELLE (1975), BERMAN (1980), RES-SEL (1985, 1988), and DIACONIS and FREEDMAN (1987, 1990). The underlying characterization of stable distributions goes back to LÉVY (1924, 1925). The general operator version in Theorem 1.37 is quoted from K(1993). Our L^p-invariance is related to a notion of "pseudo-isotropy," studied by MISIEWICZ (1990). For $p \neq 2$, every linear isometry on $L^p[0, 1]$ is essentially of the type employed in the proof of Lemma 1.39, according to a characterization of LAMPERTI, quoted by ROYDEN (1988). In the Hilbert-space case of $p = 2$, a much larger class of isometries is clearly available.

The only full-length survey of exchangeability theory, previously available, is the set of lecture notes by ALDOUS (1985), which also contain an extensive bibliography. The reader may also enjoy the short but beautifully crafted survey article by KINGMAN (1978a). Brief introductions to exchangeability theory appear in CHOW and TEICHER (1997) and in K(2002).

DE FINETTI himself, one of the founders of BAYESian statistics, turned gradually (1972, 1974–75) away from mathematics to become a philosopher of science, developing theories of subjective probability of great originality, where his celebrated representation theorem plays a central role among the theoretical underpinnings. An enthusiastic gathering of converts payed tribute to de Finetti at a 1981 Rome conference, held on the occasion of his 75th birthday. The ensuing proceedings (eds. KOCH and SPIZZICHINO (1982)) exhibit a curious mix of mathematics and philosophy, ranging from abstract probability theory to the subtle art of assigning subjective probabilities to the possible outcomes of a soccer game.

2. Conditioning and Martingales

The first use of martingale methods in exchangeability theory may be credited to LOÈVE (1960–63), who used the reverse martingale property of the empirical distributions to give a short proof of de Finetti's theorem. A related martingale argument had previously been employed by DOOB (1953) to give a simple proof of the strong law of large numbers for integrable, i.i.d. random variables. Though LOÈVE himself (1978) eventually abandoned his martingale approach to exchangeability, the method has subsequently been adopted by many text-book authors. The present characterizations of exchangeability in terms of reverse martingales, stated in Theorems 2.4, 2.12, and 2.20, appear to be new.

Martingale characterizations of special processes go back to LÉVY (1937–54), with his celebrated characterization of Brownian motion. A similar characterization of the Poisson process was dicovered by WATANABE (1964). Local characteristics of semi-martingales were introduced, independently, by JACOD (1975) and GRIGELIONIS (1975, 1977), both of whom used them to characterize processes with independent increments. The criteria for mixed Lévy processes in Theorem 2.14 (i) were obtained by GRIGELIONIS (1975), and a wide range of further extensions and applications may be found in JACOD and SHIRYAEV (1987).

The relationship between exchangeability and martingale theory was explored more systematically in K(1982), which contains the basic discrete- and continuous-time characterizations of exchangeability in terms of strong stationarity and reflection invariance, as well as a primitive version of Theorem 2.15. For finite exchangeable sequences, the relation $\xi_\tau \overset{d}{=} \xi_1$ was also noted, independently, by BLOM (1985). Further semi-martingale characterizations in discrete and continuous time, including some early versions of Theorems 2.13 and 2.15, were established in K(1988a). The paper K(2000) provides improved and extended versions of the same results, gives the basic norm relations for contractable processes in Theorem 2.23, and contains the regularization properties in Theorems 2.13 and 2.25. Martingale characterizations of exchangeable and contractable arrays were recently obtained by IVANOFF and WEBER (1996, 2003, 2004). A property strictly weaker than strong stationarity has been studied by BERTI, PRATELLI, and RIGO (2004).

The general representation of exchangeable processes in Theorem 2.18, originating with K(1972), was first published in K(1973a). The obvious point-wise convergence of the series of compensated jumps was strengthened in K(1974a) to a.s. uniform convergence, after a similar result for Lévy processes had been obtained by FERGUSON and KLASS (1972). Hyper-contraction methods were first applied to exchangeability theory in K(2002).

The discrete super-martingale in Proposition 2.28 was discovered and explored in K(1975c), where it was used, along with some special continuous-time versions, to study stochastic integration with respect to Lévy processes. The special case of $f(x) = x^2$ had been previously considered by DUBINS and FREEDMAN (1965), and a related maximum inequality appears in DUBINS and SAVAGE (1965). The proof of the general result relies on an elementary estimate, due to ESSEEN and VON BAHR (1965). The growth rates in Theorem 2.32, originally derived for Lévy processes by FRISTEDT (1967) and MILLAR (1971, 1972), were extended to more general exchangeable processes in K(1974b). Finally, a version of Proposition 2.33 appears in K(1989b).

Pure and mixed Poisson processes were characterized by SLIVNYAK (1962) and PAPANGELOU (1974b), through the invariance of their reduced Palm distributions. The general characterization of mixed Poisson and binomial processes appeared in K(1972, 1973c), and the version for general random measures was obtained in K(1975a). Motivated by problems in stochastic geometry, PAPANGELOU (1976) also derived related characterizations of suitable Cox processes, in terms of invariance properties of the associated Papangelou kernels. Various extensions and asymptotic results were derived in K(1978a,b; 1983–86).

3. Convergence and Approximation

Central-limit type theorems for sampling from a finite population and for finite or infinite sequences of exchangeable random variables have been established by many authors, including BLUM, CHERNOFF, ROSENBLATT, and TEICHER (1958), BÜHLMANN (1958, 1960), CHERNOFF and TEICHER (1958), ERDÖS and RÉNYI (1959), TEICHER (1960), BIKELIS (1969), MORAN (1973), and KLASS and TEICHER (1987). Asymptotically invariant sampling from stationary and related processes was studied in K(1999a).

Criteria for Poisson convergence have been given by KENDALL (1967), RIDLER-ROWE (1967), and BENCZUR (1968), and convergence to more general limits was considered by HÁJEK (1960) and ROSÉN (1964). More recent developments along those lines include some martingale-type limit theorems of EAGLESON and WEBER (1978), WEBER (1980), EAGLESON (1982), and BROWN (1982), and some extensions to Banach-space valued random variables, due to DAFFER (1984) and TAYLOR, DAFFER, and PATTERSON (1985). Limit theorems for exchangeable arrays were obtained by IVANOFF and WEBER (1992, 1995).

Functional limit theorems for general random walks and Lévy processes go back to the seminal work of SKOROHOD (1957). The first genuine finite-interval results are due to ROSÉN (1964), who derived criteria for convergence to a Brownian bridge, in the context of sampling from a finite population. His results were extended by BILLINGSLEY (1968) to summation processes based on more general exchangeable sequences. HAGBERG (1973), still working in the special context of sampling theory, derived necessary and sufficient conditions for convergence to more general processes. The general convergence criteria for exchangeable sequences and processes, here presented in Sections 3.1 and 3.3, were first developed in K(1973a), along with the general representation theorem for exchangeable processes on $[0, 1]$. (The latter result appeared first in K(1972), written independently of Hagberg's work.)

The restriction and extension results for exchangeable processes, here exhibited in Section 3.5, also originated with K(1973a). In K(1982), the basic convergence criterion for exchangeable processes on $[0, 1]$ was strengthened to a uniform version, in the spirit of SKOROHOD (1957); the even stronger coupling result in Theorem 3.25 (iv) is new. The remaining coupling methods of Section 3.6 were first explored in K(1974b), along with applications to a wide range of path properties for exchangeable processes. The underlying results for Lévy processes, quoted in the text, were obtained by KHINCHIN (1939), FRISTEDT (1967), and MILLAR (1971, 1972).

Though a general convergence theory for exchangeable random measures on $[0, 1]$ or $\mathbb{R}_+$ was developed already in K(1975a), the more general results in Section 3.2, involving random measures on the product spaces $S \times [0, 1]$ and $S \times \mathbb{R}_+$, appear to be new. The one-dimensional convergence criteria of Section 3.5 were first obtained in K(1988c), along with additional results of a similar nature. Finally, an extensive theory for contractable processes was developed in K(2000), including the distributional coupling theorem and related convergence and tightness criteria.

The sub-sequence principles have a rich history, going back to some early attempts to extend the classical limit theorems of probability theory to the context of orthogonal functions and lacunary series. (See GAPOSHKIN (1966) for a survey and extensive bibliography.) For general sequences of random variables, sub-sequence principles associated with the law of large numbers were obtained by RÉVÉSZ (1965) and KOMLÓS (1967). CHATTERJI (1972) stated the sub-sequence property for arbitrary limit theorems as an heuristic principle, and the special cases of the central limit theorem and the law of the iterated logarithm were settled by GAPOSHKIN (1972), BERKES (1974), and CHATTERJI (1974a,b).

In independent developments, RÉNYI and RÉVÉSZ (1963) proved that exchangeable sequences converge in the stable sense, a mode of convergence previously introduced by RÉNYI (1963). Motivated by problems in functional analysis, DACUNHA-CASTELLE (1975) proved that every tight sequence of random variables contains an asymptotically exchangeable sub-sequence. A related Banach-space result was obtained, independently, by FIGIEL and SUCHESTON (1976). The stronger version in Theorem 3.32 is essentially due to ALDOUS (1977), and the present proof follows the approach in ALDOUS (1985).

KINGMAN (unpublished) noted the subtle connection between the two sub-sequence problems, and ALDOUS (1977), in a deep analysis, proved a general sub-sequence principle for broad classes of weak and strong limit theorems. BERKES and PÉTER (1986) proved Theorem 3.35, along with some more refined approximation results, using sequential coupling techniques akin to our Lemma 3.36, previously developed by BERKES and PHILIPP (1979). A further discussion of strong approximation by exchangeable sequences appears in BERKES and ROSENTHAL (1985).

4. Predictable Sampling and Mapping

DOOB (1936) proved the optional skipping property for i.i.d. sequences of random variables, thereby explaining the futility of the gambler's attempts to beat the odds. His paper is historically significant for being possibly the first one to employ general optional times. Modernized versions of the same result are given by DOOB (1953), Theorem III.5.2, and BILLINGSLEY (1986), Theorem 7.1, and some pertinent historical remarks appear in HALMOS (1985), pp. 74–76.

The mentioned result was extended in K(1982) to any finite or infinite, exchangeable sequence, and in K(1988a) the order restriction on the sampling times $\tau_1, \ldots, \tau_m$ was eliminated. Furthermore, the optional skipping property was extended in K(2000) to arbitrary contractable sequences. The cited papers K(1982, 1988a, 2000) also contain continuous-time versions of the same results. The optional skipping and predictable sampling theorems were extended by IVANOFF and WEBER (2003, 2004) to finite or infinite, contractable or exchangeable arrays.

The fluctuation identity of SPARRE-ANDERSEN (1953–54) was "a sensation greeted with incredulity, and the original proof was of an extraordinary intricacy and complexity," according to FELLER (1971). The argument was later simplified by FELLER (op. cit.) and others. The present proof, based on the predictable sampling theorem, is quoted from K(2002), where the result is used to give simple proofs of the arcsine laws for Brownian motion and symmetric Lévy processes.

The time-change reduction of a continuous local martingale to a Brownian motion was discovered, independently, by DAMBIS (1965) and DUBINS and SCHWARZ (1965), and the corresponding multi-variate result was proved by KNIGHT (1970, 1971). The similar reduction of a quasi-left-continuous, simple point process to Poisson was proved, independently, by MEYER (1971) and PAPANGELOU (1972). COCOZZA and YOR (1980) derived some more general reduction theorems of the same type. The general Gauss–Poisson reduction in Theorem 4.5 is taken from K(1990b).

The invariance criteria for the Brownian motion and bridge are quoted from K(1989b). Integrability criteria for strictly stable Lévy processes were first established in K(1975b), and a more careful discussion, covering even the weakly stable case, appears in K(1992a). The time-change representations in Theorem 4.24 were obtained for symmetric processes by ROSIŃSKI and WOYCZYŃSKI (1986), and independently, in the general case, in K(1992a). (Those more general results were first announced in an invited plenary talk of 1984.) The general invariance theorem for stable processes appears to be new.

Apart from the publications already mentioned, there is an extensive literature dealing with Poisson and related reduction and approximation results for point processes, going back to the seminal papers of WATANABE (1964) and GRIGELIONIS (1971). Let us only mention the subsequent papers by KAROUI and LEPELTIER (1977), AALEN and HOEM (1978), KURTZ (1980), BROWN (1982, 1983), MERZBACH and NUALART (1986), PITMAN and YOR (1986), and BROWN and NAIR (1988a,b), as well as the monograph of BARBOUR, HOLST, and JANSON (1992). Our present development in Section 4.5 is based on results in K(1990b).

5. Decoupling Identities

For a random walk $S_n = \xi_1 + \cdots + \xi_n$ on $\mathbb{R}$ and for suitable optional times $\tau < \infty$, the elementary relations $ES_\tau = E\tau\, E\xi$ and $ES_\tau^2 = E\tau\, E\xi^2$ were first noted and explored by WALD (1944, 1945), in connection with his development of sequential analysis. The formulas soon became standard tools in renewal and fluctuation theory. Elementary accounts of the two equations and their numerous applications appear in many many texts, including FELLER (1971) and CHOW and TEICHER (1997).

Many authors have improved on Wald's original statements by relaxing the underlying moment conditions. Thus, BLACKWELL (1946) established the first order relation under the minimal conditions $E\tau < \infty$ and $E|\xi| < \infty$. When $E\xi = 0$ and $E\xi^2 < \infty$, the first order Wald equation $ES_\tau = 0$ remains true under the weaker condition $E\tau^{1/2} < \infty$, as noted by BURKHOLDER and GUNDY (1970), and independently by GORDON (as reported in CHUNG (1974), p. 343). The latter result was extended by CHOW, ROBBINS, and SIEGMUND (1971), who showed that if $E\xi = 0$ and $E|\xi|^p < \infty$ for some $p \in [1, 2]$, then the condition $E\tau^{1/p} < \infty$ suffices for the validity of $ES_\tau = 0$. The ultimate result in this direction was obtained by KLASS (1988), who showed that if $E\xi = 0$, then $ES_\tau = 0$ holds already under the minimal condition $Ea_\tau < \infty$, where $a_n = E|S_n|$.

We should also mention an extension of the first order Wald identity to dependent random variables, established by FRANKEN and LISEK (1982) in the context of Palm distributions. The result also appears in FRANKEN, KÖNIG, ARNDT, and SCHMIDT (1981).

The continuous-time Wald identities $EB_\tau = 0$ and $EB_\tau^2 = E\tau$, where B is a standard, one-dimensional Brownian motion, have been used extensively in the literature, and detailed discussions appear in, e.g., LOÈVE (1978) and KARATZAS and SHREVE (1991). Here the first order relation holds when $E\tau^{1/2} < \infty$ and the second order formula is valid for $E\tau < \infty$. Those equations, along with some more general, first and second order moment identities for the Itô integral $V \cdot B$, follow from the basic martingale properties of stochastic L^2-integrals, first noted and explored by DOOB (1953).

The general decoupling identities for exchangeable sums and integrals on bounded or unbounded index sets were originally obtained in K(1989b), where the connection with predictable sampling was also noted. The tetrahedral moment identities for contractable sums and integrals were first derived, under more restrictive boundedness conditions, in K(2000).

Predictable integration with respect to Lévy processes was essentially covered already by the elementary discussion of the stochastic L^2-integral in DOOB (1953). The theory is subsumed by the more general, but also more sophisticated, semi-martingale theory, as exhibited in Chapter 26 of K(2002). A detailed study of Lévy integrals was undertaken in MILLAR (1972) and K(1975b). The quoted norm conditions, presented here without any claims to optimality, were originally derived in K(1989b), along with similar estimates in the general, exchangeable case.

6. Homogeneity and Reflections

The notions of local homogeneity and reflection invariance were first introduced and studied in K(1982), where versions can be found for random sets of the basic representation theorems. The same paper includes a discussion of the multi-state case and its connections with the Markov property, and it also contains a simple version of the uniform sampling Theorem 6.17.

Excursion theory and the associated notion of local time can be traced back to a seminal paper of LÉVY (1939). Building on Lévy's ideas, ITÔ (1972) showed how the excursion structure of a Markov process can be represented in terms of a Poisson process on the local time scale. Regenerative sets with positive Lebesgue measure were studied extensively by KINGMAN (1972a). An introduction to regenerative processes and excursion theory appears in Chapter 22 of K(2002), and a more detailed exposition is given by DELLACHERIE, MAISONNEUVE, and MEYER (1987, 1992), Chapters 15 and 20.

Exchangeable partitions of finite intervals arise naturally in both theory and applications. In particular, they appear to play a significant role in game theory. Their Hausdorff measure was studied in K(1974b), where results for regenerative sets were extended to the finite-interval case. The associated distributions also arose naturally in K(1981), as Palm measures of regenerative sets. The latter paper further contains discussions, in the regenerative case, of continuity properties for the density of the local time intensity measure $E\xi$. Similar results on a finite interval were obtained in K(1983), where a slightly weaker version of Theorem 6.16 appears.

Both CARLESON and KESTEN, in independent work, proved that a subordinator with zero drift and infinite Lévy measure will a.s. avoid any fixed point in $(0, \infty)$, a result originally conjectured by CHUNG. Their entirely different approaches are summarized in ASSOUAD (1971) and BERTOIN (1996), Theorem III.4. The corresponding result for exchangeable partitions of $[0, 1]$ was established by BERBEE (1981). Our simple proof, restricted to the regular case, is adapted from K(1983). The asymptotic probability for an exchangeable random set in $\mathbb{R}_+$ or $[0, 1]$ to hit a short interval was studied extensively in K(1999c, 2001, 2003), along with the associated conditional distribution. Some extensions of those results to higher dimensions were derived by ELALAOUI-TALIBI (1999).

BLUMENTHAL and GETOOR (1968) noted that the strong Markov property follows from a condition of global homogeneity, in a version for optional times that may take infinite values. (To obtain a true Markov process, in the sense of the usual axiomatic definition, one needs to go on and construct an associated transition kernel, a technical problem addressed by WALSH (1972).) Connections with exchangeability were noted in K(1982), and some related but less elementary results along the same lines, though with homogeneity defined in terms of finite optional times, were established in K(1987) (cf. Theorem 8.23 in K(2002) for a short proof in a special case). In K(1998), the homogeneity and independence components of the strong Markov property were shown, under suitable regularity conditions, to be essentially equivalent. A totally unrelated characterization of mixed Markov chains was noted by DIACONIS and FREEDMAN (1980a) and subsequently studied by ZAMAN (1984, 1986).

7. Symmetries on Arrays

The notion of *partial exchangeability* of a sequence of random variables—the invariance in distribution under a sub-group of permutations—goes back to DE FINETTI (1938, 1959, 1972), and has later been explored by many authors. Separately (or *row-column*) exchangeable arrays $X = (X_{ij})$, first introduced by DAWID (1972), arise naturally in connection with a Bayesian approach to the analysis of variance. Jointly exchangeable, set-indexed arrays $X = (X_{\{i,j\}})$ (often referred to as *weakly exchangeable*) arise naturally in the context of U-statistics, a notion first introduced by HOEFFDING (1948). Dissociated and exchangeable arrays were studied in some early work of MCGINLEY and SIBSON (1975), SILVERMAN (1976), and EAGLESON and WEBER (1978).

A crucial breakthrough in the development of the subject came with the first representation theorems for row-column and weakly exchangeable arrays, established independently by ALDOUS (1981) and HOOVER (1982a), using entirely different methods. While the proof appearing in ALDOUS (1981), also outlined in ALDOUS (1985) and attributed by the author to KINGMAN, is purely probabilistic, the proof of HOOVER (1982a), also outlined in HOOVER (1982b), is based on profound ideas in symbolic logic and non-standard analysis. The underlying symmetry properties were considered much earlier by logicians, such as GAIFMAN (1961) and KRAUSS (1969), who obtained de Finetti-type results in the two cases, from which the functional representations can be derived.

HOOVER (1979), in a formidable, unpublished manuscript, went on to prove the general representation theorems for separately or jointly exchangeable arrays on $\mathbb{N}^d$, $d \geq 1$, using similar techniques from mathematical logic. His paper also provides criteria for the equivalence of two representing functions, corresponding to our condition (iii) in Theorems 7.28 and 7.29. (Condition (ii) was later added in K(1989a, 1992b, 1995).) A probabilistic approach to Hoover's main results, some of them in a slightly extended form, was provided in K(1988b, 1989a, 1992b, 1995). Contractable arrays on $\mathbb{N}$ were studied in K(1992b), where the corresponding functional representation and equivalence criteria were established.

A different type of representation was obtained by DOVBYSH and SU-DAKOV (1982), in the special case of positive definite, symmetric, jointly exchangeable arrays of dimension two. HESTIR (1986) shows how their result can be deduced from Hoover's representation.

Conditional properties of exchangeable arrays have been studied by many authors. In particular, extensions and alternative proofs of the basic Lemma 7.6 have been provided by LYNCH (1984), K(1989a), and HOOVER (1989). The conditional relations presented here were developed in K(1995). Convergence criteria and martingale-type properties for exchangeable arrays have been studied extensively by IVANOFF and WEBER (1992, 1995, 1996, 2003). The relevance of such arrays and their representations, in the contexts of the analysis of variance and Bayesian statistics, has been discussed by KINGMAN (1979), SPEED (1987), and others.

An application to the study of visual perception was considered by DIACONIS and FREEDMAN (1981), who used simulations of exchangeable arrays of zeros and ones to disprove some conjectures in the area, posed by the psychologist JULESZ and his followers. Such considerations lead naturally to the statistical problem of estimating (a version of) the representation function for an exchangeable array, given a single realization of the process. The problem was solved in K(1999b), to the extent that a solution seems at all possible. A non-technical introduction to exchangeable arrays appears in ALDOUS (1985), where more discussion and further references can be found.

Motivated by some problems in population genetics, KINGMAN (1978b, 1982b) studied exchangeable partitions of $\mathbb{N}$ and proved the corresponding special case of the so-called paint-box representation in Theorem 7.38. (The term comes from the mental picture, also suggested by Kingman, of an infinite sequence of objects, painted in randomly selected colors chosen from a possibly infinite paint box, where one considers the partition into classes of objects with the same color.) The present result for general symmetries is new; its proof was inspired by the first of two alternative approaches to Kingman's result suggested by ALDOUS (1985). A continuity theorem for exchangeable partitions appears in KINGMAN (1978c, 1980). Some algebraic and combinatorial aspects of exchangeable partitions have been studied extensively by PITMAN (1995, 2002) and GNEDIN (1997).

8. Multi-variate Rotations

The early developments in this area were motivated by some limit theorems for U-statistics, going back to HOEFFDING (1948). Here the classical theory is summarized in SERFLING (1980), and some more recent results are given by DYNKIN and MANDELBAUM (1983) and MANDELBAUM and TAQQU (1984). Related results for random matrices are considered in HAYAKAWA (1966) and WACHTER (1974).

Jointly rotatable matrices may have been mentioned for the first time in OLSON and UPPULURI (1970, 1973), where a related characterization is proved. Infinite, two-dimensional, separately and jointly rotatable arrays were introduced by DAWID (1977, 1978), who also conjectured the general representation formula in the separate case. The result was subsequently proved by ALDOUS (1981), for dissociated arrays and under a moment condition, restrictions that were later removed in K(1988b). The latter paper also characterizes two-dimensional, jointly rotatable arrays, as well as separately or jointly exchangeable random sheets. It further contains some related uniqueness and continuity criteria.

Independently of K(1988b), HESTIR (1986) derived the representation formula for separately exchangeable random sheets on $\mathbb{R}_+^2$, in the special case of vanishing drift components and under a moment condition, using a characterization of jointly exchangeable, positive definite arrays from DOVBYSH and SUDAKOV (1982). Other, apparently independent, developments include a paper by OLSHANSKI and VERSHIK (1996), where a related representation is established for jointly rotatable, Hermitian arrays.

The general theory of separately or jointly rotatable random arrays and functionals of arbitrary dimension was originally developed in K(1994, 1995), with some crucial ideas adopted from ALDOUS (1981), as indicated in the main text. The paper K(1995) also provides characterizations of exchangeable and contractable random sheets of arbitrary dimension. Brownian sheets and related processes have previously been studied by many authors, including OREY and PRUITT (1973) and ADLER (1981).

9. Symmetric Measures in the Plane

The topic of exchangeable point processes in the plane first arose in discussions with ALDOUS (personal communication, 1979), and a general representation in the ergodic, separately exchangeable case was conjectured in ALDOUS (1985). The five basic representation theorems for exchangeable random measures in the plane were subsequently established in K(1990a). The extension theorem for contractable random measures was proved by CASUKHELA (1997).

Appendices

The general theory of integral representations over extreme points was developed by CHOQUET (1960), and modern expositions of his deep results may be found in ALFSEN (1971) and in DELLACHERIE and MEYER (1983), Chapter 10. The special case of integral representations of invariant distributions has been studied extensively by many authors, including FARRELL (1962), VARADARAJAN (1963), PHELPS (1966), MAITRA (1977), and KERSTAN and WAKOLBINGER (1981). The connection with sufficient statistics was explored by DYNKIN (1978), DIACONIS and FREEDMAN (1984), and LAURITZEN (1989). Our general extremal decompositions are adapted from K(2000). Ergodicity is usually defined in terms of strictly invariant sets, which may require an extra condition on the family of transformations in Lemma A1.2. The randomization and selection Lemmas A1.5 and A1.6 are quoted from K(1988b) and K(2000), respectively, and the simple conservation law in Lemma A1.1 is adapted from K(1990a).

The equivalence of tightness and weak relative compactness, for probability measures on a Polish space, was established by PROHOROV (1956). The general criteria were applied in PROHOROV (1961) to random measures on a compact metric space. Tightness criteria for the vague topology on a Polish space were developed by DEBES, KERSTAN, LIEMANT, and MATTHES (1970–71) and HARRIS (1971), and the corresponding criteria in the locally compact case were noted by JAGERS (1974). The relationship between the weak and vague topologies for bounded random measures on a locally compact space was examined in K(1975–86). The present criterion for weak tightness of bounded random measures on a Polish space is adapted from ALDOUS (1985). The associated convergence criterion may be new.

The theory of multiple integrals with respect to Brownian motion was developed in a seminal paper of ITÔ (1951). The underlying idea is implicit in WIENER's (1938) discussion of chaos expansions of Gaussian functionals. For a modern account, see Chapters 13 and 18 of K(2002). The associated hyper-contraction property was discovered by NELSON (1973), and related developments are surveyed in DELLACHERIE et al. (1992) and KWAPIEŃ and WOYCZYŃSKI (1992). Convergence criteria for Poisson and related integrals were derived by KALLENBERG and SZULGA (1989).

The characterization of completely monotone sequences is due to HAUS-DORFF (1921), and the corresponding results for completely monotone functions on $\mathbb{R}_+$ and $[0, 1]$ were obtained by BERNSTEIN (1928). Simple, probabilistic proofs of these results were given by FELLER (1971), pp. 223–225 and 439–440. BOCHNER (1932) proved his famous characterization of positive definite functions, after a corresponding discrete-parameter result had been noted by HERGLOTZ (1911). Simple proofs of both results appear in FELLER (1971), pp. 622 and 634. The fundamental relationship between completely monotone and positive definite functions was established by SCHOENBERG (1938a,b). Modern proofs and further discussion appear in DONOGHUE (1969) and BERG, CHRISTENSEN, and RESSEL (1984).

The idea of Palm probabilities goes back to PALM (1943), and the modern definition of Palm distributions, via disintegration of the Campbell measure, is due to RYLL-NARDZEWSKI (1961). The Papangelou kernel of a simple point process was introduced in a special case by PAPANGELOU (1974b), and then in general in K(1978a). NGUYEN and ZESSIN (1979) noted the profound connection between point process theory and statistical mechanics, and MATTHES, WARMUTH, and MECKE (1979) showed how the Papangelou kernel can be obtained, under the regularity condition (Σ), through disintegration of the reduced Campbell measure. The present approach, covering even the general case, is based on K(1983–86).

Bibliography

Here I include only publications related, directly or indirectly, to material in the main text, including many papers that are not mentioned in the historical notes. No completeness is claimed, and any omissions are unintentional. No effort has been made to list the huge number of papers dealing with statistical applications. There are also numerous papers devoted to symmetries other than those considered in this book.

AALEN, O.O., HOEM, J.M. (1978). Random time changes for multivariate counting processes. *Scand. Actuarial J.*, 81–101.

ACCARDI, L., LU, Y.G. (1993). A continuous version of de Finetti's theorem. *Ann. Probab.* **21**, 1478–1493.

ADLER, R.J. (1981). *The Geometry of Random Fields.* Wiley, Chichester.

ALDOUS, D.J. (1977). Limit theorems for subsequences of arbitrarily-dependent sequences of random variables. *Z. Wahrsch. verw. Geb.* **40**, 59–82.

— (1981). Representations for partially exchangeable arrays of random variables. *J. Multivar. Anal.* **11**, 581–598.

— (1982a). On exchangeability and conditional independence. In KOCH and SPIZZICHINO (eds.), pp. 165–170.

— (1982b). Partial exchangeability and $\bar{d}$-topologies. *Ibid.*, pp. 23–38.

— (1985). Exchangeability and related topics. In: *École d'Été de Probabilités de Saint-Flour XIII—1983. Lect. Notes in Math.* **1117**, pp. 1–198. Springer, Berlin.

ALDOUS, D.J., PITMAN, J.W. (1979). On the zero-one law for exchangeable events. *Ann. Probab.* **7**, 704–723.

ALFSEN, E.M. (1971). *Compact Convex Sets and Boundary Integrals.* Springer, Berlin.

ANIS, A.A., GHARIB, M. (1980). On the variance of the maximum of partial sums of n-exchangeable random variables with applications. *J. Appl. Probab.* **17**, 432–439.

ASSOUAD, P. (1971). Démonstration de la "Conjecture de Chung" par Carleson. *Séminaire de Probabilités V. Lect. Notes in Math.* **191**, 17–20. Springer, Berlin.

ATHREYA, K.B, NEY, P.E. (1972). *Branching Processes.* Springer, Berlin.

BALASUBRAMANIAN, K, BALAKRISHNAN, N. (1994). Equivalence of relations for order statistics for exchangeable and arbitrary cases. *Statist. Probab. Lett.* **21**, 405–407.

BALLERINI, R. (1994). Archimedean copulas, exchangeability, and max-stability. *J. Appl. Probab.* **31**, 383–390.

BARBOUR, A.D., EAGLESON, G.K. (1983). Poisson approximations for some statistics based on exchangeable trials. *Adv. Appl. Probab.* **15**, 585–600.

BARBOUR, A.D., HOLST, L., JANSON, S. (1992). *Poisson Approximation.* Clarendon Press, Oxford.

BARTFAI, P. (1980). Remarks on exchangeable random variables. *Publ. Math. Debrecen* **27**, 143–148.

BENCZUR, A. (1968). On sequences of equivalent events and the compound Poisson process. *Studia Sci. Math. Hungar.* **3**, 451–458.

BERBEE, H. (1981). On covering single points by randomly ordered intervals. *Ann. Probab.* **9**, 520–528.

BERG, C., CHRISTENSEN, J.P.R., RESSEL, P. (1984). *Harmonic Analysis on Semigroups.* Springer, New York.

BERKES, I. (1974). The law of the iterated logarithm for subsequences of random variables. *Z. Wahrsch. verw. Geb.* **30**, 209–215.

— (1984). Exchangeability and limit theorems for subsequences of random variables. In: *Limit Theorems in Probability and Statistics* (ed. P. RÉVÉSZ), pp. 109–152. North-Holland/Elsevier, Amsterdam.

BERKES, I., PÉTER, E. (1986). Exchangeable random variables and the subsequence principle. *Probab. Th. Rel. Fields* **73**, 395–413.

BERKES, I., PHILIPP, W. (1979). Approximation theorems for independent and weakly dependent random vectors. *Ann. Probab.* **7**, 29–54.

BERKES, I, ROSENTHAL, H.P. (1985). Almost exchangeable sequences of random variables. *Z. Wahrsch. verw. Geb.* **70**, 473–507.

BERMAN, S. (1980). Stationarity, isotropy, and sphericity in l^p. *Z. Wahrsch. verw. Geb.* **54**, 21–23.

BERNSTEIN, S. (1928). Sur les fonctions absolument monotones. *Acta Math.* **52**, 1–66.

BERTI, P., PRATELLI, L., RIGO, P. (2004). Limit theorems for a class of identically distributed random variables. *Ann. Probab.* **32**, 2029–2052.

BERTI, P., RIGO, P. (1997). A Glivenko–Cantelli theorem for exchangeable random variables. *Statist. Probab. Lett.* **32**, 385–391.

BERTOIN, J. (1996). *Lévy Processes.* Cambridge University Press, Cambridge.

— (2001a). Eternal additive coalescents and certain bridges with exchangeable increments. *Ann. Probab.* **29**, 344–360.

— (2001b). Homogeneous fragmentation processes. *Probab. Theor. Rel. Fields* **121**, 301–318.

— (2002). Self-similar fragmentations. *Ann. Inst. H. Poincaré, Sec. B* **38**, 319–340.

BIKELIS, A. (1969). On the estimation of the remainder term in the central limit theorem for samples from finite populations (Russian). *Studia Sci. Math. Hungar.* **4**, 345–354.

BILLINGSLEY, P. (1968). *Convergence of Probability Measures.* Wiley, New York.

— (1986). *Probability and Measure*, 2nd ed. Wiley, New York.

BLACKWELL, D. (1946). On an equation of Wald. *Ann. Math. Statist.* **17**, 84–87.

BLOM, G. (1985). A simple property of exchangeable random variables. *Amer. Math. Monthly* **92**, 491–492.

BLUM, J.R. (1982). Exchangeability and quasi-exchangeability. In KOCH and SPIZZICHINO (eds.), pp. 171–176.

BLUM, J.R., CHERNOFF, H., ROSENBLATT, M., TEICHER, H. (1958). Central limit theorems for interchangeable processes. *Canad. J. Math.* **10**, 222–229.

BLUMENTHAL, R.M., GETOOR, R.K. (1968). *Markov Processes and Potential Theory.* Academic Press, New York.

BOCHNER, S. (1932). *Vorlesungen über Fouriersche Integrale.* Reprint ed., Chelsea, New York 1948.

BOES, D.C., SALAS, L.C.J.D. (1973). On the expected range and expected adjusted range of partial sums of exchangeable random variables. *J. Appl. Probab.* (1973), 671–677.

BOREL, E. (1914). *Introduction géométrique à quelques théories physiques.* Gauthier-Villars, Paris.

BRETAGNOLLE, J, DACUNHA-CASTELLE, D., KRIVINE, J.L. (1966): Lois stables et espaces L^p. *Ann. Inst. H. Poincaré B* **2**, 231–259.

BRETAGNOLLE, J., KLOPOTOWSKI, A. (1995). Sur l'existence des suites de variables aléatoires s à s indépendantes échangeables ou stationnaires. *Ann. Inst. H. Poincaré B* **31**, 325–350.

BROWN, T.C. (1982). Poisson approximations and exchangeable random variables. In KOCH and SPIZZICHINO (eds.), pp. 177–183.

— (1983). Some Poisson approximations using compensators. *Ann. Probab.* **11**, 726–744.

BROWN, T.C., IVANOFF, B.G., WEBER, N.C. (1986). Poisson convergence in two dimensions with application to row and column exchangeable arrays. *Stoch. Proc. Appl.* **23**, 307–318.

BROWN, T.C., NAIR, M.G. (1988a). A simple proof of the multivariate random time change theorem for point processes. *J. Appl. Probab.* **25**, 210–214.

— (1988b). Poisson approximations for time-changed point processes. *Stoch. Proc. Appl.* **29**, 247–256.

BRU, B., HEINICH, H., LOOTGITIER, J.C. (1981). Lois de grands nombres pour les variables échangeables. *C.R. Acad. Sci. Paris* **293**, 485–488.

BÜHLMANN, H. (1958). Le problème "limit centrale" pour les variables aléatoires échangeables. *C.R. Acad. Sci. Paris* **246**, 534–536.

— (1960). Austauschbare stochastische Variabeln und ihre Grenzwertsätze. *Univ. Calif. Publ. Statist.* **3**, 1–35.

BURKHOLDER, D.L., GUNDY, R.F. (1970). Extrapolation and interpolation of quasi-linear operators on martingales. *Acta Math.* **124**, 249–304.

CASUKHELA, K.S. (1997). Symmetric distributions of random measures in higher dimensions. *J. Theor. Probab.* **10**, 759–771.

CHANDA, C. (1971). Asymptotic distribution of sample quantiles for exchangeable random variables. *Calcutta Statist. Assoc. Bull.* **20**, 135–142.

CHATTERJI, S.D. (1972). Un principe de sous-suites dans la théorie des probabilités. *Séminaire de probabilités VI. Lect. Notes in Math.* **258**, 72–89. Springer, Berlin.

— (1974a). A principle of subsequences in probability theory: The central limit theorem. *Adv. Math.* **13**, 31–54.

— (1974b). A subsequence principle in probability theory II. The law of the iterated logarithm. *Invent. Math.* **25**, 241–251.

CHERNOFF, H., TEICHER, H. (1958). A central limit theorem for sums of interchangeable random variables. *Ann. Math. Statist.* **29**, 118–130.

CHOQUET, G. (1960). Le théorème de représentation intégrale dans les ensembles convexes compacts. *Ann. Inst. Fourier* **10**, 333–344.

CHOW, Y.S., ROBBINS, H., SIEGMUND, D. (1971). *Great Expectations: The Theory of Optimal Stopping.* Houghton Mifflin, New York.

CHOW, Y.S., TEICHER, H. (1997). *Probability Theory: Independence, Interchangeability, Martingales,* 3rd ed. Springer, New York (1st ed. 1978).

CHUNG, K.L. (1974). *A Course in Probability Theory,* 2nd ed. Academic Press, New York.

COCOZZA, C., YOR, M. (1980). Démonstration d'un théorème de F. Knight à l'aide de martingales exponentielles. *Lect. Notes in Math.* **784**, 496–499. Springer, Berlin.

DABONI, L. (1975). Caratterizzazione delle successioni (funzioni) completamente monotone in termini di rappresentabilità delle funzioni di sopravvivenza di particolari intervalli scambiabili tra successi (arrivi) contigui. *Rend. Math.* **8**, 399–412.

— (1982). Exchangeability and completely monotone functions. In KOCH and SPIZZICHINO (eds.), pp. 39–45.

DABROWSKI, A.R. (1988). Strassen-type invariance principles for exchangeable sequences. *Statist. Probab. Lett.* **7**, 23–26.

DACUNHA-CASTELLE, D. (1975). Indiscernability and exchangeability in L^p spaces. *Aarhus Proceedings* **24**, 50–56.

— (1982). A survey on exchangeable random variables in normed spaces. In KOCH and SPIZZICHINO (eds.), pp. 47–60.

DAFFER, P.Z. (1984). Central limit theorems for weighted sums of exchangeable random elements in Banach spaces. *Stoch. Anal. Appl.* **2**, 229–244.

DALE, A.I. (1985). A study of some early investigations into exchangeability. *Hist. Math.* **12**, 323–336.

DAMBIS, K.E. (1965). On the decomposition of continuous submartingales. *Theory Probab. Appl.* **10**, 401–410.

DARAS, T. (1997). Large and moderate deviations for the empirical measures of an exchangeable sequence. *Statist. Probab. Lett.* **36**, 91–100.

— (1998). Trajectories of exchangeable sequences: Large and moderate deviations results. *Statist. Probab. Lett.* **39**, 289–304.

DAVIDSON, R. (1974). Stochastic processes of flats and exchangeability. In: *Stochastic Geometry* (eds. E.F. HARDING, D.G. KENDALL), pp. 13–45. Wiley, London.

DAWID, A.P. (1972). Contribution to discussion of Lindley and Smith (1972), *J. Roy. Statist. Soc. Ser. B* **34**, 29–30.

— (1977). Spherical matrix distributions and a multivariate model. *J. Roy. Statist. Soc.* (B) **39**, 254–261.

— (1978). Extendibility of spherical matrix distributions. *J. Multivar. Anal.* **8**, 559–566.

DEBES, H., KERSTAN, J., LIEMANT, A., MATTHES, K. (1970–71). Verallgemeinerung eines Satzes von Dobrushin I, III. *Math. Nachr.* **47**, 183–244, **50**, 99–139.

DELLACHERIE, C., MAISONNEUVE, B., MEYER, P.A. (1983, 1987, 1992). *Probabilités et Potentiel*, Vols. **3–5**, Hermann, Paris.

DIACONIS, P. (1977). Finite forms of de Finetti's theorem on exchangeability. *Synthese* **36**, 271–281.

— (1988). Recent progress on de Finetti's notions of exchangeability. In: *Bayesian Statistics* (eds. J.M. BERNARDO, M.H. DEGROOT, D.V. LINDLEY, A.F.M. SMITH), **3**, 111–125. Oxford Univ. Press.

DIACONIS, P., EATON, M.L., LAURITZEN, S.L. (1992). Finite de Finetti theorems in linear models and multivariate analysis. *Scand. J. Statist.* **19**, 289–315.

DIACONIS, P., FREEDMAN, D.A. (1980a). De Finetti's theorem for Markov chains. *Ann. Probab.* **8**, 115–130.

— (1980b). Finite exchangeable sequences. *Ann. Probab.* **8**, 745–764.

— (1980c). De Finetti's generalizations of exchangeability. In: *Studies in Inductive Logic and Probability*, pp. 233–249. Univ. Calif. Press.

— (1981). On the statistics of vision: the Julesz conjecture. *J. Math. Psychol.* **24**, 112–138.

— (1982). De Finetti's theorem for symmetric location families. *Ann. Statist.* **10**, 184–189.

— (1984). Partial exchangeability and sufficiency. In: *Statistics: Applications and New Directions* (eds. J.K. GHOSH, J. ROY), pp. 205–236. Indian Statistical Institute, Calcutta.

— (1987). A dozen de Finetti-style results in search of a theory. *Ann. Inst. H. Poincaré* **23**, 397–423.

— (1988). Conditional limit theorems for exponential families and finite versions of de Finetti's theorem. *J. Theor. Probab.* 1, 381–410.

— (1990). Cauchy's equation and de Finetti's theorem. *Scand. J. Statist.* **17**, 235–250.

DINWOODIE, I.H., ZABELL, S.L. (1992). Large deviations for exchangeable random vectors. *Ann. Probab.* **20**, 1147–1166.

DONOGHUE, W.F. (1969). *Distributions and Fourier Transforms*. Academic Press, New York.

DOOB, J.L. (1936). Note on probability. *Ann. Math.* (2) **37**, 363–367.

— (1953). *Stochastic Processes*. Wiley, New York.

DOVBYSH, L.N., SUDAKOV, V.N. (1982). Gram–de Finetti matrices. *J. Soviet Math.* **24**, 3047–3054.

DUBINS, L.E. (1982). Towards characterizing the set of ergodic probabilities. In KOCH and SPIZZICHINO (eds.), pp. 61–74.

— (1983). Some exchangeable probabilities are singular with respect to all presentable probabilities. *Z. Wahrsch. verw. Geb.* **64**, 1–5.

DUBINS, L.E., FREEDMAN, D.A. (1965). A sharper form of the Borel–Cantelli lemma and the strong law. *Ann. Math. Statist.* **36**, 800–807.

— (1979). Exchangeable processes need not be mixtures of independent, identically distributed random variables. *Z. Wahrsch. verw. Geb.* **48**, 115–132.

DUBINS, L.E., SAVAGE, L.J. (1965), A Tchebycheff-like inequality for stochastic processes. *Proc. Nat. Acad. Sci. USA* **53**, 274–275.

DUBINS, L.E., SCHWARZ, G. (1965). On continuous martingales. *Proc. Natl. Acad. Sci. USA* **53**, 913–916.

DYNKIN, E.B. (1953). Classes of equivalent random variables (in Russian). *Uspehi Mat. Nauk (8)* **54**, 125–134.

— (1978). Sufficient statistics and extreme points. *Ann. Probab.* **6**, 705–730.

DYNKIN, E.B., MANDELBAUM, A. (1983). Symmetric statistics, Poisson point processes and multiple Wiener integrals. *Ann. Statist.* **11**, 739–745.

EAGLESON, G.K. (1979). A Poisson limit theorem for weakly exchangeable events. *J. Appl. Probab.* **16**, 794–802.

— (1982). Weak limit theorems for exchangeable random variables. In KOCH and SPIZZICHINO (eds.), pp. 251–268.

EAGLESON, G.K., WEBER, N.C. (1978). Limit theorems for weakly exchangeable arrays. *Math. Proc. Cambridge Phil. Soc.* **84**, 123–130.

EATON, M. (1981). On the projections of isotropic distributions. *Ann. Statist.* **9**, 391–400.

ELALAOUI-TALIBI, H. (1999). Multivariate Palm measure duality for exchangeable interval partitions. *Stoch. Stoch. Reports* **66**, 311–328.

ELALAOUI-TALIBI, H., CASUKHELA, K.S. (2000). A note on the multivariate local time intensity of exchangeable interval partitions. *Statist. Probab. Lett.* **48**, 269–273.

ERDÖS, P., RÉNYI, A. (1959). On the central limit theorem for samples from a finite population. *Publ. Math. Inst. Hungar. Acad. Sci. A* **4**, 49–61.

ESSEEN, C.G., VON BAHR, B. (1965). Inequalities for the rth absolute moment of a sum of random variables, $1 \le r \le 2$. *Ann. Math. Statist.* **36**, 299–303.

ETEMADI, N., KAMINSKI, M. (1996). Strong law of large numbers for 2-exchangeable random variables. *Statist. Probab. Letters* **28**, 245–250.

EVANS, S.N., ZHOU, X. (1999). Identifiability of exchangeable sequences with identically distributed partial sums. *Elec. Comm. Probab.* **4**, 9–13.

EVERITT, C.W.F. (1974). *James Clerk Maxwell*. Scribner's, New York.

FARRELL, R.H. (1962). Representation of invariant measures. *Ill. J. Math.* **6**, 447–467.

FEIGIN, P. (1979). On the characterization of point processes with the order statistic property. *J. Appl. Probab.* **16**, 297–304.

FELLER, W. (1966, 1971). *An Introduction of Probability Theory and its Applications*, Vol. **2**, 1st and 2nd eds. Wiley, New York.

FERGUSON, T.S., KLASS, M.J. (1972). A representation of independent increment processes without Gaussian component. *Ann. Math. Statist.* **43**, 1634–1643.

FIGIEL, T, SUCHESTON, L. (1976). An application of Ramsey sets in analysis. *Adv. Math.* **20**, 103–105.

FINETTI, B. DE (1929). Fuzione caratteristica di un fenomeno aleatorio. In: *Atti Congr. Int. Mat., Bologna, 1928* (ed. ZANICHELLI) **6**, 179–190.

— (1930). Fuzione caratteristica di un fenomeno aleatorio. *Mem. R. Acc. Lincei* (6) **4**, 86–133.

— (1933a). Classi di numeri aleatori equivalenti. *Rend. Accad. Naz. Lincei* **18**, 107–110.

— (1933b). La legge dei grandi numeri nel caso dei numeri aleatori equivalenti. *Rend. Accad. Naz. Lincei* **18**, 279–284.

— (1937). La prévision: ses lois logiques, ses sources subjectives. *Ann. Inst. H. Poincaré* **7**, 1–68. Engl. transl.: *Studies in Subjective Probability* (eds. H.E. KYBURG, H.E. SMOKLER), pp. 99–158. Wiley, New York, 2nd ed. 1980.

— (1938). Sur la condition d'équivalence partielle. *Act. Sci. Ind.* **739**, 5–18. Engl. transl.: *Studies in Inductive Logic and Probability, II* (ed. R.C. JEFFREY). Univ. California Press, Berkeley.

— (1959). La probabilità e la statistica nei rapporti con l'induzione, secondo i diversi punti di vista. In: *Atti Corso CIME su Induzione e Statistica*, Varenna, pp. 1–115.

— (1972). *Probability, Induction and Statistics* (collected papers in English translation). Wiley, New York.

— (1974–75). *Theory of Probability, I–II*. Wiley, New York.

FORTINI, S., LADELLI, L., REGAZZINI, E. (1996). A central limit problem for partially exchangeable random variables. *Theor. Probab. Appl.* **41**, 224–246.

FRANKEN, P., KÖNIG, D., ARNDT, U., SCHMIDT, V. (1981). *Queues and Point Processes*. Akademie-Verlag, Berlin.

FRANKEN, P., LISEK, B. (1982). On Wald's identity for dependent variables. *Z. Wahrsch. verw. Geb.* **60**, 134–150.

FRÉCHET, M. (1943). Les probabilités associées à un système d'événements compatibles et dépendants, II. *Actual. Scient. Indust.* **942**. Hermann, Paris.

FREEDMAN, D.A. (1962–63). Invariants under mixing which generalize de Finetti's theorem. *Ann. Math. Statist.* **33**, 916–923; **34**, 1194–1216.

— (1980). A mixture of independent identically distributed random variables need not admit a regular conditional probability given the exchangeable σ-field. *Z. Wahrsch. verw. Geb.* **51**, 239–248.

— (1996). De Finetti's theorem in continuous time. In: *Statistics, Probability and Game Theory* (eds. T.S. FERGUSON, L.S. SHAPLEY, J.B. MACQUEEN), pp. 83–98. Inst. Math. Statist., Hayward, CA.

FRISTEDT, B.E. (1967). Sample function behavior of increasing processes with stationary, independent increments. *Pacific J. Math.* **21**, 21–33.

GAIFMAN, H. (1961). Concerning measures in first order calculi. *Israel J. Math.* **2**, 1–18.

GALAMBOS, J. (1982). The role of exchangeability in the theory of order statistics. In KOCH and SPIZZICHINO (eds.), pp. 75–86.

GALLARDO, L. (1983). Au sujet du contenu probabiliste d'un lemma d'Henri Poincaré. *Ann. Univ. Clermont* **69**, 192–197.

GAPOSHKIN, V.F. (1966). Lacunary series and independent functions. *Russian Math. Surveys* **21/6**, 3–82.

— (1972). Convergence and limit theorems for sequences of random variables. *Theory Probab. Appl.* **17**, 379–399.

GNEDIN, A.V. (1995). On a class of exchangeable sequences. *Statist. Probab. Lett.* **25**, 351–355.

— (1996). On a class of exchangeable sequences. *Statist. Probab. Lett.* **28**, 159–164.

— (1997). The representation of composition structures. *Ann. Probab.* **25**, 1437–1450.

GRIFFITHS, R.C., MILNE, R.K. (1986). Structure of exchangeable infinitely divisible sequences of Poisson random vectors. *Stoch. Proc. Appl.* **22**, 145–160.

GRIGELIONIS, B (1971). On representation of integer-valued random measures by means of stochastic integrals with respect to the Poisson measure (in Russian). *Litov. Mat. Sb.* **11**, 93–108.

— (1975). The characterization of stochastic processes with conditionally independent increments. *Litovsk. Mat. Sb.* **15**, 53–60.

— (1977). Martingale characterization of stochastic processes with independent increments. *Litovsk. Mat. Sb.* **17**, 75–86.

GUERRE, S. (1986). Sur les suites presque échangeables dans L^q, $1 \leq q < 2$. *Israel J. Math.* **56**, 361–380.

GUGLIELMI, A., MELILLI, E. (2000). Approximating de Finetti's measures for partially exchangeable sequences. *Statist. Probab. Lett.* **48**, 309–315.

HAAG, J. (1928). Sur un problème général de probabilités et ses diverses applications. In: *Proc. Int. Congr. Math., Toronto 1924*, pp. 659–674.

HAGBERG, J. (1973). Approximation of the summation process obtained by sampling from a finite population. *Theory Probab. Appl.* **18**, 790–803.

HÁJEK, J. (1960). Limiting distributions in simple random sampling from a finite population. *Magyar Tud. Akad. Mat. Kutató Int. Közl.* **5**, 361–374.

HALMOS, P.R. (1985). *I want to be a Mathematician.* Springer, New York.

HARRIS, T.E. (1971). Random measures and motions of point processes. *Z. Wahrsch. verw. Geb.* **18**, 85–115.

HAUSDORFF, F. (1921). Summationsmethoden und Momentfolgen. *Math. Z.* **9**, 280–299.

HAYAKAWA, T. (1966). On the distribution of a quadratic form in a multivariate normal sample. *Ann. Inst. Statist. Math.* **18**, 191–201.

HAYAKAWA, Y. (2000). A new characterization property of mixed Poisson processes via Berman's theorem. *J. Appl. Probab.* **37**, 261–268.

HEATH, D., SUDDERTH, W. (1976). De Finetti's theorem for exchangeable random variables. *Amer. Statistician* **30**, 188–189.

HERGLOTZ, G. (1911). Über Potenzreihen mit positivem, reelem Teil im Einheitskreis. *Ber. verh. Sächs. Ges. Wiss. Leipzig, Math.-Phys. Kl.* **63**, 501–511.

HESTIR, K. (1986). The Aldous representation theorem and weakly exchangeable non-negative definite arrays. Ph.D. dissertation, Statistics Dept., Univ. of California, Berkeley.

HEWITT, E., SAVAGE, L.J. (1955). Symmetric measures on Cartesian products. *Trans. Amer. Math. Soc.* **80**, 470–501.

HILL, B.M., LANE, D., SUDDERTH, W. (1987). Exchangeable urn processes. *Ann. Probab.* **15**, 1586–1592.

HIRTH, U., RESSEL, P. (1999). Random partitions by semigroup methods. *Semigr. For.* **59**, 126–140.

HOEFFDING, W. (1948). A class of statistics with asymptotically normal distributions. *Ann. Math. Statist.* **19**, 293–325.

— (1963). Probability estimates for sums of bounded random variables. *J. Amer. Statist. Assoc.* **58**, 13–30.

HOOVER, D.N. (1979). Relations on probability spaces and arrays of random variables. Preprint, Institute of Advanced Study, Princeton.

— (1982a). A normal form theorem for the probability logic $L_{\omega_1 P}$, with applications. *J. Symbolic Logic* **47**.

— (1982b). Row-column exchangeability and a generalized model for probability. In KOCH and SPIZZICHINO (eds.), pp. 281–291.

— (1989). Tail fields of partially exchangeable arrays. *J. Multivar. Anal.* **31**, 160–163.

HSU, Y.S. (1979). A note on exchangeable events. *J. Appl. Probab.* **16**, 662–664.

HU, T.C. (1997). On pairwise independent and independent exchangeable random variables. *Stoch. Anal. Appl.* **15**, 51–57.

HU, Y.S. (1979). A note on exchangeable events. *J. Appl. Probab.* **16**, 662–664.

HUANG, W.J., SU, J.C. (1999). Reverse submartingale property arising from exchangeable random variables. *Metrika* **49**, 257–262.

HUDSON, R.L., MOODY, G.R. (1976). Locally normal symmetric states and an analogue of de Finetti's theorem. *Z. Wahrsch. verw. Geb.* **33**, 343–351.

ITÔ, K. (1951). Multiple Wiener integral. *J. Math. Soc. Japan* **3**, 157–169.

— (1972). Poisson point processes attached to Markov processes. *Proc. 6th Berkeley Symp. Math. Statist. Probab.* **3**, 225–239. Univ. of California Press, Berkeley.

IVANOFF, B.G., WEBER, N.C. (1992). Weak convergence of row and column exchangeable arrays. *Stoch. Stoch. Rep.* **40**, 1–22.

— (1995). Functional limit theorems for row and column exchangeable arrays. *J. Multivar. Anal.* **55**, 133–148.

— (1996). Some characterizations of partial exchangeability. *J. Austral. Math. Soc.* (A) **61**, 345–359.

— (2003). Spreadable arrays and martingale structures. *J. Austral. Math. Soc.*, to appear.

— (2004). Predictable sampling for partially exchangeable arrays. *Statist. Probab. Lett.* **70**, 95–108.

JACOD, J. (1975). Multivariate point processes: Predictable projection, Radon–Nikodym derivative, representation of martingales. *Z. Wahrsch. verw. Geb.* **31**, 235–253.

JACOD, J., SHIRYAEV, A.N. (1987). *Limit Theorems for Stochastic Processes.* Springer, Berlin.

JAGERS, P. (1974). Aspects of random measures and point processes. *Adv. Probab. Rel. Topics* **3**, 179–239. Marcel Dekker, New York.

JIANG, X., HAHN, M.G. (2002). Empirical central limit theorems for exchangeable random variables. *Statist. Probab. Lett.* **59**, 75–81.

JOHNSON, W.B., SCHECHTMAN, G., ZINN, J. (1985). Best constants in moment inequalities for linear combinations of independent and exchangeable random variables. *Ann. Probab.* **13**, 234–253.

KALLENBERG, O. (1972). Ph.D. dissertation, Math. Dept., Chalmers Univ. of Technology, Gothenburg, Sweden.

— (1973a). Canonical representations and convergence criteria for processes with interchangeable increments. *Z. Wahrsch. verw. Geb.* **27**, 23–36.

— (1973b). A canonical representation of symmetrically distributed random measures. In: *Mathematics and Statistics, Essays in Honour of Harald Bergström* (eds. P. JAGERS, L. RÅDE), pp. 41–48. Chalmers Univ. of Technology, Gothenburg.

— (1973c). Characterization and convergence of random measures and point processes. *Z. Wahrsch. verw. Geb.* **27**, 9–21.

— (1974a). Series of random processes without discontinuities of the second kind. *Ann. Probab.* **2**, 729–737.

— (1974b). Path properties of processes with independent and interchangeable increments. *Z. Wahrsch. verw. Geb.* **28**, 257–271.

— (1975a). On symmetrically distributed random measures. *Trans. Amer. Math. Soc.* **202**, 105–121.

— (1975b). On the existence and path properties of stochastic integrals. *Ann. Probab.* **3**, 262–280.

— (1975c). Infinitely divisible processes with interchangeable increments and random measures under convolution. *Z. Wahrsch. verw. Geb.* **32**, 309–321.

— (1975–86). *Random Measures*, 1st to 4th eds. Akademie-Verlag and Academic Press, Berlin and London.

— (1978a). On conditional intensities of point processes. *Z. Wahrsch. verw. Geb.* **41**, 205–220.

— (1978b). On the asymptotic behavior of line processes and systems of non-interacting particles. *Z. Wahrsch. verw. Geb.* **43**, 65–95.

— (1981). Splitting at backward times in regenerative sets. *Ann. Probab.* **9**, 781–799.

— (1982). Characterizations and embedding properties in exchangeability. *Z. Wahrsch. verw. Geb.* **60**, 249–281.

— (1983). The local time intensity of an exchangeable interval partition. In: *Probability and Statistics, Essays in Honour of Carl-Gustav Esseen* (eds. A. GUT, L. HOLST), pp. 85–94. Uppsala University.

— (1983–86). *Random Measures*, 3rd and 4th eds. Akademie-Verlag and Academic Press, Berlin and London.

— (1987). Homogeneity and the strong Markov property. *Ann. Probab.* **15**, 213–240.

— (1988a). Spreading and predictable sampling in exchangeable sequences and processes. *Ann. Probab.* **16**, 508–534.

— (1988b). Some new representations in bivariate exchangeability. *Probab. Th. Rel. Fields* **77**, 415–455.

— (1988c). One-dimensional uniqueness and convergence criteria for exchangeable processes. *Stoch. Proc. Appl.* **28**, 159–183.

— (1989a). On the representation theorem for exchangeable arrays. *J. Multivar. Anal.* **30**, 137–154.

— (1989b). General Wald-type identities for exchangeable sequences and processes. *Probab. Th. Rel. Fields* **83**, 447–487.

— (1990a). Exchangeable random measures in the plane. *J. Theor. Probab.* **3**, 81–136.

— (1990b). Random time change and an integral representation for marked stopping times. *Probab. Th. Rel. Fields* **86**, 167–202.

— (1992a). Some time change representations of stable integrals, via predictable transformations of local martingales. *Stoch. Proc. Appl.* **40**, 199–223.

— (1992b). Symmetries on random arrays and set-indexed processes. *J. Theor. Probab.* **5**, 727–765.

— (1993). Some linear random functionals characterized by L^p-symmetries. In: *Stochastic Processes, a Festschrift in Honour of Gopinath Kallianpur* (eds. S. CAMBANIS, J.K. GHOSH, R.L. KARANDIKAR, P.K. SEN), pp. 171–180. Springer, Berlin.

— (1994). Multiple Wiener–Itô integrals and a multivariate version of Schoenberg's theorem. In: *Chaos Expansions, Multiple Wiener–Itô Integrals and their Applications* (eds. C. HOUDRÉ, V. PÉREZ-ABREU), pp. 73–86. CRC Press, Boca Raton.

— (1995). Random arrays and functionals with multivariate rotational symmetries. *Probab. Th. Rel. Fields* **103**, 91–141.

— (1998). Components of the strong Markov property. In: *Stochastic Processes and Related Topics: In Memory of Stamatis Cambanis, 1943–1995* (eds. I. KARATZAS, B.S. RAJPUT, M.S. TAQQU), pp. 219–230. Birkhäuser, Boston.

— (1999a). Asymptotically invariant sampling and averaging from stationary-like processes. *Stoch. Proc. Appl.* **82**, 195–204.

— (1999b). Multivariate sampling and the estimation problem for exchangeable arrays. *J. Theor. Probab.* **12**, 859–883.

— (1999c). Palm measure duality and conditioning in regenerative sets. *Ann. Probab.* **27**, 945–969.

— (2000). Spreading-invariant sequences and processes on bounded index sets. *Probab. Th. Rel. Fields* **118**, 211–250.

— (2001). Local hitting and conditioning in symmetric interval partitions. *Stoch. Proc. Appl.* **94**, 241–270.

— (2002). *Foundations of Modern Probability*, 2nd ed. Springer, New York (1st ed. 1997).

— (2003). Palm distributions and local approximation of regenerative processes. *Probab. Th. Rel. Fields* **125**, 1–41.

KALLENBERG, O., SZULGA, J. (1989). Multiple integration with respect to Poisson and Lévy processes. *Probab. Th. Rel. Fields* **83**, 101–134.

KARATZAS, I., SHREVE, S. (1991). *Brownian Motion and Stochastic Calculus*, 2nd ed. Springer, New York.

KAROUI, N., LEPELTIER, J.P. (1977). Représentation des processus ponctuels multivariés à l'aide d'un processus de Poisson. *Z. Wahrsch. verw. Geb.* **39**, 111–133.

KELKER, D. (1970). Distribution theory of spherical distributions and a location-scale parameter generalization. *Sankhyā* A **32**, 419–430.

KENDALL, D.G. (1966). Branching processes since 1873. *J. London Math. Soc.* **41**, 385–406.

— (1967). On finite and infinite sequences of exchangeable events. *Studia Sci. Math. Hung.* **2**, 319–327.

KERSTAN, J., WAKOLBINGER, A. (1981). Ergodic decomposition of probability laws. *Z. Wahrsch. verw. Geb.* **56**, 399–414.

KHINCHIN, A.Y. (1932). Sur les classes d'événements équivalents. *Mat. Sbornik* **33**, 40–43.

— (1939). Sur la croissance locale des processus stochastiques homogènes à accroissements indépendants (in Russian with French summary). *Izv. Akad. Nauk SSSR, Ser. Mat.* **3**, 487–508.

— (1952). On classes of equivalent events (in Russian). *Dokl. Akad. Nuak SSSR* **85**, 713–714.

KIMBERLING, C.H. (1973). Exchangeable events and completely monotonic sequences. *Rocky Mtn. J.* **3**, 565–574.

KINGMAN, J.F.C. (1972a). *Regenerative Phenomena*. Wiley, London.

— (1972b). On random sequences with spherical symmetry. *Biometrika* **59**, 492–493.

— (1978a). Uses of exchangeability. *Ann. Probab.* **6**, 183–197.

— (1978b). The representation of partition structures. *J. London Math. Soc.* **18**, 374–380.

— (1978c). Random partitions in population genetics. *Proc. R. Soc. Lond. A* **361**, 1–20.

— (1979). Contribution to discussion on: The reconciliation of probability assessments. *J. Roy. Statist. Soc. Ser. A* **142**, 171.

— (1980). *The Mathematics of Genetic Diversity*. SIAM, Philadelphia.

— (1982a). The coalescent. *Stoch. Proc. Appl.* **13**, 235–248.

— (1982b). Exchangeability and the evolution of large populations. In KOCH and SPIZZICHINO (eds.), pp. 97–112.

KLASS, M.J. (1988). A best possible improvement of Wald's equation. *Ann. Probab.* **16**, 840–853.

KLASS, M.J., TEICHER, H. (1987). The central limit theorem for exchangeable random variables without moments. *Ann. Probab.* **15**, 138–153.

KNIGHT, F.B. (1970). An infinitesimal decomposition for a class of Markov processes. *Ann. Math. Statist.* **41**, 1510–1529.

— (1971). A reduction of continuous, square-integrable martingales to Brownian motion. *Lect. Notes in Math.* **190**, 19–31. Springer, Berlin.

— (1996). The uniform law for exchangeable and Lévy process bridges. In: *Hommage à P.A. Meyer et J. Neveu. Astérisque* **236**, 171–188.

KOCH, G, SPIZZICHINO, F. (eds.) (1982). *Exchangeability in Probability and Statistics*. North-Holland, Amsterdam.

KOMLÓS, J. (1967). A generalization of a problem of Steinhaus. *Acta Math. Acad. Sci. Hungar.* **18**, 217–229.

KRAUSS, P.H. (1969). Representations of symmetric probability models. *J. Symbolic Logic* **34**, 183–193.

KRICKEBERG, K. (1974). Moments of point processes. In: *Stochastic Geometry* (eds. E.F. HARDING, D.G. KENDALL), pp. 89–113. Wiley, London.

KURITSYN, Y.G. (1984). On monotonicity in the law of large numbers for exchangeable random variables. *Theor. Probab. Appl.* **29**, 150–153.

— (1987). On strong monotonicity of arithmetic means of exchangeable random variables. *Theor. Probab. Appl.* **32**, 165–166.

KURTZ, T.G. (1980). Representations of Markov processes as multiparameter time changes. *Ann. Probab.* **8**, 682–715.

KWAPIEŃ, S., WOYCZYŃSKI, W.A. (1992). *Random Series and Stochastic Integrals: Single and Multiple*. Birkhäuser, Boston.

LAURITZEN, S.L. (1989). *Extremal Families and Systems of Sufficient Statistics*. *Lect. Notes in Statist.* **49**. Springer, Berlin.

LEFÉVRE, C., UTEV, S. (1996). Comparing sums of exchangeable Bernoulli random variables. *J. Appl. Probab.* **33**, 285–310.

LETAC, G. (1981). Isotropy and sphericity: some characterizations of the normal distribution. *Ann. Statist.* **9**, 408–417.

LÉVY, P. (1924). Théorie des erreurs. La lois de Gauss et les lois exceptionelles. *Bull. Soc. Math. France* **52**, 49–85.

— (1925). *Calcul des Probabilités*. Gauthier-Villars, Paris.

— (1937–54). *Théorie de l'Addition des Variables Aléatoires*, 1st and 2nd eds. Gauthier-Villars, Paris.

— (1939). Sur certain processus stochastiques homogènes. *Comp. Math,* **7**, 283–339.

LOÈVE, M. (1960–63). *Probability Theory*, 2nd and 3rd eds. Van Nostrand, Princeton, NJ.

— (1978). *Probability Theory*, Vol. **2**, 4th ed. Springer, New York.

LYNCH, J. (1984). Canonical row-column-exchangeable arrays. *J. Multivar. Anal.* **15**, 135–140.

MAITRA, A. (1977). Integral representations of invariant measures. *Trans. Amer. Math. Soc.* **229**, 209–225.

MANDELBAUM, A., TAQQU, M.S. (1983). Invariance principle for symmetric statistics. *Ann. Statist.* **12**, 483–496.

MANDREKAR, V., PATTERSON, R.F. (1993). Limit theorems for symmetric statistics of exchangeable random variables. *Statist. Probab. Lett.* **17**, 157–161.

MATTHES, K., KERSTAN, J., MECKE, J. (1974–82). *Infinitely Divisible Point Processes*. Wiley, Chichester 1978 (German ed., Akademie-Verlag, Berlin 1974; Russian ed., Nauka, Moscow 1982).

MATTHES, K., WARMUTH, W., MECKE, J. (1979). Bemerkungen zu einer Arbeit von Nguyen Xuan Xanh and Hans Zessin. *Math. Nachr.* **88**, 117–127.

MAXWELL, J.C. (1875). *Theory of Heat*, 4th ed. Longmans, London.

— (1878). On Boltzmann's theorem on the average distribution of energy in a system of material points. *Trans. Cambridge Phil. Soc.* **12**, 547.

McGINLEY, W.G., SIBSON, R. (1975). Dissociated random variables. *Math. Proc. Cambridge Phil. Soc.* **77**, 185–188.

McKEAN, H.P. (1973). Geometry of differential space. *Ann. Probab.* **1**, 197–206.

MERZBACH, E., NUALART, D. (1986). A characterization of the spatial Poisson process and changing time. *Ann. Probab.* **14**, 1380–1390.

MEYER, P.A. (1971). Démonstration simplifiée d'un théorème de Knight. *Lect. Notes in Math.* **191**, 191–195. Springer, Berlin.

MILLAR, P.W. (1971). Path behavior of processes with stationary independent increments. *Z. Wahrsch. verw. Geb.* **17**, 53–73.

— (1972). Stochastic integrals and processes with stationary independent increments. *Proc. 6th Berkeley Symp. Math. Statist. Probab.* **3**, 307–331. Univ. of California Press, Berkeley.

MISIEWICZ, J. (1990). Pseudo isotropic measures. *Nieuw Arch. Wisk.* **8**, 111–152.

MOIVRE, A. DE (1718–56). *The Doctrine of Chances; or, a Method of Calculating the Probability of Events in Play*, 3rd ed. (post.) Reprint, F. Case and Chelsea, London, NY 1967.

MORAN, P.A.P. (1973). A central limit theorem for exchangeable variates with geometrical applications. *J. Appl. Probab.* **10**, 837–846.

MOSCOVICI, E., POPESCU, O. (1973). A theorem on exchangeable random variables. *Stud. Cerc. Mat.* **25**, 379–383.

NAWROTZKI, K. (1962). Ein Grenzwertsatz für homogene zufällige Punktfolgen (Verallgemeinerung eines Satzes von A. Rényi). *Math. Nachr.* **24**, 201–217.

NEKKACHI, M.J. (1994). Weak convergence for weakly exchangeable arrays. *C.R. Acad. Sci. Ser. Math.* **319**, 717–722.

NELSON, E. (1973). The free Markov field. *J. Funct. Anal.* **12**, 211–227.

NEUTS, M., RESNICK, S. (1971). On the times of birth in a linear birth process. *J. Austral. Math. Soc.* **12**, 473–475.

NGUYEN X.X., ZESSIN, H. (1979). Integral and differential characterizations of the Gibbs process. *Math. Nachr.* **88**, 105–115.

OLSHANSKY, G.I., VERSHIK, A.M. (1996). Ergodic unitarily invariant measures on the space of infinite Hermitian matrices. In: *Contemporary Mathematical Physics*, 137–175. *AMS Transl. Ser.* 2, **175**, Providence, RI.

OLSHEN, R.A. (1971). The coincidence of measure algebras under an exchangeable probability. *Z. Wahrsch. verw. Geb.* **18**, 153–158.

— (1973). A note on exchangeable sequences. *Z. Wahrsch. verw. Geb.* **28**, 317–321.

OLSON, W.H., UPPULURI, V.R.R. (1970). Characterization of the distribution of a random matrix by rotational invariance. *Sankhyā Ser. A* **32**, 325–328.

— (1973). Asymptotic distribution of eigenvalues of random matrices. *Proc. 6th Berkeley Symp. Math. Statist. Probab.* **3**, 615–644.

OREY, S., PRUITT, W.E. (1973). Sample functions of the N-parameter Wiener process. *Ann. Probab.* **1**, 138–163.

PALM, C. (1943). Intensitätsschwankungen in Fernsprechverkehr. *Ericsson Technics* **44**, 1–189. Engl. trans., *North-Holland Studies in Telecommunication* **10**, Elsevier 1988.

PAPANGELOU, F. (1972). Integrability of expected increments of point processes and a related random change of scale. *Trans. Amer. Math. Soc.* **165**, 486–506.

— (1974a). On the Palm probabilities of processes of points and processes of lines. In: *Stochastic Geometry* (eds. E.F. HARDING, D.G. KENDALL), pp. 114–147. Wiley, London.

— (1974b). The conditional intensity of general point processes and an application to line processes. *Z. Wahrsch. verw. Geb.* **28**, 207–226.

— (1976). Point processes on spaces of flats and other homogeneous spaces. *Math. Proc. Cambridge Phil. Soc.* **80**, 297–314.

PATHAK, P.K. (1974). An extension of an inequality of Hoeffding. *Bull. Inst. Math. Statist.* **3**, 156.

PATTERSON, R.F. (1989). Strong convergence for U-statistics in arrays of rowwise exchangeable random variables. *Stoch. Anal. Appl.* **7**, 89–102.

PATTERSON, R.F., TAYLOR, R.L., INOUE, H. (1989). Strong convergence for sums of randomly weighted, rowwise exchangeable random variables. *Stoch. Anal. Appl.* **7**, 309–323.

PHELPS, R.R. (1966). *Lectures on Choquet's Theorem.* Van Nostrand, Princeton, NJ.

PITMAN, J.W. (1978). An extension of de Finetti's theorem. *Adv. Appl. Probab.* **10**, 268–270.

— (1995). Exchangeable and partially exchangeable random partitions. *Probab. Th. Rel. Fields* **102**, 145–158.

— (2002). Combinatorial stochastic processes. In: *École d'Été de Probabilités de Saint-Flour. Lect. Notes in Math.* Springer, Berlin (to appear).

PITMAN, J.W., YOR, M. (1986). Asymptotic laws of planar Brownian motion. *Ann. Probab.* **14**, 733–779.

POINCARÉ, H. (1912). *Calcul des Probabilités.* Gauthier-Villars, Paris.

PROHOROV, Y.V. (1956). Convergence of random processes and limit theorems in probability theory. *Th. Probab. Appl.* **1**, 157–214.

— (1961). Random measures on a compactum. *Soviet Math. Dokl.* **2**, 539–541.

PRUSS, A.R. (1998). A maximal inequality for partial sums of finite exchangeable sequences of random variables. *Proc. Amer. Math. Soc.* **126**, 1811–1819.

RAO, C.R., SHANBHAG, D.N. (2001). Exchangeability, functional equations, and characterizations. In: *Handbook of Statistics* **19** (eds. D.N. SHANBHAG, C.R. RAO), 733–763. Elsevier/North-Holland, New York.

REGAZZINI, E., SAZONOV, V.V. (1997). On the central limit problem for partially exchangeable random variables with values in a Hilbert space. *Theor. Probab. Appl.* **42**, 656–670.

RÉNYI, A. (1953). On the theory of order statistics. *Acta Math. Acad. Sci. Hung.* **4**, 191–231.

— (1963). On stable sequences of events. *Sankhyā, Ser. A* **25**, 293–302.

RÉNYI, A., RÉVÉSZ, P. (1963). A study of sequences of equivalent events as special stable sequences. *Publ. Math. Debrecen* **10**, 319–325.

RESSEL, P. (1985). De Finetti-type theorems: an analytical approach. *Ann. Probab.* **13**, 898–922.

— (1988). Integral representations for distributions of symmetric stochastic processes. *Probab. Th. Rel. Fields* **79**, 451–467.

— (1994). Non-homogeneous de Finetti-type theorems. *J. Theor. Probab.* **7**, 469–482.

RÉVÉSZ, P. (1965). On a problem of Steinhaus. *Acta Math. Acad. Sci. Hungar.* **16**, 310–318.

RIDLER-ROWE, C.J. (1967). On two problems on exchangeable events. *Studia Sci. Math. Hung.* **2**, 415–418.

ROBINSON, J., CHEN, K.H. (1989). Limit theorems for standardized partial sums of weakly exchangeable arrays. *Austral. J. Statist.* **31**, 200–214.

ROGERS, L.C.G., WILLIAMS, D. (1994). *Diffusions, Markov Processes, and Martingales*, **I**, 2nd ed. Wiley, Chichester.

ROMANOWSKA, M. (1983). Poisson theorems for triangle arrays of exchangeable events. *Bull. Acad. Pol. Sci. Ser. Math.* **31**, 211–214.

ROSALSKY, A. (1987). A strong law for a set-indexed partial sum process with applications to exchangeable and stationary sequences. *Stoch. Proc. Appl.* **26**, 277–287.

ROSÉN, B (1964). Limit theorems for sampling from a finite population. *Ark. Mat.* **5**, 383–424.

— (1967). On an inequality of Hoeffding. *Ann. Math. Statist.* **38**, 382–392.

ROSIŃSKI, J., WOYCZYŃSKI, W.A. (1986). On Itô stochastic integration with respect to *p*-stable motion: inner clock, integrability of sample paths, double and multiple integrals. *Ann. Probab.* **14**, 271–286.

ROYDEN, H.L. (1988). *Real Analysis*, 3rd ed. Macmillan, New York.

RYLL-NARDZEWSKI, C. (1957). On stationary sequences of random variables and the de Finetti's [sic] equivalence. *Colloq. Math.* **4**, 149–156.

— (1961). Remarks on processes of calls. In: *Proc. 4th Berkeley Symp. Math. Statist. Probab.* **2**, 455–465.

SAUNDERS, R. (1976). On joint exchangeability and conservative processes with stochastic rates. *J. Appl. Probab.* **13**, 584–590.

SCARSINI, M. (1985). Lower bounds for the distribution function of a *k*-dimensional *n*-extendible exchangeable process. *Statist. Probab. Lett.* **3**, 57–62.

SCARSINI, M., VERDICCHIO, L. (1993). On the extendibility of partially exchangeable random vectors. *Statist. Probab. Lett.* **16**, 43–46.

SCHOENBERG, I.J. (1938a). Metric spaces and completely monotone functions. *Ann. Math.* **39**, 811–841.

— (1938b). Metric spaces and positive definite functions. *Trans. Amer. Math. Soc.* **44**, 522–536.

SCOTT, D.J., HUGGINS, R.M. (1985). A law of the iterated logarithm for weakly exchangeable arrays. *Math. Proc. Camb. Phil. Soc.* **98**, 541–545.

SENETA, E. (1987). Chuprov on finite exchangeability, expectation of ratios, and measures of association. *Hist. Math.* **14**, 249–257.

SERFLING, R.J. (1980). *Approximation Theorems of Mathematical Statistics*. Wiley, New York.

SHAKED, M. (1977). A concept of positive dependence for exchangeable random variables. *Ann. Statist.* **5**, 505–515.

SILVERMAN, B.W. (1976). Limit theorems for dissociated random variables. *Adv. Appl. Probab.* **8**, 806–819.

SKOROHOD, A.V. (1957). Limit theorems for stochastic processes with independent increments. *Theory Probab. Appl.* **2**, 138–171.

SLIVNYAK, I.M. (1962). Some properties of stationary flows of homogeneous random events. *Th. Probab. Appl.* **7**, 336–341; **9**, 168.

SLUD, E.V. (1978). A note on exchangeable sequences of events. *Rocky Mtn. J. Math.* **8**, 439–442.

SMITH, A.M.F. (1981). On random sequences with centered spherical symmetry. *J. Roy. Statist. Soc. Ser. B* **43**, 208–209.

SPARRE-ANDERSEN, E. (1953–54). On the fluctuations of sums of random variables, I–II. *Math. Scand.* **1**, 263–285; **2**, 193–194, 195–223.

SPEED, T. (1987). What is an analysis of variance? *Ann. Statist.* **15**, 885–910.

SPIZZICHINO, F. (1982). Extendibility of symmetric probability distributions and related bounds. In KOCH and SPIZZICHINO (eds.), pp. 313–320.

STAM, A.J. (1982). Limit theorems for uniform distributions on high dimensional Euclidean spaces. *J. Appl. Probab.* **19**, 221–228.

SUN, Y. (1998). The almost equivalence of pairwise and mutual independence and the duality with exchangeability. *Probab. Theor. Rel. Fields* **112**, 425–456.

SZCZOTKA, W. (1980). A characterization of the distribution of an exchangeable random vector. *Bull. Acad. Polon., Ser. Sci. Math.* **28**, 411–414.

TAKÁCS, L. (1967). *Combinatorial Methods in the Theory of Stochastic Processes.* Wiley, New York.

TAYLOR, R.L. (1986). Limit theorems for sums of exchangeable random variables. *Int. Statist. Symp.* **1986**, 785–805.

TAYLOR, R.L., DAFFER, P.Z., PATTERSON, R.F. (1985). *Limit Theorems for Sums of Exchangeable Random Variables.* Rowman & Allanheld, Totowa, NJ.

TAYLOR, R.L., HU, T.C. (1987). On laws of large numbers for exchangeable random variables. *Stoch. Anal. Appl.* **5**, 323–334.

TEICHER, H. (1960). On the mixture of distributions. *Ann. Math. Statistist.* **31**, 55–73.

— (1971). On interchangeable random variables. In: *Studi di Probabilita Statistica e Ricerca in Onore di Giuseppe Pompilj*, pp. 141–148.

— (1982). Renewal theory for interchangeable random variables. In KOCH and SPIZZICHINO (eds.), pp. 113–121.

TRASHORRAS, J. (2002). Large deviations for a triangular array of exchangeable random variables. *Ann. Inst. H. Poincaré, Sec. B* **38**, 649–680.

TROUTMAN, B.M. (1983). Weak convergence of the adjusted range of cumulative sums of exchangeable random variables. *J. Appl. Probab.* **20**, 297–304.

VAILLANCOURT, J. (1988). On the existence of random McKean–Vlasov limits for triangular arrays of exchangeable diffusions. *Stoch. Anal. Appl.* **6**, 431–446.

VARADARAJAN, V.S. (1963). Groups of automorphisms of Borel spaces. *Trans. Amer. Math. Soc.* **109**, 191–220.

WACHTER, K.W. (1974). Exchangeability and asymptotic random matrix spectra. In: *Progress in Statistics* (eds. J. GANI, K. SARKADI, I. VINCE), pp. 895–908. North-Holland, Amsterdam.

WALD, A. (1944). On cumulative sums of random variables. *Ann. Math. Statist.* **15**, 283–296.

— (1945). Sequential tests of statistical hypotheses. *Ann. Math. Statist.* **16**, 117–186.

WALSH, J.B. (1972). Transition functions of Markov processes. *Séminaire de Probabilités VI. Lect. Notes in Math.* **258**, 215–232. Springer, Berlin.

WATANABE, S. (1964). On discontinuous additive functionals and Lévy measures of a Markov process. *Japan. J. Math.* **34**, 53–79.

WATSON, G.S. (1986). An ordering inequality for exchangeable random variables. *Adv. Appl. Probab.* **18**, 274–276.

WAUGH, W.A.O'N. (1970). Transformation of a birth process into a Poisson process. *J. Roy. Statist. Soc. B* **32**, 418–431.

WAYMIRE, E.C., WILLIAMS, S.C. (1996). A cascade decomposition theory with applications to Markov and exchangeable cascades. *Trans. Amer. Math. Soc.* **348**, 585–632.

WEBER, N.C. (1980). A martingale approach to central limit theorems for exchangeable random variables. *J. Appl. Probab.* **17**, 662–673.

WIENER, N. (1938). The homogeneous chaos. *Amer. J. Math.* **60**, 897–936.

YOR, M. (1985). Inégalités de martingales continues arrêtées à un temps quelconque, I. *Lect. Notes in Math.* **1118**, 110–171. Springer, Berlin.

ZABELL, S.L. (1995). Characterizing Markov exchangeable sequences. *J. Theor. Probab.* **8**, 175–178.

ZAMAN, A. (1984). Urn models for Markov exchangeability. *Ann. Probab.* **12**, 223–229.

— (1986). A finite form of de Finetti's theorem for stationary Markov exchangeability. *Ann. Probab.* **14**, 1418–1427.

Author Index

Aalen, O.O., 470
Accardi, L., 465
Adler, R.J., 475
Aldous, D.J., v, 9, 16, 18, 27, 100, 147, 163, 308, 325, 339, 364, 366–7, 370, 446, 464, 466, 469, 473–6
Alfsen, E.M., 443, 475
Anis, A.A., 477
Arndt, U., 471
Assouad, P., 472
Athreya, K.B., 465

Bahr, B. von, 467
Balakrishnan, N., 477
Balasubramanian, K., 477
Ballerini, R., 477
Barbour, A.D., 470
Bartfai, P., 478
Bayes, T., 466
Benczur, A., 468
Berbee, H., 286, 472
Berg, C., 459, 476
Berkes, I., 9, 166, 469
Berman, S., 466
Bernoulli, J., 52, 345
Bernstein, S., 5, 53, 68, 110–2, 458, 465, 476
Berti, P., 467
Bertoin, J., 472
Bichteler, K., 145
Bikelis, A., 468
Billingsley, P., 144, 468–9
Blackwell, D., 471
Blom, G., 467
Blum, J.R., 468
Blumenthal, R.M., 297, 472
Bochner, S., 68, 458, 476
Boes, D.C., 479
Borel, E., 25, 57, 465
Bretagnolle, J., 6, 62, 466
Brown, T.C., 468, 470
Bru, B., 479
Bühlmann, H., v, 2, 4–5, 44, 465, 468
Burkholder, D.L., 87, 471

Campbell, N.R., 124, 459–60, 476

Carleson, L., 472
Casukhela, K.S., 403, 475
Cauchy, A.L., 61, 192
Chanda, C., 479
Chatterji, S.D., 469
Chen, K.H., 493
Chentsov, N.N., 392
Chernoff, H., 468
Choquet, G., 442–3, 475
Chow, Y.S., 466, 470–1
Christensen, J.P.R., 476
Chung, K.L., 471–2
Cocozza, C., 470
Courrège, P., 231
Cox, D., 44–5, 132, 265, 272–3, 280–3, 428–9
Cramér, H., 58–60, 68, 175, 250–1, 447

Daboni, L., 465
Dabrowski, A.R., 480
Dacunha-Castelle, D., 6, 9, 62, 163, 466, 469
Daffer, P.Z., 468
Dale, A.I., 464
Dambis, K.E., 11, 470
Daniell, P.J., 68
Daras, T., 480
Davidson, R., 46, 465
Davis, B.J., 87
Dawid, A.P., 18, 356, 473–4
Debes, H., 475
Dellacherie, C., 145, 215, 472, 475–6
Diaconis, P., 464–6, 472, 474–5
Dinwoodie, I.H., 481
Doléans, C., 11, 200
Donoghue, W.F., 459, 476
Doob, J.L., 10, 92, 169, 466, 469, 471
Dovbysh, L.N., 473, 475
Dubins, L.E., 11, 103, 464, 467, 470
Dynkin, E.B., 443, 464, 474–5

Eagleson, G.K., 468, 473
Eaton, M.L., 465
Elalaoui-Talibi, H., 472
Erdös, P., 468
Erlang, A.K., 45

Esseen, C.G., 467
Etemadi, N., 482
Evans, S.N., 482
Everitt, C.W.F., 465

Farrell, R.H., 443, 475
Feigin, P., 465
Fell, J.M.G., 285
Feller, W., 458–9, 465, 470, 476
Ferguson, T.S., 467
Figiel, T., 469
Finetti, B. de, v, 1–2, 4–5, 8, 14, 24–5, 30,
    126, 267, 302, 308, 464–6, 473
Fortini, S., 483
Fourier, J.B.J., 150, 276, 278, 288, 459
Franken, P., 471
Fréchet, M., 483
Freedman, D.A., v, 5, 18–9, 55, 58, 103,
    464–7, 472, 474–5
Fristedt, B.E., 9, 108, 161, 467–8

Gaifman, H., 473
Galambos, J., 484
Gallardo, L., 465
Gaposhkin, V.F., 469
Getoor, R.K., 297, 472
Gharib, M., 477
Gnedin, A.V., 474
Gordon, L., 471
Griffiths, R.C., 484
Grigelionis, B., 7, 84, 466–7, 470
Guerre, S., 484
Guglielmi, A., 484
Gundy, R.F., 87, 471

Haag, J., 464
Haar, A., 312–3
Hagberg, J., 144, 468
Hahn, M.G., 486
Hájek, J., 144, 468
Halmos, P.R., 469
Harris, T.E., 475
Hartman, P., 160
Hausdorff, F., 5, 53, 68, 112, 458, 465, 476
Hayakawa, T., 474
Hayakawa, Y., 485
Heath, D., 485
Heinich, H., 479
Herglotz, G., 476

Hermite, C., 19, 475
Hestir, K., 473–4
Hewitt, E., v, 29, 424, 464
Hilbert, D., 19–20, 350, 359–63, 449–53
Hill, B.M., 485
Hirth, U., 485
Hoeffding, W., 32, 464, 473–4
Hoem, J.M., 470
Hölder, O., 391
Holst, L., 470
Hoover, D.N., v, 16–7, 308, 325, 330–1,
    473
Hsu, Y.S., 485
Hu, T.C., 485, 494
Hu, Y.S., 485
Huang, W.J., 485
Hudson, R.L., 485
Huggins, R.M., 493

Inoue, H., 492
Itô, K., 14, 19, 43, 264–5, 350, 451, 476
Ivanoff, B.G., 33, 465, 467–8, 470, 473

Jacod, J., 466–7
Jagers, P., 475
Janson, S., 470
Jiang, X., 486
Johnson, W.B., 486
Jordan, C., 84
Julesz, B., 474

Kallenberg, O., 465–76
Kaminski, M., 482
Karatzas, I., 471
Karoui, N., 470
Kelker, D., 465
Kendall, D.G., 68, 465, 468
Kerstan, J., 465, 475
Kesten, H., 472
Khinchin, A.Y., 9, 161, 464, 468
Kimberling, C.H., 465
Kingman, J.F.C., v, 17, 343, 346, 464–6,
    469, 471, 474
Klass, M.J., 467–8, 471
Klopotowski, A., 479
Knight, F.B., 11, 470
Koch, G., 466
Kolmogorov, A.N., 59, 61, 68, 308, 310,
    392, 455

Komlós, J., 469
König, D., 471
Krauss, P.H., 473
Krickeberg, K., 489
Krivine, J.L., 6, 62, 466
Kuritsyn, Y.G., 489
Kurtz, T.G., 470
Kwapień, S., 476

Ladelli, L., 483
Lamperti, J., 466
Lane, D., 485
Laplace, P.S. de, 52, 150, 459
Lauritzen, S.L., 475
Lebesgue, H., 19, 254, 461
Lefévre, C., 490
Lepeltier, J.P., 470
Letac, G., 465
Lévy, P., 4, 9, 12, 43–5, 60, 110–1, 192, 233–40, 466, 472
Liemant, A., 475
Lisek, B., 471
Loève, M., 464, 466, 471
Lootgitier, J.C., 479
Lu, Y.G., 465
Lynch, J., 473

Maisonneuve, B., 472
Maitra, A., 443, 475
Mandelbaum, A., 474
Mandrekar, V., 490
Marcinkiewicz, J., 159
Markov, A.A., 15–6, 56, 290, 293–4
Matthes, K., 46, 55, 465, 475–6
Maxwell, J.C., 57–8, 465
McGinley, W.G., 473
McKean, H.P., 465
Mecke, J., 465, 476
Melilli, E., 484
Merzbach, E., 470
Meyer, P.A., 11, 470, 472, 475
Millar, P.W., 9, 108, 161, 467–8, 471
Milne, R.K., 484
Minkowski, H., 99, 452
Misiewicz, J., 466
Moivre, A. de, 464
Moody, G.R., 485
Moran, P.A.P., 468

Moscovici, E., 491

Nair, M.G., 470
Nawrotzski, K., 55, 465
Nekkachi, M.J., 491
Nelson, E., 452, 476
Neuts, M., 465
Ney, P.E., 465
Nguyen X.X., 476
Nikodým, O.M., 124
Nualart, D., 470

Olshansky, G.I., 475
Olshen, R.A., 29, 464
Olson, W.H., 474
Orey, S., 475

Paley, R.E.A.C., 101, 455
Palm, C., 7, 69, 111–6, 124, 459, 476
Papangelou, F., 11, 112, 123–4, 459–60, 467, 470, 476
Pathak, P.K., 464
Patterson, R.F., 468
Péter, E., 9, 166, 469
Phelps, R.R., 475
Philipp, W., 469
Pitman, J.W., 464, 470, 474
Poincaré, H., 465
Poisson, S.D., 14–5, 21–3, 43, 55, 68, 112, 134, 173, 279, 401, 416ff, 432, 449, 453–5
Popescu, O., 491
Pratelli, L., 467
Prohorov, Y.V., 126ff, 445–8, 475
Pruitt, W.E., 475
Pruss, A.R., 492

Radon, J., 124
Rao, C.R., 492
Regazzini, E., 483, 492
Rényi, A., 9, 68, 162, 465, 468–9
Resnick, S.I., 465
Ressel, P., 464, 466, 476
Révész, P., 162, 469
Ridler-Rowe, C.J., 468
Riesz, F., 362
Rigo, P., 467
Robbins, H., 471
Robinson, J., 493

Rogers, L.C.G., 442–3
Romanowska, M., 493
Rosalsky, A., 493
Rosén, B., 144, 464, 468
Rosenblatt, M., 468
Rosenthal, H.P., 469
Rosiński, J., 198, 470
Royden, H.L., 466
Ryll-Nardzewski, C., v, 1, 4, 24–5, 126, 303, 464, 476

Salas, L.C.J.D., 479
Saunders, R., 493
Savage, L.J., v, 29, 424, 464, 467
Sazonov, V.V., 492
Scarsini, M., 493
Schechtman, G., 486
Schmidt, V., 471
Schoenberg, I.J., 5, 59, 68, 351, 356, 459, 465, 476
Schwarz, G., 11, 470
Scott, D.J., 493
Seneta, E., 493
Serfling, R.J., 474
Shaked, M., 493
Shanbhag, D.N., 492
Shiryaev, A.N., 467
Shreve, S., 471
Sibson, R., 473
Siegmund, D., 471
Silverman, B.W., 473
Skorohod, A.V., 9, 101, 137ff, 155, 468
Slivnyak, I.M., 7, 69, 112, 467
Slud, E.V., 494
Smith, A.M.F., 465
Sparre-Andersen, E., 171, 470
Speed, T., 474
Spizzichino, F., 466
Stam, A.J., 465
Stieltjes, T.J., 254
Stone, M.H., 135
Su, J.C., 485
Sucheston, L., 469
Sudakov, V.N., 473, 475
Sudderth, W., 485
Sun, Y., 494

Szczotka, W., 494
Szulga, J., 454, 476

Takács, L., 494
Taqqu, M.S., 474
Taylor, R.L., 468
Teicher, H., 466, 468, 470
Trashorras, J., 494
Troutman, B.M., 494

Uppuluri, V.R.R., 474
Utev, S., 490

Vaillancourt, J., 494
Varadarajan, V.S., 443, 475
Verdicchio, L., 493
Vershik, A.M., 475

Wachter, K.W., 474
Wakolbinger, A., 475
Wald, A., 12, 209, 470
Walsh, J.B., 472
Warmuth, W., 476
Watanabe, S., 466, 470
Watson, G.S., 495
Waugh, W.A.O'N, 465
Waymire, E.C., 495
Weber, N.C., 33, 465, 467–8, 470, 473
Weierstrass, K., 135
Wiener, N., 19, 350, 451, 476
Williams, D., 442–3
Williams, S.C., 495
Wintner, A., 160
Wold, H., 58–60, 68, 175, 250–1, 447
Woyczyński, W.A., 198, 470, 476

Yor, M., 465, 470
Yule, G.U., 465

Zabell, S.L., 481
Zaman, A., 472
Zessin, H., 476
Zhou, X., 482
Zinn, J., 486
Zorn, M., 376
Zygmund, A., 101, 159, 455

Subject Index

absolutely
  continuous, 460
  monotone, 457, 465
absorption, 297
adjoint operator, 353, 362, 450
allocation sequence, 250
analysis (of)
  non-standard, 473
  variance, 474
analytic extension, 151–2
approximation of
  function, 402
  integral, 236, 246
  local time, 274
  process, 134, 144, 158f, 236, 246ff
  sequence, 31, 166
  set, 177
arcsine laws, 470
asymptotically
  equivalent, 144
  exchangeable, 163
  invariant, 9, 128
augmented
  array, 319, 328
  filtration, 77–8, 145
average, 190, 216, 226, 362
avoidance function, 52, 111

Bayesian statistics, 466, 474
BDG (Burkholder–Davis–Gundy), 87
Bernoulli sequence, 52, 345
binary array, 386
binomial process, 5, 22, 52–5
birth process, 68, 465
Borel space, 25
Brownian
  bridge, 7, 90, 136, 189–90, 355
  invariance, 189
  motion, 19, 43, 189, 355
  sheet/sail, 19, 355, 395, 475

Campbell measure, 459–60
Cauchy process, 192
centering decomposition, 392, 396
central limit theorem, 468–9

change of basis, 380
chaos expansion, 476
characteristic(s), 43, 143, 211
  function, 5, 276, 458–9
  processes, 146
closed on the right/left, 261, 279
closure properties, 150
CLRF, 19, 351, 451
coding, 22, 310
combined
  arrays, 323, 333, 368
  functionals, 376, 389
compact operator, 450
compactification, 139, 285, 447
comparison of
  moments, 32, 149
  norms, 96
compensator, 6–7, 78, 83, 173, 202, 206
complete(ly)
  exchangeable, 422
  excursion, 56
  monotone, 5, 53, 68, 112, 457–9
composition, 39, 405, 441
concave, 103–10, 161
conditional(ly)
  distribution, 162–3, 266, 335–8, 356,
    409, 443–4
  exchangeable, 69, 426–7
  expectation, 361, 367
  i.i.d., 4, 25, 70
  independence, 29, 294, 297, 305, 307,
    335, 368, 406, 428
  integrable, 78
  invariance, 123
  Markov, 290
  martingale, 78–9, 82, 91
  moment, 218, 221, 449
  regenerative, 261
  urn sequence, 71
contiguous intervals, 268
continuous/continuity of
  random functional, 19, 351, 451
  martingale component, 6, 83
  stochastic integral, 177, 246
  paths, 150

contractable
  array, 16, 301, 318–9, 329
  increments, 36
  integral, 223, 243
  measure, 46, 148, 403
  partition, 342, 345
  point process, 46
  process, 4, 36, 146, 188
  sequence, 2, 24, 126
  series, 241
  sheet, 20, 398
contraction, 3, 10, 27, 36, 42
control measure, 63
convergence in/of
  distribution, 126ff, 447
  integrals, 454
  partitions, 346
  processes, 137–42, 150
  random measures, 130–4, 148, 447
  random sets, 285
  restriction, 150, 285
  sequences, 126–8, 162–4
  series, 90, 110, 210, 421, 438
convex
  function, 32, 103–10, 149, 161
  set, 408, 441
convolution semigroup, 43
coupling, 9, 32, 146, 158–9, 166, 184, 305,
      307, 369
covariance, 43, 152, 173, 356
covariation, 83, 136, 173, 205, 230
Cox process, 45, 139, 265, 272, 428
cyclic stationarity, 287–9

decoupling, 12
  identity, 12–3, 217ff, 229, 234, 241ff
  inequality, 455
de Finetti's theorem, 25, 302
density of
  compensator, 7, 78, 80
  endpoint distribution, 287
  local time intensity, 277–8, 285–9
  nested arrays, 332–3
  partition class, 346
determining sequence, 166
diagonal (space), 385, 422–3, 453
  decomposition, 213, 390, 397
  extension, 379, 387

diagonalization, 354, 450
diffuse random measure, 46, 52, 273
directing
  element, 8, 272, 291, 408
  measure, 8, 28, 346–7
  triple, 8, 44, 136, 139, 142
discounted compensator, 11, 200–4
disintegration, 459–60
dissociated array/sheet, 339, 393
Doléans equation/exponential, 11, 200
dual projection, 201, 203, 239, 362

eigen-value/vector, 450–1
elementary integral, 185–6
embedding, 181
empirical
  distribution, 6, 8, 31, 74, 126–7
  measure, 82
endpoint distributions, 287
equivalent
  partitions, 382
  representations, 329–31
  σ-fields, 29
  symmetries, 37, 42
ergodic, 338–9, 369, 393, 440–1
  decomposition, 443
estimation problem, 474
exchangeable
  array, 16, 300, 318, 325, 429
  excursions, 265
  extension, 17, 319, 399
  increments, 36
  integral, 224–6, 229
  measure, 5, 45–56, 130–4, 148, 401ff
  partially, 473
  partition, 17, 342, 345–6
  point process, 7, 45, 269, 429
  process, 4, 36, 44, 136ff, 176
  row-column, 473
  sequence, 3, 24, 127, 170
  set, 14–5, 268, 271–8, 282–90
  sheet, 20, 355, 391, 395
  σ-field, 29
  sum, 217
  weakly, 473
excursion, 14, 56, 256, 472
  infinite, 282
exponential

distribution, 15
Doléans, 200
moments, 250
scaling, 282
truncation, 283–5
extended
array, 306, 319, 372, 385
invariance, 42
measure, 403
process, 150, 282f, 387–96
representation, 340
extreme/al, 28, 35, 47, 93, 150, 217, 275, 349, 408, 441
decomposition, 442

factorial measure, 30
$\mathcal{F}$-contractable, 6, 70, 75–6, 83
-exchangeable, 6, 70, 75–6
-extreme, 93, 218
-homogeneous, 255, 261
-i.i.d., 70
-independent, 93
-integrable, 83
-Lévy, 75–6, 192
-martingale, 70
-prediction, 70
-reflection property, 72, 76–7
-stationary, 70, 75
-urn sequence, 71–2
Fell topology, 285
filtration, 6, 69, 75, 78, 92, 94, 100, 145
finite-dimensional convergence, 136, 147, 150
FL (Fourier–Laplace) transform, 150
fluctuation theory, 171, 470
FMP, 25
Fourier inversion, 276, 288
functional representation, 302, 318, 325, 333
fundamental martingale, 202

games/gambling, 13, 469, 472
G-array/process, 351, 354, 451
Gaussian
approximation, 57
distribution, 5, 57
integral, 451
process, 63, 351

reduction, 173, 205
representation, 358–9
generating
function, 52, 458
point process, 268
genetics, 474
globally homogeneous, 15, 293
grid process, 21, 386–90, 402
group (of)
action, 311–3
transformations, 349, 441
growth rate, 96, 108, 161

Haar measure, 312
Hermite/ian
array, 475
polynomial, 19, 354
Hilbert space, 19, 350, 449
hitting probabilities, 286, 289, 472
Hölder continuous, 391
homogeneous, 255, 294, 297
globally, 15
locally, 14
hyper-contraction, 96, 449, 452
hyper-geometric, 52

ideal, 335, 341, 367, 370
i.i.d.
sequence, 29
sum, 210, 220
uniform, 31, 49–50, 52, 90, 279, 302
increments, 36–7
independent, 54
elements/entries, 58, 309, 407
increments, 63, 424
index of regularity, 157, 275–6
indicator array, 343–6
induced filtration, 265
infinitely divisible, 175
integral
contractable, 223, 243
exchangeable, 224–6, 229
Gaussian, 451
Lévy, 211, 234
mixed multiple, 353, 453
Poisson, 453–4
representation, 442–4, 475
selection, 186

integration by parts, 214–5, 245
interval sampling, 41
invariant
    function, 312
    Palm measures, 112, 116
    set, 338, 440
    σ-field, 25, 29, 128, 272, 406, 443
inverted representation, 315–7
iso-metric, 65, 358, 373, 383, 466
    -normal, 63, 351, 451
iterated logarithm, 159, 161
Itô representation, 264–5, 354

J-rotatable, 367
J_1-topology, 125, 129
jointly
    contractable, 16, 301, 318, 398
    exchangeable
        array, 16, 300, 325, 337
        measure, 401, 411–2, 433
        processes, 93
        sheet, 354–5, 395
    Gaussian, 358
    rotatable, 18, 352, 378, 385
    symmetric/ergodic, 407
jump point process, 83, 91, 103–5, 175

kernel criterion, 409

lacunary series, 469
λ-preserving, 42, 328–31, 402, 411
    -randomization, 49
    -symmetric, 115–6
Laplace transform, 5, 52, 151, 283, 458f
lattice component, 431
law of (the)
    iterated logarithm, 161, 469
    large numbers, 28, 292, 466, 469
lcsc (loc. compact, 2nd countable), 151
Lebesgue
    decomposition, 461
    measure, 46, 53, 385, 411f, 432f
left-stationary, 358
Lévy
    integral, 211, 234
    measure, 43, 175, 192
    process, 12, 43, 157
Lévy–Itô representation, 43, 139
linear

birth process, 68, 465
isometry, 65
random functional, 19, 62, 351, 451
localization, 183
local(ly)
    characteristics, 7, 83, 87
    dichotomies, 256, 260
    growth, 108, 161
    homogeneous, 14, 255, 262ff, 290f
    intensity, 278
    martingale, 103–5, 173, 205
    prediction, 73
    time, 14–5, 264, 273, 472
    uniform, 155
L^p/l^p-invariance, 5, 62
L^p-bounded, 361

marked
    optional time, 11
    partition, 343–6
    point process, 45, 173, 430
Markov property, 15, 290, 293–4
martingale, 6
    criterion, 84, 86, 466
    measure-valued, 70, 73–4, 76, 82
    product, 230
    reduction, 175
    reverse, 6, 74, 82, 91
    uniformly integrable, 202
maximal/maximum
    components, 388
    distribution, 171
    inequality, 96, 102, 210–1, 226
    kernel, 460
measurable
    CLRF, 360
    random element, 353, 359
    selection, 444
measure-preserving, 10, 176, 328ff, 402
measure-valued
    martingale, 70, 73–4, 76, 82
    sequence/process, 70, 73, 75, 163
minimal representation, 376, 450
mixed/mixture of
    Bernoulli, 52
    binomial, 5, 46, 52, 112
    ergodic/extreme, 442–4
    i.i.d., 4, 25

integral, 353, 453
Lévy, 4, 44, 158–9
Markov, 290, 472
Poisson, 52, 55, 112
urn sequence, 5, 30
mixing random measure, 166
moment
    comparison, 32, 149, 455
    estimate, 26
    existence, 363
    identity, 217ff, 229, 234, 241f, 454
    sequence, 458
moving average, 359
multiple integral
    Gaussian, 19, 352–3, 451
    mixed, 355, 385, 390f, 396ff, 453
    Poisson, 432–3, 453–4
multi-variate
    invariance, 199
    sampling, 172

natural compensator, 200
nested arrays, 332–3
noise, 63
non-decreasing process, 159, 161
non-negative definite, 5, 150, 152, 458f
non-standard analysis, 473
norm
    comparison, 202, 211, 452
    estimate, 7, 31, 95–6, 102, 210, 216
normal distribution, 57
nowhere dense, 256, 260

occupation measure, 71
off-diagonal, 379, 381, 387, 396, 399
ONB (ortho-normal basis), 350, 450–1
one-dimensional
    convergence, 150
    coupling, 32, 146
optional
    continuity, 100
    projection, 215, 239
    reflection, 258–61, 269
    sampling, 239
    shift, 6, 70, 75–6, 255
    skipping, 10, 169, 252
    time, 10–1, 177, 200–5, 265
ordered partition, 214, 353, 378

order statistics, 68, 465
orthogonal
    functions, 364, 450–1, 469
    martingales, 175, 203
    matrix, 18, 385–6
    optional times, 203
    processes, 358
    projection, 362
over an element, σ-field, 29, 33, 244

paint-box representation, 17, 343
Palm measure, 7, 111–2, 459–60, 471–2
Papangelou kernel, 123, 460
parametric representation, 410
partial exchangeability, 473
partition, 217, 342–9, 352–3, 395, 398
path properties, 9, 44, 90, 100, 108, 161
perfect set, 256, 260
permutation, 3, 380
pinned Brownian sheet, 355, 395
Poisson
    approximation, 134
    integral, 453–4
    process, 21ff, 53ff, 112, 279, 416, 432f
    reduction, 173, 206
    sampling, 279
population genetics, 474
positive
    definite, 473, 475, 458–9
    stable, 59
predictable
    contraction, 185, 188, 253
    embedding, 181
    mapping, 10, 176, 250
    process, 10ff, 105, 173ff, 211, 223ff
    projection, 201, 203, 239
    sampling, 10, 170, 249
    sequence, 170
    set, 177
    sum/integral, 210, 217, 241–2
    time, 10, 73, 169, 179
prediction
    process, 75–6
    sequence, 6, 70
preservation laws, 338, 441
product
    martingale, 230
    moment, 203, 217ff, 229, 234, 241f

symmetry, 406–7
progressively measurable, 255, 261
projection, 60, 362
pseudo-isotropic, 466
p-stable, 5, 60
purely discontinuous, 83, 105–6, 175

quadratic variation, 6, 83, 86, 96
quasi-left-continuous, 11, 106, 173

random
    distribution, 25, 163
    element, 353, 359
    functional, 19, 62, 351, 451
    isometry, 373
    matrix, 474
    measure, 21, 446–7
    partition, 17, 342–9
    sheet, 20, 354
    walk, 470
randomization, 49, 444, 449
ratio limit laws, 291
rcll (right-continuous, left limits), 7
rectangular index set, 340
recurrent, 255
reduction of/to
    CLRF, 394
    Gaussian, 173, 205
    independence, 204
    Palm measure, 7, 112, 116, 460
    Poisson, 173, 206
reflectable, 36
reflection, 35, 387, 422–8
    invariance, 14, 72, 258–61
    operator, 71
regenerative, 14, 261, 472
regular(ity)
    index, 157, 275–6
    exchangeable set, 286
regularization, 44, 83, 100, 145
renewal theory, 470
Rényi stable, 162
representable, 325
representation of
    array, 318, 325, 370, 385
    excursions, 265, 269
    functional, 370, 376, 378
    measure, 5, 45ff, 411f, 416, 432f

point process, 46, 49, 53–6
process, 44, 90, 145
sequence, 25, 30
sheet, 391, 395, 398
restriction, 40, 54, 150, 282–5, 388
reverse martingale, 6, 74, 82, 91
rotatable
    array, 338, 370, 385
    decomposition, 393
    functional, 19, 351, 370, 378
    process, 351
    sequence, 3, 58, 356
row-column exchangeable, 473

sampling (from)
    asymptotically invariant, 128
    equivalence, 31
    finite population, 464, 468
    intervals, 41
    Poisson, 279
    uniform, 231, 279
scaling, 356–7, 385
selection
    integral, 186
    measurable, 444
self-adjoint, 450
    -similar, 191, 385
semi-continuous, 275, 278
    -martingale, 6–7, 83, 145–6, 214
        special, 83–4, 100, 105
        variations, 213, 235
separately
    contractable, 16, 29, 172
    exchangeable
        array, 16, 304, 308f, 325, 331, 335
        measure, 401–2, 411, 416, 432
        processes, 93
        sequences, 29–30, 172, 217
        sheet, 355, 391, 395
        rotatable, 18, 352, 370, 391
        symmetric/ergodic, 405–6
separating
    class, 341
    sites and marks, 429
    subspace, 64
    variables, 364
sequential
    analysis, 470

coupling, 164
shell σ-field, 308
shift, 6, 69, 75, 162–4, 358
   optional, 6, 70, 75–6, 255
   -invariant σ-field, 25, 29, 128
simple
   point process, 21, 46, 123
   predictable set, 177–9
Skorohod topology, 125, 129, 285
spacing, 55–6
special semi-martingale, 83–4, 100, 105
spherical symmetry, 57–8
spreadable, 2
stabilizer, 311
stable
   convergence, 162
   integral, 192
   invariance, 196
   Lévy process, 11, 191–9
   noise, 63
   positive, 59
   Rényi, 162
   strictly, 11, 191–9
   symmetric, 5, 11, 59, 196–8
   weakly, 191–9
standard extension, 94, 205–6
stationary, v, 1
   extension, 335–8, 356
   independent increments, 84, 424
   process, 128
   random set, 15
   strongly, 6
statistical mechanics, 465, 476
stochastic geometry, 467
stochastic integral, 10ff, 85ff, 106, 173ff,
     211, 223ff
   existence, 192, 211, 223–4,
   representation, 224, 226
strictly
   invariant, 440
   stable, 191–9
strong(ly)
   Markov, 290, 293
   orthogonal, 175, 203
   reflection property, 14, 71, 259f, 269
   stationary, 6, 70, 76
subjective probability, 466
subordinator, 161, 235, 268, 272

sub-sequence principle, 9, 162–6
sufficient statistics, 475
summation process, 8, 95, 137, 140, 144
super-martingale, 103–5
supporting
   line, 416, 428
   measure, 112, 459
   set, 388
   subspace, 450
symbolic logic, 473
symmetric
   coding, 311
   function, 302, 314–5, 318, 330
   partition, 342–9
   random element, 405, 440–1
   stable, 5, 59, 196–8
tail
   estimate, 454
   process, 200–1
   σ-field, 29
tensor product, 20, 350–2, 361, 449–51
term-wise
   exchangeable, 94
   integration, 224
tetrahedral
   decomposition, 214, 238
   moment, 241, 243
   region, 214, 398, 403
thinning, 141
tight(ness), 445
   at infinity, 140
   criterion, 137, 140, 147, 446, 449
time-change
   reduction, 11, 173, 206
   representation, 11, 198
time reversal, 258
time-scale comparison, 144
total variation, 31, 57, 128, 134
totally rotatable, 367–8
trans-finite extension, 193
transition kernel, 295, 472
transposition, 4, 36, 402
truncation, 236, 239, 248

U-array/process, 302, 411–2
U-statistics, 473–4
uniform(ly)

convergent, 90, 110, 155
distribution, 5, 15
integrable, 83, 202
randomization, 49, 205–6
sampling, 279
topology, 155
truncation, 283–5
uniqueness, 28, 88, 373, 383, 396
unitary, 19, 351, 385
urn sequence, 5, 30

vague topology, 130, 447
variations of semi-martingale, 213, 235
velocity distribution, 465
visual perception, 474

Wald identity, 12, 217, 220, 229, 470–1
weak(ly)
    compact, 163, 445
    ergodic, 440–1
    exchangeable, 473
    stable, 191–9
    tight, 445–6
    topology, 126, 285, 346–7, 445–7
Wiener–Itô integral, 19, 352, 451

Yule process, 465

zero–one law, 29, 297, 310
zero set, 255–60, 265, 269, 273

Symbol Index

A_J, A_J^*, 362
A_h^τ, a_h^τ, 297
$\mathcal{A}$, $\mathcal{A}_1$, 25, 422
α, 43ff, 50, 90, 137, 143, 272
α^h, α_π^*, 139, 142f, 353

B, 43, 90
$\mathcal{B}(\cdot)$, 25
β, β_k, 49f, 90, 143, 272
β_k^J, β^J, B_J, 229
$\hat{\beta}^1$, 133

C, C', 459
C_A, $C_{a,b}$, 3, 10, 36, 42, 185
$\hat{C}_+(\cdot)$, 445
$\mathbb{C}_+$, 150

D, D_1, D_π, D_π', 385, 388, 412, 422
D_0, D_h, 256, 267
$D(\cdot,\cdot)$, 44, 129
$\overset{d}{=}$, $\overset{d}{\to}$, $\overset{d}{\sim}$, 3, 8, 33, 144
Δ, Δ^n, Δ_h^n, 440, 457
Δ_d, $\Delta(\cdot)$, $\Delta\mathcal{I}_d'$, 214, 243, 398f, 403
ΔX_t, $\Delta\xi^d$, $\Delta\bar{\eta}_{s,j}$, 83, 200, 315
$\Delta\widetilde{\mathbf{N}}_d$, $\Delta\overline{\mathbf{N}}_d$, $\Delta\mathbf{Q}_d$, 309, 317, 325
δ_t, $\tilde{\delta}_t$, $\delta(\cdot)$, δ_{ij}, 5, 18, 158, 287

$E^{\mathcal{F}}$, 171
$\mathcal{E}$, $\mathcal{E}_\xi$, 29
$\mathrm{ex}(\cdot)$, 443
η, $\bar{\eta}$, $\bar{\eta}^c$, 200
$\eta^{\otimes n}$, 352, 451

$\mathcal{F}$, $\overline{\mathcal{F}}$, $\mathcal{F}_n$, $\mathcal{F}_t$, 6, 12, 69, 77, 145
$\hat{f}$, $\hat{f}_J$, 314, 362, 438
f^{-1}, 108
$\overset{fd}{\longrightarrow}$, 136, 149

G, G_s, 311
γ, 136f, 143
γ^h, 139, 142f

H, $H^{\otimes n}$, $H^{\otimes \pi}$, 20, 350, 353, 450

I_k, I_{nk}, 270, 385
$\mathcal{I}$, $\mathcal{I}_X$, $\mathcal{I}_\xi$, 25, 86, 128

$\mathcal{I}^d$, $\mathcal{I}_d'$, $\mathcal{I}_{01}^d$, 386f, 396

$J \circ I$, 302

$\mathcal{K}(\cdot)$, 445
$\tilde{k}$, k_J, 353, 367
$k \circ I$, $k \circ r$, 302, 326

L, 14, 270, 273, 279
$L^0(\cdot)$, $L_c^2(\cdot)$, $L_F^p(\cdot)$, 63, 351, 385
$L(\cdot)$, $\overline{L}(\cdot)$, $L\log L$, 192
$\mathcal{L}(\cdot)$, 4, 24
$l(\cdot)$, 116, 150, 256, 268, 272
$\log_2$, 159
λ, λ_i, 5, 10, 312, 453
λ_D, λ^π, 385, 389, 412

M, M_i, 418, 427
$M(\cdot,\cdot)$, 111
$\mathcal{M}$, $\mathcal{M}'$, $\mathcal{M}_1$, 25, 45
$\hat{\mathcal{M}}$, $\mathcal{M}_c$, $\hat{\mathcal{M}}_c$, 445
m_J, m_π, 219f
$\hat{\mu}$, 132
μ^∞, $\mu^{(n)}$, 4, 30, 72
μ_k, μ_t, μ_x, 70, 75f, 290, 293

N_j, N_J, 340, 362
$N(0,1)$, 57
$\mathcal{N}(\cdot)$, 130
$\mathbf{N}$, $\hat{\mathbf{N}}$, $\overline{\mathbf{N}}$, $\overline{\mathbf{N}}^\pi$, 3, 17, 301
$\widetilde{\mathbf{N}}_d$, $\overline{\mathbf{N}}_d$, $\hat{\mathbf{N}}_d$, 309, 325, 335
$\mathbf{N}_d'$, $\mathbf{N}_J'$, $\hat{\mathbf{N}}_d'$, 340
$n^{(k)}$, $(n/\!/k)$, 53, 78, 240, 278, 458
ν, ν_p, ν_J, 43ff, 211, 234, 272, 346
$\hat{\nu}^1$, 132

$\mathcal{O}$, $\mathcal{O}_J$, $\mathcal{O}_d$, $\hat{\mathcal{O}}_d$, 18, 353, 378, 385, 395
Ω, 25

P_j, P_J, 362
$\overset{P}{\to}$, 19
$\mathcal{P}$, $\mathcal{P}_d$, $\mathcal{P}_{\mathcal{J}}$, $\mathcal{P}_J$, 186, 341, 375
$\mathcal{P}_d^2$, $\hat{\mathcal{P}}_d$, $\widetilde{\mathcal{P}}_d$, 385, 391, 398
p, p_t, $p_{s,t}$, $p_n^\pm$, 276f, 287
p_J, 234
$p^{-1}R$, 342
π_B, π_k, π^c, 25, 128, 385, 388, 391

ψ, ψ_t,   416, 454

Q_k, Q_s, Q'_s,   7, 71, 76, 112, 115, 459f
$\mathbf{Q}_+, \mathbf{Q}_{[0,1]}, \mathbf{Q}_I$,   6, 36, 177
$\tilde{\mathbf{Q}}, \tilde{\mathbf{Q}}_d$,   305, 309

R, R_a, R_τ,   14, 35, 258, 460
$R_J, \underline{R_\pi}$,   217
$\mathbf{R}_+, \overline{\mathbf{R}^d}$,   2, 139
r_+,   325
ρ, ρ', ρ_X,   43, 90, 157, 275

$S, \bar{S}, \mathcal{S}, \mathcal{S}_+$,   25, 111, 447, 459
$S_1, S_n, S_\pi, S_\infty$,   115, 277, 379
$S_J, S_\infty, S_{J_1,\ldots,J_m}$,   217, 219f, 241
$\mathcal{S}(\cdot)$,   308
(Σ),   123, 460
σ,   43, 90
$\sigma_t^\pm, \tilde{\sigma}_t^\pm$,   287

T,   268, 270, 272
T_p, T'_q,   380, 383
$T_{a,b}, T_{a,b,c}$,   4, 36f
$\mathcal{T}, \overline{\mathcal{T}}, \mathcal{T}_t, \mathcal{T}_\xi$,   27, 29, 74, 82, 92
$\hat{t}_\pi$,   395
τ_r,   255
θ_n, θ_t,   6, 69, 75f, 358

$U(0,1), U\{\cdot\}$,   5, 31
$\overline{U}_t$,   189f, 226
$U^{\otimes d}$,   352
$(U \cdot X) \circ V^{-1}$,   196
$\xrightarrow{u}, \xrightarrow{ud}$,   155

V_d^+,   150
$\hat{V}^j, V^J, \hat{V}^J$,   229, 234, 243
$V \cdot \lambda, V \cdot X$,   11, 192, 211, 223
$\xrightarrow{v}, \xrightarrow{vd}$,   77, 130, 447

W, W_1,   422f
$\tilde{w}$,   101, 156
$\xrightarrow{w}, \xrightarrow{wd}$,   9, 126, 445, 447
$\xrightarrow{wP}$,   164

$\hat{X}$,   6, 83, 326
$\tilde{X}, X^\pi$,   301, 387
X^c, X^d,   83
$X^J, X_h^J, X^{\mathcal{J}}, X^{a,b}$,   40, 335, 367
$X(\cdot), X_A, X_\pi$,   177, 185f, 388
$X\varphi$,   353, 360f
$X \circ p$,   16, 300, 304
$X \circ V^{-1}$,   10, 176
$x_J, \hat{x}_J, \hat{x}'_J$,   328
$\Xi, \tilde{\Xi}, \Xi^h, \Xi^\tau$,   14f, 255, 274, 282, 287
$\Xi \cdot \lambda$,   256
$\hat{\xi}, \xi_d, \xi_t$,   6, 78, 115, 357
$\xi^d, \hat{\xi}^d$,   309, 315
$\xi^J, \hat{\xi}^J, \xi^{\mathcal{J}}$,   217, 309, 335
$\bar{\xi}_J, \hat{\xi}_J, \bar{\xi}_k$,   302, 362
$\tilde{\xi}, \tilde{\xi}_t, \bar{\xi}, \bar{\xi}_s$,   422, 426
$\xi \circ p$,   9, 29, 163

Z,   200
$\mathbf{Z}_+, \mathbf{Z}_-, \tilde{\mathbf{Z}}_-$,   6, 319
$\hat{\mathbf{Z}}_d, \overline{\mathbf{Z}}_d$,   325, 335
$\zeta, \bar{\zeta}$,   200

$\mathbf{0}, \mathbf{1}, 1\{\cdot\}$,   7, 385f
$1_B\xi$,   460
$2^d, 2^J$,   235, 302, 335
$\prec, <, \subset, \ll$,   40, 217, 332, 388, 460f
$\sim, \asymp, \lesssim$,   7, 27, 302, 325, 342f
$\perp\!\!\!\perp, \perp\!\!\!\perp_\eta$,   27, 32
$|\cdot|$,   5, 302, 353, 416
$\|\cdot\|, \|\cdot\|_A$,   31, 57, 134, 166
$[\cdot], [\cdot]^J$,   6, 201, 213, 382
$[\cdot, \cdot], \langle\cdot, \cdot\rangle$,   83, 351
$(\cdot)^*$,   7, 87, 280, 353, 362, 450
$\int_a^b, \int f$,   106, 238
$\vee, \wedge$,   294
$\otimes$,   18f, 113, 350ff, 357, 361, 373, 453
$\rightarrow$,   289

Foundations of Modern Probability
Second Edition

O. Kallenberg

This new edition contains four new chapters as well as numerous improvements throughout the text. There are new chapters on measure theory-key results, ergodic properties of Markov processes and large deviations.

2002. 638 p. (Probability and its Applications) Hardcover ISBN 0-387-95313-2

An Introduction to the Theory of
Point Processes
Volume I: Elementary Theory and Methods

D.J. Daley and D. Vere-Jones

Point processes and random measures find wide applicability in telecommunications, earthquakes, image analysis, spatial point patterns, and stereology, to name but a few areas. The authors have made a major reshaping of their work in their first edition of 1988 and now present their *Introduction to the Theory of Point Processes* in two volumes with sub-titles *Elementary Theory* and *Models and General Theory and Structure*. Volume One contains the introductory chapters from the first edition, together with an informal treatment of some of the later material intended to make it more accessible to readers primarily interested in models and applications. The main new material in this volume relates to marked point processes and to processes evolving in time, where the conditional intensity methodology provides a basis for model building, inference, and prediction.

2003. 469 p. (Probability and its Applications) Hardcover ISBN 0-387-95541-0

CPSIA information can be obtained at www.ICGtesting.com
Printed in the USA
BVOW06s0221051016

464206BV00009B/71/P